# REMOTE SENSING FOR A CHANGING EUROPE

# Remote Sensing for a Changing Europe

Proceedings of the 28th Symposium of the European Association of Remote Sensing Laboratories, Istanbul, Turkey, 2–5 June 2008

Edited by

Derya Maktav

*Istanbul Technical University (ITU), Istanbul, Turkey*

IOS
Press

Amsterdam • Berlin • Tokyo • Washington, DC

ISBN 978-1-58603-986-8
Library of Congress Control Number: 2009922570

Published by IOS Press under the imprint Millpress.

*Publisher*
IOS Press BV
Nieuwe Hemweg 6B
1013 BG Amsterdam
Netherlands
fax: +31 20 687 0019
e-mail: order@iospress.nl

*Distributor in the UK and Ireland*
Gazelle Books Services Ltd.
White Cross Mills
Hightown
Lancaster LA1 4XS
United Kingdom
fax: +44 1524 63232
e-mail: sales@gazellebooks.co.uk

*Distributor in the USA and Canada*
IOS Press, Inc.
4502 Rachael Manor Drive
Fairfax, VA 22032
USA
fax: +1 703 323 3668
e-mail: iosbooks@iospress.com



PRINTED IN THE NETHERLANDS

Published for EARSeL – European Association of Remote Sensing Laboratories

*Remote Sensing for a Changing Europe*
*D. Maktav (Ed.)*
*IOS Press, 2009*



# Preface

## The 28th EARSeL Symposium and Workshops "Remote Sensing for a Changing Europe" Istanbul, Turkey, 2–7 June 2008

Merhaba,

After Voss (Norway, 1981), Igls (Austria, 1982), Brussels (Belgium, 1983), Guildford (UK, 1984), Strasbourg (France, 1985), Lyngby (Denmark, 1986), Noordwijk (Netherlands, 1987), Capri (Italy, 1988), Helsinki (Finland, 1989), Toulouse (France, 1990), Graz (Austria, 1991), Eger (Hungary, 1992), Dundee (UK, 1993), Göteborg (Sweden, 1994), Basel (Switzerland, 1995), Malta (1996), Lyngby (Denmark, 1997), Enschede (Netherlands, 1998), Valladolid (Spain, 1999), Dresden (Germany, 2000), Paris (France, 2001), Prague (Czech Republic, 2002), Ghent (Belgium, 2003), Dubrovnik (Croatia, 2004), Porto (Portugal, 2005), Warsaw (Poland, 2006), and Bolzano (Italy, 2007), the EARSeL family met in a city where the continents meet. The '*28th EARSeL Symposium and Workshops*' with the title '*Remote Sensing for a Changing Europe*' took place in Istanbul, Turkey on 2–7 June 2008. Both the symposium and the accompanied two workshops were hosted by the Remote Sensing Division of the Istanbul Technical University (ITU) at the Süleyman Demirel Convention Center in the ITU Maslak Campus. The Symposium was chaired by Prof. Dr. Derya Maktav, Head of ITU Remote Sensing Department and co-chair of EARSeL SIG Urban Remote Sensing, and Prof. Dr. Rudi Goossens, Head of EARSeL. The Turkish Chamber of the Cadastre and Mapping Engineering, and the Scientific and Technological Research Council of Turkey (TÜBİTAK) also supported the meeting.

Technical presentations were on all fields of geoinformation and remote sensing, and especially on:

- Geoinformation and remote sensing
- New sensors and instruments
- Image processing techniques
- Time series analysis, data fusion
- Imaging spectroscopy
- Urban remote sensing, land use and land cover
- Radar remote sensing, LIDAR
- Land degradation and desertification
- Hydrology, land ice & snow, coastal zone
- Forestry, agriculture
- 3D spatial analysis
- World heritage

The meeting welcomed 220 registered participants from all over the world. Eighty papers were presented during the symposium (2–5 June) at the 21 oral sessions including the three special sessions. As well, 60 papers were presented at the poster sessions which were also presented during two 'oral communications' sessions where the authors had the opportunity to present their poster papers for 4–5 minutes and invite the participants to visit their posters for detailed information. The ITU Süleyman Demirel Convention Center with one conference hall, one senate hall and four workshop rooms served for the event.

In addition to these topics, the symposium also included three special sessions:

The special session 'ASTER', jointly held by ASTER and EARSeL, included applications of ASTER, such as the usefulness of thermal remote sensing images in the study of wet permafrost, and crop and water monitoring at the scale of a small agricultural region from ASTER data.

"SPOT" special session, jointly held by SPOT and EARSeL, focused on the new ASTRO-TERRA mission; a global database designed to build consistent and accurate geospatial datasets; assessing agri-environmental impact in the French West Indies and French Guyana; and the operational use of SPOT imagery for population and housing census in Africa.

The third special session 'Seismic Geohazards' chaired by Freek van der Meer, chair of the SIG Geological Applications, integrated valuable presentations on ASTER and geohazards; evaluation of the damaged provoked by seismic events through teledetected imagery; application of an integrated airborne hyperspectral and lidar dataset in resolving the frequency and intensity of earthquakes; predicting topographic aggravation of seismic ground shaking by applying geospatial tools; and tropical volcanic islands, coastal landslides and tsunami risk.

To encourage the establishment of two new SIG's, "*Remote Sensing for Archaeology and Cultural Heritage*" (3 June) and "*Thermal Remote Sensing*" (4 June), in addition to the already existing 14 SIG's, two lunch meetings were organized where the chairmen had the opportunity to advertise their new groups.

The 'SIG on Remote Sensing for Archaeology and Cultural Heritage', co-chaired by Dr. Rosa Lasaponara and Dr. Nicola Masini (Italy), addressed the researchers interested in the application of active and passive remote sensing technologies (ground, aerial and satellite) and in the information technologies for archaeological investigation, protection and valorization of cultural heritage.

Dr. Claudia Kuenzer of the German Remote Sensing Data Center, DFD of the German Aerospace Center, DLR presented the newly founded 'SIG on Thermal Remote Sensing'. About 20 participants joined the lunch meeting and presentation, during which the goals of this new SIG were introduced. SIG-TRS envisages bringing the thermal community among the EARSeL laboratories closer together, and promoting a platform for exchange about methods, applications, new sensors and in-situ approaches in the field of thermal remote sensing.

The ITU-Center for Satellite Communications and Remote Sensing (ITU-CSCRS) where we had the technical visit, is located at the ITU Maslak Campus and is one of the foremost institutions around the world with a highly capable ground receiving station unit. It is the first center established in Turkey to conduct application oriented projects in remote sensing and satellite communications technologies and to serve national/international civil/military companies in their research, development, and educational activities. CSCRS has the capabilities of acquiring images from remote sensing satellites, processing data, and sending the products via satellite links to resident and foreign users. The station can receive images of the Earth's surface within a radius of 3000 km, which covers from Sweden to Sudan, and England to Kazakhistan. In the center the data acquired from SPOT-2, SPOT-4, RADARSAT-1, ERS-2, NOAA-11, NOAA-14, METEOSAT satellites are archived, formatted and processed with state-of-the-art technology.

After the technical visit at ITU-CSCRS on 2 June the icebreaker party of the symposium was organized in the same place on the same day.

In addition to the technical meetings, EARSeL-bureau meeting, Council meeting, co-editors meeting, and General Assembly took place on different dates during the event.

The participants of the symposium discovered the wealth of impressive sights all along the shores in a Bosphorus boat trip along Istanbul's famous waterway dividing Europe and Asia.

The symposium dinner took place at a restaurant close to the Fatih Sultan Mehmet Bridge from where a beautiful view of the Bosphorus could be enjoyed.

In conjunction with the symposium two workshops ran parallel and after the end of the symposium:

1st Workshop "Earth Observation From Research to Teaching in Schools and Universities" of the "Special Interest Group (SIG): Education and Training" of EARSeL chaired by Mario Hernandez, UNESCO; Rainer Reuter, University of Oldenburg, Germany; and Alexander Siegmund, University of Education Heidelberg, Germany, on 6 June, aimed at making the results obtained from more than 250 EARSeL member institutes available to the public.

The workshop further addressed the Global Environment and Security (GMES) programme of the European Commission (EC) and the European Space Agency (ESA).

The topics of the workshop included: earth observation for kids; science education in schools, high schools and universities; applications in biology, chemistry, geography, physics and mathematics curricula; training activities in GMES; the Global Earth Observation System of Systems (GEOSS) and other international programmes; and public outreach of environmental sciences and global change.

On 7 June, an open meeting of the EARSeL project "Science Education through Earth Observation for High Schools (SEOS)" followed the workshop, which was an initiative for using remote sensing in science education curricula in high schools funded under the 6th Framework Programme of the EC. Eleven different partners from several European countries in cooperation with the ESA have so far implemented the project.

The second workshop of the symposium was organized as the 4th Workshop of the EARSeL SIG on Developing Countries (chaired by Gürcan Büyüksalih, Turkey; Richard Sliuzas, Holland; and Peter Lohmann, Germany), in conjunction with the 8th workshop of the GIS in Developing Countries network (GISDECO 8), with the title "Integrating GIS and Remote Sensing in a Dynamic World" on 4–7 June 2008.

This workshop brought together experts from the EARSeL and GISDECO networks for the first time. Especially for developing countries, integration of remote sensing and GIS offers unique access to primary data on the status of land surfaces, as well as possibilities for analysis, visualisation and development of possible solutions to problems associated with dynamic changes of nature and humanity. Global urbanisation, climate change and its effects on natural and human systems, land use and land cover changes, and salinisation are imminent dangers. The workshop provided a forum for presenting and discussing results, and for exchanging expertise and experience among researchers and users engaged in solving the problems of developing countries.

The topics of this workshop included:

Adapted sensor and mapping methods (TerraSAR-X, ALOS, ASTER, IKONOS, QUICKBIRD, dynamics of urban development, biomass), environmental monitoring (land degradation, desertification, erosion), model development and integration (landuse and cover models, biosphere model, effects of climate change), DEM generation for developing countries (SRTM, Cartosat-2, ALOS-PRISM), innovative remote sensing and GIS education (distance learning, professional development), theory and practice of partipatory GIS (case studies on community mapping and PGIS), GIT and poverty alleviation (food and water security, resilience), hazards and risk mitigation (measuring risks and adaptive planning systems), and managing global urbanisation (slum mapping, sustainable transport systems).

I would like to thank the members of the scientific committee who have contributed to the abstract review process, to the artist Beygü Gökçin who artistically combined space and music with piano; the band of the Turkish Air Force Academy for their exceptional performance at the opening session, to my colleagues from my department, and, of course, to my students for their great efforts before, during and after the symposium.

Finally, I wish good luck and success to my Greek colleauges who will organize the 29th EARSeL Symposium in Crete, Greece in 2009. See you in Crete.

With my best regards,
November, 2008, Istanbul

Prof. Dr. Derya Maktav
EDITOR
Istanbul Technical University
Remote Sensing Division
34469 Maslak, Istanbul, Turkey
http://www.ins.itu.edu.tr/deryamaktav
dmaktav@ins.itu.edu.tr

# Contents

*Remote Sensing for a Changing Europe*
*D. Maktav (Ed.)*
*IOS Press, 2009*


*doi:10.3233/978-1-58603-986-8-1*

# Precipitation in the Antarctic gradually decreased

L.N. Vasiliev
*Institute of Geography Russian Academy of Sciences, Moscow, Russia*

Keywords: precipitation rate, cycles, wavelet power spectrum, cumulative sum

ABSTRACT: Despite forecasts that global warming will increase Antarctic snowfall, there is some evidence that exactly the opposite is true. Analysis of the Global Precipitation Climatology Project (GPCP) data indicates a decrease in the accumulation on the Antarctic ice sheet over the last 29 years. The GPCP satellite remote sensing data, spanning the years 1979 to 2007, exhibits 6-year cycles in precipitation rate. Precipitation in these cycles from 1982 to 2005 has gradually decreased by 15%. This reduction is statistically significant.

## 1 INTRODUCTION

There is no consensus at present on how to determine the mass balance of the Antarctic ice sheet. Although most researchers Chen *et al.* (2006), Ramillien *et al.* (2006), Rignot & Thomas (2002), Wingham *et al.* (2006), Zwally *et al.* (2005) consider that the ice sheet in East Antarctica is growing and that in West Antarctica is contracting, the current balance of these two parts of the ice sheet, and its changes in the past have not been established. Basic distinctions lie, as a rule, in different assessments of the influx of matter onto the surface of the Antarctic ice sheet Giovinetto & Zwally (2000), Kotlyakov (1966), Kotlyakov (2000), Vaughan *et al.* (1999). Accumulation assessments restricted to 10-year periods and climatic models of reanalysis cannot characterize the present-day oscillations of the snowfall volume with necessary precision. Uncertainty in the prediction of the behaviour of the Antarctic ice sheet may be decreased by long-term series of precipitation measurements and elucidation of causes of its oscillations. Satellite measurements of global precipitation carried out within the framework of the Global Precipitation Climatology Project (GPCP) made it possible to assess accumulation in Antarctica with different spatiotemporal resolutions for the period of 1979–2007.

Initial data were obtained by combining precipitation measurements from several space systems. Microwave measurements are carried out from satellite sun-synchronous low Earth orbit using a Special Sensor Microwave Imager (SSM/I). The infrared precipitation estimates are obtained primarily from geostationary satellites and secondarily from polar orbiting satellites. The latter are then calibrated by precipitation values from microwave measurements. Additional precipitation estimates are obtained based on Television and Infrared Observation Satellite (TIROS) Operational Vertical Sounder (TOVS). Precipitation values obtained from the combination of space measurements at different frequencies are corrected based on the rain gauge data network. Primary data are represented in two versions: 1) monthly precipitation on a grid of $2.5° \times 2.5°$ latitude – longitude spanning the years 1979–2007 Adler *et al.* (2003) and 2) daily measurements on a grid of $1° \times 1°$ from January 1997 to 2007 Huffman *et al.* (2001).

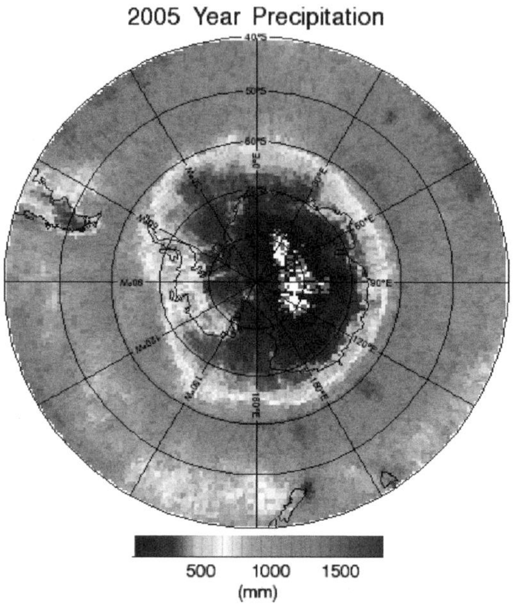


Figure 1. Spatial distribution of annual precipitation in Antarctica in 2005 with resolution 1° latitude 1° longitude. One of the examples for the period of 1979–2007.

## 2 METHODOLOGY

Spatiotemporal series of monthly and daily precipitation can be regarded as a succession of points in the phase space. To obtain a clear interpretation of the dynamic structure, it is necessary to reduce the phase space dimension without loosing essential properties in the process. In this work, we have begun with a primary representation of the spatial distribution of annual precipitation in the Antarctic (Fig. 1). Comparison of successive images similar to Fig. 1 for the last 29 years shows that satellite measurements in combination with ground-based precipitation data yields a relatively stable spatial structure of accumulation. The time series analysis of annual accumulation is presented in Fig. 2.

Errors in the assessment of accumulation stem from both random errors in the algorithm of precipitation based on brightness temperature, and random measurement errors. Random errors in the elementary 2.5° × 2.5° cell depend on the temporal density of its coverage by space measurements and the spatial density of rain gauge data. The precipitation value for each cell has therefore been accompanied by the calculated variance for that cell. Error files exhibit a structure similar to primary data, from which variance of the sum of monthly precipitation $\sigma^2$ is calculated. Assessment of the relative error of annual accumulation will be proportional to $\frac{\sigma n^{-1/2}}{P}$ where $P$ is the annual sum of monthly precipitation and $n$ is the number of months in a year. Calculated in this way, the average relative error of annual accumulation has 7.4%. It is apparent that the relative error of 10-years accumulation will decrease to 2.4%. Error values of monthly precipitation in each 2.5° × 2.5° cell are assessed by the variance, $\sigma^2$, which depends nonlinearly on the precipitation intensity. Although the probability distribution of the error it not certain, it approximates a logarithmic normal distribution. However, a final assessment of accuracy using sparse data on Antarctic snow measurements will require further investigation.

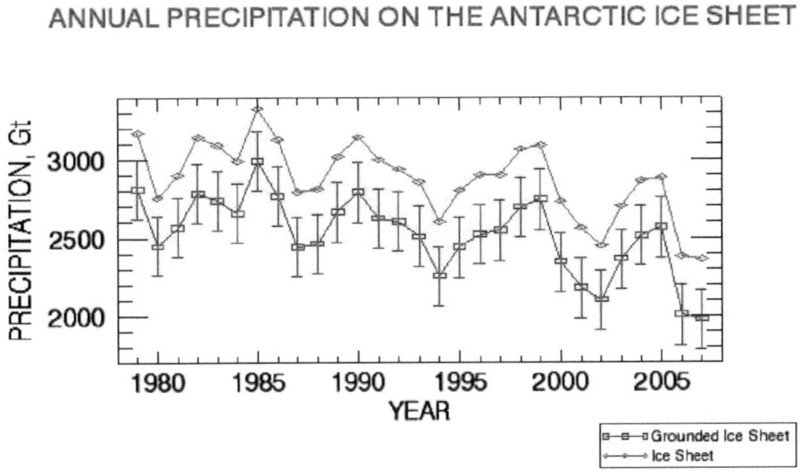


Figure 2. Annual accumulation in the Antarctic grounded ice sheet (□) and with the ice shelf (◊) over the period 1979–2007. Confidence level for points is determined by the error bars 190 Gt.

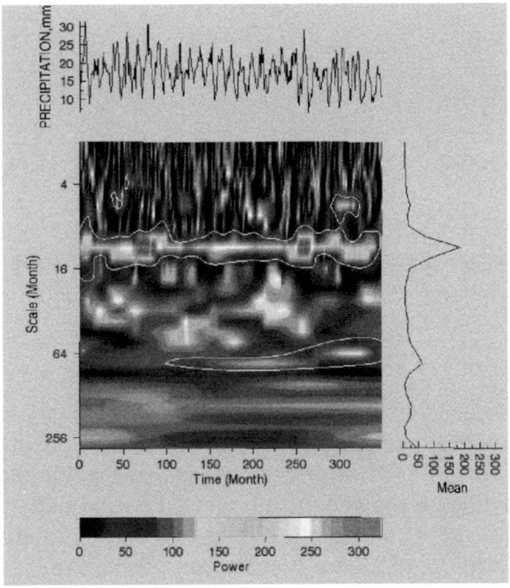


Figure 3. The Morlet wavelet transform of the time series of monthly precipitation in Antarctica, 1979–2007. (A) Time series of monthly precipitation. (B) Distribution of local values of the wavelet power spectrum for the time series of monthly precipitation (light contour lines correspond to boundaries of 90%-confidence level for local values of power spectrum. (C) Time-averaged values of wavelet power spectrum are proportional to the sum of local values at each period. Interpretation of local values of the power spectrum with a period exceeding 100 months with a time series of only 348 months is not recommended.

The comparison of the data obtained with snow measurements at polar stations for the period of 1997–2006 is presented in Table 1. Precipitation values were determined for the $1^\circ \times 1^\circ$ cell using daily space measurements. Differences between the two measurements are mainly related to spatial nonuniformity inside the cell. This is particularly prominent at Station Bellingshausen. At an average variance $\sigma^2$ of monthly precipitation, the relative error of accumulation at low $P$ values will of

Table 1. Comparison of precipitation measured by satellite systems and snowfall in water equivalent on polar surface stations for the period of 1997–2006

| Station | Duration of measurements (year) | Satellite measurements (mm) | Snowfall measurements, we (mm) |
|---|---|---|---|
| Mirny | 10 | 5640 | 5148 |
| Novolazarevskaya | 9 | 2212 | 2131 |
| Bellinshausen | 10 | 7661 | 6397 |
| Vostok | 7 | 232 | 116 |

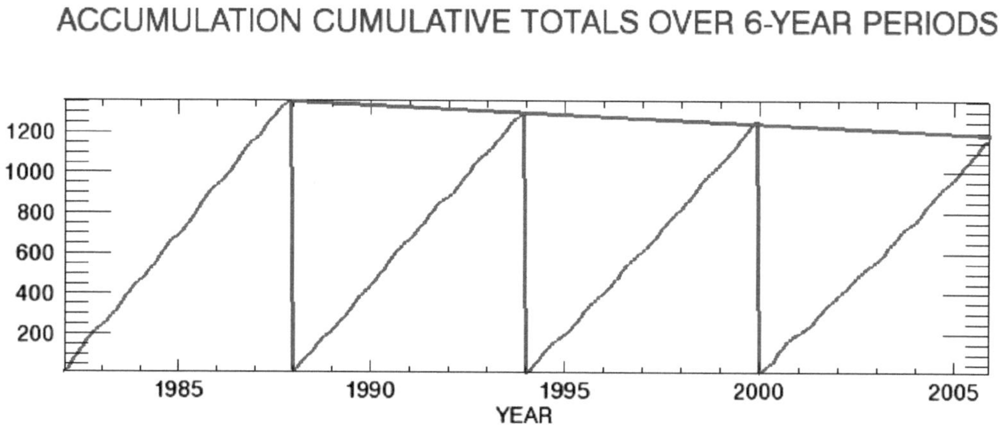


Figure 4. Cumulative totals of precipitation in Antarctica over 6-year cycles show a slow decrease in accumulation since 1982.

course increase drastically. This is observed at Station Vostok. In other cases, relative errors vary from 4 to 20%.

## 3 RESULTS

Accumulation is determined by two regimes of precipitation behaviour: inter-annual low-frequency variations, which characterize large-scale changes (Fig. 2), and intra-annual seasonal oscillations (periodicity). The latter can arbitrarily be attributed to high-frequency oscillations. To establish the time scale of accumulation variability, we used the time series of monthly precipitation in Antarctica and analyzed it with a continuous wavelet transform. Choice of the analyzing wavelet plays a crucial role. To establish actual criteria of regularity and to explain them, it is necessary to use an analyzing wavelet that corresponds best to the time series under study. The main interest lies in elucidation of criteria of periodicity and their associated time scales. The Morlet wavelet fits this objective since it precisely determines values of periods, but not the location of maximums and minimums. This corresponds to our requirements. Figure 3 presents the result of application of the Morlet wavelet. The wavelet power spectrum clearly localizes a strict periodicity along the period axis at 12-months, which is related to intra-annual seasonal behaviour of precipitation. The low-frequency region at a period of about 72–84 months includes a second (weaker) maximum, which indicates the existence of a 6 to 7-year cycle. It is worth noting that the precipitation volume in 6-year cycles from 1982 to 2005 gradually decreases by 3.5% per cycle on average. This is illustrated by variation in the cumulative sum of precipitation in four successive 6-years periods from 1982 (Fig. 4). The precipitation volume decreased by 15% during 2000–2005 compared to the period of 1982–1987. This inference quantifies the behaviour of annual precipitation that may be perceived qualitatively from Fig. 2. Of course, the inference should be treated with caution since the time series is only 29 years long and so contains only 4 cycles.

## 4 CONCLUSIONS

Our work has revealed a statistically significant variation in accumulation in Antarctica over the last 29 years, although no trend in world-wide precipitation over the last 50 years is indicated by the same data. Progress has been achieved in determining precipitation accumulation by the application of modern satellite remote sensing. The existence of a 6 to 7-year cycle is indicated but not completely confirmed. This cycle and its explanation require further study. Nevertheless, one might expect that the stability of annual periodicity in combination with low-frequency (6 to 7-year) variations can be used to refine the dynamic mass balance model of the Antarctic ice sheet. Application of modern methods for measurement of global precipitation significantly improves the determination of mass balance as a result of on-line data collection, improved accuracy, and spatiotemporal reliability.

## ACKNOWLEDGEMENTS

The one-degree daily precipitation data set and monthly precipitation in 2.5° × 2.5° grid cell were provided by the NASA/Goddard Space Flight Center's Laboratory for Atmospheres, which develops and computes the data as a contribution to the GEWEX Global Precipitation Climatology Project. We thank G. Huffman for GPCP data sets. This work was supported by the Presidium of the Russian Academy of Sciences (program 16, part 2, project 3.3, subprogram "Study and Investigation of Antarctica"), the Federal Targeted Program "World Ocean" (project no. 5), and the Russian Foundation for Basic Research (project 05-05-64168).

## REFERENCES

Adler, R.F., Huffman, G.J. et al. 2003. The version-2 Global Precipitation Climatology Project (GPCP) monthly precipitation analysis (1979 – present). *Journal of Hydrometeorology*, 4, 1147–1167.

Chen, J.L., Wilson, C.R., Blankenship, D.D., Tapley, B.D. 2006. Antarctic mass rates from GRACE. *Geophysical Research Letters*, 33, L11502.

Giovinetto, M., Zwally, J. 2000. Spatial distribution of net surface accumulation on the Antarctic ice sheet. *Annals of Glaciology*, 31, 171–178.

Huffman, G.J., Adler, R.F. et al. 2001. Global Precipitation at one-degree daily resolution from multi-satellite observations. *Journal of Hydrometeorology*. 2, 36–50.

Kotlyakov, V.M. 1966. *The Snow Cover of the Antarctic and Its Role in the Present-day Glaciation of the Continent (Jerusalem),* 246 pp.

Kotlyakov, V.M. 2000. *Selected Works: Volume 1. Glaciology of Antarctica,* Nauka, Moscow [in Russian], 423 pp.

Ramillien, G., Lombard, A., Cazenave, A. 2006. Interannual variations of the mass balance of the Antarctica and Greenland ice sheets from GRACE. Global Planet Change, 53 (3), 198–208.

Rignot, E. & Thomas, R.H. 2002. Mass balance of polar ice sheets. *Science,* 297, 1502–1506.

Vaughan, D.G., Bamber, J.L. et al. 1999. Reassessment of net surface mass balance in Antarctica. *Journal of Climate*, 12 (4), 933.

Wingham, D.J., Shepherd, A. et al. 2006. Mass balance of the Antarctic ice sheet. *Philosophical Transactions of the Royal Society* A, 364, 1627–1635.

Zwally, H.J., Giovinetto, M.B., Li, J. et al. 2005. Mass changes of the Greenland and Antarctic ice sheets and shelves and contributions to sea-level rise. *Journal of Glaciology*, 51, 509–527.

*Remote Sensing for a Changing Europe*
*D. Maktav (Ed.)*
*IOS Press, 2009*

*doi:10.3233/978-1-58603-986-8-6*

# 3D-Remote sensing, Status Report 2008

K. Jacobsen
*Leibniz University Hannover, Germany*

Keywords: optical satellites, airborne cameras, oblique images, laser scanning, DEM

ABSTRACT: Since the 27th EARSeL symposium 2007 in Bolzano with WorldView-1 and Cartosat-2A two more very high resolution optical satellites of a new group have been launched. Images with a ground sampling distance (GSD) of 0.45 m, which will be distributed with 0.5 m GSD, are operational available. In August 2008 GeoEye-1 shall be launched, having even 0.42 m GSD. With CBERS 2B an additional high resolution is in space. The system of 5 RapidEye satellites shall follow. The high number of new satellites gives a strong push to the 3D-Remote sensing. The wide spread of the stereo systems Cartosat-1 and ALOS/PRISM has improved the possibility of generating detailed and accurate digital elevation models based on space images. An overview of the new optical systems, but also a short information about new radar satellites and in near future planned missions together with the influence to the practical application is given. Only systems available for civilian use are respected in this overview.

As new trend in aerial application we have with mid format digital aerial cameras and the combination of vertical and oblique images like from Pictometry and Multivision. The digital mid format are now completed by combinations of cameras, closing the gap to the large format digital cameras.

Digital height models can be generated by interferometric synthetic aperture radar (IfSAR) or by automatic image matching. The images used for image matching should be taken within a short time interval to avoid changes of the object and different shadows. Stereo systems generating stereo combinations in general have some advantages.

## 1 INTRODUCTION

Detailed earth observation determining or respecting the three-dimensional shape of the earth surface and usable for mapping purposes is dominated by the used sensors. Low or medium resolution systems like Landsat cannot be used for this task. The first civilian use of satellite images for 3D-mapping came with SPOT 1 in 1986, but with 10m ground sampling distance (GSD) the accuracy and identification of object details is limited. With IRS-1C in 1996 the GSD was improved to 5.8m, but the real break through of the use of space images for detailed information extraction came with IKONOS in 1999. Today we have several high and very high resolution optical satellites. The number is permanently growing and also the resolution is improving, just now images taken by WorldView-1 are distributed with 0.5 m GSD. The Indian space organisation ISRO plans Cartosat-3, to be launched in 2011, with 0.35 m GSD and the US company GeoEye made a proposal for GeoEye-2 with 0.25 m GSD. The last requires a change of the US governmental limitation of the distribution of space images to currently at least 0.50m GSD. With stereo satellites like Cartosat-1 and ALOS/PRISM optimal conditions for the generation of digital elevation models (DEM) exist. In this presentation only civilian or dual use systems are respected. Most of these sensors are used for military and civilian application. Without military or secret service application we would not have such a variety of reconnaissance satellites, which we can use now also for civilian projects. Only a very limited number of systems are only based on civilian use like the German TerraSAR-X and the announced RapidEye. Sensors with medium resolution (above 10m GSD) are mainly used for land-use monitoring, they are also not included here. A special position has ASTER, allowing the generation of height models by the stereo system with 16m GSD. This sensor belongs to the overlap of the land-use monitoring and the mapping.

The optical space images are competing with aerial images. Of course the highest resolution is only possible from air, here a clear tendency to digital cameras exist. Beside the large format digital cameras also digital mid format cameras are becoming more important, supported by CCD-arrays with more pixels. In addition combinations of mid-format cameras are closing the gap to large format cameras. As platform of light weight cameras unmanned aerial vehicles are used more often, supported by partially automatic steering of these devices.

Beside very high resolution optical space images, now also very high resolution synthetic aperture radar (SAR) images are available with TerraSAR-X having the advantage of imaging through clouds. With interferometric SAR from aircraft large areas have been covered by DEMs for private companies. Such a competition to survey administrations is growing.

Laser scanning from air, also named LIDAR is used more and more for very precise and high resolution DEMs, also in not too dense forest areas.

## 2 SENSORS

### 2.1 *Optical imaging sensors*

#### 2.1.1 *Optical satellites*

Figure 1 gives an overview about the launch of the high and very high resolution optical satellites and their best ground resolution. All these systems are imaging with CCD-lines, generation the scenes by the movement and/or angular movement of the satellites. Today the majority of high and very high resolution optical satellites are flexible satellites, able to change the view direction fast and precise based on reaction wheels or control moment giros. The change of the view direction usually is precise enough to use it also during imaging without loss of accuracy.

The optical satellites today are usually equipped with positioning systems like GPS, gyros for the attitude determination and star trackers for the absolute determination of the satellite attitude (Fig. 2). Based on this equipment the exterior orientation of the sensor can be determined. In the case of IKONOS this is possible with a standard deviation of 4m without use of control points, for WorldView-1 Digital Globe talks about 2.5 m. The other sensors are not so accurate, requiring in any case control points for the correct geo-reference, but this is also recommended for IKONOS and WorldView-1 at least for a check and for a correct handling of the national net datum – the relation between the national coordinates and WGS80 in ITRF (international terrestrial reference frame).

As obvious in Fig. 1, the number of launches of such systems stays on a high level, the number of very high resolution optical sensors is growing permanently and with WorldView-1 0.45 m GSD has been reached (Fig. 3). Because of a legal restriction in the USA only images with 0.5 m GSD are distributed by US companies. This limitation will be used also for GeoEye-1, scheduled for launch in August 2008. GeoEye-1 will have 0.42 m GSD for nadir view. In addition in 2009 WorldView-2 shall be launched, it will have the same panchromatic resolution like WorldView-1, but also 8 spectral channels. In addition to the classical blue, green, red and near infrared a costal channel, yellow, red edge and a short-wave-infrared channel will be available.

The high resolution systems Resourcesat and RapidEye have a good spectral resolution with 4, respectively 5 spectral bands. They are more directed to land use monitoring, but with 5.8 m and 6.5 m GSD they are at the limit for mapping purposes.

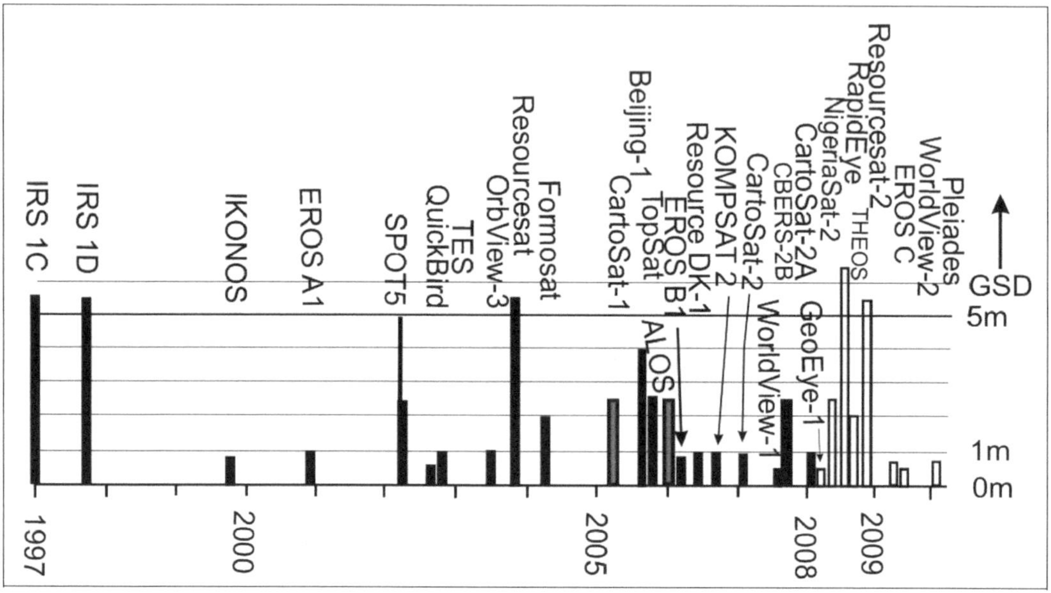


Figure 1. High and very high resolution optical satellites available for civilian use in sequence of launch with their best GSD.

Figure 2. GeoEye-1; Typical hardware components of a high resolution optical satellite courtesy: GeoEye.

Figure 3. Dolmabace palace, Istanbul in WorldView-1 image 0.5m GSD.

The development of the optical sensors is continuing as shown in Table 2. The resolution will be improved and more systems will come, which will not just replace old systems. Table 2 is not a complete list, in addition the proposals may be changed and very often the launch is postponed. Also additional countries, like Turkey, will launch very high resolution optical satellites, but no clear specifications are available yet.

Table 1. High and very high resolution optical satellites available for civilian use

| Sensor | Country | GSD (nadir) [m] | Swath [km] | Pointing in-track | Pointing across |
|---|---|---|---|---|---|
| SPOT 1-4 | France | 10/20 | 60 | – | +/–27° |
| SPOT 5 | France | 5 (2.5)/10 | 60 | – | +/–27° |
| SPOT 5 HRS | France | 5 x 10 | 120 | +20°, –20° | – |
| MOMS-02 / -P | Germany | 5.8/16.5 | 37/78 | –27°, 0°, 27° | |
| IRS-1C/1D | India | 5.8/23.5 | 70/142 | – | +/–26° |
| Resourcesat | India | –/5.8 | 70 | – | +/–26° |
| KOMPSAT | S. Korea | 6.6 | 17 | – | +/–45° |
| Terra ASTER | Japan | 15/(30, 90) | 60 | 0°, 27.2° | |
| IKONOS | USA | 0.82/3.2 | 11.3 | free view direction | |
| EROS A | Israel | 1.8 | 12.6 | free view direction | |
| QuickBird | USA | 0.61/2.44 | 16.4 | free view direction | |
| *OrbView 3* | *USA* | *1/4* | *8* | *free view direction* | not active |
| EROS B | Israel | 0.7 | 14 | free view direction | |
| FORMOSAT 2 | Taiwan | 2/8 | 24 | free view direction | |
| IRS-P5 Cartosat-1 | India | 2.5 | 30 | 26° fore, 5° after | free view to side |
| TopSat | UK | 2.5/5 | 15/10 | free view direction | |
| Beijing-1 | China | 4/32 | /600 | free view direction | |
| ALOS | Japan | 2.5 | 35 (70) | –24°, 0°, +24° | free view to side |
| KOMPSAT-2 | S. Korea | 1/4 | 15 | free view direction | |
| Resource DK1 | Russia | 1/3 | 28 | free view direction | |
| IRS Cartosat-2 | India | <1 | 9.6 | free view direction | |
| WorldView-1 | USA | 0.45 | 15.8 | free view direction | |
| CBERS-2B | China/ Brasil | 2.5/20 | 27/120 | free view direction | |
| IRS Cartosat-2A | India | 1 | 10 | free view direction | |

There is a clear tendency to smaller satellites, reducing the overall cost. The electronic components are getting smaller and most new systems are limited to just one camera. From Table 1 TopSat, Beijing-1, the not more active OrbView-3 and the EROS-satellites are belonging to the group of satellites with a weight below 500 kg, while ALOS has a weight of 4 tons and Resource DK1 even 6.8 tons.

The Indian space organisation ISRO has the most complete program for earth observation satellites from very high up to low resolution systems. It has a dense program for the development

Table 2. Proposed high resolution optical satellites available for civilian use

| Sensor | Country | Proposed launch | GSD (nadir) [m] | Swath [km] | Remark |
|---|---|---|---|---|---|
| GeoEye-1 | USA | August 2008 | 0.42/1.64 | 15 | |
| RapidEye | Germany | July 2008 | –/6.5 | 78 | 5 satellites |
| Resourcesat-2 | India | 2008 | 5.8/(23.5) | 70/140 | +/–26° |
| NigeriaSat-2 | Nigeria | 2009 | 2.5/5 | | |
| WorldView-2 | USA | 2009 | 0.45/1.8 | 15 | pan + 8 spectral |
| THEOS | Thailand | 2009 | 2/15 | 22/90 | |
| EROS C | Israel | 2009 | 0.7/2.8 | 11 | |
| Pleiades | France | Begin 2010 + 11 | 0.7/2.8 | 20 | Follow on of SPOT |
| Cartosat-3 | India | 2011 | 0.35/– | | |
| ResourceSat-3 | India | 2011 | 2.5/5.8 | 25 | |
| GeoEye-2 | USA | | 0.25/ | | proposal |

of improved optical and radar satellites and replaces also old systems. From 1995 up to 2007 ISRO launched successful 9 earth observation satellites and this program will be continued with 3 more systems in the next 2 years and 4 satellites in 2011.

The distribution of the satellite images is an important factor, causing a sharing of different satellite products to major distributors. For example SPOT Image is also distributing Formosat-2 and KOMPSAT-2 images and enlarging so the product group over 5 m, respectively 2.5 m GSD (for supermode) panchromatic SPOT images to 2 m and 1 m GSD. For longer time it was difficult to get images from the Disaster Monitoring Constellation (DMC), constructed by Surrey Satellite Technologies Ltd (SSTL), UK. SSTL has build satellites with 32 m GSD for the multispectral range for Algeria, China, Nigeria, Turkey and the UK, Spain will follow soon. Beijing-1 belongs to this group, but it is equipped in addition with a 4m panchromatic camera. The Images are now distributed by DMC International Imaging, UK. SSTL has build also the TopSat satellite having 2.5 m panchromatic GSD. All these satellites are belonging to the group with less than 500 kg weight. The DMC constellation will be continued with the Deimos Satellite for Spain, having an improved resolution of 22 m instead 32 m like the group before; it shall be launched 2008. For Nigeria a second satellite shall be launched 2009 with 2.5 m GSD for the panchromatic band.

To manage large disasters like caused by earthquakes, floods, forest fires and oil spills, with social and economic relevance for Earth and mankind, the International Charter "Space and Major Disasters" was initiated at the UNISPACE III in Vienna in 1999. The major earth observation organizations are participating and after official declaration of a disaster, they are trying to get images of the affected area, so that organizations like the Center of Satellite Based Crisis Information of the DLR are able to generate very fast ortho-maps of the affected regions as support for the local operating help organizations.

### 2.1.2 *Aerial cameras*

New aerial cameras are digital cameras. The large format digital cameras Z/I Imaging DMC, Vexcel Imaging UltraCamD and UltraCamX like also the line scan camera Leica ADS40 are well established and are taking over the imaging from standard photogrammetric film cameras. The change from film to digital systems is faster than expected few years ago. As new development we have the growing use of mid format digital cameras, equipped with just one CCD-array, using a Bayer pattern for generating colour images. With the new Kodak CCD-array having 5412 x 7216

| combination of 2 IGI DigiCam | Rolleimetric AICx4 | Track'Air MIDAS camera system |
|---|---|---|

Figure 4. Combinations of digital mid-format cameras.

pixels a capacity of 39 Mpixels has been reached. This CCD-array is also used in the UltraCamX, leading to the improved number of pixels against the UltraCamD.

For metric application fix-focus systems are required like the Applanix DSS, IGI DigiCam and Rolleimetric AIC. All these cameras are equipped with the just mentioned Kodak CCD-array. The gap between the mid-format and the large size digital cameras is closed by combinations of more than one mid-format system. DIMAC is using like IGI (Fig. 4) a combination of 2 cameras with shifted principal point, allowing the merge of 2 parallel used single camera images to a homogenous larger virtual image. Rolleimetric is offering beside a combination of 2 AIC-cameras also the use of 4 oblique mounted AIC (Fig. 4). The combination of vertical and oblique images got very fast a widespread use by companies like Pictometry and Multivision. The required images can be taken by cameras like Track'Air MIDAS, viewing to the nadir and into 4 oblique directions, partially not with overlap to the nadir image.

## 2.2 *Laser scanner (LIDAR)*

The use of laser scanners became standard for very precise and detailed digital elevation models in several countries. The range for the flying height has been enlarged, limiting the possible pulse rate because of the time of flight, but the first systems have reduced this problem by handling more than one pulse during the time interval required for sending up to receiving again the same pulse. Of course the point spacing is influenced by the flying height, but this can be fitted to the requirements.

For very high accuracy specifications in the range of 10 cm standard deviation of the height, the orientation of the point clouds for every flight line has to be improved by a procedure similar to the former model block adjustment.

## 2.3 *Synthetic Aperture Radar (SAR)*

SAR is used from space and from aircraft. With the German TerraSAR-X and the Canadian Radarsat-2 the possible ground resolution from space has been improved drastically. The ground resolution and the swath of SAR are depending upon the imaging mode. TerraSAR-X has in the SpotLight mode 1 m GSD, in StripMap mode 3 m GSD and in ScanSAR 16 m GSD, while Radarsat-2 has in total 11 different modes from Ultra-Fine with 3m GSD up to ScanSAR Wide with 100 m GSD. Corresponding to the GSD the swath width of Radarsat-2 varies between 20 km and 500 km.

The information contents of SAR-images having thc same ground resolution like optical images cannot be compared directly (Fig. 5). At first the imaging is totally different – the object reflection of the spectral range used by optical images is mainly depending upon the chemical situation of the object, while radar is mainly depending upon the physical situation of the surface. The

TerraSAR-X SpotLight image

Optical image

Figure 5. Comparison of SAR-image with optical image – both with 1m GSD; main building University Hannover.

Figure 6. Multi polarization (HH/VV) TerraSar-X SpotLight image reduced by factor 2 city of Dresden.

interpretation of SAR-images requires a special training and understanding of SAR-imaging. Special problems exist in build up areas with the radar lay over – the return signals from the facades and partially roofs are mixed with the return signal from the ground in front of the buildings. In addition total reflections by corners are causing problems like also the speckle. Of course the speckle can be reduced by filtering, but this causes also a loss of information. On the other hand some special objects like railroads with the metallic rails can be seen clearer like in optical images. The main advantage of SAR images is still the penetration through clouds – for SAR-images the partially painful waiting for cloud free weather does not exist.

The information contents of SAR-images can be improved by multi polarization like shown in Fig. 6. By the combination of HH and VV polarization the image interpretation is improved.

An important advantage of SAR-images is the combination of 2 images, taken with a small base length, to interferometric SAR (IfSAR), allowing the determination of digital height models. From space this has been done with the Shuttle Radar Topography Mission (SRTM), generating a nearly worldwide coverage (Passini et al. 2007). The height models of SRTM are available free of charge in the internet, having a spacing of 3 arcsec, corresponding to ~92 m at the equator. The standard deviation of the height values is in the range of 4 m up to 12 m depending upon the terrain inclination and the land cover – the highest accuracy is reached in flat and open areas. The used C-band of SRTM cannot penetrate the vegetation, so not digital elevation models with the height of the bare ground, but digital surface models (DSM) with the height of the visible surface are generated.

TerrSAR-X will be completed by a second, identical satellite to a tandem configuration in 2009. Both SAR-satellites are joint together to the TanDEM-X configuration, where both satellites will have a base between 0.5 km and 2 km. Based on this configuration worldwide height models shall be generated with 12 m point spacing and a standard deviation of 2 m (Pitz 2006).

The Canadian company Intermap has generated height models of the USA, Great Britain and Western Europe as NEXTMap USA, NEXTMap Britain and NEXTMap Europe by IfSAR from aircrafts operating in 6km up to 9 km flying height. The NEXTMap products have a point spacing of 12 m and shall have a vertical standard deviation of 1m and a horizontal accuracy of 2 m. Intermap generated these height models without contract from governments. The sold height models in most countries are less expensive than height models from survey administrations.

## 3 HEIGHT MODELS FROM OPTICAL IMAGES

Height models are a basic requirement for geo information systems (GIS). They have to be used for the most often generated product, the ortho image. If the available height models are not accurate enough, have not sufficient point spacing or if existing height models are too expensive or not distributed, height models have to be generated. One possibility is the generation of height models by automatic image matching. This requires stereo models where both images have to be taken under similar conditions. Optimal is the imaging of both used scenes during the same path, avoiding changes of the object and different illumination conditions. With the today dominating flexible satellites in most cases the acquisition of a stereo pair from the same orbit is possible. Nevertheless only a limited number of stereo pairs are available in the image archives because of economic reasons; this is different for the stereo systems like ASTER, SPOT-5 HRS, Cartosat-1 and ALOS/PRISM. Based on 2 or 3 optics, they are generating permanently stereo models. The images taken by SPOT-5 HRS cannot be ordered, SPOT Image only likes to distribute height models based on it. The SPOT-5 HRS height models over forest areas should be handled with care because of the limited spectral range of the images between 0.48 µm and 0.70 µm. This is not leading to sufficient image contrast in forest areas (Büyüksalih et al. 2008). ASTER stereo pairs are taken in the near infrared spectral range, optimal also for matching in forest areas, but the 15m GSD limits the vertical accuracy to approximately 15 m (Sefercik et al. 2007). ASTER/PRISM images have some problems with the image quality, but this seems not to influence the results of the image matching. In addition the orientation based on sub-images requires more control points than the orientation of a full scene. Very good results have been achieved with Cartosat-1 stereo pairs. The images are covering a spectral range from 0.50 up to 0.85µm, including the near infrared, leading to good contrast in forest regions. Even in difficult regions a good coverage by matched points has been reached (Jacobsen, 2007, Büyüksalih et al. 2008, Jacobsen et al. 2008).

The gaps in matching Cartosat-1 scenes (Fig. 7) are caused in the case of Mausanne by missing contrast in fields without vegetation, in Warsaw by areas covered by snow, in Istiranca by clouds and in Jordan by lakes and fields without vegetation. No other optical satellite could lead to completer matching results. The automatic image matching leads to the height of the visible objects, that means to digital surface model. If a mixture of points located on the ground and located on objects, like trees and buildings, is given, the points not belonging to the bare surface can be filtered. By filtering the standard deviations of the height values have been improved in any case (Table 3).

| Mausanne Jan. | Mausanne Feb. | Warsaw | Istiranca | Jordan |
|---|---|---|---|---|

Figure 7. Cartosat-1 – overlay of matched points (white) to after scenes.

Table 3. Accuracy of Cartosat-1 height models checked by precise reference DEMs

| | | SZ | bias | SZ as F (terrain inclination $\alpha$) |
|---|---|---|---|---|
| Mausanne January | open areas | 4.02 | −0.51 | 3.91 + 1.64*tan $\alpha$ |
| | open areas filtered | 3.30 | 0.48 | 3.17 + 3.14*tan $\alpha$ |
| Mausanne February | open areas | 4.13 | −1.16 | 3.96 + 3.06*tan $\alpha$ |
| | open areas filtered | 3.39 | −0.58 | 3.22 + 1.97*tan $\alpha$ |
| Warsaw | open areas | 3.23 | −0.54 | 3.16 + 1.19*tan $\alpha$ |
| | open areas filtered | 2.43 | 0.44 | 2.39 + 8.80*tan $\alpha$ |

As it can be seen in Table 3, the height accuracy is depending upon the terrain inclination. For flat and open terrain after filtering root mean square differences of the DEMs based on Cartosat-1 against reference height models are 3.17 m, 3.22 m and 2.39 m. For 2.5 m GSD and the base to height relation of 1.6 this corresponds in the average to a standard deviation of the x-parallax of 0.7 GSD, this is a very good result for DEMs. If a DEM is analysed against check points, the results would be too optimistic because check points have usually a good object contrast and are not so much affected by terrain inclination. For getting realistic information about the accuracy of height models, reference height models have to be used. As point spacing 3 GSD, identical to 7.5 m, are justified, so detailed and precise DEMs can be generated by automatic image matching of Cartosat-1 images.

## 4 CONCLUSION

Caused by the improving ground resolution the competition between aerial and space images is growing. Medium scale maps can be generated based on aerial or space images. With WorldView-1 images, maps in the scale 1:5000 can be generated by on-screen digitizing of ortho-images. The DEM required for such ortho-images can be achieved by automatic matching of Cartosat-1 images if the satellite elevation does not exceed 20° under the condition of a required geometric accuracy of 0.2mm in the map scale. Not in any case such a large map scale is requested, so also space images with larger GSD can be used. As rule of thumb for mapping, 0.1mm GSD in the map scale is required, corresponding to 1m GSD for the map scale 1 : 10 000, but the GSD should not be quite larger than 5m to allow the identification of objects which have to be shown in any small scale map.

The high number of high and very high resolution space systems is improving the situation for mapping – in most cases images are available in the archive, avoiding a waiting time for acquisition. Only if the latest object information is required, new images have to be ordered. The variety of space images also has a positive influence to the financial side, the images just required for the actual project can be ordered, avoiding higher price for a higher ground resolution.

Aerial images are taken more and more by digital cameras. The digital mid format cameras are playing an important role for several projects. The gap between the mid format and the large format cameras is closing by combinations of 2 and 4 mid-format cameras.

SAR images have growing importance, supported by the very high resolution TerraSAR-X and Radarsat-2. Of course the information contents of SAR-images is not on the same level like the information contents of optical images, nevertheless today a mapping based on SAR images taken from space is possible. A big advantage of radar is the penetration through clouds, so images can be taken in any case at the planned time without respecting the weather conditions. Only very strong rain may degrade the SAR image quality. A further advantage is the use of IfSAR for the generation of height models. This can be done from space and air. The coming TanDEM-X constellation is announced with 2m vertical standard deviation, while the NextMAP, taken from air, is specified with 1m standard deviation. DEMs also can be generated without problems based on optical space models taken from the same orbit.

The actual overview over the optical space systems is mainly based on internet information. This is a little painful – any information has to be confirmed seriously because there is still a lot of not precise, partially wrong information about the systems in the WEB. Even several information shown by operating organizations, is not updated if some parameters have been changed in the meantime.

## REFERENCES

Büyüksalih, G., Jacobsen, K., 2008: Digital Height Models in Mountainous Regions based on Space Information, EARSel Workshop Remote Sensing – New Challenges of High Resolution, Bochum + http://www.ipi.uni-hannover.de (May 2008).

Jacobsen, K., 2007: Digital Height Models by Cartosat-1, ISPRS Hannover Workshop 2007, IntArchPhRS. Vol XXXVI Part I/W51 + http://www.ipi.uni-hannover.de (May 2008).

Jacobsen, K., Crespi, M., Fratarcangeli, F., Giannone, F., 2008: DEM generation with Cartosat-1stereo imagery EARSel, Workshop Remote Sensing – New Challenges of High Resolution, Bochum 2008 + http://www.ipi.uni-hannover.de (May 2008).

Passini, R., Jacobsen, K., 2007: Accuracy Analysis of SRTM Height Models, Annual conference of ASPRS, Tampa 2007 + http://www.ipi.uni-hannover.de (May 2008).

Pitz, W., 2006: TanDEM-X, Science Meeting no. 1, http://www.dlr.de/Portaldata/32/Resources/dokumente/tdmx/TDX_Science_Meeting_No_1.pdf (March 2008).

Sefercik, U., Jacobsen, K., Oruc, M., Marangoz, A., 2007: Comparison of SPOT, SRTM and ASTER DEMs, ISPRS Hannover Workshop 2007, IntArchPhRS. Vol XXXVI Part I/W51, http://www.ipi.uni-hannover.de (May 2008).

*Remote Sensing for a Changing Europe*
*D. Maktav (Ed.)*
*IOS Press, 2009*

*doi:10.3233/978-1-58603-986-8-16*

# Geometry of vertical and oblique image combinations

K. Jacobsen
*Leibniz University Hannover, Germany*

Keywords: Pictometry, Multivision, calibration, block adjustment, Virtual Earth

ABSTRACT: For city planning and security services the combination of vertical and oblique images like from Pictometry and Multivision recently became very popular. For larger cities a high number of such image combinations are even included in Microsoft Virtual Earth. The geo-reference of the image combinations usually is based on direct sensor orientation – the combination of GPS and inertial measurement units. The images are taken by camera systems like Track'Air MIDAS, equipped with 5 Canon EOS cameras, each with 4992 x 3328 pixels. The Canon EOS is not really a metric camera; it keeps the inner and the system orientation only stable during one photo flight. The images are also used for simple measurement purposes as well as the generation of 3D-city models. If object information without disturbing loss of accuracy against the direct sensor orientation is required, a system calibration during the day of the photo flight has to be done. The method of calibration and achieved accuracy is described as well as the characteristics and potential of the image combinations.

## 1 INTRODUCTION

The interpretation and understanding of vertical aerial or space images requires experience. For untrained persons the interpretation of terrestrial or oblique images is easy because they are closer to the standard experience in viewing to objects. In addition usually the facades are not or nearly not visible in vertical images. With the combination of vertical and oblique images totally new application for photogrammetry came.

Such images like from the Pictometry system using a camera system with one vertical and 4 oblique cameras are widely used by Microsoft Virtual Earth as "Bird's Eye" (Figure 1). In Western Europe the Blom Group is imaging all cities with a population larger than 50000, that means approximately 900 cities with together approximately 100 000 km². 12 Pictometry camera systems are in use for this. The competitor MultiVision is using a similar system.

The commercial application of the combination of vertical with oblique images like generated by Pictometry and MultiVision systems is dominated by visual inspections and simple measurement of distances for public safety and planning purposes. The use together with geoinformation systems requires the knowledge of the exterior orientation of every image together with a digital elevation model (DEM). Block adjustments are too time-consuming and complex, so the exterior orientations are based on direct sensor orientation – the use of the combination of inertial measurement units (IMU) together with relative cinematic GPS-positioning. In several countries networks of permanent GPS reference stations are available, but also with satellite based reference systems like OMNISTAR and different off-line reference systems, like JPL Final, the GPS reference is available in sub-meter accuracy. Especially for planning purposes not only the visualization is important, also some metric information is used. This requires a calibration of the imaging system and the relation to the IMU and the GPS-antenna.

Figure 1: Main building of Leibniz University Hannover in Microsoft Virtual Earth, based on Pictometry system, taken by Blom Group

Oblique view 13cm GSD

The use of oblique aerial images or the combination of vertical and oblique images is not a new invention. Prior to 1938 in the USA, UK, Germany, France, Italy and Switzerland single and multiple lens cameras have been produced for oblique or combined configuration (Manual of Photogrammetry, 2nd edition) and Moffit 1967 (figure 2). Scheimflug invented an eight-lens camera in 1900, viewing oblique into 8 directions, Aschenbrenner developed in 1920 a 9-lens camera with 8 oblique and a nadir view, similar to the USGS 9 lens camera (figure 2).

In military reconnaissance oblique images are in use since long time. They combine the advantage of a view to facades and other vertical objects, together with imaging from distance. The orientation and calibration of such systems, partially with extreme long focal length, is known since longer time (Jacobsen 1988). The combination of a nadir, 2 low and 2 high oblique images taken by the Z/I Imaging KS-153 can be seen in figure 3. Today oblique images even can be taken by the very high resolution satellites like WorldView-1 (fig. 4).

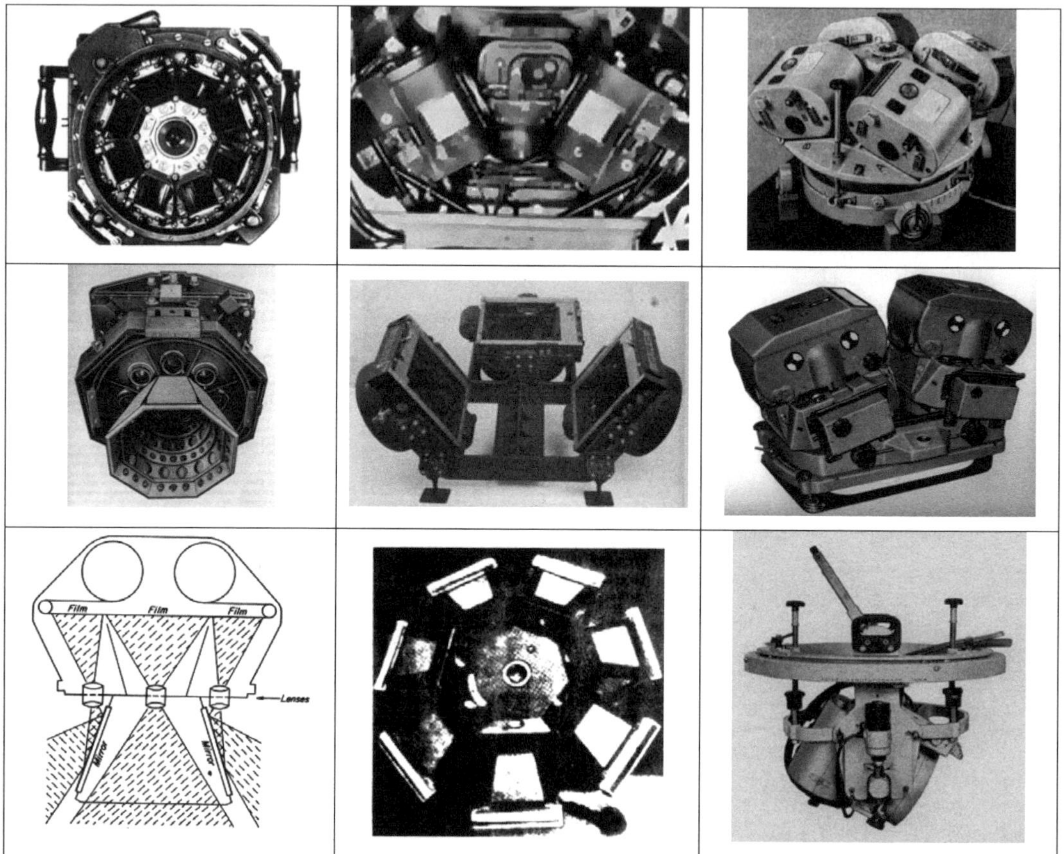


| | | |
|---|---|---|
| above: Aschenbrenners 9-lens camera 1920<br>centre and below: USGS 9-lens camera | 3 above figures: Trimetrogon<br>below: Scheimpflug's 8-lens camera (1900) | Zeiss oblique camera arrangements |

Figure 2: historic multiple lens and multiple camera arrangements for oblique aerial images

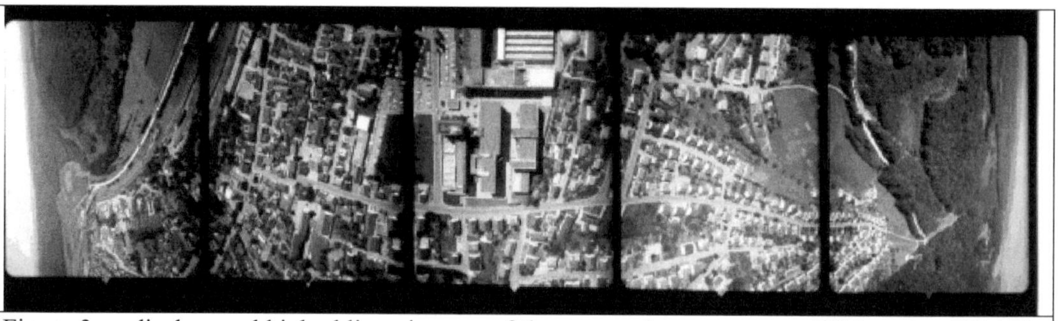

Figure 3: nadir, low and high oblique images of the reconnaissance camera Z/I Imaging KS-153, viewing from horizon to horizon

Figure 4: Dolmabahce palace, Istanbul in WorldView-1 image with 29.3° incidence angle

Today for the combination of nadir and oblique images dominating digital medium format cameras are used for civil application, allowing smaller systems which can fit to any standard aerial camera cone. Different camera combinations are in use. A typical camera system for such an application is the Track'Air (The Netherlands) MIDAS camera system. It has a combination of 5 of the shelf Canon EOS small format cameras with a focal length of 23.8mm for the nadir view and 51mm for the 4 oblique views (figure 5). The vertical camera has a field of view of 71.92° x 51.69° and the oblique cameras 38.8° x 26.4°, viewing approximately with 45° nadir angle, covering a nadir angle from 32° up to 58° (footprint arrangement – see figures 7 and 8). All 5 cameras have a CCD-array of 4992 x 3328 pixels with 7.2μm pixel size. The Canon EOS-cameras of the MIDAS system have a limited geometric stability, requiring more often a check of the calibration.

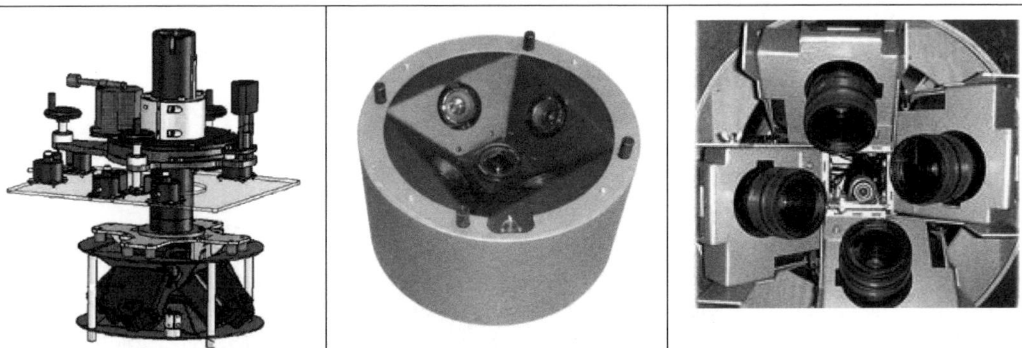

Figure 5: Track'Air MIDAS camera system – combination of 5 Canon EOS cameras

## 2 BUNDLE ORIENTATION OF VERTICAL AND OBLIQUE IMAGES

A data set with a combination of standard aerial wide angle photos (153mm focal length) and digital middle format images has been oriented by bundle block adjustment. The 3 different digital cameras have the same CCD-array like the Canon EOS, described above and focal length of 84.3mm, 169.5mm and 169.8mm. The oblique images have nadir angles in the range of 50°.

The image scale for the vertical images is approximately 1:4600 (9cm ground sampling distance (GSD) based on 20μm pixel size), for the centre of the oblique images 1:9400 (7cm GSD), 1:9900 (7cm GSD) and 1:26000 (19cm GSD). Because of the unusual arrangement and the different type of images no automatic block adjustment was possible with standard commercial software for automatic image matching, so the tie points had to be measured manually. Only 13 ground control points and no direct sensor orientation has been used, but for such a block with an extreme number of ties, with up to 16 images per ground point, this is quite enough. The bundle block adjustment did not cause any problem, only the number of blunders was higher than usual; but this was expected because of the quite different image scale, different view directions and the combination of digitized analogue photos with digital images. The sigma0 value of 20µm was satisfying for the required purpose. At the ground control points for the horizontal component root mean square errors of 2cm and for the height 7cm have been reached. The orientation by traditional bundle adjustment is possible, but the manual measurement of tie points is time-consuming. That means for such an image configuration the direct sensor orientation (use of GPS + inertial measurement system (IMU)) is a must. Only for the calibration a manual measurement of the tie and control points of such a block is justified, but after the first initial calibration, the preceding calibration can be used for support of the measurement.

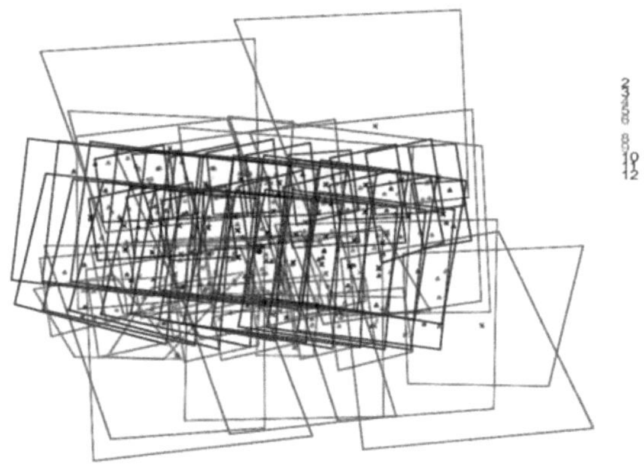

Figure 6: footprints of combination of vertical and oblique images

Colour of footprint = camera

## 3 CALIBRATION OF TRACK'AIR MIDAS CAMERA SYSTEM

A larger area has been flown with the Track'Air MIDAS camera system, organized by MultiVision. The camera with the nadir view has a focal length of 23.8mm and the oblique cameras approximately 51mm. This corresponds for the vertical view to 17cm GSD and for the oblique images to 10cm x 11cm up to 15cm x 29cm. For oblique images the GSD in the view direction is the GSD across the view direction divided by the cosine of the nadir angle, so it is not a square size.

3 flight lines, each with 4 vertical images, and the corresponding oblique views have been used for the bundle block adjustment. This should lead to a configuration like shown in figure 8c, but some of the oblique images are not well connected with the block, so they are not supporting the calibration significantly. By this reason some of these images have not been used (figure 7a). The centre flight line has been flown from north to south, while the other both have been flown in the opposite direction, improving the calibration configuration.

The data acquisition was made with LPS. LPS had no problems to match the vertical images (figure 8a) and overlapping images in the image space, but for the oblique images the matching is limited to the connection of 2 neighboured images. An automatic image matching of the whole test block, even if it was supported by initial image orientation of the first bundle block adjustment, failed. LPS has not been developed for automatic aero triangulation of such a configuration, starting with the problem of different focal length. So a combination between manual pointing and matching of 2 images was used. The automatic matching of overlapping different image combinations often leads to the extraction of corresponding object points. The used Hannover program system BLUH for bundle block adjustment is able to identify such points based on their similar object coordinates und can rename corresponding points to the same point names, leading to a better block tie. Some extension for program system BLUH was necessary, starting with the automatic exclusion of object points from the adjustment located only in images taken from the same projection centre. Such points do not allow the computation of object coordinates. As control information an orthoimage with 1m GSD and a DEM was used. For the vertical images this was leading to a sigma0 of 11µm and root mean square discrepancies at the control points of 16cm for X and Y and 1.6m for Z (table 1).

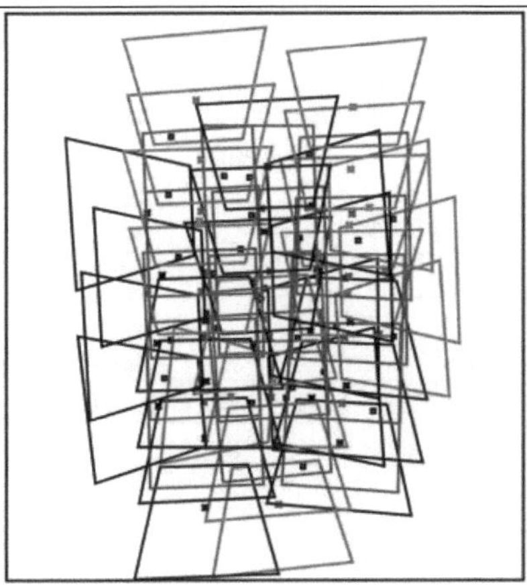

Figure 7a: footprints of calibration block, 3 flight lines, colour = camera

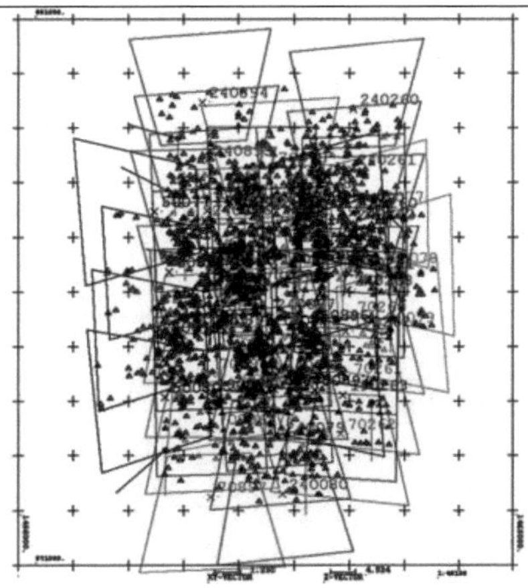

Figure 7b: footprints together with tie points

By the adjustment of all images of the test-block (figure 7) the system geometry and the camera geometry have been determined. The dominating systematic image errors are the radial symmetric components with values up to 100µm. This is usual for the used optics. The radial symmetric distortions, as well as the overall systematic image errors, are similar for the oblique sub-cameras, using the same type of lens system. For the vertical sub-camera equipped with a different lens system it varies from the other. The systematic image errors, or in other words – the difference between the mathematical model of perspective geometry and the real image geometry, can be respected in the application software system for handling the vertical and oblique images, or it can be used for resampling the images to strict perspective geometry.

The sigma0 (accuracy of image coordinates) of the vertical images of 11µm corresponds to 1.4 pixels. With the average image scale of 1:11 000 it corresponds to 12cm on the ground, not far away from the reached root mean square discrepancies at the horizontal components of the control points. The larger discrepancies at the control point height only can be explained by the limited accuracy of the DEM used for the determination of the control point heights. In an adjustment of all images together the sigma0 is with 33µm larger, caused by the different view direction, but partially also by the limited accuracy of the control point heights (table 1).

Supported by relative cinematic GPS-positions of the projection centres, also the inner orientation has been improved. Of course there is a strong correlation between a shift of the GPS-positions, especially in Z-direction, and the inner orientation, but the block configuration together with fixing the corresponding projection centres together, supports the determination of the inner orientation.

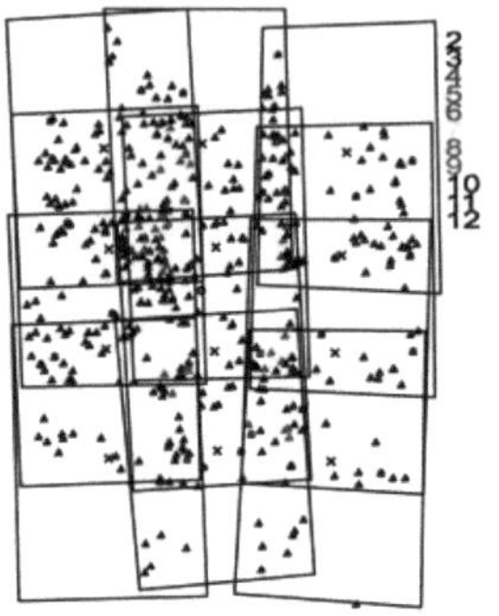

Figure 8a:only vertical images with tie points, colour coded corresponding to number of images/point

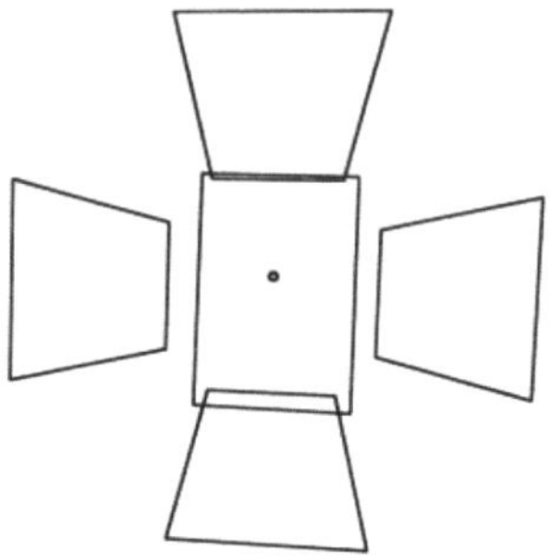

Figure 8b: footprints of one image combination

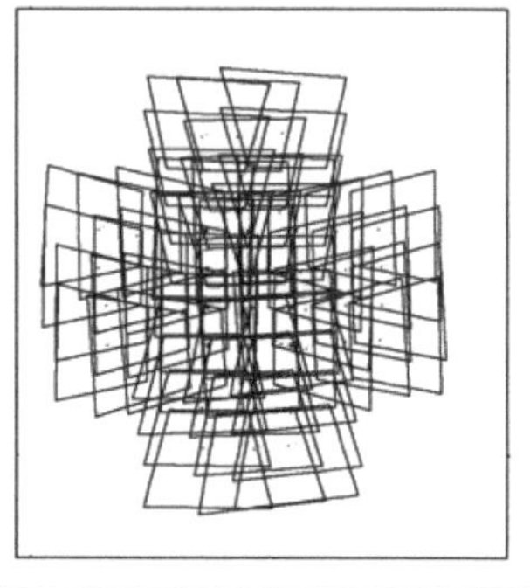

Figure 8c: maximal number of images for test block configuration

| | sigma0 | RMSX / RMSY control points | RMSZ control points |
|---|---|---|---|
| only 12 vertical images | 11 μm | 15 cm | 160 cm |
| all images | 33 μm | 38 cm | 122 cm |

Table 1: accuracy of block adjustment of control-block 32 control points

Based on the improved inner orientation by combined block adjustment with GPS-coordinates of the projection centres, joining also the corresponding projection centres of the vertical and oblique images together, the relation of the oblique cameras to the vertical camera have to be determined (table 2). The image orientations taken from each projection centre have to be rotated by multiplying the rotation matrixes by the inverse rotation matrix of the vertical image. This leads to the rotation values for the nadir image of 0° for all 3 rotations. The averaged relative orientations of the oblique images in relation to the nadir view are identical to the internal system calibration.

Based on the orientation of the combined adjustment, the boresite calibration can be computed by comparing the orientation from the controlled bundle block adjustment with the inertial orientations. The boresite values are related to the roll, pitch and yaw-system, the required transformations are made within the Hannover program GPSCOR. The so computed boresite calibration values can be used in the same run or a separate run of GPSCOR for the correction of the inertial data. The corrected inertial data are corresponding to the orientation of the vertical camera. With the values of the internal system calibration (rotations of the oblique cameras in relation to the vertical camera – see table 2), with program ROTOR the exterior orientation of all images of a project can be computed. Of course the so computed orientations, within the calibration sub-block, should be close to the orientations from the controlled bundle clock adjustment.

The kappa-values of the orientation of the sub-cameras in relation to the vertical reference camera (table 2) show, that the sub-cameras are always oriented into the oblique view direction. The oblique angles vary from 49.1 grads up to 51.3 grads (44.2° to 46.2°). With orientation of the sub-cameras in relation to the nadir images, the orientations have been computed by ROTOR. With these orientations, object coordinates have been computed by combined intersection, resulting at the control points to RMSX=0.62m, RMSY=0.60m and RMSZ=1.63m, satisfying the expectation.

| | phi [grads] | omega [grads] | kappa [grads] |
|---|---|---|---|
| sub-camera 5 | -.0261 | 49.9609 | -.3585 |
| sub-camera 7 | -49.1002 | -.5065 | -99.8977 |
| sub-camera 8 | 1.7375 | -51.1537 | 199.2017 |
| sub-camera 19 | 51.3505 | .4023 | 99.6800 |

Table 2: orientation of sub-cameras against nadir reference camera

## 4 CONCLUSON

The combined use of vertical and oblique cameras like in Track'Air MIDAS camera system requires an internal system calibration. The orientation of an image block, taken with such a combination leads to a strong overlap of images with up to 12 images per object point. Standard commercial programs are not able to handle such a block by automatic image matching and a manual measurement is very time consuming. So a direct sensor orientation with a combination of relative cinematic GPS-positioning together with an inertial measurement system is required. The direct sensor orientation together with the system calibration leads to the orientation of all sub-cameras.

With a sub-block of 3 flight lines and 4 vertical images in every flight line, with good connected oblique images, a complete system-calibration is possible. The used Canon EOS cameras have systematic image errors, dominated by the radial symmetric distortion, in the range up to 100µm or 14 pixels. The influence of the systematic image errors can be respected in a geo-coded interpretation and measurement system like from Pictometry or MultiVision by the application software or by generating perfect perspective images based on a resampling of the images using the systematic image errors.

The reached absolute accuracy within the calibration block of approximately 0.6m in X and Y is sufficient for the purposes of MultiVision applications. The relative accuracy is better than this. Of course it is depending upon the direct sensor orientation and the stability of the sub-cameras and the camera system including the misalignment. The camera geometry under usual conditions is stable within the block and the direct sensor orientation is dominated by the used hardware components. Images taken by MultiVision or Pictometry usually are only used as single images, allowing a geo-coding only by means of digital elevation models. These DEMs are an important limitation – because of the inclined views, the height errors of the DEMs are causing dislocations approximately in the same size like the height error.

## REFERENCES

Höhle, J., 2008: Photogrammetric Measurements in Oblique Aerial Images, Photogrammetrie Fernerkundung Geoinformation, 2008, issue 1, pp 7 - 14

Jacobsen, K., 1988: Handling of Panoramic and Extreme High Oblique Photographs in Analytical Plotters, ISPRS Kyoto 1988 IntArchPhRS. issue XXVII, B2, pp 232 - 237

Manual of Photogrammetry, 2nd edition, chapter II, part1, Aerial cameras and accessories, ASPRS, 1952

Moffit, F.H., 1967: Photogrammetry 2nd edition, chapter 2 Aerial Cameras International Textbook Company

*Remote Sensing for a Changing Europe*
*D. Maktav (Ed.)*
*IOS Press, 2009*

*doi:10.3233/978-1-58603-986-8-24*

# Spatial information sciences for transportation decisions

Dogukan Toraman
*Istanbul Technical University, Institute of Science and Technology, Department of Geomatics, Istanbul, Turkey; dtoraman@gmail.com*

Hande Demirel
*Istanbul Technical University, Department of Geodesy and Photogrammetry, Division of Photogrammetry, Istanbul, Turkey; hdemirel@ins.itu.edu.tr*

Nebiye Musaoglu
*Istanbul Technical University, Department of Geodesy and Photogrammetry, Division of Remote Sensing, Istanbul, Turkey; nmusaoglu@ins.itu.edu.tr*

Keywords: Transportation, Land Use, Environmental Impacts, Multi Criteria Decision Making

ABSTRACT: This paper presents a comprehensive framework for determining the interaction between transport, land-use and environmental impacts, where the developed concepts were tested using a case study. As the study area, two bridges connecting Europe and Asia and the planned third bridge was selected, where the constructed transportation infrastructures had drastically changed the land-use profile and still have negative impacts on environment. After exploring the interaction systematically and identifying the land cover/use pattern specific for the Istanbul Metropolitan Area, possible alternatives routes for the new bridge were by means of multi-criteria decision making approach and potential prediction of impacts were performed. The former land-use, transport infrastructure data were integrated with Landsat TM satellite images retrieved in 1987, 1997 and 2005. In order to identify the pattern of interaction, the land use analyses of the Asian side of the Istanbul Metropolitan area were compared within a selected box buffering the current and planned highways. For the planned third bridge alternative scenarios were selected, where alternatives were environmental friendly, cost-efficient and compatible with the current infrastructure. The results show that, easy accessibility caused by the development in transportation infrastructures created an attraction in this region and urban areas expanded rapidly. By means of the shortly described methodology and achieved results, the proposed framework aids authorities and decision-makers to better facilitate sustainable transportation.

## 1 INTRODUCTION

The interaction between transport, land-use and their impacts to the environment is prioritized at the research agenda, since the necessity of integrated policies for sustainable development is highly emerged (Rajaram *et al.* 2001; Rodier *et al.*, 2002; Shaw and Xin, 2003; Parsons and Salter, 2003; Geurs and Wee, 2004). Transport system creates new accessibility levels that encourage changes in land use patterns. Changes in land use system, on the other hand, can modify the travel demand patterns and induce changes in the transportation system (Wegener, 2001). The adverse affects of transportation on environment can be listed as air pollutant emissions, greenhouse gas emissions, noise, and diverse affects on habitat, ecosystems, endangered species. Additionally, it affects water quality through consumption and causes fragmentation and replacement of natural cover with impervious surfaces. In parallel, the urban form affects travel behavior, resulting air pollution, global climate change and noise. Residential and commercial development was indirectly effected including the distribution of employment opportunities. The transportation options available to link residential and commercial locations influence household travel behavior, including trip frequency, trip lengths, and mode of choice. Vehicle travel, in turn, generates air pollutant emissions, greenhouse gas emissions and noise (EPA, 2001). The complexity and variety of different components making up the urban environment, and the interactions among them, are the most pronounced in the mega cities. In this context, it is important to develop integrated frameworks and systems in order to detect changes and monitor dynamics, where looking at one component of change in isolation could lead to decisions being taken on the basis of insufficient information. The integration will certainly be facilitated be new technologies, tools, and expertise as land-use/cover changes are quite difficult to grasp as they occur incrementally (Kaya and Curran, 2006; Musaoglu *et al.*, 2006). Since the interaction between transport, land-use and environment has spatial and temporal characteristic, Spatial Information Sciences, involving Photogrammetry, Remote Sensing and Geographic Information Systems (GIS), provide mature solutions for integrated policies.

In order to emphasize the efficiency of Spatial Information Sciences for integrated policies, a spatial framework was designed and spatial and temporal analyses were conducted in Istanbul. The Istanbul Metropolitan Area is the only mega-city in the world that sits astride two continents – Europe and Asia, where the transportation system has been unable to keep pace with the rapid growth and changing urban structure. Two bridges connect Europe and Asia, the Bosphorus Bridge (1973, part of the inner-city traffic ring) and the Fatih Sultan Mehmet Bridge (1988, part of the outer-city traffic ring) span the Bosphorus. The city proper covers an area of 2000 km$^2$, while the metropolitan area occupies 5700 km$^2$. The research is conducted at the Asian side of the Istanbul Metropolitan Area. The study area was illustrated in Figure 1.

*Figure 1. The study area – The Asian side of the Istanbul Metropolitan Area.*

The Istanbul Metropolitan area had a population about 4.7 Million according to 1980 census data which has more than doubled and reached about 10 millions in only two decades. According to the estimations performed by State Statistical Institute, population of Istanbul may reach about 20 million in 2032 (DIE, 2008). The growth rate in Istanbul has had a significant negative effect on the remaining green area. Rapid motorization is generating serious congestion and air pollution in Istanbul due to the high population density and the lack of supporting infrastructure. According to the State Statistical Institute reports, in 2006, every fifth citizen of Istanbul owns a vehicle. Rapid motorization is generating serious congestion and air pollution in Istanbul due to the high population density and the lack of supporting infrastructure. Some negative side effects were observed, such as valuable agricultural lands and forest were cut through new routes and dense urbanization along the transportation network was settled (Demirel, 2006). The proposal for a third highway bridge by the Ministry of Public Works has been heavily criticized by transportation and city planners and non-governmental organizations. This strong opposition have resulted evaluation of other alternatives, such as railway tunnel, however the proposal of constructing a third highway bridge is still on the agenda of policy-makers.

The main objectives of the study are to discuss and evaluate the land-use changes which may be attributed to transport and monitoring the affects on environment. In order to achieve these goals, the objective of this study is to develop a comprehensive spatial framework for detecting the relationship, where the main focus will be (i) to characterize the major spatial-temporal changes of land use, which may be attributed to transportation activities, (ii) to identify the land cover/use pattern specific for the Istanbul Metropolitan Area, (iii) to explore possible alternatives routes for the new bridge by means of multi-criteria decision making approach, (iv) possible impacts of the planned highway, (v)impacts of transport and land-use on environment at a local scale.

## 2 DATA AND METHODOLOGY

The interaction between transport, land-use and environment is a dynamic process that involves changes over spatial and temporal dimensions between the three systems. Transportation, land use and environment interaction involves the time element (when), the location element (where) and the attribute element (what) that are interrelated with each other. The critical dimensions of the designed framework were the consistent definition of spatial information and time. Effects of transportation system changes on other systems and vice versa, occur at varying spatial and temporal scales. A highway construction project may introduce a traffic increase shortly after its completion and lead to some land-use changes in its immediate vicinity. However, its impacts on the land-use patterns at the regional scale may be insignificant in both short and long-terms, due to the relatively small changes to the accessibility levels on other parts of the metropolitan area.

In order to model this spatio-temporal interaction Spatial Information Technologies were used including digital image processing of satellite images and spatial data modeling, spatial analysis, information presentation and dissemination on both spatial and time domain. The methodology of this work is constituted upon Spatial Information Sciences, where photogrammetry, remote sensing and GIS forms the components. The great strength of remote sensing is that it can provide spatially consistent data sets that cover large areas with both high detail and high temporal frequency, including historical time series and remote sensing data provide a means of monitoring change in urban land cover over time (Kaya and Curran, 2008). GIS are composition of tools and methods for collecting, storing, retrieving at will, transforming, and displaying spatial data for a particular set of purposes. Given the spatial nature of many environmental impacts, GIS can have a wide application in all Environmental Impact Assessment (EIA) stages, acting as an integrative framework for the entire process, from the generation, storage, and display of the thematic information relative to the vulnerability/sensitivity of the affected resources, to impact prediction and finally their evaluation for decision (Antunes *et al.*, 2001). The designed framework for exploring the interaction is presented in Figure 2. The contribution of the spatial information sciences to the decision making process was marked in grey. The frame-work involves data acquisition, analyses, assessment, monitoring and information to the public, where the integrated policies for transport, land-use and environment were emphasized. This will ensure the integration of currently loosely coupled information and increase the efficiency of the decision making process.

In this study various spatial information were retrieved from satellite images, aerial photographs and orthophoto maps. Landsat images acquired in 1987, 1997 and 2005, were used to monitor the development of transportation infrastructures and their impacts to the land-cover/use. 1:25000 scaled topographic maps were used for rectification, where the achieved accuracy was 0.5 pixel root mean square (RMS). The first degree polynomial transformation and nearest neighborhood method were used. In order to detect the dynamics of the interaction, digital land cover classification was performed. The purpose was to establish the link between the spectral characteristics of the image to a meaningful information class value. In this study, the Iterative Self Organizing Data Analysis Technique (ISODATA) unsupervised classification algorithm was used (*Tou et al.,* 19974). From satellite images retrieved 1987 and 1997 for the Istanbul Metropolitan area the Trans

European Motorway (TEM) which was being constructed in 1987 was retrieved and from the centerline of the highway a band having a width of 4 km was created. The same process was performed for the year 1997 and 2005, in order to determine the temporal change. Since the aim of the study is to explore the interaction between transport, land-use and environment, five spectrally separable, land cover classes identified by ISODATA were; (i)settlements (involving impervious surfaces such as urban area and industrial areas), (ii) road, (iii)barren land, (iv) green area (involving forest areas, semi-natural vegetation etc.) and (v) water.

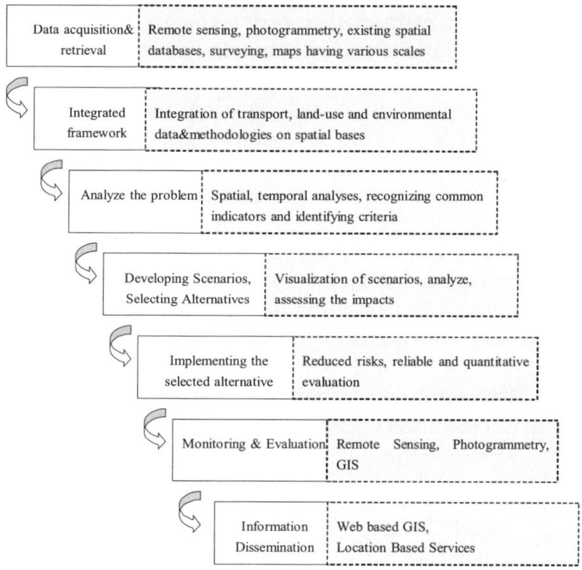


*Figure 2. The designed framework.*

The classification accuracy for the year 1987 image was 82%, for the year 1997 was 82%, where for the year 2005 the accuracy of 84% was achieved. In order to explore the interaction several data sources are being required, where within the scope of this study density of the population, district borders of the Istanbul Metropolitan Area, highways and Digital Terrain Model (DTM) were being used. Density of the population, which is attributed to district borders, is a powerful indicator for exploring the urban dynamics. Additionally, DTM of the study area was used in order to select the alternatives of the newly planned third bridge, where multi-criteria decision making approach was followed. Multi-criteria analysis (MCA) is one of the most utilized techniques for performing evaluation studies dealing with the ranking of alternatives in the presence of conflicting objectives (Malczewski, 1999). The critical aspect of spatial multi-criteria analysis is that it involves evaluation of geographical events based on criterion values and the decision maker's preferences with respect to a set of evaluation criteria. In a vector-based GIS context, attributes of geographic features may serve as decision criteria while in a raster-based system, different raster datasets would represent decision criteria. GIS requires that the evaluation criteria have a spatial reference. In this study various alternatives for the third bridge were evaluated, where multi-criteria decision analyses approach is conducted. The weighted linear combination decision rule was applied, where it allows the user to specify a set of weights representing the relative importance of criteria according to the user's preferences. The weight of a criterion defines its impact on the compensatory aggregation. By default, criterion weights are set to 1/n to represent n equally important criteria. Mathematically, the score of alternative i is calculated as $s_i = \Sigma w_j x_{ij}$, where $w_j$ is the weight of criterion j, and $x_{ij}$ is the standardized attribute value of alternative i for criterion j.

## 3 RESULTS

The developed frame-work was used to monitor the changes experienced in 1987, 1997 and 2005 within the study area. For the study area, urban areas sharply increase between 1987 and 1997, likewise the increase in the total length of road network. Hence, development of transportation infrastructures created an attraction in this region and urban areas expanded rapidly. At the satellite image of 1987 the construction site of the Trans European Motorway, which connects the European and Asian side, can be detected. At the satellite image of 1997, it can be observed that, TEM was completed and new roads were constructed in order to access TEM. However, the complete network begins to enlarge with new branches, which helps people to access the main network easily. The development is mainly observed at the northern side. Impact of developing infrastructures to the environment was analyzed by means of classification results of the satellite images. As a result of easy accessibility caused by the development in transportation infrastructures, several industries have moved to these areas and resulting unplanned and uncontrolled expansion in urban area.

In order to detect the spatial and temporal patterns of the bridges, the second bridge is selected as an example, where similar results and the same trend was achieved for the first bridge, illustrated in Figure 3. The land cover/use results within the years 1987 and 1997 of the second bridge were compared. Within only 10 years the settlement class is increased 283% only at the selected band for the second bridge. The total area classified was 13600 ha, where in 1987, 8% of the area was determined as settlements. Between years 1987-1997, the settlement was 1075 ha in 1987 and an extra 3044 ha has been gained by year 1997, only within ten years period. In contrast, crop and bare soil areas decreased 29%, indicated as barren lands in Figure 3. The green areas were also decreased 26%. The classification results and the comparison between years 1987 and 1997 were presented in Figure 3.

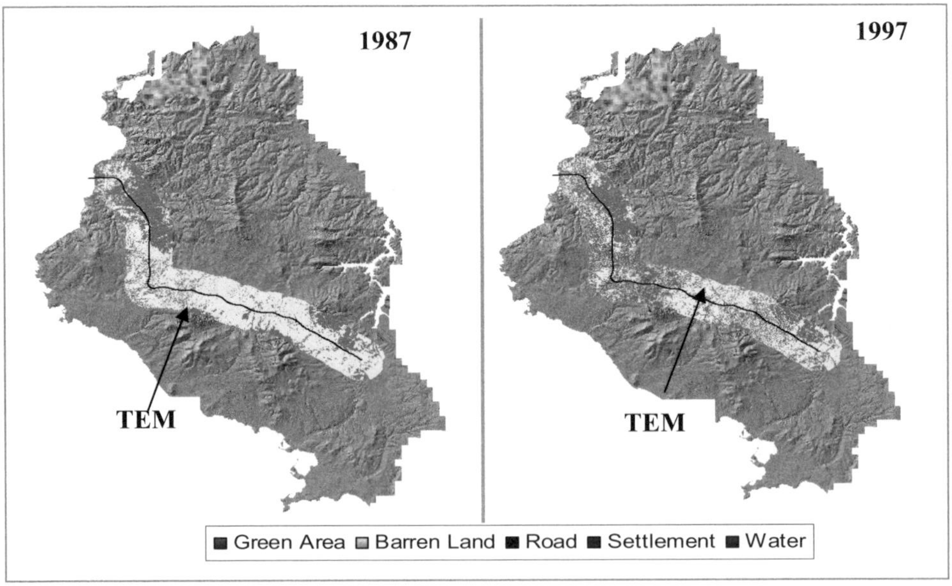


Figure 3. Classification results of the selected band within years 1987 and 1997.

According to the analyze results, there is a tendency of conversion of barren land to settlement and green area to barren land. The achieved classification results between years 1987-1997, support the recognized tendency clearly. The road class is 20 times within only ten years. The population density within the districts, where the TEM was passing through, was increased 211% within the years 1990 and 2000. The population density within the Asian side is increased 42% (DIE, 2001).

For the alternative design, criteria were selected being social, environmental, economic, topography and compatibility with the current transportation infrastructure. Four alternatives were being evaluated namely, environment friendly (alternative 1), cost-efficient (alternative 2), compatible with the current transportation infrastructures network (alternative 3) and an equal weighted option (alternative 4). There is no limit for the alternatives, where the important issue is the decision maker's preferences with respect to a set of evaluation criteria. All weights for alternatives are subjective decisions of authors. However, a detailed questionnaire, query and analyze is required for the optimum weights of an optimum alternative. The given weights for the alternatives are as follows; for alternative 1: social-0.10, environmental-0.60, economic-0.10, topography-0.10, network compentence-0.10; for alternative 2: social-0.10, environmental-0.10, economic-0.60, topography-0.10, network compentence-0.10; for alternative 3: social-0.10, environmental-0.10, economic-0.10, topography-0.10, network compentence-0.60; for alternative 4: social-0.20, environmental-0.20, economic-0.20, topography-0.20, network compentence-0.20. Total weights for the alternatives should be equal to 1. Each evaluation criteria should have a spatial reference. For each evaluation criterion, an appropriate spatial data was selected for the analysis. For the criteria "social"; land use, population density and urban plans in small scale (1:25.000) were being selected. For the criteria "environmental"; air pollution, geological maps, earthquake risk maps were used. For the criteria "economic"; land value, consumption of energy. For the criteria "topography"; slope and aspect, which were retrieved from DTM, where for the criteria "network competence"; accessibility is chosen. The used multi-criteria approach is performed by ESRI, ArcGIS 9.2, where model and extensions used are as follows: Spatial Analyst and Geo-processing. The obtained results were presented in Figure 4, where Alternative 1 is green, Alternative 2 is yellow, Alternative 3 is white and Alternative 4 is red, which is illustrated in Figure 4. For all alternatives, the initial point was selected as the junction connecting European Motorway (E-5) and TEM, which was on the TEM. The reason for selecting TEM is that, the E-5 motorway is highly used for intercity traffic and TEM is regulated as national and international freight transportation. The ending point of the alternatives is the possible foot of the third bridge, as announced in media. The Alternative 1 was passing through dense settlements, where house values on the basis of compulsory purchase rates will increase the cost of this alternative.

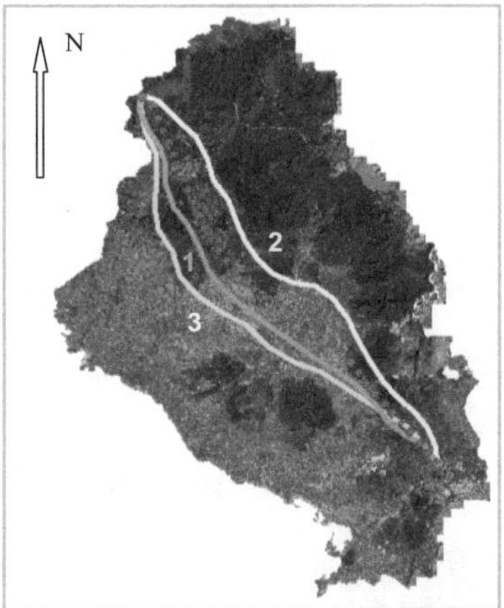

Figure 4. Alternatives for the third bridge.

The Alternative 2, was passing through a dense forestry area at the northern part of Asian-side. All alternatives were enforced not to pass through watershed areas, since adverse affects of transportation on water is known. Additionally, environmental impact assessments of the newly planned transportation infrastructures are regulated by the Ministry of Environment and Forestry, Turkey. However, system was not enforced not to pass through dense forestry areas since there is an ongoing debate on alternatives, where several alternatives were proposed passing through dense forestry area. The alternative 3, is the most compatible with the current transportation network route for the third bridge. The fourth alternative is illustrated in red in Figure 4. From the above mentioned alternatives the fourth alternative was being selected, since this alternative is "business-as-usual". All criteria have equal weights. The settlement, road, barren land, green area and water patterns and their trends were found and the second bridge spatial and temporal changes for the years 1987 and 1997 were illustrated in Figure 3. The same 10 year period was selected for the fourth alternative being 2005 and 2015 and the specific patterns and trends for the Istanbul Metropolitan Area were implemented in order to simulate the results of the given decision. The same 4 km band was being applied. For the years 1987 and 1997, the settlement class is increased 283%, where the population for the selected band was increased 211%. The land cover/use results within the years 2005 and 2015 of the third bridge alternative were compared and illustrated in Figure 5. The settlement was 3848 ha in 2005 and in ten years period it is foreseen that the settlement will be 14738 ha. The road class for the selected band was 205 ha in 2005, where in year 2015, the same class will be 4075 ha. The population density within the districts of the selected band was 273638 in 2005, where with 211% increase; it can be predicted as 851009.

These results illustrated that development of transportation infrastructures created an attraction in the region and urban areas expanded rapidly. With the selected case study area, further studies are going to be conducted. In order to determine spatial and temporal pattern and trend of the interaction other several remote sensing images are required, especially the ones belonging to former years of the Istanbul Metropolitan area, such as 60's and 70's. High resolution images can also be used for detailed analyses. In order to better facilitate decision makers, criteria applied should be determined upon the relevant documents, regulations, project results and discussions with the transportation decision makers. For determining the criteria and importance of them, questionnaires, workshops, discussions with the decision makers are required. There is no limit for the applicable alternatives of the designed framework. Alternatives are going to be determined with the decision makers and user friendly interfaces are going to designed for the implemented system. A three-dimensional city model along the transportation infrastructures is going to be integrated into the system, in order to determine the interaction between urban class retrieved from satellites and transportation. In order to support strategic planning, the data model will involve analyses for studying impacts and estimate further years interaction between transport, land-use and environment. Results will be simulated and visualized, in order to increase the awareness on the interaction and support sustainable policies.

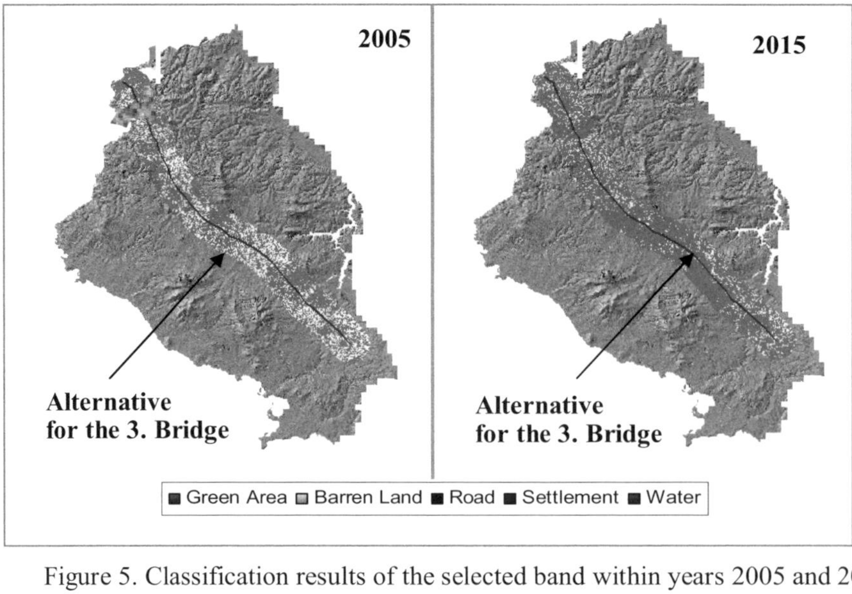


Figure 5. Classification results of the selected band within years 2005 and 2015.

## 4 CONCLUSIONS

Spatial information sciences served quite well in detecting the interaction between transport, land-use and environment. The spatial and temporal analyses indicated the rapid conversion of bare land to urban areas, where transportation activities increase. At the local scale the efforts of connecting Europe to Asia, which is inevitable, contributed to the drastic affects on urban development and environment. The distribution of accessibility in space influences location decisions and therefore results in changes of the land use and adversely affects environment. There is a tendency that urbanization sprawl will continue. With the designed framework several alternatives of a newly planned transportation infrastructure can be analyzed and future trends, depending upon former experiences can be simulated. The research results indicate how spatial information technologies can be utilized for detecting the dynamic changes in urban and environment, which may be attributed to transport.

## REFERENCES

Antunes, P., Santos, R., Jorda, L., 2001. *The application of Geographical Information Systems to determine environmental impact significance.* Environmental Impact Assessment Review, 21: 511 –535.

Demirel, H., 2006. *A Novel Approach for Determining the Impacts of Road Transportation on Air Quality.* Fresenius Environmental Bulletin (FEB). 15. 8b, 891-897.

DIE, State Statistical Institute, Turkey, retrieved on May 10, 2008 from http:// www.die.gov.tr

Geurs, K. T., Wee, B., 2004. *Accessibility evaluation of land-use and transport strategies: review and research directions.* Journal of Transport Geography, 12: 127–140.

Kaya, S., Curran, P.J., 2006. *Monitoring urban growth on the European side of the Istanbul metropolitan area: a case study.* International Journal of Applied Earth Observation and Geoinformation, 8: 18-25.

Malczewski, J., 1999.*GIS and Multicriteria Decision Analysis.* New York, John Wiley pp.392

Musaoglu, N., Gurel, M., Ulugtekin, N., Tanik, A., Seker, D.Z., 2006. *Use of Remotely Sensed Data for Analysis of Land-use Change in a Highly Urbanized District of Mega City, Istanbul.* Journal of Environmental Science and Health Part A, 41:2057-2069.

Parsons, B., Salter, L., F., 2003. *Air quality effects of traffic in a canyon-like street (Falmouth, U.K.).* Environmental Monitoring and Assessment, 82:63–73.

Rajaram, B., Nagarajan, R. and Khire, M.V., 2001. *Synergistic use of orbital satellite image and ground-based information in environmental monitoring of railway alignment construction- a case study.* International Journal of Remote Sensing, 22,16: 3187- 3201.

Rodier, C.J., Johnston, R.A., Abraham, J.E., 2002. *Heuristic policy analysis of regional land use, transit, and travel pricing scenarios using two urban models.* Transportation Research Part D. 7: 243–254.

Shaw S., Xin X., 2003. *Integrated land use and transportation interaction: a temporal GIS exploratory data analysis approach.* Journal of Transport Geography, 11:103-115.

Tou, J.T., Gonzalez, R.C., 1974. *Pattern Recognition Principles.* Addision – Wesley, Reading: Massachusetts, 1974 pp.

United States Environmental Protection Agency (EPA), 2001. *Our Built and Natural Environments, A technical review of the interaction between land-use, transportation and environmental quality.* EPA 231_R-01-002, retrieved on May 10, 2008 from http://www.smartgrowth.org/library/built.html.

Wegener, M., 2001. *New Spatial Planning GIS.* JAG, 3, 3: 224-237.

*Remote Sensing for a Changing Europe*
*D. Maktav (Ed.)*
*IOS Press, 2009*

*doi:10.3233/978-1-58603-986-8-32*

# Multisensor study of wind patterns and algal blooms in near-coastal gyres of the Mediterranean sea

Vittorio Barale
*Institute for Environment and Sustainability, Joint Research Centre, European Commission*
*TP 272, Via E. Fermi 2749, I-21027 Ispra (VA), Italy; vittorio.barale@jrc.it*

Martin Gade
*Zentrum für Meeres- und Klimaforschung, Institut für Meereskunde, Universität Hamburg*
*Bundesstraße 53, D-20146 Hamburg, Germany; martin.gade@zmaw.de*

Keywords: Mediterranean Sea, SeaWiFS, QuikScat, algal blooms, sea winds, gyres

ABSTRACT: QuikScat and SeaWiFS data (2000–2007) covering the Mediterranean Sea were used for a multisensor study of the coupling between wind patterns and algal blooms in the Gulf of Lion and the Rhodes-Ierapetra gyre systems. In these near-coastal hotspots, atmospheric forcing creates (albeit with different mechanisms) surface conditions that cause convective processes and consequent nutrient upwelling from deeper layers. As phytoplankton growth in the otherwise oligotrophic Mediterranean basin is always nutrient-limited, the blooming triggered by these processes reflects the prevailing wind patterns. Highly dynamic features recur systematically in the pigment field of both regions, in the same periods (January to May).

## INTRODUCTION

The combined use of sundry remote sensing tools, to assess complementary environmental data, provides novel opportunities for the assessment of coastal and marine processes. A comparison of QuikScat and of Sea-viewing Wide Field-of-view Sensor (SeaWiFS) data, for the period 2000–2007, was used to study the coupling between wind patterns and algal blooms in two regions of the Mediterranean Sea: the Ligurian-Provençal Sea, from 40 to 44°N, 2 to 8°E, and the Levantine Basin, from 32 to 38°N, 24 to 32°E (Fig. 1).

The Mediterranean Sea is predominantly oligotrophic and behaves like a sub-tropical basin, where the light level is never a limiting factor, so that its decrease in winter does not inhibit algal growth, but the nutrient level always is. Blooming occurs in the colder, windy and wet (winter) season, and is related to the biological enrichment of surface waters due to cooling, vertical mixing and continental runoff, but not in the warmer, calm and dry (summer) season, when the water column is strongly stratified and the nutrient supply, from coastal zones or deeper layers, is much reduced (i).

Oligotrophy increases from west to east, except for near-coastal hotspots, corresponding to point sources of runoff (often affected by a severe anthropogenic impact) or to areas where atmospheric forcing creates particular conditions that favour the blooming. This is the case of near-coastal features such as the Lions Gyre and the Ierapetra-Rhodes Gyre system (see Fig. 2). In both areas, bloom dynamics reflect enhanced air-sea interactions due to the prevailing wind patterns (i.e. the Mistral and the Etesian winds, respectively). In the north-western basin, the northerly wind increases the density of surface waters through intense evaporation and cooling, particularly in the

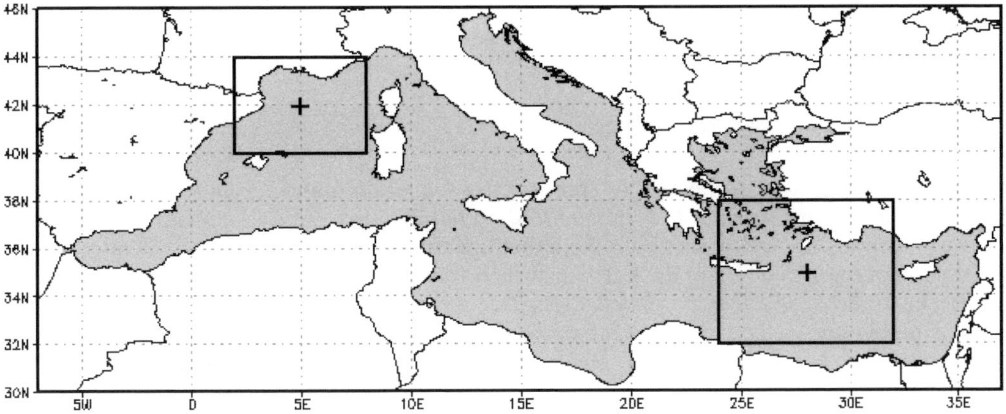


Figure 1. Mediterranean Sea, areas of interest: Ligurian-Provençal Sea (upper left box, centred at 42°N, 5°E) and Levantine Basin (lower right box, centred at 35°N, 28°E).

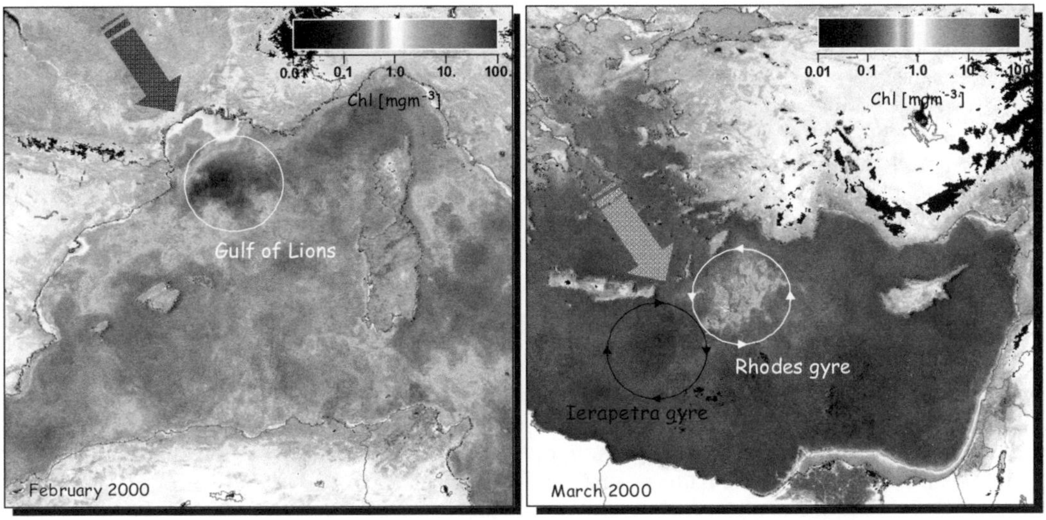


Figure 2. Monthly means of SeaWiFS-derived (iv) chlorophyll-like pigment concentration [mg/m$^3$], in the Ligurian-Provençal Sea (left panel), February 2000, and in the Levantine Basin (right panel), March 2000. Colour coding over land is based on a vegetation index, i.e. the Fraction of Absorbed Photosynthetically Active Radiation (FAPAR). Both images cover an area of about 1000 km × 1000 km, with 2 km spatial resolution. The circles indicate sites of deep convection, in the Gulf of Lion, and of upwelling/downwelling, in the Rhodes-Ierapetra gyre system. The large arrows recall the prevailing wind patterns.

Gulf of Lions, and generates deep convection processes (ii). This can lead to a complete overturning of the water column, thus ventilating the deepest parts of the basin and bringing deep nutrient-rich waters to the surface. When the water column stabilizes, this favours the onset of intense algal blooms. In the south-eastern basin, northerly winds blowing along the axis of the Aegean Sea funnel through the straits between the Islands of Crete and Rhodes, generating a (quasi) permanent vortex pair (ii). This comprises the cyclonic Rhodes Gyre, north-east of the straits, and the anticyclonic Ierapetra Gyre, south-west of the straits – often entrained in the complex basin-wide surface circulation of the eastern Mediterranean (iii). In the first case, the divergence due to Ekman pumping is linked to the upwelling of colder, nutrient-rich waters; in the second, the convergence is linked to the accretion and downwelling of warm, nutrient-poor surface waters.

## THE QuikScat AND SeaWiFS DATA SET

The wind speed data were collected by the microwave scatterometer SeaWinds, which was launched on the QuikBird satellite in June 1999. The QuikBird SeaWinds, later dubbed QuikScat, is the third in a series of NASA scatterometers that operate at 14 GHz (Ku-band). Scatterometers transmit microwave pulses down to the Earth's surface and then measure the power that is scattered back to the instrument, and that is related to surface roughness. For water surfaces, the surface roughness is highly correlated with the near-surface wind speed and direction (v, vi). Hence, wind speed and direction at a height of 10 meters over the ocean surface can be retrieved from measurements of the scatterometer's backscattered power. The 2000–2007 data used here, for comparison with the SeaWiFS data, are mapped to a 0.25 degree grid. Overlaid to the colour-coded maps of wind speed are black arrows denoting the mean surface wind direction, on a 0.5 degree grid. The wind speeds represent scalar averages of daily data, while the wind directions are vector averages of the same data[1].

The chlorophyll-like pigment concentration (*Chl*) data were collected by the visible and near infrared radiometer SeaWiFS, which was launched on the SeaStar satellite in September 1997. The original imagery was processed[2] to correct top-of-the-atmosphere radiances from atmospheric noise, to derive normalized water-leaving radiances, and then to compute from these a series of derived parameters (including *Chl*). Except for the pictures shown in Fig. 2 (see Figure caption for details), each daily image of the 2000–2007 dataset was treated using the SeaDAS algorithm set (vii) and re-mapped on a common equal-area grid, with a grid cell, or "bin", of 9 km × 9 km. For the creation of the composite data products, all valid pixels of a given time period and grid cell were compiled in the same bin, and their weighted mean was generated. The weight was based on the number of valid pixels used in the binning process (viii). Single images at full resolution (~1.2 km at nadir) display a larger amount of details, but are often incomplete over areas of the size considered here, due to imaging geometry, algorithm failure or cloud cover, even with daily overpasses of a wide-swath sensor. Hence the choice of utilizing composite images to assess the main *Chl* patterns from a statistical point of view. Because the number of valid pixels increases with larger and longer binning intervals, the compositing process generates complete, cloud-free images. As will be seen in the following, though, the short-term variability is averaged out too, and the composites retain only those patterns that persist over significant areas and periods of time.

## WIND PATTERNS AND ALGAL BLOOMS

The seasonal and multi-annual trends of the average wind speed and *Chl* values, computed in the Ligurian-Provençal Sea and the Levantine Basin domains[3] shown in Fig. 1, during the period considered (January 2000 to April 2007), are shown in Fig. 3. The wind speed record shows that maxima are reached systematically in the fall-winter period, i.e. between October and March. Conversely, the *Chl* record peaks in January-February, for the Levantine Basin, and even later, in March-April, for the Ligurian-Provençal Sea.

---

[1] QuikScat data are produced by Remote Sensing Systems and sponsored by the NASA Ocean Vector Winds Science Team. Data available at: http://www.remss.com/.

[2] Source: National Aeronautics and Space Administration (NASA); SeaWiFS Ocean Reprocessing 5.1 (2005); for details see: http://oceancolor.gsfc.nasa.gov/REPROCESSING/SeaWiFS/R5.1/.

[3] For the Levantine Basin, the two areas, for which the average wind speed and *Chl* values were calculated, are slightly displaced, in order to compensate for the displacement of the driving wind field and the Rhodes Gyre.

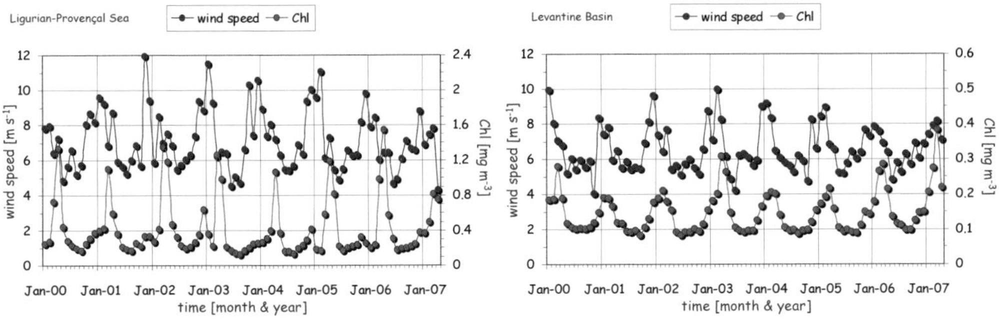


Figure 3. Monthly mean wind speed and *Chl* in the Ligurian-Provençal Sea (left panel) and in the Levantine Basin (right panel). Note the different *Chl* scale, on the right vertical axis.

Examples of the surface wind fields in the two areas considered are shown in panels (a) and (c) of Fig. 4. The corresponding surface pigment fields are shown in panels (b) and (d) of the same Figure. In both cases, the image series comprises monthly averages from January through May, for the year 2003. This period was chosen to capture the development of the classical "blue hole" and the ensuing spring bloom in the Gulf of Lion area, and the somewhat more episodical development of an analogous, but much smaller, spring bloom in the Rhodes Gyre (see trends in Fig. 3). The surface features recurring in these near-coastal hotspots represent the main blooming events of the Mediterranean Sea, respectively in the western basin and in the eastern basin, that are not driven directly by continental (e.g. fluvial) runoff[4].

In the Ligurian-Provençal Sea, the highest wind speeds (around 10 m $s^{-1}$, or more) occur in January and February. This corresponds to low pigment concentrations (*Chl* < 0.5 mg $m^{-3}$), and often to the formation of a distinct "blue hole" (*Chl* < 0.2 mg $m^{-3}$), in the Gulf of Lion – although, at the monthly scale considered, the link is not always evident and unambiguous, in some of the years considered (not shown here). Lower wind speeds (between 5 and 10 m $s^{-1}$, or less), from March onwards, correspond to strong blooming (*Chl* > 1.0 mg $m^{-3}$).

A simple regression analysis, carried out using the 2000–2007 monthly mean values of the Ligurian-Provençal Sea (as plotted in Fig. 3), is shown in Fig. 5. Various time lags, from 0 to 6 months, were considered, in order to take into consideration the effect of two environmental factors: first, the biological response of the ecosystem cannot be instantaneous, after the set up of conditions favourable to algal growth (i.e. the increase in nutrient concentration due to vertical mixing); second, when the wind is strongest, the continuing deep convection can prevent blooming and lead instead to the formation of a "blue hole" in the pigment field. With no time lag (i.e. when the *Chl* values are matched with the wind speed of the same month; see Fig. 5a), the linear fit shows essentially no correlation ($R^2 \sim 0$), while the second order polynomial fit ($R^2 = 0.05$) suggest that *Chl* is lower at lower wind speeds, and that it grows with wind speed, but only up to a point; then, as wind speed becomes higher and higher, the *Chl* values drop again (i.e. corresponding to the occurrence of deep convection). As the time lag increases, so does the correlation. The maximum correlation is reached after 4 months (i.e. when the *Chl* values are matched with the wind speed of 4 months before), with $R^2 \sim 0.3$ for both the linear and the polynomial fit (see Fig. 5b). At higher time lags, the correlation decreases again to very low values ($R^2 \sim 0.01, 0.02$).

The lack of a significant correlation, when wind speed and *Chl* are matched with no time lag, is due essentially to the fact that the ecosystem of the Ligurian-Provençal Sea responds in a very distinct manner to the extreme atmospheric forcing typical of winter months. In fact, if the regression analysis of monthly mean values is restricted to the first 4 months of the year, the correlation is more significant ($R^2 \sim 0.28$, both in the linear and in the polynomial case), but

[4] Note that the Rhone plume, in the Gulf of Lion area, presents separate dynamics.

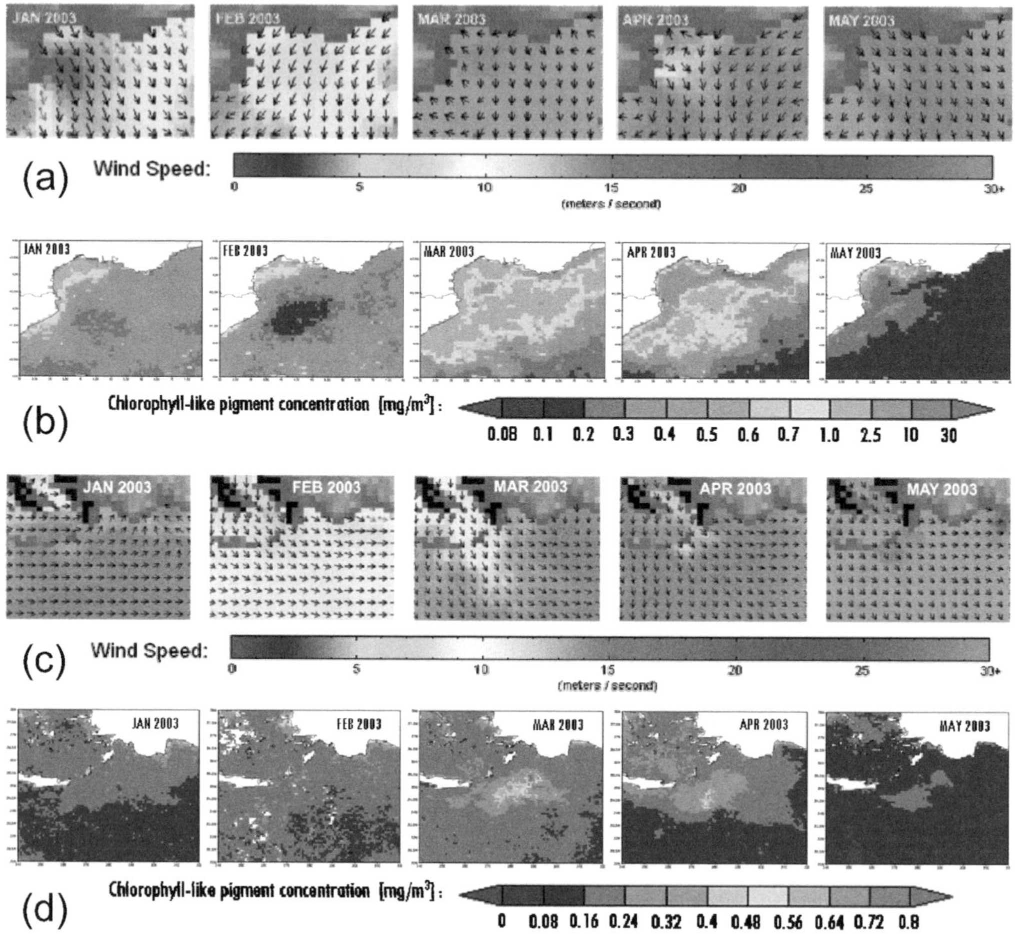


Figure 4. Monthly means, January to May 2003, of wind speed (panels a and c) and of *Chl* (panels b and d), in the Ligurian-Provençal Sea (panel a and b), February 2000, and in the Levantine Basin (panels c and d). The images cover the areas shown in Fig. 1.

negative, as shown in Fig. 6a. In this period, then, high *Chl* values occur only at lower wind speed. When the wind speed increases, and triggers the deep convection, the algae are mixed down to great depth in the water column, and cannot regain the surface to start over their photosynthetic activity. Conversely, for the remainder of the year, i.e. from May to December, the correlation is much lower ($R^2 \sim 0.07$, both in the linear and in the polynomial case), but positive, as shown in Fig. 6b.

Somewhat weaker winds are measured in the Levantine Basin. Although the pattern is not as clear as in the Ligurian-Provençal Sea, wind speed is still higher in winter months (but well below the 10 m $s^{-1}$ mark, in general), only to decrease slightly in spring months. Blooming events (*Chl* > 0.5 mg $m^{-3}$) in the Rhodes Gyre trail the periods of high wind speed. However, blooms just as intense as, if not more intense than, that of spring 2003 (see Fig. 4), are seen to occur also in other years, following winter months when the wind field did not present the same high values (i.e. with wind speed ranging between 5 and 10 m $s^{-1}$).

The regression analysis, carried out using the 2000–2007 monthly mean values of the Levantine Basin, is shown in Fig. 7. Once again, various time lags were considered, this time mainly to take

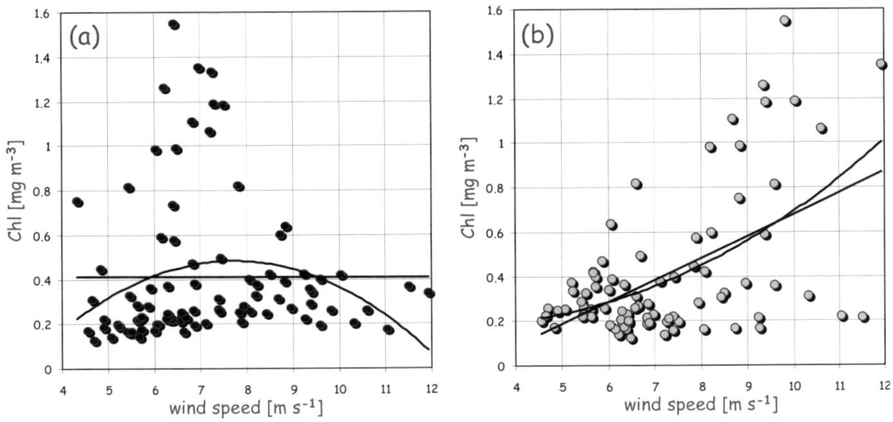


Figure 5. Regression analysis of monthly mean wind speed and *Chl* in the Ligurian-Provençal Sea, at two time lags, respectively of 0 (panel a) and 4 months (panel b).

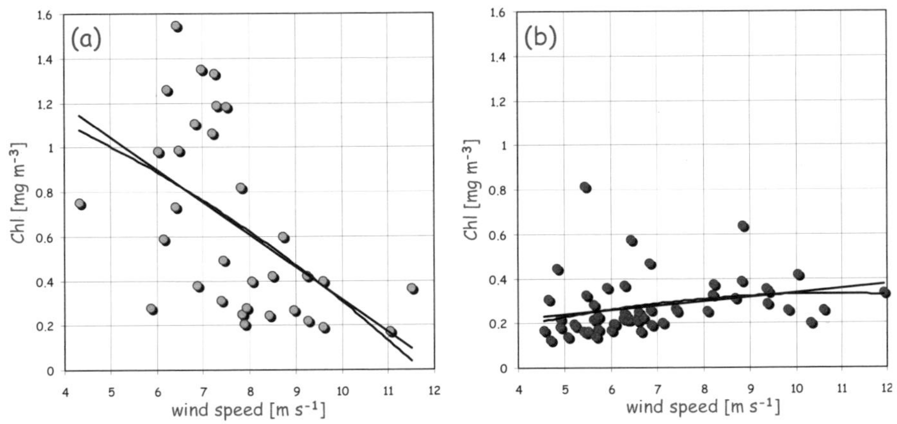


Figure 6. Regression analysis of monthly mean wind speed and *Chl* in the Ligurian-Provençal Sea, at 0 months time lag, for January to April (panel a) and May to December (panel b).

into consideration the delay of the biological response to the set up of conditions favourable to algal growth (since in this case the increase in nutrient concentration is essentially due to Ekman pumping within the cyclonic Rhodes Gyres, and not to sustained deep convection, continuing for long periods of time). With no time lag (*Chl* and wind speed of the same month; Fig. 7a), both linear and polynomial fit show a positive correlation (with $R^2 = 0.33$ and $R^2 = 0.35$ respectively). The maximum correlation is reached with a time lag of 1 month (*Chl* matched to wind speed of the preceding month, Fig. 7b), so that $R^2 = 0.53$ for both linear and polynomial fit. At higher time lags, the correlation decreases quickly to very low values, with $R^2 \sim 0.3$ at 2 months, $R^2 \sim 0.1$ at 3 months, and $R^2 \sim 0$ at 4 months.

Once again, the first 4 months of the year present a rather unique situation, in the Levantine basin, as already seen for the Ligurian-Provençal Sea. The correlation is still low, in this period, when wind speed and *Chl* are matched with no time lag ($R^2 = 0.03$ in the linear case and $R^2 = 0.13$ in the polynomial case), but becomes much better with a time lag of 1 month ($R^2 = 0.38$ in the linear case and $R^2 = 0.48$ in the polynomial case), as shown in Fig. 8a. From May to December, the

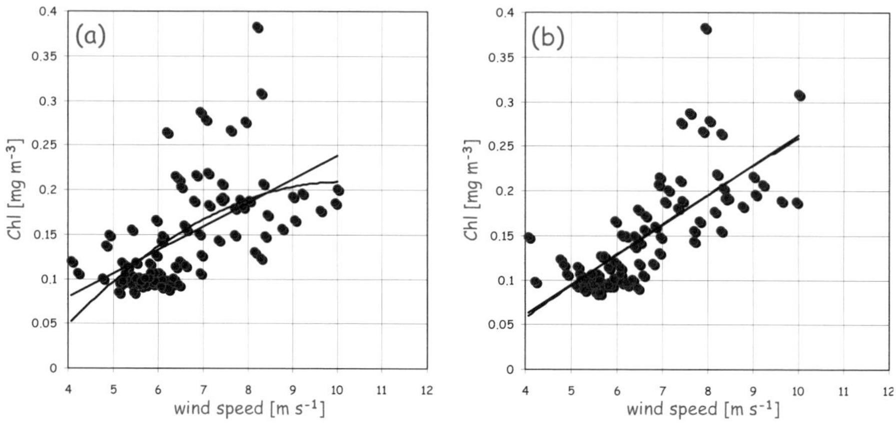


Figure 7. Regression analysis of monthly mean wind speed and *Chl* in the Levantine Basin, at two time lags, respectively of 0 (panel a) and 1 month (panel b).

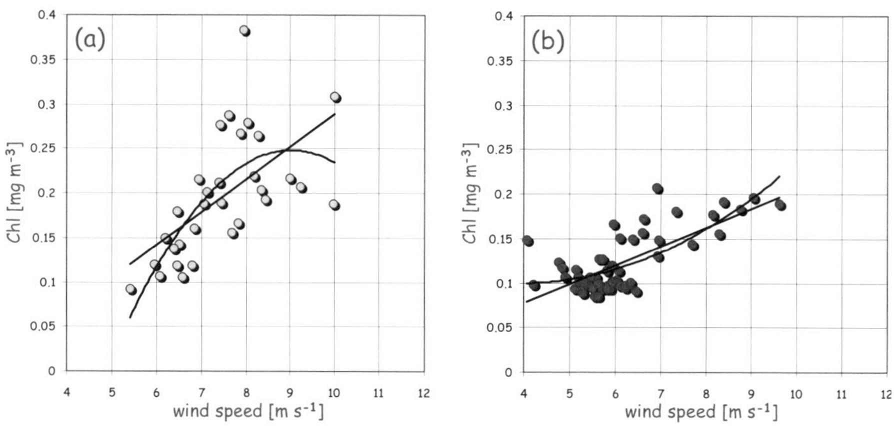


Figure 8. Regression analysis of monthly mean wind speed and *Chl* in the Levantine Basin, at 1 month time lag, for January to April (panel a) and May to December (panel b).

correlation is high, both with no time lag ($R^2$ = 0.39 and $R^2$ = 0.51) and with 1 month time leg ($R^2$ = 0.51 and $R^2$ = 0.55), as shown in Fig. 8b.

## CONCLUSIONS

Algal growth in the Mediterranean Sea is always nutrient-limited, with consequent low biomass and primary production. The fertilization of the basin, supporting localized algal blooming, is ruled mainly by the impact of (coastal interactions and) atmospheric forcing in broad near-coastal zones, and then by the ensuing thermohaline processes.

In the Ligurian-Provençal Sea, the winter winds lead to convective processes that promote nutrient upwelling and then a sustained spring blooms, once the wind regime relaxes and surface stratification sets in. In the Levantine Basin, upwelling due to Ekman pumping, within the cyclonic component of the wind-generated Rhodes-Irapetra vortex pair, also results in similar, but weaker, algal blooms. However, other factors appear to be critical, in this second case (due to the different physical processes taking place), as the pattern of high wind regime followed by blooming has a much shorter time scale and does not recur systematically.

The examples presented here demonstrate that an improved interpretation of remote sensing data sets, and of the bio-geo-physical processes responsible for the observed phenomena, can be achieved through the combination of multi-sensor techniques. However, it is obvious that knowledge of the variability of pigment concentrations and surface winds is not sufficient for a full understanding of ecological dynamics in the two regions of interest.

Modeling of the correlation between *Chl* and wind speed, planned for the future development of this line of work, might be able to provide more clues on the coupling between algal blooms and wind patterns in the Mediterranean Sea. It should be noted, however, that much longer data sets would be required for a long-term analyses of the Mediterranean trends, particularly for the case of local changes in times of global warming.

## REFERENCES

i. Barale V., 2005. Satellite observations as indicators of the health of the Mediterranean Sea. In: The Mediterranean Sea, edited by A. Saliot (The Handbook of Environmental Chemistry, Vol 5 Water Pollution, Part K, Springer-Verlag, Berlin, Heidelberg), 387–408.

ii. Millot C. & I. Taupier-Letage, 2005. Circulation in the Mediterranean Sea. In: The Mediterranean Sea, edited by A. Saliot (The Handbook of Environmental Chemistry, Vol 5 Water Pollution, Part K, Springer-Verlag, Berlin, Heidelberg), 31–66.

iii. Hamad N., C. Millot & I. Taupier-Letage, 2006. The surface circulation in the eastern basin of the Mediterranean Sea. Scientia Marina, 70(3): 457–503.

iv. Mélin F., B. Bulgarelli, N. Gobron, B. Pinty & R. Tacchi, 2000. An integrated tool for SeaWiFS operational processing (European Commission Publication no. EUR 19576 EN, Ispra, Italy) 34 pp.

v. Schroeder C., H. Boggs, G. Dome, M. Halberstam, L. Jones, J. Pierson & J. Wentz, 1982. The relationship between wind vector and normalized radar cross section used to derive Seasat-A satellite scatterometer winds. Journal of Geophysical Research, 87: 3318–3336.

vi. Donelan A. & W. J. Pierson, 1987. Radar scattering and equilibrium ranges in wind-generated waves with application to scatterometry. Journal of Geophysical Research, 92: 4971–5029.

vii. Baith K., R. Lindsay, G. Fu & C. R. McClain, 2001. SeaDAS, a data analysis system for ocean-color satellite sensors. EOS Transactions AGU, 82: 202.

viii. Campbell J. W., J. M. Blaisdell & M. Darzi, 1995. Level-3 SeaWiFS Data Products: Spatial and Temporal Binning Algorithms. SeaWiFS Technical Report Series, Volume 32, NASA Technical Memorandum 104566. edited by S. B. Hooker, E. R. Firestone & J. G. Acker (NASA Goddard Space Flight Center, Greenbelt), 23 pp.

*Remote Sensing for a Changing Europe*
*D. Maktav (Ed.)*
*IOS Press, 2009*

*doi:10.3233/978-1-58603-986-8-40*

# Determination of flood risks in the yeniçiftlik stream basin by using remote sensing and GIS techniques

İrfan Akar
*University of Atatürk, Institute of Social Sciences, Erzurum, Turkey*

D. Maktav & C. Uysal
*Dept. of Civil Engineering, Istanbul Technical University, Istanbul, Turkey*

Keywords: Remote sensing, GIS, yeniçiftlik stream, flood, hec-ras, hec-georas.

ABSTRACT: Floods are among the most devastating natural hazards in Turkey and worldwide, causing the largest amount of casualties and property damage. GIS and remote sensing methods are very attractive, fast, and reliable tools for various flood applications and management. In this study, we investigated floods which occurred and are likely to occur in a study area in Istanbul, Turkey, to determine the potential use of these tools with respect to these floods. Floods which caused loss of life and property in the Yeniçiftlik stream basin located within the boundaries of Beykoz, a suburb of Istanbul, attracted our attention due to their negative impact on human life and activities, and this was selected as the study area. Many geographical parameters such as vegetation, topographic and geologic features, precipitation, and land use features play a significant role in the occurrence of flood related disasters. Data used were topographic, soil, vegetation, and geological maps at scale 1:25000, IKONOS pan-sharpened imaging (02.03.2008), as well as aerial photographs taken in 2006. Using the Arcinfo 9.2 Spatial Analyst module, flood risk maps were created, assigning different weights to vegetation, geologic and land use features, and other morphometric features such as slope, aspect, and so on. Land use and vegetation features were determined by applying a supervised classification technique to IKONOS data. All data were processed using HEC-GeoRAS (in ArcGIS) and HEC-RAS software. The results indicate that the precision and diversity of the data used greatly affects the precision of these risk maps.

## 1 INTRODUCTION

Floods are one of the significant natural disasters that substantially impact the economic and social lives of people in many regions and countries (CEOS, 2003). Various computer models have been developed in order to understand floods that occur and demonstrate their impacts. It is possible to classify the basic components of these models into four parts: hydrological models, hydraulic models, flood mapping, and generation of spatial data for use in the model (Snead, 2000). Together with developments in GIS technologies and the increase in resolution of Digital Terrain Models (DTM), applications for research related to hydrology, water resources, and the environment have increased. Integration of GIS with hydrologic and hydraulic modeling software has been realized for various purposes. Among these, HEC-GeoRAS is an ArcGIS extension program that has been produced for generating spatial data. With this extension it is possible to extract and process geometric data pertinent to river basins over the existing DTM, read the database pertinent to HEC-RAS hydraulic software, and map water levels and floods (Özdemir, 2007). HEC-RAS is a one-dimensional hydraulic program in which water surfaces related to continuous flows can be calculated and discontinuous streams can be modeled (USACE, 2002). It has data storage and management capacity and is widely used throughout the world . With the utilization of GIS and development of this technology, hydraulic models of river systems can be generated more easily (Maidment, 2000 and 2002). Thus by using flood modelings performed with GIS and hydraulic software, it is possible to calculate the distribution area of water surfaces of potential floods and to make relevant volume and depth calculations. The study field was located within the borders of the Beykoz district of the Istanbul province between UTM/ED_1950/Zone35N, Max. West: 681127-436675, Max. East: 686377-436675, Max. North: 4561198-271090, and Max. South: 4549558-271090 coordinates. Yeniçiftlik stream basin has a perimeter of 33 km and a surface area of 31.1751 km$^2$; the north–south length of the basin is 12.555 km, and the east–west length is 3.75 km. The research field comprised a 3.66 km$^2$ area of this basin (Figure 1).

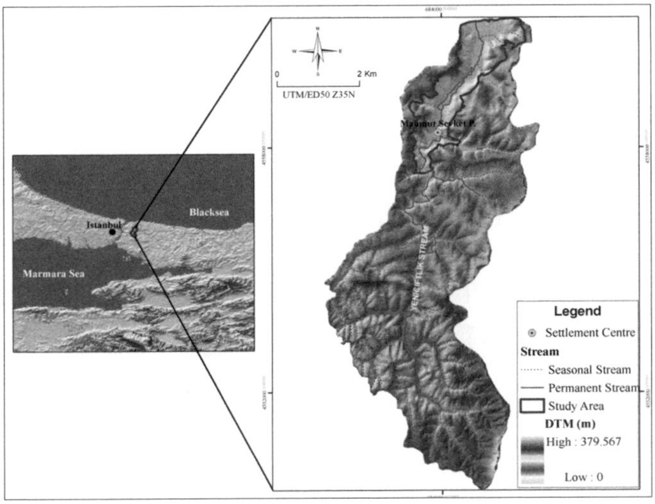


Figure 1. Study Area.

## 2 DATA AND METHOD

Data resources used in the study were topographical data at scale 1:25000, IKONOS pan-sharpened (02.03.2008) satellite imaging, aerial photographs (2006), and digital soil, geological, and vegetation data at scale 1:25000. Two different methods were applied to determine the flood areas. The first of these methods was assessed using HEC-GeoRAS and HECRAS, using the data generated. Applications related to this method are shown in Figure 2. Since there was no high resolution DTM for the study field, stages of the study were compiled under four main headings: Triangular Irregular Network (TIN) generation, processes performed using HEC-GeoRAS, processes performed using HEC-RAS, and generation of flood maps. One of the most significant stages in flood mapping is the generation of high resolution DTMs (Sugumaran and Davis et. al., 2000). For this the TIN model, which best reflects the linear layers from geometric data, was used (USACE, 2005). TIN were generated using data generated by 5–10 m contours on topographic maps at scale 1:25000. The geometric data to be used in flood mapping were generated using HEC-GeoRAS. IKONOS pan-sharpened satellite imaging and aerial photographs were used for generating these. Data completion for geometric data was performed using HEC-RAS.

The second method used was Multi-Criteria Decision Analysis (MCDA). In this method the parameters are taken into consideration one by one in order to determine the areas of risk. Basic elements of the basin's flood risk comprised the geographical properties of the basin and its flood characteristics. A flow chart detailing the method is shown in Figure 3. Applying this method, the parameters pertinent to flood risk in the study field were assessed and weight values were assigned to each.

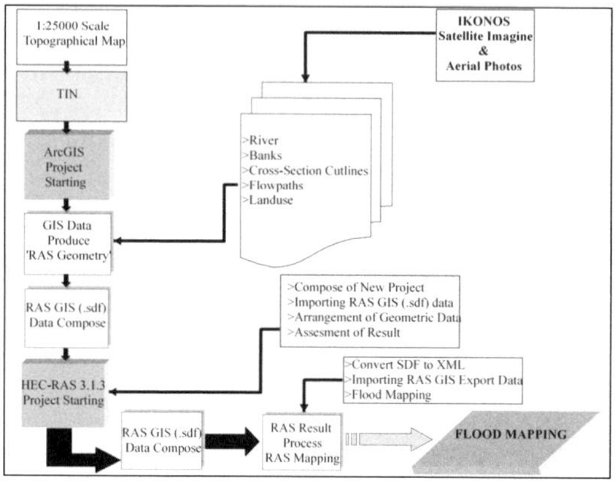


Figure 2. Flow chart of Hec-GEORAS and Hec-RAS Applications.

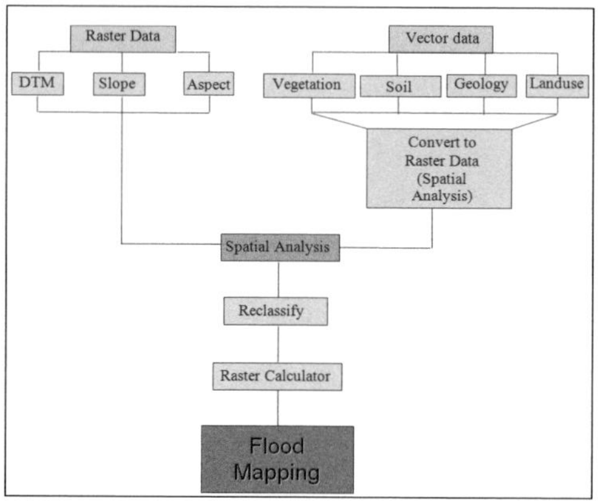


Figure 3. Flow chart of MCDA method.

## 3 HYDRAULIC MODELING AND FLOOD MAPPING

### 3.1. *Creation of geometric data*

In order to conduct hydraulic modeling of rivers, geometric data for river basins are required (Özdemir, 2007). These data generally comprise establishing the connection of river systems, bed cross-sections, defining connection points, data pertinent to hydraulic structures, and cross-section interpolations. Geometrical data used in the study included rivers, riversides, flow paths, cross-sections and land use. Only data from the 3.66 $km^2$ study area of the Yeniçiftlik stream basin were used. In generating geometric data, the generation of TIN models as base data was performed as a priority. Digitalization of geometric data in the area to be modeled after TIN generation and data entry processes were performed (Figure 4).

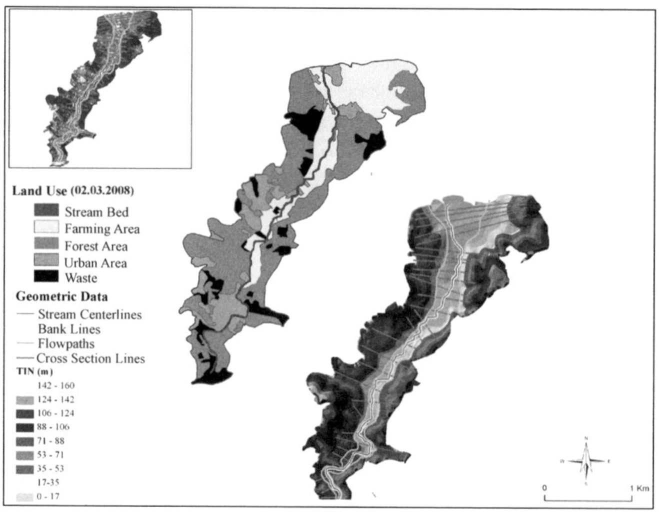


Figure 4. Using geometric data in flood mapping.

| Land Use Type | Manning's N |
|---|---|
| Stream Bed | 0.050 |
| Farming Area | 0.060 |
| Urban Area | 0.055 |
| Forest Area | 0.150 |
| Waste | 0.070 |

Figure 5. Manning's value.

In digitalizing these data, the TIN model, IKONOS pan-sharpened satellite imaging, and aerial photographs were used. Some characteristics emphasized in the digitalization of the geometric data and data entry were as follows:

- Rivers, river coasts, and stream paths were digitalized towards the flow direction.
- Definitions of river paths were conducted (such as right, left and center).
- Bed cross-section lines were digitalized towards river stream direction, from left coast to right coast.
- Cross-section lines intersected the river and stream paths once and both lines intersected with each other.

A land use layer was generated using IKONOS pan-sharpened satellite imaging. For types included in this layer Manning's N value, which is accepted as an indicator of the action demonstrated against the stream hydraulically, was assigned (Figure 5; Fleckkenstein, 1998; Özdemir, 2007).

### 3.2 *Generation of the flood model*

HEC-RAS is one-dimensional software in which the steady and unsteady streams are modeled (Merwade, 2006). Geometric data for the river basin and data for the stream have to be entered in order to conduct the modeling (Özdemir, 2007). Geometric data pertinent to the study field were generated using HEC-GeoRAS and other data were completed in this section (Figure 7). Subsequently the data was transferred to HEC-RAS software. After completing the entry of geometric data, stream data for the river in the area in which modeling was to be conducted were entered. Stream data for the repetition frequencies of floods were accepted as Steady Flow data. Since the gradient value of the study field was not very high, subcritic was selected for stream calculation. Subsequently, models for 10, 50, and 100 years were generated (Figure 6).

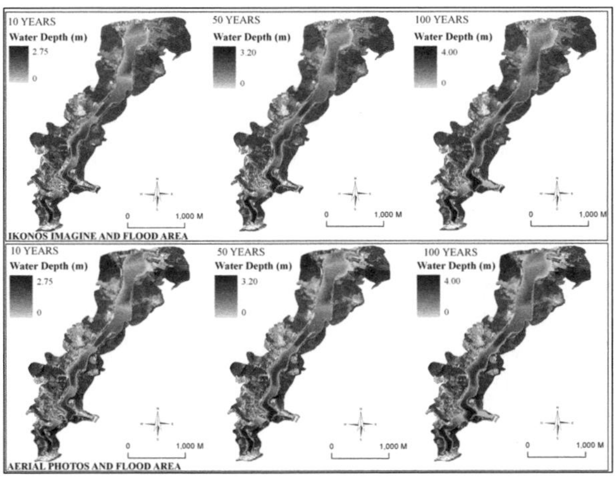


Figure 6. Models pertinent to different flood repetition frequencies.

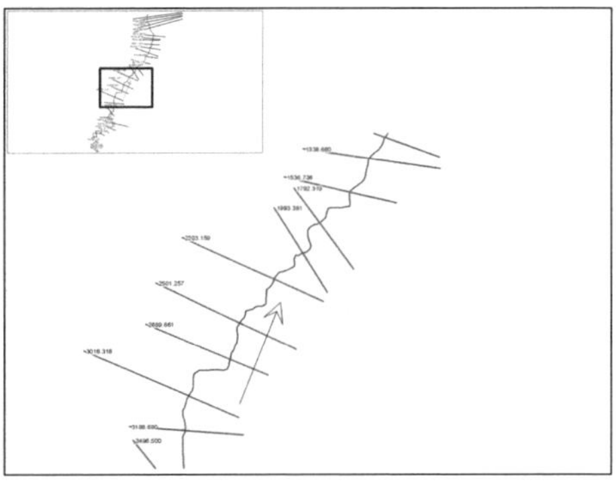

Figure 7. Geometric data generating from Hec-RAS.

### 3.3. *Multi-criteria decision analysis (MCDA)*

This is the process of assessing options of a definite number using numerous criteria which are generally weighed, contradict one another, do not use the same unit of measurement unit, and may even take qualitative values for the purposes of selection, sorting, classification, prioritization, or screening (Yoon ve Hwang, 1995). In the method applied, parameters were taken into consideration one by one while areas of risk were determined. Basic elements relevant to the flood risk conducted for the study field comprised the geographical properties of the basin and its flood characteristics. The flow chart of the method is detailed in Figure 2. Since a critical stage is the generation of basic base maps whilst carrying out the risk analysis of the study field, base data for topographic, vegetative, soil, and geographical characteristics were generated from data for the basin digitalized in the digital media. Land use characteristics and geological and soil data were converted into raster data structure for analytical purposes. In the method applied, the parameters for flood risk in the study field were assessed and weight values were assigned to each (Figure 8). Weight values ranged between one and 10. Those closer to one have the least risk and those closer to 10 have the highest risk. The reclassify module of the Spatial Analyst module is used for applications at this stage. Due to the fact that the impacts of parameters on relevant disasters show varying rates, different values are assigned to each by means of this module. The overlay process is realized following this value assignment. This process was performed using the Raster Calculator module of the Spatial Analyst module. As the method was applied, the flood risk map of the study field was determined (Figure 9). Five different degrees were selected for flood risk. Examination of the flood risk map shows in particular that the risk is quite high in parts where the gradient values are low and the settlement and agricultural areas are intense (Figure 9 and 10).

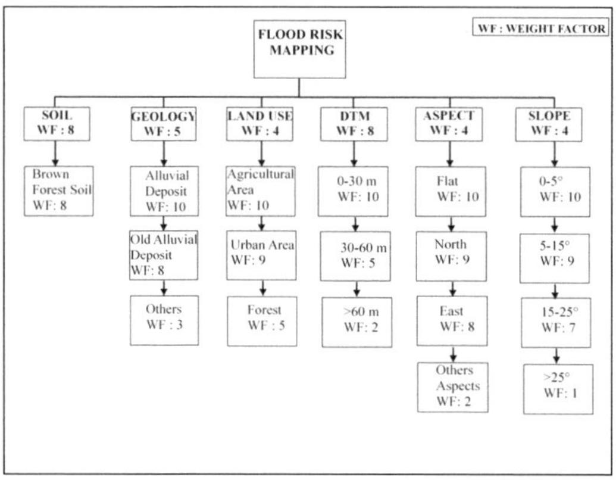


Figure 8. Decision hierarchies for flood hazard index ranking.

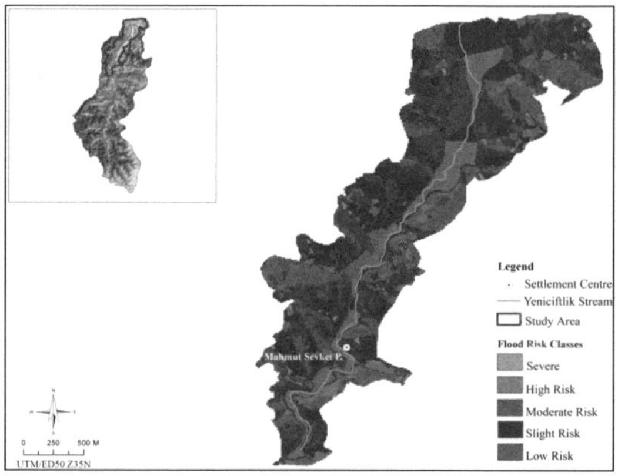


Figure 9. Flood risk mapping generating from MCDA method.

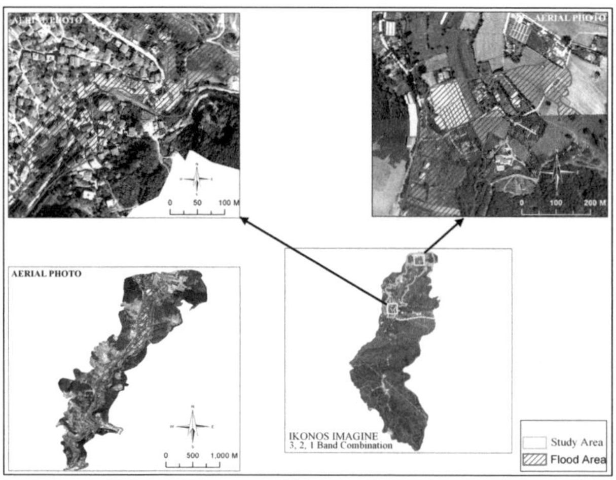


Figure 10. Displaying aerial photos of flood area determined by MCDA method.

## 4 CONCLUSION

In this study an area of 3.66 $km^2$ in the Yeniçiftlik stream basin was examined. Flood risk analysis was performed using Remote Sensing and GIS. Remote Sensing data were used in determining land use, demonstrating changes in land use, determining the river basin characteristics of the study field, and creating geometrical data for the floods. As the study field is a basin with flood risk potential any flood taking place may significantly impact the settlement areas, industrial zones, and productive agricultural fields. In the hydrologic and hydraulic modeling conducted for flood risk studies, flood rates for 10, 50, and 100 years were calculated using maximum flood rates for 33 years, and flood risk maps were generated through hydraulic modeling using the values obtained. A TIN model was used for the geometric data generated in the hydraulic model application, and the river's central line was digitalized over stream paths and cross-sections, satellite imaging, aerial photographs, and the TIN model. Land use factors between the cross-sections were determined, flood models were created according to different repetition frequencies, and a 100 yearly potential flood model was utilized for determining the flood areas. In MCDA, another method used, criteria for risk analysis were determined using the Analytic Hierarchical Method. In the study, the flood risk was considered according to five parameters: the DTM of the basin, gradient, aspect, geological characteristics, and land use. As a result of the model generated it was determined that the flood potential of the study field is high. Due to the risk of flooding in the coming years it will be necessary to take relevant precautions in this area and implement them rapidly. It has also been determined that in order to conduct a more realistic modeling using HEC-GeoRAS and HEC-RAS hydraulic software, further data would be required.

## REFERENCES

Sugumaran, R., C.H. Davis, J. Meyer & T. Prato, 2000. High resolution digital elevation model and a web-based client-server application for improved flood plain management, IEEE Geoscience & Remote Sensing Symposium IGARSS 2000.

Committee on Earth Observation Satellites (CEOS), 2003. The Use of Earth Observing Satellites for Hazard Support: Assessments and Scenarios. Final Report of the CEOS Disaster Management Support Group (DMSG). http://www.ceos.org/pages/DMSG/2001Ceos/Reports/flood.html (accessed 14 April 2008).

Fleckkenstein, J., 1998. Using GIS to derive Velocity Fields and Travel Times to Route Excess Rainfall in a Small-Scale Watershed, Univ. of California Davis. http://scs.ucdavis.edu. (accessed 08. 03.2008).

Maidment, D., Djokic, D., 2000. Hydrologic and Hydraulic Modeling Support with GIS, ESRI Press, USA.

Maidment, D.R., 2002. ArcHydro GIS for Water Resources, Esri Press, California.

Snead, D.B., 2000b. Development and Application of Unsteady Flow Models Using Geographic Information Systems, Departmental Report, Master of Science in Engineering, The University of Texas at Austin, USA.

Özdemir, H., (2007), Özdemir, H., 2007, Floodplaın Mappıng Usıng Hec-Georas And Hec-Ras: A Case Study Of Havran Rıver (Balıkesir). http://www.cbs2007.ktu.edu.tr/bildiri/S_33.pdf (accessed 15.05.2008).

Yoon K P and Hwang C L, 1995, Multiple Attribute Decision Making: An Introduction, Sage Publications, Thousand Oaks.

USACE (US Army Corps Engineers), 2002. HEC-RAS River Analysis System, User's Manual, Davis, USA.

USACE (US Army Corps Engineers), 2005. HEC-GeoRAS GIS Tools for Support of HEC-RAS Using ArcGIS, User's Manuel, Davis, USA.

Merwade, V., 2006. Tutorial on using HEC-GeoRAS with ArcGIS 9.1 Prepared by School of Civil Engineering, Purdue University web.ics.purdue.edu/~vmerwade/education/georastutorial.pdf (accesed 24.04.2008).

*Remote Sensing for a Changing Europe*
*D. Maktav (Ed.)*
*IOS Press, 2009*

*doi:10.3233/978-1-58603-986-8-46*

# ERS data for rainfall-runoff models

L. Halounová, M. Švec
*CTU in Prague, Faculty of Civil Engineering, Prague 6, Czech Republic*

J. Horák , J. Unucka, M. Hanzlová, L. Juříkovská
*TU Ostrava Faculty of Mining and Geology, Ostrava-Poruba, Czech Republic*

Keywords: rainfall-runoff model, ERS, GIS, soil moisture changes

ABSTRACT: The presented project being a part of rainfall-runoff modeling is focused on determination of relations between soil moisture changes detected in radar data and precipitations. One catchment area of– the Olse River in the northern Moravia is used for. The catchment´s land cover classes derived from Thematic Mapper data are decisive for creation of homogenous land cover patches. The changes are determined from ERS-2 data. The radar data were chosen according to the flood precipitations date in the catchments covering either before and after flood period or only after a flood one. Radar image data processing is performed by following working steps: image data filtering, band subtraction, delineation and selection of the same land cover patches; these areas were subdivided according to their change values derived from the radar data subtraction. These subareas were evaluated in GIS to find relations between the spatial distribution of the radar data changes and hydrological, morphological, and soil conditions.

## 1 INTRODUCTION

Rainfall-runoff modeling should predict flood occurrence in case of high precipitations in various catchments. The project whose preliminary results of relations between changes in soil moisture and changes in ERS radar data are summarized in the paper is focused on spring drainage basins – areas representing the first part of catchments. Their flood conditions are derived and caused only by their precipitation and water absorption capacity of the region. Therefore the rainfall-runoff hydrological model covers the whole spring part of the river including the Polish part.

The practical usage of the final model is a quick processing of the immediate rainfall and the water stream discharge situation, deciding if the state is in fact a flood state which should be managed as the real flood situation including all warranty activities.

Similar demands to rainfall–runoff condition assessment and more accurate weather predictions are required both for extreme events, and for ecological and water management purposes.

This modeling and weather forecast are processed by the CHMI (Czech Hydrometerological Institute) and watershed managers in the Czech Republic. Modeling is also performed by other authors on individual catchment scale with more research orientation. Eichler, Kulhavy, Dolezal and Soukup analyzed drainage runoff, agriculture impact and hydrological balance models (Kulhavy 2000), including simulation approaches for drainage runoff assessment (Dolezal 2000). Neruda (2002) used artificial intelligence network approach for the rainfall–runoff modeling, predicting the following day flow using historical flow and precipitation data .

Relations between anthropogeneous landscape changes and flood intensity were assessed by Langhammer (within the Evaluation of natural environment changes impact to flood development grant project), methodology of rainfall–runoff conditions in urbanized and industrial areas assessment using combination of classical hydrological methods, mathematical simulations and GIS processing and visualization were studied by Rapantova (2004). Usage of GIS for watershed models, ecological aspects, erosion impacts and sediment transport modeling were studied by Vozenilek (1999) and Pavelka (1996). Runoff conditions are affected by various factors, above all by geomorphologic characteristics, soil profile character and condition, cover type; evapotranspiration level and snow cover condition. Their error can reach tens of per cent, i.e. of snow cover or evapotranspiration. Individual factors are inter-related and cannot be evaluated individually. For example, snowmelt modeling can be performed by various methods (mostly used is degree/day method) (Beven 2001, Chow 1964, Hradek 2002) and Anderson´s method (Anderson 1968), whose results are influenced by altitude, terrain morphology, exposure and vegetation cover.

The modeling in the project combines rainfall forecast and hydrological model of water surface outflow. One input to the hydrological model is soil moisture. The analyzed watersheds are evaluated for historical events well documented by detailed discharge measurements, however, without soil moisture measurement.

The surface outflow is a spatial phenomenon and therefore soil moisture should be also used as a spatial data layer. However, remote sensing imagery is the only data source describing spatially the characteristic. The problem was that no ground truth data were available for the real soil moisture calibration. Having no soil moisture input information, the only possibility were to use only relative values for a further model calibration. Changes calculated between two time levels close to the high rainfall were applied. Their values formed one GIS layer from the system comprising land cover, soil types, slopes, distance to water courses, and rainfalls.

## 2 INPUT DATA

The input data means data used for the watershed GIS model describing relations between hydrological conditions on one side and "local conditions" on the other side.

### 2.1 *The study area*

The project is focused on the Olse River catchment. The Olse River flows from Istebny in Poland into the Odra River in the flat Ostrava region forming the frontier between Poland and the Czech Republic after leaving Poland.

Table 1. Geographical characteristics of the Olse River watershed

| River Catchment | | Olse |
|---|---|---|
| Length | all | 99 km |
| | Czech Republic | 83 km |
| | Poland | 16 km |
| Area | all | 1118 km2 |
| | Czech Republic | 639 km2 |
| | Poland | 479 km2 |

### 2.2 *Meteorological data*

The meteorological data in the Olše River watershed were measured by 6 rain gauges (Bílý Kříž, Jablunkov, Lučina, Hradiště, Třinec, Lysá hora) and delivered in one hour time intervals. Their values were processed by kriging for the regular grid networks. The grid values were used for the precipitation GIS layer. Only one layer as a sum of all precipitations was created due to relatively similar time distribution of the rainfalls.

The final rainfall GIS layer values were sums in individual pixels for the whole before flood period between the radar image pair's dates. This simplification was used since important rains and the rain curve mode appeared only with several hours′ delays.

## 2.3 *Land cover of the catchment*

Land cover of the catchment is formed by 3 main areas: urban (72) sq km, agricultural (318,6 sq km) and forest (241,4 sq km) according to the land use/land cover CORINE data from 2000 in the Olse River. The land cover was compared to the 1990 CORINE data and only less than one per cent of difference was found. The same comparison was prepared from Thematic Mapper data and land cover changes formed nearly 20 per cent of the area. Therefore more accurate land cover information is used from the optical data.

## 2.4 *Soil types*

The map of soil classes of the Czech Republic is formed by homogenous regions having its coded description. The feature class attributes are embedded in 8-numbered code characterizing climate region, main soil unit, slope gradient, soil skeleton, boulderity, and soil depth. The maps are processed at 1 : 5000 scale. Four "soil" type areas have different 3-number coding - bare land, water, quarry, and forest. The final detailed soil types comprise tens of classes. However, they were aggregated according to their skeleton content and soil classes according to following characteristics were formed.

Table 2a. Soil type classification according to the skeleton content occurring in the region

| Class | description |
|---|---|
| 0 | Without skeleton, with skeleton less than 10 % |
| 1 | Low skeleton less than 25 % |
| 2 | Mid skeleton less than 50 % |
| 3 | High skeleton higher than 50% |

Table 2b. Soil type classification according to the soil depth

| Class | description |
|---|---|
| 0 | Deep – depth more then 60 cm |
| 1 | Mid – depth 30 – 60 cm |
| 2 | Shallow – depth less than 30 cm |

## 2.5 *DEM of both catchments*

Digital elevation model was derived from 1: 25 000 scale contour lines using TIN and processed into the 25 x 25 m grid format.

## 2.6 *Remote sensing data*

Three remote sensing data types were used for. The first detailed data processing was performed from ETM+ Landsat data, a pair of SPOT data and a pair of ERS-2 data formed the whole group of remote sensing data.

### 2.6.1 *SPOT image data*
The data list is in table 2. After preprocessing, the data were georeferenced and used for the land cover change detection between 2002 and 2004.

### 2.6.2 *ERS-2 image data*
The regular water level measurements showed individually the flood dates on both rivers. The ERS-2 data were selected to cover the period of these events (see table 2); one flood (2005 year) is imaged by a pre- and post-imagery and one by two post flood imageries (2002 year).

Table 2. List of SPOT and ERS data for individual flood events

| Flood event date | SPOT | | ERS acquisition date |
|---|---|---|---|
| | acquisition date | MS/P | |
| 19/07/2002 | 19/07/02 | SPOT 4, MS - HRV1 | 22/07/2002 |
| | 21/07/02 | SPOT 5, MS, P -HRG2 | 26/082002 |
| 24/08/2005 | | | 27/07/2005 |
| | | | 31/08/2005 |

## 3 METHODOLOGY

The soil moisture change analysis was processed in GIS. The soil moisture depends on many phenomena. Several of them are time independent – soil type, orientation, and slope, etc., however, some of them are time dependent – rainfall, temperature and evapotranspiration (with hours′ changes), land cover (with months′ changes).

The soil moisture changes were supposed to be expressed by radar data changes. The radar data depend on material type, roughness and moisture. The changes were determined from ERS data measured within 34 days. The period is generally quite short for the roughness changes on one side, and it is the moisture which the radar data are sensitive for in case of steep incident angle as the ERS data have. Therefore the hypothesis that only moisture caused the changes was used.

The changes were calculated as two data difference – subtraction of the younger image from the older one. The subtraction was performed after filtering by 5 x 5 average filter. The resulting data were simplified and therefore grouped into ten classes comprising data with the same digital value ranges. Negative values (classes 1 – 7 for 2002 and 1 – 6 for 2005) represented decreasing of the radar intensity what means decreasing of the soil moisture; positive values (classes 9 – 10 for 2002 and 8 – 10 for 2005) depicted increasing soil moisture. The $8^{th}$ or $7^{th}$ classes with both negative and positive values near zero were handled as areas with no changes.

It was supposed that the changes were caused by rainfalls between acquisitions of the ERS data pair. Dependency of soil moisture on previous rainfalls, land cover, slope, and soil type was analyzed in the first phase.

Non-stationarity can be found at land cover data (months′ scale) and hydrological data.(hours′scale, evapotranspiration is calculated in days′ scale in 90 places in the Czech Republic). Their input must be prepared in a spatial form by an interpolation method to get a spatial layer.

The changes were analyzed in two main land cover classes – forest and agricultural areas where the rainfall, slope, and soil type are shown. The characteristics were evaluated for land cover homogenous areas.

The relations showed influence of time invariant characteristics on the soil moisture changes. The relations need to be studied as a combination of characteristics. Their separated evaluation showed that all detailed classes of all characteristics occur in all change classes.

## 4 RESULTS

The graph (fig. 1) of forest change classes (1 – 10) in 2002 showed that the largest area (45 per cent – column 3 - 6) have lower soil moisture at the end of the period between two ERS measurements with extreme change areas covering less than 5 per cent (column 1, 2). Positive changes occurred at 33 (column 8, 9, 10) per cent of forest areas. All slope classes are found in all change classes in a similar ratio. No direct simple influence is possible to be derived from the slope and soil moisture mutual relation. Similar results were determined for all other characteristics.

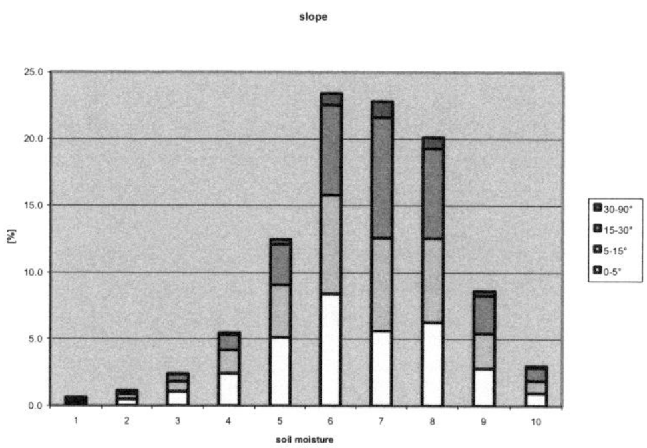


Fig. 1. Slope classes in forest soil moisture classes

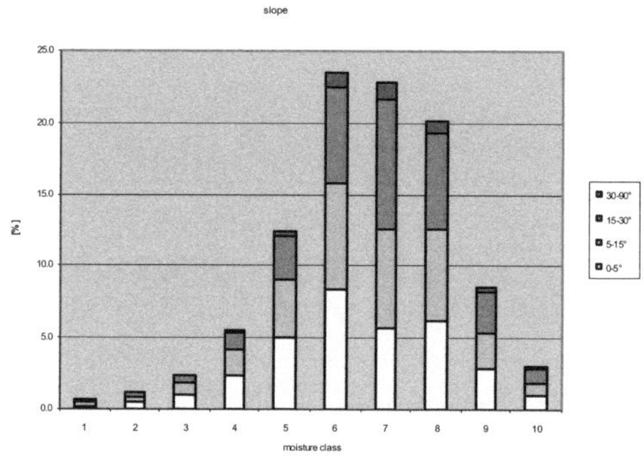


Fig 2. Slope classes in agricultural area soil moisture classes

The graph (fig. 2) of agricultural area change classes proved that the soil moisture decreases more quickly if compared with forest areas. Dryer area covers more than 50 per cent, positive areas around 25 per cent. Mutual occurrence ratio of all slope classes shows again no close direct dependency on slopes.

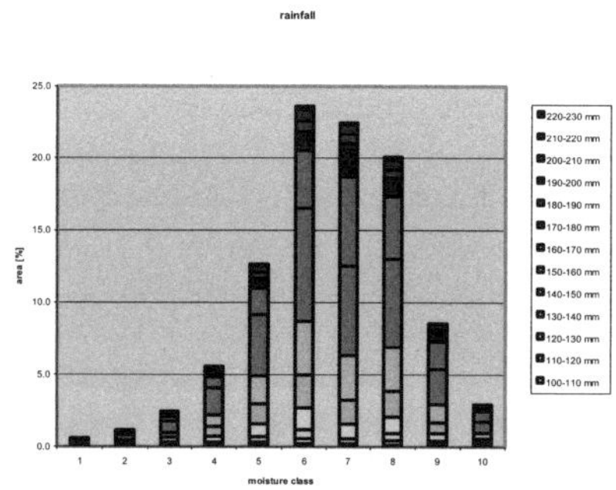


Fig. 3. Rainfall classes in soil moisture classes

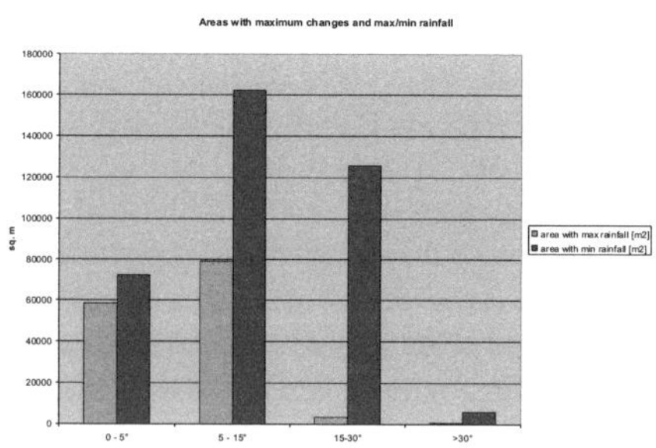


Fig 4. Moisture change classes in areas with highest and lowest rainfall distributed according to slope terrain

Fig 3 showing rainfall classes is another prove of a complex relation among individual characteristic and moisture changes. The precipitation distribution is really an invariant moisture feature. However, comparing slope, rainfall and moisture change shows that moisture changes caused by different rainfall occur in areas with different slope (see table 3)

Table 3. Comparison of areas with high and low rainfalls in the highest moisture changes in forest areas

| | extremely high rainfall | | | extremely low rainfall | | |
|---|---|---|---|---|---|---|
| slope (°) | number of soil classes | area [m²] | % | number of soil classes | area [m²] | % |
| <5° | 3 classes* | 58482 | 41.2 | 1 class* | 72282 | 19.7 |
| 5 - 15° | 2 classes | 79214 | 55.8 | 3 classes | 162179 | 44.3 |
| 15°- 30° | | 3564 | 2.5 | 1 class | 125680 | 34.3 |
| >30° | | 617 | 0.4 | | 6196 | 1.7 |

- classes whose area is larger than 1 hectare

Further analysis (table 3) compared areas with extremely high moisture changes and with extreme (low and high) rainfalls related to soil type and slope. 97 per cent areas with the highest changes and precipitation occurs in areas with slope 0 – 5° and 5 – 15° with 4 different soil classes (one occurring in both slope classes). Areas with the highest change and lowest precipitation have half percentage in the lowest slope (compared to the highest precipitation area), and one third in areas with slope up to 30° which nearly do not occur in wettest areas.

## 5 CONCLUSIONS

The results showed that the relation between the changes found in radar data and other characteristics cannot be studied individually. It was only the last table – table 3 which showed a certain correlation – only few soil classes are decisive for the surface changes; slope where most changes were found is lower than 15 ° for extremely high rainfall and up to 30 % for low rainfall.

The partial results lead the solution to a multicriterial analysis. The analysis will be focused on determination of rainfall and land cover influence to other (positionally constant) characteristics where the evapotranspiration will be used in the form of a layer which is subtracted from the rainfall layer. The rainfall will be a time dependent value and land cover time independent in the analysis with above mentioned data.

The analysis will be oriented to determination of weights in the equation (1)

$$\textit{Change (digital value)} = (\textit{precipitation} - \textit{evapotranspiration} - \textit{surface outflow}) \cdot w_{time} \cdot w_{soil} \cdot w_{slope} \cdot w_{land\ cover} \cdot w_{temperature} \quad (1)$$

where $w_{time} = f\,(time)$,
$w_{temperature} = f\,(time)$,

others $w$ are constant for the place in case of unchanged land cover
*surface outflow* - calculated using CN curves.

The equation will be evaluated for existing area and further calibration of soil moisture will be performed in the surface runoff model (Hanzlova et al. 2008).

## ACKNOWLEDGEMENT

The project is financially supported by the Czech Agency grant 205/06/1037 Application of Geoinformation Technologies for Improvement of Rainfall-Runoff Relationships
SPOT data provided by the CNES OASIS project Application of Geoinformation Technologies for Improvement of Rainfall-Runoff Relationships
ERS data provided by the European Space Agency - Category 1 project No 3912

## REFERENCES


Anderson, E.A., 1968. Development and testing of snow pack energy balance equations. Water Resources Research 3.

Beven, K. J. 2001. *Rainfall-runoff modelling*. The Primer. John Wiley & Sons Chichester.

Dolezal F., Kodesova R., Kulhavy Z., Soukup M. 2000. Study of hydrological function of agricultural soils drainage system via simulation modeling. Hydrological days Plzen. IX.2000.

Chow, V.T., 1964. *Handbook of applied hydrology*. McGraw-Hill New York..

Hanzlova et al 2008. CN derivation from LANDSAT ETM+data and LC changes for rainfall-runoff models. Proceedings of the 28th EARSeL symposium.

Hradek F.et Kurik P. 2002. *Hydrology. University mimeographed,* CZU Prague.

Kulhavy Z., Kovar P. 2000. Hydrological balance models utilization for small watersheds. Metodological help – User output of project NAZV - EP 7062: Evaluation of agricultural farm impact to runoff rate in small watersheds via mathematical modeling. VUMOP Prague, II.2000.

Neruda M., Neruda R. 2002. Comparsion of quantitative and qualitative properties water properties via neural networks. Proceedings of the conference *Extreme hydrological phenomena in floods*, 12. 11. 2002, Ceska vedeckotechnicka vodohospodarska spolecnost, Prague.

Pavelka, K.,1996. New Quality of Remotely Sensed Data in the Next Ten Years. Conference GIS Frontiers in Business and Science. Brno: Masaryk University, 1996, p. 20-41. ISBN 80-210-1330-3.

Rapantova, N., Grmela, A ., Vojtek D., Tylcer, J. 2004. Assessment of natural and artificial recharge in the urban and industrial areas. XXXIII. Congress IAH. Mexico.

Vozenilek, V. 1999: Time and Space in Network Data Structures for Hydrological Modelling. In Craglia M., Onsrud H.: *Geographic Information Research - Trans-Atlantic Perspectives*. Olomouc: Univerzita Palackeho Olomouc, pp. 189-202.

Vysusilova, T, 2007. Bachelor thesis – Comparison of time changes of GIS layer from different data sources. CTU in Prague.

*Remote Sensing for a Changing Europe*
*D. Maktav (Ed.)*
*IOS Press, 2009*


*doi:10.3233/978-1-58603-986-8-53*

# Risk assessment of forest fires by using satellite data with remote sensing techniques

Emre Ozelkan* and Cankut Ormeci
*Informatics Institute, Satellite Remote Sensing and Communication Program, Istanbul Technical University, Maslak-34469, Istanbul, Turkey*
*emre.ozelkan@gmail.com, cankut@itu.edu.tr*

Keywords: Landsat, SPOT, GIS, forest fire, forest fire risk map, Kibriz stream canyon, remote sensing, analysis, disaster, normalized difference vegetation index

ABSTRACT: One of the most important events that cause anomalies in nature's balance is forest fires. Forest fires are also very serious threats in Turkey, which is located in the Mediterranean Region. While complete prevention of a forest fire is impossible, it is possible to reduce the damages of forest fires by constructing a forest fire risk map. In this study, we analyze a forest fire that took place at Kibriz Stream Canyon near the city of Antalya. The most important aspect of this fire was that the interference was very difficult due to the canyon's harsh landscape. Here, we show that a fire risk map can be especially beneficial for this area, and for other areas where fire interference is very difficult due to the unfriendly topography. Remote Sensing and Geographic Information System (GIS) were applied in this study for the assessment of the situation before and after the fire and for forming the fire risk map. Landsat TM (01.08.1990), Spot XS (24.06.2007) imageries and 1/25000 scale topography maps were used to generate a digital terrain model and to establish the land use classes by means of unsupervised and supervised classification algorithms. In addition, the normalized difference vegetation index (NDVI) was computed to compare the classified imageries before and after the fire. Comparison of the NDVI values helped us to determine the vegetation pattern change after the fire.

## 1 INTRODUCTION

One of the most important events that cause anomalies in nature's balance is forest fires. Forest fires are also very serious threats in Turkey, which is located in the Mediterranean Region. In the past sixty years, more than 1.5 million hectare of forest area had disappeared in Turkey due to forest fires. Forest fires influence directly all living things in that area, affecting all natural balance directly or indirectly. While complete prevention of a forest fire is impossible, it is possible to reduce the damages of forest fires by constructing a forest fire risk map. (Hernandez, P.A., 2006)

A forest fire risk map shows us how to take preventive measures before and during the fires. (Coskun, O., 1998) A fire risk map is constructed based on slope, aspect direction, vegetation pattern, altitude and distance from roads of the forest. Based on the fire risk map, the higher risk areas are detected for special preventive treatment and for quick interference during a fire to minimize the damage.

In this study, we analyze a forest fire that took place at Kibriz Stream Canyon near the city of Antalya. This particular fire started in August 18 2006 and lasted more than one week. At the fire location the depth of the canyon is nearly 700–800 meters. More than 450 hectare forest area was destroyed. The most important aspect of this fire was that the interference was very difficult due to the canyon's harsh landscape. Here, we show that a fire risk map can be especially beneficial for this area, and for other areas where fire interference is very difficult due to the unfriendly topography.

Remote Sensing and Geographic Information System (GIS) are applied in this study for the assessment of the situation before and after the fire and for forming the fire risk map. Landsat TM (01.08.1990), Spot XS (24.06.2007) imageries and 1/25000 scale topography maps were used to generate a Digital Terrain Model (DTM) and to establish the land use classes by means of unsupervised and supervised classification algorithms. In addition, in order to compare the classified imageries before and after the fire, the Normalized Difference Vegetation Index (NDVI) values were computed. Comparison of the NDVI values helped us to determine the vegetation pattern change after the fire.

The main contribution of this research are 1) identification of the actual damage in the Kibriz Stream Canyon using remote sensing and GIS data and 2) illustration of how a fire risk map can help in forest fire prevention, and fire interference with an application in Kibriz Stream Canyon.

## 2 STUDY AREA

The area of study, Kibriz Stream Canyon, which is between 36°–37° northern parallels, 29°–30° eastern meridians, is located 210 km far away from the city Antalya.

### 2.1 *Vegetation type*

Most of the area is covered by Turkish Pine. A low height area, where the Turkish Pine is rare, is mostly covered by Maki types. Turkish Pine trees may be found in the area of the Maki vegetation area. These kinds of vegetations are especially susceptible to fire.

### 2.2 *Climatic conditions*

While in general, a Mediterranean climate is observed in this area, terrestrial climate is effective the inner land. Some of the meteorological variables' mean values are:

Annual Rain 782.9 mm
Annual Temperature 19.6 C
Annual Relative Humidity 50%
Annual Wind Speed 2.4 m/sec
Annual Insolation Time 8.11 hour/day

The mean insolation time in August is almost 11.2 hours and the faster wind directions are West-South West, North-North East and East-North East.

### 2.3 *Topography*

Topography, which influence all risk parameters (insolation, humidity, distance from road, aspect and slope) is an important physiographic factor. So the high slope and high elevation structure of the canyon increase the risk of the forest fire. Also wind is very critical parameter for spreading the fire because of the bosphorus structure of Kibriz Stream Canyon.

### 2.4 *Distance from roads*

Most of the root-causes of the fires are related to human activities. Roads are vehicles that help humans to reach the forest areas. So "distance to roads" is an important factor that may increase the fire risk, but on the other hand, roads are also necessary to reach the fire area. Because of the absence of roads, the interference to Kibriz Stream Canyon is very difficult.

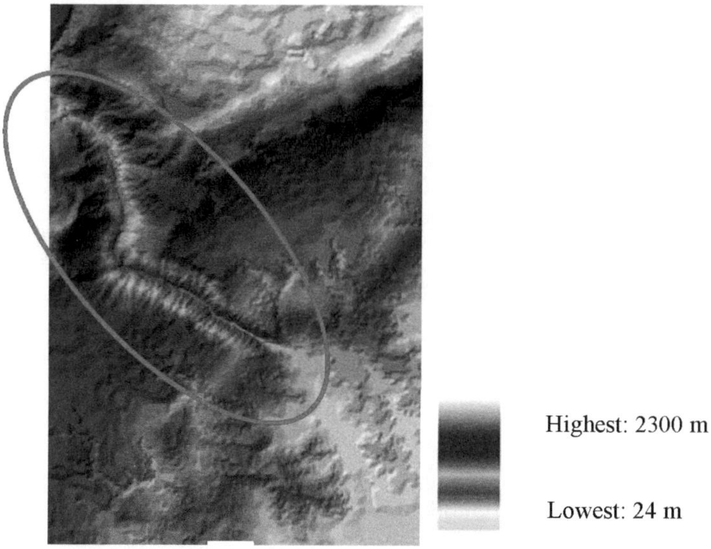


Figure 1. DTM of the study area.

## 3 MATERIALS AND METHODS

### 3.1 *Data*

The image of Landsat TM in 01.08.1990 before the fire and the image of Spot XS in 24.06.2007 after the fire were used to show the effect of the forest fire. Also topographic maps were used to constitute DTM and according to DTM slope, aspect maps were formed.

### 3.2 *Methods*

1/25000 scale topographic maps were rectified according to the WGS84 and a UTM projection and digitized for producing DTM for the study area (Fig. 1).

Also satellite images were geo-coded by the help of rectified topographic maps according to the WGS84 and a UTM projection to combine and analyze with the remaining data.

The purpose of the digital classification is selecting and examining the reliable control areas to exhibit the landscape characteristics (Erten, E., 2005). In this study, the Maximum Likehood supervised and unsupervised classification algorithms were used to classify satellite images (Ormeci, C., 1987). Classification of the satellite images one by one is the best way to compare before and after fire situations. The statistical results of the separate classification of the different date images introduce the affect of the fire.

The risk map of the study area was constructed to show the risk area by classifying the areas from major risk to high risk areas. Fire risk map and the risk classes were constructed based on the following factors: slope, aspect direction, vegetation pattern, and distance from roads of the forest.

Since the vegetation type is an important determinant of risk, they were classified using the Normalized Difference Vegetation Index (NDVI). The high NDVI values indicate the high moisture, healthy vegetation, and the low values indicate the low moisture, unhealthy and vegetation-free areas (Parmiggiani, F.).

The other factor is slope. High slope areas naturally do not sustain high risk but after flame expansion of the fire becomes very fast (Akar I., 2006).

Aspect takes a determinative role on area's humidity, precipitation, wind, insolation time and intensity. Because of the orientation with respect to the sunlight, the vegetation which takes

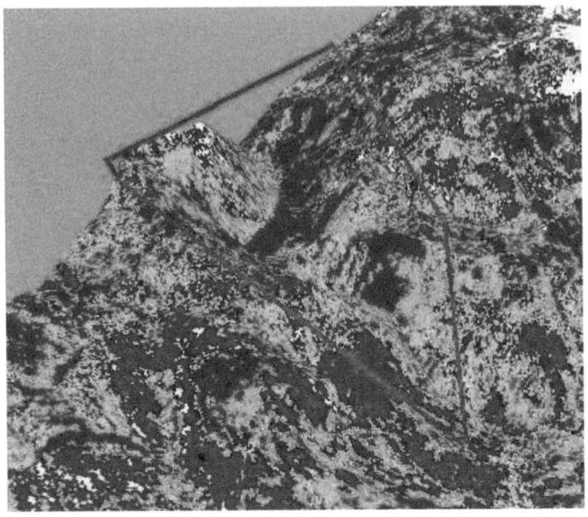

Figure 2. Landsat image classification.

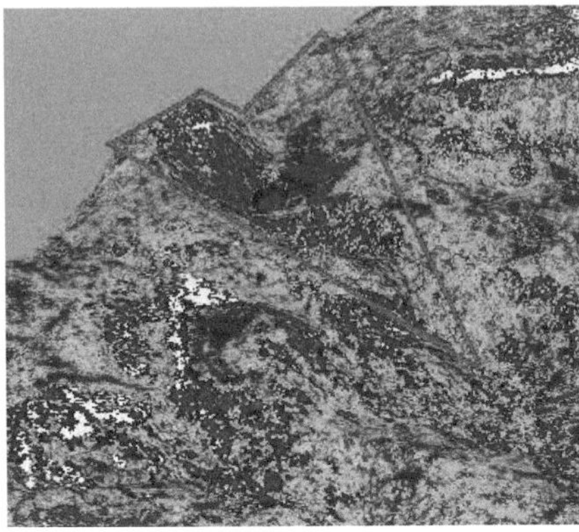


Figure 3. Spot image classification.

sunlight from South faces water shortage problem (Bilgin, 2001). The fire risk probability is high at these kinds of areas. As indicated earlier, distance to roads increases the possibility human source fire risk. For the areas, which are far from the roads, the fire probability is taken low, and vice versa. All risk parameters, weights, classes and factors values are given in Table 1. The multi-criteria equation used in GIS to determine forest fire risk areas is as follows:

$$RC = NDVI*0.9 + Slope*0.7 + Aspect*0.8 + Distance*0.5 + Altitude*0.5$$

## 4 RESULTS

### 4.1 *Classification and NDVI*

Landsat TM and SPOT XS images were used to identify and indicate the forest risky area by the way of classification and NDVI. 7-4-3 band combination for Landsat TM and 3-1-2 band combination for SPOT XS were used for classification and making NDVI analyses. The images were classified into 6 classes as forest 1, forest 2, forest 3, non forest area, river and snow by using maximum like hood algorithm (Figs 2, 3). The images show the effect of the fire clearlyindicating a destruction of more than 450 hectare forest area.

Addition to the classification analyses, NDVI results were used for understanding the forest differentiation after fire. (Parmiggiani, F., 2006) Also NDVI values were classified into 6 classes and 1990 Landsat image NDVI values were used as a risk factor for risk analyses to establish the risk map before the fire.

### 4.2 *GIS results*

GIS results as slope, aspect, distance to roads, altitude, and vegetation index are the factors that form the infrastructure of the risk map. This step of this study exhibits the risk of forest fire at our study area.

#### 4.2.1 *Slope*

In the study area, slope values are changing between 0–90 degrees. And in the study area the slope values are very high, so this situation causes very high risk (Fig. 4).

#### 4.2.2 *Aspect*

The aspect directions are directly proportional to vegetation humidity values. For example; the humidity of vegetation decreases, if the vegetation look to the South direction and it is dry

Table 1. Risk parameters, weights, classes and factors values

| Parameter | Weight | Class | Factor |
|---|---|---|---|
| Vegetation (NDVI) | 0.9 | Few or no vegetation | 1 |
| | | Very dry | 10 |
| | | Dry | 8 |
| | | Moist | 6 |
| | | Fresh like | 4 |
| | | Fresh | 2 |
| Slope | 0.7 | 64–89 | 10 |
| | | 45–64 | 9 |
| | | 36–45 | 8 |
| | | 29–36 | 7 |
| | | 23–29 | 6 |
| | | 17–23 | 5 |
| | | 11–17 | 4 |
| | | 3–11 | 3 |
| | | 0–3 | 2 |
| Aspect | 0.8 | South | 10 |
| | | South West | 9 |
| | | South East | 8 |
| | | Flat | 7 |
| | | West | 6 |
| | | East | 5 |
| | | North west | 4 |
| | | North east | 3 |
| | | North | 2 |
| Distance to roads | 0.6 | 0–308 m | 10 |
| | | 308–616 m | 8 |
| | | 616–925 m | 6 |
| | | 925–1233 m | 4 |
| | | 1233–1541 m | 2 |
| | | 1541–1850 m | 1 |
| Altitudes | 0.5 | 1790–2300 | 10 |
| | | 1425–1790 | 8 |
| | | 1069–1425 | 6 |
| | | 730–1069 | 4 |
| | | 425–730 | 2 |
| | | 24–425 | 1 |

vegetation area which causes high risk. (Fig. 5) The risk order of the directions is: South>South West>South East>Flat>West>East>North West>North East>North.

In the study area, aspect directions which mean insolation values are mostly south, southwest, and southeast and west directions, which cause very high risk. These values are coherent with the NDVI values which show the vegetation humidity potential.

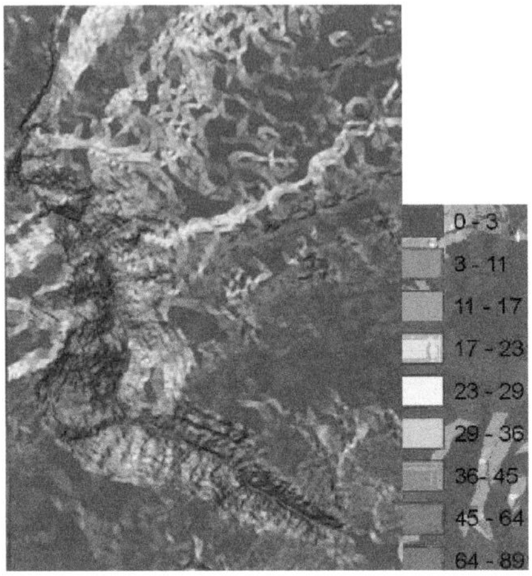


Figure 4. Slope map of the study area.

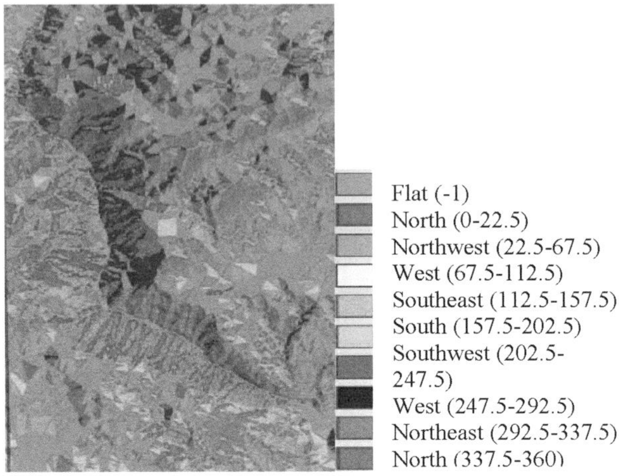


Figure 5. Aspect map of the study area.

Figure 6. Road map of the study area.

### 4.2.3 *Distance to Roads*

The roads which are the linear element of the map, is very important to determine the fire risk. The roads were drown on the 1/25000 scale topographic maps and the distance of each pixels to all roads were calculated. (Fig. 6)

### 4.2.4 *Altitude*

For the humidity and temperature values, high altitude areas are more dangerous than the low altitude areas. With increasing altitude, the humidity and temperature values decrease, resulting in a quick spread of the fire. As shown in the Digital Terrain Model, the altitude of the study area is very high.

### 4.2.5 *NDVI*

As it was mentioned in section 4.1, NDVI values were used for fire risk parameter and also the NDVI application of different date images supply the chance of comparison vegetation changes before fire and after fire (Figs 7, 8) (Talesca, L, Lasaonaro, R., 2005).

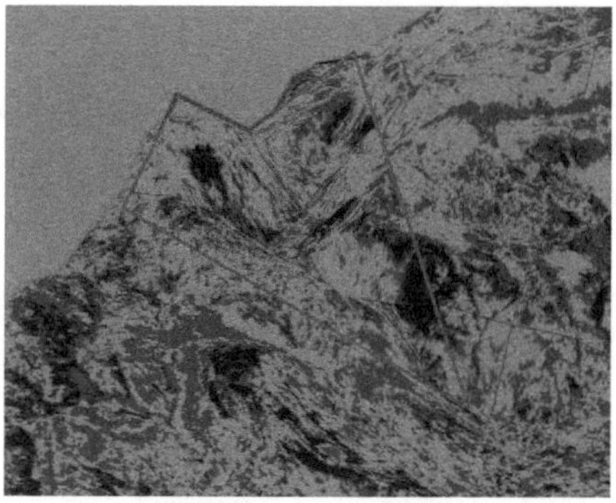

Figure 7. NDVI application before the fire.

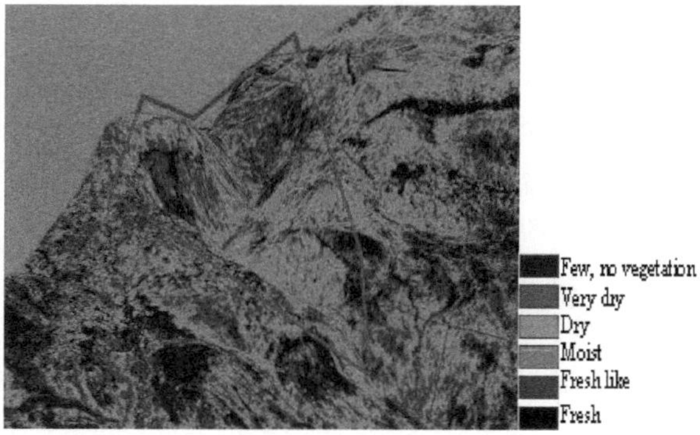


Figure 8. NDVI application after the fire.

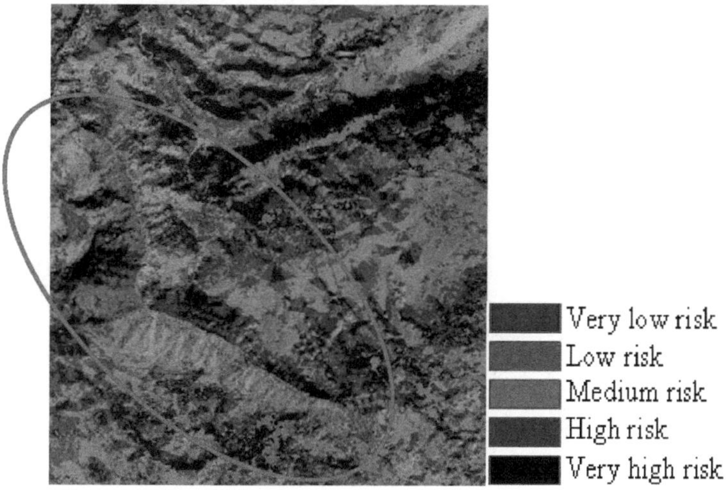


Figure 9. Fire risk map of the study area.

### 4.2.6 *Fire risk map*

Considering all fire risk parameters (slope, aspect, distance to roads, altitude and NDVI values), a fire risk map is established using the risk formula given in section 3.2 and the weight and factor values in Table 1. As shown in Fig. 9, the risk map indicates the fire risk potential of the Kibriz Stream clearly.

## 5 CONCLUSIONS

Fire risk map is a very important instrument to pinpoint and interfere the fire and to decrease the impact of the fire. Here, topographic maps were used to analyze the fire risk parameters (slope, aspect, distance to roads, and altitude), and satellite images were used to analyze NDVI values. Classification and NDVI analyses exhibit the effect of the fire and helped us establish the fire risk map of the study area. We would like to conclude by emphasizing that the fire risk maps can be beneficial especially for those areas such as Kibriz Stream Canyon where interference is very difficult because of the hard topographic conditions.

## REFERENCES

Akar, I., Ozdemir, S., Ozdemir, H., 2006, Using GIS in Geomorphologic Studies: A Case Study of Kasatura Bay and Vicinity, Book of Statements for Information Technology Days, Fatih University, Istanbul, Turkey. 13–16 September, pp. 527–534.

Beldaa, F.,* Melda, J., 2000, Relationships between climatic parameters and forest vegetation: application to burned area in Alicante (Spain); *Forest Ecology and Management.*

Bilgin, T., 2001, General Cartography II. Filiz Bookstore, Istanbul.

Coskun, O, 1998 Forest Fire Analyze by Using Remote Sesning Data, *Master Thesisi,* ITU Library, Istanbul.

Erten, E., Kurgun, V., Musaoglu, N., 2005, Forest Fire Risk Zone Mapping From Satellite Imagery and GIS a Case Study.

Hernandez, P.A.,* Arbelo, M, Gonzales Calvo, A., 2006, Fire Risk Assessment Using Satellite Data, ELSEVIER.

Ormeci, C., 1987, Basic Princiles and Percaption Analyses of Remote Sensing, Volume 1, ITU Library, Istanbul.

Parmiggiani, F., Quarto, G., MARRA, G., and Dario Conte, 2006, NDVI fluctuations from 1995 to 2006 in South Italy and North Africa: a search for or a climate change indicator.

Talesca, L., Lasaonaro, R., 2005, Vegetational patterns in burned and unburned areas investigated by using the detrended fuctuation analysis.

*Remote Sensing for a Changing Europe*
*D. Maktav (Ed.)*
*IOS Press, 2009*


*doi:10.3233/978-1-58603-986-8-61*

# Assessment of different topographic corrections in MODIS data for mapping effective snow covered areas in mountainous terrain

Z. Akyürek & A. Ü. Şorman
*Middle East Technical University, Civil Engineering Department, Ankara, Turkey, zakyurek@metu.edu.tr, sorman@metu.edu.tr*

Keywords: Snow, mountainous terrain, MODIS, topographic normalization

ABSTRACT: Topography and its derivatives (altitude, slope and aspect) have an effect on satellite-measured radiances. For mountainous areas the sun zenith and azimuth angles, as well as direction of observation relative to these are more limiting factors. In this paper four topographic normalization methods were used to correct the reflectance values of medium spatial resolution satellite data, namely MODIS. The performance of the topographic normalization methods is examined for snow covered areas of the study area located in the eastern part of Turkey. Modeling of snow-covered area in the mountainous regions of Eastern Turkey has significant importance in order to forecast snowmelt discharge especially for optimum use of water in energy production, flood control, irrigation and reservoir operation optimization. MOD09GKM data, which have the land surface reflectance having atmospheric correction, digital elevation model (DEM) and the geo-location files (MOD03) were used. It is obtained that statistical empirical correction method worked better compared to the other methods in removing the terrain effects for snow covered areas. The importance of topographic normalization in mapping the effective snow covered area in snowmelt modeling is also discussed and the early findings of Satellite Application Facilities on Hydrology (H-SAF) project, which is financially supported by EUMETSAT, is presented. Turkey is a part of the H-SAF project, both in product generation (e.g. snow recognition, fractional snow cover and snow water equivalent) for mountainous regions for whole Europe, cal/val of satellite-derived snow products with ground observations (synoptic, automated weather stations and snow courses) and impact studies with hydrological modeling in the mountainous terrain of Europe.

## 1 INTRODUCTION

Optical and near-infrared sensors can distinguish between snow-covered and snow-free ground. This is based on the very high reflectance of snow in the visible (0.5~0.7 μm) wavelengths compared to other natural targets (Warren, 1982). The Nasa Earth Observing System (EOS) moderate-resolution imaging spectroradiometer (MODIS) provides global automated binary maps of snow cover by using a normalized difference snow index (NDSI) with threshold tests. The approach used for obtaining the snow-covered area (SCA) maps from MODIS images employs the advantage of the fact that snow reflectance is high in the visible (0.5~0.7 μm) wavelengths and has low reflectance in the shortwave infrared (1~4 μm) wavelengths (Hall et al., 2001).

Snow in mountainous is challenging to map with optical remote sensing techniques because of the relief or topography may affect the image radiometry. This effect causes a high variation in the reflectance response for similar target types: shaded areas show less than expected reflectance, whereas in sunny areas the effect is the opposite. Akyurek and Şorman (2002a) analyzed the effect of topography on snow depletion curves derived from NOAA-AVHRR images for the Karasu Basin located in the eastern part of Turkey. Srinivasulu and Kulkarni (2004) used a model to estimate spectral reflectance for the Himalayan terrain using the LISS-III data from Indian Remote Sensing satellites. They suggest the use of terrain slope and aspect in estimating the accurate spectral reflectances over the Himalayan terrain. Chaponniere et al. (2005) used a combined high and low spatial resolution approach for mapping snow covered areas in the Atlas Mountains, where monitoring the snow dynamics can not be performed routinely due to management costs and difficult access. Richter (1998) developed a method (ATCOR3) for the radiometric and atmospheric correction of satellite images over mountainous terrain. The proposed model utilize the small angle approximation, i.e. the sensor's swath angle is less than $+\!-8^{\circ}$ with respect to the image center and the solar zenith angle is assumed constant for a recorded scene. These scenes are provided by high spatial resolution satellite sensors such as Landsat TM, SPOT, Quickbird, and Ikonos. Therefore, the process of topographic normalization may be critical in areas for rough terrain, as a preliminary step for determining the fractional snow cover area from images having moderate spatial resolution.

Turkey can be described as a country with abundant snow cover on mountainous regions in the eastern part of the country where Euphrates River basin is located. This basin is largely fed from snow precipitation whereby nearly two-thirds occur in winter and may remain in the form of snow for half of the year. The river feeds large reservoirs with great volume of runoff. The concentration of discharge mainly from snowmelt during spring and early summer months causes not only extensive flooding, inundating large areas, but also the loss of much needed water required for irrigation and power generation purposes during the summer season. Accordingly, modeling of snow-covered area in the mountainous regions of Eastern Turkey, as being one of the major headwaters of Euphrates–Tigris basin, has significant importance in order to forecast snowmelt discharge especially for energy production, flood control, irrigation and reservoir operation optimization. There is not yet a well established operational snow monitoring system in the country. Therefore comparison of satellite derived snow maps and snow course ground measurements is vital for improvement of the existing mapping algorithms. The snow cover area validation was performed by Tekeli et al. (2005) using the MODIS images during the accumulation and ablation periods of 2002–2003 water year and as well during the winter period of 2003–2004. Over the ablation period of 2004, daily snow albedo values retrieved from MODIS Terra were compared with ground-based albedo measurements (Tekeli et al., 2006) for the same area. MODIS-8 day snow products (MOD10A2) were used to model the snowmelt runoff process in the eastern part of Turkey (Tekeli et al., 2006). They found out that the hydrograph shape might have been affected by the elevation bias of the MODIS snow mapping algorithm, which may under represent the snow cover in lower elevation regions and over represent the snow in the higher elevation ranges. This leads the need of fractional snow-covered area determination from the MODIS-data. There are several remote sensing approaches that have been applied for estimating the fractional snow cover within a sensor pixel (Vikhamar and Solberg, 2003; Kaufman, Kleidman, Hall, Martis and Barton, 2002; Barton, Hall and Riggs, 2001; Salomonson and Appel, 2004; Metsamaki et al., 2005).

In this study widely used topographic normalization methods, available in the literature, are applied to MODIS data, having medium spatial resolution, in order to correct the reflectance values of the satellite data due to terrain effect. The performance of these methods in finding the fractional snow cover area is examined.

## 2 DATA

The study area is located in the eastern part of Turkey called Karasu basin (Fig. 1) which is the headwater of the Euphrates River. The basin has a drainage area of 10 216 $km^2$ controlled by the stream gauging station with an elevation ranging from 1125 m to 3487 m. The topographic map of the basin (DEM) and the location of the AWOS stations are shown in Fig. 2.

The dataset consists of the Terra MODIS images, a digital elevation model (DEM), land cover/use map derived from Landsat image and validated by ground survey, snow measurements and meteorological observations. The MODIS Level 1B radiance files at 500 m resolution (MOD02HKM), land surface reflectance files at 500 m resolution (MOD09GHK) and geolocation

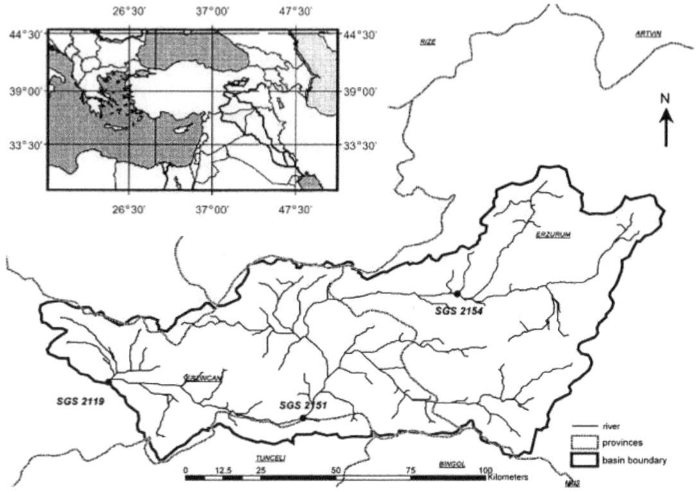


Figure 1. Location of Karasu Basin (upper Euphrates River) in Turkey and the location of stream gauges in the basin.

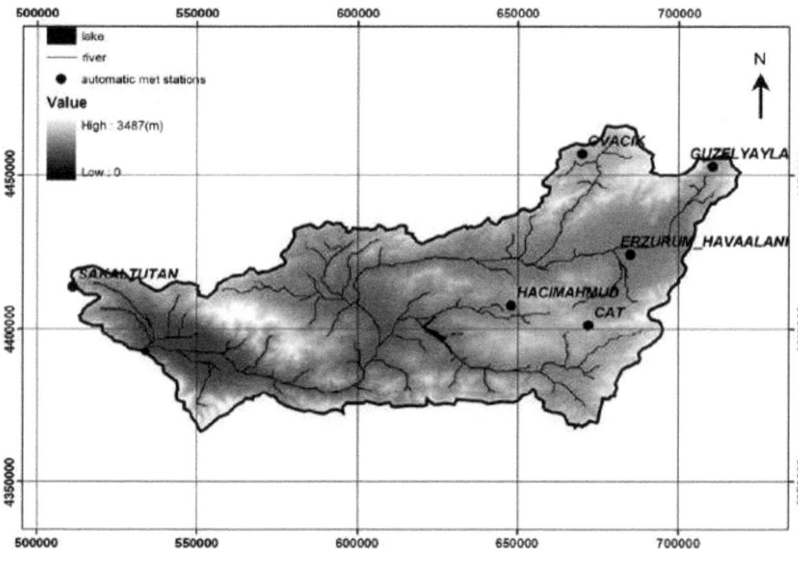


Figure 2. Location of automatic stations on DEM of Karasu Basin.

files (MOD03) acquired on 14 March 2006 and 30 March 2006 were downloaded from the internet site http://daac.gsfc.nasa.gov/MODIS/data access.html. Those two dates were selected for the analysis, since the snow water equivalence values measured at AWOS were at their max value for those dates, in April the starting of melting was observed. The land cover information was obtained from a previous study (Akyurek and Şorman, 2002b), where the land cover types were obtained from high spatial resolution image, using a multi-level classification technique based on knowledge-based segmentation of Landsat images. The level of the land use/cover classification was determined according to CORINE land cover classification scheme. The field data (snow depth, meteorological observations) were obtained from the AWOS at the site (Fig. 2) for the same dates.

## 3 METHODS

Digital values (DVs) derived from satellite optical systems cannot be used for geophysical measurements and multitemporal studies, because of including effects derived from sensor calibration, atmospheric and topographic interferences. The image metadata provides the calibration values to transform DV to sensor radiance. The model to obtain true reflectance from sensor radiance may be expressed as in the Equation 1 (Richter, 1998).

$$\rho_k = \frac{K\pi\left(\dfrac{(L_{sen,k} - L_{a,k})}{\tau_{k,0}}\right)}{E_{0,k}\cos\theta_z \tau_{k,i} + E_{d,k}} \tag{1}$$

where $K$ is a correction factor of the annual variations of earth-sun distance, computed from Julian day; $L_{sen,k}$, is the sensor radiance in band $k$; $L_{a,k}$ is the atmospheric upwelling radiance scattered at the sensor for the same band; $\tau_{k,0}$ and $\tau_{k,i}$ are the path atmospheric transmittances of the upwelling (ground surface-sensor path) and downwelling (sun-ground surface path) flows, respectively; $E_{o,k}$ is the solar irradiance at the top of the atmosphere; $\theta_z$ is the sun zenith angle; and $E_{d,k}$ is the diffuse irradiance at the surface.

### 3.1 *Atmospheric correction*

Satellite measurements of the Earth's surface are contaminated by atmospheric effects. Various problems arise during interpretation of remotely-sensed data due to this atmospheric contamination. The top of the atmosphere (TOA) reflectance derived from radiance measured by sensors may be increased or decreased when compared to the surface reflectance as a function of the reflectance of the target and of its environment. The radiative transfer in clear atmosphere is now well understood and accurate models exist (Rahman and Dedieu, 1994). However there are still the problems of developing a fast method for the inversion of surface reflectances from remotely-sensed measurements and to determine the atmospheric parameters for large areas and long time periods.

### 3.2 *Topographic correction*

Topography has an effect on satellite-measured radiances. For mountainous areas the sun zenith and azimuth angles, as well as direction of observation relative to these are more limiting factors. For mountainous terrain three effects that caused by the topography can be listed: 1) Some areas receive exclusively diffuse irradiance due to cast shadows; 2) Shielding of the sky hemisphere reduces the diffuse irradiance; and 3) surrounding terrain reflects irradiance towards the observed ground area (Proy et al., 1989). The shadowed areas become smaller on slopes facing the sun, while they increase on slopes oriented away from the sun (Gemmel, 1998). Topographic correction should affect the denominator parameters of Equation 1, since changing illumination conditions affect the actual solar irradiance received at a single pixel.

Several methods have been developed so far with the purpose of removing terrain effects from the measured pixel radiance and available in the literature. Widely used methods are the Lambertian cosine correction, the statistical-empirical correction, the C-correction and the Minnaert correction (Vikhamar et al., 2004; Riano et al., 2003). The general approach of these methods is to normalize the observed radiance from inclined surfaces ($L_T$) to flat (horizontal) surfaces ($L_H$) by modelling the local incidence angle to the terrain surface cos (i) for each pixel (Table 1). Illumination angle; IL ($\gamma$i) is defined as the angle between the surface normal and the solar beam (Fig. 3). Using information about the solar position at the acquisition time for the satellite image and the local terrain relief, il can be calculated for a pixel by the formula (Smith et al., 1980):

Table 1. Methods for topographic correction of satellite measured radiances

| Method | Equation |
|---|---|
| Lambertian cosine correction | $L_H = L_T\left(\frac{\cos(sz)}{\cos(i)}\right)$ |
| C-correction | $L_H = L_T\left(\frac{\cos(sz)+C}{\cos(i)+C}\right)$ |
| Minnaert correction | $L_H = L_T\left(\frac{\cos(sz)}{\cos(i)}\right)^k$ |
| Statistic-empirical correction | $L_H = L_T - \cos(i)m - b + \overline{L_T}$ |

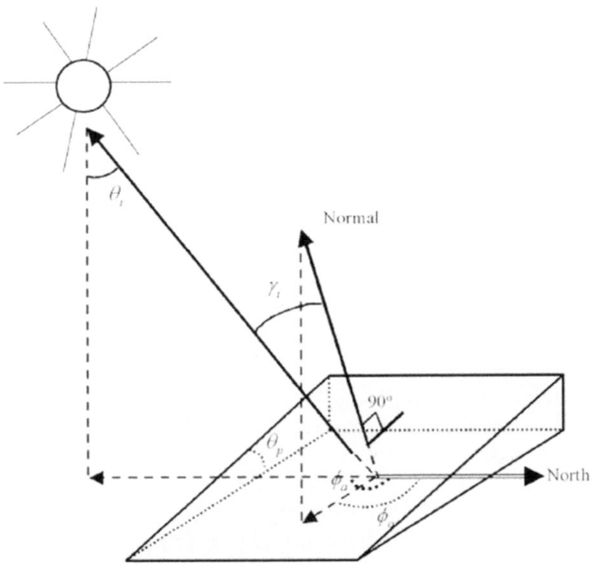


Figure 3. Angles involved in the computation of illumination angle (il) (After Riano et al., 2003).

IL = Cos i = cos sz cos tz + sin sz sin tz cos (sa-ta) (2)

where sz(θi) is the solar zenith angle, sa (ϕo )is the solar azimuth angle, tz(θp) is the surface normal zenith angle or the terrain slope and ta (ϕa) is the terrain azimuth angle.

The most critical points for the radiometric correction are bidirectional reflectance effects, atmospheric correction, the spatial resolution of the DEM, and the calculation of slope and aspect from the digital terrain data as topographic factors.

### 3.3 *Fractional snow cover modeling*

Taking advantage of the fact that snow reflectance is high in the visible (0.5–0.7 μm) wavelengths and has low reflectance in the shortwave infrared (1–4 μm) wavelengths to enable distinguishing snow fromclouds and other non-snow-covered conditions, Hall et al. (2002) have used the normalized difference snow index (NDSI) to develop an automated approach to provide daily, global observations of snow cover. In this study the fractional snow cover area (SCA) algorithm is based on a sub-pixel reflectance model. Pixel reflectance is modelled as a linear mixture of snow and snow free bare ground. The original model has previously been developed by Vikhamar et al. (2004). In that model pixel reflectance is modelled as a linear mixture of snow, individual tree species and snow-free bare ground (e.g. rock, soil, low vegetation):

$$\overline{R}_G = A_p R_p + A_s R_s + A_B R_B + A_{SW} R_{SW} + A_{BG} R_{BG} \tag{3}$$

where $\overline{R}_G$ is the modelled pixel reflectance for a given wavelength and $A_p + A_{ss} + A_B + A_{SW} + A_{BG} = 1$ A represents area fractions of a pixel and $R$ is reflectance. The subscripts $P$, $S$ and $B$ refer to several different trees ($P$: pine, $S$: spruce, $B$: birch), SW and BG refer to snow and bare ground respectively. For the mountainous areas the Equation 3 consists of snow and bare ground, since the trees are not available at high altitudes in the study area. Therefore the area of the pixel having fractional snow cover can be represented as $A_{SW} + A_{BG} = 1$.

## 4 RESULTS

### 4.1 *Atmospheric correction*

Atmospheric correction is important to obtain the true reflectance of the Earth features. However rather than applying an atmospheric correction method to the downloaded data, atmospherically corrected data are downloaded. In this part MOD02HKM data, this is the calibrated Earth view data at 500 m resolution, and MOD09GHK, which is the land surface reflectances having atmospheric correction, were used.

Salomonson and Appel (2004) used the radiances observed at the top of the atmosphere in estimating the fractional snow cover from MODIS using the normalized difference snow index. They underlined the need of showing the effect of atmospheric correction on the reflectances derived from MODIS observations for snow index computation. In this study the NDSI was calculated both for MOD02HKM and MOD09GHK data (Fig. 4). It is seen from this figure that NDSI computed from the reflectance values (MOD09) for snow covered pixels have larger scattering compared to the NDSI values computed from the radiances (MOD02). This indicates that for fractional snow cover determination atmospherically corrected satellite data can give better results.

### 4.2 *Topographic effect*

In using the MODIS data artifacts might arise when the DEM resolution cell is not adequate for the pixel size of the imaging satellite sensor (Richter, 1998). DEM having a spatial resolution of 0.25 times the pixel size or better is recommended to be desirable for the topographic correction. In this study DEM having a resolution of 100 m (Fig. 2) was used.

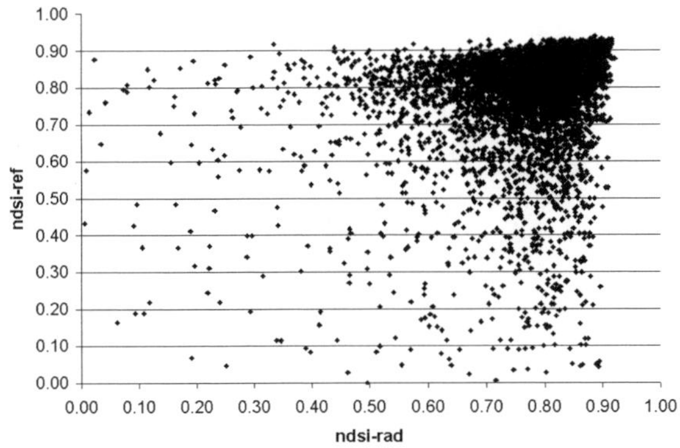


Figure 4. NDSI computed from MOD02 radiance (x-axis) and MOD09 reflectance (y-axis) values for March 14, 2006.

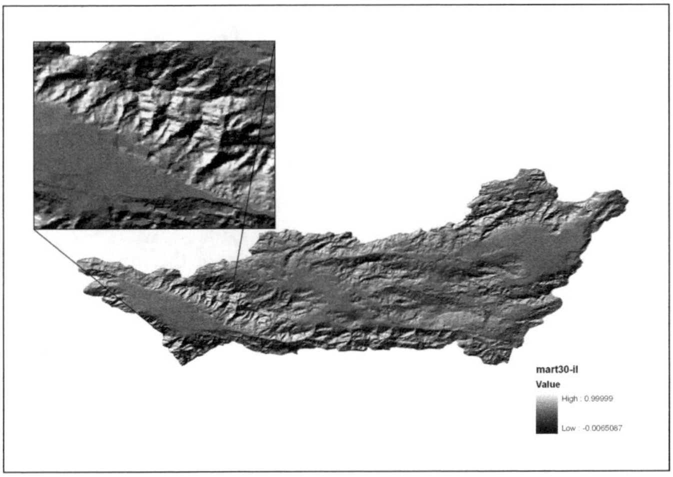


Figure 5. Illumination (il) map of March 30 2006.

There is no topographic correction method generally accepted by the scientific community for all applications. As it is summarized in Table 1, four commonly used topographic correction methods (Lambertian Cosine correction, C-correction, Minnaert correction and Statistic-empirical correction) were applied for the optic band (Band 4) of MOD09GHK data. MODIS view angles, the sun zenith and azimuth angles were obtained from MOD03 data for March 14, 2006 and March 30, 2006, and the illumination maps for March 14 and March 30 were obtained by using Equation 2. The illumination map of March 30 is depicted in Fig. 5. From the analysis no illumination effect on the infrared band was seen. Although atmospherically corrected images were used in the analysis, still reflectances for band 6 greater than 100% were observed in the fully snow covered pixels of the images. In theory, reflectance values should lie within the 0–100% range and the NDSI ratio should lie within the −1.0 to +1.0 ranges. These violations suggest that error or other anomalies may have crept into the input data and indicate that further investigation may be warranted to uncover the causes (MODIS ATBD, 1999). Summary statistics are kept within the algorithm for pixels that exceed violations of these limits; however the test for topographic normalization is done regardless of violations of these limits.

The effect of the applied correction methods are given in Fig. 6 and Fig. 7. Considering the correction for two dates (March 14, March 30), the results obtained from the statistic-empirical correction method are comparatively good.

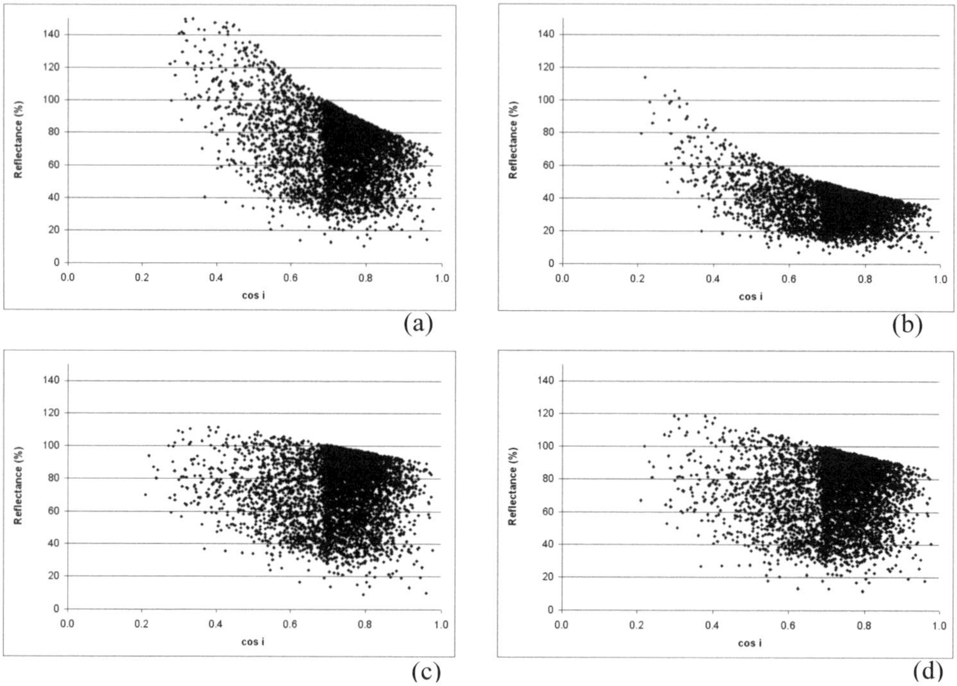


Figure 6. The effect of different topographic correction methods on the reflectances for the image acquired on March 14, 2006: a-cosine correction, b-Mineart correction (k = 0.5), c-statistic-empirical correction, d-C correction.

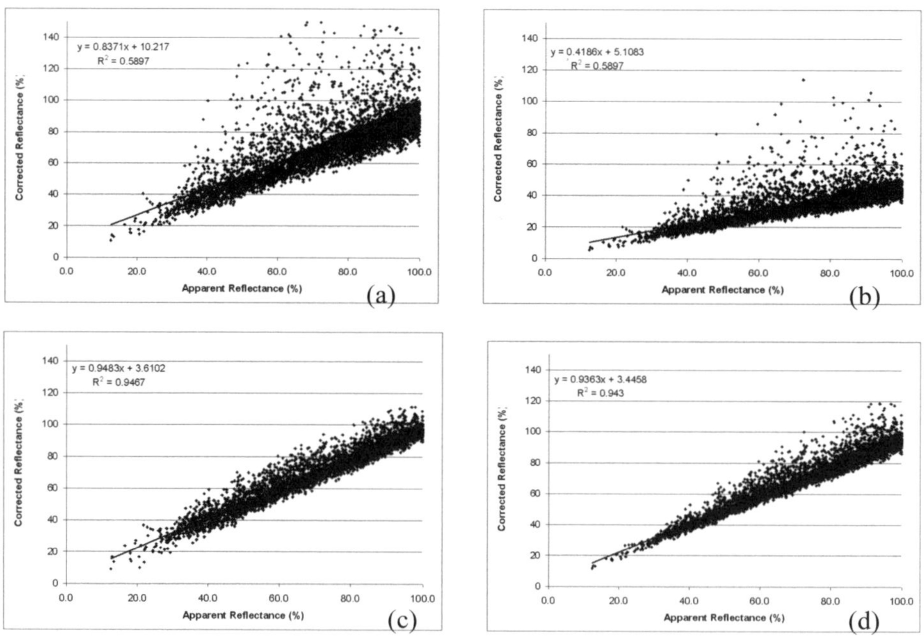


Figure 7. The corrected and apparent reflectance from March 14, 2006: a-cosine correction, b-Mineart correction (k = 0.5), c-statistic-empirical correction, d-C correction.

The Lambertian cosine correction over-corrects significantly for low cos(i) values due to near zero division (Fig. 6a). Teillet et al. (1982), Meyer et al. (1993), Ekstrand (1996) and Vikhamar et al. (2004) described this observation. Areas with cos(i) values less than zero are located in shadows and in this study these areas were excluded from the analysis in order to avoid the over-correction. The Minnaert correction is a modification of Lambertian cosine correction method by indroducing k parameter depending on the land cover types. For snow covered areas Minneart correction with different k values were applied, results obtained with k = 0.5 are given in Fig. 6b. However the results of this correction were generally weak, with some improvements over the Lambertian cosine correction for snow covered area. The C-correction is a modification of the Lambertian cosine correction, which displaces the over-correction towards lower cos (i) values (Fig. 6d). Statistic-empirical correction gave the best results for the snow covered areas (Fig. 6c).

### 4.3 *Fractional snow cover area mapping*

The fractional snow cover area (SCA) algorithm which is based on a sub-pixel reflectance model was applied for the study area. Since the land cover in the study area is not complex and it is composed of bare ground and pasture, the pixel reflectance is modeled as a linear mixture of snow and snow free bare ground. The bare ground and pasture classes are grouped into bare ground class. NDSI was used as an indicator between pure snow and non-snow classes located on the horizontal terrain in the study area. It is known that this grouping would bring error in determining the pure reflectances of the snow free area in the late spring months (e.g. April and May) where pasture will grow up and the greeness of the pasture will force to make three classes (bare ground, pasture, snow) rather than two (bare ground, snow). An accurate estimation of pure spectral reflectances (end-members) of the land cover classes is crucial to a successful application of the linear mixture model. The end members can be determined from the data (image end-members), from a spectral library or from field reflectance measurements (reference end-members) (Roberts et al., 1998). In this study, since the land cover is not complex in the study area, the pure spectral reflectances of snow and bare ground were determined from the image with the help of a land cover map.

Figure 8 shows an example of snow cover map over the study area in the east part of Turkey (Karasu basin). The ground observations from the AWOS are also marked on the figure indicating the fractional SCA values at the observations, the values for Hacımahmut and Guzelyayla stations are also given in Table 2.

## 5 CONCLUSIONS

The results obtained in this paper indicate the needs of topographic normalization for rough terrains. To map the snow in mountainous terrain, the primary factors, snow free bare ground, wavelength and topography should be accounted for. The performance of four extensively used topographic correction methods was examined for snow covered areas in mountainous terrain. Statistical empirical correction method worked better compared to the other methods in removing the terrain effects for snow covered areas. Better performance of statistical empirical correction method was also obtained by Vilkhamar et al. (2004) for snow-covered forest in hilly areas.

The FSCA maps obtained by sub-pixel reflectance algorithm give comparatively better results than binary snow products. In order to improve the algorithm the reflectance values (end-members) of different land cover classes are started to be determined and the sub-pixel reflection model will be implemented to cover other land classes like; snow, bare ground, pasture and forest. Thresholding approach was used to determine the end-members from the images, some other techniques like principle component analysis (PCA), independent component analysis (ICA) can be also implemented to determine the end-members of different land cover classes from the image.

The algorithm is under development in retrieving fractional snow cover maps from NOAA-AVHRR data on a daily basis within the H-SAF project.

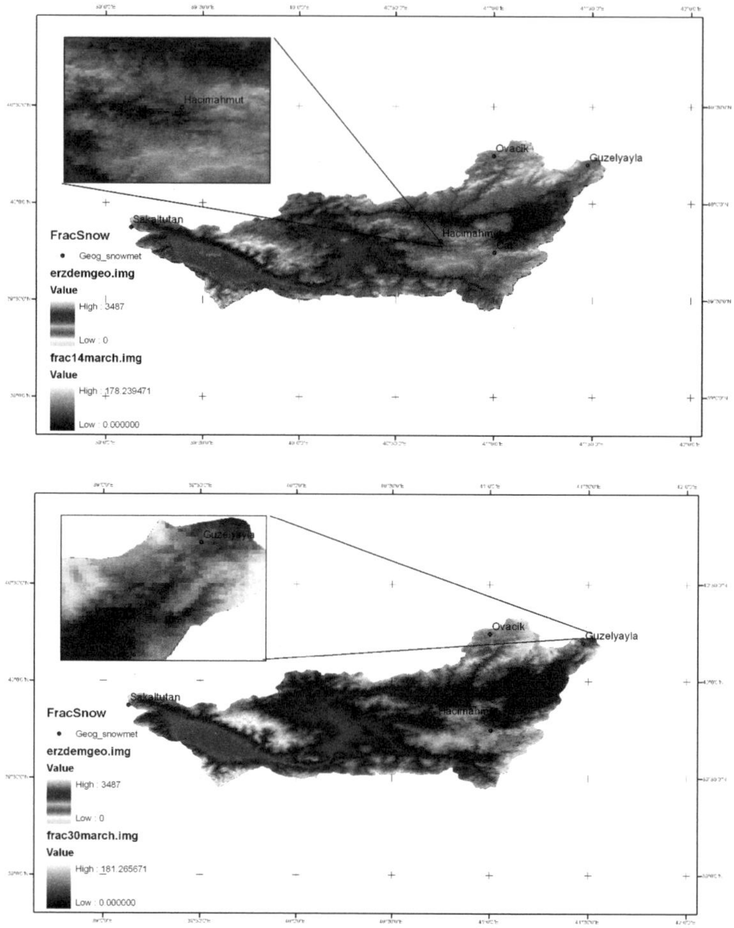


Figure 8. Snow mask maps in 500 × 500 m grid cells for Karasu Basin (14th of March 2006 (top) and 30th of March 2006 (bottom)). In the maps, SCA coverage is presented as running colours describing SCA percentage.

Table 2. Comparison of the fractional snow cover values with NDSI values at AWOS

| **AWOS** | | **14.March.2006** | | | | **30.March.2006** | | | |
|---|---|---|---|---|---|---|---|---|---|
| Name | Elev. (m) | SD (cm) | SWE (mm) | SCA (%) S-pixel | NDSI | SD (cm) | SWE (mm) | SCA (%) S-pixel | NDSI |
| Hacımahmut | 1965 | 14.5 | 58.0 | 73.9 | 100 | – | – | – | – |
| Guzelyayla | 2065 | 53.0 | 168.6 | 99.1 | 100 | 20.41 | No obs | 74.8 | 100 |
| Ovacik | 2130 | 67.7 | 209.2 | 100 | 100 | 64.8 | 245.2 | 94.9 | 100 |
| Sakaltutan | 2150 | 83.06 | 199.34 | 100 | 100 | No obs | No obs | 95.3 | 100 |
| Cat | 2340 | 93.75 | 335.8 | 100 | 100 | 91.89 | 349.2 | 95.9 | 100 |

SD: snow depth, SWE: snow water equivalence, SCA: snow cover area, S-pixel: sub-pixel

## ACKNOWLEDGEMENTS

This paper was prepared within the framework of H-SAF project supported by EUMETSAT. We kindly thank to H-SAF snow team for their valuable contributions.

## REFERENCES

Akyurek, Z., & Şorman, A. U. (2002a). Monitoring snow covered areas using NOAA-AVHRR data in the eastern part of Turkey. *Hydrological Sciences Journal*, 47(2), 243–252.

Akyurek, Z., & Şorman, A.U., (2002b). Accuracy assessment of a Landsat assisted land-cover mapping. Case study: City of Erzurum and its vicinity—Turkey. *3rd International Symposium Remote Sensing of Urban Areas*, Istanbul, Turkey, 11 – 13. June, 2002, 2 (pp. 529–536).

Barton, J. S., Hall, D. K., & Riggs, G. A. (2001). Remote sensing of fractional snow cover using Moderate Resolution Imaging Spectroradiometer (MODIS) data. *Proceedings of the 57th Eastern Snow Conference*, May 17–19, 2000, Syracuse, NY, 171–183.

Chaponniere E., Maisongrande, P., Duchemin, B., Haich, L., Boulet, G., Escadafal, R., & Elouaddat, S. (2005). "A combined high and low spatial resolution approach for mapping snow covered areas in the Atlas mountains". *International Journal of Remote Sensing*, 26(13), 2755–2777.

Ekstrand, S. (1996). Landsat TM-based forest damage assessment correction for topographic effects. *Photogrammetric Engineering and Remote Sensing*, 62(2), 151–161.

Gemmel, F. (1998). An investigation of terrain effects on the inversion of a forest reflectance model. *Remote Sensing of Environment*, 65, 155–169.

Hall, D. K., Riggs, G. A., & Salomonson, V. V. (2001). Algorithm theoretical basis document (ATBD) for the MODIS snow and sea ice mapping algorithms, Available at: http://modis.gsfc.nasa.gov/data/atbd/atbd_mod10.pdf.

Hall, D. K., Riggs, G. A., Salomonson, V. V., DiGirolamo, N. E., & Bayr, K. J. (2002). MODIS snow cover products. *Remote Sensing of Environment*, 83, 181–194.

Kaufman, Y. J., Kleidman, R. G., Hall, D. K., Martins, J. V., and Barton, J. S. (2002). "Remote Sensing of subpixel snow cover using 0.66 and 2.1 µm channels". *Geophysical Research Letters*, 29(16), 1781 (doi:10.1029.2001GLO13580).

Metsamaki, S., Anttila, S.T., Markus, H.J., & Vepsalainen, J.M. (2005). A feasible method for fractional snow cover mapping in boreal zone based on reflectance model. *Remote Sensing of Environment*, 95, 77–95.

Meyer, P., Itten, K.I., Kellenberger, T., Sandmeier, S., & Sandmeier, R. (1993). Radiometric corrections of topographically induced effects on Landsat TM data in an Alphine environment. *ISPRS Journal of Photogrammetry and Remote Sensing*, 48(4), 17–28.

Modis ATBD (1999). Atmospheric Correction Algorithm: Spectral Reflectances (MOD09) http://modissr.ltdri.org/MAIN_SURFACE_PRODUCTAND%20USER%20GUIDE/atbd_mod08.pdf.

Proy, C., Tanre , D., & Deschamps, P. Y. (1989). Evaluation of topographic effects in remotely sensed data. *Remote Sensing of Environment*, 30, 21–32.

Rahman H. & Dedieu, G., (1994). SMAC: a simplified method for the atmospheric correction of satellite measurements in the solar spectrum. *Int. J. Remote Sensing*, 15(1), 123–143.

Riano, D., Chuvieco, E., Salas, J., & Aguado I. (2003). Assessment of different topographic corrections in Landsat-TM data for mapping vegetation types. *IEEE Transactions on Geoscience and Remote Sensing*, 41(5), 1056–1061.

Richter R., (1998). Correction of satellite imagery over mountainous terrain. *Applied Optics*, 37, 18, 4004–4015.

Salomonson, V.V. & Appel, I., (2004). Estimating fractional snow cover from MODIS using the normalized difference snow index. *Remote Sensing of Environment* 89, 351–360.

Simith, J. A., Lin, T.L., & Ranson, K.J. (1980). The Lambertian assumption and Landsat data. *Photogrammetric Engineering and Remote Sensing*, 46(9), 1183–1189.

Srinivasulu, J. & Kulkarni, A.V. (2004). Estimation of spectral reflectance of snow from IRS-1D LISS-III sensor over the Himalayan terrain. Proc. *Indian Acad. Sci. (Earth Planet Sci.)*, 113, 1, 117–128.

Teillet, P.M., Guindon, B., & Goodenough, D. G. (1982). On the slope-aspect correction of multispectral scanner data. *Canadian Journal of Remote Sensing*, 8(2), 84–106.

Tekeli E., Z. Akyurek, A.A. Şorman, A. Şensoy, & A.Ü. Şorman, (2005). Using MODIS Snow Cover Maps in Modelling Snowmelt Runoff Process in the Eastern Part of Turkey, *Remote Sensing of Environment*, 97, 216–230.
Tekeli E., A. Şensoy, A.A. Şorman, Z. Akyurek, & A.Ü. Şorman, (2006). Accuracy assessment of MODIS daily snow albedo retrievals with in situ measurements in Karasu basin, Turkey. *Hydrological Processes*, 20(4), 705–721.
Vikhamar D., Solberg R., & Seidel K. (2004). Reflectance Modelling of Snow covered Forests in Hilly Terrain. *Photogramettric Engineering and Remote Sensing*, 70, 9, 1069–1079.
Vikhamar, D., & Solberg, R. (2003). Subpixel mapping of snow cover in forests by optical remote sensing. *Remote Sensing of Environment*, 84(1), 69–82.
Warren, S. G. (1982). Optical properties of snow. *Reviews of Geophysics and Space Physics*, 20(1), 2–52.

*Remote Sensing for a Changing Europe*
*D. Maktav (Ed.)*
*IOS Press, 2009*

*doi:10.3233/978-1-58603-986-8-73*



# Classification and information extraction in very high resolution satellite images for tree crops monitoring

B. Mougel, C. Lelong
*CIRAD/UMR TETIS – Montpellier- FRANCE*

J.-M. Nicolas
*Telecom Paris - FRANCE*

Keywords: Classification; Fourier; Tree; Groves, Very high resolution.

ABSTRACT:
Recent access to Very High Spatial Resolution (VHSR) Satellite Images allows vegetation monitoring at metric and sub-metric scale, with individual trees now detectable. Therefore, it discloses new applications in precision agriculture for orchards and other tree crops.
In this paper, we present some methodological directions for classification, and extraction of specific agricultural information from these images. Aims are tree crop detection, plot mapping, species identification, and cropping-system characterization. This latter includes for instance row management (e.g. grid vs. line pattern, width of rows and inter-rows, row orientation), crown shape, and crown size estimation.
In this paper, we skip the segmentation step and consider that we have got a precise delimitation of plots that have a homogeneous content. To classify these plots, we have used expert knowledge in agronomy combined with image information in a decision tree.
Classification criteria were based on parameters resulting from the Fourier transform analysis or vegetation indices, derived as one single descriptor for the whole plot.
As a conclusion, the proposed methodology was found capable of classification and characterization of tree crops, provided the trees are clearly seen from above, and their planting is regular enough to give a response with Fourier analysis.

## INTRODUCTION

This study is part of the ORFEO methodological program led by CNES (the French Space Agency) and several research institutes for the development of algorithms dedicated to image processing of the future Very High Spatial Resolution (VHSR) *Pléiades* sensor. This part of the project aims at developing automatic tools for the recognition of landscapes elements, such as groves and other tree plantations.

Since 2001 a new generation of satellite sensors delivers more accurate details and information of the Earth surface. We are now able to distinguish individual trees in VHSR satellite images. They should allow a better identification of the landscape units based on their content like, for the groves, the identification of the species or the crop system. Although current processing of this type of images goes back to airborne photographs and are based on visual photo-interpretation, this technique is time consuming and it is thus necessary to develop new tools based on computer processing for more automatic extraction of spatial information.

We propose to develop a multi-step VHSR-image processing sequence in an attempt to improve current practices in this field. The first step consists in a segmentation of the image in homogeneously textured units using tools previously published in the literature. The second step corresponds to an independent classification of the obtained units into a set of pre-defined classes; the segmentation preparing the best data set to be efficiently classified. The third step allows deriving some information about the grove structure and cropping practices.

In this paper, we present the second step of this treatment, based on an expert system grouping Fourier analysis, vegetation indexes comparison, and a decision tree. The objective is to classify each plot as a single unit, in a typology of height types of land use and structure.

## 1. MATERIAL

We focus in this study on a test-area in the South of France (Department of Gard), located south of Nîmes and north of St Gilles. It includes of a large variety of land use types, including several types of groves, orchards and forests. It thus gives the opportunity to test our method on a wide range of plot structures.

### *1.1 IMAGE DATA*

We analyze a Quickbird [1] image acquired in July 2005 with a spatial resolution of 0.7 m in the panchromatic mode and 2.5 m in the multispectral mode (three spectral bands in the visible (blue, green and red) and near-infrared domains). These data were merged using a Brovey transform [2] to provide a multispectral image at 0.7 m spatial resolution, similar to that of the future *Pléiades* products. We finally extracted a subset of this image of 2411 pixels x 2122 pixels (2.5 km$^2$) to allow quick computing.

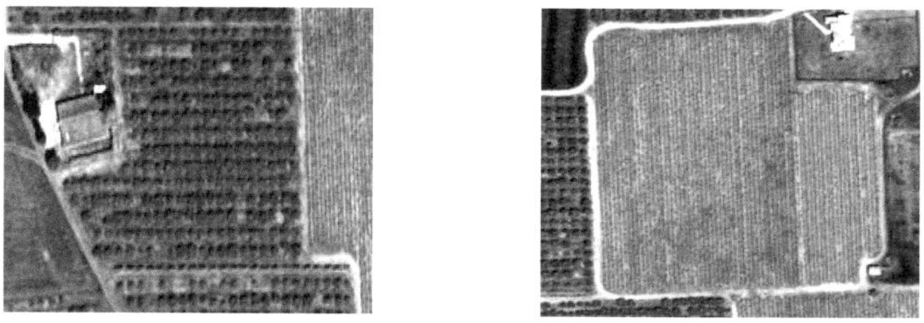

***Figure 1: Test-image samples: peach grove (left) and vineyards (right).***

Based on a field survey combined to photointerpretation, we produced a ground-truth image delineating and classifying each plot in a typology composed of height classes:

1. Fields (large crops in continuous cover: cereals, oilseeds…)
2. Market-garden (small crops in discontinuous cover: vegetables, strawberries…)
3. Vineyards
4. Young grid-groves (small trees with faint coverage density)
5. Adult grid-groves (big trees with dense cover)
6. Trellis-groves (row-structure of tree plantation due to wire-support)
7. Fallow lands with random isolated trees
8. Forests (natural tree canopy structure)

Inside this typology, we can observe that the Fields class (class 1) is the only one without any texture, while Fallow lands (class 7) and Forests (class 8) do not correspond to structured plantations. These three classes will thus be unprocessed when we will focus on the structured groves only.

**Table 1: Number of plots in the ground-truth image for each class of interest**

| Class number | 1 | 2 | 3 | 4 | 5 | 6 | 7 | 8 |
|---|---|---|---|---|---|---|---|---|
| Number of plots | 28 | 14 | 16 | 15 | 15 | 3 | 23 | 17 |

### *1.2 RESULT OF SEGMENTATION*

Our choice of methodology is to classify each plot as a single object, assuming it has been already correctly segmented. Several algorithms can be used for this purpose [e.g. 3-4-5-6-7-8-9-10], that we won't detail here. We have though tested some of them, and evaluated their potential to provide plots of homogeneous land use systems with specific indicators of efficiency at the plot level of segmentation [11] and the operational point of view. As a result, we have finally applied the SxS hierachical segmentation [5-6] to extract image-subsets of each plot. The total number of individual subsets is 673, containing plots that correspond to one of the classes of interest, and others to untreated objects (roads, buildings, gardens, hedges, undefined landscape units…). Only the 131 plots related to classified ground-truth were processed for classification.

## 2. METHOD

The general principle of the proposed classification is based on a decision tree, including agronomical knowledge at each step of decision. Indeed, each different plot structure that can be observed is strongly correlated with the cropping practices for each kind of grove, often specific to the tree species (e.g.: vine, fruit, olive, pine…). This expert system is then fed by descriptors derived from the image processing, which give information about the structure (e.g. Fourier parameters, texture indices) or the density of the canopy (e.g. vegetation index). This section will first introduce and explain these descriptors, and then it will present the decision tree in its whole.

### *2.1 FOURIER PARAMETERS*

Our approach is based on the Local Fast Fourier Transform [12]. We thus calculate the FFT on small windows that fit the inner area of each plot. The window size was fixed to 64 pixels, to reach enough repeatability of the planting structure and allows FFT. If the plot size is too small to contain such a window, its dimension is decreased to 32 pixels but the result will be of lower accuracy. To avoid singular sampling, at least three windows were used for each plot. The FFT is then derived for each of the different spectral bands of the multispectral image, and the final FFT spectrum is the result of the cumulative spectrum over these different bands. Finally, the Fourier spectrum for the whole plot is the mean of the spectrum derived for these three windows.

FFT spectrum of periodic features is characterized by the presence of peaks of strong intensity that give information about the corresponding image structure (cf. *Figure 2*). Indeed, if there is no peak, there is no periodic structure and the plot belongs to one of the classes "Fields", "Fallow lands" or "forests". If there are one to two peaks, the structure is linear. If there are four or more peaks, the structure is a grid arrangement of individual objects. First step of the analysis is thus to detect the exact number of peaks in the Fourier spectrum of each plot and classify it in the three main classes characterizing plot pattern: "rows", "grid", or "unstructured", this latter corresponding to the class untreated by the next steps. The expert knowledge is here important to discriminate between the different possible crops in each class. For instance, vineyards and orchards have different width or rows and inter-rows. Moreover, vine-stocks are fixed on wires and pruned to fit only one meter width, thus displayed as only one pixel line in the image. It is thus represented in the Fourier space by one single peak corresponding to the inter-row of about 2m [13]. Besides, trellis-orchards, like apple crops, can have a canopy width of two meters or more, thus displayed by a strip of several pixels in the image, and represented in the Fourier space by two different peaks corresponding respectively to the tree-row and the inter-row.

The FFT pattern of these two classes "rows" and "grid" is also characterized by the angle separating two different peaks: for linear structures, like vineyards or trellis-groves, the angle between the two principal peaks is null, while it is close to 90° for simple grid-groves or to 45° for complex grid-groves (cf. *Figure 2*).

In any case, the distance from one given peak to the centre of Fourier spectrum corresponds to the period of the structure, and to the width of the plantation pattern: rows or inter-rows for trellis-grove, and grid dimensions for grid-groves (cf. Table 2).

Finally, the association of these different observations on the Fourier patterns with expert knowledge allows an easy discrimination of the unstructured and the structured plots, and, in this latter case, plot classification in one of the four following classes: grid-grove, trellis-grove, vineyard, and market-garden.

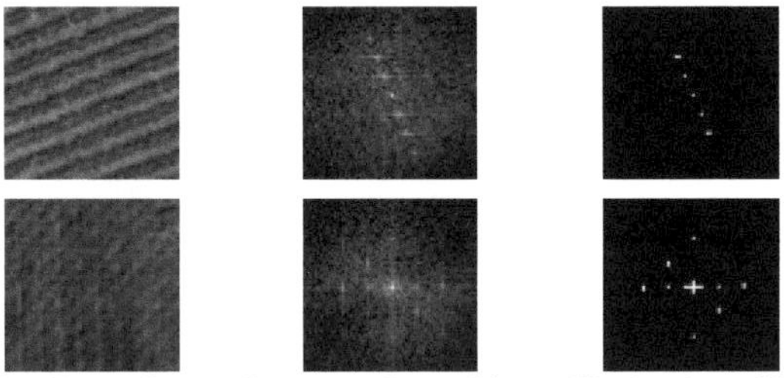

***Figure 2: Fourier Transform of groves planted in trellis (top) and in grid (bottom)***

**Table 2: Relationships between Fourier parameters and class characteristics:**

| | Peaks | Angle between peaks | Distance to center (Plantation period) |
|---|---|---|---|
| Vineyards | 1 | 0° | 2m |
| Trellis-groves | 2 | 0° | 4m |
| Market-garden | 1 or 2 | 0° | 0.5 to 1m |
| Grid-groves | 4 to 6 | 45 or 90° | 6-8m |

### *2.2 TEXTURAL INDICES*

Textural indices based on the cooccurrence matrix derivation [13] can be particularly useful for discriminating plots of Fields, Fallow lands and Forests [14-15-16]. As an example, the Entropy index (cf. *Figure 3*) displays values close to zero for large fields, and different kinds of grey for the plots containing trees. Nevertheless, we have not analyzed in more detail this discrimination ability in the present study, which concerns mainly the managed groves, not the natural or fallow areas.

***Figure 3: Panchromatic image (left) and Entropy index (right) of a subset of the studied image.***

### *2.3 VEGETATION INDEX (NDVI)*

The NDVI [17] is a vegetation index derived from multispectral images strongly correlated to the canopy biomass. It is thus efficient to distinguish different densities of groves having the same structure (cf. Figure 4). Indeed, the mean NDVI of a given plot will be higher if trees have a larger and denser crown. We will thus use this indicator to discriminate the young from the adult grid-groves. It can add some criteria for vineyard and other trellis-grove discrimination too. Finally, the NDVI also gives an indication of the canopy soil coverage; market-garden, which have very small covering ratios, will thus be easily distinguished from other row-crops (eg. vineyard, trellis-grove).

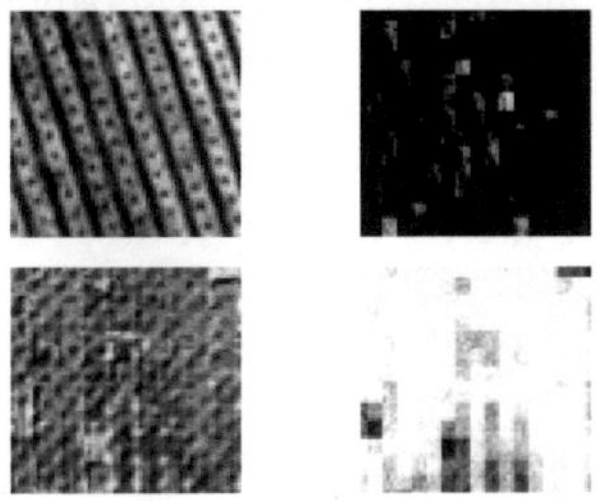

**Figure 4: False colour composite image (left) and NDVI (right) of young (top) and adult (bottom) groves.**

### *2.4 DECISION TREE*

Based on the synthesis of these different observations on relevant descriptors of plot structure and content, we propose the following decision tree as the classification tool for the extracted plots (*Figure 5*).

1. The first analyzed parameter is the presence of peak in the Fourier spectrum. If there is no peak, the plot is not processed through the decision tree. It can possibly be treated afterwards with textural indexes-related classification if needed.
2. The angle between the different peaks in the Fourier pattern allows separating the plots in two groups, corresponding respectively to row and grid structures.
   a) In the case of grid-crops, the NDVI is used to determine the global age of the grove.
   b) In the case of row-crops, the number of peaks allows separating the plots as possible vineyards and possible trellis-grove, but still containing the market-gardens. Afterwards, the NDVI value helps extracting from these two groups the plots actually corresponding to market garden.

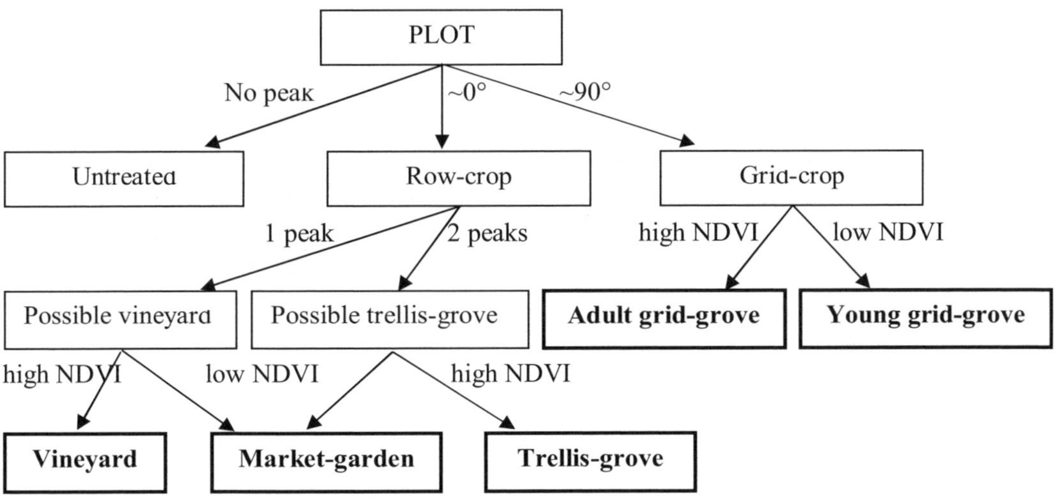


***Figure 5: Sketch-map of the classifier decision tree.***

Finally, we check that the planting period of the plots corresponds well to the specifications of the class it has been assigned to. If not, the plot is processed again with a lower filtering of the Fourier transform.

## 3. RESULTS AND DISCUSSION

Initially in this study, we wanted to distinguish height different classes. However, given the method based on the analysis of planting frequency, we cannot discriminate plots of Fields / Fallow lands / Forests at this stage. The confusion matrix of final classification of the 131 plots in the six resulting classes is given Table 1. At first sight, it shows that no big confusion is made between the large classes Row-crop and Grid-crop, and that the classification result is very good.

Even for adult grid-groves having a closed canopy, an FFT-peak can still be obtained, leading to good performances of the method.

However we can note that Market-gardens are poorly classified (only 64% of good occurrence). This class displays a great variability which may be due to confusion with vineyards. This occurs mainly after irrigation of the market-gardens which then display a high NDVI value similar to those of vineyards, and when vegetables cover a large row reaching vine-stock width.

About 14% error is committed on Fields, that are confused with vineyard or grove at an early development phase, when only partial vegetation cover does not hide completely the ploughing pattern in soil. When recently irrigated, they can even be mistaken with Market-gardens.

| | Un-structured | Market-garden | Vineyards | Young grid-groves | Adult grid-groves | Trellis-groves |
|---|---|---|---|---|---|---|
| Fields | 22 | 2 | 3 | 0 | 0 | 1 |
| Fallow lands | 23 | 0 | 0 | 0 | 0 | 0 |
| Forests | 17 | 0 | 0 | 0 | 0 | 0 |
| Market-garden | 2 | 9 | 2 | 1 | 0 | 0 |
| Vineyards | 0 | 2 | 14 | 0 | 0 | 0 |
| Young grid-groves | 1 | 2 | 0 | 10 | 2 | 0 |
| Adult grid-groves | 1 | 0 | 0 | 1 | 13 | 0 |
| Trellis-groves | 0 | 0 | 0 | 0 | 0 | 3 |

***Table 3: Confusion Matrix***

NDVI has also a limited effect for discriminating young-groves from adult ones when intercrops are not bare. Especially, 16% error is committed on the Young Grid-groves which understorey vegetation increases the total plot NDVI to value similar to those of adult-plots.

## CONCLUSION AND PERSPECTIVES

As a conclusion, the method works well provided it uses lots of a priori knowledge on the study area. It allows discriminating between different tree crop structures and ages, with a good accuracy in most of the cases. Some classes would still need more efficient descriptors to be fully discriminated, and deeper analysis of NDVI, textural indices, and other descriptors will certainly improve the results.

Among perspectives of this study, it could be interesting to compare this classifier results with other methods relevant for processing complex data architectures, such as, for instance, the Support

Vector Machine [18]. It should also be more automatic to allow an operational use. Finally, it should be tested on several other types of groves to analyse its generic abilities.

ACKNOWLEDGEMENTS

This work was funded by the French National Space Agency (CNES) and National Geographic Institute (IGN).

## REFERENCES

[1] http://www.digitalglobe.com/about/quickbird.html

[2] C. Pohl and J.L. Van Genderen. Multisensor image fusion in remote sensing: Concepts, methods and applications. International Journal of Remote Sensing, 19 (5), 823-854 (1998).

[3] http://www.definiens-imaging.com/ecognition/

[4] C. Burnett and Thomas Blaschke. A multi-scale segmentation/objet relationship modelling methodology for landscape analysis. Ecological Modelling 168, 233-249 (2003).

[5] L. Guigues, H. Le Men and J.-P. Cocquerez. Scale-sets image analysis. In Proc. of IEEE Int. Conf. on Image Processing (ICIP'03), Barcelona, Spain, September 2003.

[6] L. Guigues. Modèles Multi-Echelles pour la Segmentation d'Images. PhD thesis of *Cergy-Pontoise University*, 287p. (2003).

[7] D. Mumford & J. Shah, Optimal approximation by piecewise smooth functions and associated variational problems, Comm. Pure Appl. Math, 42, 577–685 (1989).

[8] P. LI, X. Xiao. Multispectral image segmentation by a multichannel watershed-based approach. International Journal of Remote Sensing, 28, 4429-4452 (2007).

[9] http://orfeo-toolbox.sourceforge.net/Docs/OTBSoftwareGuide.pdf, page 164-170 (2006).

[10] http://www.itk.org

[11] B. Mougel, C. Lelong et J.-M. Nicolas. Comparison of three segmentation methods for groves recognition in very high resolution satellite images. Proc. of "PSIP 2007 conference - Physics in Signal and Image Processing", 31 jan.-2 feb. 2007, Mulhouse, France, 5pp.

[12] R.A. Schowengerdt. *Remote Sensing: Models and methods for image processing.* Academic Press, Elsevier (Third Edition) 2007.

[13] R. M. Haralick, K. Shanmugam, I. Dinstein. Textural features for image classification. IEEE Transactions on Systems, Man, and Cybernetics, 3 (6), 610-621 (1973).

[14] C. Delenne, S. Durrieu, G. Rabatel, M. Deshayes, J.S. Bailly, C. Lelong, and P. Couteron. Textural approaches for vineyard detection and characterization using very high spatial resolution remote-sensing data. *International Journal of Remote Sensing, 29 (4), 1153-1167 (*2008**)**.

[15] C. Lelong and A. Thong-Chane. Application of textural analysis on very high resolution panchromatic images to map coffee orchards in Uganda. Proc. of IGARSS, II, 1007-1009, Toulouse (FR), 21-25 juillet 2003.

[16] C. Lelong, C. Lesponne, N. Lamanda, G. Lainé and E. Malézieux. Understanding the Spatial Structure of Agroforestry Systems using Very High Resolution Remote Sensing: an Application to Coconut-based Systems in Melanesia", Proc. of 1st World Congress of Agroforestry, Orlando (USA), 28 juin - 2 juillet 2004.

[17] Cihlar, J., St-Laurent, L.; Dyer, J. Relation between the normalized vegetation index and ecological variables. *Remote Sensing of Environment, 35*, 279-298 (1991).

[18] G. Foody, & A. Mathur. A relative evaluation of multiclass image classification by support vector machines. IEEE Transactions on Geoscience and Remote Sensing, 42, 1335–1343 (2004).

*Remote Sensing for a Changing Europe*
*D. Maktav (Ed.)*
*IOS Press, 2009*

*doi:10.3233/978-1-58603-986-8-80*

# Determination of minimum burning area on active fire detection by means of MSG-SEVIRI

A. Calle, J.L. Casanova, J. Sanz & D. de la Fuente
*Remote Sensing Laboratory. LATUV. University of Valladolid. Spain. e-mail: abel@latuv.uva.es*

F. González-Alonso
*CIFOR-INIA. Ministry of Science and Innovation. Madrid. Spain.*

Keywords: PSF, MTF, MSG-SEVIRI, Active fire detection

ABSTRACT: In order to improve operation on active fire detection, geostationary satellites as GOES, MSG and MTSAT are already operative and they have led the international community to think that the global observation network in real time may become a reality. The implementation of this network is the aim of the Global Observations of Forest Cover and Land Cover Dynamics (GOFC/GOLD) FIRE Mapping and Monitoring program, focused internationally on taking decisions concerning the research of the Global Change. The estimation of the minimal burning area detected by MSG-SEVIRI is analyzed in this work; it's a very relevant issue since it implies the sensor's operational availability in active fire detection. This minimum size is conditional on the fire location inside the pixel, as shown before from the theoretical point of view, among other factors such as atmospheric conditions, fire temperature, among others. A spatial analysis has been carried out, analyzing several parameters referred to spatial performances of MSG-SEVIRI sensor as pixel footprint in the 3.9 microns spectral band. In order to obtain results based on real data, a large data base focused on the active fire detection in the Iberian Peninsula has been studied. Statistical results obtained from a large data base of fires detections, show that the probability to detect a minimal size of 10ha is 90% and 4ha with 50%.

## 1 INTRODUCTION

The MTF shape in the frequency domain and, consequently, the Point Spread Function (PSF) in the spatial domain has not a special relevance when the surface observed shows a homogeneous distribution of radiance; nevertheless, when there are heterogeneous distribution of radiance inside the pixel, as is frequently the case of forest fires, PSF and deconvolution processes must be considered. Markham, 1985, has analyzed the spatial characterization of sensors for Multispectral Scanner (MSS) and Thematic Mapper (TM), by means of a sensor model in order to define their MTF and PSF functions accurately. In the case of Advanced Very High Resolution Radiometer (AVHRR), Mannstein and Gesell, 1991, have shown several aspects concerning data deconvolution. Rojas *et al.*, 2002 have shown the spatial characterization in orbit for Moderate Resolution Imaging Spectroradiometer (MODIS) based on the comparison with Advanced Spaceborne Thermal Emission and Reflection Radiometer (ASTER) images and Huang, *et al.* 2002, have studied the impact of the sensor's PSF on land cover characterization in the optical spectrum. Finally, with regard to finer spatial resolution sensors such as Ikonos and QuickBird, Helder *et al.* 2006, used high-contrast reflectance plastic-panels, from 3.6% to 52.1%, for MTF estimation. Concerning geostationary satellites, Zhang et al. 2006, have analyzed the impact of PSF on infrared radiance by means of a parameter related with the fraction of energy coming from the central part of pixel: the Ensquared Energy (EE). EE is the ratio of energy detected from the nominal Ground Field Of View

(GFOV) divided by the energy detected from all GFOVs, being the nominal GFOV the threshold ground sample distance, which is the centroid-to-centroid adjacent spatial samples on the Earth's surface as measured at the satellite sub-point. This quantity is known as Threshold Ground Sample Distance (TGSD). The conclusions obtained from the EE analyzed are related to the PSF shape. Unlike Zhang *et al.* 2006, who worked with simulated PSF, this paper shows results by using real MTF functions of the Spinning Enhanced Visible and Infrared Imager (SEVIRI) sensor on board of Meteosat Second Generation (MSG).

On the other hand, many thermal parameters in remote sensing are estimated by solving multi-spectral processes, such as the estimation of the temperature using split-window procedures or the estimation of thermal parameters in hot-spots through Dozier's method (Dozier, 1981; Matson and Dozier, 1981). In these estimations, it is assumed that the pixels of the bands involved correspond to the same spatial target and contribute with the same sensitivity to the radiance measurement. However, even in the case of a perfect co-registration between bands, this assumption would not be true since each band has a different PSF. This is one of the problems mentioned by Wooster *et al.* 2005, in order to propose a single-channel method to estimate the fire temperature instead of applying a bi-spectral method. In addition, the influence of the PSF has been highlighted as responsible for the differences in the Fire Radiative Power (FRP) when different sensors are compared (Wooster *et al.,* 2003). Concerning geostationary satellites, MSG is providing operational results in fire detection and biomass burning in Africa (Wooster et al. 2005) and Mediterranean countries (Calle *et al.*, 2006) and Geostationary Operational Environmental Satellites (GOES) are used operationally in South-Central-North-America (Prins and Menzel, 1992, 1994).

## 2 METHODOLOGY

The function mostly used to express a sensor's spatial resolution is the MTF. This function is expressed in the frequency domain and it defines the sensor capacity to solve details at different spatial frequencies. The higher frequency beyond which the sensor is unable to distinguish detail is known as Nyquist frequency. Beyond this limit, aliasing occurs in the image. The relation between the PSF and the MTF is established through Fourier transform so that the MTF is the module of Fourier transform. That is:

$$MTF(u,v)=\left|\Im\{PSF(x,y)\}\right| \tag{1}$$

where (u,v) are the coordinates in the frequency domain. The use of MTF for spatial characterization of sensors is useful because, in the frequency domain, the sensor's final MTF is the product of the MTF's from affecting components: optic lens, detector, and electronic system. (Note that multiplying several functions is easier than convoluting them.). The MTF functions of the SEVIRI sensor were provided by European Organisation for the Exploitation of Meteorological Satellites (EUMETSAT). Since the MTF only contains information on the Fourier transform module, the PSF function cannot be determined from the MTF by means of the inverse Fourier transform. The PSF is determined by applying Hankel transform (Gaskill, 1978):

$$PSF(x) = \int_0^\infty u \cdot MTF(u) \cdot J_0(x \cdot u)\, du \tag{2}$$

where $J_0(x)$ is the Bessel function of first order.

The objective of this paper is to estimate the impact of the PSF-SEVIRI's shape on the estimation of minimal burning area detected by sensor. This assessment is related to several issues as the overlapping area between pixels. Figure 1 shows the typical cross shape when fire is detected, due to PSF effects and overlapping between pixels. If two or more neighbour pixels are affected by fire, the overlapping of pixels may produce several cross shapes, or other geometric figures, as figure 1 right shows, since the cross shape is the basic-single structure. In addition, the discrepancies in the results obtained in the application of bi-spectral calculations, as a consequence of the non-coincidence between the PSF functions of the bands involved, will be also analysed.

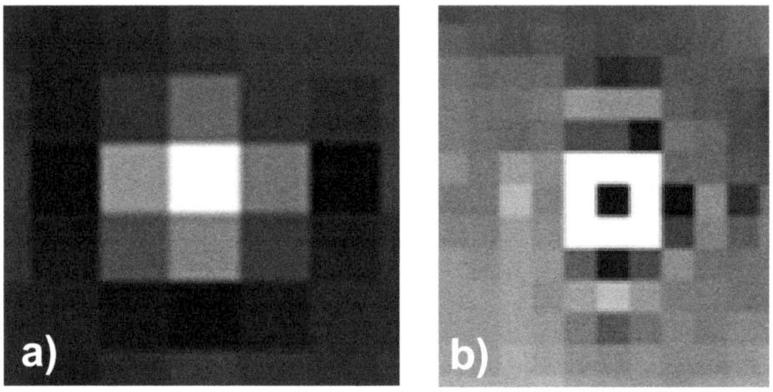


Figure 1: Impact of overlapping PSFs on pixels affected by fire. a) Fire appears as a cross shape as basic structure. b) Combination of several affected pixels can give place to squares.

Concerning to the analysis of database of fire detections, it was applied a statistical distribution in order to estimate the probability of fire size detection. The statistical distribution applied to the data base of fires was the continuous probability distribution of Weibull, with the probability density function given by:

$$f(x;k,A)=\frac{k}{A}\left(\frac{x}{A}\right)^{k-1}e^{-\left(\frac{x}{A}\right)^{k}} \tag{3}$$

for $x \geq 0$ and $f(x; k, a) = 0$ for $x < 0$, where $k > 0$ is the shape parameter and $a > 0$ is the scale parameter of the distribution. Its complementary cumulative distribution function is a stretched exponential. Weibull distribution is flexible because it can mimic the behaviour of other statistical distributions such as the normal and the exponential (Weibull, 1951).

## 3 ANALYSIS OF RESULTS

Figure 2 shows a three-dimensional graph where the Brightness Temperature in the 3.9μm band (vertical axis) is shown versus the fire temperature and the distance from the pixel centre (PSF impact), where the background temperature is 300K and the one-dimensional burning area is 50m. Saturation plane is shown in the figure. Note that for low fire temperatures (below 450K, taking into account that we are talking of flaming and smoldering mixed phases, the PSF impact is not noticeable. However, large differences in Brightness Temperature are found in hotter fires. In order to explain the importance of a 10K-difference in the 3.9μm band, note that if a contextual detection algorithm is applied (Calle *et al.*, 2006) the detection will be lost when the standard BT deviation around the pixel is higher than 3K.

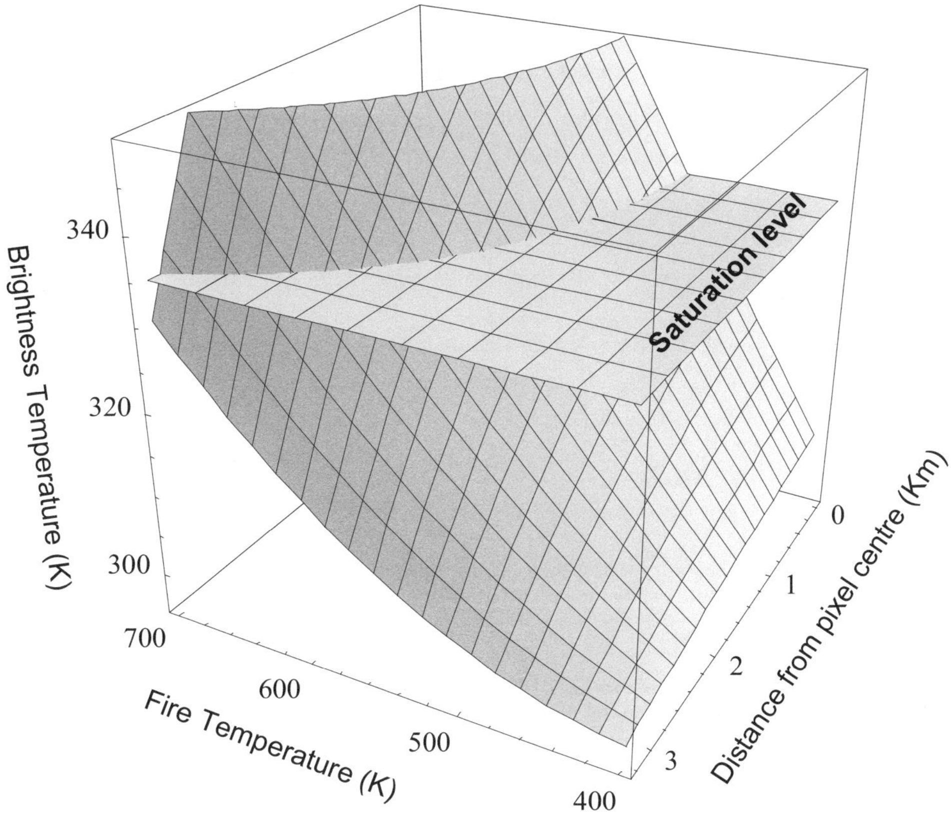


Figure 2: Brightness temperature in the 3.9μm band (vertical axis) versus fire temperature and distance from pixel centre (PSF impact). One-dimensional burning size: pxy=50m and background temperature 300K

The estimation of the minimal burning area detected by MSG-SEVIRI is a relevant topic since it implies the sensor's operational availability in active fire detection. This minimum size is conditional on the fire location inside the pixel, as shown before from the theoretical point of view, among other factors such as atmospheric conditions, fire temperature, etc.

In order to obtain results based on real data, a large data base focused on the active fire detection in the Iberian Peninsula has been studied. The period of study has been July and August, 2006 and March to November, 2007. Only records with solution in the bi-spectral equations system (Dozier, 1981) were used in order to apply a statistical distribution of the burning area size; in addition, several outlier situations were filtered in order to improve the quality of the data used.

Taking into account this cumulative distribution function, the minimal burning area detected is obtained. Figure 3 shows the probability to detect a fire versus the burning area. As this figure shows, there is 50% probability to detect a fire with a burning area of 4ha whereas a fire of 10ha is detected by SEVIRI with a probability of 90%. Weibull statistical distribution was applied to the active fire detection database with $r^2$=0.96.

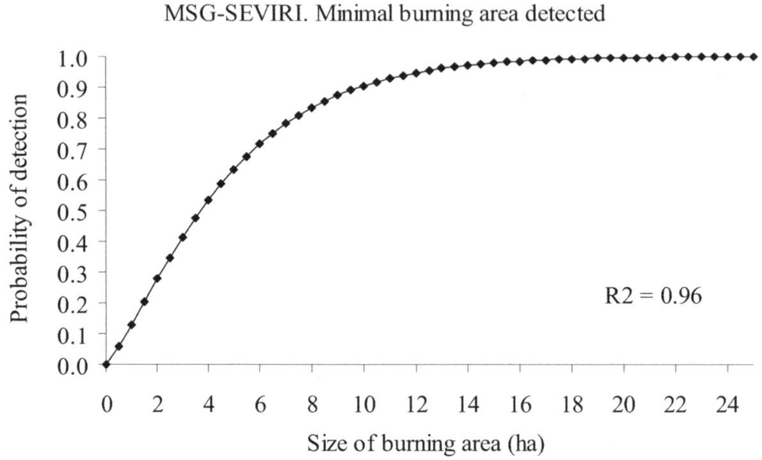


Figure 3: Fire detection probability versus size of burning area, for MSG-SEVIRI sensor, predicted by a Weibull's statistical distribution applied to real detections.

On the other hand, taken into account that active detection is based on contextual analysis based on brightness temperature in the MIR spectral band (3.9μm) and difference MIR-TIR (3.9μm-10.8μm), the knowledge of these values on real fire detection will provide the best thresholds to be applied to detection algorithm. Figure 4 shows the histogram of frequencies of difference brightness temperature MIR-TIR in the hot-spots detected by MSG-SEVIRI.

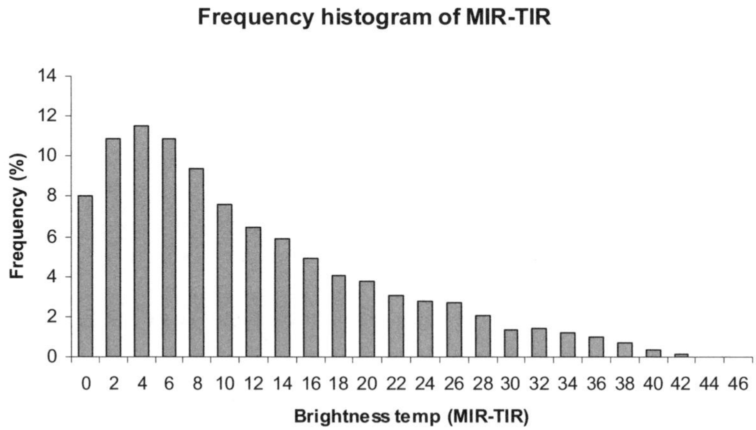


Figure 4: Histogram of frequencies of difference brightness temperature MIR-TIR in the hot-spots detected by MSG-SEVIRI. This result provides the threshold to be used in the contextual algorithm of detection.

## 4 CONCLUSIONS

The influence of the PSF shape, and its non-coincidence between bands, should not produces large errors when the radiance is homogenously distributed in the surfaces analysed. However, the opposite is true in wildfires, where the spatial location of a HTE inside the pixel may influence the results obtained. The brightness temperature provided by the sensor, in the 3.9μm band, is the main magnitude involved in fire detection, since this value is compared with a threshold to decide the existence of forest fires. As has been shown, the location of a particular fire provides large differences of brightness temperature values depending on the distance at which it is located from the pixel centre. It could even be responsible for the saturation or non-saturation of the sensor and for the loss of detection that sometimes occurs (omission errors).

## 5 ACKNOWLEDGEMENTS

This paper is part of the project PREVIEW, ref PREVIEW-D3100-1-SER-ASTRIUM-002 founded by Sixth Framework Programme priority 4 Aeronautics & Space – Space 2004. The authors wish to thank these Institutions for their support in making it possible.

## 6 REFERENCES

Calle, A., Casanova, J.L. and Romo, A. 2006. Fire detection and monitoring using MSG Spinning Enhanced Visible and Infrared Imager (SEVIRI) data. *Journal of Geophysical Research*, 111, No. G4, G04S06, doi 10.1029/2005JG000116

Dozier, J. 1981, A method for satellite identification of surface temperature fields of subpixel resolution. *Remote Sensing of Environment*, 11, 221-229.

Gaskill, D., 1978. Linear Systems, Fourier Transforms, and Optics, *Ed. John Wiley & Sons, New York*, 1978.

Helder, D., Choi, J. and Anderson, C. 2006. On-orbit Modulation Transfer Function (MTF) Measurements for IKONOS and QuickBird. *Proceedings of 2006 Civil Commercial Imagery Evaluation workshop*, Laurel, Maryland, March 14-16 USA, Sponsored by Joint Agency Commercial Imagery Evaluation (JACIE) Team.

Huang, C., Townshend, J.R.G., Liang, S, Kalluri, S.N.V. and DeFries, R.S. 2002. Impact of sensor's point spread function on land cover characterization: assessment and deconvolution. *Remote Sensing of Environment*, 80, 203-212.

Mannstein, H. and Gesell, G 1991. Deconvolution of AVHRR data. *Proccedings of 5th AVHRR data users' meeting.* Tromso, Norway, 25-28 June 1991. EUM P 09 (Darmstadt-Eberstadt:EUMETSAT), 53-58.

Markham, B.L. 1985. The Landsat sensors' spatial responses. *IEEE Transactions on Geoscience and Remote Sensing*, 23 (6), 864-875

Matson, M. and Dozier, J. 1981, Identification of sub-resolution high temperatures sources using a thermal IR sensor. *Photogrametric Engineering and Remote Sensing*, 47 (9), 1311-1318.

Prins, E.M. and Menzel, W.P. 1992. Geostationary satellite detection of biomass burning in South America. *International Journal of Remote Sensing*, 13, 2783-2799.

Rojas, F., Schowengerdt, R.A. and Biggarb, S.F. 2002. Early results on the characterization of the Terra MODIS spatial response. *Remote Sensing of Environment*, 83, 50–61.

Weibull, W. 1951. A statistical distribution function of wide applicability. *Journal of Applied Mechanics Transactions.* ASME 18 (3), 293-297

Wooster, M.J., Zhukov, B and Oertel, D. 2003. Fire radiative energy for quantitative study of biomass burning: derivation from the BIRD experimental satellite and comparison to MODIS fire products. *Remote Sensing of Environment*, 86, 83-107.

Wooster, M.J., Roberts, G., Perry, G.L.W and Kaufman, Y.J 2005. Retrieval of biomass combustion rates and totals from fire radiative power observations: FRP derivation and calibration relationships between biomass consumption and fire radiative energy release. *Journal of Geophysical Research*, 110, D24311, doi: 10.1029/2005JD006318.

Zhang, P., Li, J., Olson, E., Schmit, T.J., Li, J. And Menzel, P. 2006. Impact of Point Spread Function from geostationary satellites. *IEEE Transactions on Geoscience and Remote Sensing*, 44 (8), 2176-2183.

*Remote Sensing for a Changing Europe*
*D. Maktav (Ed.)*
*IOS Press, 2009*

*doi:10.3233/978-1-58603-986-8-86*

# Real time snow recognition from msg satellite for mountainous areas

S. Sürer
*Middle East Technical University, Geodetic and Geographic Information Technologies Department, Ankara, Turkey, serdarsurer@gmail.com*

O. Gökdemir
*Hacettepe University, Hydrogeological Engineering Dept., Ankara, Turkey, orhan.gokdemir@gmail.com*

Ö. Beşer, Z. Akyürek & A. Ü. Şorman
*Middle East Technical University, Civil Engineering Department, Ankara, Turkey, obeser@gmail.com, zakyurek@metu.edu.tr, sorman@metu.edu.tr*

A. G. Ertürk
*Turkish State Meteorological Services, Remote Sensing Department, Ankara, agerturk@meteoroloji.gov.tr*

Keywords: Snow, MSG-SEVIRI, NOAA AVHRR, SAFNWC, multi-temporal

ABSTRACT: In Hydrology Satellite Application Facilities (HydroSAF) Project which is a financially supported project by EUMETSAT, the use of snow products retrieved from satellite images in hydrological applications and to observe the impact of the products are key issues. Turkey is a part of the HydroSAF project, both in developing satellite derived snow products (snow recognition, effective snow cover, and snow water equivalent) for mountainous areas, cal/val of satellite-derived snow products with ground observations and impact studies with hydrological modeling in the mountainous terrain of Europe. The snow recognition product for mountainous regions is evaluated in this paper. An algorithm has been developed for snow recognition over mountainous areas of Europe. The proposed algorithm uses Satellite Application Facility to support Nowcasting and Very Short Range Forecasting's (SAFNWC) cloud products. Two main validation processes have been applied for the snow cover product belonging to 19th January 2008 produced with SEVIRI data. First, the comparison of the SEVIRI snow cover product with the snow cover product produced from a single NOAA AVHRR data for the same date 19th January 2008 has been performed. Second, SEVIRI snow cover product was validated with 43 synoptic weather stations distributed over Europe.

## 1 INTRODUCTION

The distribution of snow covered area (SCA) on the land surface highly affects the Earth's energy balance and hydrological cycle. Monitoring and estimating the snow water equivalent for discharge predictions gains more importance during melting periods since snow is one of the major renewable water resources. The most common parameters required for distributed snow models are: snow water equivalent, grain size, SCA, albedo, snow temperature profile and some meteorological parameters like solar radiation. Because snow appears continuously in the mountainous terrain in mid-latitudes and the accessibility difficulties of such terrains makes the use of satellite images and/or aerial photographs in monitoring and estimating the snow parameters more valuable. Remote sensing data has been used for better comprehension of SCA information (Painter et al. 1998, Cline et al. 2003). Several satellite sensors have been used for SCA mapping such as: AVHRR, MODIS, MERIS, and ASTER (Harrison & Lucas 1989, Hall et al. 2003, Tampellini et al. 2004).

Due to its high temporal resolution, SEVIRI provides with better discrimination capacity between quickly changing phenomena like ice clouds and snow. Within the framework of HydroSAF Project, a snow cover mapping algorithm has been developed and it is generating SC products on real time bases. Algorithm uses Meteosat Second Generation satellite (MSG) instrument Spinning Enhanced Visible and Infra-Red Imager (SEVIRI) data that are acquired in every 15 minutes through whole day. The snow recognition algorithm uses 4 different channels: channel-1 (visible 0.64 μm), channel-3 (near-infrared 1.6 μm), channel 4 (infrared 3.9 μm), and channel-9 (infrared 10.8 μm). The proposed algorithm uses cloud products of Satellite Application Facility to support Nowcasting and Very Short Range Forecasting's (NWCSAF) (Derrien & Gléau, 2003). After generation of SC products, the testing of the algorithm has been done with the SC product produced from a better spatial resolution satellite sensor AVHRR data and with snow depth measurements of synoptic weather stations.

## 2 DATA

Eight successful launches of Meteosat satellites followed the first Meteosat satellite which was in 1977. In 21st December 2005, Meteosat-9 also named as MSG-2 was launched that has significantly enhanced services and products of Meteosat satellites. MSG-2 provides imagery with 15 minutes temporal resolution and 3 km spatial resolution (at nadir) through 12 spectral channels for the whole hemisphere of the Earth. We have been receiving SEVIRI sensor data of MSG-2 satellite for two years from Turkish State Meteorological Services (TSMS) in the framework of HydroSAF Project as Level 1.5 high rate information transmission (HRIT) data format. The received SEVIRI data has its own projection that needs additional processes. We have developed our own software for converting HRIT to Hierarchical Data Format (HDF) and calculating reflectance values from radiance values in order to make the data ready for the proposed thresholding algorithms. Among 12 spectral channels, channel-1, channel-3, channel-4, and channel-9 were used for SC products generation from SEVIRI images. SEVIRI images between 08:00-15:45 GMT time intervals of day (32 individual images for each day) have been used in order to keep the Sun at the highest possible elevation for reducing the negative effects of shadowing that may occur due to rough terrain. In the framework of HydroSAF project we have studied on a subset domain of SEVIRI which is within 25° N to 75° N latitudes and 20° W to 45° E longitudes. In addition to the SEVIRI data, two adjacent scans (07:59 and 09:39 GMT) of NOAA AVHRR sensor for 19$^{th}$ January 2008 have been used. The NOAA AVHRR channels used for SC product generation were: channel-1 (0.58 μm – 0.68 μm), channel-3A (1.58 μm – 1.64 μm), and channel-4 (10.30 μm – 11.30 μm). A digital elevation model (DEM), GTOPO DEM which has 1 km spatial resolution obtained from Eurosat/GISCO has been used for elevation based analysis. The ground observations of Europe as synoptic measurements were also gathered from TSMS in order to be used in accuracy assessment test of satellite derived products.

## 3 METHODS

For the proposed snow recognition algorithm, spectral thresholding methods were applied on sub pixel scale of SEVIRI image. The main idea was using the different spectral characteristics of cloud, snow and land. Discrimination of snow and cloud was the most challenging part of snow recognition algorithm development. Before going into further investigation for snow pixels, discrimination of cloud has been done and thereafter only cloud free pixels considered for snow and land discrimination. In order to get rid of cloud covered pixels, cloud mask (CMa) and cloud type (CT) products of NWCSAF have been used (User Manual for the Cloud Products of the SAFNWC/MSG 2007). After detailed analysis on CMa and CT products of NWCSAF, we have generated a combination of these products in order to integrate into the proposed snow recognition algorithm as the cloud recognition part. The accuracy of the generated cloud map from NWCSAF cloud products was investigated on a preceding study and acquired high accuracy results (Sürer et al. 2008). Snow cover maps using SEVIRI data have been produced for each 15 minutes cycle between 8:00–15:45 GMT that makes 32 individual images in a day. All individual 15 minute images acquired during a day are subjected to a series of thresholding tests. First, the high visible reflectance of snow was considered and pixels having reflectance values higher than 0.45 are collected. Then, we used one of the spectral indices similar to the (Dozier 1989) has used, called snow index (SI), by dividing NIR1.6 to VIS0.6. The pixels having NIR1.6 / VIS0.6 values higher than a fixed threshold value 0.6 have been collected. Then, pixels having low sun zenith angle (sza) are discarded by a filter accepting pixels only higher than 5°. A final test for covering all cold pixels below freezing point has been applied and pixels having temperature lower than 275 K° on channel-9 are accepted considering that the temperature of snow can not exceed the freezing point (Romanov et al. 2003). After finding snow cover maps for each individual 15 minutes image, we have generated daily snow cover map by accepting pixels having at least 3 snow hits among 32 images during a day. Finally, a daily thematic map has been generated which is consisting of 3 different classes: snow, cloud and land.

## 4 RESULTS

In order to test the performance and accuracy of the proposed snow recognition algorithm, 19th January 2008 product has been chosen. In Fig. 1 a representative RGB image from that particular day at 12:00 GMT is displayed. The combination of the bands used for RGB image generation was: red (0.64 μm), green (1.6 μm), and blue (3.9 μm – 10.8 μm). For making snow pixels more apparent the high visible reflectance, low medium infrared reflectance and low brightness temperature difference of the snow have been considered making snow pixels bright red color on RGB image.

Figure 1. MSG-SEVIRI, RGB composite image of Europe for January 19th, 2008 at 11:00 GMT. Red = 0.64 μm, green =1.6 μm, blue = 3.9 μm – 10.8 μm.

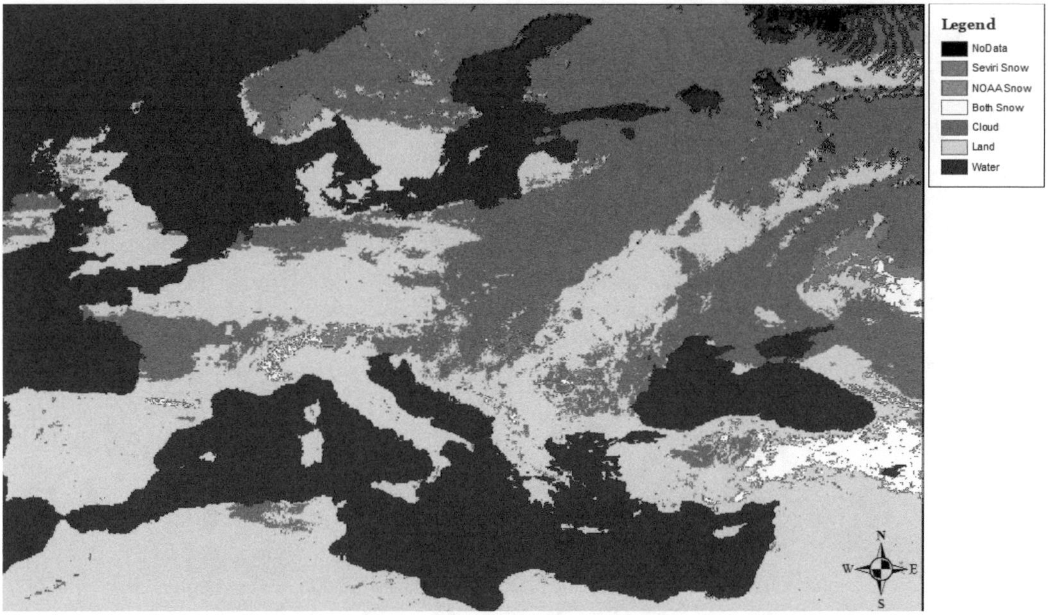


Figure 2. Overlay of snow cover products of SEVIRI and AVHRR for 19th January 2008.

Using this popular RGB combination in snow mapping, the difference between main classes also became more apparent. Water clouds appear as white due to showing high values on all of the three color combinations. Some cold clouds appear as purple and even red, showing similar spectral characteristics with snow and making the discrimination between cloud and snow pixels difficult. At this particular time of 19th January 2008, it is possible to test the snow recognition algorithm especially on cloud clear parts of the image such as the snow covered areas of Eastern Turkey. The performance of the SC map produced from SEVIRI data was tested by three analyses.

## 4.1 *Overlay analysis of SC products obtained from SEVIRI and AVHRR*

An overlay analysis has been applied to SC products generated from both SEVIRI and AVHRR sensors in order to have better comprehension about the snow detection abilities of two sensors, as displayed in Fig. 2. The same SC detection algorithm was applied for both of the images and also their cloud coverages were obtained from the same NWCSAF CMa and CT products. It should be remembered that the SC product generated from SEVIRI sensor uses 32 sequential images from 08:00 – 15:45; on the other hand the SC product of NOAA AVHRR uses only 2 adjacent images mosaiced to cover the same domain as MSG-SEVIRI provides. In Fig. 2, while red color is showing the snow pixels that were detected only by SEVIRI, the color magenta is representing the snow pixels detected only by AVHRR sensor, and the white color was assigned to the snow pixels that were detected by both of the sensors. From the number of pixels information for each class on this classified image, it has been observed that the 50% of the areal coverage of the total SC was detected only by SEVIRI sensor, 10% of the SC was detected only by AVHRR sensor, and the resting 40% of the total snow cover was detected by both of the sensors. The low temporal resolution of the AVHRR sensor causes it to detect less number of snow pixels than SEVIRI detects. As can be seen in Fig. 2, red pixels (snow detected only by SEVIRI) are mostly located at upper part and left-hand side of the image, this is mainly due to AVHRR having lack of adequate reflectance values information because of high viewing zenith angles at that instant of time. In addition, it was

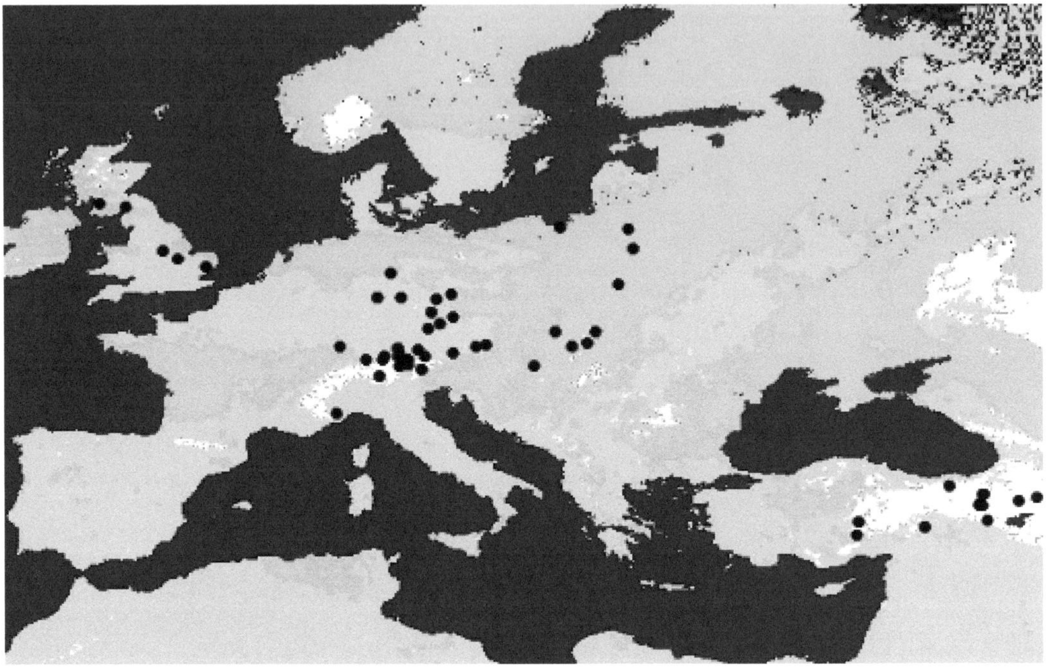

Figure 3. Black dots representing the distribution of 54 synoptic weather stations used for validation of MSG-SEVIRI SC product of 19th January 2008 on thematic snow cover map (white=snow, cyan=cloud, green=land, blue=water).

observed that SEVIRI also detects more or less the same amount of snow pixels while using only one image other than 32 composite images for that particular time of the day. Moreover, because AVHRR has very low temporal frequency of measurement, it is difficult to detect snow pixels lying under moving clouds with SC product obtained from AVHRR data, it is more obvious when comparing Fig. 1 and Fig. 2, the Eastern part of the Alps is covered with clouds (in Fig. 1) and the same area is assigned with red pixels (in Fig. 2) which are the snow pixels detected only by SEVIRI. Thus, SEVIRI is able to detect 90% of the total SC, this is mostly because of its high temporal frequency. So, it can also detect snow pixels lying under moving clouds for some part of the day.

## 4.2 *The correlation of SC products with elevation*

An additional analyzes with the same image in order to interpret the relation of snow detection capabilities of SEVIRI and AVHRR with elevation has been done with a DEM. GTOPO DEM which has 1 km spatial resolution obtained from Eurosat/GISCO has been used for acquiring elevation values of classified pixels in Fig. 2. It was observed that while the mean elevation of the red pixels (snow detected only by SEVIRI) was around 1300 m, the mean elevation of magenta (snow detected only by AVHRR) pixels was below 1200 m, indicating that AVHRR has better ability to catch snow pixels on lower elevations. On a region of interest covering only the Eastern Turkey, it was also observed that the snow pixels detected only by SEVIRI were nearly 300 m higher mean elevation than the snow pixels detected only by AVHRR.

It can easily be seen from Fig. 2 that the magenta pixels on Eastern Turkey are mostly located around the mountains which are located at lower elevation zones. The reason why AVHRR SC product detects more snow pixels than SEVIRI SC product on lower elevations (possible fractional snow) is most probably due to its better spatial resolution than SEVIRI.

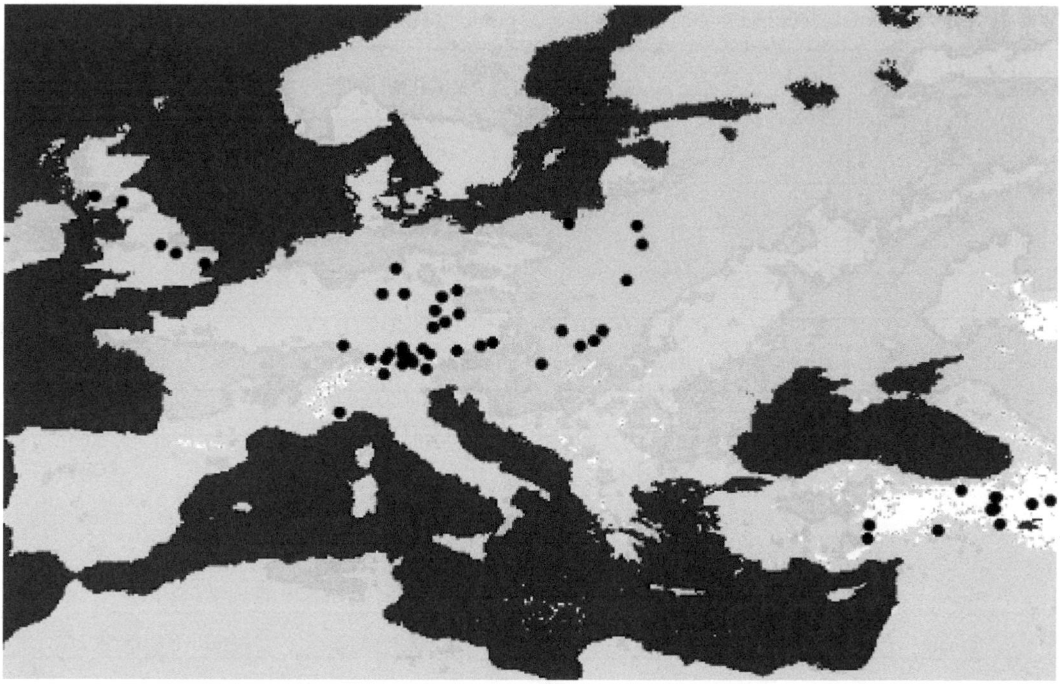

Figure 4. Black dots representing the distribution of 54 synoptic weather stations used for validation of NOAA AVHRR SC product of 19th January 2008 on thematic snow cover map (white=snow, cyan=cloud, green=land, blue=water).

### 4.3 *The validation of SC products with measurements of synoptic weather stations*

Besides comparison of SEVIRI SC product with another SC product obtained from a sensor data which has better spatial resolution, the validation of SC products (SEVIRI and AVHRR) has been done with 54 synoptic stations distributed over Europe (Fig. 3 and Fig. 4).

Out of available 54 synoptic stations that have measured snow depth in 19th January 2008, we were able to use 43 of them since 11 stations were falling over the cloud covered regions in $19^{th}$ January 2008. The 1.1 km spatial resolution AVHRR SC product was resampled to 5 km which is the spatial resolution of SEVIRI SC product in order to have an agreement between pixels of two thematic SC maps. The snow data measured by synoptic weather stations are on a point scale. On the other hand, resulting SC maps would have 25 $km^2$ (5×5) areas for one single pixel. Having these facts in mind, the two classes of SC products: snow and land, are processed for an accuracy test with snow depth measurements of synoptic stations. According to the accuracy test with synoptic stations, 74.4% overall accuracy for SEVIRI SC product (Table 1), and 58.1% overall accuracy for AVHRR SC (Table 2) product has been determined.

## 5 CONCLUSIONS

A multispectral thresholding technique has been used for snow cover mapping over mountainous terrain of Europe domain with MSG-SEVIRI imagery. The outcome results in this paper showed that however SEVIRI SC product has 5 times coarser spatial resolution than AVHRR SC

Table 1. Accuracy assessment test results of SEVIRI SC product with synoptic stations

| | Synoptic Stations | | | | |
|---|---|---|---|---|---|
| SEVIRI SC Product | Snow | Land | Total | Users Accuracy | Producers Accuracy |
| Snow | 20 | 0 | 20 | 100% | 62.5% |
| Land | 12 | 11 | 23 | 52.1% | 0 |
| Total | 32 | 11 | 43 | Overall Avg. = 74.4% | |

Table 2. Accuracy assessment test results of AVHRR SC product with synoptic stations

| | Synoptic Stations | | | | |
|---|---|---|---|---|---|
| AVHRR SC Product | Snow | Land | Total | Users Accuracy | Producers Accuracy |
| Snow | 14 | 0 | 14 | 100% | 43.75% |
| Land | 18 | 11 | 29 | 38% | 0 |
| Total | 32 | 11 | 43 | Overall Avg. = 58.1% | |

product, SC derived from SEVIRI shows more reliable snow detection with the advantage of its high frequency scanning. The accuracy assessment tests made with synoptic stations also supported this result. It was also observed that the SC product derived from AVHRR data has the ability to detect snow pixels present on the lower elevations near snow line boundary (snow/no-snow), because of its better spatial resolution than SEVIRI. The most important contribution to the success of snow recognition algorithm is surely having observations from geostationary satellite imagery SEVIRI which provides a high temporal frequency measurement with spectral information required for snow mapping. This information helped us to identify the number and location of snow pixels which have stayed under clouds for most of the daytime.

## REFERENCES


Cline D.W., Bales R.C., Dozier J., 1998. Estimating the spatial distribution of snow in mountain basins using remote sensing and energy balance modeling. Water Resources Research, 5:1275–1285.

Derrien, M., H. Le Gléau, 2003. SAFNWC/MSG SEVIRI Cloud Product, Proceeding of the EUMETSAT Meteorological Satellite Conference, 191–198.

Dozier J, 1989. Spectral signature of alpine snow cover from Landsat Thematic Mapper. Remote Sensing of Environment, 28:9–22

Hall D.K., Riggs G.A., Salomonson V.V., DiGirolamo N.E., Bayr K.J., 2002. MODIS snow-cover products. Remote Sensing of Environment, 83:1:181–194.

Harrison A.R., Lucas R.M., 1989. Multi-spectral classification of snow using AVHRR AVHRR imagery. International Journal of Remote Sensing, 10:4:5:907–916.

Painter T. H., Dozier J., Roberts A.D., Davis R.E. & Green R.O., 2003. Retrieval of subpixel snow-covered area and grain size from imaging spectrometer data. Remote Sensing of Environment, 85: 64–77.

Romanov P., Tarpley D., Gutman G. & Carroll T.R, 2003. Mapping and monitoring of the snow cover fraction over North America. Journal of Geophysical Research, 108:D16:8619.

Sürer S., Gökdemir O., Akyürek Z., Şorman A.Ü., Beşer Ö., 2008. Snow cover mapping over the mountainous areas in europe with MSG-SEVIRI. EARSeL eProceedings, in-review.

Tampellini M.L., Brivio P.A., Carrara P., Fantoni D., Gnocchi S., Ober G., Pepe M., Rampini A., Ratti R., Nodari F.R., Strozzi T., 2004. Monitoring snow cover in alpine regions through the integration of MERIS and AATSR ENVISAT satellite observations. Proceedings of MERIS User Workshop (ESA SP-549), 10–13.

User Manual for the PGE01-02-03 v1.3 (Cloud Products) of the SAFNWC/MSG: Scientific part SAF/NWC/IOP/MFL/SCI/SUM/01, 2007. Issue 1, Rev. 3.

*Remote Sensing for a Changing Europe*
*D. Maktav (Ed.)*
*IOS Press, 2009*

*doi:10.3233/978-1-58603-986-8-94*

# Evaluation of the damages provoked by seismic events through teledetected imagery

V. Baiocchi
*Dipartimento Idraulica Trasporti e Strade, Area di Geodesia e Geomatica, "Sapienza" University of Rome, Italy, valerio.baiocchi@uniroma1.it*

R. Brigante & F. Radicioni
*Dipartimento di Ingegneria Civile e Ambientale, University of Perugia, Italy, topos@unipg.it*

Keywords: DSM, earthquake, Ikonos, Umbria

ABSTRACT: The availability of high resolution satellite stereopairs also for the civil users opens new possible fields of application among which the automatic extraction of digital models of the surface, the stereoscopic restitution, as well as the possibility to appraise changes and transformations of areas following catastrophic events, as for instance seismic events. In this last case, obviously, as for all the monitoring studies, it is necessary to effect a comparison with the situation before the event.
Photogrammetry can fortunately also use older aerial acquisitions, very useful when substantial changes have occurred in the territory; normally, such historical aerial frames are easily available at a very reasonable cost.
In this paper is illustrated an experimentation including a first series of tests to evaluate the real possibilities of use of high resolution satellite images, acquired by the Ikonos satellite, to estimate changes caused by a seismic event happened in September 1997 in central Italy, causing relevant damages to a lot of buildings in the zone. It is interesting to evaluate the different capabilities of change detection tecniques on single images and the difference of DSMs extracted by aerial and satellite stereopairs. As reference have been assumed ortophoto maps and aerial frames of the same zone, acquired in years antecedent to the seismic event. From the comparisons, it is possible to underline the variations in the urban areas as the presence of new constructions, changes of the roads, areas with different use of the ground, etc.

## 1 INTRODUCTION

The investigated area presents higher mountains, mainly in the eastern part, hilly areas and lowlands near Topino river and its tributary Menotre, with heights above sea level varying from 300 to 1200 meters. The area includes some small urban agglomerations, mainly constituted by buildings of two-three floors, and some industrial zones with ample sheds. On September 26$^{th}$, 1997 a seismic crisis started, lasting up to the first months of 1998, and interesting different municipalities, beginning mainly in Colfiorito and Cesi, with various replicas of notable intensity, particularly that of October 14$^{th}$, 1997 that has interested the territory of Sellano and Preci. The areas interested by this seismic event cover a big part of the regional territory of Umbria, with damages especially in the zones of the Apenninic chain, in the municipalities of Colfiorito, Tesina, Sostino, La Franca. Following such events a joint project of CNR/GNDT, National Italian Seismic Service and Umbria and Marche Regional Administrations, was started also with the collaboration of the Geologists Professional Orders of the of the involved Regions. The main goal of the project was to individualize, in the town circle, areas with homogeneous characteristics regarding the local seismic response, considering geologic, geomorphologic, hydrogeologic and sismostratigraphic conditions.

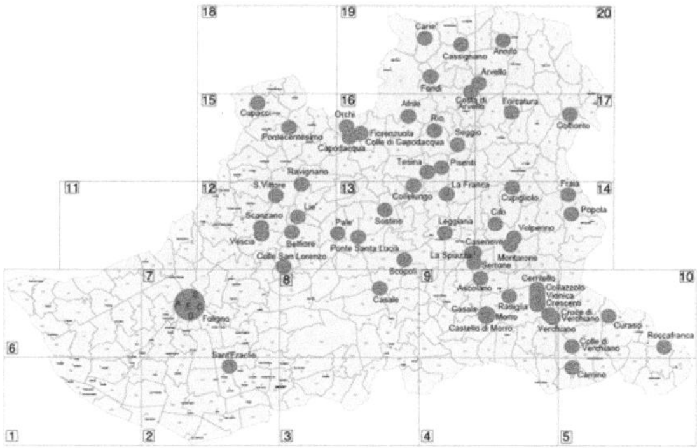


Figure 1. Foligno Municipality map with most damaged areas.

Figure 2. Encircled in red the study area (from Google Earth©).

In this way nine different zones (E1 ÷ E9) were individuated assigning different coefficients of seismic amplification (Fa).

In the municipalities of La Franca, Lie, Pontecentesimo, Tesina, Sostino, Colfiorito the unstable or potentially unstable zones has been individuated, where indeed the greatest damages have been observed. For each of the interested municipalities, after detailed surveys, integrated programs of recovery have been compiled, foreseeing the remaking of the destroyed buildings and the restructuring of those damaged.

With such job we wanted to try to put in evidence the modifications happened in the municipality territory, urban and extra urban, following the 1997 seismic events, exploiting some different photogrammetric techniques, particularly comparing satellite images acquired by Ikonos satellite in 2006, that represent the situation post seismic event, with aerial images of the same zone acquired in 1977. A planimetric comparison based on change-detection techniques was performed. Then an evaluation of the possibilities to detect the same changes comparing DSM extracted automatically by the same image was performed.

Figure 3. Kinemtic tracks (red) and GCPs (green)

Figure 4. 2006 DSM.

## 2 IMAGES AND METHODOLOGIES UTILIZED

The situation of the area in 2006 has been described through the Ikonos panchromatic stereo images of part of the zone interested by the seismic event (an area of around 150 $km^2$), characterized by a geometric resolution of 1 m, 11 bit of gray depth, standard geometric corrections, and pseudo epipolar geometry. An automatic DSM extraction was performed through the rigorous photogrammetric model implemented in the *PCI Geomatica OrthoEngine* software. The GCPs, necessary for photogrammetric orientation of the images, have been measured through a GPS survey in NRTK and rapid static post-processing modes.

The obtained DSM has been validated through comparisons with some checkpoints, with height spot points derived from the Regional Technical Map (CTR) and with the results of some kinematic surveys performed during the transfers for measuring the GCPs. Such comparisons reported an accuracy around 2-4 m in height. From the images an orthophoto map was also derived; its planimetric accuracy is less or equal to pixel dimension, therefore inferior to one meter.

The morphology before the seismic event has been derived from 1977 archive aerial images at 1:33000 scale (grey scale), and 1:13000 scale (colour), acquired with a *Wild RC10* camera (with known calibration data), digitalized with a photogrammetric scanner *Vexcel UltraScan 5000*, at a 1200 dpi resolution. Using the software *Socet Set* by *Bae System* the photogrammetric model has been estimated, then it has been possible to extract a DSM representative of the area of study as it was thirty years ago. Using the same GCPs to orient the images and the model, an orthophoto map was realized. The estimated accuracy of these products is around 40 cm planimetrically and 80 cm vertically.

To check possible errors of interpretation an orthophotomap of 1999 was also studied. At that time most of the damages caused by the aforesaid seismic events were still visible, because the destroyed buildings had not been reconstructed yet.

The tests on monoscopic images have been executed through the software *Socet Set*, performing a *change detection* between the orthophoto of 2006 (Ikonos 1 m) and 1977 (aerial 1:13000). It has been necessary to perform a series of pre-elaborations of the two images (matching of the histograms of grey scale, variation of resolution) this to make them radiometrically as similar as it was possible. This step of the elaboration was performed in the *Envi* package. The change detection, performed through *Socet Set*, according to its documentation, is based on an algorithm of pixel for pixel subtraction; the most relevant changes are underlined therefore with a very clear or very dark coloration, while what is unchanged gets an intermediate gray tone. The Ikonos DSM was extracted using *PCI Orthoengine 10.1* with the Toutin photogrammetric model (Toutin 2004), while the aerial images DSM was obtained using the *Socet set* package. Different software may led to different results whose effects can mask actual changes. Being this only a first test, we wanted to use the most standard package for each sort of image. DSM differentiation was operated in GIS environment using "subtractions" DEM commands.

Figure 5. 1977 Orthophoto.

Figure 6. 1977 DSM.

Figure 7. Detail of 1999 orthophoto with underlined changed areas.

In any case, it is important to consider that the comparisons were performed on images different in terms of geometric and radiometric resolution and with a wide time gap, but this can be very similar to the actual operating situation of these techniques in emergency situations.

## 3 RESULTS

The monoscopic change detection comparison has been the one that better evidenced changes, both caused by the seismic event and due to the strong urbanistic expansion of the zone in study, especially in the last ten years. In fact, as shown in the figures 8 and 9, the buildings of new construction, the variations of the road tracks and the changes on the utilization of the ground are well visible, particularly in the rural areas. At the same time buildings present in 1977 and destroyed by the 1996 event are easily detectable (Fig. 10). The mentioned documentation released by the municipal administration of Foligno has confirmed the results obtained by the *change detection*, mainly for completely destroyed and not reconstructed buildings. As known, a similar algorithm of change detection is very influenced by chromatic variations that necessarily point out not only true changes in the characteristics of the territory. The different insolation, for instance, creating different shadings can determine notable differences in the tones of grey among the two images compared that don't correspond to real variations.

Figure 8. 1977 - 2006 change detection: new buildings underlined in black on the right.

Figure 9. 1977 - 2006 change detection: new road underlined in black on the right.

Figure 10. 1977 - 2006 change detection: destroyed building underlined in white on the right.

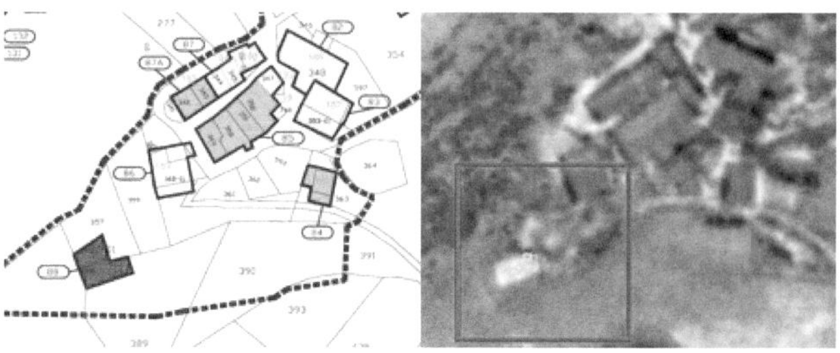

Figure 11. Destroyed building (in red) present on the municipality map (to the left) and on the change detection (in white to the right).

Figure 12. "False" change detection due to different color of the soil

Also the different state of cultivation of a ground is able to trigger the algorithm as a notable change, but in reality no change of interest has happened. Whereas the change detection doesn't notice meaningful changes, can be useful to analyze the difference of DSMs. The results of the volumetric comparison are in this case to be considered as indicative, also because the expected precision of the 2006 DSM, obtained by satellite images with a 1-meter resolution, is of 3-6 m as on specific bibliography. This accuracy is very similar to the height of the buildings present in the area of study so that their detection can be masked by errors of the model. We've also to keep in mind that height variations we'll observe could be caused by other factors and not exclusively from the event, this because the period of time passed between the two acquisitions (aerial and satellite) it's very wide.

It's necessary again to remember that the two DSMs used for the comparison are very different from each other for the different acquisition method and the elaboration with two different software of the most standard type for the different images. Anyway, some interesting results were observed: as an example in this urbanized area (Fig. 13) of the municipality of Colfiorito on which the change detection didn't reveal particular changes.

Figure 13. area of Colfiorito city where change detection doesn't reveal variations.

Figure 14-15. the same area on 1977 aerial frame and on Ikonos 2006 image

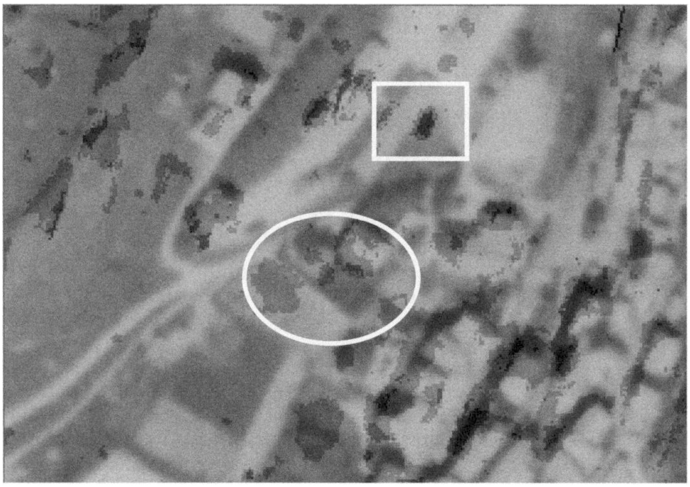

Figure 16. DSM variation (red positive, blue negative, light blue unchanged) over 1977 frame, in the ellipse changes described in the text, in the rectangles probable "false" changing.

In the same area the comparison of DSM underlines positive and negative volumetric variations, where as positive we consider the buildings existing before the event and no longer existing after. In the figure can be noticed the red zones that indeed represent buildings destroyed by the event, immediately nearby to them the blue color underlines buildings constructed later. The blue colour on existing buildings before and after the event suggests a variation of heights on the building itself. You can notice besides some positive variations in areas that from the photos don't seem modified; this could be due to the observed (Baiocchi 2004) sensibility of some algorithms for extraction of DSM to zones with particular chromatic homogeneity or different shadowing. The check of the real correspondence can be however achieved only with an on site survey.

## 4 CONCLUSIONS

These first step experimentation showed that the algorithms of change detection individuate many variations happened in the urbanized areas, keeping in mind that post-earthquake effects after about 10 years from the event, are not sufficiently representative of the immediate consequences of the event; many buildings have been in fact reconstructed. The job has allowed to show however that the variations of the urban and road tracks can be easily detected especially using the change detection algorithm on single orthophoto images. Surely better results would have been achieved, using a photogrammetric documentation of the real situation pre and post-earthquake. The same considerations can be made for the DSM comparison. The comparison of DSM has allowed anyway to easily underline variations not detectable using only the change detection algorithms. Regarding this, it has to be investigated the influence of the different algorithms of extraction of DSM used with the aerial and satellite images.

It's however important to underline that it was possible to compare very different initial data, separately elaborated with different software, getting important information on the changes happened during the years and above all following the seismic period. Surely, it has been demonstrated that the satellite images can be used for the generation of ortophotos of wide areas, with the possibility of update the existing cartography, or, as in this case, to supply an useful tool for the individualization of not anymore present buildings.

## ACKNOWLEDGEMENTS

The authors want to thank the municipal administration of Foligno for putting at our disposition the recovery programs of all the villages interested by the 1996 earthquake, and the Umbria Regional administration for the 1977 archive aerial photos.

## REFERENCES

Baiocchi, V., Crespi, M., De Vendictis, L. & Giannone, F. 2004. A new rigorous model for the orthorectification of synchronous and asynchronous high resolution imagery, *In Proceedings of the 24th EARSeL Symposium*, Dubrovnik (Croatia): 461-468.

Jacobsen, K., 2002. Mapping with IKONOS images. *In Proceedings of the 22th EARSeL Symposium,* Prague,: 98-104.

Fraser, C.S., Baltsavias, E. & Gruen, A. 2002. Processing of Ikonos imagery for submetre 3D positioning and building extraction. *Journal of Photogrammetry & Remote Sensing 56, February*: 177– 194.

Toutin, T., Chénier, R. & Carbonneau, Y., 2002. 3D models for high resolution images: examples with Quickbird, Ikons and EROS. *Proceedings of ISPRS Commission IV Symposium, Joint International Symposium on Geospatial Theory, Processing and Applications, Ottawa, July 8-12-2002*: 547-551.

Toutin, T., 2004. Geometric processing of remote sensing images: models, algorithms and methods (review paper). *International Journal of Remote Sensing, 25 (10)*: 1893-1924.

Zhang L., 2005. Automatic Digital Surface Model (DSM) Generation from Linear Array Images. *PhD Dissertation, Institute of Geodesy and Photogrammetry.* ETH Zurich, ISBN 3-906467-55-4

*Remote Sensing for a Changing Europe*
*D. Maktav (Ed.)*
*IOS Press, 2009*

*doi:10.3233/978-1-58603-986-8-102*

# Remote sensing monitoring to preserve ancestral semi-natural mountain meadows landscapes

I. Pôças[1], M. Cunha[2], A.R.S. Marçal[3] & L.S. Pereira[4]
[1]*Faculdade de Ciências, Universidade do Porto and Agricultural Engineering Research Center, Portugal, ipocas@mail.icav.up.pt*
[2]*Faculdade de Ciências, Universidade do Porto and Centro de Investigação em Ciências Geo espaciais, mcunha@mail.icav.up.pt*
[3]*Faculdade de Ciências, Universidade do Porto, DAM and Centro de Investigação em Ciências Geo espaciais andré.marcal@fc.up.pt*
[4]*Institute of Agronomy, Technical University of Lisbon and Agricultural Engineering Research Center, Portugal, lspereira@isa.utl.pt*

Keywords: Ancestral Landscapes, Mountain meadows, Vegetation indices, remote sensing, SPOT, Landsat

ABSTRACT:

"Lameiros" are ancestral semi-natural meadows, essential elements of mountain landscapes in Northern Portugal. In the "lameiros" a traditional irrigation system is used and water is applied all year around. They are mainly used for forage production for autochthonous bovine feeding, but they are also important for the water and nutrients cycle regulation, erosion control and as barrier to forest fires propagation. Although recognized for their economical, environmental, landscaping, cultural and genetic value, the perpetuation of these "lameiros" could be at risk, at medium term, due to human desertification in the mountain regions and to the announced constraints in use of water resources. To preserve these landscapes it is essential to know them better and to better characterize them. Therefore a monitoring program using remote sensing tools is now being developed to evaluate different patterns of "lameiros", and their spatial extent and evolution. Two important questions are determinant in this program: the selection of the most appropriate spatial resolution for monitoring "lameiros", and the availability of satellite historical data. In this context, NDVI were compared in two selected test sites, with and without full irrigation. Data were derived from several field campaigns with a spectroradiometer and using different sensors: i) Landsat 5 and Landsat 7 (30m pixel), ii) SPOT 4 and SPOT 2 (20m pixel), iii) SPOT 5 (10m pixel). The NDVI temporal series produced were evaluated considering "lameiros" management and weather data. Results obtained so far indicate that the SPOT images provide data at the most adequate scale.

## 1 INTRODUCTION

The landscape of North of Portugal is characterized by several elements typical of the mountains traditional agriculture, in which "lameiros" present an important role. The "lameiros" are permanent semi-natural meadows originated several hundred years ago (Moreira et al., 2001; Pires et al., 1994). They mainly occur above 700m elevation, in regions with high water availability. "Lameiros" are irrigated all year around, depending upon the water availability: i) in winter and early spring the irrigation aims at promoting a thermal regulation at the soil and vegetation level, to avoid or minimize the freezing risks for the vegetation (Gonçalves, 1985) and then irrigation is called "lima" irrigation; ii) in summer, which is a dry season, the irrigation aims to satisfy the crop water requirements.

In this context, these meadows can be classified as irrigated meadows, imperfect or deficit irrigated meadows and non-irrigated meadows, according to their proximity to water streams and springs and consequently to water availability all over the year (Pires et al., 1994; Pôças et al., 2008a; 2008b; Teles, 1970). The traditional irrigation system is constituted by a network of earth channels and small contour ditches that allows the water to spread over the pasture after being derived from water courses or springs. This ingenious system increases the infiltration opportunity time and reduces the runoff time of concentration, consequently favouring a more efficient use of water in the mountain fields and decreasing the peak flows downstream.

These semi-natural meadows represent the main forage resource for the autochthones bovine races produced in the mountain regions of the north of the country. This livestock production constitutes the first economical input for the local farmers and is a fundamental key for the rural development of regions dealing with an increasing human desertification and aging of population. Besides its economical, rural development and water regulation functions, several other benefits are attributed to "lameiros": i) prevention of erosion risks in sloping areas because the soil is permanently and full covered and runoff flow is slow down when spreading over the forage crop; ii) conservation of the biodiversity related to a variety of plant and animal species; iii) preservation of landscape values, creating an integration of several elements as forest, agricultural fields, communitarian pastures; iv) prevention of the propagation of forest fires because acting as buffer zones.

Despite their multifunctionality, the sustainability of "lameiros" may be compromised by the human desertification and the population aging in the mountain regions and by restrictions in water use. To prevent this situation, a monitoring program is being developed by remote sensing in order to better know and understand the ecological behaviour of these semi-natural meadows, their dynamics in the traditional agriculture and in featuring the mountain landscapes of North of Portugal. The information derived from this monitoring program might help improving the "lameiros" management, mainly water and irrigation management, and support the definition of economical, agricultural and ecological strategies and water and land use policies in the region.

The remote sensing applicability is recognized for biotope monitoring (Bock, 2003), land use monitoring (Marçal and Wright, 1997; Marçal et al., 2005; Sawaya et al., 2003), evapotranspiration mapping (Allen et al., 2007), and crop yield forecasting (Salazar et al., 2008), among others. Vegetation indices computed from red and near infra-red bands are often used to quantify canopy attributes, considering their relationships with various plant biophysical parameters. The Normalized Difference Vegetation Index (NDVI), one of the most used vegetation indices, is a sensitive indicator of the vegetation cover and condition (Rouse *et al.* 1973).

The success on using remote sensing data for monitoring "lameiros" dynamics is, however, conditioned by the spatial resolution adequacy due to the small size of the "lameiros" fields and the limited availability of satellite historical data to study the evolution of these meadows in the mountain landscape during the last decades, including in periods of water scarcity. This study is focused on these two issues.

## 2 MATERIAL AND METHODS

### *Test sites*

The study area was selected among the mountainous regions of north Portugal, in Salto (Montalegre), where the principal agricultural activity – livestock production – is sustained by "lameiros". Two agricultural fields were selected in this site, having different water regimes, an irrigated

meadow (IM) and a non-irrigated one (NIM). Each selected field was divided in three plots, thus a total of six plots are considered: IM1, IM2, IM3 and NIM1, NIM2, NIM3.
For the evaluation of the most appropriate spatial resolution for "lameiros" monitoring, the following data were used: i) reflectance measurements obtained at field level using a spectroradiometer and ii) satellite image data, from SPOT and Landsat sensors.

*Ground field radiometric measurements*

To obtain reflectance measurements at field level, a portable spectroradiometer FieldSpec UV/VNIR was used, with a conic IFOV around 25° and a reflectance data capture between 325 nm and 1075 nm of the electromagnetic spectrum.

In each plot, measurement points representative of the field conditions were established: i) four measurement points in the IM1; ii) three measurement points in NIM1, NIM3 and IM2; iii) 2 measurement points in NIM2 and IM3. At each measurement point, the spectroradiometer was set to an integration time of 136 ms and ten reflectance files were saved. These reflectance files were automatically compared with a white reference measurement taken over a reference panel (in which the reflectance all over the electromagnetic spectrum is close to 100%).

Seven campaigns of reflectance measurements were carried out in the period July-December 2007, sampling different vegetation stages – maximum vegetation development, vegetation cut (hay cut), and the initial development after cut, and vegetation dormancy. All measurements were performed in sunny days – 10 July 2007, 20 July 2007, 8 August 2007, 11 September 2007, 20 October 2007, 23 November 2007 and 27 December 2007 – between 11-13 hours.

*Satellite data*

For the evaluation of the spatial resolution adequacy for the "lameiros" monitoring, images from SPOT and Landsat sensors were used. Data relative to the images used are presented in Table 1.

Table 1 – Satellite images used: sensors, spatial resolution and acquiring dates

| Sensors | Spatial Resolution | Red band | Infrared band | Acquiring Dates |
|---|---|---|---|---|
| SPOT 2 | 20m | 0,61 - 0,68μm | 0,78 – 0,89μm | 4 July 2005 |
| | | | | 8 August 1991 |
| SPOT 4 | 20m | 0,61 - 0,68μm | 0,78 – 0,89μm | 8 July 1998 |
| SPOT 5 | 10m | 0,61 - 0,68μm | 0,78 – 0,89μm | 31 July 2002 |
| | | | | 20 August 2003 |
| | | | | 6 September 2006 |
| Landsat 5 | 30m | 0,63 - 0,69μm | 0,76 – 0,90μm | 15 September 1998 |
| | | | | 20 October 1997 |
| | | | | 25 December 1994 |
| Landsat 7 | 30m | 0,63 - 0,69μm | 0,78 – 0,90μm | 24 June 2000 |
| | | | | 21 November 2001 |

The satellite images were processed with the PCI Geomatica v8.2 software. The processing included geometric correction, calibration and radiometric correction, and atmospheric correction. The geometric correction was based on the manual identification of Ground control points (GCP). The GCP were identified in each satellite image using a 0.5 m spatial resolution georeferenced database (ortophotomap 1/10000). Georeferenced satellite images with 5x5m pixels were created for the study area, using a first order polynomial transformation and bilinear resampling

The meadows fields of interest were identified in the ortophotomap and then converted to 5m pixel resolution, in order to match the georeferenced satellite images.

The digital numbers of the georeferenced Landsat images were converted to spectral radiance using the following equation (Chander and Markham, 2003):

$$L_\lambda = G_{rescale} \times Q_{cal} + B_{rescale} \quad (1)$$

where $L_\lambda$ is the spectral radiance at the sensor's aperture (W/(m$^2$ sr µm)), $G_{rescale}$ and $B_{rescale}$ (W/(m$^2$ sr µm)/DN) are band specific rescaling factors (sensor gain and offset, respectively), and $Q_{cal}$ is the quantized calibrated pixel value in digital numbers.

The digital numbers of the georeferenced SPOT images were converted into spectral radiance, according to

$$L_\lambda = (DN - offset)/Gain \quad (2)$$

where $L_\lambda$ is the spectral radiance at the sensor's aperture (W/(m$^2$*sr*µm)), DN is the digital number, Gain represents the sensor gain and Offset is the sensor offset (equal to zero).

The spectral radiance data were then converted into reflectance at the top of the atmosphere through the following equation (Chander and Markham, 2003):

$$\rho = (\pi \cdot L_\lambda \cdot d^2)/(ESUN_\lambda \cdot \cos\theta_s) \quad (3)$$

where ρ is the planetary reflectance (dimensionless), d is the earth-sun distance in astronomical units, $ESUN_\lambda$ is the mean solar exoatmospheric irradiance, and $\theta_s$ is the solar zenith angle (degrees). The assumption that the water reflectance at the near infrared (NIR) band should be around zero was used to establish a simplified atmospheric correction process. The reflectance values over a water surface measured in the NIR band were assumed to be the contribution of atmosphere, and subtracted from the reflectance values of all image bands.

*Data analysis*

NDVI data were computed with the reflectance data obtained in field measurements, considering the red and near infrared band widths of the sensors SPOT2, SPOT4, SPOT5, Landsat5 and Landsat7 (Table1). The NDVI is computed as

$$NDVI = \frac{\rho_{NIR} - \rho_{red}}{\rho_{NIR} + \rho_{red}} \quad (4)$$

where $\rho_{NIR}$ is the reflectance at the near infrared wavelength band and $\rho_{red}$ is the reflectance at the red wavelength band. The mean NDVI values calculated for each measurement point were therefore obtained, and a mean NDVI value was computed for each agricultural field (IM and NIM) and each plot (IM1, IM2, IM3, NIM1, NIM2 and NIM3)

A one-way ANOVA was applied to the NDVI data in order to evaluate the potential of the different sensors to distinguish between irrigated and non-irrigated meadows. Results are expressed as F values and significance level (p value).

The Pearson correlation was applied to evaluate the correlation between the NDVI results obtained through satellite images and spectroradiometer. However, since the NDVI data obtained through the different sensors represent different years, with variable weather conditions, it was necessary to correct the NDVI values to eliminate the weather variation effect. In order to set that correction a NDVI corrected was defined (NDVIcorr $sensor_x$) as

$$NDVI_{corr\ sensor\ 1} = NDVI_{sensor\ 2} \cdot \frac{ETr_2}{ETr_1} \qquad (5)$$

where $ETr_x$ is the relative evapotranspiration, for the spectroradiometer measurements and satellite image dates (1 and 2, respectively). The relative evapotranspiration was calculated using the ratio between the reference evapotranspiration at a 10-days period covering the dates of satellite image or the spectroradiometer measurements (five days before and five days after the date of the spectroradiometer measurement and satellite image) and the 17 years (1991-2007) mean reference evapotranspiration for the same dates. The reference evapotranspiration, which represents the climate effect over a crop, was computed using the FAO56 methodology (Allen et al., 1998).

## 3 RESULTS AND DISCUSSION

In each satellite image 787 pixels in the irrigated meadow and 690 in the non irrigated meadow were used, distributed for each plot according to field conditions of vegetation development, water distribution and slope and aspect variation – 285 pixels in IM1, 329 in IM2 and 173 in IM3, and 176 pixels in NIM1, 297 in NIM2 and 217 in NIM3.

The mean NDVI values were set for each pair "satellite image – spectroradiometer measurement" – and analysed, as presented in Table 2.

In the three first months of the period considered (July - September) all the satellite sensors used in this study were able to distinguish between irrigated and non-irrigated meadows. The results of the October-December period did not distinguish meadows with different water regimes. These results on the last 3 months reflect the generalized water availability due to precipitation, over both study fields.

Table 2 – Comparison of mean NDVI values (n=3), obtained by different satellite sensors and spectroradiometer (Sp), in irrigated (IM) and non irrigated meadows (NIM).

| | | Sensors | | | | | | | | | |
|---|---|---|---|---|---|---|---|---|---|---|---|
| | | | SPOT(10m) | | | SPOT20m | | | Landsat5 | | Landsat7 |
| Month | | Sp | D1 | D2 | Sp | D1 | D2 | Sp | D1 | Sp | D1 |
| | IM | | | | 0,82 | 0,84 | 0,76 | | | 0,83 | 0,71 |
| Jul | NIM | | | | 0,64 | 0,61 | 0,56 | | | 0,59 | 0,50 |
| | *Sig* | | | | *0,000* | *0,000* | *0,000* | | | *0,001* | *0,004* |
| | IM | 0,65 | 0,68 | 0,63 | 0,65 | 0,61 | | | | | |
| Aug | NIM | 0,27 | 0,37 | 0,45 | 0,27 | 0,38 | | | | | |
| | *Sig* | *0,003* | *0,015* | *0,001* | *0,003* | *0,001* | | | | | |
| | IM | 0,70 | 0,72 | | | | | 0,71 | 0,77 | | |
| Sep | NIM | 0,49 | 0,49 | | | | | 0,48 | 0,43 | | |
| | *Sig* | *0,001* | *0,000* | | | | | *0,008* | *0,000* | | |
| | IM | | | | | | | 0,77 | 0,69 | 0,77 | 0,73 |
| Oct | NIM | | | | | | | 0,60 | 0,66 | 0,59 | 0,64 |
| | *Sig* | | | | | | | *0,046* | *0,173* | *0,008* | *0,021* |
| | IM | | | | | | | | | 0,44 | 0,59 |
| Nov | NIM | | | | | | | | | 0,50 | 0,65 |
| sig | Sig | | | | | | | | | 0,542 | 0,180 |
| | IM | 0,43 | 0,62 | 0,65 | | | | 0,66 | 0,42 | | |
| Dec | NIM | 0,43 | 0,63 | 0,60 | | | | 0,64 | 0,41 | | |
| | *sig* | *0,928* | *0,818* | *0,313* | | | | *0,748* | *0,942* | | |

Sig. - significance of F test performed by ANOVA

The Pearson correlation was computed for each pair "satellite image – spectroradiometer measurement", considering the corrected and non-corrected NDVI values over the "lameiros", as presented in Table 3. When compared with the results at field level (represented by the spectroradiometer), both SPOT and Landsat images showed a very interesting behaviour for the monitoring of the studied "lameiros", as reflected by the correlation values. The results for November and December were less consistent than those from the other months, probably reflecting the unfavourable light conditions during this period, which affect the NDVI measurements.

Although SPOT images with higher spatial resolution (10m) presented better results, it seems that the 20m and 30m spatial resolution images are also relevant for "lameiros" monitoring. This is of particular interest because it allows the use of a broader historical archive of satellite images to characterize the spatial evolution and distribution of these meadows since 1977.

Table 3 – Pearson correlation (r) between spectroradiometer and different satellite sensors for NDVI and NDVIcorr values for each month.

| Month | | SPOT (10m) | SPOT (20m) | Landsat5 (30m) | Landsat7 (30m) |
|---|---|---|---|---|---|
| Jul | NDVI | | 0,74*** | | 0,94*** |
| | NDVIcorr | | 0,83*** | | 0,73* |
| | Obs. number | | 12 | | 12 |
| Aug | NDVI | 0,95*** | 0,95*** | | |
| | NDVIcorr | 0,96*** | 0,79* | | |
| | Obs. number | 12 | 6 | | |
| Sept | NDVI | 0,79* | | 0,77* | |
| | NDVIcorr | 0,77* | | 0,60 | |
| | Obs. number | 6 | | 6 | |
| Oct | NDVI | | | 0,60 | 0,78* |
| | NDVIcorr | | | 0,78* | 0,90** |
| | Obs. number | | | 6 | 6 |
| Nov | NDVI | | | | 0,04 |
| | NDVIcorr | | | | -0,62 |
| | Obs. number | | | | 6 |
| Dec | NDVI | -0,64** | | -0,15 | |
| | NDVIcorr | -0,40 | | 0,73* | |
| | Obs. number | 12 | | 6 | |

Correlation coefficients significance: *$p<0.10$; ** $p<0.05$; ***$p<0.01$

## 4 CONCLUSIONS

The results revealed very interesting perspectives for the use of SPOT and Landsat sensors for monitoring the "lameiros" dynamics, particularly for the July-September period. However, the study was not conclusive about the potential of both sensors to distinguish between irrigated and non-irrigated meadows all year around. The study described is to be continued, with spectroradiometer measurements performed during 2008, and with satellite image acquisitions planned.

An important aspect of this investigation was the fact that the 30m resolution Landsat images are adequate to monitor "lameiros", which opens good perspectives in terms of historic coverage using the Landsat thermal band, which might be useful for evapotranspiration mapping to support assessing water management issues for these semi-natural mountain meadows.

## ACNOWLEDGEMENTS

The first author wishes to acknowledge the "Fundação para a Ciência e a Tecnologia" for the PhD grant. The authors acknowledge the OASIS European Program for making available the SPOT image data used in this study through the project, ref. 307 and the "Fundação para a Ciência e a Tecnologia" for funding of the PTDC/AGR-AAM/67182/2006 scientific project.

## REFERENCES

Allen, R.G., Tasumi, M., Morse, A., Trezza, R., Wright, J.L., Bastiaanssen, W., Kramber, W., Lorite, I. 6 Robinson, C.W. 2007. Satellite-based energy balance for mapping evapotranspiration with internalized calibration (METRIC) – Aplications. *Jounal of Irrigation and Drainage Engineering.* July/August 2007: 395-406

Allen R.G., Pereira L.S., Raes D. & Smith M., 1998. Crop Evapotranspiration. Guidelines for Computing Crop Water Requirements. *FAO Irrig. and Drain. Paper 56*, FAO, Rome, 300p.

Bock, M. 2003. Remote sensing and GIS-based techniques for the classification and monitoring biotopes. Case examples for a wet grass and moor land area in Northern Germany. *Journal for Nature Conservation* (11) 145-155

Chander, G. & Markham, B. 2003. Revised Landsat-5 TM radiometric calibration procedures and postcalibration dynamic ranges. *IEEE Transaction on Geosciences and Remote Sensing* 41

Gonçalves, D. A. 1985. A rega de lima no interior de Trás-os-Montes. Instituto Universitário de Trás-os-Montes e Alto Douro. Vila Real.

Marçal, A.R.S. & Wright, C.G. 1997. The use of 'overlapping' NOAA-AVHRR NDVI maximum value composites for Scotland and initial comparisons with the land cover census on a Scottish Regional and District basis. *International Journal of Remote Sensing* (18) 491-503

Marçal, A.R.S; Borges, J.S.; Gomes, J.A. & Pinto da Costa, J.F. 2005. Land cover update by supervsed classification of segmented ASTER images. *International Journal of Remote Sensing* (26) 1347-1362

Moreira, N.; Aguiar, C. & Pires, J.M. 2001. Lameiros e outros prados e pastagens de elevado valor florístico. Pastagens de Montanha. *Direcção Geral de Desenvolvimento Rural.* LIsboa

Pires, J.M.; Pinto, P.A. & Moreira, N.T. 1994. Lameiros de Trás-os-Montes. Perspectivas de futuro para estas pastagens de montanha. Série Estudos Escola Superior Agrária. Edição do Instituto Superior de Agronomia.

Poças, I.; Pereira, L.S.& Cunha, M. 2008a. Pastagens como factor de conservação da água em zonas de montanha, os Lameiros. *Pastagens e Forragens.* (28) (In press)

Poças, I.; Cunha, M. & Pereira, L.S. 2008b. Los lameiros, pastizales seminaturales de regadio de montaña: sistemas ancestrales en el paisaje rural portugués del siglo XXI. In: J. Palerm, T. Martinéz, M. Castro, L.S. Pereira (eds.) El pequeño riego y herencias de culturas hídricas en Iberoamérica. COLPOS, Montecillo, México (In press).

Rousse, W.; Haas, R.; Scheel, J. & Deering, W. 1973. Monitoring vegetation systems in great plain with ERST. In *Proceedings of the Third ERST Symposium, NASA SP-3511*, US Government printing office, Washington DC, 309-317

Sawaya, K.E.; Olmanson, L.G.; Heinert, N.J.; Brezonik, P.L. & Bauer, M.E. 2003. Extending satellite remote sensing to local scales: land and water resources monitoring using high-resolution imagery. *Remote Sensing of Environment* (88) 144-156

Teles, A.N. 1970. Os lameiros de montanha do norte de Portugal. Subsídios para a sua caracterização fitossociológica e química. *Separata da Agronomia Lusitana – Vol. XXXI.* Tomo I – II

*Remote Sensing for a Changing Europe*
*D. Maktav (Ed.)*
*IOS Press, 2009*


*doi:10.3233/978-1-58603-986-8-109*

# Current state of terrestrial and earth observation techniques for biomass estimation in Europe

C. Kalaitzidis & D. Zianis
*Mediterranean Agronomic Institute of Chania, Chania, Greece*

Keywords: biomass estimation, terrestrial methods, bioenergy, carbon sequestration

ABSTRACT: Methods of estimating vegetation biomass are invaluable, as they allow the forecast of potential bioenergy production, the assessment of carbon sequestration and the sustainable planning of natural resources. Traditionally, biomass estimation procedures have been employing in situ measurements of plant characteristics, which accurately determine the current and predict potential biomass production. As remote sensing platforms and sensors evolve, earth observation is developing in a reliable source of information that can be used towards the same purpose.

Each European member-state, and particularly those in Eastern Europe and the Balkan peninsula, employ a variety of individual or combination of biomass estimation methods. This study focuses on the differences in the present status of terrestrial methods and earth observation techniques for biomass estimation in Greece and Europe. Information regarding the methods of biomass estimation was acquired through personal interviews with the directors of regional Greek Forest Service departments and supplemented by a thorough review of scientific and 'gray' literature. A comparison was also made with the state-of-the-art terrestrial and earth observation methods, applied in other European countries, for the same purpose. Recommendations on harmonizing remote sensing techniques to terrestrial methods, in order to achieve low cost and increased accuracy of biomass estimation, are finally proposed in this article.

## 1 INTRODUCTION

The increasing demand for energy on a global level, is putting pressure on the nations to turn their attention to alternative sources of fuel. With the supply of oil short of the worldwide requirements, biofuel produced from vegetation biomass appears as a viable alternative. Extracting fuel from vegetation biomass turns almost any type of biomass into a renewable source of fuel.

In 2008, the European Council has proposed to increase the share of renewable energy consumption in Europe to a level of 20% to 30% by the year 2020 (EC, 2008). The aim for this proposal was to increase energy security for the European Union and reduce its dependency on imported oil and gas, as well as reducing the cost of those imports. This requires the orientation of each member-State towards the production of biomass, for use as an alternative energy source. For the long term reliance on the production of bioenergy from vegetation biomass to cover a substantial portion of a country's energy requirements, it is essential to develop a management plan of vegetation species that can be used for biomass production.

Some European countries are already very active in the exploitation of biomass for the production of fuel and energy. The methods of assessing biomass production potential and future yield vary greatly, depending on the vegetation type used in each case. The development of such methods within a country, are based on previous scientific research, usually conducted within that particular country. As a result those methods are focused on particular vegetation species, and are optimised for the environmental and socio-economic conditions of that country.

Traditionally, methods of estimating biomass and biomass potential relied on field data collection through sampling. Recently, with the advent of remote sensing, earth observation data acquired from a distance have been shown to be able to provide information regarding the amount of biomass present, as well as make estimates on future biomass production. The extent to which, methodologies falling under those two categories are employed, is also a characteristic that varies between the European Union states.

In a pan-European effort to achieve greater degree of energy independence, it is imperative that information exchange regarding biomass estimation methods is promoted, concurrently with the efforts for the development of biomass production. Harmonisation of the methodologies of biomass potential estimation across the European Union, employing either or both field and earth observation measurements, is a necessity.

This study is initially identifying the current status of biomass estimation methods, using either or both terrestrial and earth observation data, at a scientific level. Having established the potential methods that can be employed, the results of a survey of the Greek national authorities involved in biomass estimation for bioenergy, is presented. The survey concerned the data and methods used by those bodies, with respect to monitoring, assessing and predicting current and future biomass potential.

## 2 CURRENT METHODS OF BIOMASS ESTIMATION FOR BIOENERGY

The initial methods of measuring the biomass of vegetation relied on visual observations and quantitative measurements of certain vegetation properties. The number and type of those properties depended on factors such as the vegetation species, the amount and area covered and the quantitative model that those measurements were used as input. Such methods are still used today in many occasions and are accepted as reliable and producing estimates of adequate accuracy.

The introduction of remote sensing as a source of earth observation data for a large variety of applications, has led to the proposal of the use of such data in the estimation of biomass and biomass potential. Remote sensing has certain advantages and disadvantages over the traditional terrestrial measurement methods and the performance of methodologies employing earth observation data varied greatly between cases. Having been evaluated comprehensively in the recent years, such methodologies have gradually moved from a pure scientific to the application level, in some countries and to a certain extent. On the other hand, development of new sensors able to provide data of higher quality, are opening doors for the formulation of more elaborate methods of biomass estimation, initiating a process requiring laborious testing and evaluation before their value is determined.

### 2.1 *Terrestrial methods of biomass estimation*

Vegetation biomass is produced in the form of woody and green biomass of forest trees species, as well as agricultural crops, residues and byproducts. According to the Terminology of Forest Science (1971), biomass is generally defined as “the total quantity, at a given time, of living organisms of one or more species per unit area (species biomass) or of all species in a community (community biomass)”.

Direct determination of forest biomass by destructive harvest of all trees growing in a specific area is a time consuming and labour intensive procedure and has rarely been implemented (Brown and Iverson, 1992). Thus, the following three approaches have been developed for indirect estimation of forest biomass:

1) The regression analysis, which is the most widely used method and requires destructive harvest of sample trees. The obtained empirical model is subsequently applied to the stand scale and predictions of biomass are therefore obtained. Detailed information on the procedures was reported by Ogawa and Kira (1977), Parde (1980), and Satoo and Madgwick (1982).
2) The mean tree method, where some 'average' trees are sampled and the mean biomass of these trees is determined. Thus, the stand biomass is calculated as the product between the number of the trees and the obtained mean value (Ovington, 1956).
3) The unit area method where the total biomass of sample plots is determined and the average biomass per unit of ground area is determined. Finally, this value is multiplied by the total forested area and the stand biomass is obtained (Satoo and Madgwick, 1982; Chave *et al.*, 2001).

Recently, aboveground biomass expansion factors (ABEF) values have been developed for estimating forest biomass from readily available datasets of forest inventories (Lehtonen *et al.*, 2004). ABEF is calculated as the ratio between aboveground biomass and stem volume. Living branch biomass expansion factors (BBEF) and leaf biomass expansion factors (LBEF) could also be estimated by dividing the leaf and branch biomass by stem volume, respectively. The main drawbacks in this approach are mainly two: i) the values of the expansion factors change according to species, site and ontogenetic stage and ii) wood density is usually used as parameter which relates stem volume values to biomass estimates. Wood density fluctuates both within the same tree as well as between different trees in a stand and hence, using wood density to estimate expansion factors inevitably increases the variability of BEF values within a stand, and in turn within a forest.

Undoubtedly, the most commonly used mathematical model for estimating tree biomass is the allometric equation (obtained through approach 1 above) with the following power form:

$$Y = aX^b \qquad (1)$$

where $a$ and $b$ are scaling parameters that vary with the variables under investigation; $Y$ is the total tree biomass or one of its components (foliage, branches, roots, etc) and $X$ a tree dimension variable (*i.e.* $D$, $D^2$, $D^2H$, $DH$, *etc*; $D$ stands for stem diameter measured at 1.3 m above ground and $H$ is the tree height).

The standard method to obtain estimates for $a$ and $b$ values in the allometric equations is by the least-square regression of log-transformed data of $Y$ and $X$ pairs measured from destructively sampled trees that represent the size range of the stands under investigation (Ter-Mikaelian and Korzukhin, 1997; Eamus *et al.*, 2000; Zianis and Mencuccini, 2004; Zianis *et al.*, 2005). This method has been applied to either one species (*e.g.*, Bartelink, 1996; Bartelink, 1997) or to different tree species growing at the same site (*e.g.*, Martin *et al.*, 1998). Additionally, generalized equations that were developed by Pastor *et al.* (1983/1984) for six tree species growing in northeast U.S.A., for beech trees growing in Europe (Zianis and Mencuccini, 2003), and for several species growing in European forests (Muukkonen, 2007) may be used to estimate forest biomass at regional or national level. Most of the published allometric equations use $D$ as the only independent variable while introducing a second independent dimension variable (e.g., $H$, crown length, crown radius, diameter at crown base, *etc*) leads to the multicollinearity problem and unequivocally to biased biomass estimates. Parresol (1999) developed new statistical techniques and procedures to estimate woody biomass at the stand scale and reviewed several statistics for evaluating and comparing biomass models. Cannell (1983) has compiled biomass and production data from forest ecosystems across the globe for different tree compartments, while Parde (1980) thoroughly reviewed several aspects of forest biomass studies. In their textbook, Satoo and Madgwick (1982) presented a quite extensive methodology for biomass estimation while an effort was made to focus on the major ecological variables affecting primary productivity and standing biomass of forest resources.

## 2.2 *Biomass estimation using earth observation data*

Remote sensing is already being used in many countries for the recording of national forest inventories, instead of or in conjunction with field measurements, as a fast, accurate and cost efficient method of data collection. Data from remote sensors are used for the calculation of the forest area or canopy volume, as well as mapping of the forested areas (McRoberts & Tomppo, 2007). There are two type of remote sensors employed in the estimation of biomass. The passive sensors, dealing with the 400 – 2500 nm region of the electromagnetic spectrum and relying on the sun to provide the energy, and the active sensors, which produce the electromagnetic signal themselves before recording the response.

Passive sensors usually employ vegetation indices, derived from specific wavebands related with vegetation spectral characteristics. Early studies developed empirical relationships between vegetation indices extracted from remotely sensed data and physiological or biochemical properties of the vegetation in question, which could be related to biomass (Dong *et. al.*, 2003; Todd *et. al.*, 1998). Alternatively regression analysis is used to select the optimal wavebands that can be used for biomass estimation, employing high spectral resolution sensors. The impediment of these particular relationships is the lack of broad applicability to landscapes other than those from which they were developed. On the other hand, non-parametric methods also show some degree of success in estimating canopy volume and have the advantage of being generally more applicable, as the parameters of the exact quantitative relationship are not pre-defined and are calculated for each individual case. An example of non-parametric method is the *k*-Nearest Neighbour (*k*-NN), the efficiency of which increases if followed by empirical optimization (Chirici *et. al.*, 2008). Vegetation indices have been used for the estimation of both forest species and agricultural crops (Christensen & Goudriaan, 1993; Gilabert *et. al.*, 1996; Leblon *et. al.*, 1991; Zheng *et. al.*, 2004).

Two of the issues that the estimation of biomass through remote sensing has to face, are those of scale and homogeneity. These two factors define the spatial resolution and area coverage requirements of the imagery. For large scale applications, a high degree of spatial resolution is not required. In those cases, such as global biomass estimation studies, simple indices like the NDVI, or combinations of red and near-infrared bands in regression models, perform with a satisfactory degree of efficiency (Dong *et. al.*, 2003; Muukkonen & Heiskanen, 2007). Smaller scale studies however, require finer resolution and imagery from satellites such as Landsat or ASTER are essential. Muukkonen and Heiskanen (2005 and 2007) have used ASTER data, employing multiple regression and neural networks methods of analysis, estimating standwise biomass with relatively high accuracy with the red and near-infrared bands.

Active sensors used for biomass estimation primarily include lidar and radar sensors. Lidar is an instrument that measures distance at a very high spatial resolution. This instrument can be used on an airborne platform to derive canopy height, or even on the ground to capture the canopy profile and canopy metrics (tree height, crown size), which can be accurately related to above-ground biomass (Drake *et. al.*, 2002; Kimes *et. al.*, 2006).

Synthetic Aperture Radar (SAR) data capture the manner that forests scatter the emitted pulse and that signal can be used to make estimates on the canopy volume and hence, above-ground biomass. Previous studies have suggested that the choice of waveband depends on the species in question, but propose the use of longer wavebands (P- and L-band) with cross-polarization (Kasischke *et. al.*, 1997). Others have taken advantage of the effect that vegetation moisture has on the scattering of the microwave signal and have used radar to estimate total canopy moisture as a parameter that can indicate the amount of biomass present (Wigneron *et. al.*, 1999).

Some of the principles of radar can also be applied to certain passive sensors, with the capability of collecting data at an off-nadir view angle. Prime example is the Multiangle Imaging Spectro-Radiometer (MISR), which can provide data at nine different angles, but only in four bands in the visible and near-infrared region (Diner *et. al.*, 1998). The capability of multi-angle data collection can be used in conjunction with geometric models to provide fractional crown cover, mean canopy height and predict the vertical structure of forest canopies, leading to the estimation of aboveground biomass, in addition to the use of the red and near-infrared bands that supply valuable vegetation-related information (Chopping *et. al.*, 2008; Kimes *et. al.*, 2006).

## 3 BIOMASS ESTIMATION FOR BIOENERGY IN GREECE

Despite the fact that a voluminous amount of biomass allometric relationships has been developed for different tree species growing in several European countries, a limited number of empirical equations were built for trees growing in Greece. Tsiouvaras (1978) developed allometric models which relate the dry biomass of Kermes oak (*Q. coccifera* L) shoots to their diameter and/or length, in order to study forage productivity in northern Greece. Alifragis *et al.* (2001) estimated above-ground biomass for Aleppo pine (*P. halepensis* Mill.) trees growing in Chalkidiki peninsula in northern Greece, to estimate the effect of stand age on nutrient allocation. Allometric equations for several tree components for beech species (*F. moesiaca* Cz.) were built by Zianis and Mencuccini (2003) to study productivity potential across an elevation gradient in northern Greece. Recently, Mitsopoulos and Dimitrakopoulos (2007) estimated the crown fuel weight of Aleppo pine in central Greece through allometric equations built from 40 trees. All of these endeavours were made at a scientific level.

From interviews with the Forest Directors, it emerged that wood products of Greek ecosystems are estimated through the use of volumetric measurements units, since forest managers are only interested in tree stem production. Silvicultural residues either remain on the site or are removed by rural population for heating purposes. Forest management plans are based on stem volume estimations and standard statistical procedures are followed (usually systematic sampling on several plots) to acquire the appropriate datasets. The Forest Directorate is based on remotely sensed data and digital image processing techniques in order to produce ortho-photomaps (scale 1:20000), which describe the vegetation type and ground coverage for the majority of Greek forests. Based on field surveys ortho-photomaps are amended with stand density and standing volume values but the accuracy of the provided estimates is questioned since these maps are rather ‘static’ in time and are not very often updated. No other remote sensing product appears to be used in the domain of forest biomass estimation.

In the domain of agricultural crops, Greece is slowly beginning to cultivate crops for their exclusive use of producing biodisel. These efforts were initiated following a directive of the European Union, which states that by 2010, 5.75% of the annual diesel and petrol consumption of each EU state, must originate biological sources. This corresponds to a total of 370,000 hectares being used for bioenergy crops. Sunflowers and rape will be used on approximately 55% of those fields, with the remaining being used for sorghum, sugar, wheat and corn for the production of bioethanol.

However, no methods seem to be in place for the estimation of biomass of those crops and early and accurate estimation of the yield. Currently the estimations are made based on the size of the plots used for the cultivation of the crop and the variations of these estimates are between ±15% and ±35% of the mean (personal communication).

## 4 HARMONISING REMOTE SENSING AND TERRESTRIAL TECHNIQUES

Terrestrial methods of estimating biomass are accepted as the most accurate and reliable. In order for remote sensing data to be used towards the same goal, it must be shown that such data can indeed improve on these accuracies.

The weakness of the terrestrial methods of estimating biomass, lie within the statistical extrapolations made from the samples to the plot/field and the bias in the selection of representative samples. These statistical errors are known and are considered as acceptable. However, this is the area where remotely sensed data can assist in the improvement of the accuracies of those biomass estimations. The synoptic view provided by remote sensing provides average unbiased measurements of the plot or field in question. The use of vegetation indices and the employment of appropriate bands (Lu, 2006), can provide with accurate estimations of biomass. The choice of such indices must however, be made carefully, since many are weakly or insignificantly correlated with biomass (Foody *et. al.*, 2001). Terrestrial methods are by no means replaced, as they provide invaluable ground information that can be used to calibrate and validate the models employing remotely sensed data. Regression methods, whether parametric or non-parametric will still form the backbone of the methodology.

Even though terrestrial measurements will still form an integral part of a biomass estimation methodology, remote sensing can provide an immense amount of data at short temporal intervals and at a comparably lower cost. The final obstacle that remains, is the identification of the appropriate remote sensing data, in terms of spatial, spectral and temporal resolution, that can be used for biomass estimation of a particular vegetation type, in specific environmental conditions and geographic location.

## 5 CONCLUSIONS

Science has provided with methods for estimation biomass of vegetation that can or will be used for the production of biofuel. Depending on the case a combination of terrestrial and earth observation data can supply relatively accurate estimates of biomass. The technological evolutions of remote sensors and platforms is promising that in the future data availability will allow the improvement of the current methods and the development of new ones, that will provide even more accurate estimates.

As it is evident that the reliance of humanity on mineral fuel must be restricted, the extraction of bioenergy from biomass appears as a viable solution. Appropriate management of the cultivation of crops or forests, requires methods of estimating biomass at en early stage, prior harvesting. Some European countries are already routinely employing methods for mapping and monitoring the areas covered by vegetation with potential for use for bioenergy. In Greece, such monitoring is very scarce and in biomass estimation is almost non-existent. However, this is a situation that will soon change, as the European Union is promoting and enforcing the use of crops for the extraction of biofuel.

## ACKNOWLEDGMENTS

We would like to thank the forest directors of Imathia and Pella prefectures, for the valuable information provided. This work was undertaken within the frame of an FP7 funded programme with the acronym CEUBIOM, and full title “Classification of European Biomass Potential for Bioenergy Using Terrestrial and Earth Observations”.

## REFERENCES

Alifragis, D., Smiris, P, Maris, F., Kavvadias, V., Konstantinidou, E. and Stamou, N. 2001. The effect of stand age on the accumulation of nutrients in the aboveground components of an Aleppo pine ecosystem, *Forest Ecology Management*, 141: 259-269

Bartelink, H.H. 1996. Allometric relationships on biomass and needle area of Douglas-fir. *Forest Ecology Management* 86:193-203

Bartelink, H.H. 1997. Allometric relationships for biomass and leaf area of beech (Fagus sylvatica L), *Annals of Forest Science* 54: 39-50

Brown, S. & Iverson, L.R. 1992. Biomass estimates for tropical forests, *World Resource Review* 4: 366-384

Cannell, M.G.R. 1983. *World forest biomass and primary production data*, Academic press, London

Chave, J., Riera, B., Dubois, M.A. 2001. Estimation of biomass in a neotropical forest of French Guiana: spatial and temporal variability, *Journal of Tropical Ecology* 17: 79-96

Chirici, G., Barbati, A., Corona, P., Marchetti, M., Travaglini, D., Maselli, F. & Bertini, R. 2008. Non-parametric and parametric methods using satellite images for estimating growing stock volume in alpine and Mediterranean forest ecosystems, *Remote Sensing of Environment* 112: 2686-2700

Chopping, M., Moisen, G.G., Su, L., Laliberte, A., Rango, A., Martonchik, J.V. and Peters, P.C. 2008. Large area mapping of southwestern forest crown cover, canopy height, and biomass using the NASA Multiangle Imaging Spectro-Radiometer (MIRS), *Remote Sensing of Environment* 112: 2051-2063

Christensen, S. & Goudriaan, J. 1993. Deriving light interception and biomass from spectral refletance ratio, *Remote Sensing of Environment* 43: 87-95

Diner, D.J., Beckert, J.C., Reilly, T.H., Bruegge, C.J., Conel, J.E., Kahn, R., Martonchik, J.V., Ackerman, T.P., Davies, R., Gerstl, S.A.W., Gordon, H.R., Muller, J-P., Myneni, R., Sellers, R.J., Pinty, B. & Verstraete.,M.M. 1998. Multiangle Imaging SpectroRadiometer (MISR) description and experiment overview. IEEE Transactions on Geoscience and Remote Sensing, 36: 1072-1087

Dong, J., Kaufmann, R.K., Myeni, R.B., Tucker, C.J., Kauppi, P.E., Liski, J., Buermann, W., Alexeyev, V. & Hughes, M.K. 2003. Remote sensing estimates of boreal and temperate forest woody biomass: carbon pools, sources, and sinks, *Remote Sensing of Environment* 84: 393-410

Drake, J.B., Dubayah, R.O., Knox, R.G, Clark, D.B & Blair, J.B. 2002. Sensitivity of large-footprint lidar to canopy structure and biomass in a neotropical rainforest, *Remote Sensing of Environment* 81: 378-392

Eamus, D., McGuinness, K. & Burrows, W. 2000. Review of allometric relationships for estimating woody biomass for Queensland, the Northern Territory and Western Australia. In: *National Carbon Accounting System Technical Report 5b*, Australian Greenhouse Office, Canberra

EC. 2008.20 20 by 2020 – Europe's climate change opportunity, COM(2008)0030, Brussels, Belgium

Foody, G.M., Cutler, M.E.J., McMorrow, J., Pelz, D., Tangki, H., Boyd, D.S. & Douglas, I. 2001. Mapping the biomass of Bornean tropical rain forest from remotely sensed data, *Global Ecology & Biogeography* 10: 379-387

Gilabert, M. A., Gandia, S. & Melia, J. 1996. Analyses of spectral-biophysical relationships for a corn canopy, *Remote Sensing of Environment* 55: 11-20

Kasischke, E.S., Melack, J.M. & Dobson, M.C. 1997. The use of imaging radars for ecological applications – A review, *Remote Sensing of Environment* 59: 141-156

Kimes, D.S., Ranson, K.J., Sun, G. & Blair, J.B. 2006. Predicting lidar measured forest vertical structure from multi-angle spectral data, *Remote Sensing of Environment* 100: 503-511

Leblon, B., Guerif, M. & Baret, F. 1991. The use of remotely sensed data in estimation of PAR use efficiency and biomass production of flooded rice, *Remote Sensing of Environment* 38: 147-158

Lehtonen A., Mäkipää R., Heikkinen J. & Sievänen R. Liski J. 2004. Biomass expansion factors (BEFs) for Scots pine, Norway spruce and birch according to stand age for boreal forests, *Forest Ecology and Management* 188: 211-224

Lu, D. 2006. The potential and challenge of remote sensing-based biomass estimation, *International Journal of Remote Sensing* 27: 1297-1328

Martin, J.G., Kloppel, B.D., Schaefer, T.L., Kimbler, D.L. and McNulty S.G. (1998). Aboveground biomass and nitrogen allocation of ten deciduous southern Appalachian tree species. *Canadian Journal of Forest Resesearch* 28: 1648-1659

McRobetrts, R.E & Tomppo, E.O. 2007. Remote sensing support for national forest inventories, *Remote Sensing of Environment* 110: 412-419

Mitsopoulos, I.D. & Dimitrakopoulos, A.P. 2007. Allometric equations for crown fuel biomass of Aleppo pine (*Pinus halepensis* Mill.) in Greece, *International Journal of Wildland Fire* 16: 642-647

Muukkonen, P. 2007. Generalized allometric volume and biomass equations for some tree species in Europe, *European Journal of Forest Research* 126: 157-166

Muukkonen, P. & Heiskanen, J. 2005. Estimating biomass for boreal forests using ASTER satellite data combined with standwise forest inventory data, *Remote Sensing of Environment* 99: 434-447

Muukkonen, P. & Heiskanen, J. 2007. Biomass estimation over a large area based on standwise forest inventory data and ASTER and MODIS satellite data: A possibility to verify carbon inventories, *Remote Sensing of Environment* 107: 617-624

Ogawa, H. & Kira, T. 1977. Methods for estimating forest biomass. In: *Primary productivity of Japanese forests-Productivity of terrestrial ecosystems*, Shidei, T. and Kira, T. (eds), JIBP Vol 16 pp 15-25

Ovington, J.D. 1956. The form, weights and productivity of tree species grown in close stands, *New Phytologist* 55: 289-304

Parde, J. 1980. Forest biomass, *Forestry Abstracts* 41: 343-362

Parresol, R.B. 1999. Assessing tree and stand biomass: a review with examples and critical comparisons, *Forest Science* 45: 573-593

Satoo, T. & Madgwick, H.A.I. 1982. *Forest biomass*. Martinus Nijhoff, London, pp 152

Ter-Mikaelian, M.T. & Korzukhin, M.D. 1997. Biomass equations for sixty-five North American tree species, *Forest Ecology Management* 97: 1–24

Todd, S.W., Hoffer, R.M. & Milchunas, D.G. 1998. Biomass estimation on grazed and ungrazed rangelands using spectral indices, *International Journal of Remote Sensing* 19: 427-438

Tsiouvaras, C.N. 1978. Twig diameter-length-weight relations of Quercus coccifera L. *Annals of Science*, Vol KA. Forestry section, Aristotelian University

Wigneron, J.-P., Ferrazzoli, P., Olioso, A., Bertuzzi, P. & Chanzy, A. 1999. A simple approach to monitor crop biomass from C-band radar data, *Remote Sensing of Environment* 69: 179-188

Zheng, D., Rademacher, J., Chen, J., Crow, T., Bresee, M., Le Moine, J. & Ryu, S.-R. 2004, Estimating aboveground biomass using Landsat 7 ETM+ data across a managed landscape in northern Wisconsin, USA, *Remote Sensing of Environment* 93: 402-411

Zianis, D. & Mencuccini, M. 2003. Aboveground biomass relationships for beech (*Fagus moesiaca* Cz.) trees in Vermio Mountain, Northern Greece, and generalised equations for *Fagus* spp. *Annals of Forest Science*, 60: 439-448

Zianis, D. & Mencuccini, M., 2004. On simplifying allometric analyses of forest biomass. *Forest Ecology Management* 187: 311–332

Zianis, D., Muukkonen, P., Mäkipää, R. & Mencuccini, M. 2005. Biomass and stem volume equations for tree species in Europe. *Silva Fennica Monographs* No 4

*Remote Sensing for a Changing Europe*
*D. Maktav (Ed.)*
*IOS Press, 2009*

*doi:10.3233/978-1-58603-986-8-117*


# Use of ASTER-data for a soil erosion risk model application, Chios island (Greece)

S. Veraverbeke, R. Goossens & T. Vanderstraete
*Department of geography, Ghent University, Ghent, Belgium*

Keywords: ASTER, soil erosion risk model application, Chios

ABSTRACT: ASTER-satellite data provide useful information for a soil erosion risk model application. An orthorectified FCC and a DEM are the results of a digital photogrammetric restitution. An elevation model can be used to calculate slope angles and the drainage area of each location. So the sensitivity for erosion due to the topographic situation is estimated. In the Mediterranean, rainfall is often of an orographic origin. A linear regression equation taken from literature with the height above sea level as independent variable, and the mean annual rainfall as dependent variable has been adapted for the Chios island. Then the erosivity of the rainfall is estimated. The orthorectified bands are used to make a LULC-classification by means of the maximum likelihood classifier. Before classifying, the image is corrected for differential illumination effects to improve the classification. The different LULC classes have a different sensitivity towards erosion. One non-satellite based information source has been added to the dataset, namely a generalised lithologic map. This map allows tracking of the parent material, which is an important factor determining the soil formation. Different lithologic formations have different erodibility characteristics. When the rainfall erosivity and the sensitivity for erosion due to the topographic situation and the lithology are combined, the potential soil erosion risk is estimated. The potential soil erosion risk is the erosion risk in a virtual world without vegetation. When the sensitivity for erosion of the land cover is taken in account, the actual soil erosion risk is estimated; this is the current soil erosion risk like. At the island of Chios a high potential soil erosion risk is prevalent for most of the land. However the actual soil erosion risk is relatively low for most parts of the island. Vegetation tempers soil erosion. Nevertheless the Mount Epos area is badly degraded and has high actual soil erosion risks.

## 1 INTRODUCTION

In the Mediterranean land degradation has been recognised as an important environmental problem (Thornes 1996; Geeson et al. 2002). Land degradation has several definitions. An appropriate definition is "the reduction and loss of the biological or economic productivity caused by land use change, from a physical process, or a combination of the two (Thornes 1996)". In the Mediterranean the subject has been labelled "Mediterranean Desertification" (Brandt and Thornes 1996; Geeson et al. 2002). In Mediterranean Europe soil erosion due to water erosion (especially sheet erosion) is the main erosion process. This is a consequence of the Mediterranean climate characteristics: dry and hot summers alternated with wet winters.
Although soil erosion models have been constructed for a long time, each model has some specific limitations. Wischmeier and Smith (1978) were the first who constructed a soil erosion model

(Universal Soil Loss Equation USLE) by combining various factors that influence erosion. Their model is designed for plot scale measurements. CORINE (1992) and Kosmas (1999) used the USLE principle for the suite in Mediterranean climate conditions.

Remote sensing can offer opportunities for data gathering in and monitoring of land degradation (e.g. Cannizzaro et al. 2002; Harrison et al. 1996). Its synoptic nature makes it a consistent and continuous data source. ASTER-data provide topographic information (after photogrammetric restitution of the nadir and backward near-infrared bands) and information of terrain features (e.g. land use, vegetation) that can be used in a soil erosion model.

## 2 STUDY AREA

The Greek island Chios is situated in the east of the Aegean Sea. It lies off the Turkish coast and south of the Greek island Lesbos (Figure 1). The climate is typical Mediterranean with a total annual rainfall average of 543 mm and most of the precipitation falling in the winter months (Hellenic National meteorological Service, unpublished data).

*Figure 1: study area (www.maproom.psu.edu)*

## 3 METHODS

### *3.1. DEM and orthoimage*

The near-infrared band of the ASTER-sensor is acquired nadir and backward in an along-track stereomodel. A digital photogrammetric restitution is performed using VirtuoZo software. The relative orientation runs fully automatically in VirtuoZo. Ground control points were collected in August 2006. Absolute orientation was performed using 34 ground control points. When the epipolar images are resampled, the image matching algorithm calculates the height value. In doing so the DEM and the orthoimage are created.

### *3.2. LULC classification*

Differences in terrain illumination can result in differences in reflectance values of similar terrain features, leading to a possible misclassification. To overcome this problem a topographic correction for differential illumination effects is applied.

*3.2.1. Topographic correction for differential illumination effects*

The correction method that has been chosen is C-correction. The basic C-correction formula is (McDonald et al. 2000):

$$Ln = L \times \left( \frac{cos(\Theta_{sz}) + c}{cos(i) + c} \right)$$

Where $Ln$ is the radiance observed from horizontal surface (corrected image value), $L$ is the radiance observed over sloped terrain (uncorrected image value), $\Theta_{sz}$ is the sun's zenith angle and $i$ is the sun's incidence angle in relation to the surface normal direction, $c$ is the estimated C-correction coefficient. The value of $cos(i)$ is derived using the following equation (Law and Nichol 2004):

$$cos(i) = cos\Theta_n \times cos\Theta_{sz} + sin\Theta_n \times sin\Theta_{sz} \times cos(\Phi_s - \Phi_n)$$

Where $i$ is the sun's incidence angle in relation to the surface normal direction, $\Theta_n$ is the terrain's slope angle, $\Theta_{sz}$ is the solar zenith angle, $\Phi_s$ is the solar azimuth and $\Phi_n$ is the aspect of the terrain.

The C-value is derived by linear regression with the radiance value $L$ as the dependent variable and the cosine of the sun's incidence angle $cos(i)$ as independent variable (McDonald et al. 2000):

$$L = m \times cos(i) + b$$

The parameter C is then defined as (McDonald et al. 2000):

$$c = \frac{b}{m}$$

*3.2.2. Multispectral image classification*

After their correction for differential illumination effects, the green, red and NIR ASTER-band are used in a multispectral image classification in order to classify the land use and land cover of the island. The maximum-likelihood algorithm was chosen to perform the classification. The different classes that were spectrally distinguishable are: river bank vegetation-irrigated agriculture, phrygana, maquis, pine forest, polyculture olives-grain, naked rock, fire spot, phrygana on fire spot.

### *3.3. Soil erosion model application*

Modelling is always done by simplifying the reality. The model that is applied here is an empirical grey-box model. That implies that the model is based on statistical significant relationships, though not all details of the system are fully known (Morgan 2005). The results of the model application are expressed on an ordinal scale. The model application has as goal to estimate soil erosion by water, more specified sheet erosion. Wischmeier and Smith (1978) were with the Universal Soil Loss Equation the first who combined various erosion influencing factors to estimate erosion. The USLE is constructed for use in North American conditions on plot scale. This is an important difference than when using satellite raster data.

*3.3.1. Erosion influencing factors*

The erosivity of the rainfall R, the sensitivity for erosion due to the topographic situation T, the sensitivity for erosion due to the lithology L and the sensitivity due to erosion of the land use and land cover are taken in account. For the factors R, T and L a scaling between 1 and 2 is conducted. The factor LU is scaled between 0.1 and 1 (more information about the scaling in 3.3.2.). These factors are all partly correlated; however their influence on erosion is modelled separately.

The relation between annual rainfall and altitude, expressed in a linear regression with altitude, expressed in a linear regression with altitude as independent variable and mean annual rainfall as dependent variable, was found for Crete (Naoum and Tsanis 2003) and modified for the Chios isl-and using climate records. Then the erosivity of the rainfall R is estimated using the formula of Re-nard et al. (1995):

$$ER = 0{,}0048P^{1.61}$$

Where ER is the erosivity of the rainfall and P is the mean annual precipitation. The value is then rescaled between 1 and 2 resulting in the value of the factor R.

To estimate the sensitivity for erosion due to the topography T, the slope angle as well as the drai-nage area for each location is taken in account. The resulting formula is (Warren et al. 2005):

$$T = A^m (\sin \beta)^n$$

Where A is the drainage area and $\beta$ is the slope angle. For sheet estimating sheet erosion, the pa-rameters m and n are set on 1. Finally the factor T is rescaled between 1 and 2.

The sensitivity for erosion due to the lithology L is determined after Yassoglou (1995) and Kosmas et al. (1999) (Table 1).

*Table 1: Sensitivity for erosion for the different lithologies and deposits of Chios (after Yassoglou 1995 and Kosmas et al. 1999)*

| Lithology or deposit | Index |
|---|---|
| Alluvium, colluvium | 1 |
| Basic volcanic rocks, flysch | 1.5 |
| Sands and gravels, sandstones | 1.7 |
| Marl, carbonate rocks, acid volcanic rocks | 2 |

The sensitivity for erosion of the land cover is determined after Kosmas et al. (1999) (Table 2).

*Table 2: Sensitivity for erosion for the different land uses and land covers of Chios (after Kosmas et al. 1999)*

| Land Use Land Cover | Index |
|---|---|
| Maquis | 0.1 |
| Pine forest | 0.2 |
| River bank vegetation-irrigated agriculture | 0.4 |
| Polyculture olives-grain | 0.6 |
| Phrygana | 0.8 |
| Naked rock | 1 |

### *3.3.2. Potential and Actual Soil Erosion Risk*

An overview of the methodology is given in Figure 2.

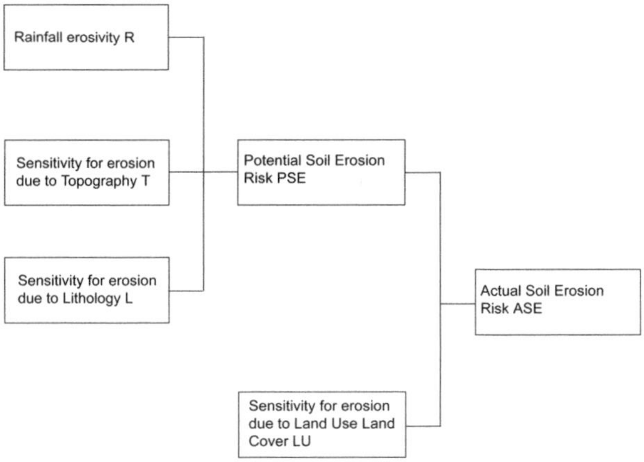


*Figure 2: Methodology of the application of the soil erosion model*

The potential soil erosion risk PSE is calculated as the soil erosion risk in a virtual world without vegetation. The factors R, L and T are taken in account. The factor LU is not taken in account. Where the other factors are relatively stable on a long term, vegetation processes are more dynamic (Aforestation, forest fires, overgrazing,…). The PSE is calculated by:

$$PSE = \sqrt[3]{R \times T \times L}$$

The resulting PSE values are in the range between 1 and 2. Values near 2 indicate a high potential soil erosion risk, whether values near 1 indicate a low potential soil erosion risk.
By calculating the Actual Soil Erosion Risk ASE the factor LU is taken in account:

$$ASE = LU \times PSE = LU \times \sqrt[3]{R \times T \times L}$$

The resulting ASE values are in the range between 0.1 and 2. Normally the ASE value should be less than the corresponding PSE value as vegetation tempers erosion. Only by the class "naked rock" the potential soil and actual soil erosion risk are equal. A class discretisation is performed to visualise the results.

## 4 RESULTS

### *4.1. DEM and orthoimage*

Hirano et al. (2003) consider a RMS-deviation between 7 and 15 meter as a good result when working with ASTER-data. The resulting RMS-deviation of the study is 11.4 m. A resulting height map is displayed in Figure 3.

### *4.2. LULC classification*

In order to make an evaluation of the correction for differential illumination effects, a test spot of pixels of the same land cover, consisting of both illuminated pixels and "shadow" pixels has to be defined. By regressing the $cos(i)$-values as x-values against the DN-values as y-values for the uncorrected and for the corrected image bands two regression equations are obtained per band (one for the uncorrected image band and one for the corrected image band). The DN-values of uncorrected image should normally vary more than the DN-values of the corrected image band. As a consequence the slope of the regression equation should have a lower value by the corrected image band. An ideal correction for differential illumination effects should result in a slope value of zero (McDonald et al. 2000). The regression equations of the green, red and NIR band are given in Table 3. The results of the correction for differential illumination effects are satisfying. Though in some areas the C-correction method seems to overcorrect.

*Table 3: Regression equations for the uncorrected and corrected image bands ($cos(i)$ as x, DN as y)*

| Green band | Green band corrected | Red band | Red band corrected | NIR band | NIR band corrected |
|---|---|---|---|---|---|
| y = 0.1704x + 66.491 | y = -0.0209x + 63.794 | y = 0.2236x + 45.16 | y = -0.0078x + 44.001 | y = 0.209x + 29.179 | y = 0.0203x + 32.21 |

The result of the LULC-classification conducted on the for differential illumination effects corrected image bands is presented in Figure 4.

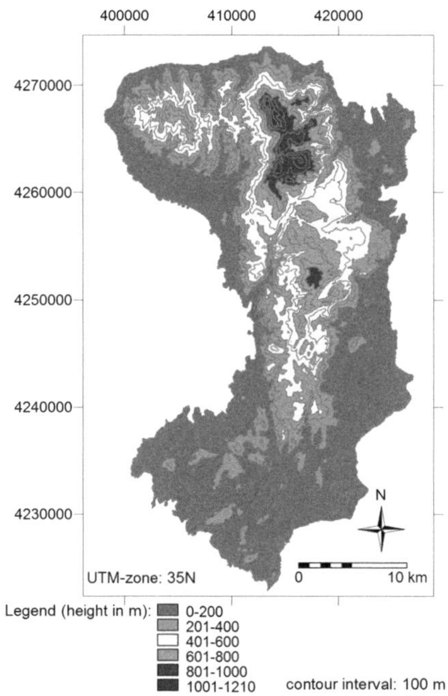


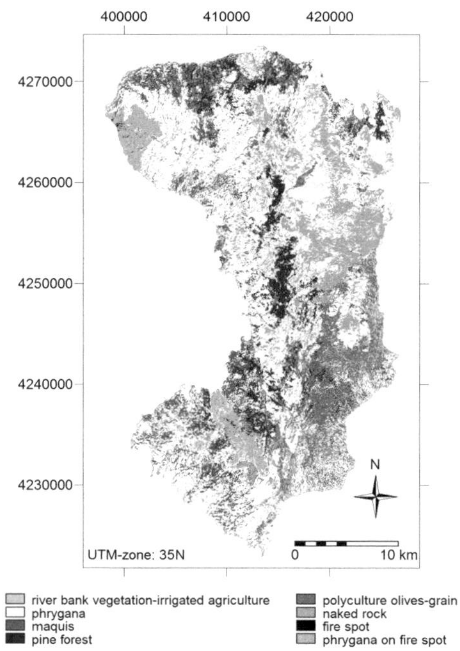


*Figure 3 (Left): height map of Chios*

*Figure 4 (Right): Land Use Land Cover Classification of Chios*

### *4.3. Soil erosion model application*

The resulting Potential Soil Erosion risk map and the resulting Actual Soil Erosion risk map are presented in respectively Figure 5 and Figure 6. The classes high and very high soil erosion risk are abundant in the potential soil erosion risk map. This is normal because the potential soil erosion risk considers the situation of maximal erosion. This map shows which areas that are more sensible for soil erosion. When the Actual Soil Erosion risk is calculated, the role of the vegetation as an erosion restraining factor is emphasized. Most parts of the island have a relatively low actual soil erosion risk. Nevertheless the Mount Epos area (location on Figure 6) is an area with high actual soil erosion risks. This is because the Mount Epos area is locked in a vicious fire circle (Grove and Rackham 2001). Shepherd set fires to use the burned areas as rangeland for their animals (Margaris en Koutsidou 2002). These activities make the Mount Epos area a degraded spot.

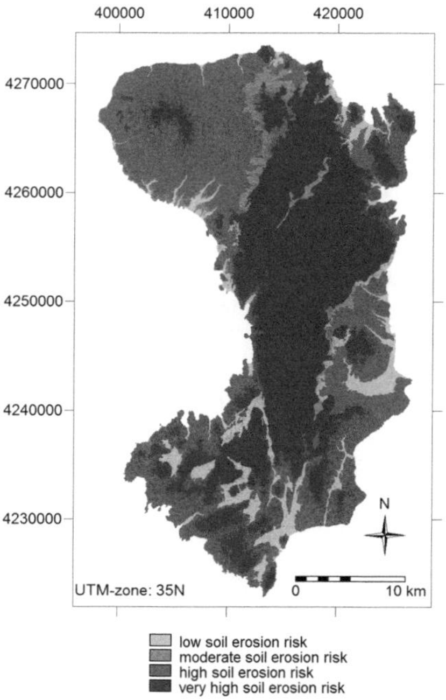


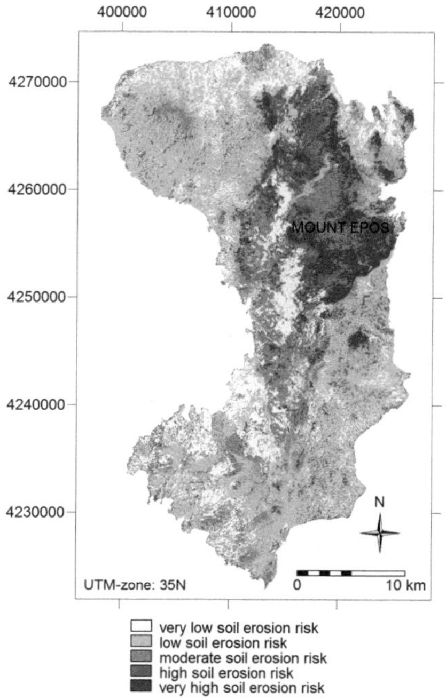


*Figure 5 (Left): Potential Soil Erosion Risk map*

*Figure 6 (Right): Actual Soil Erosion Risk map*

## 5 CONCLUSIONS

ASTER-data provide useful information to monitor soil erosion risk. ASTER-data provide topographic information and information about the terrain features. This information can be used to construct a relative simple soil erosion model by combining different data layers. It was not the purpose of the paper to suggest an elaborated methodology for soil erosion modelling. Though the use of satellite data combined with the existing knowledge of soil erosion modelling, gives opportunities for soil erosion monitoring in areas where other datasets are not existent or difficult to access.

The soil erosion model application was constructed for the Greek island Chios. A digital photogrammetric restitution provided a DEM. The LULC classification was improved by correcting the image bands for differential illumination effects. On most parts of the island the actual soil erosion risk is relatively low; nevertheless the Mount Epos area has high actual soil erosion risks.

## REFERENCES

Brandt C., Thornes J. (eds.) 1996. Mediterranean Desertification and Land Use (Wiley), 554 p.

CORINE, 1992. Soil Erosion Risk and Important Land Resources in the Southern Regions of the European Community (EUR 13233 EN), 97 p.

Geeson N., Brandt C., Thornes J. (eds.) 2002. Mediterranean Desertification, A Mosaic of Processes and Responses (Wiley), 440 p.

Grove A., Rackham O., 2001. The Nature of Mediterranean Europe. An Ecological History (Yale University Press), 384 p.

Kosmas C., Kirkby M., Geeson N., 1999. The MEDALUS Project. Mediterranean Desertification and Land Use. Manual on Key Indicators of Desertification and Mapping Environmentally Sensitive Areas to Desertification. 87 p.

Law, K.H., Nichol, J. 2004. Topographic Correction for Differential Illumination Effects on IKONOS Satellite Imagery. XXth ISPRS Congress, Istanbul.

Margaris N., Koutsidou, E., 2002. Landscape Protection from Grazing and Fire. In: Mediterranean Desertification: A Mosaic of Processes and Responses (eds. Geeson N., Brandt C., Thornes J.)(Wiley), pp. 83-92

McDonald E., Wu X., Caccette P., Campbell N., 2000. Illumination Correction of Landsat TM Data in South East NSW, Proceedings of the Tenth Australasian Remote Sensing and Photogrammetry Conference, Adelaide, pp. 1-13

Morgan R., 2005. Soil Erosion and Conservation (Blackwell Publishing), 304 p.

Naoum S., Tsanis I., 2003. Temporal and Spatial Variations of Annual Rainfall on the Island of Crete, Greece. In: Hydrological Processes 17, 1899-1922

Renard G., Lane, L., Foster, G., Laflen, J. 1995. Soil Loss Estimation. In: Soil Erosion, Conservation and Rehabilitation. (Ed. Agassi, M.). pp. 169-202

Thornes J., 1996. Introduction. In: Mediterranean Desertification and Land Use, Edited by C.J. Brandt, J.B. Thornes (Wiley), pp. 1-11

Warren S., Mitasova H., Hohmann M., Landsberger S., Iskander F., Ruzycki T., Senseman G. 2005. Validation of a 3-D Enhancement of the Universal Soil Loss Equation for Prediction of Soil Erosion and Sediment Deposition. Catena 64, pp. 281-296

Wischmeier W., Smith D. 1978. Predicting rainfall erosion losses. A guide to conservation planning, (US Dept. Agric. Agricultural Handbook), 537 p.

Yassoglou N., 1995. Land and Desertification. In: Desertification in a European Context: Physical and Socio-economic Aspects (eds. Fantechi, R., Petr, D., Balabanis, P. and Rubio, J.L.).European Commission, DG XII EUR 15415 EN:, pp. 35-56

*Remote Sensing for a Changing Europe*
*D. Maktav (Ed.)*
*IOS Press, 2009*

*doi:10.3233/978-1-58603-986-8-125*

# Evaluation of data fusion methods for agricultural monitoring based on synthetic images

A. S. Rodrigues, A. R. S. Marçal & M. Cunha
*Centro de Investigação em Ciências Geo-Espaciais, Faculdade de Ciências, Universidade do Porto, Portugal*

Keywords: synthetic images, data fusion, NDVI, agricultural monitoring.

ABSTRACT: There are several data fusion methods widely used to produce a high resolution multi-spectral image from a pair of images - a panchromatic high resolution and a multi-spectral lower resolution image. Although the fused images can be visually satisfactory, it is not clear whether they provide additional information for quantitative measurements made from satellite images. A methodology to evaluate data fusion algorithms is proposed, based on the production of synthetic images that reproduce real satellite images. An experiment was conducted testing the performance of six data fusion methods in the production of NDVI values for land parcels from SPOT HRG and Landsat TM data. The fusion methods evaluated were: Brovey, IHS Hexcone, IHS Cylinder, PCA, Wavelet IHS and Wavelet Single Band. The best data fusion method overall was found to be Wavelet IHS, although better results were obtained by using directly the lower resolution multi-spectral data instead. The software tools developed and a number of test images datasets are freely available at the SITEF website (www.fc.up.pt/sitef).

## 1 INTRODUCTION

The number of satellite sensors acquiring high and very high resolution images of the Earth has been steady increasing in the last few years. Most of these sensors use two complementary image modes – a multi-spectral image (M) and a higher spatial resolution panchromatic image (P). Examples of such sensors include IKONOS (multi-spectral image with 4m pixel and panchromatic image with 1m pixel), QuickBird (2.4m / 0.6m) and SPOT (10m / 5m, or 20m / 10m). The objective of data fusion in this context is to generate a multi-spectral image with both high spatial and spectral resolutions (Ranchin & Wald 2000). The effective application of a data fusion algorithm produces a high resolution multi-spectral (fused) image that is usually satisfactory for visual perception or cartographic applications. However, it is not clear whether the fused image provides valuable additional information when the aim is to make quantitative measurements from the satellite image.

The purpose of this work is to investigate if the use of data fusion improves the information provided by satellite images for quantitative measurements in a practical application (agriculture monitoring), and to evaluate the most widely used data fusion methods using synthetic images.

## 2 DATA FUSION

There are several well established image data fusion methods, such as Intensity Hue Saturation (IHS), Principal Components Analysis (PCA), Brovey and Wavelet (Wang *et al* 2005b).

One of the simplest methods is the Brovey fusion. The fused image is simply obtained by a normalization of M, which is then multiplied by P (Wehrmann *et al* 2005). In the IHS method, a RGB color composite produced from M is mapped to the IHS color space. The I component is then replaced by P, after histogram matching, and the resulting IHS image converted back to the RGB color space. There are three IHS models – cylinder, triangular and hexcone - providing slightly different results in the RGB colors produced. The IHS method is simple and effective, but it can only be applied to multi-spectral images with 3 bands. An alternative is to use the PCA method, which can be applied to images with any number of bands, and works in a somehow related way. Initially the principal components of M are computed. The first component (which has the most information) is replaced by P, after histogram matching, and then the principal components are converted back to the initial space (Wald 2002). The wavelet method is the most recent one and harder to implement. Four images are produced from P: one of rough detail and three of high resolution corresponding to the horizontal, vertical and diagonal components. The rough image of P is replaced by a histogram matched version of M, after which the inverse transform is computed to produce the final fused image (Balcik & Sertel 2007). The wavelet fusion can be used in one of three models: Single Band, IHS and PCA.

The quality of the fused image depends on the method used and on the data itself. The evaluation is usually based on visual analysis, or sometimes on statistics that examine the similarity or the discrepancies between the fused and original products, on a pixel by pixel and band by band basis. The quantitative evaluation of an image produced by data fusion is not a straight forward task, as there is usually no reference (high resolution multi-spectral image) to be used for comparison. The common approach is to degrade the fused image to the spatial resolution of M, and to compare it with M (Meenakshisundaram & Couloigner 2005). Several parameters have been used for this task, such as Euclidean Distance (Wang *et al* 2005a), Coefficient of Correlation (Scheunders & Backer 2001) and Root Mean of Square Error (Wehrmann *et al* 2005). All these evaluation parameters focus on the differences between individual pixels and bands. However, for many remote sensing applications the objective is to extract information based on a combination of bands for areas with several pixels. One common example is the production of Normalized Difference Vegetation Index (NDVI) values for land parcels (Jensen 2000).

The NDVI is computed using the reflectance values in the red ($\rho_R$) and near infrared ($\rho_{NIR}$) bands, using (1). Two NDVI ranges were used: -1 to 1 and 0 to 1, by setting to zero all negative values. The interval 0 to 1 was used as the standard range.

$$NDVI = \max\left\{0, \frac{\rho_{NIR} - \rho_R}{\rho_{NIR} + \rho_R}\right\} \tag{1}$$

The calibration of image band *i* to reflectance ($\rho_i$) is done using (2), where $R_i$ is the radiance, $E_{0i}$ the equivalent solar spectral irradiance and $\theta_s/\theta_v$ are the solar/viewing zenith angles (FIFE 2004). The radiance is obtained directly from the recorded Digital Numbers (DN) using a linear relation $R_i = \alpha_i DN + \beta_i$, where $\alpha_i$ and $\beta_i$ are the calibration coefficients (FIFE 2004, GAEL 2003).

$$\rho_i = \frac{\pi R_i}{E_{0i}\cos(\theta_s)\cos(\theta_v)} \tag{2}$$

A suitable evaluation of the quality of a data fused product would be to compare the value of NDVI computed using the data fused image ($NDVI_{FUS}$) with the NDVI obtained directly using a high resolution multi-spectral image ( $NDVI_{MH}$ ). The absolute error in NDVI (δ) is thus computed from (3). However, this approach is difficult to implement using real satellite images, as there is usually no high resolution multi-spectral image available to use for validation of the fused image.

$$\delta = |NDVI_{MH} - NDVI_{FUS}| \quad (3)$$

## 3 METHOD

The data fusion evaluation was done using synthetic images produced with the Synthetic Image Testing Framework (SITEF).

### 3.1 *Synthetic Image Testing Framework (SITEF)*

The SITEF provides images with controlled spatial and spectral characteristics, which simulate real multi-spectral satellite images by making use of a reference image where training areas are identified (Marçal & Rodrigues 2008). The SITEF software is freely available at www.fc.up.pt/sitef, where some test images datasets are also available. The objective is to simulate land parcels of various sizes with different land cover types. Initially, the number of land cover types (c), the size of the smallest unit (u), the range of sizes (s) and a repetition parameter (r) are used to produce a base image. As an example, figure 1 shows four base images, all with u=4, s=4, r=2, and with c=4, 5, 6, 7. The smallest squares on the top left section of these images have 4 by 4 pixels, while the largest ones on the lower right have 16 by 16 pixels (s=4). In this case there are 4 parcels of each size (r=2). The whole images are 80 by 80 pixels, with a total of 64 parcels. There are 4 single unit parcels (4 by 4 pixels), and generally 4 parcels of i by j units, with i,j=1,2,3,4. The classes are assigned to parcels assuring that two neighboring parcels always belong to different land cover types.

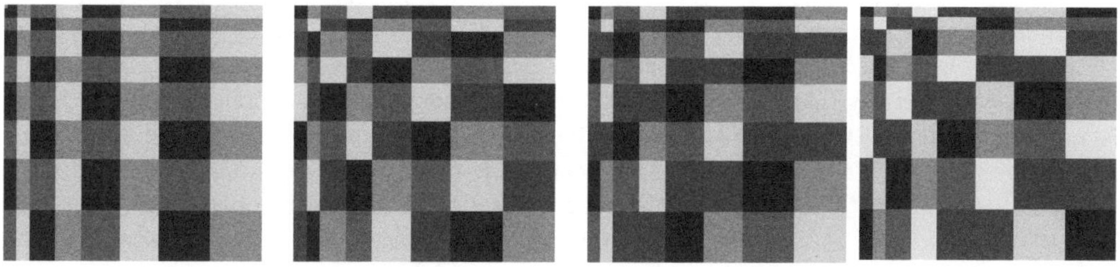

Figure 1. Synthetic base images with c=4, 5, 6, 7 (left to right), all with u=4, s=4, r=2.

The base image and a reference satellite image are used to produce a multi-spectral synthetic image. Reference areas are previously established in the multi-spectral satellite image, one for each land cover class considered (1, …, c). The multi-spectral synthetic image produced (MH) will have the same number of bands as the reference satellite image, with the pixel values of each class in the base image replaced by random vectors from the reference areas. A lower resolution version (ML) of this image is also produced (reduced in size by a factor of 2), as well as a high resolution panchromatic image (PAN). The process is presented schematically in figure 2.

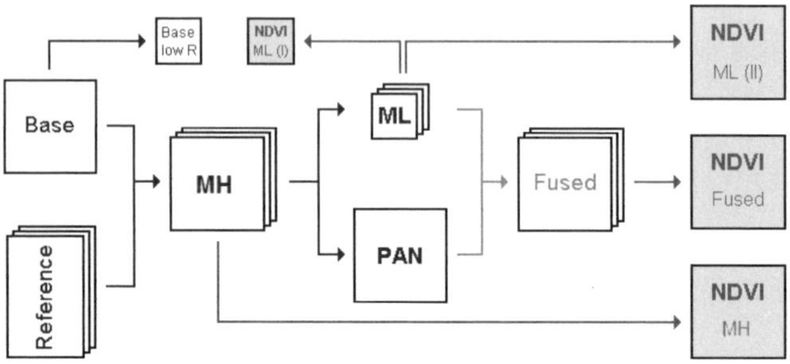


Figure 2. Schematic representation of the data fusion evaluation experiment: MH - multi-spectral high resolution, ML - multi-spectral low resolution and PAN - panchromatic high resolution.

## 3.2 *Test images*

Two satellite images, from Landsat and SPOT, both covering a mountainous area around Montalegre (Portugal) were used as reference. The SPOT 5 HRG image was acquired in 2005 and the Landsat 5 TM image in 1997. Two test images were produced using the SPOT (TIS) and Landsat (TIL) satellite images as reference. A total of six land cover classes were considered: irrigated (1) and non-irrigated permanent semi-natural mountain meadows (2), evergreen forest (3), deciduous forest (4), communitarian pastures (5) and annual crops (6).

The multi-spectral synthetic images TIS and TIL were created with u=3, s=8, r=5 and c=6, corresponding to an image size of 540 by 540 pixels. Although both SPOT HRG and Landsat TM sensors provide multi-spectral images with more than three bands, the synthetic images were both produced with three bands, as only the red and near infrared bands are used to compute the NDVI and the spectral coverage of the images only overlap three bands. Three versions were prepared for each test image: MH - multi-spectral high resolution, ML - multi-spectral low resolution and PAN - panchromatic high resolution. The MH synthetic images are presented in figure 3, with histogram linear enhancement. The lower resolution multi-spectral images were created by averaging every 2 by 2 pixel block into a single pixel. The panchromatic versions were created by combining the three bands with weights that reproduce the spectral response of the satellite sensors. For SPOT 5 HRG these weights are 0.617, 0.383, 0.000 for bands 1,2,3. As Landsat 5 TM does not have a panchromatic band, the characteristics of ETM panchromatic band were used instead. The relative contributions from Landsat 5 TM bands 2,3,4 to reproduce the ETM panchromatic band are: 0.333, 0.335, 0.332. Alternative versions of the test images used here are available at SITEF (www.fc.up.pt/sitef) with 8 and 5 classes – the Montalegre dataset.

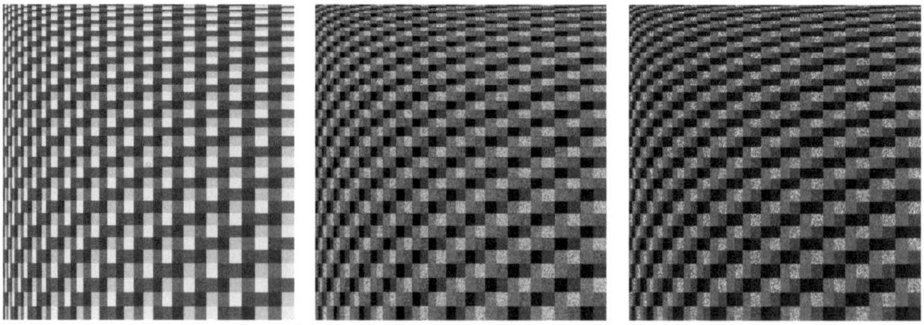

Figure 3.Base image (left) and synthetic test images TIS (center) and TIL (right), with histogram linear enhancement.

An evaluation of the spatial resolution degradation process was also made. A Gaussian filter of size 19 and standard deviation 5 was applied to the SPOT MH image, resulting in an image without high frequencies (MHGauss). From the MHGauss image two versions were created: MLGauss and PANGauss.

## 4 RESULTS

The data fusion experiment was performed using PCIGeomatica (PCI Geomatics 2005) and ERDAS Imagine (Erdas Imagine 2007) software. The versions ML and PAN of the test images were used as input for the data fusion methods tested, resulting in six fused images: (1) Brovey, (2) PCA, (3) IHS model Cylinder, (4) IHS model Hexcone, (5) Wavelet model IHS and (6) Wavelet model Single Band.

A total of nine NDVI images were produced for each test image using the original MH and ML images (two versions), and using the six fused images. These NDVI images are all high resolution except one of the NDVI image produced from ML. The overall process is illustrated schematically in figure 2.

### 4.1 *Evaluation of data fusion methods*

The average NDVI was computed for each of the 1600 parcels of the base image, for all seven high resolution NDVI images. The NDVI of each parcel obtained from the fused images was compared with the NDVI computed with the original MH data. As an illustration, figure 4 shows a plot of all NDVI values obtained from the Brovey (left) and Wavelet IHS (right) fused data versus NDVI values obtained from the original MH data, for TIS. The plots for the other fusion methods exhibit roughly the same behavior, both for TIS and TIL.

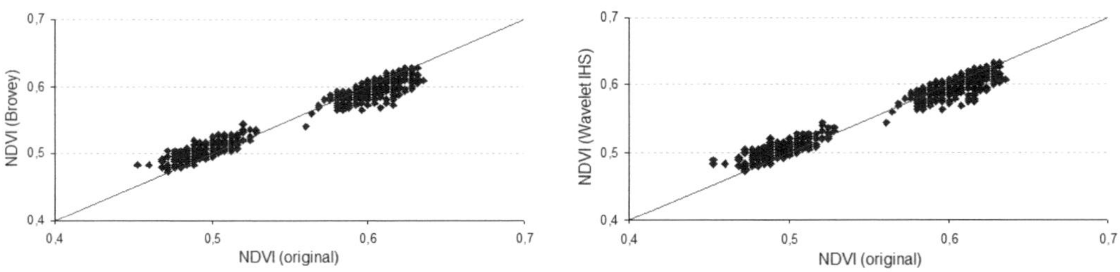


Figure 4. Average NDVI values for all 1600 parcels, computed with Brovey (left) and Wavelet IHS (right) fused data versus the original MH data, for SPOT test image (TIS).

The average absolute errors in NDVI (δ) were computed for all parcels and for all fused images. Average values of δ were calculated for parcels of the same size within each image. These results ($\delta \times 1000$) are presented in table 1 for the square parcels of TIS. In bold is the best result for each scale tested. Overall the best methods were Brovey, Wavelet IHS and IHS Hexcone, all with low average absolute errors (6.5, 6.7 and 7.2). The values of δ tend to become smaller with the increase of parcel size, as the influence of neighboring parcels becomes less important.

Table 1 – Average absolute error in NDVI (δ x1000) for six data fusion methods for SPOT test image (TIS).

| Parcel size (units) | Brovey | IHS Cylinder | IHS Hexcone | PCA | Wavelet IHS | Wavelet Single Band |
|---|---|---|---|---|---|---|
| 1 | 14.0 | 30.7 | 13.9 | 31.2 | **13.8** | 32.8 |
| 2 | 17.3 | 37.4 | **15.4** | 31.0 | 16.2 | 36.5 |
| 3 | 7.7 | 25.1 | **6.9** | 25.9 | 7.7 | 16.0 |
| 4 | **0.8** | 19.7 | 4.3 | 21.8 | 2.6 | 6.2 |
| 5 | 2.4 | 19.5 | 3.8 | 23.5 | **2.1** | 10.1 |
| 6 | 5.9 | 25.9 | **4.5** | 24.2 | 5.6 | 17.3 |
| 7 | 3.2 | 21.0 | 3.8 | 23.5 | **3.0** | 10.2 |
| 8 | **1.0** | 20.3 | 4.6 | 22.6 | 2.2 | 6.2 |
| All | **6.5** | 25.0 | 7.2 | 25.5 | 6.7 | 16.9 |

A similar experience was carried out with the synthetic data for Landsat (TIL). The results obtained from the fused data are presented in table 2. In bold is the best result for each scale tested. There are four fusion methods that provide NDVI values with low errors (IHS Cylinder, PCA, Wavelet IHS and Wavelet Single Band) while the others two methods (Brovey and IHS Hexcone) do not provide satisfactory NDVI results for all parcel sizes tested.

Table 2 – Average absolute error in NDVI (δ x1000) for six data fusion methods for Landsat test (TIL).

| Parcel size (units) | Brovey | IHS Cylinder | IHS Hexcone | PCA | Wavelet IHS | Wavelet Single Band |
|---|---|---|---|---|---|---|
| 1 | 131.5 | 30.6 | 66.2 | **22.9** | 24.2 | 24.8 |
| 2 | 109.3 | 29.3 | 55.8 | **22.6** | 23.4 | 23.8 |
| 3 | 122.1 | 14.6 | 60.0 | 12.6 | **12.2** | 13.6 |
| 4 | 123.5 | 7.2 | 58.2 | **6.9** | 9.9 | 11.2 |
| 5 | 128.3 | **9.8** | 59.4 | 11.7 | 11.8 | 13.3 |
| 6 | 118.4 | 11.2 | 57.1 | 11.7 | **9.9** | 14.7 |
| 7 | 124.8 | **8.8** | 57.8 | 11.0 | 10.6 | 12.3 |
| 8 | 132.2 | **7.0** | 58.1 | 7.7 | 10.2 | 11.8 |
| All | 123.8 | 14.8 | 59.1 | **13.4** | 14.0 | 15.7 |

An alternative approach is to obtain the NDVI values of each parcel directly from the low resolution multi-spectral image (ML). As the parcels are located in the base image, which is only available in high resolution, this can be achieved by one of two modes: (I) reducing the resolution of the base image, to match the lower resolution multi-spectral image, or (II) increasing the resolution of the ML image to match the resolution of the base image (see figure 2). Each of these modes provides estimates of the NDVI values for each parcel, which can be compared with the reference NDVI obtained directly from the high resolution original test data (MH). The results for TIS and TIL, presented in table 3, are about the same as the best ones from the fused data, with a very slight advantage of mode II. The values of δ are again generally much lower for large parcels.

Table 3 - Average absolute error in NDVI (δ x1000) using the original lower resolution TIS and TIL.

| Parcel size (units) | TIS | | TIL | |
|---|---|---|---|---|
| | Mode I | Mode II | Mode I | Mode II |
| 1 | 13.9 | 13.6 | 28.8 | 21.8 |
| 2 | 19.4 | 17.1 | 26.1 | 22.1 |
| 3 | 6.1 | 7.5 | 7.7 | 9.6 |
| 4 | 0.8 | 0.8 | 5.4 | 5.4 |
| 5 | 2.7 | 2.2 | 10.2 | 8.0 |
| 6 | 6.1 | 5.9 | 9.4 | 8.5 |
| 7 | 3.0 | 2.9 | 5.1 | 6.2 |
| 8 | 1.0 | 1.0 | 5.1 | 5.1 |
| All | 6.6 | 6.4 | 12.2 | 10.8 |

A similar analysis was performed using the evaluation parameters Euclidean Distance (ED), Coefficient of Correlation (R) and Root Mean of Square Error (RMSE) instead of δ. The results obtained for TIS are almost identical to the evaluation based on δ. For TIL, the IHS Cylinder is less rated with these parameters than it was for δ, particularly for R. Otherwise the results for TIL are also consistent with the evaluation based on δ.

4.2 *Evaluation of the spatial degradation process*

The image pairs MLGauss / PANGauss and MLGauss / PAN were used to produce two fused images (fusGauss and fusNew), using the IHS Hexcone fusion method. Average values of δ were calculated for parcels of the same size within each image. The results obtained for the two fused images based on MLGauss were compared with those obtained with the fused image obtained directly from ML and PAN, without using a Gaussian filter (FUSED). Figure 5 shows the values of δ for the NDVI images produced from these three fused images, using a range of 0 to 1 (left) and -1 to 1 (right) for the NDVI. The use of a Gaussian filter in the degradation stage results in better NDVI values in the fused images. The range -1 to 1 was found to be better for all methods and scales.

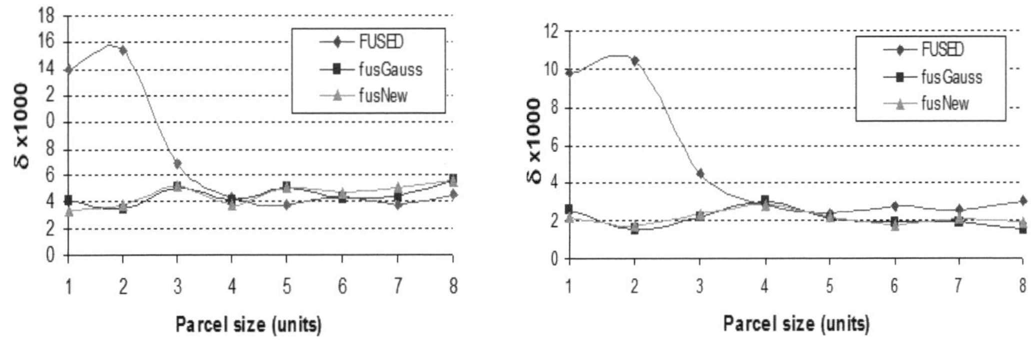


Figure 5: Average absolute error in NDVI (δ x1000) obtained from the fused images with a NDVI range of 0 to 1 (left) and -1 to 1 (right)

## 5 CONCLUSIONS

The methodology proposed, based on the production of synthetic images reproducing real satellite images, proved to be effective for the evaluation of data fusion results. The test images were produced with the Synthetic Image Testing Framework (SITEF), which is available at www.fc.u.pt/sitef. The SITEF software and test datasets can be used to test not only data fusion methods, but also image segmentation and image classification.

The experiment carried out performed an evaluation of six data fusion methods with simulated SPOT HRG and Landsat TM data. For SPOT HRG, the best fusion methods were Brovey, Wavelet IHS and IHS Hexcone, while for Landsat TM the best results were produced from PCA, Wavelet IHS and IHS Cylinder. For the two test datasets combined, the best fusion method was found to be Wavelet IHS. This is in line with the results reported by various authors, such as Vijayaraj *et al* 2004, Karathanassi *et al* 2007 and Zhou *et al* 1998, for pixel and bands based tests. Overall, both the best fusion methods and the lower resolution multi-spectral images produce NDVI average values that are very close to those obtained from the high resolution multi-spectral images (differences in NDVI below 0.02), except for the very small parcels. However, all the evaluation parameters on all images indicate that the smallest error in the computation of NDVI average for land parcels is obtained using the lower spatial resolution multi-spectral image directly, with the parcel location by mode II. This indicates that for quantitative measurements, such as NDVI, there is no benefit from using fused data instead of the original lower resolution multi-spectral data. It is nevertheless worth noting that the best results from the fused data are only marginally worse than those obtained from the lower resolution data.

The evaluation of the spatial degradation process indicates that the use of a Gaussiam filter and the range -1 to 1 for NDVI provides better results when using the fused data to compute average NDVI values per parcel. The spatial degradation mode could be further improved, by using a Modulation Transfer Function filter (Aiazzi *et al* 2006). Other fusion methods, besides the six tested here could perhaps produce better results, such as the modified version of the IHS method proposed by Tu *et al* 2004, and Context-Based Decision which was voted as the best fusion method in a recent contest (Alparone *et al* 2007). However, even with these improvements the best results would still likely be obtained using the low resolution multi-spectral images, as data fusion methods are not intended to improve spectral information but rather the visual interpretation.

## ACKNOWLEDGMENTS

This work is part of the FCT LameSat_XXI project (PTDC/AGR-AAM/67182/2006). The authors wish to thank the OASIS programme (nº307) for providing the SPOT satellite image, and Isabel Pôças for the identification of land cover types.

## REFERENCES

Aiazzi, B., Alparone, L., Baronti, S., Garzelli, A. & Selva, M. 2006. *MTF-tailored multiscale fusion of high-resolution MS and Pan imagery*. Photogrammetric Engineering and Remote Sensing, vol.72, no.5, 591-596.

Alparone, L., Wald, L., Chanussot, J., Thomas, C., Gamba, P. & Bruce, L. 2007. *Comparison of Pansharpening Alghorithms: Outcome of the GRS_S Data Fusion Contest.* IEEE Transactions on Geoscience and Remote Sensing, vol.45, no.10.

Balcik, F.B. & Sertel, E. 2007. *Wavelet_Based Image Fusion of Landsat ETM Images: A Case Study for Different Landscape Categories of Istanbul* (Presented Conference Paper style). Conference on Information Extraction from SAR and Optical Data, with Emphasis on Developing Countries. Istanbul 2007.

Erdas Imagine. 2007. http://gi.leica-geosystems.com/default.aspx . Consulted in 2007-12-19.

FIFE. 2004. http://*dac.ornl.gov/FIFE/guides/Satellite_SPOT_Extracted_Data.html#4.1*. *Satellite SPOT Extracted Data.* Consulted in 2007-12-12.

GAEL 2003. http://earth.esa.int/pub/ESA_DOC/landsat_FAQ/. *Inventory of Landsat product anomalies*, Issue 1, Revision 0, 5-11. Consulted in 2007-11-07.

Jensen, J.R., 2000. *Remote Sensing of The Environment – An Earth Resource Perspective*. Prentice Hall.

Karathanassi, V., Kolokousis, P. & Ioannidou, S. 2007. A *comparison study on fusion methods using evaluation indicators*. International Journal of Remote Sensing, vol. 28, no. 10, 2309-2341.

Marçal, A.R.S. & Rodrigues, A.S. 2008. *A Framework for the Evaluation of Multi-Spectral Image Segmentation.* Submitted to the Proceedings of GEOBIA 2008, August 2008, Calgary, Canada.

Meenakshisundaram, V. & Couloigner, I. 2005. *Quality assessment of Fusion Methods for High Resolution Images*. 3rd International Symposium Remote Sensing and Data Fusion Over Urban Areas. USA 2005.

PCI Geomatics. 2005. http://www.pcigeomatics.com/ .Consulted in 2007-12-19.

Ranchin, T. & Wald, L. 2000. *Fusion of High Spatial and Spectral Resolution Images: The ARSIS Concept and its Implementation*. Photogrammetric Engineering & Remote Sensing, vol. 66, no.1, 49–61.

Scheunders, P. & Backer, S.De. 2001. *Fusion and merging of multispectral images using multiscale fundamental forms*. Journal of the Optical Society of America A, vol.18, no.10, 2468-2477.

Tu, T.M., Huang, P.S., Hung, C.L. & Chang, C.P. 2004. *A Fast Intensity-Hue-Saturation fusion technique with spectral adiustment for IKONOS imagery*. IEEE Geoscience and Remote Sensing Letters, vol.1, no.4, 309-312.

Vijayaraj, V., O'Hara, C. G. & Younan, N. 2004. *Quality analysis of Pansharpened Images*. Geoscience and Remote Sensing Symposium. IEEE International, vol.1, 85- 88.

Wald, L. 2002. *Data Fusion: Definitions and Architectures – Fusion of images of different spatial resolutions*. Paris: Les Presses de l´École des Mines.

Wang, L., Zhang, Y. & Feng, J. 2005a. *On the Euclidian Distance of Images*. IEEE Transactions on Pattern Analysis and Machine Intelligence, vol.27, no.8, 1334- 1339.

Wang, Z., Ziou, D., Armenakis, C., Li, D. & Li, Q. 2005b. *A Comparative Analysis of Image Fusion Methods*. IEEE Transactions on Geoscience and Remote Sensing, vol. 43, no.6, 1391-1402.

Wehrmann, T., Colditz, R.R., Bachmann, M., Steinnocher, K. & Dech, S. 2005. *Evaluation of Image Fusion Techniques*. Remote Sensing & GIS Environmental Studies, vol.113, 296-302.

Zhou, J., Civco, D.L. & Silander, J.A. 1998, *A wavelet transform method to merge Landsat TM and SPOT panchromatic data*. International Journal of Remote Sensing, vol. 19, no. 4, 743-757.

*Remote Sensing for a Changing Europe*
*D. Maktav (Ed.)*
*IOS Press, 2009*

*doi:10.3233/978-1-58603-986-8-134*

# Improved classification of urban areas with pixel-based measures of object size

Rik Bellens & Sidharta Gautama
*Ghent University – TELIN – IPI – IBBT*
*St. Pietersnieuwstraat 41, B-9000, Ghent, Belgium*
*Email: rik.bellens@telin.ugent.be*

Keywords: urban classification, shape, morphological profiles

ABSTRACT: Meter to sub-meter resolution satellite images have generated new interests in extracting man-made structures in the urban area. However, classification accuracies for such purposes are far from satisfactory. Spectral characteristics of urban land cover classes are so similar that they cannot be separated using only spectral information. As a result, there is an increased interest in incorporating geometrical information. In current literature, this is achieved by using an object-based approach. This requires a segmentation process. However, the complex objects in urban remote sensing images make this process very difficult. In this paper, we propose a method to measure the minimum and maximum dimension of an object, without however performing a segmentation. This method is based on morphological profiles (MP). Previous work on MP's have shown the potential for improving classification results. However, an MP contains many values for each pixel, which can lead to problems of dimensionality. Feature extraction algorithms could reduce the dimensionality, but the resulting features are no longer interpretable. In this paper we use MP's to derive a measure of minimum and maximum object dimension. These two measures allow to differentiate between long (roads) and more compact objects (buildings). We show that these new features improve the classification substantially.

## 1 INTRODUCTION

Recent advances in earth observation technology have led to an increased availability of data products at very high spatial resolutions. These may open up new areas in the application of satellite imagery. Land-use mapping in complex settings such as urban and suburban environments is one of the domains for which the new Very High Resolution (VHR) data will offer new possibilities. Of particular interest herein is the detection and identification of a variety of man-made structures such as roads and buildings. The difficulty of using VHR imagery such as IKONOS and QuickBird is that the classification accuracy for such purposes is far from satisfactory (Zhang and Couloigner, 2006). Spectral characteristics of urban land cover classes such as road surfaces, parking lots and open areas are so similar that they cannot be separated using only spectral information. As a result, there is an increased interest in incorporating geometrical information in image classification.

Automated land-cover classification can be accomplished using either pixel-based or object-based approaches (Liu et al., 2005; Marangoz et al., 2004; Hofmann, 2001). Pixel-based methods classify each pixel individually, while object-based methods first group together pixels in a meaningful way by image segmentation. The object-based approach provides a straightforward method to incorporate geometrical information. Different shape characteristics of objects can easily be calculated on the segments. However, the segmentation process is a very difficult task. Good parameters for the process highly depend on the image data and the classification task (Hofmann,

2001). In pixel-based approaches, it is less straightforward to incorporate geometrical information. Consequently, most pixel-based approaches only rely on spectral and possibly textural information.

One possibility to incorporate geometrical information in pixel-based methods is the use of mathematical morphology (Serra, 1982; Soille and Pesaresi, 2002; Soille, 2003). Image features defined by their morphological characteristics are reportedly useful to improve accuracies in urban classifications using VHR data. In Chanussot et al. (2006); Benediktsson et al. (2003) a multiscale approach is used. Morphological profiles (MP) are generated from a panchromatic image. A morphological profile is a composition of morphological openings and closings with increasing size of the structuring element. Each opening and closing results in a transformation of the original image in which objects smaller than the structuring element are deleted. As such, the morphological profile carries information about the size and the shape of objects in the image. In Bellens et al. (in press) this approach is extended to a directional morphological profile, which contains information about the maximal dimension of objects.

To improve the classification of satellite images in urban areas, we can include morphological profiles as extra bands in the classification process. However, an MP typically consist of more than 10 values for each pixel. This might cause problems of dimensionality. In Benediktsson et al. (2003) two approaches to reduce the dimensionality are discussed. The first method uses feature extraction algorithms. This leads to a reduced set of features with high information content. However, these features are not interpretable. Therefore, it is not easy to analyse and/or fine-tune these features. Moreover, feature extraction algorithm can be computationally intensive. The second approach is an attempt to extract the size information from the MP. However, the method used is very sensitive to different parameters and does not result in a good classification. In this paper we will propose a more robust method to derive a measure of minimal and maximal object size from the regular and the directional MP. We will show that these features will improve the classification of man-made objects substantially.

## 2 MORPHOLOGICAL PROFILE

Mathematical morphology (Serra, 1982; Soille, 2003) is a popular tool in image processing and can be used in very diverse tasks such as feature detection, image segmentation, image denoising, image sharpening and many more. Successful applications of mathematical morphology have been reported in remote sensing image processing as well(Soille and Pesaresi, 2002).

Many morphological operators exist. Two are of particular importance for this paper: opening and closing. Here we will give an informal description of these operators. For a formal definition we refer to Serra (1982); Soille (2003). Basically, a morphological opening (resp. closing) removes objects from the image which are brighter (resp. darker) than their background and which are smaller than a certain structuring element (SE). An object is smaller than the SE if the SE does not fit into that object. The SE can have different sizes and shapes. Figure 1 shows two closings with disk shaped SE's of different sizes. With the smallest sized SE none of the objects are removed, with the larger SE, the smallest objects have disappeared. We use the term object to clarify to the reader the behaviour of morphological transformations. Note, however, that openings and closings are local operations with no explicit notion of objects and object boundaries. As a result, objects do not only disappear, but also the shape of objects are changed. This is an unwanted side-effect which could be solved by using morphological reconstruction. However, as shown in Bellens et al. (in press, 2007), this often leads to worse results.

Besides disk shaped SE's, also other shapes can be used. Figure 2 shows three closings with linear SE's with different orientations of the line. This time objects are removed if their length in the orientation of the line is shorter than the length of the line. Taking the maximum over different orientations but with the same length results in images like shown in Fig. 3. Here objects are removed if their maximum dimension is smaller than the length of the line.

By increasing the size of the SE, more and more objects disappear. We will use the term scale of an opening or closing to refer to this size. The vector containing the pixel values of pixel x in

(a) original (b) R = 3 (c) R = 6

Figure 1. An image (a) and its closings with disk shaped SE's of size R = 3 (b) and size R = 6 (c). Objects with a width smaller than 2 ·R are deleted from the image. As a side effect, borders of objects are smoothed.

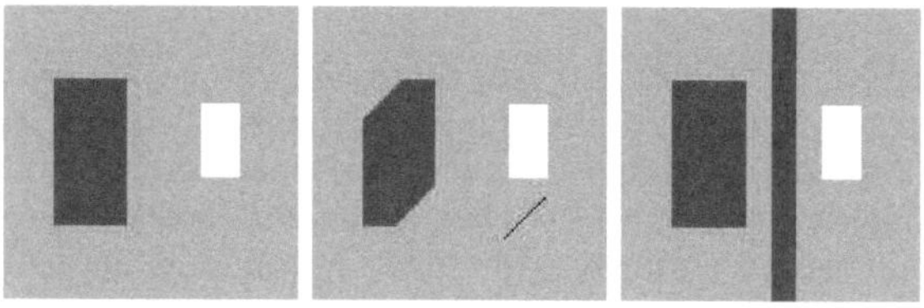

(a) L = 25, *6 = 0* (b) L = 25, *6* = 7r/4 (c) L = 25, *6* = 7r/2

Figure 2. Three closings with linear SE's of length *L* and different orientations. Dark objects with small dimensions in all directions are removed in every closing. Dark objects with a large dimension in at least one direction remain visible in some closings.

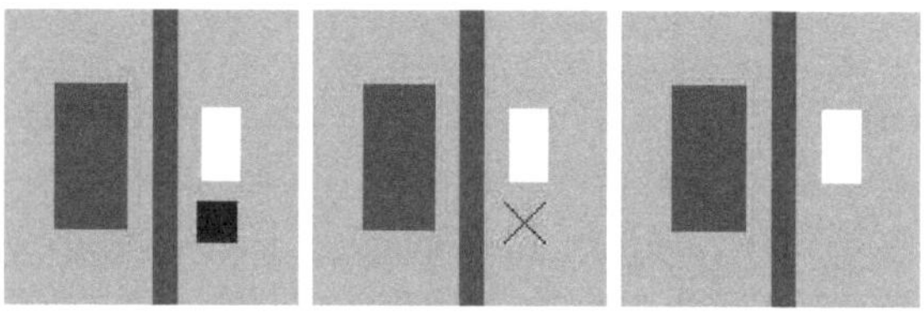

(a) L = 15 (b) L = 25 (c) L = 35

Figure 3. The minimum of closings with linear SE's for three different lengths. Objects are filtered out if their maximum dimension is smaller than L. Rectangular objects with length smaller than L and diagonal size larger than L are partly removed.

openings and closings of different scales is called the morphological profile (MP). The differential morphological profile (DMP) is the vector containing the differences between subsequent values in the MP. A large value in the DMP at scale s is an indication of the presence of an object of size s. The DMP thus contains information about the size of objects.

An MP based on disk shaped SE's(Pesaresi and Benediktsson, 2001; Chanussot et al., 2006; Benediktsson et al., 2003) gives information about the minimum dimension of objects, while an MP based on linear SE's (or a directional MP)(Bellens et al., in press) contains information about the maximum dimension of objects. Together they are able to distinguish between compact objects like buildings and elongated structures like roads. In these papers it is shown that the MP's are able to improve the classification substantially.

## 3 MEASURE OF SIZE

The morphological profile is an indirect indication of the size of objects. It consists of a large number of dimensions which makes it very impractical to analyse and could lead to problems of

(a) IKONOS image (b) scaleofmaximum changeinMP

Figure 4. An IKONOS image of a suburban area in Ghent, Belgium (a) and the scale of maximum change in the MP (b).

dimensionality when classifying. Therefore, Benediktsson et al. (2003) proposed the use of feature extraction algorithms. However, this leads to uninterpretable features, which makes it difficult to analyse and are computationally intensive. In the same paper these features were compared to a size feature derived from the MP. This size feature equals the scale in the MP at which the highest change occurred. However, this feature is very sensitive to different imperfectons and properties of the image as we will see later. In this paper we propose a method to derive a stable and relevant measure of object size.

Let us first have a closer look to how the pixel value changes in openings and closings as the size of the SE increases. Suppose we are dealing with a pixel in a bright object with a darker background. In this case the relevant information is in the openings. For dark objects in bright backgrounds, the behaviour is analogously, but now the relevant information is in the closings. For small SE's, smaller than the object, the pixel value in the opening will be almost equal to the original pixel value. A small decrease can occur due to variations within the object and possibly noise. When the size of the SE increases, the MP will remain almost constant until the SE reaches the size of the object (this can be the minimum dimension for disk shaped SE's or the maximum dimension for linear SE's). At that point, the MP will show a substantial decrease in pixel value. This decrease might be sudden or more gradual in case of blurry edges. The value of the MP at this scale will be the intensity value of one of the neighbouring objects. We could say that the object is fused with that neighbouring object. If we further increase the scale, the MP remains constant for a while, until it reaches the size of this joined object. At that point, the MP decreases again to the intensity value of another neighbouring object. This process repeats until you decide to stop investigating larger SE's or until there are no more darker objects.

When using the maximum difference between subsequent scales as an indication of the scale of objects, as in Benediktsson et al. (2003), it can easily be seen that this will not always lead to the desired result. For example, when edges are blurry, the change in the MP at the scale of the object might be gradual, while due to some noise or other effects, the change in the MP at other scales might be larger. Especially, when using a dense sampling of scales, this might be problematic. Also, we are interested in the first scale where the MP changes substantially as this is the scale of the object and not that of joined objects. However, the change might be larger at the second, third, ...change. Figure 4(b) illustrates this problem. The figure shows the size feature for a suburban area with disk shaped SE's. Most of the houses in this image are assigned a value which is in relation to the object size. However, some whole areas are covered with much higher values.

In this paper we introduce a method to derive a more stable measure of object size from the MP. A disk based MP will lead to a measure of minimum dimension (width), while a directional MP leads to a measure of maximum dimension (diagonal length). First, the MP, consisting of a number of openings at different scales and a number of closings at different scales, is transformed in a new vector with one value for each scale. This value is equal to the difference between the original pixel value and the pixel value in the opening or closing at that scale, whichever is highest. This value is thus an indication of whether or not the pixel belongs to an object that is smaller or larger than the SE. When this value is small, it is likely due to some variation within the object. Once this value becomes larger, the size of the SE is probably larger than that of the object. If it becomes even larger, this does not give any more information about the size of individual objects. We now translate this qualitative reasoning into a fuzzy membership function. This membership function takes the difference *($d_s$)* between opening or closing at scale *s* and the original pixel value and translates it into a number between 0 and 1, indicating how true it is that the size of the object is smaller than the size of the SE. We can choose different functions as long as for small values of $d_s$, the function value is small and increases when $d_s$ becomes larger. Once $d_s$ is larger than a certain limit, the function value should be very close to 1 and only increase slightly as $d_s$ further increases. In this paper we use the following function:

$$\mu_s = 1 - exp(-\frac{d_s^2}{2 \cdot \sigma^2}) \tag{1}$$

This function has one parameter *a* that has to be set according to the typical contrast between objects. We now are able to take the weighted mean of the scales, where the weights are equal to the difference between the fuzzy membership values at subsequent scales. This weighted mean will be the measure of size. Besides the weighted mean, it would also be possible to derive other measures indicating the quality or uncertainty of the measure.

## 4 EXPERIMENTAL RESULTS

Figure 5 shows the results on an IKONOS image. For the minimum dimension, roads and houses have similar values, while larger buildings and fields have much higher values. For the maximum dimension roads clearly have much higher values than houses. Also the fields and the larger buildings have high values. Both measures of size now allow us to differentiate between roads and other objects. By simply thresholding on the minimum and maximum dimension we get the result in Fig. 5(c). This could be further improved by also using the spectral information, using more advanced classification algorithms or doing some postprocessing.

A second test area consist of a QuickBird image of a more difficult area in the city of Ghent, Belgium (Fig. 6). Here we are interested in classifying each pixel in one of nine classes. Table 1 shows the accuracies of this classification. If only using spectral information, the classification of man-made objects are far from satisfactory. Also shadow is not well classified. By also including the size measures, the overall accuracy increases by almost 5%, but especially the improvement on the man-made classes and the shadow classes are apparent.

## 5 CONCLUSION

Past research has showed promising results for including geometrical information in the classification process using morphological profiles. However, including the full morphological profile in a classification could lead to problems of dimensionality. In this paper we propose a method to derive a relevant and meaningful feature from a morphological profile which is easy to calculate. This feature is a measure of the minimum dimension of an object when using regular

(a) minimum dimension

(b) maximum dimension

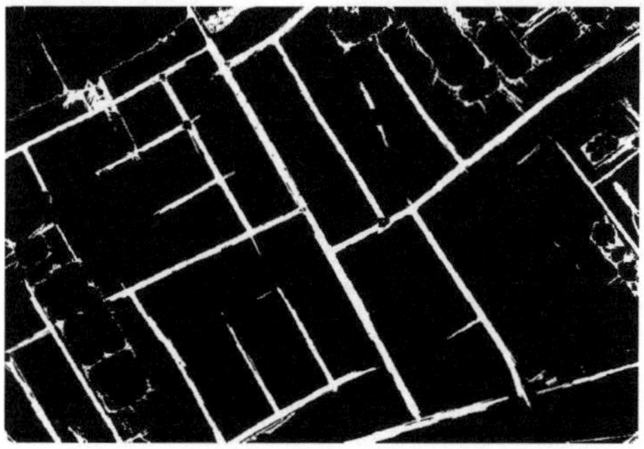

(c) road detection by threshold

Figure 5. The measure of minimum (a) and maximum (b) dimension and a road detection result (c) obtained by simply thresholding both measures of size.

Table 1. Comparison of producer's accuracies when only using spectral information or using both spectral information and the measures of size.

| class | spectral only | spectral and shape features |
|---|---|---|
| water | 92.62 | **96.75** |
| grass | **71.41** | 64.96 |
| trees | 83.42 | **87.35** |
| dark roof | 37.80 | **54.33** |
| red roof | 52.57 | **60.68** |
| white roof | **86.43** | 83.91 |
| road | 55.48 | **71.93** |
| other man-made | **54.05** | 52.57 |
| shadow | 41.03 | **61.81** |
| Overall Accuracy | 72.15 | **76.93** |

(a) QuickBird image

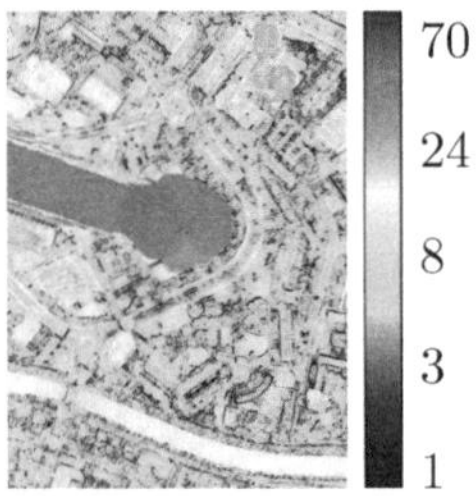


(b) minimum dimension

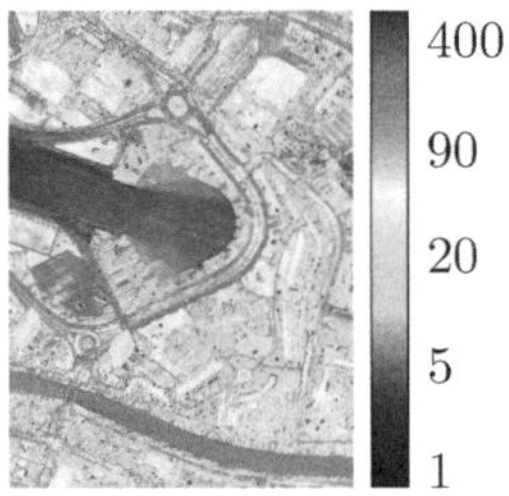


(c) maximum dimension

Figure 6. A QuickBird image located in the city of Ghent, Belgium and the measures of minimum (b) and maximum (c) dimension.

MP's or a measure of the maximum dimension when using directional MP's. Experimental results have shown that these features are relevant and are able to substantially improve classifications in urban areas.

## ACKNOWLEDGMENT

This work was funded by a scholarship of the Institute for the Promotion of Innovation through Science and Technology in Flanders (IWT-Vlaanderen).

## REFERENCES

R. Bellens, L. Martinez-Fonte, S. Gautama, J. C.-W. Chan, and F. Canters. Potential problems with using reconstruction in morphological profiles for classification of remote sensing images from urban areas. In Proceedings of IEEE International Geoscience and Remote Sensing Symposium (IGARSS) 2007, Barcelona, Spain, July 2007.

R. Bellens, S. Gautama, L. Martinez-Fonte, W. Philips, J. C.-W. Chan, and F. Canters. Improved classification of vhr images of urban areas using directional morphological profiles. IEEE Trans.Geosci. Remote Sens., in press.

J. A. Benediktsson, M. Pesaresi, and K. Arnason. Classification and feature extraction for remote sensing images from urban areas based on morphological transformations. IEEE Trans. Geosci.Remote Sens., 41(9):1940 – 1949, 2003.

J. Chanussot, J. A. Benediktsson, and M. Fauvel. Classification of remote sensing images from urban areas using a fuzzy possibilistic model. IEEE Geosci. Remote Sensing Letters, 3(1):40–44, 2006.

P. Hofmann. Detecting urban features from ikonos data using an object-oriented approach. In Proceedings of RSPS2001, pages 79–91, 2001.

Z.J. Liu, J. Wang, and W.P. Liu. Building extraction from high resolution imagery based on multi-scale object oriented classification and probabilistic hough transform. In Proceedings of IEEE International Geoscience and Remote Sensing Symposium (IGARSS) 2005, pages 2250– 2253, Seoul, South Korea, 2005.

A. M. Marangoz, M. Oruc, and G. Buyuksalih. Object-oriented image analysis and semantic network for extracting the roads and buildings from ikonos pan-sharpened images. In Proceedings of the ISRPS 2004 Annual Conference, Istanbul, Turkey, July 19–23, 2004.

Martino Pesaresi and Jon Atli Benediktsson. A new approach for the morphological segmentation of high-resolution satellite imager. IEEE Trans. Geosci. Remote Sens., 39(2), February 2001.

J. Serra. Image Analysis and Mathematical Morphology. Academic, London, U.K., 1982.

P. Soille. Morphological Image Analysis: Principles and Applications, Second Edition. Springer, Berlin,Germany, 2003.

Pierre Soille and Martino Pesaresi. Advances in mathematical morphology applied to geoscience and remote sensing. IEEE Trans. Geosci. Remote Sens., 40(9), September 2002.

Q. Zhang and I. Couloigner. Automated road network extraction from high resolution multi-spectral imagery. In Proceedings of ASPRS 2006 Annual Conference, Reno, Nevada, May 1–5, 2006.

*Remote Sensing for a Changing Europe*
*D. Maktav (Ed.)*
*IOS Press, 2009*

*doi:10.3233/978-1-58603-986-8-142*

# Distributed information system in support of aerospace research

H. Nikolov[1], R. Kancheva[1], D. Petkov[1], G. Georgiev[1]
I. Kibardina[2], V.P. Savorskiy[2], Yu.G. Tishchenko[2], M.T. Smirnov[2]

[1] *Solar-Terrestrial Influences Laboratory (STIL BAS), Bulgaria, e-mail: rumik@abv.bg, hristo@stil.bas.bg*
[2] *Institute of Radioengineering and Electronics (IRE CPSSI), Russia, e-mail: tishchen@ire.rssi.ru*

ABSTRACT: Modern remote sensing technologies for Land Use/Land Cover applications rely on the integration of a big variety of data from both airborne and ground-based instruments. The final product highly depends on the proper and successful exploitation of as much data as possible. This is the reason why a unified data integration and management system should be at the disposal of researchers from different science fields. One possible approach is to develop an information system (IS) based on an open source technology providing an easy data access. Also a standard way of seamless addition of new data should be considered. The approach adopted in this research study is based on a distributed database management system with simple web interface to the data and models. A single cluster contains the following components: a subsystem for data archiving and exchange; a subsystem for specific data preprocessing and calibration; a subsystem for metadata management and integration; a subsystem for the whole catalogue management; a subsystem in support of external users.

The above-described IS was developed in the framework of a joint Russian-Bulgarian project for distributed system infrastructure for aerospace and in-situ data. The system consists of a set of archived data and is supported by hardware/software facilities allowing the exchange of catalogue and ancillary information in an on-line mode. The infrastructure allows to carry out general and detailed data search and to prepare orders for data delivery. The focus at this stage of the work was put on the refinement of the system specification, details of interchange protocols and archive formats, the development of software prototypes ensuring metadata exchange between the system's clusters and access to the information resources of the system. Along with a description of the IS the paper presents a detailed operational scheme of a single standard cluster.

## 1 INTRODUCTION

In the last years the need for geoinformation services raises at different levels from municipality administration to ministerial officers. This information has various uses – decision making in urban and agricultural planning, elaborating ecological strategies for water basins management, atmospheric conditions estimation and much more. The data needed for producing it come from diverse sources having its own geographical coordinates and file formats. At that step data integration, also called data synergy, becomes vital but not an easy task. For this reason some standard must be set up and all data must follow it otherwise the integration is not possible. Another key element of this structure is the manner the end user receives and interprets the processed data. Our experience has shown that this should be done in the most general and common way the end user got accustomed.

Having in mind the abovementioned researchers from IRE CPSSI and STIL BAS initiated creation of distributed Earth observation information system (EOIS) providing data access to both scientific and national authorities. Besides admission of the preprocessed data in the EOIS data manipulation and modeling is also foreseen. This system is in full relevance with the strategy of the

EC for development of shared environmental information system (SEIS) whose aim is to bridge the gap between several initiatives such as GMES, INSPIRE, WISE, EIONET.

## 2 METHODOLOGY

During the development phase the team decided all software tools to be open source thus widening the scope of potential users. The main components of the EOIS include distributed database for data archiving and access; set of server-side scripts for data processing, visualization data exchange and modeling; web server for user interaction handling. Special care is paid on every part of EOIS since it has to have high level of reliability of data storage and efficiency in providing users with space/air borne data products and supporting services. Along with data processing, archiving, and cataloguing facilities, EOIS provides the end users with simple, comfortable, and easy-to-learn interface that conforms all user requirements concerning geodata search, select, order, and procurement.

The basic integrated system integrates the design of:

- space data archive,
- catalogue system,
- user access interfaces,
- data distribution and exchange system.

Key element is the practical implementation of the physically distributed database between both contributors to the project. One of major achievements is that the database could be filled up simultaneously and independently with data by all members of the team thus providing more data (space/air borne, field, and laboratory) for one specific area of interest on the earth surface. Here should be mentioned that databases synchronization mechanism which runs on specified time intervals is also included. For this purpose special authentification mechanism has been elaborated for user identification on the system entry. Based on the user rights given by the administrators of EOIS every single user is allowed set of specific actions to be taken. For instance a group of regular users is assigned the right to list the archive of images and other auxiliary data while other group of users can run simple operations on those data such as building graphical (graphs) representations for single dates or whole periods.

On the figure below is described one possible scenario for using EOIS in joint use of data from multiple sources (optical and thermal imagers, passive radar together with field and laboratory data) in the decision support and decision making for regional planning of agricultural test sites located in the central part of Bulgaria.

After successful fusion of the raw data several objects corresponding to different LC classes are created and then they are inserted in the database. On the nest step the decision-maker could require more data (in our case multispectral images) of the same area, but from other dates thus receiving more information on historical development of the region. This is helpful in determination of the best soil-vegetation performance thus cost effective farming.

One of the users' priorities is the reduction of time consume on data searching, selecting, ordering, and receiving operations. To fulfill these requirements user access subsystem included into EIOS provides users with friendly designed interfaces that help to select and order data sets in the most convenient and well-guided way. To this end our data access subsystem is equipped with user interface which central benefit in the provision of the end users with integrated interface that allows to search, select, browse, and order data from single access point. In addition the data access subsystem is designed in way that reveals actual distributed archive content, i.e. despite physical location of data the end user receives information on products available in the EOIS. This functionality is provided by a powerful search engine (which composes the core part of user access subsystem) that addresses directly to catalogue systems of both IRE CPSSI and STIL BAS parts of

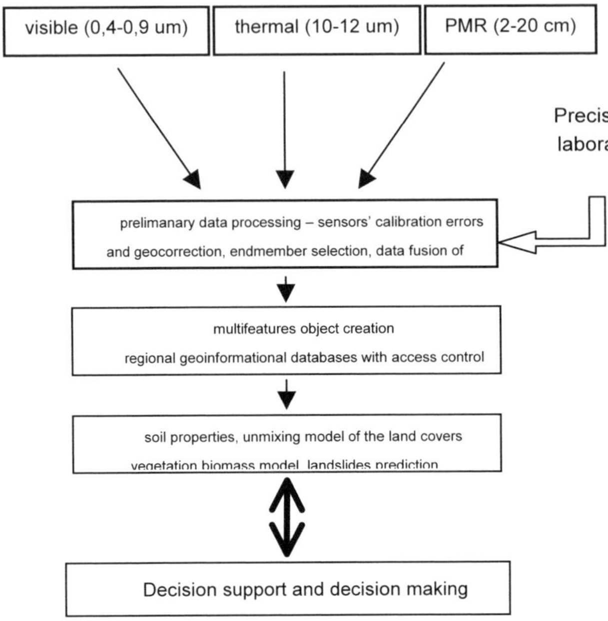


Figure 1. Information flow in the decision making process.

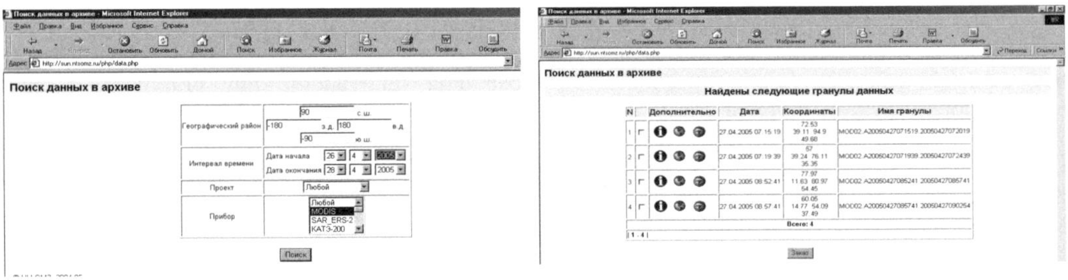


Figure 2. User interface for data search and ordering.

the EOIS. As well catalogue system is equipped with metadata exchange which used to synchronize common part of catalogues once per day.

## 3 CONCLUSIONS

Presented was the prototype of an EOIS developed mutually by IRE CPSSI and STIL BAS which is in full relevance with aims of the creation of shared database for environment data initiated by European Commission. Its main features are end user interaction easiness and environmental modeling based on multiple source data. Given is a practical implementation of this database into decision making process for precision farming purposes.

## ACKNOWLEDGEMENTS

This research was supported by the NSFB under contracts NZ-1410/04 and INI-12/05 and by the joint IRE-RAS – STIL-BAS project "New technologies in remote sensing of the Earth".

## REFERENCES

Interoperable Catalogue System, ICS Valids, ver. 1.1, 2002, CEOS/WGISS/PTT/Valids.

Interoperable Catalogue System (ICS) System Design Document (SDD). CEOS/WGISS/PTT/SDD, Issue 1.4, June 1998, Committee of Earth Observation Satellites/Protocol Task Team.

Petkov D., R. Kancheva, A. Shutko, T. Coleman, A. Krumov, H. Nikolov, D. Borisova. Synergetic in remote sensing technology – joint use of multispectral and microwave data. 36th COSPAR Scientific Assembly, Beijing, China, 16 – 23 July 2006. COSPAR2006-A-01258, JASR2006, (in press).

Tishchenko Yu.G., V.P. Savorskiy, M.T. Smirnov,H.Nikolov, R.Kancheva, D.Petkov, G.Georgiev, Information System for Multipurpose Aerospace Research: Structure fnd Functional Features, 57th International Astronautical Congress, Valencia, Spain, 02–06 Oct. 2006.

*Remote Sensing for a Changing Europe*
*D. Maktav (Ed.)*
*IOS Press, 2009*

*doi:10.3233/978-1-58603-986-8-146*

# Tracing soil pollution dynamics near mining dump site lakes, Mirkovo flotation plant

H. Nikolov, D. Borisova
*Solar-Terrestrial Influences Laboratory, Bulgarian Academy of Sciences, Bulgaria, hristo@stil.bas.bg / dborisova@stil.bas.bg*

Keywords: land cover change detection, vegetation indices, soil line concept, soil reclamation

ABSTRACT: Mining plants are one of the factors having major negative impact on the area where they are situated. In the case of Mirkovo floatation plant, located in the outskirts of Stara Planina Mountain in the middle of Bulgaria, the pollution comes from two major sources – dust from milling shop and waste water from floatation shop. The investigations are carried out deal with determination of the impact on the soils and vegetation in the neighborhood areas using reflectance information from multispectral data and supporting hyperspectral in-situ measurements. During the research preliminary information about mineral content of the ore material coming from the mine and soil type is also considered.
Numerous studies have analyzed the variance of spectral reflectance of rocks, soils and vegetation in response to their cover using remote sensing. The goal of the study is to show land cover changes detected through vegetation indices as NDVI, RVI, SAVI and the soil line concept in remote sensing. On the next step change detection methods are used to support local authorities in preparation of short-term reclamation plans and as well to recommend farmers in planting suitable vegetation spices in assisting the rehabilitation of the top soils. In this research the data from Landsat TM/ETM+ combined with in-situ measured data are used. The obtained results show that the analyzed data and the implemented approach are useful in environmental monitoring and economically attractive for the company responsible for the ecological state of the region.

## 1 INTRODUCTION

Traditional classification techniques are "hard" in the sense that a single pixel is assigned to a single land cover class. At coarse resolutions pixels predictably become mixed. For these mixed pixels "soft" classifiers can be used, which assign a pixel to several land cover classes in proportion to the area of the pixel that each class covers.

Several researchers have addressed this sub-pixel mixture problem. Among the most popular techniques for sub-pixel classification are mixture modeling (Kerdiles and Grondona, 1996) and supervised fuzzy c-means classification (Foody,, 1994).

These techniques aim at estimating the proportions of specific classes that occur within each pixel. The result is a number of fraction images, one for each land cover class concerned. This information describes the class composition, it does not provide any indication as to how this is spatially distributed within the pixel. The result is thus quite different from the classic classification algorithms, where a single land cover map, containing all classes, is produced. One way to go from fraction images to a classic land cover map would be to use "hardeners", where rules can be formulated to determine which class dominates the pixel. The main disadvantage is that information is lost much in the same way as happens with classic hard classifiers. Atkinson (1997) has formulated an idea for an alternative approach. It consists of converting raster data to vector data by threading the vector boundaries through the original image pixels (instead of between pixels, as classic raster-to-vector conversion algorithms would do). This process is called sub-pixel mapping. The key problem is determining where the relative proportions of each class are most likely to occur.

## 2 METHODS

### 2.1 *Sub-pixel Classification*

The spectral unmixing technique that will be applied is based upon Linear Mixture Modeling. The basic hypothesis is that the image spectra are the result of mixtures of surface materials and that each of these components is linearly independent of the other. The mixtures in the image are expressed as linear combinations of their respective spectra in the image. Spectral response of each image pixel in every spectral band can be considered as a linear combination of the response of each component (end-member) present in the mixture. Therefore, every pixel contains information about the proportion (or fraction) and spectral response of each component.

Hence the spectral reflectance r, for every image pixel in any band, can be modeled as follows:

$$r_\Sigma(\lambda_i)=p_1r_1(\lambda_i)+p_2r_2(\lambda_i)...+p_mr_m(\lambda_i)+\varepsilon=\Sigma p_jr_j(\lambda_i)+\varepsilon \qquad (1)$$

where $p$ = fraction cover,
$r$ for pure component reflectance
$\lambda$ is wavelength
and $\varepsilon$ is an error term.

This equation can become a linear system of n equations and k unknowns in two ways:

- applying the equation to every pixel in a single band (n pixels)
- applying the equation to every band of a single pixel (n bands)

In order to solve this set of equations, the number of unknowns cannot be larger than the number of equations. In case the linear mixture model is applied to every pixel in a single band, this condition is not likely to be limiting, as the number of pixels will normally largely exceed the number of end-members. In the other case, where the model is applied to every band for a given pixel, this condition is much more likely to be constraining. Thus the spectral dimensions of the image data set will limit the total number of end-members. In hyperspectral remote sensing this limitation generally is no restriction. In multispectral image analysis however this is a serious difficulty. The problem can be addressed by using a multitemporal data set, where the maximum number of end-members equals the number of spectral bands multiplied by the number of dates for which images are available. The first application has been widely used to calculate pure end-member reflectance, given the end-member fractions, while the second application is typically being used to derive end-member fractions, given the pure components reflectance.

In "soft" classifications the estimated variables (the fractions or proportions of each land cover class) are continuous, ranging from 0 to 100 percent coverage within a pixel. Consequently the above mentioned technique is not suitable for accuracy assessment. Nevertheless it is possible to apply statistical techniques that use a comparable approach and yield similar results.

Purposely dividing the known land cover proportions derived from the data with high spectral resolution into two data sets – the first one is used for unmixing model creation and the second one is for validation.

The validation could be achieved by calculating the sample correlation coefficient R, which is the estimator of the correlation coefficient ρ, between the known land cover proportions from the validation set and the soft classification output. In the case of perfect agreement, the correlation coefficient would be 1 and the linear regression line would have a zero intercept and 45 degrees slope.

### 2.2 *Sub-pixel Mapping*

The major challenge in sub-pixel mapping is determining the most possible locations of the fractions of each land cover class within the pixel. Creating unmixing model from data with high spectral resolution can solve this. A simple representation of the problem and a possible solution is given in Figure 1 (adapted from Atkinson, 1997).

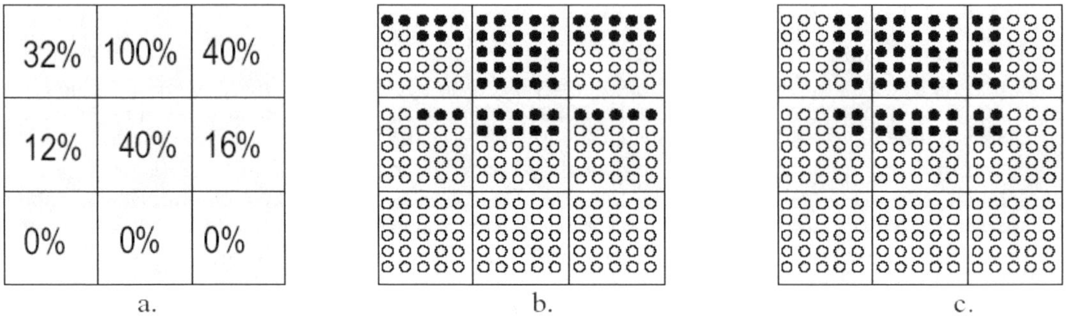


Figure 1. A raster grid of 3 by 3 coarse pixels, each discretised into 5 by 5 sub-pixels

## 3 STUDY AREA AND DATA

The geological exploration of the copper-bearing rocks in the Sredna gora region, located in the middle of Bulgaria, started in the late 50-ies of 20-th century. As a result the mining plant "Elacite" was built who started its production 1986. The main activity of this plant is production of copper concentrate from chalcopyrite. Two dump sites from the flotation process were established near village Benkovski. They are considered as a main source of contamination of neighboring agricultural areas.

As a source for airborne data with moderate spectral resolution for the spectral reflectance of the land cover was used data sets from Landsat TM/ETM+ instrument (GLCF, 2005). The acquisition data is in first decade of June which guarantees good illumination conditions.

From this dataset the two target areas (two open dumps) were extracted by their visual discrimination from the surroundings. This way the dataset necessary for sub-pixel procedure was created.

Laboratory measurements of ore minerals were performed with laboratory spectroradiometer. The instrument provides data in the spectral range 0,5 – 2,55 um with bandwidth less than 0.01 um. The statistic reliability is guaranteed by integrating 10 spectra per sample per measurement. The obtained spectral reflectance of the samples from the dump sites were compared with similar ones taken from spectral library (Clark, 2003)

## 4 RESULTS

Since in our case the number of mixed pixels (which is the case for pixels from the borders of the areas) is not very big but it is crucial to find every correct pixel. This was achieved by implementing the methodology of the unmixing theory on the data from the dump slag. We also found a correct proportion for the boundary pixels between the slag and the water and between the slag and the surrounding agricultural areas. This helped us to map correctly (determining the surfaces of the dumps) the dump areas thus assessing their environmental impact. The surfaces determined this way allowed us to monitor the rate of reclamation activities for the older dump. These activities are summarized in the Table 1.

Table 1. Reclamation activities in Elacite mine plant dump sites

| Activities | Reclama-tion as planne | Reclama-tion as identifie |
|---|---|---|
| Pilot reclamation of dump boundary ($m^2$*1000) | 50 | 30 |
| Reclamation of dump boundary and neighboring agricultural areas ($m^2$*1000) | 180 | 150 |
| Biological purification of the water (NTUs) | 80 | 140 |

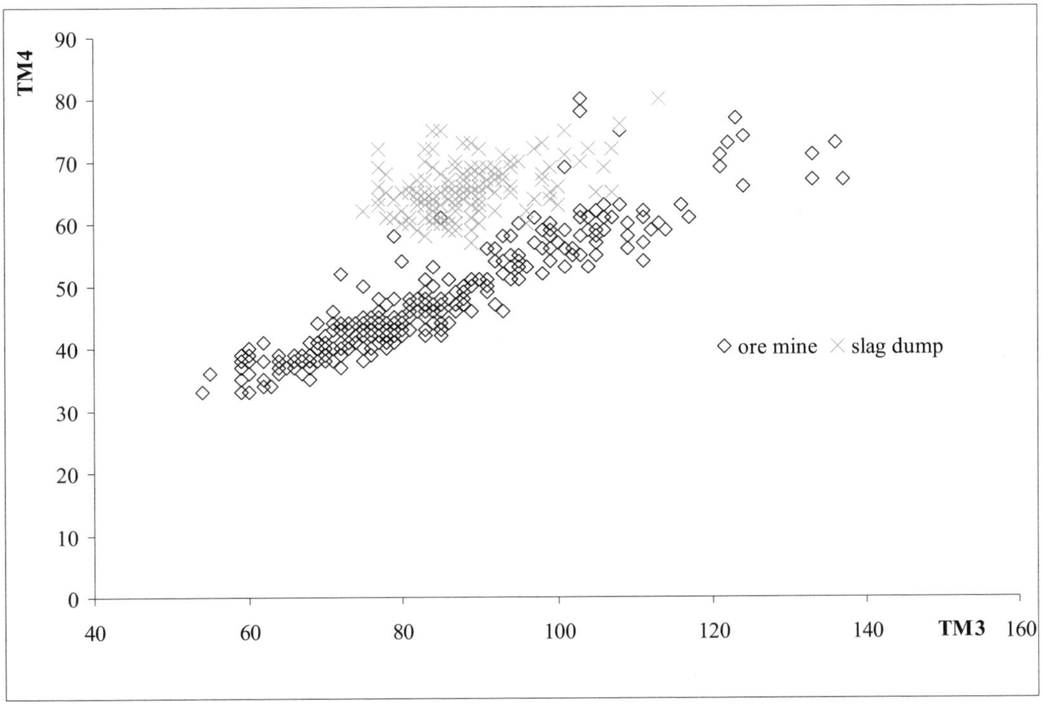


Figure 2. DN corresponding to red and NIR for open mine and slag dump areas

In Figure 2 the minor overlapping of two areas (open pit mine and slag dump) show that two end-members are easily recognized. The values for dump slag were used in process of unmixing before the sub-pixel mapping was done.

If the ground resolution for the Landsat data is considered to be 28.5 m instead of 30 m and above sub-pixel unmixing is applied on the data the discrepancy between the calculated (see Table 1) and in-situ measured surfaces is less than 2%. This proves that the applied methodology for surface determination of the slag dumps is suitable for similar type of applications.

## 5 CONCLUSIONS

The above results prove that methodology for sub-pixel classification is capable of estimating the end-member fractions quite accurately.

The main advantage of the presented technique is that mixed pixels are used during the training phase. Compared to these other techniques, the present one is simple, cheap and objective.

The results of the sub-pixel mapping exercise indicate that the technique can be useful to increase the resolution while keeping the classification accuracy high.

## ACKNOWLEDGEMENTS

This research was supported by the NSFB under contracts MUNZ-1502/05, NZ-1410/04 and INI-12/05 and by the joint IRE-RAS – STIL-BAS project "New technologies in remote sensing of the Earth".

## REFERENCES

Atkinson P. M., M. E. J. Cutler, H. Lewis, 1997. Mapping sub-pixel proportional land cover with AVHRR imagery. International Journal of Remote Sensing, vol. 18, no.4, pp. 917-935.

Clark R., G.Swayze, A.Gallagher, T.King, W.Calvin. 2003. The U. S. G. S., Digital Spectral Library Ver. 05: Version 1: 0.2 to 3.0 um. U. S. Geological Survey, http://speclab.cr.usgs.gov (accessed 03 February 2005)

Foody G. M., D. P. Cox, 1994. Sub-pixel land cover composition estimation using a linear mixture model and fuzzy membership functions. International Journal of Remote Sensing, vol. 15, no. 3, pp. 619-631.

Global Land Cover Facility - University of Maryland, http://glcfapp.umiacs.umd.edu:8080/esdi/index.jsp (accessed 06 April 2006)

Kerdiles H., M. O. Grondona, 1996. NOAA-AVHRR NDVI decomposition and sub-pixel classification using linear mixing in the Argentinean Pampa. International Journal of Remote Sensing, vol. 16, no. 7, pp. 1303-1325.

*Remote Sensing for a Changing Europe*
*D. Maktav (Ed.)*
*IOS Press, 2009*

*doi:10.3233/978-1-58603-986-8-151*



# Colorimetrical analysis in vegetation state assessment

R. Kancheva, D. Borisova
*Solar-Terrestrial Influences Laboratory, Bulgarian Academy of Sciences, Bulgaria, rumik@abv.bg*

Keywords: spectral reflectance, colorimetrical analysis, vegetation parameters

ABSTRACT: Remote sensing is an established technique in environmental studies. First of all, this concerns soil-vegetation ecosystems where the availability of means for vegetation monitoring, stress detection and state assessment is of great importance. A significant amount of research has been performed to develop efficient methods for monitoring of vegetation dynamics. A prevailing part of the works is devoted to the use of multispectral data transformations (vegetation indices) such as spectral bands ratios and linear combinations in order to estimate vegetation parameters. The dependence of vegetation spectral features in the visible and near infrared bands on plant biomass, chlorophyll content, canopy cover, etc. lies at the root of this approach. In this paper we report some results of the colorimetrical analysis of vegetation spectral data. The work was conducted in order to reveal plant senescence effects due to plant growth or stress factors and the impact of the soil background on vegetation reflectance. One of the goals of the study was to evaluate the potential of various colorimetric features for vegetation assessment. Another objective was to compare this approach to the results of the implementation of vegetation indices for plant bioparameters retrieval from multispectral data. The integration of both methods was examined as well showing good predictive capabilities.

## 1 INTRODUCTION

The problem of mixed classes is essential in remote sensing and concerns most aspects of data processing and interpretation. It is associated with spectral mixture decomposition and classification. Soil-vegetation covers are a typical example and a most common case of mixed classes where the determination of the components' proportions is of a particular interest. The objective of the paper is to study the influence of different soil background and plant senescence on vegetation reflectance and color features as well as to reveal the potential of colorimetric analysis technique for mixture decomposition focusing on the accuracy of green cover estimation which is related to vegetation state assessment.

## 2 MATERIALS AND METHODS

In-situ and green-house reflectance measurements of various vegetation types (alfalfa, wheat, spring barley, peas, carrots, grass) with different soil background and degree of senescence were performed in the spectral range 400-820 nm.

The soil diversity was presented by dark soils (black, brown, reddish laterite) and light soils (grey forest, alluvial, meadow) with different properties (mineral composition, organic and moisture content, surface roughness and salinity). The spectral reflectance curves of some of these soils are given in Fig.1 illustrating the large range of soil reflectance signatures.

The variety of green and dry vegetation amounts (the latter related to chlorophyll decrease in mature plants or plant stress conditions) was achieved during plant development as well as simulated from bare soil and vegetation full-canopy cover reflectance using the theory of mixed classes (Mishev 1991):

$$r_{\Sigma}(\lambda) = \sum_i p_i r_i(\lambda) \quad (1)$$

where $r_{\Sigma}(\lambda)$ are the spectral reflectance signatures of the mixed class, $r_i(\lambda)$ - the reflectance of the components (classes) composing the mixture, $p_i$ - the relative amounts of the pure classes in the mixture.

In the case of soil and green vegetation mixtures, for instance, considering that $\sum_i p_i = 1$:

$$r_{sv} = p_v(r_v - r_s) + r_s \quad (2)$$

The spectral reflectance curves of pasture with different vegetation cover are shown in Fig.2 where the impact of the soil type (dark black soil – 1 and light alluvial-meadow soil – 2) is illustrated. As seen, this impact is much stronger in case of sparse vegetation cover (1a and 2a).

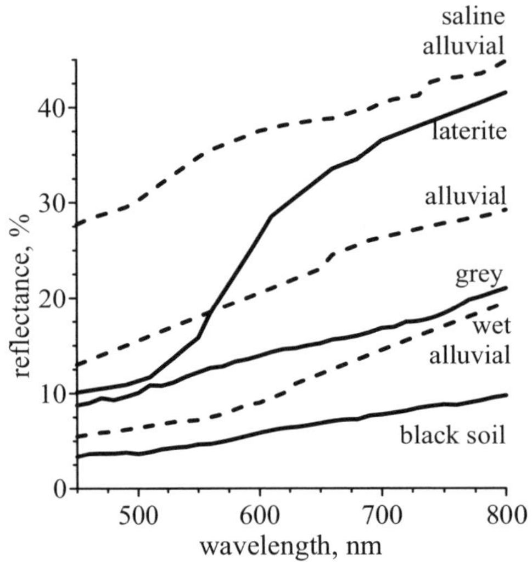


Figure 1. Reflectance spectra of different soils

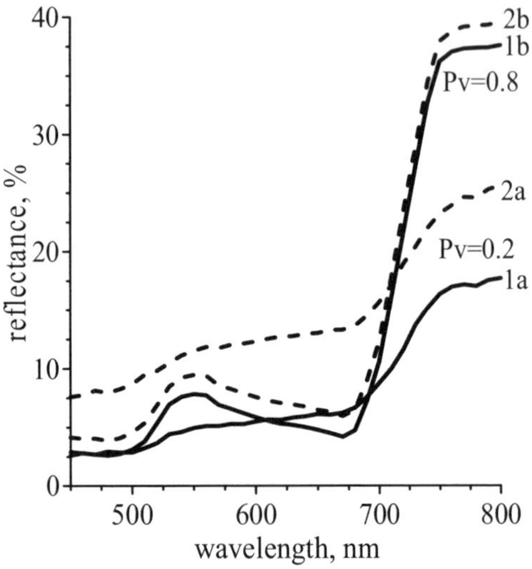


Figure 2. Soil-vegetation mixture reflectance

Colorimetric characteristics such as tristimulus values *X,Y,Z*, chromaticity coefficients *x,y,z* and dominant wavelength $\lambda_d$ of the measured objects and simulated mixtures were computed in the spectral range 450-750 nm according to the CIE1964 methods and $D_{65}$ light source.

In correspondence with the theory of mixed classes (the same being true for *Y, Z* and *W=X+Y+Z*) (Mishev 1992, Kancheva & Borisova 2003):

$$X_{sv} = \sum_{\lambda} D_{65}\left[p_v(r_v - r_s) + r_s\right]\bar{x}\Delta\lambda \quad (3)$$

$$X_{sv} = p_v(X_v - X_s) + X_s \quad (4)$$

$$x_{sv} = \frac{p_v(X_v - X_s) + X_s}{p_v(W_v - W_s) + W_s} \quad (5)$$

Some spectral ratio indices (Table 1) commonly used in vegetation studies were calculated in the green (G)–550 nm, red (R) – 670 and near infrared (NIR) – 800 nm bands.

Table 1. Vegetation indices

| № | VI | № | VI | № | VI | № | VI |
|---|---|---|---|---|---|---|---|
| 1 | ND (NIR,R) | 3 | ND (G,R) | 5 | (NIR-R)/R | 7 | TVI(NIR,R) |
| 2 | NIR/R | 4 | G/R | 6 | (G-R)/R | 8 | G+NIR-2R |

Statistical data processing was carried out to investigate the variance of vegetation reflectance and color features due to different soil background and plant canopy (green and dry) fraction. Correlation and regression analysis was applied to the experimental and simulated data sets to reveal the relationships between soil-vegetation reflectance (in terms of *VI*), color characteristics (presented here in terms of $\lambda_d$) and the proportions of mixture components ($p$).

## 3 RESULTS AND DISCUSSION

Strong correlations were found between the green canopy cover $p_v$, the dominant wavelength $\lambda_d$ and the *VI* values of the soil-vegetation mixtures. The correlation coefficients are given in Table 2 separately for dark soils (black and wet brown) and light soils (alluvial and dry grey forest) as well as for all soils. The vegetation samples include wheat, barley, peas, grass and alfalfa.

Table 2. Correlation matrix of the green vegetation cover ($p_v$), dominant wavelength ($\lambda_d$) and spectral indices (*VI*) for dark and light soils

| | | VEGETATION INDICES | | | | | | | |
|---|---|---|---|---|---|---|---|---|---|
| | $\lambda_d$ | 1 | 2 | 3 | 4 | 5 | 6 | 7 | 8 |
| $\lambda_d$ | dark soils | -0.91 | -0.91 | -0.97 | -0.93 | -0.91 | -0.94 | -0.9 | -0.94 |
| $p_v$ | -0.96 | 0.95 | 0.91 | 0.93 | 0.88 | 0.91 | 0.88 | 0.94 | 0.94 |
| $\lambda_d$ | light soils | -0.93 | -0.89 | -0.95 | -0.91 | -0.89 | -0.91 | -0.92 | -0.92 |
| $p_v$ | -0.93 | 0.98 | 0.88 | 0.94 | 0.85 | 0.88 | 0.85 | 0.98 | 0.94 |
| $\lambda_d$ | all soils | -0.7 | -0.76 | -0.89 | -0.85 | -0.76 | -0.85 | -0.68 | -0.87 |
| $p_v$ | -0.88 | 0.92 | 0.88 | 0.92 | 0.86 | 0.88 | 0.86 | 0.91 | 0.94 |

In Fig.3 the green vegetation cover fitting models by $\lambda_d$ are presented. The correlation and the accuracy of $p_v$ retrieval appeared to be higher for dark soil mixtures (standard error of estimation 0.07 and 0.09 for the dark and light soil case respectively). An explanation could be the greater $\lambda_d$ range for mixture samples with dark soils. Fig.4 shows the relationship between *VI 1* and the green canopy fraction (SEE=0.073).

If the soil non-differentiated model is applied the green fraction values $p_v$ ($\lambda_d$) are overestimated for light soil mixtures and underestimated for dark soils (Table 3). Just the opposite is the situation with *VI* where $p_v$ retrieval from models developed. regardlessly of the soil type lead to underestimating of green vegetation fraction in case of light soils. This provides a possibility for verifying the $p_v$ values estimated through $\lambda_d$ and/or *VI*. The accuracy of mixture colorimetric decomposition through a soil-combined model increases with increased density of the vegetation cover. One reason is the smaller difference between $\lambda_d$ of mixtures with equal $p_v$ and different soil type (light or dark) component (Table 4).

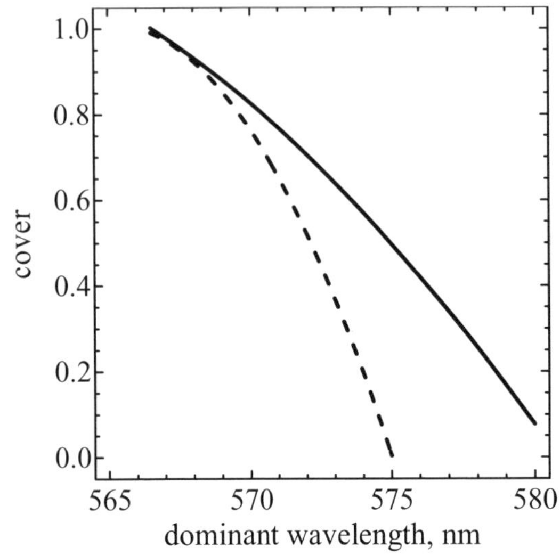


Figure 3. Green cover fitting models $p_v=f(\lambda_d)$ for dark (—) and light (- - -) soil-vegetation mixtures

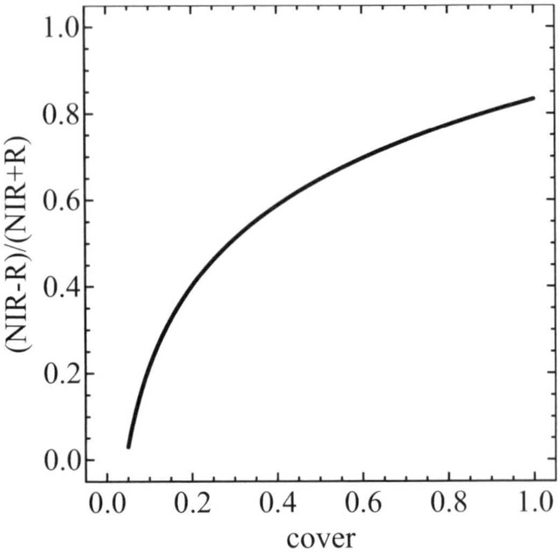


Figure 4. Green cover $p_v$ fitting model by *VI 1 (NIR-R)/(NIR+R)*

Table 3. Relative errors of green cover prediction using the soil non-differentiated $p_v=f(\lambda_d)$ model

| green cover | estimated (light soils) | error % | estimated (dark soils) | error % |
|---|---|---|---|---|
| 0.2 | 0.4 | 100 | 0.1 | -50 |
| 0.4 | 0.54 | 37 | 0.32 | -20 |
| 0.6 | 0.67 | 11 | 0.53 | -11 |
| 0.8 | 0.80 | 0 | 0.80 | 0 |
| 1.0 | 0.95 | 5 | 1.0 | 0 |

Table 4. Differences between $\lambda_d$ for dark and light soil-vegetation mixtures

| for $p_v$ | 0 | 0.2 | 0.4 | 0.6 | 0.8 | 1.0 |
|---|---|---|---|---|---|---|
| $\Delta\lambda_d$, nm | 5.5 | 4.4 | 3.3 | 2.2 | 1.0 | 0 |

Using two- or three-factor inverse dependencies of canopy cover $p_v$ on $\lambda_d$ and *VI* a very good correspondence was achieved between the actual and estimated vegetation fraction. The SEE and MAE (mean absolute prediction error) are lower than any of the *VI* and $\lambda_d$ green cover-fitting models (Table 5).

Table 5. Correlation and errors of green fraction ($p_v$) retrieval using $\lambda_d$, *VI* and joint models

| y (x) | model | r | SEE | MAE |
|---|---|---|---|---|
| $p_v$ | $a+b\lambda_d$ | 0.93 | 0.078 | 0.062 |
| $p_v$ | a+bVI 3 | 0.94 | 0.088 | 0.073 |
| $p_v$ | $a+b\lambda_d$+cVI 3 | 0.95 | 0.066 | 0.05 |
| $p_v$ | $a+b\lambda_d$+cVI 3+dVI 8 | 0.98 | 0.054 | 0.044 |

The studies did not reveal significant model differences associated with the vegetation type. This is explained by Fig.5 where the narrow full-vegetation canopy cluster on the color locus is seen (565.5-568 nm) as well as the wider band of bare soils (light and dark including dry and wet samples with smooth and rough surface) location (575-580.5 nm). This is another reason for the worse accuracy of sparse vegetation cover assessment, especially when using not-differentiated models. The wider $\lambda_d$ range of the soil cluster within the color locus and the narrower one of different full-cover green canopies suppose bigger errors in assessment of sparse vegetation covers if soil type and properties are not taken into account.

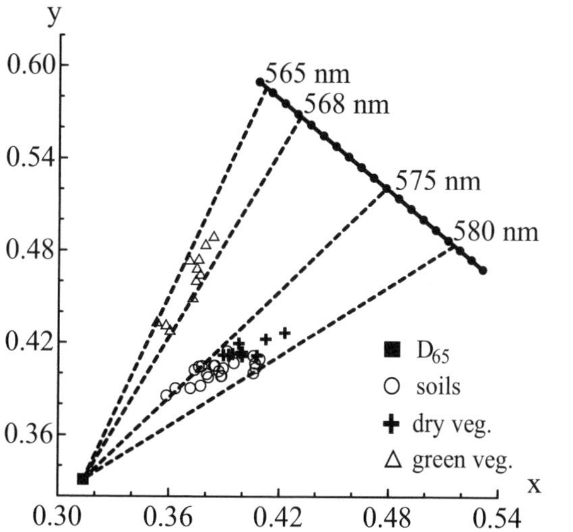


Figure 5. Dominant wavelengths of various soil types and full-cover vegetation canopies

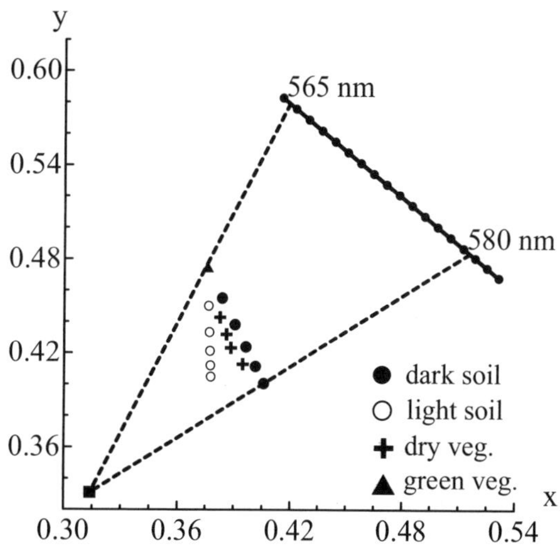


Figure 6. Dominant wavelengths of 2- and 3-component mixtures

One important thing to point out in Fig. 5 is that dry vegetation canopies and their soil mixtures lie within the soil $\lambda_d$ band and have an average dominant wavelength of 577 nm which is the same as the average $\lambda_d$ of the soil cluster. The senescence effects on vegetation reflectance are quite similar to the soil impact. This means that the senescence influence on plant spectral and color features can be consider similar to a proportionally increased soil impact. Evidence is Fig.6 where the position on the color locus of two and three component mixtures of soil (dark or light), green and dry vegetation is presented.

## 4 CONCLUSIONS

High correlation was found between the dominant wavelength $\lambda_d$ of soil-vegetation mixtures and plant canopy cover. Regression models describing this relationship were derived using large data sets from different soil and vegetation types. Green cover retrieval by the calculated from reflectance measurements $\lambda_d$ values has satisfying accuracy with mean absolute error 0.06. What should be further done is to verify the predictive potential of color models over various vegetation types and growth stages, and to test the models sensitivity toward varying external factors (illumination conditions). In this respect some vegetation indices have a proven stability although in our experiment most of them showed a big variance to soil surface structure, humidity and especially color as a whole. Best results were obtained by the use of multiple regression models for retrieving green fraction values. These results provide grounds for considering the idea of wider experimental study of plant color characteristics and their potential in vegetation monitoring.

The advantage in using color features is that the visible spectral range is closely related to biophysical parameters of the objects and that the whole reflectance curve is used normalized on the spectral distribution of the incident radiation. Besides, $\lambda_d$ allows the comparison of slightly differing color stimuli. A disadvantage is the sensitivity of the regression coefficients in $p_v=f(\lambda_d)$ retrieving models. But as a whole the obtained results are an encouraging confirmation of the potential of mixture analysis for vegetation state monitoring using both techniques - spectral reflectance transformation and colorimetric analysis. Future work is intended in precising the dependences of *VI* and $\lambda_d$ on green canopy cover by using larger data sets and different experiment conditions including airborne spectral data.

## ACKNOWLEDGEMENTS

This research was supported by the NSFB under contracts NZ-1410/04 and INI-12/05 and by the joint IRE-RAS – STIL-BAS project "New technologies in remote sensing of the Earth".

## REFERENCES

Mishev, D. 1991. Spectral characteristics of mixed classes of natural formations, *Acta Astr. 25 (8/9)*, 443-446.

Mishev, D. 1992. Colour coordinates of a mixed class. *Compt. rend. Acad. bulg. Sci. 45 (3)*, 51-54.

Kancheva, R., Borisova, D. 2003. Two techniques for spectral classes decomposition from their mixture reflectance. *Compt. rend. Acad. bulg. Sci. 56 (2)*, 43-48.

Kancheva, R., Mishev, D. 2000. Colorimetrical characteristics for detection of plant chlorophyll variations. *Compt. rend. Acad. bulg. Sci., 53 (4)*, 43-46.

*Remote Sensing for a Changing Europe*
*D. Maktav (Ed.)*
*IOS Press, 2009*


*doi:10.3233/978-1-58603-986-8-157*

# Land use/cover mapping using multisensor image fusion technique

S. Abdikan & F. Balik Sanli
*Yildiz Technical University, Civil Engineering Faculty, Department of Geodesy and Photogrammetry Engineering, Turkey*

M.T. Esetlili & Y. Kurucu,
*Ege University, Faculty of Agriculture, Department of Soil Science, Turkey*

Keywords: ASAR, PALSAR, SPOT, fusion, landuse/cover

ABSTRACT: Remote Sensing is an important technique for mapping land use and land cover in the vast acreages. In this sense, the fusion of optical and radar remote sensing data offers the opportunity to combine complementary sensors with different features. In this study, beside the capability of the combined multi source imagery, the contribution of SAR images to the optical images for identifying land use/cover types was investigated. For this purpose, using the synergy between SAR and Optical data, the improvement in the classification accuracy was analyzed. The study area, covering urban and agricultural areas, lies in the Menemen Plain to the west of Gediz Basin in the Aegean Region of Turkey. The satellite data used in this study are multispectral SPOT, ENVISAT-ASAR, and ALOS-PALSAR images. The 3-2-1 band combination of a SPOT-2 image was fused with C band ASAR imagery and with the new mission L band PALSAR imagery. The land use/cover types were defined from both of the fused images. In this case, since the SAR images have different bands (C band and L band) the penetration property is the key factor to see the affects on extracting information from fused images. Before the fusion application, the speckle reducing filter techniques were used for the preprocessing of SAR images. For the filtering of SAR images, kernel windows with different size were tried. Then the SPOT image was registered to SAR images. For the registration of SAR images, image to image registration method was used with a root mean square error of less than 1 pixel. A pixel based fusion method was carried out. Both of the fused images (SPOT-ASAR and SPOT-PALSAR) were classified to determine the land use/cover map. The results were compared with a classified SPOT image, which is commonly used to define land cover types. While processing the classification, the training areas were selected covering a large portion of the individual fields and were away from the field boundaries to reduce the mixed pixels. The ground truth data were used for the accuracy assessment process.

## 1 INTRODUCTION

Remote Sensing is an important technique for mapping land use and land cover in the vast acreages. In this sense the fusion of optical and radar remote sensing data offers the opportunity to combine complementary sensors with different features. Since the response of radar is more of a function of geometry and structure while optical wavelengths occur as surface reflection, their role in landuse cover mapping can be viewed as complementary.

Usually conventional multispectral classification methods, which make use of spectral response of ground objects, are used for thematic mapping. Basically, the spectral vector of a pixel is used to classify the pixel by using a classifier. These per pixel classification approaches often result in mixed pixel classification for low resolution images such as Landsat and Spot ones. Since a wide spectrum of data can be available for the same site observed, additional sources may provide complementary data, and fusion of different information can produce a better understanding of the observed site. Information provided from an individual sensor can be incomplete, imprecise and inconsistent. Therefore fusion of multispectral, multipolarization and multitemporal data may decrease the uncertainty related to the single data sources (Simone et al. 2003).

In this study, we aim to see the influence of fusion approaches on classification accuracy using different algorithms and different images for the same area. For this purpose, contribution of SAR images to optical images in defining land use types (natural products and textures) was investigated. The 1,2,3 band combination of SPOT-2 images were fused with each PALSAR and ASAR SAR data. The fusion results were classified using Maximum Likelihood Classification method. Classification results were compared with the classified SPOT images.

## 2 STUDY AREA

The study area is situated on the Menemen (Izmir) Plain to the west of the Gediz Basin, and covers about 400 square km. The Aegean Sea lies on the west side of the study area, and Manisa Province lies to the North. The area is bordered by Bakircay Basin in the North, and Izmir Bay in the South (Fig. 1). On the Menemen Plain, a semi-dry and low humidity mesothermal climate is dominant, and the main period for rain (which is 616 mm / m2) is the winter season. Thus, the summer season is drier, with excessive sun light and evaporation. Relative humidity is around 50%. Although all crop types can be grown on the plain, the main products are cotton, corn, wheat, barley, cereals and viniculture. The coastal areas which have become very salty due to poor drainage conditions are used as pastures. The texture of the soil is mostly medium and the soil profile is very deep. Gediz river floodings have created plain fields, levees and geomorphologic depression formations in its surroundings. Soil texture and distribution of the soil moisture on the Menemen Plain is coherent with these geomorphological formations. While clay texture is dominant in depression fields, levee lands are sandy and the flood plains have a loamy texture. The area has a smooth micro relief with an approximate slope of 1 %. In the large part of the study area, where the fields were prepared for cotton and corn farming, the actual planting (seeding) started at the beginning of May 2006. The study area was unplanted except for winter crops which are wheat and barley. The surface roughness of the study area is homogeneous. In this season, since the area had not received enough rain until the beginning of May, the soil moisture levels varied usually according to its water holding capacity.

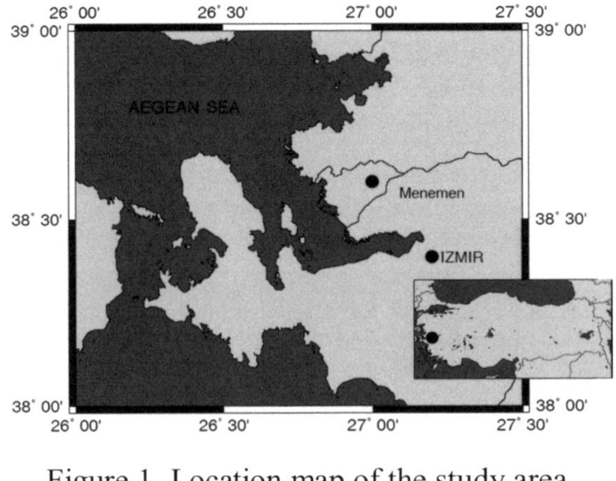


Figure 1. Location map of the study area

# 3 DATA USED AND METHODOLOGY

## 3.1 *Data*

In order to validate the above concept, the image data used in this study were acquired by the ENVISAT-ASAR, ALOS-PALSAR and SPOT-2. SPOT-2 satellite image of 14 May 2006, ENVISAT-ASAR satellite image of 11 June 2006 and ALOS-PALSAR satellite image of 10 June 2006 that include agricultural fields of Menemen Plain were used. The ASAR (Advanced Synthetic Aperture Radar) operates in the C band. It can acquire images in both single and dual polarizations as HH, HV, and VV. The incidence angles vary between 15 ° and 45° (Baghdadi et al. 2006). For this study a VV polarized image with the resolution of 12.5 m x 12.5m was selected. The PALSAR (Phased Array L-band Synthetic Aperture Radar) operates in L band and acquires images in five observation modes (Fine Beam Single (FBS), Fine Beam Dual (FBD), Direct Transmission (DT), ScanSAR and Polarimetry). The incidence angles range between 18° and 55°. It can acquire data in four polarizations (Rosenqvist et al. 2004). Fine beam mode with 6.25m x 6.25m resolution was selected from the five different beam modes. The detailed description of the SAR data used is given in (Table 1).

| | **ASAR** | **PALSAR** | **SPOT-2** |
|---|---|---|---|
| **Date** | 11/06/2006 | 10/06/2006 | 14 /05/2006 |
| **Sensor** | ASAR/IM | PALSAR/FBS | HRV/HRG |
| **Pixel Spacing** | 12.50 m. | 6.25 m. | 20m |
| **Orbit** | 2384 | 2010 | |
| **Flight direction** | Ascending | Ascending | |
| **Processing** | PRI | L1.5 | |
| **Polarization** | V/V | H/H | |
| **Swath** | IS3-82 km | 80 km | 60 km |
| **Incidence angle** | 26-31.4 | 41.5 | L29.6 |

Table 1. Properties of the SAR data

Cadastral maps in 1/5000 scale and topographic maps in 1/25000 scale were used for the rectification of SPOT-2 data as the ancillary data. The orthorectified SPOT image and SRTM data were used for the orthorectification of the SAR data.

## 3.2 *Methodology*

### 3.2.1 *Image Fusion*

Image fusion is an application dependent framework for making use of two or more complementary images/spectral bands of the same or different sensors for the same area, to get more information which is not achieved using each image/band alone in order to enhance the quality of image interpretation. Hall (1992), Genderen & Pohl (1994), Pohl (1996), Pohl & Genderen (1993, 1995, 1998), Wald (1998) and Simone et al. (2003) describe image fusion and its techniques in detail and Kuplich et al. (2000) and Sun et al. (2003) assess the contribution of fusion techniques to rural and urban land use classification.

In this study, two fusion methods have been examined for gamma filtered SAR images namely Intensity, Hue and Saturation (IHS), and High Pass Filter (HPF).

IHS method separates the Intensity, Hue and Saturation components of a RGB image. Three bands of an MS image are converted from RGB colour to IHS colour. Spatial frequency related I component is replacing the high resolution image and back transformation IHS to RGB is required (Pohl & Van Genderen 1998). The HPF fuses both spectral and spatial information with the band-addition approach. Edge information is extracted from high resolution image and added pixel by pixel basis to the low resolution one. High frequency component of the high resolution image is concerned to spatial information. High pass filter of the high resolution image corresponds to the high frequency component. In conclusion adding filter to the low resolution band, spatial information content of the high resolution image replace and will be seen in the fused image (Bethune et al., 1998).

In this study pre-processed images were fused. Three bands of SPOT images were merged with each ASAR and PALSAR images. Resulting fused images were resampled to the resolution of SAR images, i.e 8m. x 8m. for PALSAR and 25m. x 25m. for ASAR.

### 3.2.2 *Statistical comparison of fused images*

In order to evaluate the advanced spectral quality of the fused images, SPOT XS image is compared with the produced fused images. Assessment analyses for each of the PALSAR-SPOT fused image and ASAR-SPOT fused image were performed. Performing the increased interpretation capabilities, the spectral relationship between the original SPOT image and each of the fusion results were compared to see the advances in the spectral quality. Correlation coefficient (CC), Standard deviation (SD) and mean values were computed and compared in Table2.

| **Image** | **Mean** | SD | CC |
|---|---|---|---|
| SPOT | | | |
| B1 | 96.8084 | 37.2656 | |
| B2 | 76.9506 | 21.4078 | |
| B3 | 106.549 | 19.7607 | |
| IHS-(PALSAR +SPOT) | | | |
| B1 | 71.1392 | 61.5873 | 0.576661 |
| B2 | 50.8029 | 51.6274 | 0.240118 |
| B3 | 79.5988 | 50.0148 | 0.188911 |
| IHS-(ENVISAT +SPOT) | | | |
| B1 | 103.24 | 56.707 | 0.675132 |
| B2 | 82.3085 | 47.7533 | 0.486662 |
| B3 | 111.423 | 45.9042 | 0.275237 |
| HPF-(PALSAR +SPOT) | | | |
| B1 | 96.234 | 34.7693 | 0.956516 |
| B2 | 76.4309 | 20.1752 | 0.950098 |
| B3 | 106.016 | 18.5624 | 0.949144 |
| HPF-(ENVISAT +SPOT) | | | |
| B1 | 96.2826 | 36.6429 | 0.976182 |
| B2 | 76.4424 | 21.0979 | 0.973563 |
| B3 | 106.037 | 19.4426 | 0.972617 |

Table 2. Statistical Comparisons of the fusion results

### 3.2.3 *Land cover classification*

Classification is the process of grouping pixels of images into patterns of varying gray tones or assigned colors that have similar spectral values to transfer data into information for determining earth resources (Jansen 2004). In this study, supervised classification is applied to each fusion result using maximum likelihood classification. Intentionally, only the spectral information was used. Same training areas were applied on each fusion result. It is aimed to see the spectral differences caused by SAR images influencing the classification results. Additionally, 450 test samples for each classified image were collected and treated separately. Due to the characteristics of the study, fifteen landcover classes were defined, including wheat, summer crop, built-up, road, well developed (WD) pasture, weak pasture, scrub land, shadow scrub land, water surface, salt crust, moist soil, dry soil, shallow sea, deep sea and meadow.

## 4 RESULTS

### 4.1 *Accuracy Assessments of Classifications*

The overall classification accuracy and the overall Kappa statistics results of the classified fusion results and SPOT are given in Table 2. For the accuracy assessment process, 30 samples were chosen for each land cover categories. The overall accuracy assessment of the images for SPOT, SPOT-PALSAR fusion and SPOT-ASAR fusion were 67.11 %, 72.89 % and 64.89 % respectively. Kappa statistics were calculated as 0.65, 0.71 and 0.62 respectively for the above order (Table 3).

The mixing of classes is considered to be the main implementation problem affecting the accuracy of the classification process. As clearly seen from the variety of the land cover categories, this study area has a heterogeneous structure (Table 3).

Although the salt crusts are classified with higher accuracy from the SPOT image, the borders of the saltpans are recognized better in the PALSAR fused image.

The heights of the newly seeded summer crops are about 5-10 cm. These fields were classified better in the fused images. The PALSAR fused image gave better results (about 10 %) than that of the others.

In all the classification results, it was observed that well developed pasture areas were mixed with the meadow and summer crops. Weakly developed pasture areas having low coverage percentages were mixed with the bare soil in all the classified images. In both fused images, weakly developed pasture areas were classified better (about 7 %) than the SPOT image. Scrublands not lying on the hilly and shadow areas were classified about 14 % better with the PALSAR fused image than the SPOT image.

In this study, the best classification was achieved using the fusion results of the PALSAR image. The main contribution using the fused PALSAR image is in the recognition of the parcel borders such as roads, canals, salt pans etc. Although road and built–up areas were mixing, both classes were better defined in the PALSAR fused image. In addition, these objects were better recognized by visual interpretations.

While the meadow which grows in wetlands was classified with the same accuracy using the SPOT and PALSAR fused images, the water surfaces having different depths were classified better in both PALSAR fused and ASAR fused images (i.e., 100 % and 87 % respectively).

The soil was classified in the two groups depending on the moisture content. Both the moist and dry soils were classified better with the fused images. Due to the L band of PALSAR image, PALSAR fused images give best results for soil classes.

| Class Name | Reference | Classified | Number Correct | Producers Accuracy | Users Accuracy |
|---|---|---|---|---|---|
| Wheat | 53 | 30 | 29 | 54.72% | 96.67% |
| Summer crop | 20 | 30 | 14 | 70.00% | 46.67% |
| Weak Pasture | 21 | 30 | 20 | 95.24% | 66.67% |
| WD Pasture | 31 | 30 | 20 | 64.52% | 66.67% |
| Scrub | 18 | 30 | 17 | 94.44% | 56.67% |
| Shadow scrub | 23 | 29 | 23 | 100.00% | 79.31% |
| Road | 16 | 30 | 10 | 62.50% | 33.33% |
| Built-up | 18 | 30 | 16 | 88.89% | 53.33% |
| Meadow | 25 | 30 | 16 | 64.00% | 53.33% |
| Water surface | 14 | 30 | 12 | 85.71% | 40.00% |
| Shallow sea | 39 | 30 | 25 | 64.10% | 83.33% |
| Deep sea | 23 | 30 | 22 | 95.65% | 73.33% |
| Moist soil | 62 | 30 | 26 | 41.94% | 86.67% |
| Dry soil | 41 | 30 | 24 | 58.54% | 80.00% |
| Salt crust | 46 | 30 | 28 | 60.87% | 93.33% |
| Total | 450 | 450 | 302 | | |
| **SPOT** | | | Overall Classification Accuracy = 67.11%<br>Overall Kappa Statistics = 0.6477 | | |
| Wheat | 60 | 30 | 29 | 48.33% | 96.67% |
| Summer crop | 21 | 30 | 17 | 80.95% | 56.67% |
| Weak Pasture | 24 | 30 | 22 | 91.67% | 73.33% |
| WD Pasture | 29 | 30 | 20 | 68.97% | 66.67% |
| Scrub | 26 | 30 | 21 | 80.77% | 70.00% |
| Shadow scrub | 16 | 30 | 16 | 100.00% | 53.33% |
| Road | 22 | 30 | 18 | 81.82% | 60.00% |
| Built-up | 21 | 30 | 16 | 76.19% | 53.33% |
| Meadow | 25 | 30 | 16 | 64.00% | 53.33% |
| Water surface | 39 | 30 | 30 | 76.92% | 100.00% |
| Shallow sea | 32 | 30 | 24 | 75.00% | 80.00% |
| Deep sea | 23 | 30 | 21 | 91.30% | 70.00% |
| Moist soil | 45 | 30 | 28 | 62.22% | 93.33% |
| Dry soil | 40 | 30 | 26 | 65.00% | 86.67% |
| Salt crust | 27 | 30 | 24 | 88.89% | 80.00% |
| Total | 450 | 450 | 328 | | |
| **PALSAR+SPOT** | | | Overall Classification Accuracy = 72.89%<br>Overall Kappa Statistics = 0.7095 | | |
| Wheat | 56 | 30 | 28 | 50.00% | 93.33% |
| Summer crop | 21 | 30 | 16 | 76.19% | 53.33% |
| Weak Pasture | 24 | 30 | 22 | 91.67% | 73.33% |
| WD Pasture | 35 | 30 | 18 | 51.43% | 60.00% |
| Scrub | 14 | 30 | 12 | 85.71% | 40.00% |
| Shadow scrub | 21 | 30 | 20 | 95.24% | 66.67% |
| Road | 20 | 30 | 13 | 65.00% | 43.33% |
| Built-up | 19 | 30 | 13 | 68.42% | 43.33% |
| Meadow | 26 | 30 | 10 | 38.46% | 33.33% |
| Water surface | 45 | 30 | 26 | 57.78% | 86.67% |
| Shallow sea | 33 | 30 | 22 | 66.67% | 73.33% |
| Deep sea | 20 | 30 | 20 | 100.00% | 66.67% |
| Moist soil | 49 | 30 | 27 | 55.10% | 90.00% |
| Dry soil | 36 | 30 | 22 | 61.11% | 73.33% |
| Salt crust | 30 | 30 | 23 | 76.67% | 76.67% |
| Total | 450 | 450 | 292 | | |
| **ENVISAT+SPOT** | | | Overall Classification Accuracy = 64.89%<br>Overall Kappa Statistics = 0.6239 | | |

Table 3. Accuracy assessments of classifications

## 4.2 *Visual comparisons*

Among the three classification results, the fusion of PALSAR and SPOT gives better visual interpretation results which also contribute to the accuracy of classification. Roads and built-up areas are recognized better. Especially canals and roads between the agricultural fields are observed better (Fig. 2). In the PALSAR fused image, classification of the parcel borders were defined accurately (Fig. 3). Although the salt crusts are classified with higher accuracy from the SPOT images, the borders of the saltpans are recognized better in PALSAR fused image (Fig. 4).

Figure 2. Visual interpretations of fusion results

Figure 3. Parcel borders in classified images

Figure 4. Recognition of salt crust in classified images

## 5 CONCLUSIONS

In this study, we aimed to see the influence of fusion approaches on the classification accuracy using different algorithms and different images for the same area. For this purpose, contribution of SAR images to optical images in defining land use types (natural products and textures) was investigated. Since HPF fusion technique gave better result statistically and visually than IHS, HPF fusion technique was chosen. Than 1,2,3 band combination of SPOT-2 images were fused with each PALSAR and ASAR SAR data. The fusion results were classified using the Maximum Likelihood Classification method.

The results showed that the PALSAR fused image gave better classification accuracy among all. Roads, canals and built-up areas were classified with a higher accuracy which is useful for defining parcel borders. In addition, water surface was classified with 100% accuracy with the PALSAR fused image. These are important contributions since they help defining agricultural inventories. Furthermore, both the moist and dry soils were better classified with the fused images. Due to the L band of PALSAR image, PALSAR fused images gave best results for the soil classes.

## 6 REFERENCES

Baghdadi, N., Holah, N. & Zribi, M. 2006. Soil moisture estimation using multi-incidence and multi-polarization ASAR data. *International Journal of Remote Sensing* 27, 1007-1920.

Bethune, S., Muller, F. & Donnay, J.P. 1998. Fusion of multispectral and panchromatic images by local mean and variance matching filtering techniques. *Proc. Fusion of Earth Data, Sophia Antipolis, France, 28-30 Jan 1998.*

Genderen, J. L. Van & Pohl, C. 1999. Image Fusion: Issues techniques and Applications. *Proc. EARSeL Workshop on intelligent image fusion*, Strasburg, France: 18-26.

Hall, D. L. 1992. *Mathematical Techniques in Multisensor Data Fusion*. (Norwood: Arteech House Inc.)

Jensen, J. R. 2004. *Digital Change Detection. Introductory Digital Image Processing: A Remote Sensing Perspective.* New Jersey' Prentice-Hall: 467– 494.

Kuplich, T.M., Freitas, C.C. & Soares, J.V. 2000. The study of ERS-1 SAR and Landsat TM synergism for landuse classification. *International .Journal of Remote Sensing*, 21, 10, 2101-2111.

Pohl, C. & Genderen, J. L. Van. 1993. Geometric Integration of Multi-image Information. *Proc. 2nd ERS-1 Symposium-Space at the Service of our Environment.* Hamburg, Germany), ESA SP-361., : 1255-1260.

Pohl, C. & Genderen, J. L. Van. 1995. Image Fusion of Microwave and Optical Remote Sensing Data for Map Updating in The Tropics. *Proc. SPIE Conference EUROPTO'95-Image and Signal for Remote Sensing II Processing.* Paris, France, Spie proceeding series: 2570: 2-10.

Pohl, C. 1996. *Geometric Aspects of Multisensor Image Fusion for Topographic Map Updating in the Humid Tropics.* ITC Publications, 39. ISBN 90 6164 121 7.

Pohl, C. & Van Genderen, J. L. 1998. Multisensor image fusion in remote sensing: concepts, methods and applications, *International Journal of Remote Sensing*, 19, 823-854.

Rosenqvist, A., Shimada, M., Watanabe, M., Tadono, T. & Yamauchi, K. 2004. Implementation of Systematic Data Observation Strategies for ALOS PALSAR, PRISM and AVNIR-2. *Geoscience and Remote Sensing Symposium, 2004. IGARSS '04. 20-24 Sept.2004,* Vol7: 4527- 4530.

Simone, G., Farina, A., Morabito, F.C., Serpico, S.B. & Bruzzone, L. 2003. Image Fusion techniques for remote Sensing Applications. *Information Fusion*, 3, 1: 3-15.

Sun, W., Heidt, V., Gong, P. & Xu, G. 2003. Information fusion for rural land-use classification with high resolution satellite imagery. *IEEE Transactions on Geoscience and Remote Sensing*, 41, 4: 883-890.

Wald, L. 1998. Data Fusion: A Conceptual Approach for an efficient Exploitation of Remote Sensing Images, *EARSeL Conference on Fusion of Earth Data, Sophia Antipolis, France, 28-30 Jan 1998.*

*Remote Sensing for a Changing Europe*
*D. Maktav (Ed.)*
*IOS Press, 2009*


*doi:10.3233/978-1-58603-986-8-165*

# Urban growth pattern of Didim

H. Esbah
*Istanbul Technical University, College of Architecture, Landscape Architecture, Turkey.*

B. Kara & B. Deniz
*Adnan Menderes university, Faculty of Agriculture, Landscape Architecture, Turkey.*

B. Kesgin
Ege University, Faculty of Agriculture, Landscape Architecture, *Turkey.*

Keywords: LULC change, ASTER, SPOT, change detection, urban growth.

ABSTRACT: Didim peninsula is the fattest growing urban area in the Aydin province, Turkey. Since 1990, the Town of Didim has changed significantly after discovered by domestic and international tourist. In spite of the recent rapid LULC change, Didim has not been spoiled compared to other big touristic towns of Turkey. Didim has been announced as "Tourism hot spot" in 2000, thus its planning is overseen by the Ministry of Culture and Tourism. Monitoring of the Didim's development is necessary to guide the Ministry in promoting sustainable planning guidelines. The present work aims to determine the characteristics and the amount of urban growth in Didim by using remote sensing and GIS technology. Already rectified Aster (dated 04/27/2005) and Spot 2X (03/02/1994) images were used as well as the population information, aerial photographs, city plans and thematic maps from previous studies. Object oriented classification technique is employed. Some complementary information is extracted from aerials and maps by on-screen digitization. Total of 16 LULC categories are defined. After, putting all information in the GIS database, the pattern of landscape change in Didim is described by using selected landscape metrics. The case study of the Town of Didim offers a good example of the impact of national policies on land use dynamics at local landscape scale. The findings indicate three simultaneous key trends: loss of coniferous forests, the thinning of the maqui vegetation cover, and intensification of urban areas on valuable class II type of soils. Identified trends have significant consequences in terms of the response that ecosystems have given to these anthropogenic landscape alterations. A strategy to promote sustainable land use management should be generated timely manner.

## 1 INTRODUCTION

The present work aims to understand the urban growth pattern of Didim and its ongoing effects on the local landscape characteristics by using remote sensing technology. Surrounded by very sensitive ecosystems, the Town of Didim is the fastest growing urban area in the Aydin province, Turkey. Monitoring of its growth is necessary to take actions to mitigate the adverse effects on these ecologically important areas.

Examples of using remote sensing technology to understand the land use /land cover dynamics are gaining popularity among the scientists in Turkey. Doygun (2005) investigated the consequences of the urban development in the fourth biggest metropolitan area of Turkey, and found out that Adana and its surroundings experienced 107.58% urbanization between 1984 and 2000, and agricultural lands decreased 34% due to this development. Another research on the coastal areas of the same geography displayed that the expansion of agricultural practices over ecologically sensitive natural areas was evident, and urban settlements continued to grow mostly at the expense of prime agricultural land in their close vicinity (Alphan & Yılmaz 2005). In a rapidly developing; heavily industrialized city of Bursa, Aksoy & others (2004) found out that 3006.1 ha of agricultural land was converted to urban uses corresponding with a 73.9% change between 1984 and 1998.

Maktav & others (2002) investigated the urban growth in the Buyukcekmece, Istanbul area between 1984 and 1998 by using satellite images, ancillary data and remote sensing technology. Their results indicated approximately 288.2% overall population increase in 14 administrational units of the central district. In this period, due to the migration that is triggered by industrialization, the urban area expanded approximately 20% as a detriment of the open systems and agricultural lands.

There has been no previous attempt to understand the landuse/landcover (LU/LC) change dynamics of rapidly growing Didim area. Obviously, remote sensing and GIS technology could be an effective tool in this endeavor. Spreading to almost 300 square kilometers, Didim is shaped as a peninsula surrounded by Mugla and large inlet of Akbuk on the east, Aegean Sea on the west and south, and legally protected Bafa Lake Nature Park on the north east, and the Big Meander Delta National Park on the north (Fig. 1).

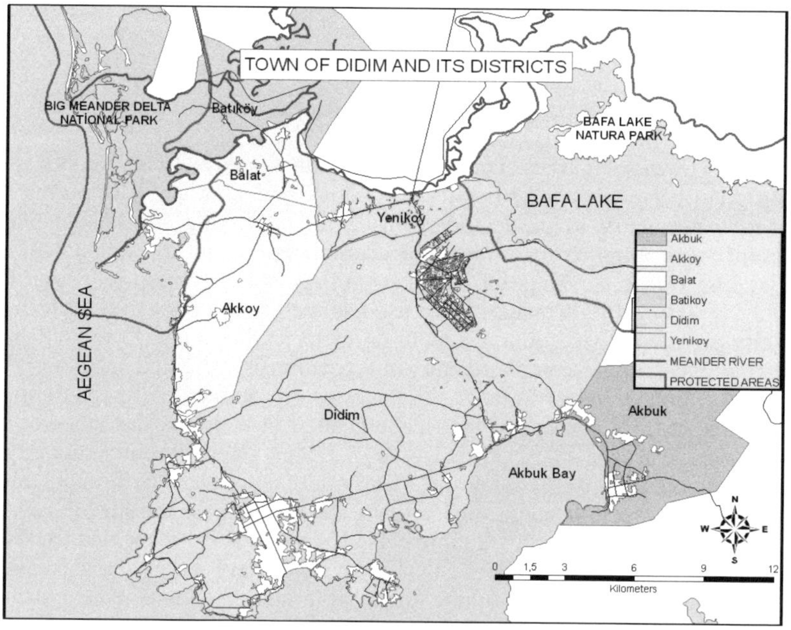


Figure 1. Study area, Town of Didim

Didim was a small village for a long time whose economy relied on agriculture that was conducted on relatively unfertile soils of the peninsula. Since 1990, Didim has changed significantly after discovered by international tourists who seek a modest vacation in nice climate and beaches. The blue flag holder Altinkum beach, various ancient remnants such as Temple and Oracle of Apollo, and comfortable weather especially soothing for asthma patients are some of the major reasons why Didim became an attractive place for tourists.

When the Altinkum beach has started to be famous around the country in 1980s, people mainly from Ankara and other Anatolian cities rushed here to have summer houses. When Turkey's economy started to decline those people found very hard to survive in the big cities, most of the summer house owners, who were mainly retired people, have decided to move into Didim. This has initiated the development. Subsequently, agricultural lands have gained development status, whilst more people from larger cities hunted for opportunities to settle. Summer houses, hotels, resorts, restaurants and bars mushroomed. In late 1990s, after the regulation which enables foreigners to own property in Turkey became effective, a great number of Europeans (British mostly) started to buy houses and land in Didim, thus the real estate sector grew significantly. In 2000s, Didim has been registered as a "Cultural and Tourism Protection and Development Region" by the Ministry of Culture and Tourism, which means the Turkish Government has designated it as an investment area and there are plans for a fully operational marina, golf courses, sports centers and 5 star holiday villages. Today, Didim has a bed capacity of 15.000 in 200 facilities either certified by the Ministry or by the Municipality. Not only, are property values rising dramatically as foreign buyers follow the lead, but also the urbanization is escalating. This is changing the social, ecological, and economic characteristics of the town. Understanding and monitoring of the Didim's development is necessary to guide the Ministry in promoting sustainable planning guidelines.

## 2 MATERIALS

This study utilizes 1:25000 scale topographic maps and soil maps. Existing GIS data from the Forest service and other institutions are also used along with paper maps obtained from the Municipality. Already rectified Quickbird (dated March 05, 2006), Aster (dated April 27, 2005), and Spot2X (dated March 03, 1994) images are used for the classification and the accuracy assessment. Also, 1: 35000 scale, August 1993 dated black and white aerial photographs are used for the visual interpretations and guiding the accuracy assessment of the Spot data.

## 3 METHOD

Overall workflow of this study is five fold: Literature review, site studies, analysis of population change, objects based classification, and analysis of landscape structure. Population is a land resource impact indicator. Population analysis reveals information with regards to the density of new urbanization. Data obtained from the state statistical institute is utilized for the analysis. Also, on site interviews with the officials and the inhabitants generated further understanding of the population characteristics.

Before starting the object oriented classification, based on the literature review, site visits and scrutinizing the existing material 16 landuse/land cover (LU/LC) classes are defined: 1- Conifer (Coniferous forest), 2- Maquis 1 (dense and tall maqui cover), 3- Maquis 2 (moderately dense, %60 coverage), 4- Maquis 3 (maqui formation with %10-60 coverage), 5- Garique (discontinuous, low, bushy associations), 6- Young forest (naturally growing new forest), 7- Sparsely vegetated (steppes, tundra and badlands), 8- Croplands, 9- Groves (Olive, Citrus, Plum plantations), 10- Artificial surfaces (buildings and impermeable surfaces), 11- Salt marshes (salines), 12- Coastal vegetation (vegetated areas above the high tide line), 13- Grassland (Grasslands, moors and heath lands) 14- Bare rock, 15- Water, 16- Cloud .These classes do not represent functional meanings (as in the case of industrial, open space, park and so fort), but rather physical attributes. This approach in defining the classes seems more reasonable because information stored in the remote sensing data is about the physical characteristics (spatial and spectral) of the earth

### 3.1 *Object based classification*

Object based classification focuses on homogeneous image objects (the group of pixels) rather than individual pixels (Baatz et. *al* 2001, Platt R.V. et.*al*. 2008). It not only uses spectral information (as in the case of pixel based supervised or unsupervised classification approaches) but also utilizes spatial information. Three major steps exist during the procedure:

1- Segmentation: It is also called multi resolution segmentation. For the classification, 3 bands of VNIR in Aster data and 3 bands of Spot 2X are used along with DEM and Aspect layers First, the scale parameters, and then, the homogeneity criteria and shape ratio are identified. Scale parameter is important as it affects the size of the object polygons. Scale factor for this study is determined as 10 and 20. Attributes such as color, shape, area and so on are included in the attribute table.

2- Defining the training samples and applying classifier: Based on the ground data gathered by GPS, and Quickbird images sets of training samples are defined for each LU/LC type of the study. Standard nearest neighbor classifier has been chosen for describing the training samples.

3- Classification: After segmentation and obtaining training samples, a very straightforward classification procedure is run in the Definiens Professional 5.0 software. The preceding two steps were repeated few times with minor modifications until the acceptable accuracy (defined as 80% at the beginning of the study) was attained.

An accuracy analysis was performed by using reference data. In addition to ground data, Quickbird image and the black and white aerials were used to test the accuracy of the classifications from Aster and Spot images, respectively. An equalized random sampling method was used to generate 800 random points (50 points in each class) in the Erdas Imagine 8.7 software. Error matrices that describe the patterns of mapped class relative to the reference data were generated, from which the overall accuracies, users and producers accuracies, and Kappa statistics were derived. 81.68% (Kappa: 0.8029), and 80.14% (Kappa: 0.7857) overall classification accuracy is acquired for 1994 Spot 2X and 2005 Aster data, respectively.

### 3.2 *Landscape structure indices*

In order to further understand the characteristics of the landscape change two levels of landscape structure metrics are utilized: Class and Patch Level. At the class level, which is dealing with attributes of the each LU/LC class, Proportion of Landscape metric is utilized. Proportion of Landscape measures the proportion of a certain LU/LC type in the landscape. At the patch level, which is dealing with the patch dynamics in a certain LU/LC class, Number of patches and Mean patch size metrics are used. Size is an important parameter for species richness and survival; more available the suitable habitats, higher the chances for populations to persist in altered habitats (Groom et *al.* 2005). ArcGIS 9.2 software was used for the landscape structure analysis.

## 4. RESULTS

### 4.1 *Population change*

The population of the town tripled between 1985 and 2007 (Table 1). Today, Didim's population is 39549. In Batikoy, population has declined continuously and dramatically. Because most of the land in this district is within the borders of the national park, any kind of development is restricted hence leading people to move out to other areas of the region. Population decrease has also occurred in Balat and Akkoy districts within a 22 year period, probably due to the same reason. However, further the national park and closer to the central business district Didim, lower the rate and amount of decline.

Table 1. Population change in Didim

| *DISTRICT* | *POPULATION* | | | | |
|---|---|---|---|---|---|
| | 1985 | 1990 | 1997 | 2000 | 2007 |
| Balat | 1664 | 2002 | 1367 | 1673 | 1259 |
| Akkoy | 1281 | 1459 | 1233 | 1749 | 1116 |
| Yenikoy | 2066 | 2374 | 2425 | 2442 | 2522 |
| Akbuk | NA | 2046 | 1994 | 2965 | 2657 |
| Batikoy | 1037 | 940 | 244 | 578 | 249 |
| Didim | 5063 | 11378 | 20797 | 25699 | 31746 |

Source: State Statistical Institute (DIE) Reports, 1985 and 2008

On the other hand, the district of Didim has experienced a significant population growth accelerated after 1997. This population lives there year around, hence generating demand for different sectors such as service, government, health etc. The second major developed urban area, Akbuk, does not follow the Didim's trend, because most of its facilities are used seasonally. The population of Akbuk has increased slightly between 1990 and 2007 with ups and downs in different periods. Yenikoy is a relatively rural district with a slight but steady population increase. Here, people engage in agricultural activities more than tourism, though since 2004, the bed and breakfast type of establishments has emerged. When the marina and other investments proposed in the Ministry's plan is finished, a major population increase (almost 1000%) is anticipated.

### 4.2 *Landuse/Landcover changes*

In 1994, the landscape was dominated by garigue cover on the central part of the study area. Conifer forests and tall maqui (Maquis1) habitats was mostly on the eastern part, in the Akbuk vicinity. Agricultural areas were mostly on the northern part of the study area, where the soil is relatively more fertile and irrigation is possible from the Big Meander River.

By 2005, the garigue cover has changed to maqui cover (mostly Maquis2 and Maquis 1). This visually significant change is attributable to a major fire took place in the peninsula few decades ago: the recovering landscape had developed more of a garigue structure by 1994 and this cover evolved to be maqui type of cover by 2005. Also, to some extend, the time of the satellite data could contribute to this change. On the east part of the study area, the decline of the conifer forests is evident and the transformation from forest to garigue is striking. Fires and the edge effects generated by the urban development and lack of sustainable planning in Akbuk are the main cause of this change.

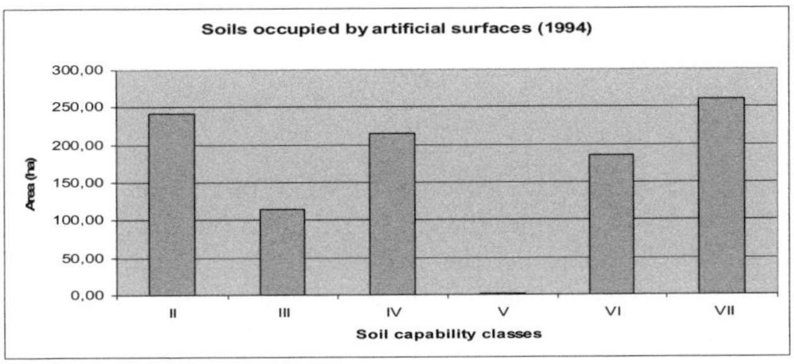


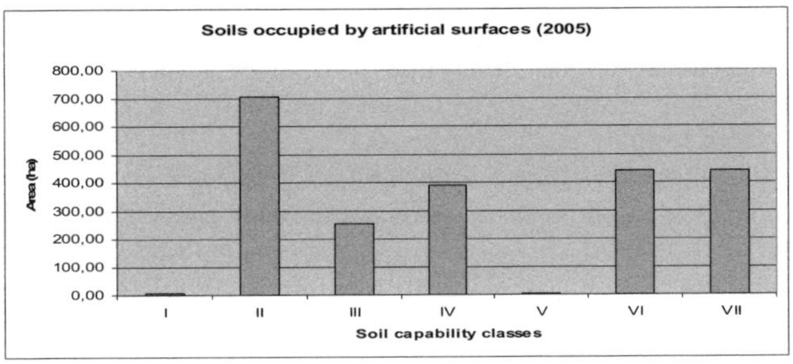


Figure 2. Distribution of artificial surfaces in relation to soil classes.

In fact, an unsustainable development prevails in the study area in general. For example, the majority of the Town of Didim is on class VI and VII soils, nevertheless, the recent development mostly has took place on the agriculturally important II class soils, which is already very scarce (corresponding only to the 5.84% of the soils in the study area) (Fig. 2).

In 1994, the landscape matrix was very heterogeneous (Table 2). Crop fields covered %18,86 of the landscape. The second biggest LU/LC was garigue followed by Maquis 1 and Maquis 2. Grasslands, which experienced the highest amount of decline by 2005, were the fifth major LU/LC class in the landscape. Other land cover classes constituted the rest of the landscape (28.06%). In this, the share of the artificial surfaces was only 3.03%.

Table 2. Proportion of landscape

| *LU/LC* | *1994* *(%)* | *2005* *(%)* |
|---|---|---|
| Conifer | 2,01 | 1,75 |
| Maquis1 | 13,80 | 9,19 |
| Maquis2 | 13,76 | 27,88 |
| Maquis3 | 5,31 | 6,43 |
| Garigue | 14,36 | 10,23 |
| Young Forest | 0,01 | 0,36 |
| Sparsely vegetated | 1,59 | 1,59 |
| Crop Land | 18,86 | 20,76 |
| Groves | 3,12 | 1,48 |
| Artificial surfaces | 3,03 | 6,30 |
| Salt Marshes | 4,72 | 4,05 |
| Coastal vegetation | 1,92 | 2,87 |
| Bare Rock | 3,37 | 0,14 |
| Water | 2,00 | 1,97 |
| Grassland | 11,17 | 5,01 |

In 2005, Crop lands were no longer the leading LU/LC type (Table 2). Instead, Maquis 2 type of habitats became the most dominant in the matrix. This is due to decline in the Maqui 1 and garigue habitats. A 33.38% and 28.76% decline has occurred in the Maqui1 and Garigue categories, respectively. Maqui1 areas, which comprise very dense and tall maqui vegetation, has turned into a less dense maqui cover during time, thus looking more like Maquis 2 type. This thinning of vegetation cover has occurred especially on the areas close to urban developments. This is a very clear sign of edge effects. Garigues, that was generated after the fire in 1980s, changed its structure as the vegetation formation became more continuous. These areas turned predominantly into Maquis2 and Maquis 3 habitats, some of them are also converted to artificial surfaces. In fact, the share of artificial surfaces doubled with in an 11 year period.

Class level analysis has showed increase in the Maqui2 type areas, not only the total area of such habitat has increased but the number of its patches and mean patch size. Even though, the share of the agricultural land has increased in the overall landscape, a slight decline occurred in their patch number but the mean patch size increased (Table 3). In the total landscape, the most fragmented habitats were Maquis3 and sparsely vegetated areas indicated by increasing numbers of patches but decreasing patch size. By looking at the numbers we can say that the shrinkage and attrition stages of land transformation are in effect for these habitats. The priority should be given to their preservation.

Table 3. Patch number and patch size

| | *Number of patches* | | *Mean patch size (ha)* | |
|---|---|---|---|---|
| TYPE | 1994 | 2005 | 1994 | 2005 |
| Conifer | 84 | 50 | 9,28 | 13,51 |
| Maquis1 | 409 | 141 | 13,07 | 25,26 |
| Maquis2 | 338 | 434 | 15,77 | 24,90 |
| Maquis3 | 167 | 253 | 12,32 | 9,84 |
| Garigue | 349 | 396 | 15,96 | 10,01 |
| Young forest | 1 | 12 | 5,05 | 11,69 |
| Sparsely vegetated | 75 | 129 | 8,21 | 4,75 |
| Cropland | 263 | 256 | 27,78 | 31,43 |
| Groves | 201 | 96 | 6,01 | 5,96 |
| Artificial surfaces | 115 | 227 | 10,22 | 10,76 |
| Salt marshes | 82 | 65 | 22,32 | 24,14 |
| Coastal vegetation | 155 | 219 | 4,80 | 5,08 |
| Cloud | 17 | 0 | 22,25 | 0,00 |
| Bare rock | 214 | 39 | 6,10 | 1,36 |
| Water | 453 | 414 | 1,70 | 1,84 |
| Grassland | 407 | 185 | 10,63 | 10,49 |

## 5. CONCLUSION

This study attempted to display the LU/LC change as a function of urban growth in the Town of Didim between 1994 and 2005. Within an 11 year period, the population has changed almost 1.4 times, but the artificial surfaces associated with urban growth doubled. Current growth pattern displays unsustainable use of the resources: Most of the development takes place in ecologically and economically important and also vary rare Class II type of soils; forest and dense maqui vegetations loose their structure transforming to a thinner cover. Fragmentation process also may be in effect in the study area as indicated by the patch number and mean patch size values.

The case study of the City of Didim offers a good example of the impact of national policies on land use dynamics at local landscape scale. Land conversion and increase in population density have been accompanied by sprawling urban pattern and fragmentation in open systems. Identified trends have significant consequences in terms of the response that ecosystems have given to these anthropogenic landscape alterations. A strategy to promote sustainable land use management should be generated timely manner.

## REFERENCES

Aksoy, E., Ozsoy, G. & Sezgin, E. 2004. Determining urbanization development and its adverse effect on soils of the alluvial plains at the Bursa Province using multi-date satellite data. *In press in Proceedings of ISC on Natural Resource Management for Sustainable Development, 7-10 June 2004. Erzurum, Turkey.*

Alphan, H. & Yilmaz, T. K. 2005. Monitoring environmental changes in the Mediterranean coastal landscape: the case of Cukurva, Turkey. *Environmental Mangement* 35 (5): 607-619.

Baatz, M., Heynen, M., Hofmann, P., Lingenfelder, I., Mimier, M., Schape, A., Weber M., & Willhauck, G. 2001. *eCognition User Guide 2.0*: Object Oriented Image Analysis. *Definiens Imaging GmbH, Munich, Germany*

DIE. 1985. *Census of Population. Report. Republic of Turkey, State Institute of Statistics, Ankara.*

DIE. 2008. *Census of Population. Report. Republic of Turkey, State Institute of Statistics, Ankara.*

Doygun, H. 2005. Urban development in Adana, Turkey, and its environmental consequences. *International Journal of Environmental Studies* 62: 391-401.

Groom, M. J., Meffe, G.K., & R. C. Carroll. 2005. Principles of conservation biology. *Third Edition. Sinauer Assoc. Inc., Sunderland*, 699pp.

Maktav, D., Erbek, F. S. & Akgun, H. 2002. Kentsel gelisimin uzaydan izlenmesi. *Metgraf Press, Istanbul*, 65pp.

Platt R.V., Rapoza L. 2008. An Evaluation of an Object-Oriented Paradigm for Land Use/Land Cover Classification, *The Professional Geographer*, 60:1, 87 – 100

*Remote Sensing for a Changing Europe*
*D. Maktav (Ed.)*
*IOS Press, 2009*


*doi:10.3233/978-1-58603-986-8-173*

# Comparison of pixel-based and object-based classification methods in detecting land use/land cover dynamics

B. Kesgin
Ege University, Faculty of Agriculture, Landscape Architecture, *Turkey.*

H. Esbah
*Istanbul Technical University, College of Architecture, Landscape Architecture, Turkey.*

Y. Kurucu
*Ege University, Faculty of Agriculture, Department of Soil Science, Turkey*

Keywords: Pixel based classification, object based classification, ASTER, land use/land cover.

ABSTRACT: Due to the complex spatial structure of the earth surface, obtaining a detailed and accurate land use/land cover (LULC) classification results with satellite data have still been problematic. The overall goal of this research is to compare the pixel based and object oriented image classification approaches in terms of the overall accuracies and robustness of the final classification product. An Aster image, dated 4/27/2005, with 3 bands from spectral regions of VNIR is used to perform the LULC classification for 16 different LULC classes. Ground truth data are collected from field surveys, available maps and Quickbird images.In pixel-based image analysis, supervised classification is performed by using maximum-likelihood classifier in Erdas 8.7. Object-oriented image analysis is conducted by utilizing Definiens Professional 5.0: The segmentation algorithm does not solely rely on the single pixel value, but also on shape, texture, and pixel spatial continuity. During the implementation, several different sets of parameters were tested for image segmentation, 20 was selected as a scale parameter and nearest neighbor was used as the classifier. At the end, the performance of pixel based and object-oriented classifications are compared based on the accuracy assessment results.

## 1 INTRODUCTION

Land use and land cover maps are very important data for landscape planners analyzing landscape dynamics. Remote sensing techniques are valuable tools to get efficient information about different LULCs. However, attaining an accurate and reliable results are challenging due to the complex structure of the earth's surface. In this paper, we compared the performances of more traditional pixel based and relatively recent object oriented classification techniques in a very heterogeneous Mediterranean landscape.

Pixel based approach is a classic approach that is based on conventional statistical techniques, such as supervised and unsupervised classification. The basic unit in the pixel based approach is a single pixel that has spectral information (Matinfar et al. 2007). Generally, earth objects like vegetation, water, buildings etc. have different spectral properties that can be assigned by this approach. But pixels within an image can represent more than one land cover on the ground. Therefore, mixed pixels problems in the images has been major handicap for obtaining accurate classification result (Thornton et al. 2006).

Object based classification classifies homogeneous image object (the group of pixel) rather than individual pixel (Baatz et al. 2001, Platt R.V. et al. 2008). The basic step of an object-oriented approach is the segmentation procedure, or grouping of neighboring pixels into meaningful, homogeneous patches (i.e. image objects). These objects might be agricultural parcels, roads, houses, water bodies, conifer cover, saline surfaces, and so on. Scale parameter is a unit-less index that describes the level of aggregation. They carry also reflectance, texture, size of object, form (spatial features) and topology information in a common attribute table (Manakos 2001, Pillai 2005, Mathieu 2007). These two approaches have been applied and compared by utilizing different satellite images (Mitri & Gitas, 2004), urban mapping (Thomas 2003, Chen et al., 2007), forest type classification (Shataee et al. 2004), mapping vegetation communities (Mathieu 2007), bare soil brightness mapping (Brodsky 2006), and coal fire research (Yan, 2003). In general, object based classification method yielded better results in comparison to pixel-based classification method. Matinfar et al., 2007 applied the object-oriented and pixel-based classification methods on Landsat ETM data for mapping LULC. The results showed that the object-oriented approach gave more accurate results compared to pixel based classification method. In this study we aim to test the performances of pixel and object based classifications by using Aster data. This work is a part of an ongoing research on understanding the LULC dynamics in the two legally protected areas and its surrounding landscapes in western Turkey

## 2 STUDY AREA

The study area is located in the south-west part of Aydin province, on the western coast of Turkey. It comprises Dilek Peninsula- Big Menderes Delta National Park and Bafa Lake Nature Park and their 10 km surrounding areas. It covers an area of approximately 152800 ha The major land cover types includes forests, shrublands, agricultural areas, olive plantations, urban areas, coastal wetlands, water bodies, etc. (Fig. 1).

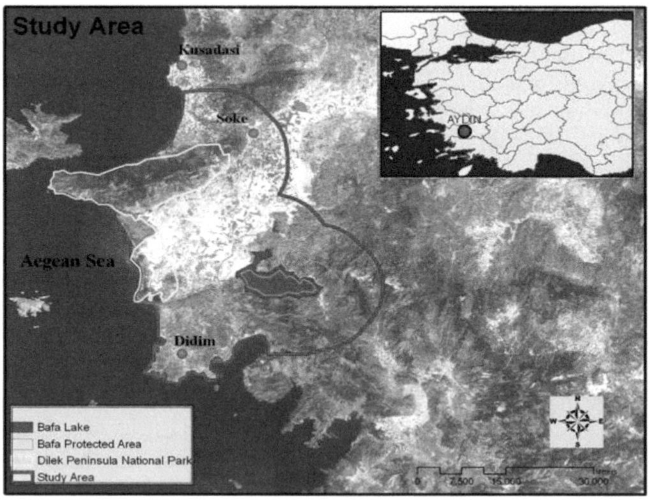


Figure 1. Study Area

## 3 DATA

Ortho-rectified Aster image, dated 4/27/2005, with 3 bands from spectral regions of VNIR is used to perform LULC classification. Normally, Aster has tree sensor system which is Visible Near Infrared (VNIR), Short Wave Infrared (SWIR), Thermal Infrared (TIR). VNIR band with 15 m. resolution provides information on common land cover features such as water, soil, and vegetation. On the other hand, SWIR bands, which have 30 m spatial resolution, are especially advantageous for discrimination of rocks and minerals and TIR bands, which have 90 m spatial resolution, are advantageous for detecting different kind of rocks such as limestone and sandstone (Yan 2003). Furthermore, Digital Elevation Map (DEM) and Aspect layers are taken into consideration to obtain better classification results. Ground control points and ground truth data are collected using Magellan GPS. For accuracy assessment, Quickbird images are used as the reference image.

## 4 METHOD

Main procedures of our work include; image pre-processing, image classification; maximum likelihood supervised classification for pixel based classification, and nearest neighborhood classification for the object base classification, and accuracy assessment.

### 4.1 *Pre- processing:*

In order to obtain more accurate results, before the classification procedures, ortho-rectified Aster image was atmospherically corrected using dark pixel subtraction method. The Aster VNIR (1, 2, 3) bands, DEM and Aspect layers are added.

### 4.2 *Classifications*

#### 4.2.*1 Land Use and Land Cover Classes*

Before classification, land cover types have been defined referring to CORINE land cover nomenclature. The details of the classes are given in Table 1.

Table 1. Defined LU/LC classes in the study

| No | Land Use/ Land Cover | Corine Classes | Explanation |
|---|---|---|---|
| 1 | **Artificial surfaces** | 1,0 | (1.1.) Urban fabric (1.1.1.), Continuous urban fabric surfaces (1.1.2.), Discontinuous urban fabric (1.2.), Industrial, commercial (1.2.1.), Industrial or commercial units and transport units (1.2.2.), Road and rail networks |
| 2 | **Agricultural areas** | 2,1 | Arable Land (Cotton, Corn,Wheat, etc...) |
| 3 | **Permanent crops** | 2,2 | Species; Citrus, Peach , Fig, Olive Species. |
| 4 | **Coniferous forest** | 3,1,2 | Pinus brutia and Pinus nigra are intensive in the area. |
| 5 | **Natural grassland vegetation** | 3,2,1 | Grasslands, moors and heath lands. |
| 6 | **Maquis1** | 3,2,3_1 | Dense Maquis, the species are about 2m height and with 70% coverage, rarely pinus species have been seen. Species; Pistacia terebinthus,Q. coccifera, Arbutus unedo, A.andrachne, Casteneum ipsicum, Ceratonia siliqua, Cercis siliquastrum, Laurus nobilis, … |
| 7 | **Maquis2** | 3,2,3_2 | Moderate dense; The species are lower than 2 m height, with 60% coverage. Species; Q.coccifera, Juniperus phoenica, J. Occicedrus, Olea europe, Phyleria latifolia. |
| 8 | **Maquis3** | 3,2,3_3 | Less dense; The species are lower than 2 m height, with 10 to 60% coverage. Species; Q.coccifera, Sparteum junceum, Cistus parvifolia. |
| 9 | **Gariques** | 3,2,3_a | Discontinuous bushy associations of the Mediterranean calcareous plateaus. Species; Q. cerris, Sarcopoterium sipinosum, Thymus , Cistus sp. |
| 10 | **Transitional woodland shrub** | 3,2,4 | Bushy or herbaceous vegetation with scattered trees. Can represent either woodland degradation or forest regeneration/colonisation. |

| | | | |
|---|---|---|---|
| 11 | **Open spaces with little or no vegetation** | 3.3 | Rocky and bare areas. |
| 12 | **Bare rock** | 3,3,2 | Scree, cliffs, rocks outcrops, including active erosion, rocks and reef flats situated above the high-water mark. |
| 13 | **Sparsely vegetated areas** | 3,3,3 | Includes steppes, tundra and badlands. Scattered high-altitude vegetation. |
| 14 | **Salt marshes** | 4,2,1 | Vegetated low-lying areas, above the high-tide line, susceptible to flooding by sea water. Often in the process of filling in, gradually being colonised by halophilic plants..Species in the area; Salicornia sp., Tamarix sp, Juncus acutus, Arundo donax, Scirpus sp.... |
| 15 | **Salines** | 4,2,2 | Salt-pans, active or in process of abandonment. |
| 16 | **Water** | 5,0 | Sea, lake, coastal lagoons, river. |
| 17 | **Cloud*** | | |

4.2.2 *Pixel Based Classification*

In pixel-based image analysis approach, supervised classification was performed using maximum-likelihood algorithm (Figure 2). ERDAS 8.7 was used for the procedure. Supervised Classification is based on a set of user-defined spectral signatures to classify an image. In our research, the spectral signatures are derived from training areas, which are evaluated according to data from land surveys such as ground truth points, photos and higher resolution images such as Quickbird images. After specifying the spectral signatures, the maximum likelihood supervised classification was performed. Maximum Likelihood (ML) is a standard pixel based approach that classifies pixel according to the multivariate probability density functions of the classes of interest (Lillesand & Kiefer 1994). To overcome the salt-pepper effect, supervised classification map has been filtered with 3x3 kernel size.

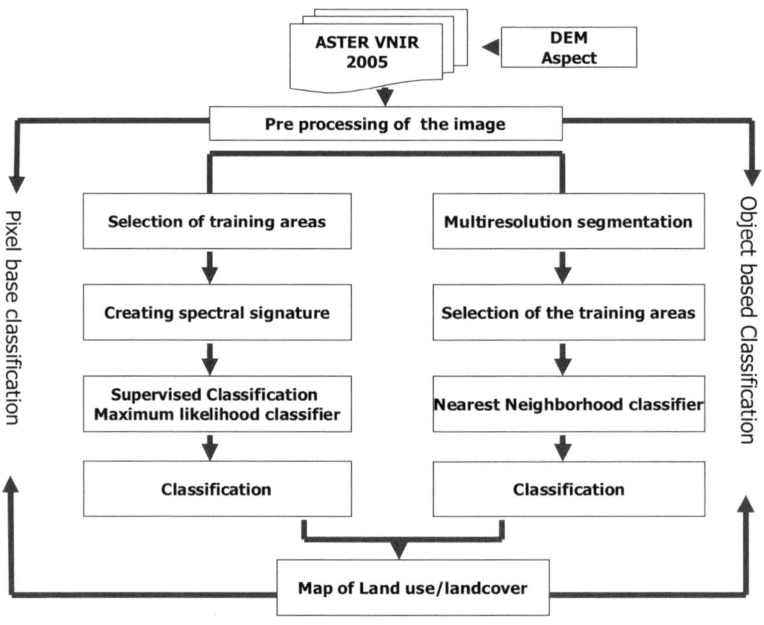


Figure 2. Flow diagram of the classification procedures

4.2.3 *Object based classification*

Object-oriented classification has at least four components (Figure 2). The first step is a multi-resolution segmentation of the image. The segmentation was applied to the 3 spectral bands and 2 extra bands including DEM and aspect information in order to create pixel based objects. After testing some parameters for segmentation, suitable two levels are applied for our study using the following parameters: for the level 1, for scale factor 10, 0.8 for color, 0.2 for shape, 0.2 for compactness, 0.8 for smoothness, for the level 2, for scale factor 20, 0.7 for color, 0.3 for shape, 0.5 for compactness, 0.5 for smoothness and a diagonal pixel neighborhood. According to spatial and spectral properties of the image, homogenous polygons are created by segmentation with given parameters.

In the second and third steps, training samples were selected according to the ground control points and the standard nearest neighbor classifier was applied. Finally, a straightforward classification was performed by the software.

4.3 *Accuracy Assessment*

An accuracy assessment of the classification results was performed using reference data created from visual interpretation of the Quickbird image data. The ASTER VNIR data were also examined during the assessment process to minimize reference dataset error. An equalized random sampling method was used to generate the random points in the software of Erdas Imagine 8.7. A total number of 850 random points were sampled, with at least 50 random points for each class. Error matrices that describe the patterns of mapped class relative to the reference data were generated, from which the overall accuracies, user's and producer's accuracies, and Kappa statistics were derived to assess the accuracies of the classification maps (Table 2).

5 RESULT

Final LULC maps are presented in Figure3. An evaluation of the overall results of the accuracy assessment clearly shows the advantage of using object based classification method: The overall accuracy for the object based classification (77.06%) is higher than that of the pixel based classification 73.06%. Also, The Overall Kappa Statistics are also higher for the object based classification 7562%).

With regards to the evaluation of each individual class, the Kappa statistics yields better results with OBC for artificial surfaces, agricultural areas, maquis1&2, Garique, Open spaces, sparsely vegetation, salt and marshes. On the other hand, PBC gives better results for permanent crops, coniferous forest, natural grassland, maquis3, and bare rock.

The result of the user's and producer's accuracy also varies by individual classes. Producer's accuracy of pixel based classification has been higher in artificial surface, coniferous forest, natural grasslands, bare rock and salt-marshes. However, it is lower in permanent crops, coniferous, Maquis1&2, open spaces, salines. Additionally, the producer accuracy of the agricultural land has been too low for both classification methods.

User's accuracy of pixel based classification has been higher in permanent crops, coniferous forest, natural grassland, maquis3, bare rock and salines, however, it is lower in artificial surfaces, agricultural areas, maguis1 &2, Gariques, open spaces, sparsely vegetated areas. Generally, user's accuracy of permanent crops, young forest and salt- marshes have been too low for both methods.

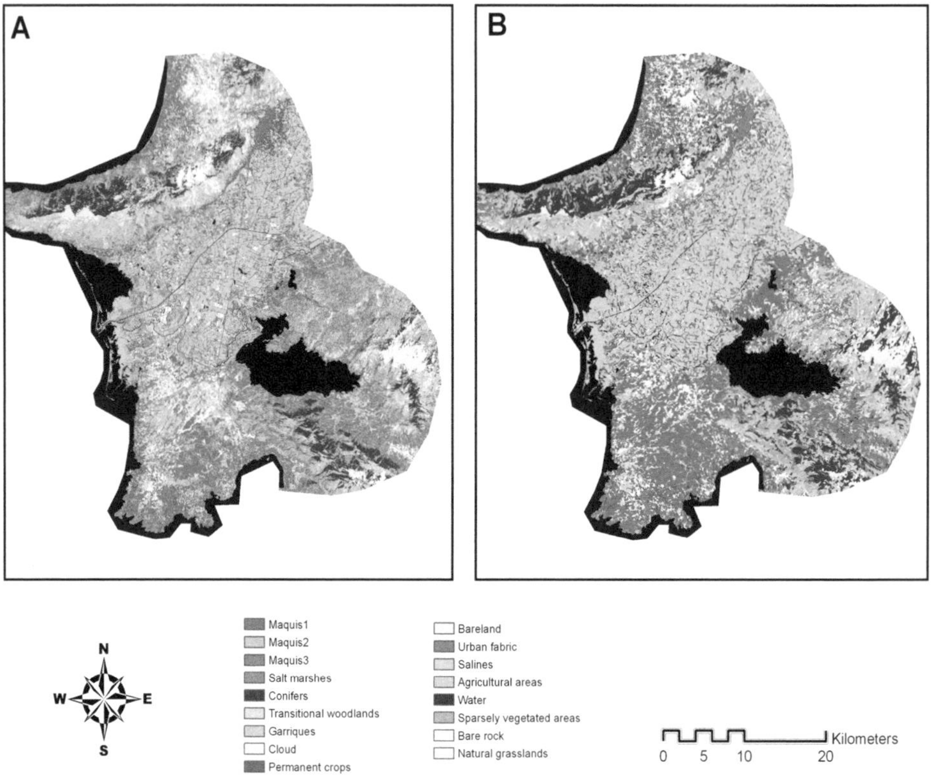


Figure 3. Pixel based classification map (A), Object based classification map (B)

Table 2. Accuracy assessment results.

| | | **Pixel Based Classification** | | | **Object Based Classification** | | |
|---|---|---|---|---|---|---|---|
| **No** | **Land Use/ Land Cover** | **Producers Accuracy** | **Users Accuracy** | **Kappa** | **Producers Accuracy** | **Users Accuracy** | **Kappa** |
| **1** | Artificial surfaces | 77.78% | 56.00% | 0.54 | 73.91% | 68.00% | 0.66 |
| **2** | Agricultural areas | 39.47% | 90.00% | 0.88 | 41.82% | 92.00% | 0.91 |
| **3** | Permanent crops | 91.43% | 64.00% | 0.62 | 92.00% | 46.00% | 0.44 |
| **4** | Coniferous forest | 100.00% | 92.00% | 0.92 | 100.00% | 84.00% | 0.83 |
| **5** | Natural grassland vegetation association | 84.62% | 88.00% | 0.87 | 75.00% | 66.00% | 0.64 |
| **6** | Maquis1 | 60.94% | 78.00% | 0.76 | 74.14% | 86.00% | 0.85 |
| **7** | Maquis2 | 50.00% | 68.00% | 0.65 | 70.49% | 86.00% | 0.85 |
| **8** | Maquis3 | 71.15% | 74.00% | 0.72 | 65.38% | 68.00% | 0.66 |
| **9** | Gariques | 68.85% | 84.00% | 0.83 | 68.18% | 90.00% | 0.89 |
| **10** | Transitional woodland shrub | 86.36% | 38.00% | 0.36 | 96.77% | 60.00% | 0.58 |
| **11** | Open spaces with little or no vegetation | 53.49% | 46.00% | 0.43 | 95.12% | 78.00% | 0.77 |
| **12** | Bare rock | 85.42% | 82.00% | 0.81 | 83.72% | 72.00% | 0.71 |
| **13** | Sparsely vegetated areas | 78.72% | 74.00% | 0.72 | 81.63% | 80.00% | 0.79 |
| **14** | Salt marshes | 96.55% | 56.00% | 0.54 | 87.18% | 68.00% | 0.66 |
| **15** | Salines | 88.89% | 80.00% | 0.79 | 94.59% | 70.00% | 0.69 |
| **16** | Water | 100.00% | 100.00% | 1.00 | 100.00% | 98.00% | 0.98 |

## 5 CONCLUSION

Attaining a high overall classification accuracy is challenging, when we are dealing with very heterogeneous landscape characteristics. Object based classification method is promising in this endeavor, if the most appropriate segmentation parameters are applied. Pixel based classification method has clear advantages when classifying coniferous forest, natural grassland, low density maquis and bare rock cover. A hybrid (supervised and unsupervised) classification method would yield more satisfying results, however this is left to the future research. Regardless of the method, the success of any classification relies on detailed knowledge of the satellite data and characteristics of the landscape in hand.

## REFERENCES

Baatz, M., Heynen, M., Hofmann, P., Lingenfelder, I., Mimier, M., Schape, A., Weber M., & Willhauck, G. 2001. *eCognition User Guide 2.0: Object Oriented Image Analysis. Definiens Imaging GmbH, Munich, Germany*

Brodsky L.& Boruvka L. 2007. Object-oriented Fuzzy Analysis of Remote Sensing Data for Bare Soil Brightness Mapping. *American-Eurasian J. Agric. & Environ. Sci.*, 2 (4): 448-456, 2007 ISSN 1818-6769

Chen, Y., Shi, P., Fung, T., Wang, J., Li, X. 2007. Object-oriented classification for urban land cover mapping with ASTER imagery. *International journal of remote sensing,* v28, Issue 20, pp 4645-4651.

EC. 1993. CORINE Land Cover: technical guide. *Report EUR 12585 (European Commission,Luxembourg).*

Lillesand, M.,T., Kiefer, W., R. 1994. *Remote Sensing and Image Interpretation, Third Edition, John Wiley & Sons, Inc., New York,* 750 pp.

Mathieu R., Aryal J., Chomg K. A. 2007. Object-Based Classification of Ikonos Imagery for Mapping Large-Scale Vegetation Communities in Urban Areas. *Sensors 2007,7,* 2860-2880, ISSN 1424-8220.

Matinfar H.,R., Sarmadian F., Alavi Panah S.K., Heck R.J. 2007. Comparisons of Object-Oriented and Pixel-Based Classification of Land Use/Land Cover Types Based on Lansadsat7, Etm+ Spectral Bands (Case Study: Arid Region of Iran), *American-Eurasian J. Agric. & Environ. Sci.*, 2 (4): 448-456, ISSN 1818-6769

Mitri, G. H., and I. Z. Gitas. 2004. A semi-automated object-oriented model for burned area mapping in the Mediterranean region using Landsat-TM imagery. *International Journal of Wildland Fire*, 13(3) 367–376

Manakos, I. 2001. eCognition and Precision Farming. *http://www.lrz-muenchen.de/~lnn/. eCognition Application Notes,* Vol. 2, No 2, April 2001.

Platt R.V., Rapoza L., 2008. An Evaluation of an Object-Oriented Paradigm for Land Use/Land Cover Classification, *The Professional Geographer*, 60:1, 87 – 100

Shataee S., Kellenberger T., Darvishsefat A.A., 2004. Forest types classification usıng ETM+ data ın the north of Iran/ comparison of object-orıented wıth pixel-based classification techniques, *XXth ISPRS Congress Commission PS, Working Group VII/1.*

Pillai, R. B., Weisberg, P. J., Lingua, E. 2005. Object-orıented classıficatıon of repeat aerıal photography for quantıfyıng woodland expansıon ın central nevada. *Proceeding of the 20th biennial workshop on aerial photography, videography and high resolution digital imagery for resourse assessment. Wescalo, Texas. American society for photogrammetry and remote sensing.*

Thornton M.V., Atkinson P.M., Holland D.A., 2006. Super-resolution mapping of rural land cover objects from fine spatial resolution satellite sensor imagery. *International journal of remote sensing,* v27, No3-4, pp. 473-491.

Thomas, N., Hendrix, C. & Congalton, R.G. 2003. A comparison of urban mapping methods using high-resolution digital imagery. *Photogrammetric Engineering and Remote Sensing*. vol 69: 963-972.

Yang, X., 2002, Satellite monitoring of urban spatial growth in the Atlanta metropolitan area. *Photogrammetric Engineering and Remote Sensing*, 68, pp. 725–734.

Yan, G. 2003. Pixel based and object oriented image analysis for coal fire rsearch. *International institute for geo-information science and earth observation enschede, the Netherlands. Master of science thesis.*

*Remote Sensing for a Changing Europe*
*D. Maktav (Ed.)*
*IOS Press, 2009*

*doi:10.3233/978-1-58603-986-8-180*

# Oil-spill detection using generalized linear model and genetic algorithm: A case study in Lebanon

C. Özkan
*Erciyes University, Engineering Faculty, Geodesy and Photogrammetry Engineering Dept., 38039 Kayseri, Turkey,* cozkan@erciyes.edu.tr

F. Sunar
*Istanbul Technical University, Civil Engineering Faculty, Remote Sensing Division, 34469 Maslak Istanbul, Turkey,* fsunar@ins.itu.edu.tr

Keywords: Oil pollution, Radarsat-1, Generalized linear model, Genetic Search

ABSTRACT: Oil spill discharges have serious biological and economic impacts upon the marine environment. One of the most hazardous land-based threats with an estimation of 110,000 barrels of oil discharges was occurred in the Lebanon coast of the eastern Mediterranean Sea recently. Because of not enough cleaning activities, the oil pollution has spread 100 miles along the coast and effected also to Syria's shoreline. Depending on the wind and current condition, the spilled oil could also be a big threat to other neighborhood countries such as Turkey, Greece, and Cyprus.

It is necessary to identify some features such as the type, areal extent, position, (nearest ship position if any), the possibility percentage for the oil spill analysis. For this purpose, satellite-based oil pollution monitoring systems are being used to take precautions and even to determine the possible polluter. Today, Synthetic Aperture Radar (SAR) satellites are often preferred to optical sensors due to the all-weather and all-day capabilities and being used to detect the oil spills discharged into the sea with sufficient accuracies.

In this study, Radarsat-1 images covering Lebanon coasts acquired by ITU-CSCRS (Istanbul Technical University – Center of Satellite Communication and Remote Sensing) during the event were employed for oil spill analysis. After some required image enhancement and transformation techniques such as spatial filtering and extraction of textural features based on gray level co-occurrence matrix (GLCM), the oil spill areas were determined by supervised classification using Generalized Linear Model (GLM). Genetic Algorithm (GA) was used to optimize the parameters of the GLM.

## 1 INTRODUCTION

The increase and development in the marine transportation began to be an important factor for the oil pollution of maritime environment both socially and economically. The amount of oil spilled annually worldwide has been estimated at more than 4.5 million tons (Bava *et al*. 2002). Illegal oil spills at sea occur during tanker and cargo ship operations, where ballast waters, tank washing residues, fuel oil sludge and machinery space bilge deliberately are discharged into the sea (Maar *et.al.* 2003; Harahsheh *et al.* 2004). The %30 of international marine trade takes place among the ports of Mediterranean and nearby seas. As the world's %28 of the petroleum transport passes through Mediterranean, the %50 of the total transportation is considered to be risky and dangerous. Every year nearly 20.000 tons of petroleum leaks to Mediterranean from the surrounded 60 oil refining plants as a result of consciously or unconsciously accidents (Günel 2004).

Therefore to detect and monitor oil spills at sea is becoming increasingly important due to the threat that such pollution poses to marine ecosystem. Today, remote sensing technology is being used for this purpose over the past decade. Synthetic Aperture Radar (SAR) satellites are often preferred to optical sensors due to the all weather and all day capabilities and being used to detect the oil spills discharged into the sea with sufficient accuracies (Solberg *et al.* 2007). Oil-Spill detection procedures in SAR data generally comprises segmentation, feature extraction and classification stages (Solberg *et al.* 2007; Brekke & Solberg 2007). The backscatter energy level for oil-spilled areas is too low since the oil dampens the capillary waves of the sea surface. But, there is a serious problem that the other natural phenomenas also dampen the short waves and create dark areas on the sea surface (Solberg *et al.* 2007). Thus, the dark image regions of which the probabilities of being either oil spill or alike are high are segmented. These segmented parts of original images are used to obtain the features of shape, contrast and textural characteristics. In terms of these features, some classification algorithms based on statistical, neural, fuzzy, rule based, boosting algorithms etc. are used for identification of the dark areas in a manner of binary classification, i.e. oil or alike (Fiscella *et al.* 2000; Frate *et al.* 2000; Solberg *et al.* 1999; Keramitsoglou *et al.* 2006; Ramalho & Medeiros 2006; Ozkan & Sunar 2007). In this study, as a different approach, GLM was used to model the relation between features and binary classes. GLM provides a general unified conceptual and methodological framework to many of the most commonly used applied statistical techniques for the analysis of both continuous and discrete data (Khuri *et al.* 2006). The unknown parameters of GLM are conventionally solved by Iterative Weighted Least Squares (IWLS) algorithm (Lindsay 1997; Dobson 2002). For parameter estimation, in stead of IWLS, GA can be employed effectively. GA is a highly efficient stochastic global optimization algorithm for problem solving. GA emulates the natural evolution process and maintains a population of potential solutions to a given problem, which are evaluated by a problem-specific fitness function. This population is evolved by selecting relatively fit members and performing genetic operations (Pham & Karaboga 2000).

In this study, oil pollution occurred recently in the Lebanon coast was examined. Due to not enough cleaning activities carried on, the oil pollution has spread 100 miles along the coast and effected also to Syria's shoreline. Depending on the wind and current condition, the spilled oil could also be a big threat to other neighbourhood countries such as Turkey, Greece, and Cyprus. Radarsat-1 images covering Lebanon coasts acquired by ITU-CSCRS (Istanbul Technical University – Center of Satellite Communication and Remote Sensing) during the event were used to test the usage of the GLM for identification of the oil-spill areas. The general methodology in this study consists of image filtering, segmentation of the dark areas, feature extraction, setting GLM, parameter estimation by GA and performance analysis. In order to be able to show the effectiveness of GLM algorithm, Bayesian Maximum likelihood and Artificial Neural Network Multi Layer Perceptron classification methods were also employed.

## 2 STUDY AREA AND DATA USED

As one of the East Mediterranean countries, Lebanon, neighboring Syria and Israel, has 225 km coastline (Fig.1). The most recent oil pollution, occurred due to the bombing of a power plant at Jieh, has approximately affected 1/3 of the whole Lebanese coastline. Depending on the weather conditions, it could be also a serious threat to the neighboring Mediterranean countries such as Turkey, Cyprus and Syria. Among them, Northern Cyprus, 162 km away from Lebanon, has a length of approximately 330 km coastline with up to 12km nautical miles of territorial waters (Sunar *et al.* 2007; Ozkan & Sunar 2007).

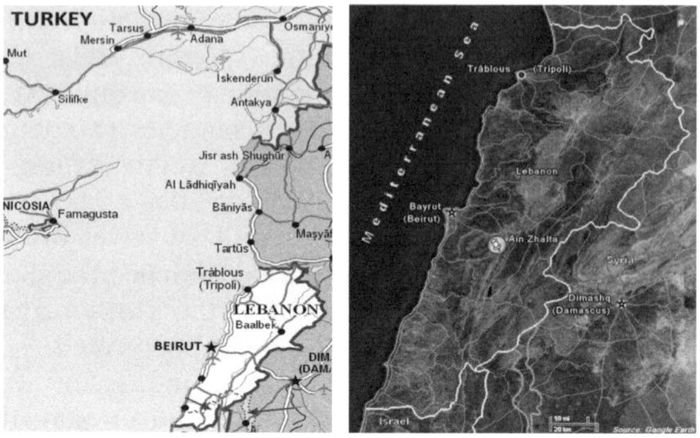


Figure 1. Map and satellite image of the study area.

In this study, as an active sensing system, 5 of total 20 Radarsat -1 images acquired by ITU-CSCRS were used in the analyses (Table 1).

Table 1. The characteristics of the radar dataset used.

| Satellite | Acquisition date | Beam mode | Spatial resolution (m) | Band | Wavelength (cm) | Pol. | Coverage (sq km) |
|---|---|---|---|---|---|---|---|
| | 2006-07-21 | Fine | 8 | | | | 50x50 |
| | 2006-08-09 | Standard | 25 | | | | 75x75 |
| RADARSAT-1 | 2006-08-17 | Wide | 30 | C | 5.3 | HH | 150x150 |
| | 2006-08-23 | Standard | 25 | | | | 75x75 |
| | 2006-10-01 | Fine | 8 | | | | 50x50 |

## 3 METHODOLOGY

Oil slicks on the sea surface can be detected because of the dampening effect of oil on capillary waves. Therefore radar images have an advantage for oil spill detection at this point where oil can be detectable as black patches on images. In general there are three different techniques for oil spill detection; *i)* manual detection; *ii)* semi-automated detection; and *iii)* fully-automated detection. In this study, semi-automated method was used and for this purpose ML, ANN MLP and GLM methods were tested.

### 3.1 *Artificial Neural Network (ANN)*

Artificial neural networks are computational systems based on the principles of biological neural systems, i.e. it is a mathematical model composed of many neurons operating in parallel. These networks have the capacity to learn, memorize and create relationships amongst data. They have some advantages such as their non-parametric and non-linear nature, arbitrary decision boundary capabilities, easy adaptation to different types of data and input structures, and good generalization capabilities over classical statistic and analytic approaches. Although the network design as a classifier is a hard task despite the increment in the performance of classification, an approach for oil spill detection based on a Multilayer Perceptron (MLP) neural network are described in recent research studies (Frate *et al.* 2000; Topouzelis *et al.* 2005).

### 3.2 *Generalized Linear Models (GLM)*

Linear models that describe a linear relationship between a response and one or more predictors can be written as (Eq.1);

$$E(Y_i) = \mu_i = x_i^T \beta \quad ; \quad Y_i \sim N(\mu_i, \sigma^2) \tag{1}$$

Where responses are independent and normally distributed with $x_i^T \beta$ mean and $\sigma^2$ variance. But generally the underlying relationship between independent and dependent variables are nonlinear. As a special class of nonlinear models, generalized linear models employ linear approaches (Matlab, 2008). There are three main parts of GLM: (i) a distribution function of the response ($Y_i$), (ii) linear predictor ( $x_i^T \beta$ ) and (iii) a link function.

Due to some mathematical requirements, the distribution function of the responses follow any probability distribution from the exponential family of distributions such as Normal, Binomial, Poisson, Multinomial, Gamma, Negative Binomial, etc. with parameters including a mean $\mu$. In that case, the expected value of a response $Y_i$ ($\mu_i$) is linked to the linear predictor through a monotone and differentiable link function as follows (Eq.2):

$$g(\mu_i) = x_i^T \beta \tag{2}$$

The link function relates the probability distribution of responses to the linear predictor of the model. The canonical link functions can be any of identity, log, logit, reciprocal, probit, loglog etc. (Table 2). So, a general linear model is a special type of the generalized linear model with a normal distribution for response and an identity function for link.

The parameters ($\beta$) of linear predictor are generally estimated by IWLS method in the manner of maximum likelihood estimation (Hardin & Hilbe 2007).

Table 2. Some link functions.

| | |
|---|---|
| Identity | $\mu = x\beta$ |
| Log | $\log(\mu) = x\beta$ |
| Logit | $\log(\mu/(1-\mu)) = x\beta$ |
| Probit | $\text{norminv}(\mu) = x\beta$ |
| Loglog | $\log(-\log(\mu)) = x\beta$ |
| Reciprocal | $1/\mu = x\beta$ |

### 3.3 *Genetic Algorithm (GA)*

The essential idea of Genetic Algorithms was put forward by Holland in 1975 (Holland 1975) as a directed random search technique. GA is a highly efficient stochastic global optimization algorithm which can find the global optimal solution in complex multi-dimensional search spaces. Two main advantages of GA are *(i)* it is not easy to slump into a local optimized region and *(ii)* it can be realized by a parallel computational method. GA has been applied to many computer vision fields, medical image registration, image segmentation, model based matching, camera calibration, object recognition (Bhanu *et al.* 1991; Grefenstette & Fitzpatrick 1985; Hill & Taylor 1992; Tsang 1997; Ji & Zhang 2001).

GA emulates the natural evolution process and maintains a population of potential solutions to a given problem through a problem-specific fitness function by performing common genetic operators such as selection, crossover and mutation (Pham & Karaboga 2000).

## 4. APPLICATION and RESULTS

In general, SAR data are subject to speckle which is a natural phenomenon generated by coherent processing of radar echoes. Therefore the radar images were firstly filtered by a 3x3 median and 5x5 low-pass convolutions, respectively for speckle noise reduction. After filtering, co-occurrence texture measurements were computed. Because of the coherent nature of SAR imaging, these measures are more suitable for classification instead of using individual pixel brightness values. In contradiction to the pixel intensity itself, texture provides information about the spatial correlation among neighboring pixels. From these measures, homogeneity, angular second moment (Assilzadeh & Mansor 2001), variance and mean were found effective in discrimination of oil spills from other features at the sea surface (Ozkan & Sunar 2007).

The patterns from which these textural values computed were totally collected from the RADARSAT-1 images used. The training areas were determined by taking into consideration both the meteorological conditions (weather state, wind etc.) and manual visual inspection (Ozkan & Sunar 2007).

Afterwards, the training patterns were sampled randomly for a reasonable number of patterns to reduce the computational burden from training areas. For each class, 5000 training and 5000 test patterns were collected separately (Ozkan & Sunar 2007).

From artificial neural network structures, Multilayer Perceptron was employed. The training algorithm used for updating the synaptic weights and bias values was selected as Resilient Backpropagation (Reidmiller & Braun 1993). This learning algorithm was found as the best from a test sequence of different learning rules such as standard gradient decent with momentum, conjugate gradient methods and one-step secant. Only one hidden layer and 50 neurons in this hidden layer were used (Ozkan & Sunar 2007).

In GLM application, binomial distribution was chosen for the response distribution since the response is binary categorical data. Loglog function was used as the link function because it gave better results than the others. For loglog function, the linear equation would become as follows (Eq.3):

$$\log(-\log(\mu_i)) = \beta_0 + \beta_1 x_{i1} + \beta_2 x_{i2} + \beta_3 x_{i3} + \beta_4 x_{i4} \quad (3)$$

The estimation of parameters was done by both IWLS and GA. Both algorithms were evaluated in Matlab platform. In the evaluation of GA, the selection, crossover and mutation functions are stochastic uniform, Scattered and Gaussian respectively.

The classification results of the test data from ML, ANN and GLM were given in Tables 3, 4, 5 and 6. In these tables, the abbreviations of PA, UA and OA mean producer accuracy, user accuracy and overall accuracy. In addition to overall accuracy, the test of hypotheses based on Z-test was done with respect to the Kappa and Z statistics values (Congalton & Green 1999). The significance level was considered as 0.05. In the first test, Kappa statistics was individually tested for evaluating the difference of the algorithmic classification from a random one. The second test was done to see how well the classification methods differ from each other. The Kappa and Z statistics values were given in Table 7 and 8, respectively.

Test 1:
$H_0$:$Kappa_1=0$ , $H_1$:$Kappa_1 \neq 0$ and the rejection region R: $|Z| \geq z_{\alpha/2}$
For $\alpha$ significance level of 0.05, R: $Z \geq 1.96$ or $Z \leq -1.96$
The test statistics, $Z = Kappa_1 / \sqrt{Var(Kappa_1)}$

Test 2:
$H_0$:$(Kappa_1-Kappa_2)=0$ and $H_1$:$(Kappa_1-Kappa_2) \neq 0$ and the rejection region R: $|Z| \geq z_{\alpha/2}$
For $\alpha$ significance level of 0.05, R: $Z \geq 1.96$ or $Z \leq -1.96$
The test statistics, $Z = (| Kappa_1 - Kappa_2 |) / \sqrt{Var(Kappa_1) + Var(Kappa_2)}$

Table 3. ML classification.

| | Oil | Alike | UA |
|---|---|---|---|
| Oil | 2714 | 733 | 78.7 |
| Alike | 2286 | 4267 | 65.1 |
| PA | 54.3 | 85.3 | **OA=69.8** |

Table 4. ANN classification.

| | Oil | Alike | UA |
|---|---|---|---|
| Oil | 3311 | 521 | 86.4 |
| Alike | 1689 | 4479 | 72.6 |
| PA | 66.2 | 89.6 | **OA=77.9** |

Table 5. GLM classification through IWLS.

| | Oil | Alike | UA |
|---|---|---|---|
| Oil | 2918 | 1085 | 72.9 |
| Alike | 2082 | 3915 | 62.3 |
| PA | 58.4 | 78.3 | **OA=68.3** |

Table 6. GLM classification through GA.

| | Oil | Alike | UA |
|---|---|---|---|
| Oil | 2535 | 587 | 81.2 |
| Alike | 2465 | 4413 | 64.2 |
| PA | 50.7 | 88.3 | **OA=69.5** |

Table 7. Kappa and Z statistics.

| | Kappa(%) | Z-statistic |
|---|---|---|
| ML | 39.62 | 45.40 |
| ANN | 55.80 | 69.16 |
| GLM (IWLS) | 36.70 | 40.21 |
| GLM (GA) | 38.96 | 45.64 |

Table 8. Pair-wise Z statistics.

| | ML | ANN | GLM (IWLS) | GLM (GA) |
|---|---|---|---|---|
| ML | - | 13.61 | 2.35 | 0.54 |
| ANN | 13.61 | - | 15.72 | 14.34 |
| GLM (IWLS) | 2.35 | 15.72 | - | 1.84 |
| GLM (GA) | 0.54 | 14.34 | 1.84 | - |

For a selected oil spilled region excluding the test patterns, totally 194376 dark patterns of textural features were extracted (Fig.2). The classification images of ANN MLP, ML and GLM GA were given in Figure 2. The cross-coherence matrixes of these classifiers through these patterns were given in Table 9 and 10.

Table 9.Cross coherence matrix for oil class.

| | MLP | ML | GLM GA |
|---|---|---|---|
| MLP | 193470 | 192277 | 181991 |
| ML | 192277 | 193024 | 182182 |
| GLM GA | 181991 | 182182 | 182251 |

Table 10.Cross coherence matrix for alike class.

| | MLP | ML | GLM GA |
|---|---|---|---|
| MLP | 906 | 159 | 646 |
| ML | 159 | 1352 | 1283 |
| GLM GA | 646 | 1283 | 12125 |

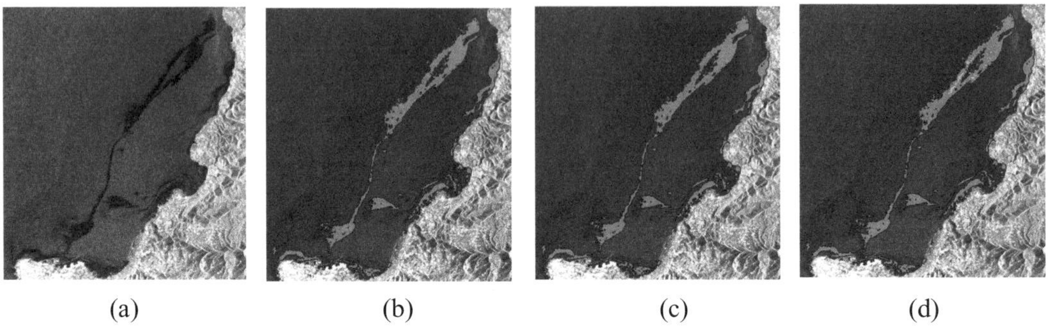


Fig 2. Classification of dark areas (a) Raw image; (b)ANN MLP (c) ML (d) GLM GA. (Red and green colors are showing oil and alike, respectively).

The classifiers are supposed to produce only oil class for these patterns. So, the table 10 can be somewhat considered as error. As GLM has lower performance, it gives low coherence with both MLP and ML for oil class. But for alike class, GLM has low coherence with MLP and high coherence with ML. The accordance between GLM an MLP for alike class is higher than ML.

## 5. CONCLUSIONS

Remote sensing and satellite data are effective tools for man-made hazards like illegal oil spill discharges that need a synoptic view when there is an emergency status and where on site surveying is not possible. Another very important aspect of remote sensing is the production of the near-real time data by operational satellite-based systems so that authorities can quickly and directly use it. One of the most critical data to be supplied is the position of the possible oil spill and its probability level (i.e., low, medium or high). Since this information will be used for the response crews in a short enough time following an oil spill incident, the accuracy of the detection has an importance.

In this study, ML, ANN, MLP and GLM classifiers were compared for capability of oil spill detection. As shown in the test 1, the classification matrixes of all classifiers are better than random, i.e. rejection of null hypothesis with respect to the significance level of 5%. ANN MLP is found far and away the winner of this competition. The performance of GLM through GA is comparable to the ML. This situation is also proved statistically by the second test, i.e. null hypothesis can't be rejected and substantiated. But, although overall accuracy of GLM GA is superior to the accuracy of GLM IWLS, statistical tests in the second group show that there is not adequate evidence to substantiate the alternative hypothesis between GLM GA and GLM IWLS, i.e. there is statistically no difference between GA optimization and IWLS parameter estimation with respect to the significance level of 5%.

Despite being a new method as a classifier, these results should encourage to use GLM for oil spill detection. Since GA intelligent optimization depends on selection, crossover and mutation functions, the effectiveness of GA will be deeply examined with different compositions in future.

## REFERENCES

Assilzadeh, H., & Mansor, S.B. 2001. Early warning system for oil spill using SAR images. *Proceedins of ACRS 2001—22nd Asian Conference on Remote Sensing,* Singapore, November 5–9: 460–465.

Bava J., Tristan, O., & Yasnikouski J. 2002. Earth observation applications through systems and instruments of high performance computer. *ASI/CONEA Training Course*, September-December, Matera, Italy.

Bhanu, B., Lee, S. & Ming, J.C. 1991. Self-optimizing image segmentation system using a Genetic Algorithm. *International Conference on Genetic Algorithms,* San Diego, CA, USA: 362-369.

Congalton, R.G. & Green, K. 1999. *Assessing the accuracy of remotely sensed data: Principles and practices.* Lewis Publishers.

Dobson, A. J. 2002. *An introduction to generalized linear models.* Chapman & Hall.

Fiscella, B., Giancaspro, A., Nirchio, F., Pavese, P. & Trivero, P. 2000. Oil spill detection using marine SAR images, *International Journal of Remote Sensing* 21(18) 3561–3566.

Frate, F.D., Petrocchi, A., Lichtenegger, J. & Calabresi, G. 2000. Neural networks for oil spill detection using ERS-SAR data, *IEEE Transactions on Geoscience and Remote Sensing* 38(5) 2282–2287.

Günel, İ. 2004. Mediterranean Sea will die. Radikal newspaper, *http://www.radikal.com.tr/haber.php?haberno=111649* (in Turkish, last accessed: 30.10.2006).

Grefenstette, J.J. & Fitzpatrick, J.M. 1985. Genetic search with approximate function evaluation. *In Proceedings of International Conference on Genetic Algorithms and their Applications*: 112–120.

Harahsheh, H., Essa, S., Shiobarac, M., Nishidaid, T. & Onumad, T. 2004. Operational satellite monitoring and detection for oil spill in offshore of United Arab Emirates, *http://www.isprs.org/istanbul2004/comm7/papers/130.pdf* (last accessed: 30.10.2006).

Hardin, J.W. & Hilbe, J.M. 2007. *Generalized linear models and extensions.* Stata Press.

Hill, A., & Taylor, C.J. 1992. Model-based image interpretation using genetic algorithm, *Image Vision Comput.* 10 295–300.

Holland, J.H. 1975. *Adaptation in natural and artificial systems.* MIT Press, Cambridge, MA.

Ji, Q. & Zhang, Y. 2001. Camera calibration with genetic algorithms, *IEEE Transactions on Systems, Man and Cybernetics* 31(2) 120-130.

Keramitsoglou, I., Cartalis, C. & Kiranoudis, C. 2006. Automatic identification of oil spills on satellite images, *Environ. Model. Softw.* 21(5) 640–652.

Khuri, A.I., Mukherjee, B., Sinha, B.K. & Ghosh, M. 2006. Design issues for generalized linear models: A review, *Statistical Science* 21(3) 376–399.

Lindsay, J.K. 1997. *Applying generalized linear models.* Springer.

Maar M., Burchard H. & Bolding, K. 2003. Simulating the fate of illegal oil discharges in the marine envronment– Oceanides deliverable report #7, *http://pta.jrc.cec.eu.int/show.gx?Object.object_id=SEC-----0000000000025AD9* , Bolding & Burchard Hydrodynamics (last accessed: 30.10.2006).

Matlab 2008. *Statistics toolbox user guide.* The Mathworks Inc.

Ozkan, C. & Sunar, F. 2007. Comparisons of different semi-automated techniques for oil-spill detection: A case study in Lebanon. *27th EARSeL Symposium*, Italy.

Pham, D.T. & Karaboga, D. 2000. *Intelligent optimization techniques.* Springer.

Ramalho, G.L.B. & Medeiros, F.N.S. 2006. Using boosting to improve oil spill detection in SAR images. *The 18th International Conference on Pattern Recognition (ICPR'06).*

Reidmiller, M. & Braun, H. 1993. A direct adaptive method for faster backpropogation learning: The rprop algorithm. *Proceedings of the IEEE Int. Conf. On Neural Networks,* San Francisco, CA: 586–591.

Solberg, A.H.S., Storvik, G., Solberg, R. & Volden, E. 1999. Automatic detection of oil spills in ERS SAR images, *IEEE Transactions on Geoscience and Remote Sensing* 37(4) 1916–1924.

Solberg, A.H.S., Brekke, C. & Husøy, P.O. 2007. Oil spill detection in radarsat and envisat sar images, *IEEE Transactions on Geoscience and Remote Sensing* 45(3) 746-755.

Sunar, F., Akkartal, A., Göral, B. & Uça Avcı, Z.D. 2007. The Threat of the oil pollution incident occurred in Lebanon to the Northern Cyprus Coasts and the importance of operational satellite monitoring system. *Int. Conference on Environment: Survival and Sustainability*, 19-24 Feb. Nicosia, Northern, Cyprus.

Topouzelis, K., Karathanassi, V., Pavlakis, P. & Rokos, D. 2005. Oil spill detection using rbf neural networks and sar data. *Proceedings on ISPRS, July: 724–729.*

Tsang, P.W.M. 1997. A genetic algorithm for affine invariant recognition of object shapes from broken boundaries, *Pattern Recognition Letters* 18 631-639.

*Remote Sensing for a Changing Europe*
*D. Maktav (Ed.)*
*IOS Press, 2009*

*doi:10.3233/978-1-58603-986-8-188*

# National and regional scale InSAR DEMs and applications

Sowmya Gopal
*Intermap Technologies GmbH, Heimeranstrasse 35, Munich, Germany 80339*

Bryan Mercer
*Intermap Technologies Corp.1200, 555 - 4th Ave. S.W.Calgary, AB Canada, T2P 3E7*

Keywords: InSAR, DEM, DSM, DTM

ABSTRACT: The purpose of this paper is to provide an update on a program, referred to as NEXTMap® to create a single uniform DEM (Digital Elevation Model) for Western Europe and similarly for the USA using the airborne InSAR (Interferometric Synthetic Aperture Radar) technology. The program is well advanced and completion of this program is slated for the end of 2009. The database is being populated country by country and is available under commercial license to the user community. The characteristics of this DEM are intermediate in detail between those provided over wide regions by satellite technology and the more localized technology of lidar and airborne photogrammetry.

The NEXTMap® program aims at being able to provide DSM (Digital Surface Model), DTM (Digital Terrain Model) and ORI (Orthorectified Image) products in a seamless fashion on a national and regional scale. This paper addresses three potential market areas that would benefit from the NEXTMap coverage (as listed below) and compares the products so created with existing solutions, as applicable.

- Intelligent Transportation Systems (ITS): This includes predictive adaptive lighting, bird view navigation systems and use of 3D road vectors to improve fuel efficiency.
- Flood Risk Solutions: Changes in the insurance market has increased the demand for the GIS solution to develop forecast estimates for natural hazards.
- Personal Navigation Devices (PNDs): Information ranging from 3D routes to cell coverage can be made available in PNDs with a very simple user interface, to serve navigation and safety as well as recreational purposes.

## 1 INSAR PRINCIPLES

The interferometric process has been widely discussed in the literature, (e.g. Zebkor and Villsenor (1992), Bamler and Hartl (1998), Rodriguez and Martin (1992)). The geometry relevant to height extraction, 'h', is illustrated in Fig. 1.

If two antennas, separated by baseline 'B', receive the back-scattered signal from the same ground pixel, there will be a path difference '$\delta$' between the two received wave-fronts. The baseline angle '$\theta_b$' is obtainable from the aircraft inertial system, the aircraft height is known from differential GPS and the distance from antenna to pixel is the radar slant range. Then it is simple trigonometry to compute the target height 'h' in terms of these quantities as shown in equations 1–3.

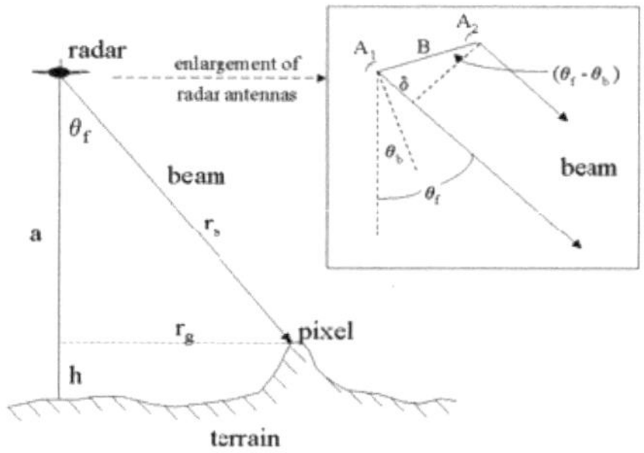


Figure 1. Schematic of Airborne InSAR Geometry.

Table 1. Selected System Parameters of Intermap's Airborne InSAR Systems (STAR Series only)

| System / Parameter | STAR-3i | STAR-4 | STAR-5 | STAR-6 |
|---|---|---|---|---|
| Operational year (Initial) | 2002 (1996) | 2004 | 2007 | 2007 |
| Platform | Lear 36 | King Air 200T | King Air 200T | Lear 36 |
| Typical speed (km/hr) | 720 | 430 | 430 | 720 |
| Wavelength / Polarization | 3 cm / HH | 3 cm / HH | 3 cm / HH | 3 cm / HH |
| Bandwidth (MHz) | 135 | 135/270 | 135/270 | 135 |
| Baseline length (m) | 0.93 | 0.96 | 0.93 | 0.94 |
| Look direction | Left or right | Left or right | Left or right | Left or right |
| Platform altitude (km) | 6 – 12 | 4 – 8.5 | 4 – 7.5 | 6 – 12 |
| Swath width* | 8 – 15 | 6 – 11 | 8 – 15 | 6 – 11 |

$$\sin(\theta_f - \theta_b) = \delta / B \quad (1)$$
$$\delta/\lambda = \varphi / (2 * \pi) + n \quad (2)$$
$$H = H - r_s \cos(\theta_f) \quad (3)$$

The path difference '$\delta$' is measured indirectly from the phase difference '$\varphi$' between the received wave fronts (eqn. 2).Because the phase difference '$\varphi$' can only be measured between 0 and $2\pi$ (modulo $2\pi$), there is an absolute phase ambiguity ('n' wavelengths) which is normally resolved with the aid of relatively coarse ground control. A "phase unwrapping" technique completes the solution. Thus the extraction of elevation is performed on the "unwrapped" phase. Often the InSAR is operated in a so-called ping-pong mode which effectively doubles the value of the geometric baseline B. These equations become the basis for sensitivity and error analysis (e.g. Rodriguez and Martin, 1996). For 'single-pass' InSAR airborne systems as described in this work, the signals are received almost simultaneously so that errors induced by temporal-decorrelation are not a factor as is the case for satellite systems such as ERS and Radarsat which operate in a 'repeat-pass' mode. Provided the baseline length, position (from DGPS) and attitude (from coupled GPS/inertial) are adequately controlled and/or measured, the dominant noise-like error source arising out of these sensitivity equations is 'phase noise $\sigma_\varphi$' so that the signal-to-noise ratio, a function of flying altitude among other system-related factors, becomes a means of (partly) controlling height error specifications. That is, other parameters being fixed, the height noise will increase as a function of flying altitude. For example, DEMs created from the STAR-3i system (Table 1), when operated at approximately 9 km altitude, have a height-noise level of about 0.5 m (1 sigma, 5 m sample spacing) at the far edge of the swath. Systematic errors, with reference to STAR-3i DEMs, are usually slowly varying and arise from a variety of sources but are limited through calibration, operational and processing procedures.

Figure 2. Clockwise from upper left: STAR-3i, STAR-4, STAR-5 and STAR-6.

Table 2. Selected Intermap Core Product specifications for InSAR DSMs and DTMs

| Product Type | ORI/CORI | | DSM | | DTM | |
|---|---|---|---|---|---|---|
| | Pixel size | RMSE horizontal | Spacing | RMSE vertical* | Spacing | RMSE vertical* |
| I | 1.25 m | 2.0 | 5.0 m | 0.5 | 5 m | 0.5 |
| II | 1.25 m | 2.0 | 5.0 m | 1 | 5 m | 1 |
| III | 1.25 m | 2.0 | 5.0 m | 3 | – | – |

*RMSE vertical accuracy is with respect to terrain that is moderately sloped, bare (DSM) and unobstructed. DTM specifications apply to areas for which the forest or other above ground cover is 'patchy' to a maximum scale of about 100 m. Details of these specifications can be found at www.intermap.com.

The schematic idealization of Fig. 1 is replaced of course by many factors in the practical implementation of InSAR. For example a complex image containing phase and magnitude information is created from the signal received at each antenna. Subsequent operations on the complex images allow three ortho-rectified products to be derived: DEM, Magnitude and Correlation. The DEM, as noted earlier, is usually referred to as a DSM in recognition that the received signal relates to the scattering surface which may be the terrain or could be an object upon the terrain, natural or otherwise. The magnitude is often referred to simply as an Ortho-Rectified Image (ORI). The ORI is fused with multi spectral imagery to create a Color Ortho-rectified Image (CORI). In relatively open urban or forest situations, it is possible to create a DTM (Digital Terrain Model) from the DSM (Wang, et. al., 2001) and is offered as a core product along with the DSM and ORI (Table 2).

## 2 CURRENT INTERMAP AIRBORNE INSAR SYSTEMS

Intermap has developed five airborne InSAR systems and currently operates four of them. Current specifications of the four STAR systems used in the NEXTMap® program are summarized in Table 1 and the platforms are shown in Fig. 2.

In principle, the acquisition rate for a single system could reach about 7,000 sq. km/hr for the Lear Jets. In practice, however, the effective acquisition rate is considerably reduced by turning times at end-of-line, side lap (terrain-dependent, so is the swath width*) and the requirement for partial re-flights due to factors which would otherwise drive the data outside specification.

## 3 PRODUCT SPECIFICATIONS

The core products available from Intermap's online store include ORI, CORI, DSM and the bare earth DTM. X-band images are available at 1.25 m resolution with similar horizontal accuracy. DSM and DTM are posted at 5 m spacing. The elevation products are available in three standard vertical accuracy specifications as illustrated in Table 2. It is worth noting that all four of the STAR family of sensors is able to achieve these product specifications despite the nuance of individual system design or platform specifics. Apart from these CORE specifications, other accuracies and image/DEM resolution can be supported to meet specific requirements. Optical/radar merged products are now also becoming available. Varying flying altitudes and operating modes enable different accuracy specifications to be achieved which may be reflected in cost.

## 4 OPERATIONAL FLOW

The operational flow consists of four major stages:

(1) Planning and acquisition
(2) Interferometric processing
(3) Editing and finishing
(4) Independent Quality Control (IQC)

Following the final acceptance from IQC, the data sets are delivered to the data base repository (the 'Intermap-Store'). The operational concept has evolved to accommodate the requirements imposed by the current NEXTMap® goals as well as custom projects. The NEXTMap® Europe and USA objectives alone require the data acquisition for an area incorporating more than 10 million sq. km, to be completed by the end of 2008. All aspects of production are managed with rigorous QC checks throughout and within the framework of the companies' ISO9000 certification.

**Planning and Acquisition:** Typically blocks of 1200 km × 400 km in size are planned. Within these blocks, parallel flight lines are planned according to the terrain. Sufficient overlap is incorporated into the plan to ensure there are no gaps in coverage between imaged swaths. Thirteen tie lines per block are flown, orthogonal to the main data passes. The tie lines are required to assist in the removal of systematic errors during the processing stage. Radar reflectors are postioned on the ground at the ends of the tie lines prior to imaging and surveyed in precisely. These ultimately fix the DEM and ORI to the reference ellipsoid.

Navigation and certain other data are processed and analysed after each mission in order to determine whether the quality limits, for the particular data type specified, have been satisfied. Data must be re-acquired if it fails to meet quality criteria. Normally this quality check is performed before the flight and ground crew demobilize.

**Interferometric Processing:** The production process is very complex as it accomodates multiple platforms, high data volumes, differing product specifications, dynamic schedule priorities and demand for high throughput. This part of the process, now performed in a single processing centre, has been automated to a considerable extent with internally developed software and procedures. The system is PC-based and much of the processing power comes from the 64-bit blade hardware. In particular, the processing architecture has been developed to be scalable in order to accommodate the increasing data volume as acquisition efficiency and platform numbers have increased. The process ingests the raw signal data ('phase history'), ancillary data and navigation data. Following ingestion, the system performs the interferometric processing to ultimately create strips of image and elevation data. Systematic errors are reduced through an internally-developed adjustment process using the controlled orthogonal tie lines that run across the data block. Output products in the form of 7.5'×7.5' tiles are created for editing.

**Editing and Finishing:** A digital mapping system referred to as Interferometric Editing System (IES) was developed, incorporating a feature rich set of editing tools within a stereo-viewing environment. The input data tiles (images and DSMs) are ortho-rectified so false stereo is created to enable the image to be viewed and manipulated in three dimensions. The operator uses the tool set to address hydrological features (e.g. flatten water bodies, ensure rivers run down hill), edit transportation features (roads, railways, airports, etc) and radar artifacts. A set of editing rules determines a consistent application of the IES tools across different geomorphological units and across different editors. Editing is the most labor intensive component of the operational activity and incorporates many quality control steps within the process. At the end of the process, 7.5'×7.5' tiles of DSM, DTM and ORI are delivered to IQC.

**IQC (Independent Quality Control):** IQC is an autonomous group, separated from the production team. The mandate of the IQC group is to review the IES output tiles for editing completeness and accuracy. The accuracy checks are usually with respect to publicly-available GCP data such as the NGS (National Geodetic Survey) network in the USA. Tiles judged incomplete at the 95% level, or whose accuracy checks are outside specification are returned to the production side for re-processing as required. The IQC approved tiles are forwarded to the Intermap-Store, for ingestion into the database. Subsequent to this stage, the data are available to the public under license. A global interface tool allows regions of interest to be displayed for availability and detail, including thumbnail presentations.

## 5 NATIONAL MAPPING PROGRAMS: NEXTMAP®

NEXTMap® is a registered trademark to describe Intermap's InSAR based national and regional mapping programs. Specifically the concept is to make DSM, DTM, ORI and CORI products generally available in a seamless fashion over national and trans-national regions where multiple applications and markets may benefit.

Forecast dates for product availability of NEXTMap® USA and of NEXTMap® Europe areas are shown in Fig. 3. Two factors will likely impact these forecasts: (1) the time it takes to obtain overflight permits vary from country to country, and (2) priorities are subject to market forces.

At the time of this writing, acquistion has been completed for NEXTMap® Europe and processing is well underway while the NEXTMap® USA acquisition is 65% complete. The data acquisition phase for Germany was completed in a 3-month period of 2006 and the completed products are available in Intermap-Store, since November 2007.

## 6 APPLICATIONS

### Intelligent Transportation Systems

Intermap is building a countrywise 3D road vector database by adding the height extracted from the DSM to the 2D road vectors. Rapid growth of the Intelligent Transportation Systems (ITS) market has been triggered by nationwide coverage. Prototype results from NEXTMap® Germany data meet the auto industry accuracy standards for 3D road vectors in North America and Europe.

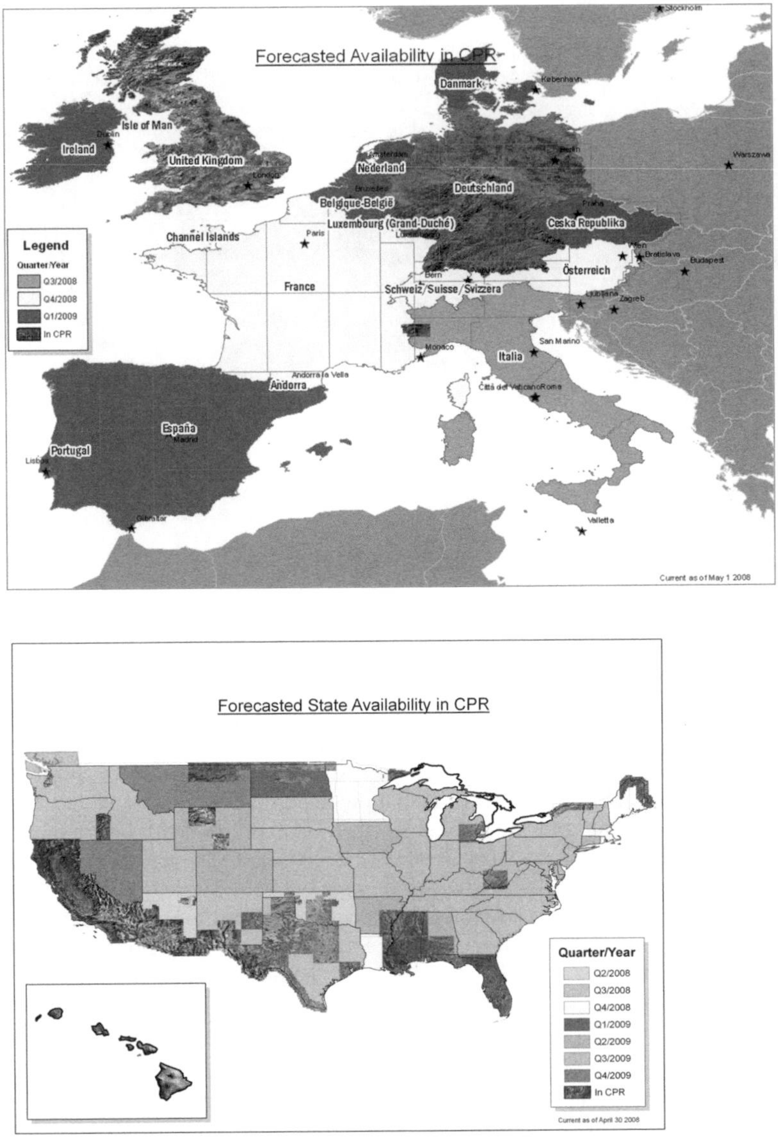


Figure 3. Forecast Dates* for Product Availability of NEXTMap® Europe and NEXTMap® USA Areas.
* This forecast may change subject to operational and other priorities.

The 3D road vectors can not only improve the accuracy of existing routing data but also provide accurate information about the 3D characteristics of the road ahead and hence serve as a 'sensor' in driver assistance systems – headlight steering, lane departure warnings, curve notification, throttle management, to mention a few. An example of headlight steering and 3D road vectors overlaying image and terrain is shown in the bottom left and right parts of Fig. 4, respectively.

## Flood Risk Solutions

Figure 5 illustrates the importance of accurate and detailed elevation data for modelling applications such as a flood induced by a storm surge. In this example, a Katrina like event is applied to Miami.

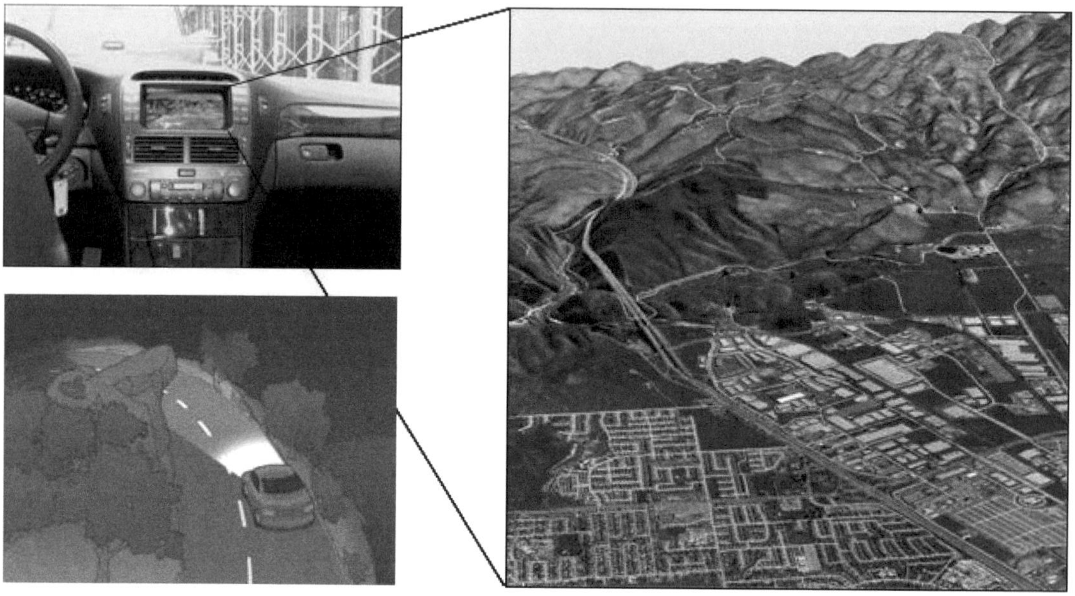

Figure 4. Predictive Adaptive Lighting (bottom left) and 3D road vectors (right).

Note the differences between simulations based on NEXTMap® data compared to the prediction prepared by FEMA based on the elevation data from another source. One area (parts of it indicated by red circles in the Figure) of the FEMA prediction overestimates the inundation by 50% while it also misses an entire area that would be inundated through the channel at the bottom of the image (yellow circle in the Figure), resulting in an unpleasant surprise for the insurance company.

Risk management solutions offer gecoding and 2D as well as 3D visualization (with house based coordinates), risk data assessment tool (for natural hazards like flood, windstorm, hail, earthquake etc.), portfolio analysis (for eg, report on what percentage of the areas are in a 50 year, 100 year or 200 year risk zones), exposure modeling (influence of exact geo location on optimum calculation of reserves to meet legal and strategic requirements) and maintenance services (for e.g., regular updates of the address database) and hence can prove to be viable candidates in the insurance market.

**Personal Navigation Devices**

Using the NEXTMap® USA data products enables the PNDs to be far more robust; providing uniformly reliable map information where data does not exist, while significantly improving the accuracy of the existing data. Users have access to clearly identified trails, paths and roads, overlaid on a 3D model. They can virtually elevate from their location to an elevation they choose. They can even navigate across the touch screen to view in 3D the surrounding terrain, public and private land boundaries, and cell-phone coverage grids, as shown in Fig. 6. While the left part of the figure depicts the GPS tracking mode, the right part shows the cell phone coverage with the gold color overlay.

## CONCLUSIONS

The NEXTMap® program, based upon airborne InSAR, is in the process of creating a 3-dimensional, homogeneous, seamless database comprising DSM, DTM and ORI products, for eighteen countries of Western Europe (in addition to Britain that is already available) and for the USA (excluding Alaska). The DSM is specified at 1 m RMSE vertical accuracy for 5 m sample spacing, while the

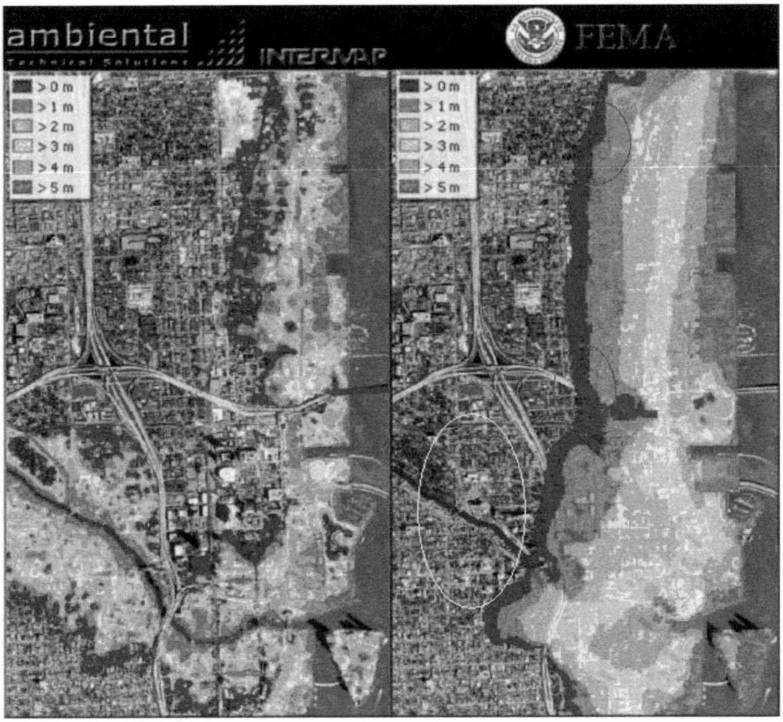


Figure 5. Forecast of Katrina like event in Miami.

Figure 6. Examples of PND features: 3D view of route and cell coverage.

ORI is specified at 1.25 m resolution with better than 2 m RMSE horizontal accuracy. For many applications the combination of detail provided, national and trans-national availability and shared pricing through license arrangements should produce an attractive user opportunity. The creation of the datasets for both areas is well underway and scheduled for completion in 2009. With respect to many market applications such as ITS, flood risk solutions, PNDs and visualization these products should occupy a solid niche, relative to alternative technologies.

## ACKNOWLEDGEMENTS

The authors would like to acknowledge their colleagues at Intermap who have made the NEXTMap® program a reality.

## REFERENCES

Bamler, R., and P. Hartl (1998), *Synthetic aperture radar interferometry*, Inv. Probl., Vol. 14, pp. R1–R54.

Mercer, B. (2004) DEMs Created from Airborne InSAR – An Update, *Proceedings of the ISPRS XXth Congress.*

Rodriguez, E., and J.M. Martin (1992). Theory and design of interferometric synthetic aperture radars. *IEEE Proceedings-F*, Vol. 139, No. 2, pp. 147–159.

Schwäbisch, M. and J. Moreira (1999). The high resolution airborne interferometric SAR AeS-1, *Proceedings of the Fourth International Airborne Remote Sensing Conference and Exhibition/21st Canadian Symposium on Remote Sensing, Ottawa, Canada.*

Tennant, J.K., and T. Coyne (1999). STAR-3i interferometric synthetic aperture radar (INSAR): some lessons learned on the road to commercialization. In: *Proceedings of the 4th International Airborne Remote Sensing Conference and Exhibition/21st Canadian Symposium on Remote Sensing*, Ottawa, Canada. 21–24.

Wang, Y., B. Mercer, V. Tao, J. Sharma, and S. Crawford (2001). Automatic Generation of Bald Earth Elevation Models from Digital Surface Models Created Using Airborne InSAR, *Proceedings of ASPRS 2001.*

Zebker, Howard A., and J. Villasenor (1992). Decorrelation in Interferometric Radar Echoes, *IEEE Transactions on Geoscience and Remote Sensing*, Vol. 30: Number 5, pp. 950–959.

*Remote Sensing for a Changing Europe*
*D. Maktav (Ed.)*
*IOS Press, 2009*

*doi:10.3233/978-1-58603-986-8-197*



# Rule-based classification of SPOT imagery using object-oriented approach for detailed land cover mapping

St. Lewinski & Z. Bochenek
*Institute of Geodesy and Cartography, Warsaw, Poland*
*stanislaw.lewinski@igik.edu.pl; zbigniew.bochenek@igik.edu.pl*

Keywords: object-oriented classification, SPOT, land cover, land use

ABSTRACT: Preparation of land cover / land use maps for large areas, based on automatic classification of high-resolution satellite data is the objective of many application programmes, e.g. GSE Land Monitoring Services. The crucial point for this kind of activity is to apply optimal classification approach, which will ensure high class recognition accuracy and classification repeatability. Among different approaches object-oriented approach seems to give the best results, as it allows to use various spectral and non-spectral features in the classification process and enables to have more control of the final map accuracy.
The presented paper demonstrates the method, developed within object-oriented approach, called rule-based classification. In this method SPOT satellite image with 20-meter resolution was analyzed in eCognition environment, assuming that Standard Nearest Neighbour classifier, based on selected sample objects, will not be applied. Instead of it, contents of SPOT satellite image was characterized by various spectral / texture parameters describing directly particular land cover / land use classes, as well as by pre-defined functions, determined on the basis of graphical analysis of feature space constructed for particular terrain objects.
The classification process begins with recognition of water class, whose objects were delineated using multi-resolution segmentation. For the remaining land cover classes new segmentation is prepared. In turn, general forest class and urban class are discriminated; in the next stage both classes are divided into more specified categories. Consequently, broad agricultural and grassland classes are determined, being next divided into more detailed information classes. In total 13 land cover / land use categories were discriminated in the presented work. Final classification map has been prepared using aggregation procedure, in order to obtain map resolution fulfilling 4ha size of Minimum Mapping Unit. Overall accuracy of all classes checked through verification process reached 89.1% with Kappa statistic 0.87.

## 1 INTRODUCTION

Concept of division of image contents into so-called objects is not new (Gonzalez & Wintz, 1977). However, only 8 years ago, when new eCognition software appeared, rapid development of object-oriented methods of image classification can be observed. New possibilities of object recognition are clearly visible, while observing improvement of tools offered to users in subsequent software versions. In many scientific centers research works aimed at preparation of optimal object-oriented methods for classification of satellite data were undertaken. These works were concentrated on two principal aspects: multi-resolution image segmentation, which would allow to delineate homogeneous land cover classes, and on comprehensive class characteristics, taking into account both spectral and textural features, as well as relation between objects. Authors of the works used multi-source satellite data (Whiteside, 2005), as well as high-resolution images (QuickBird, Ikonos), which enable to apply texture and shape features more effectively (Wei et al., 2005, Kressler et al., 2005). Standard Nearest Neighbour classifier, which is based on using training areas for particular land cover classes, was the most commonly applied method of classification (Yuan & Bauer, 2006, Hajek, 2005). Recently, taking advantage of development of eCognition software, research works on new object-oriented approach were undertaken. This approach is based on using parametric spectral / texture values for the determined classes and hierarchical classification. which applies decision tree method. Mostly these works were concentrated on the selected urban or natural land cover categories (Ehlers et al., 2006, Lucas et al., 2007). The presented work is a new proposal for preparing land cover map based on satellite data with 10 –30 meter resolution, using parametric approach for object-oriented classification.

## 2 STUDY AREA AND SATELLITE DATA

Methods of object-oriented classification were studied using multispectral SPOT 4 satellite image characterized by 20 meter spatial resolution. The image was collected on July 16, 2007. It covers area in central-north Poland, comprising Kujawy region.

Arable land is predominant type of land cover within this region; large areas of grasslands, as well coniferous and deciduous forests also appear here. There is also quite extensive drainage system (canals, lakes, Notec and Vistula rivers), as well as large urban agglomeration (Bydgoszcz).

SPOT satellite image prior to object-oriented classification was pre-processed. This stage included image rectification and atmospheric correction of raw satellite data. While rectifying Nearest Neighbour algorithm was used, whereas for atmospheric correction ATCOR module working within PCI software environment was applied.

## 3 METHODICAL APPROACH

The assumed scheme of classification is presented in figure 1. Hierarchical process starts with recognition of objects representing land cover class “water”. Next, general classes of forest, built-up land, meadows and agricultural land are discriminated. These general land cover categories are divided into subclasses in subsequent stages of classification. Then the objects, which are recognized through relations with previously discriminated classes, are classified. These are the following subcategories: green urban areas, construction sites and deforestations.

Classification flowchart starts, as usually in object-oriented approach, with segmentation. The aim of first segmentation is to create objects representing solely class “water”. This segmentation, based only on SPOT 3 band, is multistage; in the course of subsequent five cycles scale coefficient changes from 5 to 25. The assumed approach enables to delineate objects, which describe adequately shapes of lakes and rivers with various lengths and sizes. Next class “water” is recognized using threshold values in SPOT 3 and SPOT 4 bands, taking into account shape parameters, which help in identification of elongated river beds.

While recognizing class “water” image area represented by all non-classified objects was segmented once again; this segmentation was used in discrimination of the remaining classes. All four SPOT bands were applied in segmentation process; scale parameter was set to 10.

At the next stage on the basis of segmented image general forest class is delineated. (without division into sub-classes). In order to do that, non-typical solution was applied; it is based on defining equation of straight line, which separates forest objects from the remaining objects in two-dimensional space formed by SPOT 3 and SPOT 4 reflectance values. Coefficients of equation were determined on the basis of graphical analysis of scatterplot formed by SPOT 3 and SPOT 4 bands. This approach is not complicated but it requires to select appropriately not large subset of image, which includes forest areas and their surroundings. The example of above mentioned scatterplot with the defined straight line is presented in figure 2; forest areas are represented by points located above the line. In addition, in order to limit number of objects threshold values in SPOT 1, SPOT 3 and SPOT 4 bands were applied. At the next phase forest class was divided into 3 sub-classes: deciduous, coniferous and mixed forests To do that brightness parameter, representing mean values of objects in all four SPOT bands was utilized. Such an approach is similar to that presented in de Kok et al. works (2000) and Lewinski work (2006), where forest sub-classes were discriminated on the basis of analysis of reflectance values in panchromatic band.

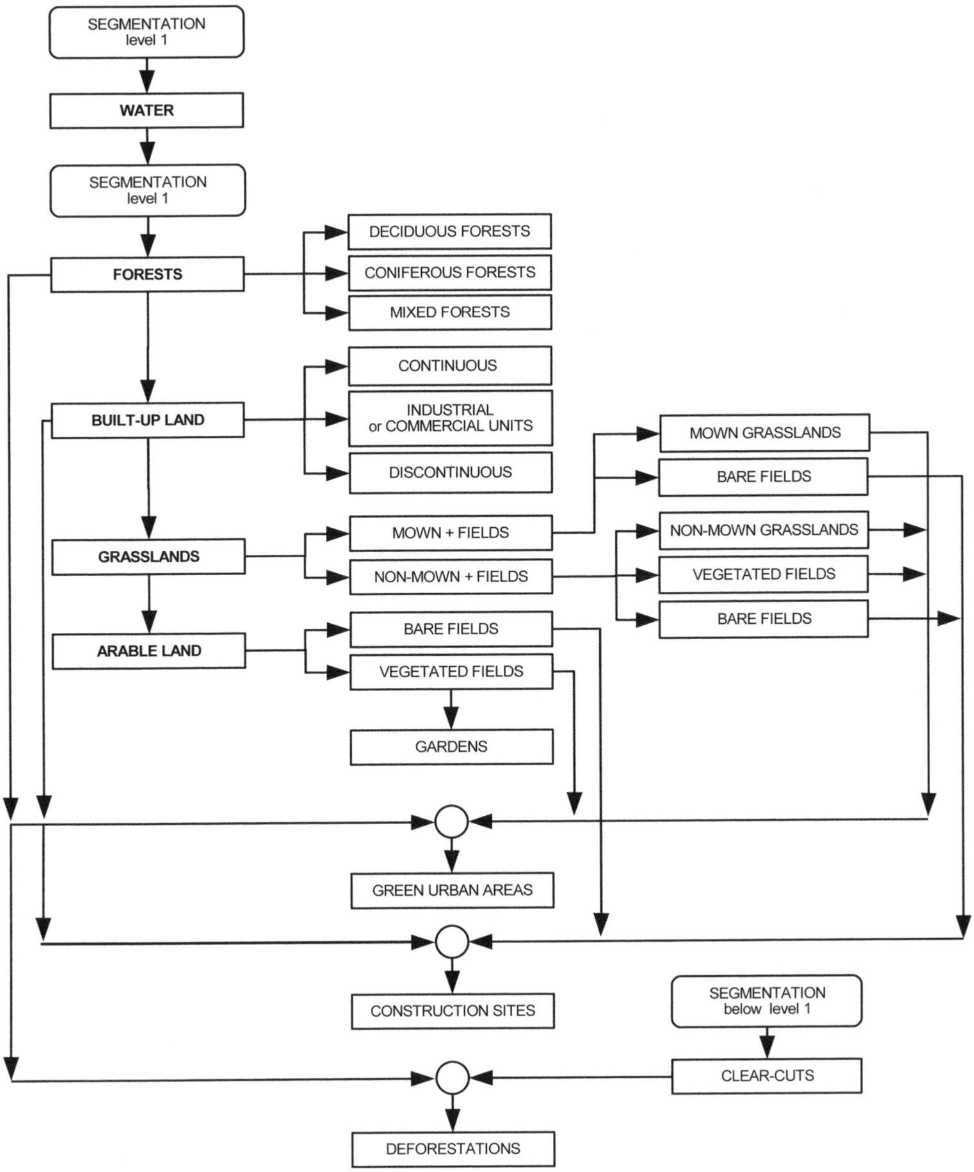


Figure 1. Classification flowchart of SPOT satellite image

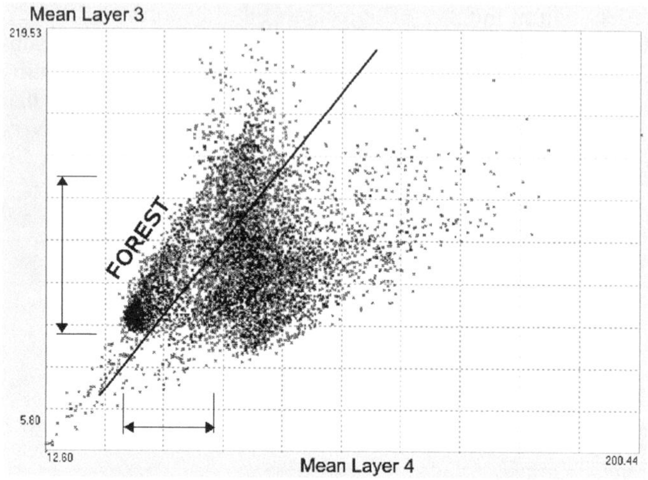


Figure 2. Principle of recognizing forest class with the use of straight line equation

At the next stage of classification non-classified objects are analyzed in order to recognize classes of built-up land . As in case of forests first general class of built-up land is defined and next it is divided into sub-classes. It was found as a result of analyses, that Haralick texture measure GLCM Contrast (Haralick et al., 1973) will be the basic classifier. However, texture values were not calculated on the basis of original spectral bands, but using specially transformed image, namely SPOT 2 band filtered with the use of Laplacian Edge Detector (Type 1) offered by PCI Geomatica software. This approach is similar to that proposed for forest recognition at de Kok et al. work (2005) . The applied transformation enhanced edges of objects representing built-up land; the transformed image is shown in figure 3. In addition, in order to increase accuracy of class recognition the following class parameters were utilized: GLCM Homogeneity texture measure based on SPOT 2 band, ratio SPOT 3 / SPOT 2 as well as threshold values in SPOT2, SPOT3 and SPOT 4 bands. Next, general class of built-up land was divided into 3 sub-classes: continuous built-up land, discontinuous built-up land, industrial or commercial units. The division was done using ZABUD2 criterion, which is based on combination of SPOT 1, SPOT 2 and SPOT 4 bands;
ZABUD2 = 10*(SPOT 2 –2*SPOT 3 + SPOT 4).

After recognizing classes of built-up land grasslands and arable land are classified. In order to do that, as in case of forests, equation of straight line was applied. The line was determined in a feature space formed by SPOT 4 reflectance values and NDVI values. Objects located above this line were attributed to general class "grasslands", while below it to general class "arable land". In turn, grasslands were divided into 2 sub-classes: mown grasslands + fields and non-mown grasslands + fields; SPOT 3 / SPOT 4 ratio was the basis of this division. At next step, from subclass "mown grasslands + fields" final subclasses "mown grasslands" and" bare fields" were discriminated, using threshold values in SPOT 4 band. Second subclass "non-mown grasslands" was delineated applying difference of mean reflectance values in SPOT 2 and SPOT 1 bands, as well as threshold value in SPOT 4 band. The remaining objects existing within subclass "non-mown grasslands + fields" were divided into vegetated fields and bare fields, utilizing threshold value in SPOT 4 band.

Objects classified preliminarily as arable land were attributed to sub-classes "bare fields" and "vegetated fields", applying threshold values in SPOT 1 and SPOT 4 bands.

At this stage of classification all objects existing on satellite image were already classified. Further action was aimed at re-classification on the basis of direct object properties and relations among particular classes.

Class "gardens" was derived from sub-class "vegetated fields". The following parameters were used for its delineation: texture measure GLCM Contrast based on SPOT 1 band, SPOT3 / SPOT 4 ratio, mean reflectance value in SPOT 4 band, standard deviation in SPOT 3 band. Also information on distance from straight line dividing SPOT 4 / NDVI feature space into grasslands and arable land was here utilized Application of quite a large number of parameters was indispensable due to spectral heterogeneity of the class and its diversified texture. In Poland class "gardens" includes clusters of small plots, ca 300 $m^2$ in size, incorporating holiday cottages, garden houses or out-buildings. Vegetation within these plots can be very diversified, mainly composed from fruit trees and shrubs.

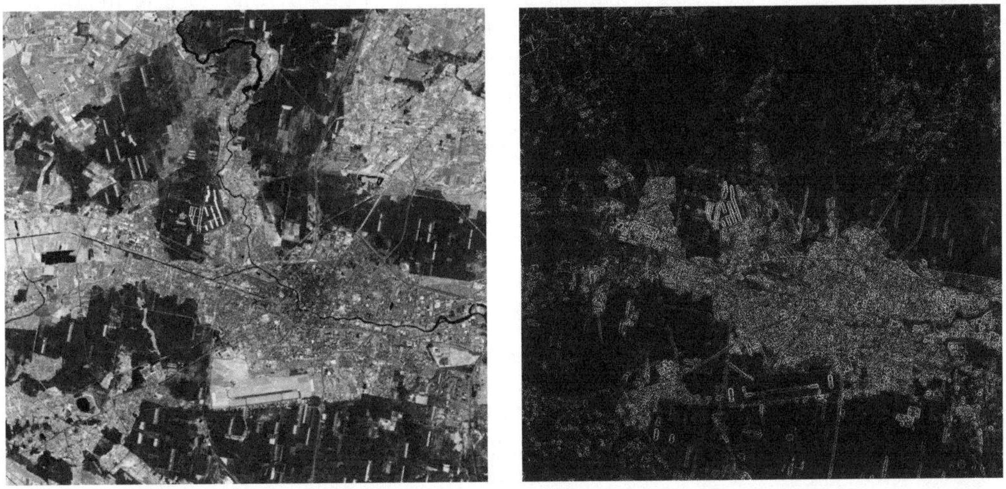

Figure 3. Example of applying Laplacian Edge Detector (Type 1) for filtering SPOT image with built-up land.

"Green urban areas" were the next class defined on the basis of already classified objects. Forests, grasslands and vegetated fields were taken here into analysis. Degree of adjacency to built-up land was the basis classification criterion in this analysis. Similar criterion was applied in case of delineating class "construction sites", which was initially classified as bare fields.

According to classification flowchart deforestations were the last class taken into consideration. Its recognition is quite complex process, which implies application of additional segmentation of satellite image. It is created below the main classification level in such a way, to be able to obtain small objects, which potentially could be clear-cuts. This segmentation was done on the basis of SPOT 4 band, using small scale coefficient. Clear-cuts were identified assuming, that they have higher reflectance values in SPOT 3 and SPOT 4 bands than the surrounding objects. Next re-classification of forest class was performed, using relation with layer where clear-cuts and small objects surrounded with forest were identified

Classes recognized separately, but representing the same (or similar) land cover category, were joined. Class "grasslands" was created from joining mown and non-mown grasslands. Similarly, in case of arable land this class was formed through joining objects representing vegetated and bare fields. In total 13 classes were discriminated; they were presented in table 1.

## 4 FINAL MAP PREPARATION AND DISCUSSION OF THE RESULTS

Getting information about land cover / land use with a pre-determined degree of spatial resolution was one of the assumptions of the presented work. To achieve this goal area of 4 ha was assumed as a Minimum Mapping Unit (MMU); it corresponds to 100 SPOT pixels. Size of MMU was carefully justified. While making visual interpretation 4 ha, i.e. square 200 x 200 m is practically the smallest area, which can be delineated at a scale of 1:50 000; at this scale 4 ha corresponds to 2 x 2 mm. Size of minimum mapping unit is also justified by classification principles, which include information derived from Haralick texture measures. These measures are reliable, when calculations are based on high enough number of pixels. De Kok (2002) found, that objects, which will be analyzed through texture measures, should have at least 30 pixels.

In order to achieve the assumed spatial resolution the classification raster image was generalized. Standard PCI Geomatica functions were used for this purpose. It was assumed, that group of pixels with the same values smaller than minimum mapping unit is incorporated into adjacent, the largest homogeneous group of pixels. Analysis of size of pixel groups was done in multi-stage process, owing to this approach number of possible "non-logical" changes of classification was minimized. Table 1 includes areas of particular land cover classes and their percentage contribution to total classified area.

Table 1. Class areas after generalizing process with the use of 4ha MMU.

| Class No | class name | ha | % |
|---|---|---|---|
| 1 | continuous built-up land | 1 432.3 | 0.40 |
| 2 | discontinuous built-up land | 4 772.8 | 1.32 |
| 3 | industrial or commercial units | 4.6 | 0.001 |
| 4 | construction sites | 71.9 | 0.02 |
| 5 | green urban areas | 561.9 | 0.16 |
| 6 | arable land | 206 097.8 | 57.01 |
| 7 | grasslands | 52 633.7 | 14.56 |
| 8 | gardens | 1 857.3 | 0.51 |
| 9 | coniferous forests | 49 899.1 | 13.80 |
| 10 | deciduous forests | 13 487.8 | 3.73 |
| 11 | mixed forests | 23 548.1 | 6.51 |
| 12 | deforestations | 487.6 | 0.13 |
| 13 | water | 6 674.6 | 1.85 |
| | total | 361 529.5 | 100 |

At the next stage accuracy assessment of the obtained classification was carried out. 1400 points were distributed randomly within the classification image, proportionally to the area of the recognized classes. Results of accuracy assessment in the form of error matrix, including producer's and user's accuracy, are presented in table 2.

Overall accuracy of classification was 89.1 %. Taking into account number of classes and total area covered with classification (over 3600 $km^2$ ) the obtained result can be found as satisfactory. This conclusion is supported by high Kappa coefficient (0.87).

The best results were achieved for class "water". Accuracy of the largest class "arable land" reached over 90 %; similar result was achieved for class "grasslands". Analyzing forest classes, coniferous forests were classified with the highest accuracy, influenced by precision of classifying mixed forests. Some objections one can have for classification of deciduous forests, reaching ca. 80 %. Main misclassifications are due to confusion with two classes: mixed forests and arable land. In first case accuracy is dependent on precise definition of mixed forests, while arable land covered with vegetation is sometimes mixed with deciduous forests due to spectral similarity of both classes.

Quite good results were obtained for three classes of built-up land. However, it must be considered, that after generalization class "industrial or commercial units" is represented only by one object having area over 4 ha. Smaller objects were incorporated to two remaining classes of built-up land.

Table 2. Accuracy assessment of classification of SPOT image.

| Class No | Reference data | | | | | | | | | | | | | Sum |
|---|---|---|---|---|---|---|---|---|---|---|---|---|---|---|
| | **1** | **2** | **3** | **4** | **5** | **6** | **7** | **8** | **9** | **10** | **11** | **12** | **13** | |
| **1** | 37 | 1 | 3 | | | 5 | | | | | | | | 46 |
| **2** | | 46 | 1 | | | 4 | | | | | | | | 51 |
| **3** | | | 46 | | | | | | | | | | | 46 |
| **4** | | 5 | 1 | 31 | 1 | 7 | | | | | | | | 45 |
| **5** | | 7 | | 1 | 38 | | | | | | | | | 46 |
| **6** | | 7 | 1 | | | 498 | 11 | | 2 | 11 | 4 | | 1 | 535 |
| **7** | | | | | | 12 | 145 | | | 1 | 1 | | | 159 |
| **8** | | | | | 1 | 17 | 4 | 17 | | | 1 | | | 40 |
| **9** | | | | | | | | | 156 | 1 | 2 | 1 | | 160 |
| **10** | | | | | | 1 | 2 | | 1 | 56 | 10 | 1 | | 71 |
| **11** | | | | | | | | | 10 | 1 | 80 | | 1 | 92 |
| **12** | | | | | | | 1 | | 3 | 2 | 1 | 33 | | 40 |
| **13** | | | | | | | 1 | | 2 | | 1 | | 65 | 69 |
| Sum | 37 | 66 | 52 | 32 | 40 | 544 | 164 | 17 | 174 | 72 | 100 | 35 | 67 | 1400 |
| Accuracy % | | | | | | | | | | | | | | |
| Producer's | 100 | 69.7 | 88.5 | 96.9 | 95.0 | 91.5 | 88.4 | 100 | 89.7 | 77.8 | 80.0 | 94.3 | 97.0 | |
| User's | 80.4 | 90.2 | 100 | 68.9 | 82.6 | 93.1 | 91.2 | 42.5 | 97.5 | 78.9 | 87.0 | 82.5 | 94.2 | |
| Overall accuracy 89.1%, Kappa coefficient 0.87 | | | | | | | | | | | | | | |

* Full class names are given in table 1

Results of accuracy assessment obtained for discontinuous built-up land are characterized by significant differences, while analyzing producer's and user's accuracy. Classification image should comprise more objects attributed to this class. Good results were achieved for classes "green urban areas" and "construction sites"; the last one was recognized excessively. The same applies to class "deforestations". The lowest accuracy was reached for class "gardens" ; although producer's accuracy is 100 %, user's accuracy reaches only 42.5 %. It means, that too many objects were attributed to this class; it can be found in table 2 that some objects representing class "arable land" were misclassified as "gardens". In case of this class the applied classification approach proved to be not sufficient. It is caused by spectral heterogeneity of class "gardens" and spatial resolution of SPOT image, which is not adequate for recognizing texture features formed by relatively small objects.

## 5 CONCLUSIONS

The presented work demonstrates object-oriented approach, which assumes recognition of land cover classes based solely on spectral / textural / relational parameters, without applying Standard Nearest Neighbour classifier. Using tools of eCognition software (version 5) land cover / land use classes are recognized in subsequent processes, applying principle of decision tree. At first stage five general classes are discriminated: water, forests, built-up land, grasslands and arable land. Next, they are divided into sub-classes and in turn the successive classifications are performed using relations existing between already recognized objects.

Innovation in this approach is to apply, beside threshold values of spectral / textural parameters describing objects, parameters defined on the basis of equations of straight lines. Owing to these lines objects are divided into classes in two-dimensional feature space. Coefficients of equations are determined on the basis of visual analysis of scatterplots.

The applied approach enabled to obtain overall accuracy of classification image of 89.1 %. This result should be considered as satisfactory, taking into account large area of the whole SPOT scene (60 x 60 km), which implies high diversity of land cover forms within particular thematic classes.

The presented approach was verified through its application for two other SPOT satellite scenes. They covered eastern and western part of Poland, different in fragmentation of land cover forms. In each case, in order to achieve high classification accuracy, modification of parameters used for identification of particular classes, was necessary. However, it did not change the principles of the proposed method. In all three cases the achieved classification accuracy was ca. 90 %; it confirms the correctness of the applied approach.

## ACKNOWLEDGEMENT

The study has been funded by Polish Ministry of Science and Higher Education, project N520 016 31/2481.

## REFERENCES

De Kok R., Buck A., Schneider T., Ammer U., 2000. Analysis of image objects from VHR imagery for forest GIS updating in the Bavarian Alps. *Proceedings of the ISPRS, Amsterdam, July 2000, Working Group III/5.*

De Kok R., Koziol K., Wezyk P., 2005. Zastosowanie klasyfikacji obiektowej wysokorozdzielczych obrazow teledetekcyjnych oraz analiz przestrzennych GIS w kartowaniu drzewostanow. *Roczniki Geomatyki, t. 3, z. 4: 99-108.*

Ehlers, M., U. Michel, G. Bohmann and D. Tomowski, 2006. Decision based data fusion techniques for the analysis of settlement areas from multisensor satellite data. *Proceedings of ASPRS 2006 Annual Conference, Reno, Nevada; May 1-5, 2006.*

Gonzalez R. C., Wintz P., 1977. *Digital image processing*. Addison-Wesley Publishing Company.

Hajek, F. 2005. Object-oriented classification of remote sensing data for the identification of tree species composition. *Proceedings of ForestSat 2005 conference, May 31 - June 3, 2005, Boras, Sweden.*

Haralick R. M., Shanmugan K., Dinstein I., 1973. Textural Features for Image Classification. *IEEE Transactions on Systems, Man and Cybernetics, Vol. SMC-3, No.6: 610-621.*

Kressler, F.P., K. Steinnocher and Y.S. Kim, 2005. Enhanced Semi-Automatic Image Classification of High-Resolution Data. *Proceedings of the IGARSS 2005 Symposium. Seoul, Korea. July 25-29, 2005.*

Lewinski St., 2006. Land use classification of ASTER image – Legionowo test site. *Proceedings of the 25th Symposium of the European Association of Remote Sensing Laboratories, Porto, Portugal, 6-9 June 2005: 441-447.*

Lucas R , Rowlands A, Brown A, Keyworth S, Bunting P., 2007. Rule-based classification of multi-temporal satellite imagery for habitat and agricultural land cover mapping. *ISPRS Journal of Photogrammetry and Remote Sensing, Volume 62, Issue 3, August 2007: 165-185*

Wei, W., X. Chen and A. Ma , 2005. Object-oriented Information Extraction and Application in High-resolution Remote Sensing Image. *Proceedings of the IGARSS 2005 Symposium. Seoul, Korea. July 25-29, 2005.*

Whiteside, T., 2005. A multi-scale object-oriented approach to the classification of multi-sensor imagery for mapping land cover in the top end. *Proceedings of NARGIS 2005 – Application in Tropical Spatial Science. 4th - 7th July 2005 Charles Darwin University, Darwin, NT, Australia.*

Yuan, F., 2006. Mapping impervious surface area using high resolution imagery: A comparison of object-based and per pixel classification. *Proceedings of ASPRS 2006 Annual Conference, Reno, Nevada; May 1-5, 2006.*

*Remote Sensing for a Changing Europe*
*D. Maktav (Ed.)*
*IOS Press, 2009*


*doi:10.3233/978-1-58603-986-8-205*

# REDD pilot project in Cameroon – method development and first results

Manuela Hirschmugl & Mathias Schardt
*Institute of Digital Image Processing, JOANNEUM RESEARCH, Graz, Austria*

Thomas Häusler & Sharon Gomez
*GAF AG, Munich, Germany*

Joseph Amagu Armathe
*Ministry of Environment, Yaounde, Cameroon*

Keywords: REDD, tropical forest, deforestation, land cover classification

ABSTRACT: This paper describes a method for monitoring the implementation of the REDD (Reducing Emissions from Deforestation and Degradation) process for a pilot project in Cameroon. A method was developed and tested in a test area in the transition zone between tropical evergreen forest and savannah in Cameroon. The method developed involved a series of processing steps which can be divided into three groups: (1) pre-processing steps including cloud masking, geometric and radiometric adjustment as well as topographic normalization; (2) forest masking including segmentation, classification and manual correction and finally (3) land cover classification including spectral signature analysis and training. The last part showed the most crucial problem encountered in this test phase: due to shifting cultivation, it is difficult to differentiate cropland in the stage, where it is left to bush fallow and natural bush land and secondary forest. The method and results were compared to similar studies already performed in Bolivia and Vanuatu and the differences are discussed. The results gained in the Cameroon test area showed, that 529.786 $km^2$ (equivalent to 3.45% of the forested area) was deforested between 1990 and 2000, further 88.547 $km^2$ (3.44% of the forested area) between 2000 and 2005. Further investigations will include more detailed studies on the issue of degradation and the applicability of the method for the roll-out on the whole country of Cameroon as well as on other countries in the Congo region.

## 1 BACKGROUND & OBJECTIVES

At the UNFCCC and Kyoto Protocol Conference of Parties (COP) meeting in Montreal, 2005, the governments of Papua New Guinea and Costa Rica, supported by Latin American and African countries, submitted a proposal for the consideration of reducing emissions from deforestation and degradation in developing countries (REDD) as a mechanism for the post-Kyoto reporting. Avoiding deforestation is considered to be a contribution to the reduction of green house gas (GHG) emissions. Parties agreed to a two-year process of evaluation of the issue by initiating REDD pilot projects to better understand the REDD process. Under the auspices of the European Space Agency (ESA) GMES Service Element on Forest Monitoring a pilot project was developed for REDD implementation and testing in Bolivia and Cameroon with user consultation and endorsement. The overall aim of the REDD pilot in Cameroon is to establish baseline projections of emissions caused by deforestation (using Earth Observation) combined with regional projections of degradation nested in a wall-to-wall approach. In order to assess deforestation and degradation at a national level a two tier remote sensing analysis to provide forest area maps and forest cover change maps at specific time periods of 1990–2000–2005 using high resolution data for the whole country was proposed.

Several methodological studies for the REDD process are currently underway in tropical rainforest countries. An example of a study that also uses high resolution satellite data is the one being undertaken in the pacific island state of Vanuatu (Herold et al. 2007). In this study, the deforestation between 1990 and 2000 was assessed using Landsat, Spot and Aster data. The approach consists of several pre-processing steps such as co-registration, image mosaicking, water and cloud masking and image normalization. The main procedure is a wall-to-wall deforestation approach on a pixel basis with manual interaction.

The second similar work has been done in Bolivia (Steininger et al. 2001; Killeen et al. 2007) also based on Landsat imagery. The time span was from the mid-1980ies until 2005 with a focus on the Santa Cruz department and thus was not a full country wall-to-wall mapping. Also, the mapping is restricted to the forest zone with a precipitation of more than 1000 mm per year. The procedure was a pixel-based mapping with a common pixel size of 60 m. A classification of different forest types and also other land cover classes was performed. Training and evaluation was based on field trips and aerial surveys. A study dealing with the topic of selective logging, a type of forest degradation has been carried out in Brazil (Souza et al., 2005). The authors used a contextual classification algorithm (CCA) to assess logging and fire-derived canopy damages.

## 2 DATA

For the REDD pilot project in Cameroon data for national coverage from three time periods was obtained. For 1990 and 2000, Landsat data (Landsat 5 and Landsat ETM+ respectively) was purchased. Due to the SLC (Scan Line Corrector) failure (Markham et al. 2004) of Landsat ETM+ in 2003, for 2005 Landsat data is not available without significant data gaps. DMC (Disaster Monitoring Constellation, http://www.dmcii.com/) data was used alternatively, since it provides a similar geometric resolution (32 m compared to 28.5 m from Landsat) and also the Near InfraRed (NIR) spectral band. The advantage of DMC data can be seen in its large swath width (600 km) and thus a reduced number of images was needed to cover the whole area. In order to cover the whole area of Cameroon, 32 Landsat scenes from 1990 and 2000 respectively were necessary. In addition, 14 partly overlapping DMC scenes have been purchased to cover the time slot of 2005.

## 3 METHOD

In order to develop the forest mask a semi-automatic processing chain was developed in this project. It consists of three types of processing steps: pre-processing steps, processing steps related to forest masks and processing steps to derive the land cover parameters for the deforested areas. The order and connection of all steps is depicted in Fig. 1.

### 3.1 *Pre-processing*

The application of optical remote sensing data in tropical areas often faces the problem of high cloud coverage. Large parts of the data sets used in this study were covered by clouds, haze or cloud shadows. From the 1990 Landsat coverage, about 26% of the image was cloud covered. In the 2000 Landsat imagery, it was approximately 16% and the DMC data had the highest cloud cover with 28%. In order to deal with the cloud cover a semi-automatic approach was chosen owing to the patchy appearance of the clouds. In the first step, the clouds were classified by spectral thresholding. In the second step, the cloud mask was treated with a morphological dilation algorithm in order to integrate also the spectrally fuzzy border areas of the clouds. In a third step, the mask was duplicated and geometrically shifted in the direction of the cloud shadows (opposite to the sun position) to cover the cloud shadows. This is necessary, because cloud shadows are not distinguishable from 'normal' land cover types based on their spectral properties. In a final step a

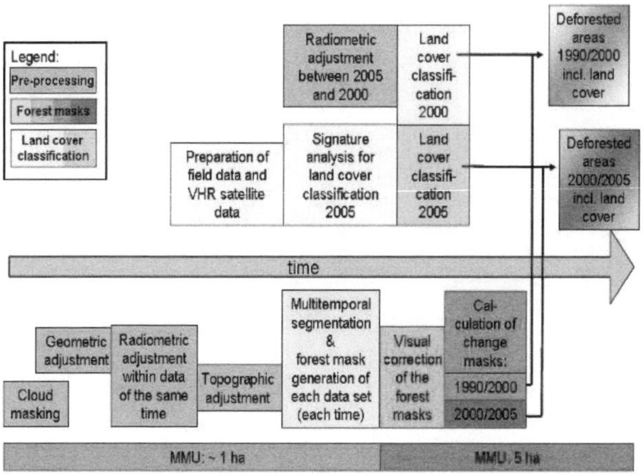


Figure 1. Overview of the developed procedure.

quality control was done by manual interaction, where cloud and cloud shadow masks are corrected and hazy areas are added. The result of these steps was a mask of all areas which cannot be used in the further processing. For these areas, other data have to be found to generate the required information, such as lower resolution optical data like MODIS or SAR data (Mayaux et al. 2002).

Strong topography causes different illumination of the north- and south-facing slopes. This effect has to be corrected by normalization procedures in all areas with mountainous terrain. The Minnaert correction (Colby 1991, Meyer et al. 1993) was used to correct this illumination differences. The Minnaert correction was chosen, because it has proven to deliver good results in earlier studies, e.g. Tottrup 2007.

The original geo-referenced scenes were overlaid to assess their geometrical congruence. Shifts of up to four pixels, especially in W/E direction could be observed in the test area. A deforestation mapping based on this data, especially of areas with a lot of small structured (riparian) forest, would result in severe erroneous deforestation on one side and wrongly classified afforestation on the other side. Therefore, the data sets have to be adjusted to ensure congruence and thus correct results. Two different fully-automatic approaches were tested to perform the geometrical adjustment: (1) Coarse registration and (2) Fine registration. The main difference between the two procedures is that fine registration performs local correction. This is important, if the images are internally distorted. For each area, the shifts between the images are calculated and then applied. During the investigation of the test site, it turned out that with the coarse registration the images could not be fitted into a congruent geometry. Therefore, the fine registration was performed resulting in a stack of three geometrically congruent image data sets.

For national coverage of Cameroon, it is more time and cost effective not to treat each of the 78 images independently. Therefore, neighboring images from similar acquisition dates were combined into blocks and made radiometrically comparable. For this radiometric adjustment, two approaches were checked in the test site: (1) Adjustment based on linear regression and (2) Histogram-matching. The results of both approaches were statistically evaluated based on the comparison of different land cover classes in the overlap area of the two neighboring images, called image $A_{orig}$ and $B_{orig}$. The same land cover area was selected in these images ($A_{orig}$ and $B_{orig}$) and in both adjustment results being linear-regression adjusted image B ($B_{linreg}$) and histogram-matching based adjusted image B ($B_{histmatch}$). The deviation in terms of digital numbers between the image $A_{orig}$ and $B_{orig}$; $A_{orig}$ and $B_{linreg}$; $A_{orig}$ and $B_{histmatch}$ respectively was evaluated. The average difference for each band calculated based on 14 reference areas located in the overlap is given in Table 1. The lowest difference for each band is highlighted in red color. Except for the fifth band, the results of the histogram-matching based adjustment performed best and thus, histogram-matching was used.

Table 1. Results of the radiometric adjustment tests

| Band | Difference between $A_{orig}$ and $B_{orig}$ | Difference between $A_{orig}$ and $B_{linreg}$ | Difference between $A_{orig}$ and $B_{histmatch}$ |
|---|---|---|---|
| 1 | 4,04 | 3,17 | 1,21 |
| 2 | 1,94 | 1,62 | 0,84 |
| 3 | 3,54 | 3,79 | 1,49 |
| 4 | 7,75 | 5,77 | 4,42 |
| 5 | 11,71 | 5,67 | 7,52 |
| 6 | 4,15 | 2,75 | 1,58 |

## 3.2 *Segmentation*

Pixel-to-pixel change detection has widely been used in the past, often with limited success only. Many different factors influence the final pixel value even if the area is mapped with the same sensor: slightly different viewing angle, geometric shifts, atmospheric effects, phenological effects and sun-position, to name a few of them. In addition, if images (pixels) from different sensors should be compared, there are yet more influencing factors: different geometric resolution (Landsat = 28.5 m, DMC = 32 m), different spectral resolution (e.g. Landsat ETM+ NIR = 0.775–0.9, DMC NIR = 0.77–0.9), different swath widths with the associated properties (Landsat swath width = 185 km, DMC swath width = 600 km). All those factors, together with still small geometric inaccuracies remaining despite the fine-registration of the images, lead to the fact that most pixel values in question are not directly comparable.

Different illumination, atmospheric effects, phenological properties or spectral resolution of the sensor cannot be changed. However, slight shifts resulting from different viewing angles, different geometric resolution etc. can be taken care of by using a segmentation approach. Segment-based approaches deliver more reliable results in this context. In addition, there are three more advantages of the segment-based approach compared to the pixel-based method: first, the object's information (e. g. area and shape) can be used to suppress objects with a size below the minimum mapping unit or very long and thin segments, which can be attributed to remaining geometric inaccuracies. Second, the scattered appearance of pixels, commonly known as "salt-and-pepper-effect" is avoided. Third, segments can easier be modified than single pixels; this saves considerable time in the control phase applying a manual correction. The segmentation procedure consists of a series of pre-processing and processing steps. These are:

a) Layer stacking

This step is necessary to include all three time frames into the segmentation. In order to avoid the Landsat data with their 7 bands to have a stronger influence than the DMC data, only three spectral bands (NIR, red and green) of the Landsat data were used. The selection of the bands goes along with the spectral properties of the DMC data.

The main advantage of using all multi-temporal data as inputs for the segmentation is that only one set of segments is used for each analysis.

b) Auto-scaling

Auto-scaling is used to scale the gray values of each band (from all three input data sets) to the same minimum and maximum (or alternatively mean and standard deviation) values. This step is a linear transformation, which changes the pixel values, but maintains the general characteristic (e.g. the relationship in gray levels between different land cover classes) of the image. This step is necessary to ensure equal conditions for each band and time. Otherwise, one band or one time frame would affect the segmentation more than another.

c) SNN (Symmetric Nearest Neighbor) filtering

The SNN filtering algorithm maintains the edges, while smoothing the area between edges. This is a very useful procedure, which allows the following segmentation algorithm to more easily find the correct edges.

d) Region growing segmentation

This algorithm is a standard region growing algorithm with a local and a global threshold to be set by the operator. With the thresholds once defined, it is expected, that they can be transferred to other regions, which show similar land cover patterns.

e) Spatial merging

After segmentation, very small regions are merged with their most similar neighbors. In the current test site, the minimum size was set to be very small (8 pixels, which is equivalent to an area of 6498 $m^2$). The reason for this small minimum size can be found in the forest pattern: the transition zone between the actual rainforest and the savannah which is characterized by thin and long-patches of forest along the rivers. These forests would not be adequately mapped with a larger minimum size in this processing step.

### 3.3 *Classification forest/non-forest*

For each region, the mean values for each time period and also the NDVIs are calculated. Based on this information, for each time, forest/non-forest thresholds are defined and preliminary forest masks are generated. The generated segments including their preliminary information about forest or non-forest was converted to vector format, because vector data is more efficient for manual interaction. First, all segments smaller than the defined minimum mapping unit (MMU) of 5 ha with a long and thin shape, mostly only one pixel width (shape index = area/perimeter < 18) are excluded from the manual correction step, because they are predominantly caused by different image structure (LANDSAT vs. DMC data) or by remaining geometrical inaccuracies. Therefore, no actual change is assumed and the automatic classification results from the 1990 images were used for all three time slots. All other segments of the preliminary forest mask were manually checked for each time and corrected if necessary.

### 3.4 *Land cover classification of deforested areas*

The results obtained in this study serve as inputs for a modeling phase using models which take into account a variety of different factors such as economic, social and policy driven factors. One important input is the land cover class in which the former forest had been turned into. In order to train the model (training phase is 1990–2000), it was necessary to obtain the land cover information for the deforested areas in 2000. For the validation of the model, also the land cover information of 2005 was necessary. Therefore both images (2000 and 2005) were classified using two different sources for the derivation of the training data: First, very high resolution (VHR) satellite data (geometrical resolution around 1m) and second, data from a field survey. This classification consists of a series of processing steps:

1. Preparation of data (maps) for field work in the country.
2. Selection and ordering of Quickbird data for deriving additional training data and for the verification of the classification results.
3. Selection of training data from the Quickbird imagery, their transfer to the DMC data (images 2005) and integration in the signature analysis.
4. Integration of the field measured data in the signature analysis for 2005.
5. Signature analysis of all data (Quickbird and field) – iterative adjustment in order to avoid mixed pixels in the training data set.
6. Classification of the DMC (2005) data.
7. Spectral adjustment of the Landsat 2000 data to the DMC 2005 data. This was again performed used the histogram-matching tool.
8. Transfer of the signatures from the DMC 2005 data to the Landsat 2000 mosaic.
9. Test of the transferability and eventually adjustment (signature analysis and iterative adjustment).
10. Classification of the Landsat 2000 data.

11. Eventually post-classification processing steps such as majority filtering, neighborhood-based corrections, etc.
12. Intersection of the deforestation maps with the respective land cover classification (change 1990/2000 with Landsat2000 classification and change 2000/2005 with DMC classification).

## 4 RESULTS FOR THE TEST AREA

In the first stage of the pilot study, a test site was defined to develop the methodology and to test its feasibility. The test site is located in the center-east part of the country and comprises an area of 45.000 km². The area is in the transitional vegetation zone between closed evergreen forests in the south and savanna in the north. The data used in the test area are two Landsat images from 1990 and 2000 respectively and one DMC image from 2005. From the test area, about 8.5% is covered by clouds, cloud shadows or haze in at least one of the time slots. These parts are not yet substituted by neighboring scenes, which is foreseen to be done in the roll-out step. Topographic adjustment was not necessary in this area, because the region is rather flat or hilly. Geometric adjustment was performed according to the fine registration procedure described above. Histogram matching was used for the radiometric adjustment of the images. The manual correction step of the forest masks turned out to be more time-consuming than estimated. This can be explained by the very small structured riparian forest in this region. The effort of the corrections was about 100 hours in total for the test area. It is expected, that the effort is less for areas either in the tropical rainforest zone as well as in the dry areas of the northern territory. However, this still has finally to be confirmed within the roll-out phase.

Based on the corrected forest masks, two change masks considering both de- and reforestation were derived. The change masks showed a deforested area between 1990 and 2000 of 529.786 km² and a reforested area for the same time span of 441.239 km². From 2000 to 2005, the deforested area is 524.925 km² and the reforested area amounts up to 583.111 km². Between 1990 and 2000, these figures lead to a net loss of 88.547 km² and between 2000 and 2005 to a net gain of 58.186 km². These net change values have to be interpreted with care, as already mentioned by Grainger (2008). Trends can be misinterpreted and a prognosis for the future is very unstable based on such data. A further critical point in GHG budgeting based on net changes only is, that secondary forest might not be binding as much green house gases as the original forest. This can be caused be lower biodiversity, lower plant density or simply a lower average tree size. Therefore, reforestation areas are mapped individually and have to be further analyzed for its contribution to REDD.

A field work campaign was initiated for the ascertainment of training data and additionally training data were collected from the interpretation of Quickbird data. Consequently, the required main classes (grassland, cropland, wetland, settlement and other land) could be successfully extracted. Grassland and settlement areas are rather easy to distinguish. However, there are still some topics to be looked at in more detail. Scrubland is a dominant class in this vegetation zone and therefore it had to be included as a separate class. A mayor spectral confusion was found between the three classes 'open deciduous scrubland', 'riparian forest' and 'cropland'. This confusion was noted in the field as well. It could be explained by the dominant agricultural system in this area: the shifting cultivation. According to FAO & World Bank (2001), in the so-called 'Forest-based Farming System', people are "*[...] clearing a new field of forest every year, cropping it for 2 to 5 years [...] and then abandoning it to bush fallow for 7 to 20 years.*"

Based on the procedure just described, the deforested areas were classified into the categories Grassland, Cropland (including burnt areas), Scrubland, Wetland (in the vicinity of water bodies only), Settlement, Other land (including bare soil, water) and Forest. The classification result was evaluated on a segment by segment basis. The majority of land cover class with a majority fraction of more than 50% was taken, the remaining segments are considered to be segments of mixed land cover under the heading "other". This leads to the percentage given in Table 2. Between 10 (1990–2000) and 20% (2000–2005) of the deforested areas still show a significant

Table 2. Deforested areas by land cover classes (segment-based evaluation on the principle of majority with more than 50% of the segment belonging to the majority class)

| Land cover class | Deforestation 1990–2000 (segment – majority) | Deforestation 2000–2005 (segment – majority) |
|---|---|---|
| Settlement | 0,12% | 3,06% |
| Grassland | 0,93% | 2,30% |
| Other (incl. water and mixed classes) | 37,76% | 31,38% |
| Shrubland | 3,51% | 12,71% |
| Forest | 10,26% | 19,25% |
| Cropland | 47,40% | 31,30% |
| Wetland | 0% | 0% |

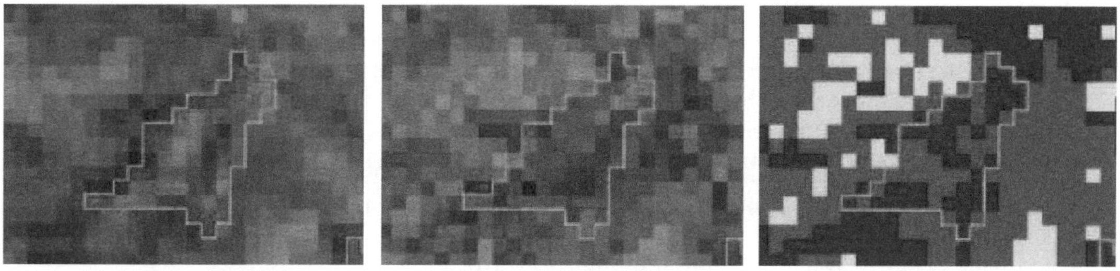

(a) Segment in 1990 (b) Segment in 2000 (c) Classification 2000

Figure 2. Example of an area largely deforested between 1990 and 2000 and the result of the land cover classification in 2000 data.

amount of forest. The reason can be found in the segment-based approach. Some segments are only partly deforested leading to remaining forest patches in the deforested segments. An example of such an area is depicted in Fig. 2. The forest patch in 1990 has been largely deforested, however, some patches of forest remained and there seems to be already secondary vegetation. These two factors influence the classification results from the 2000 images. Still more than 50% of the pixels are classified as forest. This example shows why it is important to keep these areas still as deforested areas, because otherwise large parts of deforestation and forest degradation would be suppressed. These areas can also be seen as candidates for further degradation studies.

## 5 DISCUSSION AND CONCLUSION

In conclusion, it can be stated that the first phase of the REDD pilot study Cameroon has been successfully completed. For a test site involving 5 satellite scenes, a methodology was developed and its applicability was tested. Compared to both the Vanuatu and the Bolivian study, the current project additionally involved topographic normalization of the mountainous terrain. According to Riaño et al. (2003), topographic correction methods show two advantages compared to using band ratios: firstly, diffuse irradiation is different in each band and can, therefore, not be corrected by band rationing and secondly, spectral resolution is lost, which is a drawback for multispectral classification. Killeen et al. (2007) reported for Bolivia a confusion of inundated (riparian) forest with sunny montane slopes, which could possibly be solved by using an adequate topographic normalization procedure. Co-registration was also performed in both existing studies. In Vanuatu, a coarse registration was performed. In the current study, a regional 'fine registration' was necessary. In the Bolivian case study, co-registration is mentioned, but no details about the used method are given. The novel segmentation approach proved to be advantageous compared to a pixel based method as used in the Vanuatu and also in the Bolivian study. The first advantage is clearly the improved comparability between the different time slots. This has been shown by visually comparing the results of pixel-wise and segment-based change detection between 1990 and 2000. The second advantage can be found in the easier manual interaction using segments instead of pixels. In Vanuatu, the authors used two images of the same spot and for the same time slot in many cases, which improved the quality of the result. Due to limited data availability and the large size of Cameroon with the related costs this was not possible in the current study. Land cover classification was not performed for Vanuatu, but for Bolivia. Similar problems regarding confusion of classes have been found in Bolivia and in Cameroon, such as the confusion of natural grassland and scrub vegetation with croplands. In general, more research is needed in the field of forest degradation, which was only marginally covered in the current pilot study. This could possibly be done similar to the works of Souza et al. (2005) or using the Linear Mixture Modeling (LMM) approach described in Tottrup (2007).

The classification into the land cover classes turned out to be difficult for certain land cover types. This is mainly caused by the agricultural system of shifting cultivation with very different spectral signatures for cropland in the different stages of the usage-cycle. Newly-burnt areas are overlapping with wetland and cropland left to bush fallow is hardly distinguishable from natural scrubland. More field data will be collected to better understand the land cover/ land use systems occurring in Cameroon. The results obtained in the test site are encouraging to take the study to the next step, which is the roll-out for the whole country of Cameroon. Future activities include:

- to elaborate an automatic method to mosaic the images and to get rid of the clouded-, cloud-shadow- and hazy areas
- to obtain better field data for both training and verification
- to develop a procedure for the automatic classification of forest degradation (different degrees of degradation)
- to find alternative data for the clouded and misty parts to get a full country coverage of information.

## REFERENCES

COLBY J.D., 1991: Topographic normalization in rugged terrain. Photogrammetric Engineering and Remote Sensing, Vol. 57, No. 5, pp. 531-537.

FAO AND WORLD BANK, 2001: Farming Systems and Poverty – Improving Farmers' Livelihoods in a Changing World. Rome and Washington DC. Available online at: ftp://ftp.fao.org/docrep/fao/003/y1860e/y1860e00.pdf. Last accessed: XII 2007.

GRAINGER, A. 2008: Difficulties in tracking the long-term global trend in tropical forest areas. Proceedings of National Academy of Sciences of the United States of America (PNAS), Vol. 105, No. 2, pp. 818-823.

HEROLD M., SAMBALE J., LINDNER M., URBAN M. AND WEAVER S., 2007: Satellite based monitoring of the national forest resources in the pacific island state of Vanuatu. DGPF Tagungsband 16 / 2007 – Dreiländertagung SGPBF, DGPF und OVG, pp. 391-398.

STEININGER M., TUCKER C., TOWNSHEND J.R.G., KILLEEN T.J., DESCH A., BELL V. AND ERSTS P., 2001: Tropical deforestation in the Bolivian Amazon. Environmental Conservation Vol. 28, No. 2, pp. 127-134.

KILLEEN T.J., CALDERON V., SORIA L., QUEZADA B., STEININGER M., HARPER G., SOLÓRZANO L.A. AND TUCKER C.J., 2007: Thirty years of land-cover change in Bolivia. AMBIO: A Journal of the Human Environment, Vol. 36, No. 7, pp. 600-606.

MARKHAM B.L., STOREY J.C., WILLIAMS D.L. AND IRONS J.R., 2004: Landsat sensor performance: history and current status. IEEE Transactions on Geoscience and Remote Sensing, Vol. 42, No. 12, pp. 2691-2694.

MAYAUX P., DE GRAND G.F., RAUSTE Y., SIMARD M. AND SAATCHI S., 2002: Large-scale vegetation maps derived from the combined L-band GRFM and C-band CAMP wide area radar mosaics of Central Africa. International Journal of Remote Sensing, Vol. 23, No. 7, pp. 1261-1282.

MEYER P., ITTEN K.I., KELLENBERGER T., SANDMEIER S. AND SANDMEIER R., 1993: Radiometric correction of topographically induced effects on Landsat TM data in an alpine environment. ISPRS Journal of Photogrammetry and Remote Sensing, Vol. 48, No. 4, pp. 17-28.

RIAÑO D., CHUVIECO E., SALAS J. AND AGUADO I., 2003: Assessment of Different Topographic Corrections in Landsat-TM Data for Mapping Vegetation Types. IEEE Transactions on Geoscience and Remote Sensing, Vol. 41, No. 5, pp. 1056-1061.

SOUZA JR. C.M., ROBERTS D.A. AND COCHRANE, M.A., 2005: Combining spectral and spatial information to map canopy damage from selective logging and forest fires. Remote Sensing of Environment, Vol. 98, pp. 329-343.

TOTTRUP C., 2007: Forest and Land Cover Mapping in a Tropical Highland Region. Photogrammetric Engineering & Remote Sensing, Vol. 73, No. 9, pp. 1057-1065.

*Remote Sensing for a Changing Europe*
*D. Maktav (Ed.)*
*IOS Press, 2009*

*doi:10.3233/978-1-58603-986-8-214*

# Comparison of pixel based and feature based fusion of high resolution optical and SAR imagery

G. Atay
*Karadeniz Technical University Engineering Faculty, Dept. of Geodesy and Photogrammetry, Trabzon, Turkey, gatay@ktu.edu.tr*

J.D. Wegner, P. Lohmann, P. Hofmann, U. Sörgel
*Institute of Photogrammetry and Geoinformation (IPI), Leibniz University of Hannover, Germany*

Keywords: SAR, pixel based, feature based, image fusion, ITK, OTB, coregistration

ABSTRACT: The number of high resolution imaging remote sensing systems is increasing rapidly. But, because of technologic limitations, not every system can produce images with both very good quality and all desired features. Images obtained with different sensors may have complementary information. Fusion of these data with a fine image fusion approach can provide better information about the observed scene. Two common complementary information sources are optical and Synthetic Aperture Radar (SAR) images. These two products have some advantages and disadvantages separately because of their sensor characteristics. Combining advantages of the two images may decrease disadvantages of them; moreover the process will provide extra information. The study includes two parts. First, it consists of pixel based image fusion of SAR and optical images while the second part is fusion of the same data with a feature based method. Registration strategies of both pixel based and the feature based methods were determined and they were implemented for our optical and SAR images. The results of these methods were compared. According to the results it was clear that the feature based method worked better than pixel based method.

## 1 INTRODUCTION

Lately, the number of imaging remote sensing systems has been increasing rapidly. But, because of the technologic limitations, every system cannot produce images with the desired quality and features. Therefore, especially in recent years, combining images taken by various sensors has attracted increasing attention. Image fusion is a process to obtain a more useful joint image from more than one input image. With the availability of multi-sensor, multi-temporal, multi-resolution image data, the fusion of digital image data has become an important tool in remote sensing.

The objective of this study is to register one high resolution optical image and one high resolution SAR image. A pixel based method and feature based method are developed and the different results are compared. The pixel based method accomplishes the fusion at the lowest level. It is applied on raster data in which every pixel and its grey value are examined. The values are modified by mathematical calculations and operators (Schmitz 2007). The feature based method requires the extraction of objects recognized in the optical and the SAR image. Features correspond to characteristics extracted from the initial images which are depending on their environment such as extent, shape and neighborhood (Pohl & Van Genderen 1998).

In this study a SAR and an optical image were used. The SAR image was acquired by the EMISAR sensor (Electromagnetic Institute Synthetic Aperture Radar Sensor). EMISAR is a dual frequency (L- and C-band) polarimetric SAR sensor build by the Technical University of Denmark for radar and remote sensing research. It is installed on a Danish Air Force Gulfstream G3. For this study a C-band SAR image was used (Fig.1 (right)). The optical image is a high resolution aerial image (Fig.1 (left)). Panchromatic data and color information have been captured covering the same area as the SAR image. The optical image was resampled to the pixel size of the corresponding SAR image and hence the resolution of both images is four meters.

Figure1. Optical (left) and SAR (right) images

The viewing geometry of optical and SAR images as well as the wavelength domain are different. Hence, geometric and radiometric differences occur. Geometric differences appear because SAR sensors measure distances in slant range geometry whereas optical sensors measure angles with usually small off-nadir angles. Radiometric differences occur because optical sensors capture the terrain reflectivity response to the visible sun light. SAR sensors, however, image the terrain response to actively sent microwaves. The higher the resolution of the images becomes, the greater become the differences that have to be accounted for during the registration process.

## 2 METHODOLOGY

First of all, two registration strategies, one pixel based and one feature based, were developed in this study. The steps of the strategies are shown in Figure 2. The feature based image registration algorithm strategy was developed by (Wegner 2007).

It can be observed from Figure 2 that the common steps of pixel based and feature based image fusion are ortho-rectification, preprocessing and registration. The first common step is ortho-rectification, i.e. the optical and the SAR image are projected from sensor space to object space on the ground. It is implemented in order to decrease the geometric differences between optical and SAR imagery. Both optical and SAR images were already ortho-rectified before the study. Hence, this step had not to be implemented. The second common step is preprocessing which is very important in order to reduce some undesirable signals in the images. After the images are preprocessed some extra processes are needed for the feature based registration strategy before the registration process. First, features must be extracted from both optical and SAR images. It is necessary to get features from the optical image in order to register the results with the feature image of the SAR image. In this study edges are used as features for both optical and SAR images. The best results for the optical image were achieved with the Canny edge detection algorithm (Canny 1986). Lines of the SAR image are extracted using the algorithm developed by Touzi et al 1988 followed by a threshold operation.

After extracting features in the images, distance maps of the optical and the SAR image are produced. A distance map is an image where the grey value of each pixel is the distance to the nearest pixel from a set of objects (Cuisenaire 1989). This step was done in order to transform one-dimensional lines to two-dimensional continuous information. In our study distance maps display the Euclidean distance between a background pixel and the nearest line pixel. Then the numerical value of the distance is translated to a grey value. An approach developed by Danielsson (Danielsson 1980) was chosen for the computation of the Euclidean distances.

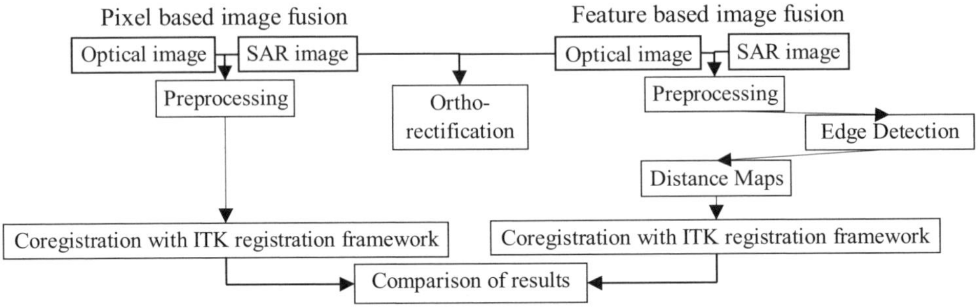


Figure 2: Comparison of the pixel based and the feature based registration strategies

For the implementation of the registration strategy, algorithms already existing in the open source library ORFEO Toolbox were used. ORFEO Toolbox (OTB) is distributed as an open source library of image processing algorithms. It is based on the medical image processing library ITK and offers particular functionalities for remote sensing image processing in general and for high spatial resolution images in particular. OTB was set up in order to prepare for the exploitation of high resolution images, derived from Pleiades (PHR) and Cosmo-Skymed (CSK) systems, the French Space Agency (CNES) (CNES 2008).

## 3 PREPROCESSING

Before the data obtained by satellite systems or digital airborne systems can be analyzed, it is often necessary to preprocess them in order to correct defects (Gibson & Power 2000). A preprocessing step is necessary for both pixel and feature based fusion methods. It is necessary to get rid of noise and to reduce the speckle effect that the images include for further steps. For this purpose two edge preserving smoothing filters were used. An anisotropic diffusion filter (Perona & Malik 1990) was applied to the optical image and a Frost filter (Frost et al. 1982) to the SAR image.

## 4 IMAGE REGISTRATION

Image registration is the task of finding a spatial transform mapping one image onto another. In ITK, registration is performed within a framework of pluggable components that can easily be interchanged. This flexibility means that a combinatorial variety of registration methods can be created, allowing users to pick and choose the right tools for their specific application. (Ibánez et al. 2005)

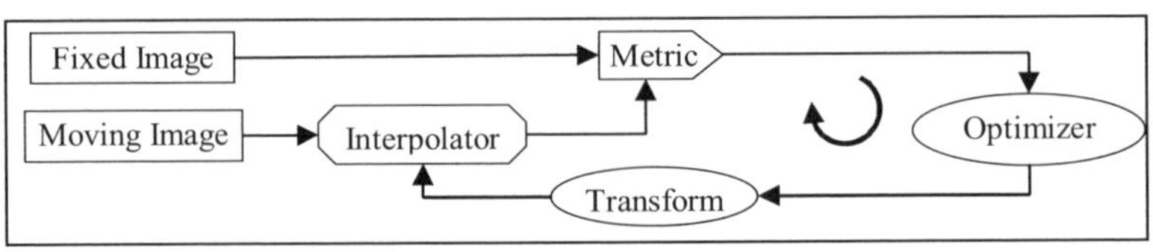


Figure 3: The ITK registration framework (Ibánez et al. 2005)

The ITK registration method requires two input images, a transform, a metric, an interpolator and an optimizer (Fig. 3). Input data to the registration framework are two images: one is called the fixed image and the other one the moving image. Registration is interpreted as an optimization problem with the goal of finding the optimum spatial mapping that will align the moving image with the fixed image. In this study the distance map of the optical image is the fixed image and the distance map of the SAR image is the moving image. The transform component represents the spatial mapping of points from fixed image space to points in moving image space. The interpolator is used to evaluate moving image intensities at non-grid positions and the metric component provides a measure of how well the fixed image is matched by the transformed moving image. This measure is the quantitative criterion to be optimized by the optimizer over the search space defined by the parameters of the transform. (Ibánez et al., 2005)

In this study a two-dimensional translation transform was used as the transformation component for simplicity reasons. It involves shifting the origin of the current coordinates system horizontally and vertically by a specific amount. For simplicity and computation cost reasons a bilinear interpolation was used as the interpolator component. It has been extensively shown that metrics based on the evaluation of mutual information are well suited for overcoming the difficulties of multi-modality registration. (Ibánez et al., 2005) Therefore, in the study a "Mutual Information Metric" was implemented as the metric component. A gradient descent optimizer was used for the optimizer component. It searches for the set of transformation parameters that maximizes the metric value by iteratively changing the transformation parameters.

In ITK there are several alternatives for these components. For example, a mean squares, normalized correlation, and mutual information can be chosen for the metric, too. In our study we used translation transform for the transform parameter, a bilinear interpolation for the interpolator, mutual information for the metric and a gradient descent optimizer for the optimizer.

## 5 RESULTS AND DISCUSSION

As mentioned before, both optical and SAR images were already ortho-rectified and coregistered and thus fit very well. So for this project their positions one to another were changed a bit. A shifting was applied to them. The amount of the translations is known. Therefore, the results of both the pixel and the feature based algorithm can be compared to the original coregistered images for evaluation reasons.

(a) (b) (c)

Figure 4: Test region of the optical image (a), Test region of the SAR image (b), Real translations between images in pixels (c)

In Figure 5 the registration result of pixel based approach is shown. The ellipses point some important features than can show the results very well. The red ellipses display the situation before the registration and the yellow ellipses show the situation after the registration. The program also gives some numerical results about the registration concerning the transformation parameters. In this study the parameters are only the translations between the two images. These computed translations were used to calculate an Euclidean distance value. The Euclidean distance between the real translation values and the computed translations was 2.98 pixels. In Figure 5 both optical and SAR images can be observed in both (a) and (b). For better visualization checkerboard scheme was used.

Figure 5: Checkerboard before the registration (a), Checkerboard after the registration (b)

Feature extraction results of Canny and Touzi algorithms and distance maps of both optical and SAR images can be observable in Figure 6.

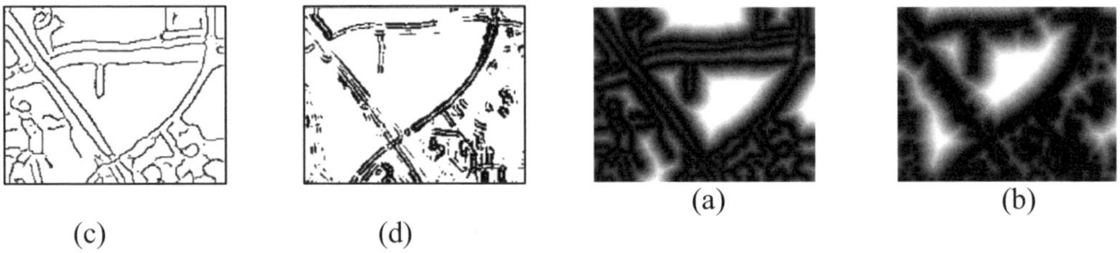


Figure 6: (a) Extracted lines from optical image (b) Extracted lines from SAR image (c) Distance map of the optical image (d) Distance map of the SAR image

Then the registration framework was implemented to these distance maps. That is to say, the main inputs of the ITK registration framework are distance maps. In Figure8 (a) and (b) the results of the registration are displayed. The Euclidean distance between real transition values and computed translations between images was 0.47 pixel.

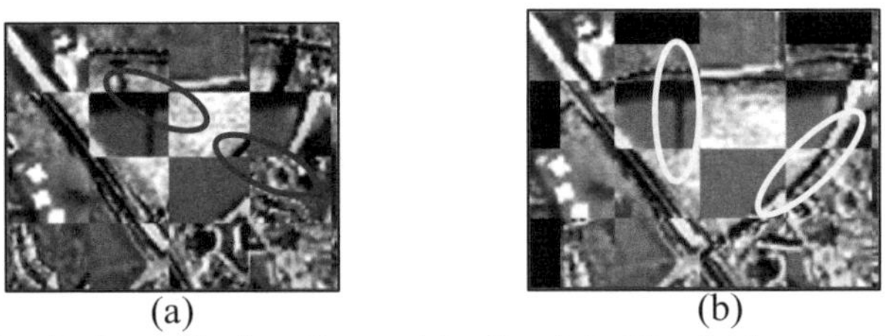


Figure 8: Checkerboard before the registration (a), Checkerboard after the registration (b)

| Axis | Real value (pixel) | Pixel based fusion (pixel) | Feature based fusion (pixel) |
|---|---|---|---|
| x | 16 | 18,3021 | 15,5427 |
| y | 17 | 15,1134 | 16,8995 |
| | **Euclidean distance** | 2,98 | 0,47 |

Table 1: Numerical values of registration processes.

The numerical values of two approaches are shown in Table 1. In Figure 9 results of a selected specific detail were shown. It is much clearer that feature based approach gives the better results than the pixel based method.

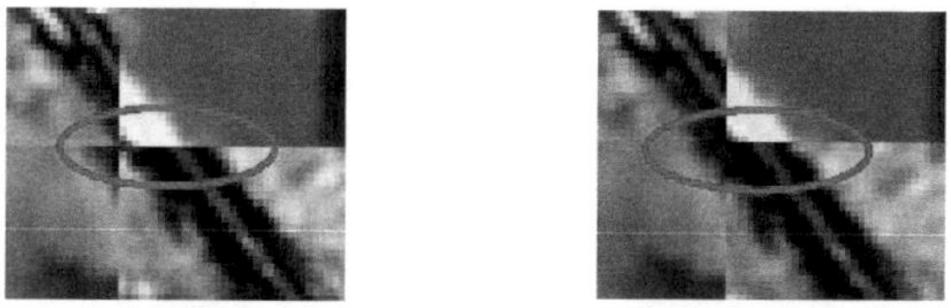

(a) (b)

Figure 9: Pixel based registration result (a), Feature based registration result (b)

## 6 CONCLUSION

In conclusion, the feature based registration approach works better on high resolution optical and SAR imagery than the pixel based approach. In terms of visual results and numerical values, the feature based registration algorithm aligns the SAR image better to the optical image. Due to high resolution and multi-modality imagery (different sensors) the pixel based method does not work as well. OTB proved to be very useful for this image analysis and registration task. For further work it is important to implement and test additional metrics, optimizing methods and also different images in order to be able to make more general statements.

## ACKNOWLEDGEMENT

The authors thank the Scientific and Technical Research Council of Turkey (TUBITAK) for supporting the stay of the first author Gülçin ATAY at the Institute of Photogrammetry and Geoinformation (IPI), Leibniz University of Hannover, Germany.

## REFERENCES

Canny, J. 1986. A Computational Approach to Edge Detection. *IEEE Transactions on Pattern Analysis and Machine Intelligence*, 8(6):679-698.

CNES.Pleiades. 2008. ORFEO Accompaniment Program. Accessed 21.05.2007. http://smsc.cnes.fr/PLEIADES/A_prog_accomp.htm

Cuisenaire, O. & Macq, B. 1999. Fast and Exact Signed Euclidean Distance Transformation with Linear Complexity. *IEEE International Conference*. 06 : 3293-3296.

Danielsson, E. 1980. Euclidean distance mapping. *Computer Graphics and Image Processing*, 14:27-248.

Frost,V. S., Stiles, J. A., Shanmugan, K. S. & Holtzman, J. C. 1982. A Model for Radar Images and Its Aplication to Adaptive Digital Filtering of Multiplicative Noise. IEEE Transactions on Pattern Analysis and Machine Intelligence. 4(2):157-166.

Gibson, P.J. & Power, C.H. 2000. Introductory remote sensing: Digital Image Processing and Applications. New York. Taylor & Francis.

Perona, P., Malik, J. 1990. Scale Space and Edge Detection Using Anisotropic Diffusion. IEEE Transactions on Pattern Analysis and Machine Intelligence. 12:629-639.

Pohl, C. & Van Genderen, J. L. 1998. Multisensor image fusion in sensing: concepts, methods and applica tions. *Int. J. Remote Sensing*. 19(5): 823-854.

Touzi, R., Lopez, A. & Bosquet, P. 1988. A Statistical and Geometrical Edge Detector for SAR Images. *IEEE Transactionson Geoscience and Remote Sensing*. 26:764-773.

Wegner, J.D. 2007. Automatic Fusion of SAR and Optical Imagery, Diploma Thesis.

Wegner, J.D., Inglada, J. & Tison C. 2008. Image Analysis of fused SAR and Optical Images Deploying Open Source Software Library OTB. Article to be published in proceedings of EUSAR 2008.

Zhang, Y., 2004. Understanding image fusion, *Photogrammetric Engineering and Remote Sensing*. 70(6): 657-661.

Ibánez, J., Schroeder W., Ng., Lydia & Cates J. 2005. The ITK software guide. 2 edition. version 2.4.

*Remote Sensing for a Changing Europe*
*D. Maktav (Ed.)*
*IOS Press, 2009*

*doi:10.3233/978-1-58603-986-8-220*

# Compilation and assessment of Pan-European land cover changes

Gerard Hazeu, Sander Mücher, Henk Kramer[1] and Felix Kienast[2]
[1]*Alterra, Wageningen University and Research Centre, Centre for Geo-Information, Droevendaalses-teeg 3, 6708 PB Wageningen The Netherlands; Tel : +31 317 481928, Fax : +31 317 419000, e-mail :gerard.hazeu@wur.nl*
[2]*Swiss Federal Research Institute WSL, Zuercherstrasse 111, CH-8903 Birmensdorf, Switzerland*

Keywords: historical land cover changes, regional analysis, Pan-European land cover

ABSTRACT: Simulating future land use changes requires knowledge about the present land use and the land use in the past. The knowledge on historical land cover changes is important for spatial planning, resource evaluation, ecological modeling etc. This paper describes the production of various land cover datasets, the assessment of the historical cover land dataset for 1960 (HISLU60) and the changes between 1960 and 1990.

## 1 INTRODUCTION

Land cover is changing in many parts of Europe at an increasing rate. The knowledge on these land cover changes is important for spatial planning, resource evaluation, ecological modeling etc. Modification of ecosystems is most visible through changing land cover. Furthermore, spatio-temporal models that describe and predict land cover change due to social and economic processes need reliable information on land cover changes in order to calibrate and validate these models (Kaufman and Seto 2001).

Since the CORINE Land Cover covers only EU member countries, high resolution land cover information is still lacking for countries as Switzerland and Norway. Therefore, the best available land cover databases has to be integrated to compile Pan-European land cover databases.

Against this background three new Pan-European land cover databases are created for the years 1960, 1990 and 2000. The aim of this paper is a) to present the compilation of Pan-European land cover databases for 1960, 1990 and 2000, and b) to provide an assessment of the 1960 database and the land cover changes between the years 1960-1990.

## 2 DATA

The data sources used are the following:

- CORINE Land Cover databases (CLC1990, CLC2000 and CLC-changes)
- Pan-European Land Cover database (PELCOM),
- Global Land Cover database for the year 2000 (GLC2000),
- National land cover databases for Switzerland and Norway,
- World Atlas of Agriculture published in 1969 (scale 1:2.500.000),
- Additional datasets for the assessment of HISLU60.

The CLC1990, CLC2000 and CLC-changes are land cover databases based on satellite imagery of Europe. A common methodology was applied by each participating country to discern the 44 different land cover types (Heymann *et al.* 1994; Bossard *et al.* 2000; Feranec *et al.* 2007; Buttner *et al.* 2004). The vector database was produced at a scale 1:100.000 with 25ha as minimum mapping unit. The 100 by 100m raster database version which is used here is downloadable from the EEA website (http://dataservice.eea.europa.eu/dataservice/metadetails.asp?id=1007).

The 1km resolution Pan-European Land Cover database PELCOM contains 16 thematic classes and was based on the interpretation of NOAA-AVHRR satellite data of 1997 (Mücher *et al.* 2001). Only the land cover classes urban areas, wetlands and water bodies were derived from ancillary data sources.

The classification of the SPOT4-Vegetation dataset (spatial resolution 1km) by 18 regional teams resulted in the global land cover database for the year 2000 (GLC2000). The legend based on the United Nations land cover classification system (LCCS) consists of 23 classes (Di Gregorio 2005; Bartholomé & Belward 2005).

The national database of Norway is a vector database for the year 2000 (AR2000). It has the following 8 land cover classes: production forest, non-production forest, agricultural land, other land with vegetation, barren land, built-up areas, glaciers and water. The database was produced with help of satellite images and different ancillary data sources. It can be seen as an update of the forest and land use map of Norwegian Atlas of 1983.

The Swiss CORINE Land Cover datasets for the years 1985 and 1997 are 100m raster datasets at CLC-level 2 nomenclature (13 classes). The Swiss Land Use Statistics were used to produce the spatial databases (Nippel & Klingl 1998).

The World Atlas of Agriculture published in 1969 (scale 1:2.500.000) is the basis for the construction of the historical land cover database (HISLU60) (WAA 1969).

Additional independent data sources used to assess the quality of the HISLU60 database are:

1. national land cover database of the Netherlands for 1960 (HGN60),
2. Swiss national statistics for 1972, and
3. 73 BIOPRESS windows of 30 by 30 km across Europe for the year 1950 (Gerard *et al.* 2006; Hazeu & Mücher 2005).

## 3 METHOD

*Production of land cover datasets for the years 1960, 1990 (PLCM1990) and 2000 (PLCM2000)*
The basis for the production of the 2000 Pan-European Land Cover Mosaic (PLCM2000) is the CLC2000 database. The 100 by 100 m grid size and the CLC-level 3 nomenclature formed the basis for the integration of the GLC2000, PELCOM and both national land cover databases with the CLC2000 database. The conversion of the original land cover classes of the GLC2000, PELCOM and national databases into CLC-level 3 classes is based on expert knowledge and previous studies (Mücher *et al.* 2004; Neumann *et al.* 2007; Hazeu *et al.* 2008).

The PLCM1990 database is a combination of CLC1990, the CLC1990 database of United Kingdom, the Swiss national database of 1985 and the PLCM2000 database for the remaining countries for which, unfortunately, we did not have information from 1990. Furthermore, the PLCM1990 database is produced under comparable conditions as the PLCM2000 database. The production of both PLCM databases and the thematic aggregation is extensively described in the ECOCHANGE report (Hazeu *et al.* 2008).

The main data source for the HISLU60 database is the World Atlas of Agriculture. The methodology used for the construction of the historical land use database is described by Kramer & Mücher (2006). The following 7 classes are discerned in the HISLU60 database: urban, arable land, grassland, forest, non-agricultural land, inland water and sea. The database has a spatial resolution of 250 by 250 m and is based on the integration of different map sheets out of the World Atlas of Agriculture.

All databases used in the assessment of land cover changes are converted to a common projection (ETRS_1989_LAEA). They are spatially aggregated into 1 by 1 km grids and thematically aggregated into the 7 HISLU60 land cover classes as the characteristics of the HISLU60 database define the level of detail for the assessment of land cover changes.

*Assessment of the HISLU60 dataset*
The assessment of the HISLU60 dataset is based on the three independent data sources mentioned in the data section. For all three data sources the land cover classes are converted into the 7 HISLU

classes. The Dutch assessment (HGN1960) covers the Dutch territory with a border of sea around. In total an area of 41739 km$^2$. The Swiss assessment deals with 41366 km$^2$ (national territory) and the BIOPRESS assessment covers 59304 km$^2$ across Europe.

*Assessment of 1960-1990 land cover changes at regional level*
An index is calculated to carry out the assessment of land cover changes for the 1960-1990 period at regional level (NUTSx). The index which is the HISLU60 area divided by the area of the reference dataset is calculated per HISLU60 class for each assessment study. On basis of this index a new normalized HISLU60 area per NUTSx region is calculated for the grassland and forest classes. The normalized 1960 values are used in the regional analysis of changes for the 1960-1990 period.

## 4 RESULTS

*Compilation of HISLU60, PLCM1990 and PLCM2000*
The compilation of the PCLM1990 and 2000 databases by integrating several data sources on thematic and spatial content resulted in homogeneous Pan-European land cover databases. All databases have the same geographical extent, i.e. EU27+2 (Norway, Switzerland), Albania, Andorra, Belarus, Bosnia-Herzegovina, Croatia, Iceland, Kosovo, Liechtenstein, Macedonia, Moldavia, San Marino, Serbia and Montenegro, part of Russia, Turkey and Ukraine.

Table 1. Area (km$^2$ and %) of 7 HISLU land cover classes for the years 1960, 1990 and 2000. The spatial resolution is 1 by 1 km (Hazeu *et al.* 2008).

| HISLU classes | HISLU60 | | PLCM1990 | | PLCM2000 | | PLCM1990-HISLU60 | PLCM2000-PLCM1990 |
|---|---|---|---|---|---|---|---|---|
| | km2 | % | km2 | % | km2 | % | km2 | km2 |
| Urban | 87057 | 0.97 | 149620 | 1.67 | 157924 | 1.76 | 62563 | 8304 |
| Arable land | 3432502 | 38.27 | 3595102 | 40.09 | 3587947 | 40.01 | 162600 | -7155 |
| Grassland | 1716263 | 19.13 | 643819 | 7.18 | 640629 | 7.14 | -1072444 | -3190 |
| Forest | 2260972 | 25.21 | 2935273 | 32.73 | 2935987 | 32.74 | 674301 | 714 |
| Non-agri.land | 1205102 | 13.44 | 1416475 | 15.79 | 1416887 | 15.80 | 211373 | 412 |
| Inland waters | 193284 | 2.15 | 219111 | 2.44 | 219956 | 2.45 | 25827 | 845 |
| Sea | 74259 | 0.83 | 8786 | 0.10 | 8769 | 0.10 | -65473 | -17 |
| Total | 8969439 | 100 | 8968186 | 100 | 8968099 | 100 | | |

* discrepancies in total areas due to small differences in extent

* HISLU60 data deviate slightly from Hazeu *et al.* (2008) due to recent improvements with DCW

A statistical comparison of HISLU60 with PLCM1990 unveils a very strong increase in urban and forest areas at the expense of grasslands. After a first assessment of the HISLU60 dataset it appears that the urban area was largely underestimated. A combination with the urban areas of the Digital Chart of the World (DCW) resulted in a better estimation of urban areas around 1960. The improved HISLU60 data are presented in Table 1 (2$^{nd}$ and 3$^{rd}$ column).

The land cover changes between PLCM1990 and PLCM2000 mainly refer to those countries that have both CLC databases available (see Hazeu *et al.* 2008). Weber (2007) and Haines-Young & Weber (2006) elaborate on these land cover changes between CLC1990 and CLC2000.

*Assessment of the HISLU60 dataset*
HISLU60/reference dataset indices, omission (accuracy) and commission (reliability) errors for the three assessments are calculated and presented in Table 2. All figures indicate limited per class accuracies and reliabilities (in most cases below 60%). The index is an indication for the extent of under- or overestimation of the surface area in the HISLU60 database.

The deviation of the index between the three assessments is small and around 1 for the HISLU classes forest and inland water. All three assessments indicate a low index value for the urban class indicating a large underestimation of urban area in the HISLU60 database. Despite the improvement of urban area due to the combination of HISLU60 data with the DCW database, it is still largely underestimated (column 7). The other indices for the three assessments vary largely and over- or underestimation of land cover are in contradiction with each other. Arable land is largely overestimated (23%) in the Netherlands but underestimated in the other case studies. Grassland is underestimated according to the Dutch case-study but largely overestimated (and in equal amounts) in the Swiss and BIOPRESS assessments. Non-agricultural land is largely underestimated according to the BIOPRESS data.

The comparison of HISLU60 data with the BIOPRESS data shows large regional differences. One important discrepancy between both datasets is the overestimation of grasslands in the Mediterranean countries on behalf of the non-agricultural class. The majority (>80%) of moors and heath lands (CLC class 322), sclerophyllous vegetation (CLC class 323) and sparsely vegetated areas (CLC class 333) are not classified as non-agricultural land. The area occupied by those classes is for more than 50% classified as grasslands in the HISLU60 database. A detailed analysis of the Dutch assessment also revealed large regional differences in index values. Only arable land (all overestimation) and urban area (all underestimation with the exception of Friesland) present the same trends for all provinces.

The overall accuracy of the HISLU60 database ranges from 38.4 – 58.1% depending on the used reference dataset. The accuracy assessment with the Swiss data resulted in the lowest overall accuracy while the BIOPRESS assessment took an intermediate position with 48.1%. These low accuracies are a clear indication to restrict the change analysis to total statistics per region, country or for the entire database. A more detailed analysis of changes does not make sense as the pixel (1 by 1 km) or locational accuracy is very low for all three assessments.

Table 2. Omission (accuracy) and commission (reliability) errors for the three HISLU60 assessments. The index (HISLU60/reference dataset) indicates under- or overestimation of a land cover class in the HISLU60 dataset.

| HISLU classes | HGN1960 | | | BIOPRESS1950 | | | SWISS1970 | | |
|---|---|---|---|---|---|---|---|---|---|
| | Om. error | Com. error | Index | Om. error | Com. error | Index | Om. error | Com. error | Index |
| Arable land (class 1) | 0.57 | 0.46 | 1.23 | 0.57 | 0.63 | 0.90 | 0.16 | 0.60 | 0.27 |
| Grassland (class 2) | 0.57 | 0.64 | 0.89 | 0.46 | 0.22 | 2.13 | 0.51 | 0.24 | 2.13 |
| Forest (class 3) | 0.28 | 0.27 | 1.05 | 0.52 | 0.55 | 0.95 | 0.37 | 0.38 | 0.97 |
| Non-agricultural land (class 4) | 0.39 | 0.37 | 1.04 | 0.11 | 0.40 | 0.28 | 0.64 | 0.56 | 1.14 |
| Inland water (class 5) | - | - | - | 0.28 | 0.25 | 1.11 | 0.61 | 0.67 | 0.91 |
| Urban (class 6) | 0.16 | 0.46 | 0.35 | 0.29 | 0.66 | 0.43 | 0.06 | 0.65 | 0.10 |
| Sea (class 7) | 0.85 | 0.86 | 0.99 | 0.91 | 0.68 | 0.46 | - | - | - |
| Overall accuracy | 0.58 | | | 0.48 | | | 0.38 | | |

* Sea and inland water are taken together in HGN1960 assessment

* HGN60 assessment with fruit trees/ orchards of HISLU recoded to Grassland

* BIOPRESS assessment with new HISLU60 dataset

*Regional land cover changes for 1960-1990*

Despite the inaccuracies in the HISLU60 database a regional analysis of changes for the 1960-1990 period is possible. Figure 1 shows the percentage of changes in forest or grassland expressed as percentage of their normalized 1960 area per NUTSx regions. The BIOPRESS and Swiss indices of Table 2 are used to calculate the normalized 1960 area as they are comparable and have the most widespread coverage. However, the general picture of Figure 1 will not deviate much if the increase or decrease of grassland and forest is derived from indices based on the HGN assessment.

The total forest area increases with 236122 km2 and the grassland area decreases with 563730 km2 for the EU27+2. These figures show a similar trend as the statistics presented in Table 2. Very strong increases of forest or grassland have to be dealt with care as the area in 1990 excessively exceeds the area in 1960.

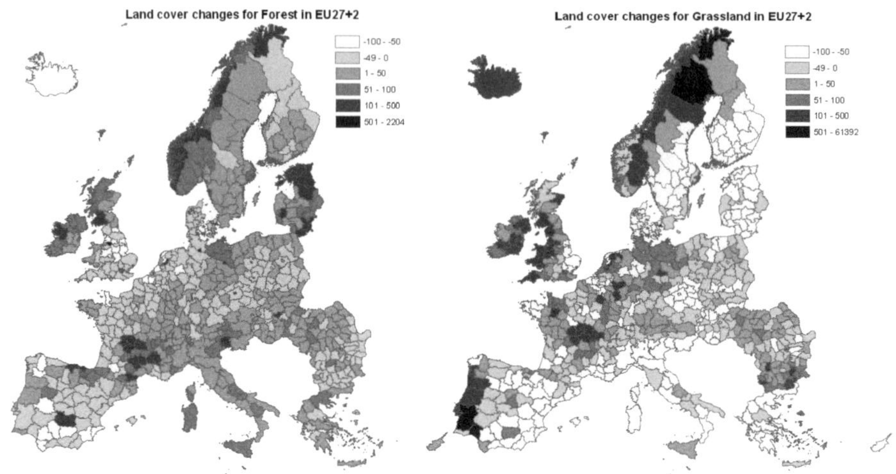


Figure 1. Spatial distribution of the amount of changes during the 1960-1990 period for forest (left) and grasslands (right) in Europe for NUTSx regions. The in- or decrease is expressed as percentage of the 1960 area.

## 5 DISCUSSION AND CONCLUSIONS

The analysis presented in this paper is restricted to land cover data with a spatial resolution of 1 by 1 km and 7 HISLU60 classes. The spatial and thematic detail is determined by the less detailed database HISLU60. The production of a Pan-European land cover map for 1960 is determined by the availability of non-digital information. Homogeneity of such maps, thematic and spatial detail are limited in today's perspectives. Improvements of the urban areas of the HISLU60 database resulted in better but still low accuracies.

The assessment of the HISLU60 database on basis of three independent data sources shows low overall locational accuracies (<60%). Therefore, a pixel comparison of changes between 1960-1990 is difficult. A statistical comparison of land cover at regional (NUTSx) level have its value in the understanding of processes (e.g. increase/decrease of forest/grassland) that determine the landscape and took place within the 1960-1990 period. However, regional differences in the HISLU60 database make sure that these indications have to be dealt with care. Besides varying the value of the HISLU60/reference database indices, a comparison with BIOPRESS 1950-1990 statistics can help to better interpret the results of the regional land cover change assessment.

Historical change analysis remains difficult and needs to be further explored. Results of a historical land cover change analysis needs to be validated and interpreted in relation to the way they are produced and used.

The PLCM databases show some shortcomings mainly due to the production of a Pan-European mosaic out of a limited number of available and suitable databases. For example, the PLCM2000 is produced on basis of 5 different databases with each a different thematic, spatial and temporal detail. In this sense the potential of new land cover datasets, such as Globcover needs to be explored. However, the availability of 'historical data' to construct land cover datasets of the past remains difficult. For example, the PLCM1990 partially coincides with the PLCM2000 database as no Pan-European coverage exist for both years. The production of future pan-European land cover datasets will be possible due to the foreseen availability of CLC2006 (extended European window, detailed land cover of 2006) and detailed soil sealing and forest layers.

## ACKNOWLEDGEMENTS

The research presented in this paper was done in the framework of the ECOCHANGE project which is an integrated FP6 project funded by EU.

## REFERENCES


Bartholomé, E. & Belward, A.S. 2005. GLC2000: a new approach to global land cover mapping from Earth observation data. *International Journal of Remote Sensing* 26 (9), 1959-1977.

Bossard, M., Feranec, J., & Otahel, J. 2000. CORINE land cover technical giude – Addendum 2000. *Technical report* 40. European Environment Agency, Copenhagen. http://terrestrial.eionet.eea.int

Büttner, G., Feranec, J., Jaffrain, G., Mari, L., Maucha, G. & Soukup, T. 2004. The CORINE Land Cover 2000 Project. in R. Reuter, (Editor). *EARSeL eProceedings*, 3, (3). EARSeL, Paris, pp. 331-346.

Di Gregorio, A. 2005. *Land Cover Classification System (LCCS): classification concepts and user manual.* FAO, Rome.

Feranec, J., Hazeu, G., Christensen, S., & Jaffrain, G. 2007. CORINE land cover change detection in Europe (case studies of the Netherlands and Slovakia). *Land Use Policy*, 24, 234-247.

Gerard, F., Thompson, A., Manchester, S., Smith, G., Wadsworth, R., Swetnam, S., Petit, S., Gregor, M., Luque, S., Huitu, H., Köhler, R., Olschofsky, K., Hazeu, G., Mücher, C.A., Halada, L., Bugár, G. & Pino, J. 2006. Land Cover Change in Europe from the 1950'ies to 2000. *Aerial photo interpretation and derived statistics from 59 samples distributed across Europe.* Edited by Raul Köhler, Konstantin Olschofsky and France Gerard. Book published by University of Hamburg, World Forestry Institute, Germany. ISBN 80-89088-46-5, 364 pp.

Haines-Young, R. & Weber, J.-L. 2006. Land accounts for Europe 1990-2000. Towards integrated land and ecosystem accounting. *EEA Report* 11. Copenhagen, European Environment Agency.

Hazeu, G.W. & Mücher, C.A. 2005. Historic land use dynamics in and around Natura2000 sites as indicators for impact on biodiversity. Wageningen, Alterra. *Alterra-report* 1077, 154p., 19 figs., 8 tables and 15 refs

Hazeu, G.W., Mücher, C.A., Kienast, F. & Zimmerman, N. 2008. *Land cover maps for environmental modeling at multiple scales.* Deliverable D01.02.01 for the ECOCHANGE project. Integrated project EU Funded Project FP6 2006 GOCE 036866.

Heymann, Y., Steenmans, Ch., Croissille, G. & Bossard, M. 1994. *CORINE land cover. Technical guide.* Luxembourg, Office for Official Publications European Communities.

Kaufman, R.K. & Seto, K.C. 2001. "Change detection, accuracy, and bias in a sequential analysis of Landsat Imagery in the Pearl River Delta, China: econometric techniques," *Agriculture, Ecosystems and Environment,* Vol. 85, 2001, pp. 95-105.

Kramer H. & Mücher, C.A. 2006. Land Use from the days before the satellites. *Proceedings of the 2nd Workshop of the EARSeL SIG on Land Use and Land Cover*, 28-30 September 2006, Center for Remote Sensing of Land Surfaces, Bonn, Germany.

Mücher, C.A., J.L. Champeaux, K.T. Steinnocher, S. Griguolo, K. Wester, C. Heunks, W. Winiwater, F.P. Kressler, J.P. Goutorbe, B. ten Brink, V.F. van Katwijk, O. Furberg, V. Perdigao & Nieuwenhuis, G.J.A. 2001. Development of a consistent methodology to derive land cover information on a European scale from Remote Sensing for environmental monitoring. The PELCOM report. 160 pp. *Alterra-report* 178, ISSN 1566-7197.

Mücher, C.A., Hennekens, S.M., Bunce, R.G.H. & Schaminée, J.H.J. 2004. Mapping European Habitats to support the design and implementation of a Pan_European Ecological Network. The PEENHAB project. *Alterra-report* 952, ISSN 1566-7197.

Neumann, K., Herold, M., Hartley, A. & Schmullius, C. 2007. Comparative assessment of CORINE2000 and GLC2000: Spatial analysis of land cover data for Europe. *International Journal of Applied Earth Observation and Geoinformation* 9 (2007) 425-437.

Nippel, T. & Klingl, T. 1998. Swiss Land Use in the European context. Integration of Swiss Land Use Statistics with CORINE Land Cover. Swiss Federal Statistical Office, Neufchâtel, Switzerland.

Weber, J.-L. 2007. Implementation of land and ecosystem accounts at the European Environment Agency. *Ecological Economics* 61, 695-707.

World Atlas of Agriculture 1969. Land utilisation maps prepared by the Committee for the World Atlas of Agriculture (Instituto Geografico De Agostini S. p. A. – Novara), 62 maps.

http://dataservice.eea.europa.eu/dataservice/metadetails.asp?id=1007

*Remote Sensing for a Changing Europe*
*D. Maktav (Ed.)*
*IOS Press, 2009*

*doi:10.3233/978-1-58603-986-8-226*

# Historic land cover changes at Natura 2000 sites and their associated landspaces across Europe

C.A. Mücher & G.W. Hazeu
*Alterra, Wageningen UR, Droevendaalsesteeg 3, 6708 PB Wageningen, NL; e-mail: sander.mucher@wur.nl*

R. Swetnam & F. Gerard
*Centre for Ecology and Hydrology, CEH, Monks Wood, Huntingdon, Cambridgeshire, UK*

S. Luque
*Cemagref, UR AMM, 2, rue de la papeterie, BP 76, F-38402, St-Martin-d'Hères, cedex, France.*

J. Pino
*Center for Ecological Research and Forestry Applications, CREAF, Bellaterra, Spain.*

L. Halada
*Institute of Landscape Ecology of the Slovak Academy of Sciences, ILE SAS, Stefanikova 3, PO Box 254, 814 99 Bratislava, Republic of Slovakia*

Keywords: land cover changes; Europe; Landscapes, Natura 2000; Habitats; biodiversity

ABSTRACT: In order to test the potential efficiency of the Natura 2000 network in protecting habitats and landscapes from land use changes that have a significant impact on biodiversity, an analysis of land use changes over the last fifty years was undertaken for 71 protected sites across Europe. The analysis showed that designation of sites limits the amount of dramatic land cover changes. Unfortunately, the speed of land cover changes outside protected sites does not slow down and causes increased isolation of protected sites and fragmentation of remaining habitats within the wide diversity of European landscapes. Therefore, the landscapes in which the remaining habitats are embedded should receive more attention for conservation strategies. This paper focuses firstly on a very detailed level (scale 1:20.000) on land cover changes at protected sites over the last fifty years, and secondly it focuses on a more regional scale (scale 1:100.000) at land cover changes within the European landscapes over the last 10 years. The impacts of the land cover changes on biodiversity are very site-specific and depends on the habitat types occurring with the specific landscape type and what the range of the specific habitat is.

## 1 INTRODUCTION

Modifications of ecosystems are most visible through changing land cover which has resulted in both positive and negative changes in landscape quality and biodiversity (Mander & Jongman 2000). Many of these changes are incremental and as such, go unnoticed until the changes become quite significant. For example, in Europe there have been significant losses in the area of agricultural land over the last fifty years (Gerard et al., 2006). These changes are most clearly indicated through the measurement and monitoring of land cover change which can be detected and quantified through the analysis of Earth Observation (EO) data. Legal frameworks to protect the environment, to ensure sustainable use of its natural resources and to maintain an acceptable level of biodiversity exist at both national and international level. Within the European Union the most significant of these include the Bird, Habitats, Nitrate and Water Directives. For the policy makers, there remains an important need to monitor these networks in order to ensure effective delivery of landscape and nature conservation targets. Information on land cover derived from EO data can provide important benchmarks against which change can be measured. However interpretation of the land cover changes in terms of impact on biodiversity require site specific information which can be partly obtained from existing environmental databases integrated over landscapes, but information from field sampling is also highly esteemed. This paper is partly based on the results of the BIOPRESS project (EVK2-2002-00178). The BIOPRESS project is strongly linked to the CORINE Land Cover project (CLC90 and CLC2000) and its methodology. The CORINE land cover project is based on the visual interpretation of high spatial resolution satellite imagery at a scale of 1: 100.000. The minimum mapping unit is 25 hectares and 5 hectares for changes. The CORINE nomenclature has three hierarchical levels, with 5 classes at level one (artificial surfaces, agricultural areas, forests and semi-natural areas, wetlands, and water bodies), 15 classes at level two, and 44 classes at level three (CEC, 1994; Feranec et al. 2007). For more details, see also the CORINE website at http://terrestrial.eionet.europa.eu/CLC2000. BIOPRESS aimed to capture overall patterns of change in Europe, with a focus on Natura 2000 sites, and to develop ways of converting this information into measures of impact on biodiversity (F. Gerard et al. 2006; Hazeu & Mücher, 2005). Main product of the BIOPRESS project was a consistent database of historical land cover changes (1950-1990-2000) based on the interpretation of georeferenced aerial photographs in and around Natura 2000 sites across Europe for 73 window samples (30 km by 30 km, scale 1:100.000) and for 59 detailed transect samples (15 km by 2 km, scale 1:20.000). This paper focuses firstly on a very detailed level (scale 1:20.000) on land cover changes at protected sites over the last fifty years, and secondly it focuses on a more regional scale (scale 1:100.000) at land cover changes within the European landscapes over the last 10 years. The impacts of the land cover changes on biodiversity are very site-specific and depends on the habitat types occurring with the specific landscape type and what the range of the specific habitat is.

## 2 BIOPRESS RESULTS

When summarizing the land cover/use changes for all transects, the following trends can be found for Europe for the period 1950-1990 and the period 1990-2000 (see Figure 1): a general net decrease in agricultural land (e.g. crops, orchards, pastures); a net increase in artificial surfaces (e.g. built- up and road networks); and a net increase in forest and semi-natural areas (e.g. natural grasslands, moors and heathlands). Although wetlands (e.g. salt marshes and intertidal flats) and water bodies (e.g. rivers and lakes) were showing local losses, their total acreage remained stable (Hazeu & Mücher, 2005; Gerard et al, 2006). Remarkable is that in the period 1990-2000 the rate of land cover changes has only slightly slowed down compared to period 1950-1990. Same trends are found for the two time periods.

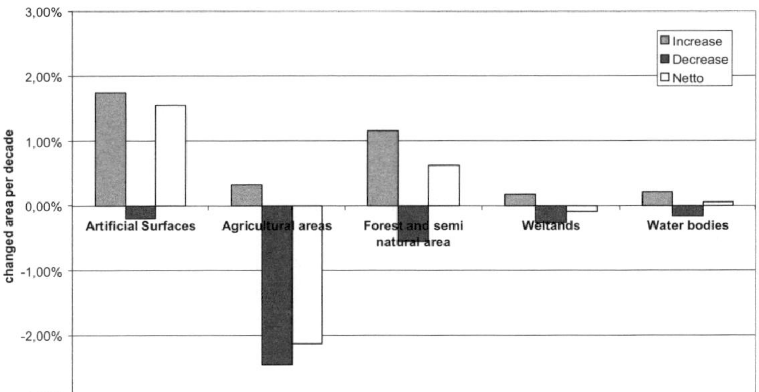


*Figure 1 Land cover changes, expressed in percentage area changed per decade, derived from historical AP interpretation for 59 transects across Europe at scale 1:20.000 for the period 1950-1990.*

## 3 LAND COVER CHANGES AROUND NATURA 2000 SITES

Land cover changes (1950-1990-2000) at a scale 1:20.000 were analyzed for 42 transects that intersected with 71 Natura 200 sites distributed over 6 countries (Belgium, Finland, Germany, Spain, the Netherlands and the United Kingdom. Most land cover changes take place outside Natura 2000 sites. However, if one considers the total area inside versus outside Natura 2000 sites the relative amount of changes is almost equal inside and outside the Natura 2000 sites (Corine level 3) . So, the perception that not much has changed within protected sites over the last fifty years is wrong. Therefore, the amount of land cover changes (in ha) should be normalized for the surface area it concerns (ha), in other words we should look at the percentual change. At CORINE level 1, so only considering the severe changes, eg. from agriculture to forest or urban areas, twice as much land cover changes took place outside (20.8 %) than inside (9,8%) the protected sites. So, designation helps to reduce the amount of severe land cover changes. Analysis of the spatial distribution of the land cover changes (at CORINE level one at various distances from the Natura 2000 borders over the period 1950-2000) indicates that the absolute amount of changes took place just outside the Natura 2000 sites and just inside the Natura 2000 sites. So, very close to the Natura 2000 borders. The further the distance from the Natura 2000 border the less land cover changes can be observed. However, to account for the varying size of the buffer zones the amount of changes (ha) have to be divided by the total area of the relevant buffer area (ha), as a measure of the relative change. In the core of the Natura 2000 sites only 0.5 % of the area has changed significantly of the last fifty years. Which means that in the core area of most Natura 2000 sites the environmental pressures are very low and so well protected. Since there is a gradient of increasing percentual change from the core of the Natura 2000 sites to a distance of 5 km outside the Natura 2000 sites, it makes sense to create bufferzones around designated sites to improve protection against severe changes. At CORINE level 3, these concern more gradual and natural changes, e.g. from heathlands to forest and from natural grasslands to sclerophyllous vegetation, the gradient is less weaker from inside to outside Natura 2000 sites.

The types of land cover changes differ very much between inside and outside the designated sites. The land cover flows inside Natura 2000 sites are especially related to natural succession and only partly to changes from agriculture. While land cover flows outside Natura 2000 sites are also related to these processes, they are even more to changes in agriculture, and to urbanization. In principle, there are 1892 land cover change (LCC) types possible at CORINE level three (44 classes). However, the realized types of land cover changes is much lower. Moreover, the amount of land cover change types at CORINE level three is much lower inside Natura 2000 sites with 376 LCC types outside with 555 LCC types. Human activities lead to a wider range in land cover flows, and is stronger outside the Natura 2000 sites. Land cover changes at CORINE level one have a stronger impact on biodiversity than land cover changes at lower levels of the CORINE hierarchy. Land cover changes at level one always concern major modifications in the ecosystem (e.g. change from agriculture to urban) while changes at level three concern slighter changes (e.g. natural succession from heathland into transitional woodland-scrub). Therefore, it is interesting to analyse the CORINE level one land cover changes.

The most important Land cover flows at CORINE level one are: from agriculture to urban with 41.0 %; from agriculture to forests with 25.1 %; from forests to agriculture with 6.8 %; from forests to urban (3-1) with 6.8 %; from wetlands to forests (4-3) with 5.8 %. So, agriculture has lost terrain at the expense of expansion of forests and urban areas over the last fifty years. The main land cover flow "from agriculture to urban", occurs for 96.8 % outside Natura 2000 sites. The change from agriculture to forests occurs for 70% outside the designated sites. Changes that occur mainly inside Natura 2000 sites are changes from forests to wetlands (with 98.8%), from wetlands to water (with 85.2%) and vice versa (75.5%), and from forests to water (62.4%) and vice versa (54.3%). These changes are partly natural, but are triggered to a large extent by changes in agriculture outside Natura 2000 sites that have for example a big influence on the groundwater table and deposition of nitrogen. Also measures by nature conservation have triggered some of these changes (e.g. from forests to wetlands). So, also here, the overall conclusion is that designation helps to restrict the amount and types of land cover changes that have the largest impact on biodiversity. So, from the analysis of historic land cover changes in and around Natura 2000 sites over the last fifty year it becomes clear that designation helps to protect biodiversity.

## 4 LAND COVER CHANGES IN RELATION TO EUROPEAN LANDSCAPES AND ASSOCIATED HABITATS

This first part of the paper focused on very detailed land cover changes (scale 1:20.000) over the last fifty years across Natura 2000 sites. It was shown that designation helps the reduce not only the amount of severe land cover changes but designation influences also the types of land cover changes. However, it must be stressed that land cover changes outside designated sites still can have a severe impact on non-protected habitats and increase the isolation of such sites across all European landscapes. Therefore, we concentrate now on a more regional scale (scale 1:100.000) at land cover changes across all European landscapes over the last 10 years. For that purpose we integrated the European Landscape Classification - LANMAP (Mücher et al., 2006) with the Natura 2000 site database and the CORINE land cover changes over the period 1990-2000. Integration of the Natura 2000 database with LANMAP indicates that diversity in habitat types is very different over the European landscapes. As we all know Europe has an enormous diversity in landscapes that are very rich in their cultural history. Europe has therefore also more cultural landscapes than natural landscapes. Nevertheless, these cultural landscapes are often richer in terms of habitats than the natural landscapes.

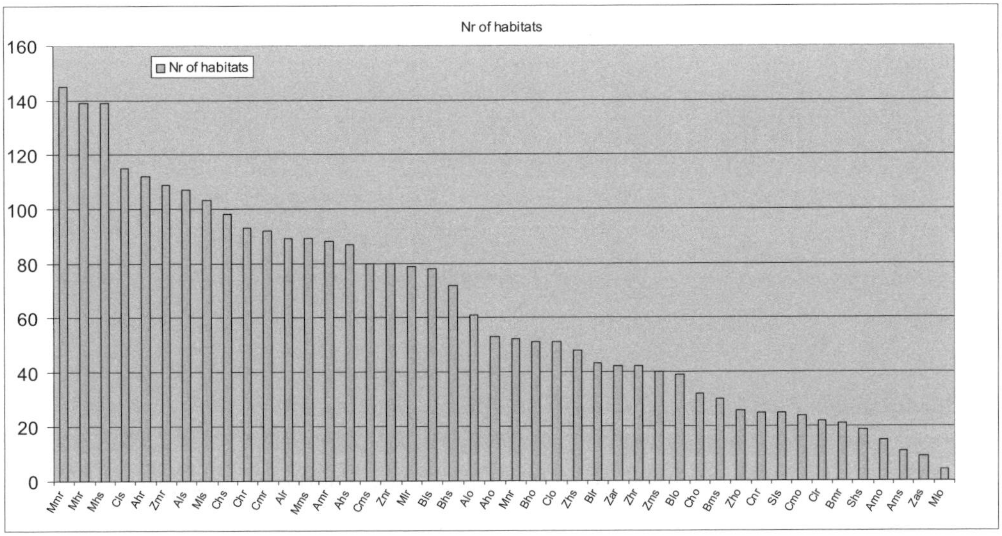


*Figure 2 Number of Annex I habitat types per European landscape type (level 3).*

For example, cultural landscapes such as the Mediterranean mountains (*Mmr*) and Mediterranean hills (*Mhr* and *Mhs*) are very rich in their number of Annex I habitat types. While more natural Boreal landscapes, like Boreal mountains (*Bms* and *Bmr*), count much less Annex I habitat types. This does not make these landscapes les valuable but it should stressed that there is a wide diversity in habitats between landscapes. At the same time we see that some habitat types, like H-6430 "Hydrophilous tall herb fringe communities", H-91E0 "Alluvial forests", H-7140 "Transition mires and bogs" and H-3150 "Natural eutrophic lakes", are present in more than 120 European landscape types and have therefore a very broad range. While other habitats such as H-1180 "submarine structures made by leaking gases", H-1610 "Baltic esker islands", H5140 "Cistus palhinhae formations on maritime wet heaths", H-91X0 """Dobrogean beeach forests", are restricted to only one landscape type. If we have a look now at the land cover changes that took place over the period 1990-2000 we see that the amount and types of land cover changes differ very much across the different European landscapes. For example, afforestation affected much more the Mediterranean mountains (*Mmr*) and hills (*Mhr*) than the Boreal hills (*Bho* and *Bhs*) and lowlands (*Blo* and *Bls*). The Mediterranean hills on rocks are very rich in natural grassland formations (15 Annex I habitat types), sclerophyllous scrub (12 Annex I types), Rocky habitats (12 Annex I habitat types) and heath and shrublands (6 Annex I habitat types) and all these vegetation formations are very sensitive to afforestation which will lead in the end to a loss of biodiversity. Urbanisation takes especially place in the Atlantic lowlands dominated by sediments (*Als*) and has a major impact on all habitat types.

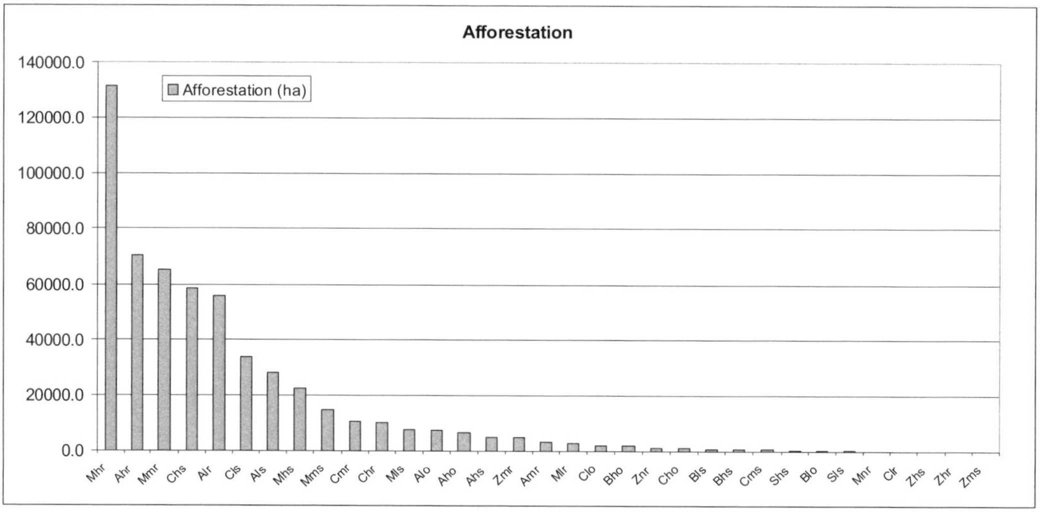


*Figure 3 Afforestation (natural regeneration and planting) expressed in ha per European landscape type (level 3) over the last ten years.*

## 5 CONCLUSIONS

The BIOPRESS project showed that Europe has been affected by an enormous amount of land cover changes of the last fifty years and unfortunately the rate of land cover changes has not really slowed down. The highest annual change rates were found in the Alpine and Atlantic zone over the last fifty years. Urban areas and forests have been expending at the expense of agricultural land. An important policy instrument to reduce the amount of severe land cover changes is the designation of sites. Analysis of land cover changes inside versus outside protected sites showed that designation indeed reduced the amount of dramatic changes. Even the distance from the borders influences the amount of severe changes. In the core of the designated sites 0.5% has changed on average, while on a 3,5 km distance outside the sites 27% has changed on average. So the amount of LC changes per ha increases with the distance outside Natura 2000 sites and decreases with the distance inside. More gradual changes, such as vegetation succession, are distributed more even inside and outside the protected sites. In general, there is much difference in the land cover flows from inside to outside Natura 200 sites. For example, changes from agriculture to urban areas took place for 97.4% outside Natura 2000 sites. While specific changes, e.g. from semi-natural to wetlands, took for 98.8 % place inside the sites. However, it must be stressed that land cover changes outside designated sites still can have a severe impact on non-protected habitats and increase the isolation of such sites. Therefore, a second step was to see how land cover changes evolved across all European landscapes. It was shown that European landscapes are very diverse in their Annex I habitat types and that the extent of the habitat types itself differ very much across the European landscapes. Land cover flows cause different pressures on biodiversity and it was shown that over the period 1990-2000 afforestation affected especially Mediterranean mountains and hills which are rich in natural grassland formations, sclerophyllous vegetation, rocky habitats and shrubs. Urbanisation affected especially the Atlantic lowlands and has a major impact on all habitat types. This paper demonstrated that LC changes and associated pressures differ very much between all European landscapes and that the biodiversity impact of these local LC changes depend very much on the habitat types present in a particular landscape and on the extent of the specific habitat type. In other words, impact of land cover changes on biodiversity are very site specific and it is recommended not only to analyse existing environmental databases but also to add information from field surveys to interpret the land cover changes in the right way.

## ACKNOWLEDGEMENTS

We would like to thank the Alternet programme and the EU projects BIOPRESS and ECO-CHANGE for supporting our research and to enable the publication of our results. In this context, we are also grateful to European Topic Centre on Land Use and Spatial Information (ETC-LUSI) and the European Topic Centre on Biological Diversity (ETC/BD) for their support.

## REFERENCES

Commission of the European Communities (1994). CORINE land cover. Technical guide. Office for Official Publications of European Communities, Luxembourg.

Feranec, J., Hazeu, G.W., Christensen, S., & Jaffrain G. (2007). CORINE Land Cover change detection in Europe (Case studies of the Netherlands and Slovakia). Land Use Policy 24, 234-247.

Gerard, F. Gregor, M., Luque S., Huitu H., Köhler, R., Olschofsky, k., Hazeu, G., Mücher, C.A.., Halada, L., Bugár, G., & Pino, J, (2006). Land Cover Change in Europe from the 1950'ies to 2000. Edited by Raul Köhler, Konstantin Olschofsky. University of Hamburg, World Forestry Institute, Germany. 364 pp.

Hazeu, G.W., & Mücher, C.A. (2005). Historic land use dynamics in and around Natura 2000 sites as indicators for impact on biodiversity. Phase I of the BIOPRESS project for the Netherlands. Alterra Report 1077, Wageningen, the Netherlands.

Mander, Ü., & Jongman, R. H. G. (2000). Consequences of land use changes. Southampton, WIT Press, 314 pages.

Mücher, C.A., Wascher, D.M, Klijn, J.A. Koomen A.J.M. and Jongman, R.H.G. (2006) A new European Landscape Map as an integrative framework for landscape character assessment. In: Landscape Ecology in the Mediterranean: inside and outside approaches, R.G.H. Bunce and R.H.G. Jongman (Eds) 2006. IALE Publication Series 3, pp. 233-243.

*Remote Sensing for a Changing Europe*
*D. Maktav (Ed.)*
*IOS Press, 2009*

*doi:10.3233/978-1-58603-986-8-232*

# Integration of Panchromatic and Multispectral Images by Local Fractal Dimension

M. Lillo-Saavedra
*Facultad de Ingeniería Agrícola, Departamento de Mecanización y Energía Av. Vicente Méndez 595, Casilla 537, Chillán, CHILE. e-mail: malillo@udec.cl*

C. Gonzalo-Martín
*Departamento de Arquitectura y Tecnología de Sistemas Informáticos, Facultad de Informática, Universidad Politécnica de Madrid, Campus de Montegancedo 28660, Madrid, ESPAÑA.*
*e-mail: consuelo.gonzalo@upm.es*

Keywords: Fusion Image, Wavelet *à trous*, Fractal dimension

ABSTRACT:
The fusion image strategies are a good solution to obtain a synthetic image with high spatial and spectral characteristics simultaneously. Some of them are based on the Wavelet Transform, computed by means of the *à trous* algorithm (AWT). Most of them do not differentiated between spectral bands. In this sense, a new approach that weights differently the spatial information integrated from the high resolution image in each of the fused image spectral bands by the optimization of the trade off between the spatial and spectral quality of the fused images, was proposed. The main problems of this approach are that a unique weighting factor for the whole spectral band is computed, and the need of indices, that separately measure the spectral and spatial quality of the fused images. In this work, a new strategy that tries to avoid the problems above mentioned is introduced. For that, it is proposed to determine a local weighting factor for each panchromatic pixel by means the fractal map, using the box-counting algorithm. Panchromatic and multispectral Quickbird images have been used to show the performances of this new methodology. The local quality of the final fused images has been evaluated by means of local quality maps of Q index. It has been proved that the proposed fusion strategy preserve the high frequency information of the panchromatic image in areas with a high detail, while in homogeneous areas the low frequency information of the multispectral image are conserved.

## 1 INTRODUCTION

Today, a great number of methods and algorithms for fusing multispectral and panchromatic images are available. Most of them are based on the use of different types of transforms. Some are very simple from a conceptual point of view, like the methods based on the Brovey transform (BT) (Hallada and Cox 1983), or the IHS transform (Intensity, Hue, Saturation) (Chavez et al. 1991). In addition, there exists a high number of methodologies based on multi-resolution analysis techniques, that essentially use the Discrete Wavelet Transform (DWT), both in its pyramidal version, by means of the Mallat algorithm (Ranchin and Wald 2000), and in its redundant version, using the *à trous* ("with holes") algorithm (Dutilleux 1987, Gonzalez-Audícana et al., 2005, Gonzalo and Lillo-Saavedra 2006). All these methods try to merge in a coherent way the spatial information of the panchromatic image with the spectral information of the multiespectral one.

A common characteristic to all fusion strategies of panchromatic and multispectral images, available today, is that they do not distinguish between different land cover areas. That means the same information retraived from the panchromatic image is integrated in the multispectral image independently of the surfaces characteristics. However, it is know that for some remote sensing application, like traditional spectral classification on pixel basis, high spatial resolution increases within-field variability and therefore may decrease the classification accuracy due to the blow-up information. It has been shown that different classes need different resolution images to be correctly identified (Huang et al. 2003).

In this sense, it can be very useful the availability of fusion strategies that apply different fusion rules dependent on land cover characteristics. Thus, areas with a high detail should preserve the high frequency information of the panchromatic image in areas with a high detail, while in homogeneous areas the low frequency information of the multispectral image should be conserved.

It is well known that natural surfaces are often erratic and present complex features at any scale. They do not reveal Euclidean shape and therefore they can not be analysed by the traditional Euclidian geometry. The ability of fractal geometry to describe irregular shapes or complex objects present in natural surfaces allows their analysis.

Most of the traditional applications of fractal techniques to image analysis are based on the estimation of fractal dimension. This is a key parameter in fractal geometry which measures the irregularity of complex objects, as well as the homogeneity of uniform surfaces. Two dimension objects have a fractal dimension greater than two and less than three. Rough surfaces have higher fractal dimension than smooth surfaces and tend to fill the 3D space. In this sense, the fractal dimension can be related to the intuitive idea of roughness (Pentland 1984) and consequently can be use to discriminate between surfaces with different textural features. It has been proved that the characterization of a whole image by a unique number (fractal dimension) can not always discriminate between very different surfaces. In this sense, it seems to be more effective estimate the local fractal dimension around each pixel and generate a fractal map to characterize the roughness of a whole image.

The main aim of this work is to investigate the ability of the local fractal dimension to improve the quality of fused images by the DWT using *á trous* algorithm (AWT).

## 2 FUSION METHODOLOGY

### 2.1 *Weighting ATW Fusion Methodology*

The algorithm proposed by Dutilleux (1987) (AWT), presents two principal characteristics, first a direction independence of the filtering process and moreover it is redundant, in the sense that, there is not any dyadic spatial compression of the original image between two successive degradation levels, thus the size of that image is maintained.

The AWT algorithm consists basically in the application of consecutive convolutions between the image under analysis and a scaling function at distinct decomposition levels. One of the most widely used scaling functions for the computation of the *à trous* algorithm is the b3-spline (Lillo-Saavedra and Gonzalo 2006, González-Audícana et al. 2005).

If the original image is represented by $I_j(x,y)$, the wavelet coefficients, $C_{j+n}(x,y)$ for the decomposition level j+n, are obtained by the difference between the corresponding two consecutive degraded images, $I_{j+n-1}(x,y)$ and $I_{j+n}(x,y)$, as it is shown in equation (1):

$$C_{j+n}(x,y) = I_{j+n}(x,y) - I_{j+n-1}(x,y) \quad (1)$$

To synthesize the image from a decomposition level j+n, an additive criterion that adds all the coefficients obtained to the last decomposition level can be applied, as it is shown in equation (2):

$$I_j(x,y) = I_{j+n}(x,y) + \sum_{k=1}^{n} C_{j+k}(x,y) \tag{2}$$

Where $I_{j+n}(x,y)$ represents a background image that contains low frequency information of the original image, and $C_{j+n}(x,y)$ their respective wavelet coefficients, which contain high frequency information. From equations (1) and (2) an image fusion strategy can be proposed, in which the low frequency information, contained in a multispectral image, can be integrated with the high frequency information, contained in the wavelet coefficients of a high resolution spatial image (panchromatic image), resulting a multispectral image with high spatial resolution. This fusion strategy does not provide control on the spatial and spectral quality of the fused images. In this sense, a weighting AWT version of a fusion strategy, based on the determination of an objective trade-off criterion between spectral and spatial quality of the fused images, has been proposed in Lillo-Saavedra & Gonzalo (2006).

$$I^{i}_{Fus}(x,y) = I^{i}_{MULTI_{j+n}}(x,y) + \alpha^{i} \sum_{k=1}^{n} C_{PAN_{j+k}}(x,y) \tag{3}$$

Where the index i represent the number of bands of the MULTI image and n the number of decompositions levels from the j level; $\sum_{k=1}^{n} C_{PAN_k}$ represents the sum of all PAN image wavelet coefficients; and $\alpha^i$ represents the weighting factors of these coefficients for each band.

Although this fusion methodology improves the AWT benefits, the $\alpha^i$ value is applied to whole spectral band, not having this methodology a strategy to discriminate the amount of high frequency information that is required to inject in the multispectral source image, in areas with different texture characteristics o roughness, in other word in different kind of land covers.

## 2.2 *Weighting ATW by local fractal dimension*

In this paper it is proposed to estimate the $\alpha^i$ (x,y) values, one for each pixel of each spectral band, by means of the information provides by the local fractal maps of the panchromatic image and the spectral band to be fused. Different relations between the local fractal maps of source images can be established for $\alpha^i$ (x,y) values estimation.

$$\alpha^{i}(x,y) = f(\mathrm{dim\_frac}(\mathrm{MULTI}^{i}(x,y)), \mathrm{dim\_frac}(\mathrm{PAN}(x,y)) \tag{4}$$

Where f(·) could be any algebraic relations between the local fractal maps of source images. These maps have been calculated by means the box-counting algorithm. Different experiments have show that the best results are reached when the information from sources images are averaged and normalized.

## 2.3 *Data and pre-processed*

The data used to evaluate the AWT fusion methodology, weighted by local fractal dimension, correspond to a scene recorded by the panchromatic and multispectral sensors transported on QUICKBIRD satellites. The multispectral image size is 512x512 pixels and consequently the panchromatic image size is 2048x2048 pixels, covering 420 Ha. The scene has been recorded on february 18, 2005, and is geographically located in the Peumo Valley, Cachapoal watershed, Chile (34º17'58'' S, 71º19'52'' O). Color compositions of the multispectral image and its corresponding panchromatic image are presented in Fig. 4 (a) and (b), respectively.

a) b)

Figure 1: Source images. a) Panchromatic. B) Multispectral color composition

Previously to the fusion process, source images should be pre-processed. The MULTI image must be resized to the PAN image size by an interpolation method, and also it should be co-registered with this last image.

## 3 RESULTS

Obtained results from this study have confirmed that local fractal dimensions maps are an efficient tool to discriminate areas with different texture or roughness at different scales. Thus it can be appreciated at figure 2, the different values of local fractal dimension for a same scene at two different resolutions: in figure 2 (a) for the panchromatic image (0.6 m) and in figure 2 (b) for the spectral band 3 of the multispectral image (2.4 m). It should be noted, that the polygons at the center of the scene, present high local fractal dimension in the high resolution image (panchromatic), where the lines crops can be identified, while for low resolution (multispectral image) it has a fractal dimension near zero, since at that resolution the mentioned lines can not be detected.

This kind of information has resulted very useful to discriminate areas where each one of the source images should have different weight in the fused image, from the point of view of certain applications of fused images like classification task. The idea already mentioned at introduction is to preserve the high frequency information of the panchromatic image in areas with a high detail and conserved the low frequency information of the multispectral image in homogeneous areas.

In this paper, fusion results for the source images included in figure 1, are displayed at figure 3. In order to do a first evaluation of the performances of the proposed fusion strategy, the standard AWT fusion strategy has been considered as reference. Thus the source images have been fused according equation 3, for $\alpha=1$ for all spectral bands (figure 3 a), and estimating the different $\alpha^i$ values for each spectral band from the local fractal dimension maps of the panchromatic image and the corresponding spectral band of the multispectral image (figure 3 b).

From the visual inspection, apparently the two fused images are very similar. However, a local analysis shows significant differences between them. In order to show these differences local quality maps of Q index (Zhou 2002) have been calculated. These maps are displayed at figure 4, for the two fused images. It should be noted that high values of Q index expresses high quality. Then, it can be appreciated that in those areas where the Q values are different for the two images,

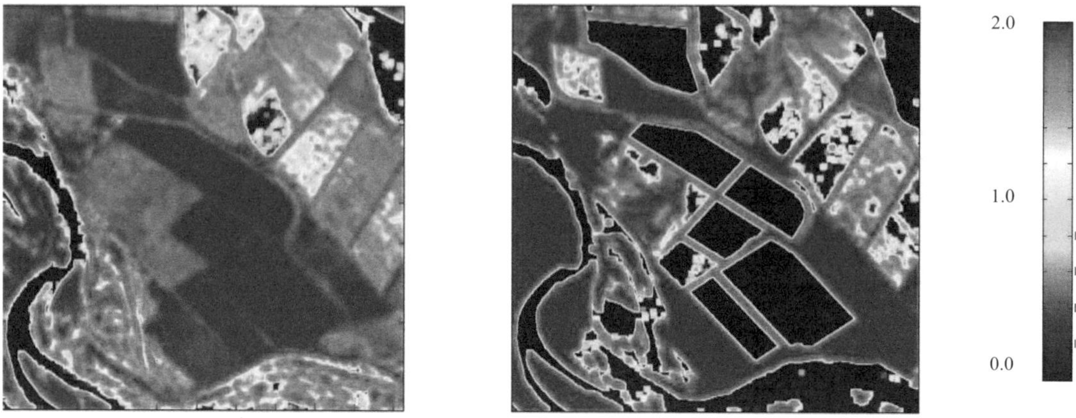


a) b)

Figure 2: Local fractal dimension maps. a) Panchromatic and b) band 3 of multispectral images

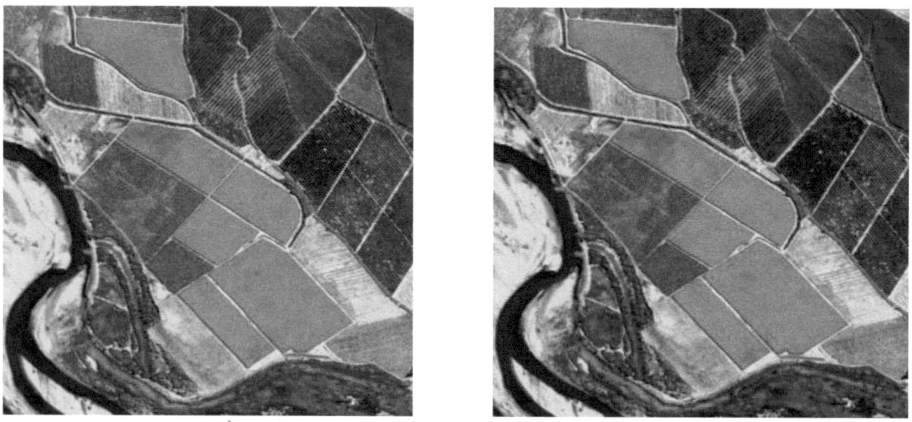

a) b)

Figure 3: Fused images. a) Standard AWT fusion method. b) Fusion method based on local fractal dimensions

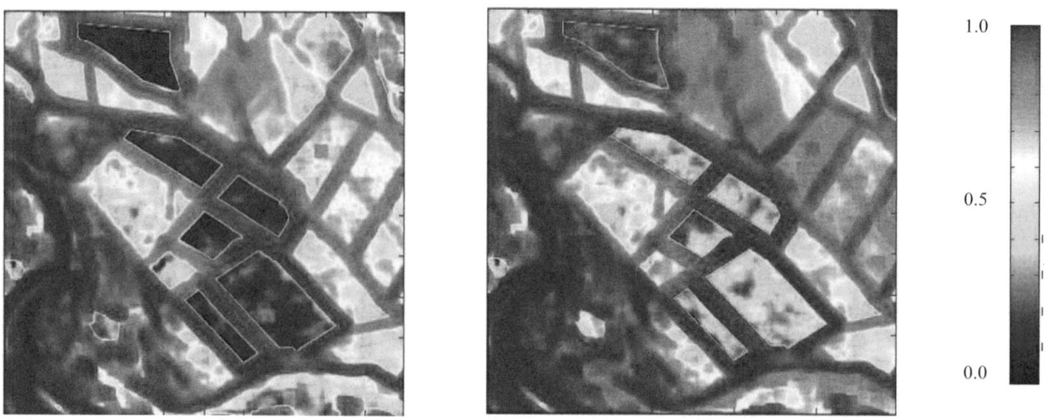


a) b)

Figure 4: Q-Maps: Standard AWT fusion method. b) Fusion method based on local fractal dimensions they are always higher for the image fused by the method based on local fractal dimension maps. Those differences are especially noteworthy in homogeneous areas, avoiding the common problem to other fusion methods that introduce artefacts in these areas.

Moreover, to show differences between the fused images, zooms of a particular area of the whole multispectral image, AWT fused image weighted by local fractal dimension and standard AWT fused image are displayed at figure 5 a), b) and c). The close up shows a zone of vineyard in the left-down diagonal (high spatial frequency) and homogeneous crop in the other side (low spatial frequency). It can be appreciated that for both fused images (figure 5 b) and c)), the spatial information has improved respect the multispectral image. Nevertheless, the smoothness is conserved for the proposed method (figure 5 b)) in the low spatial frequency zone of the multispectral image (figure 5 a)).

In order to corroborate this observation, an unsupervised classification, for the three whole images, has been carried out. The k-means parameters used has been: seven class and 5% of change threshold. Fig. 5 d), e) and f) presents the classification of zooms displayed in Fig. 5 a), b) and c). It is evident that the AWT fused image, weighted by local fractal dimension, maintains the same multispectral image classification in the homogeneous zone, but increases the class in the vineyard zone as well as that AWT fused image.

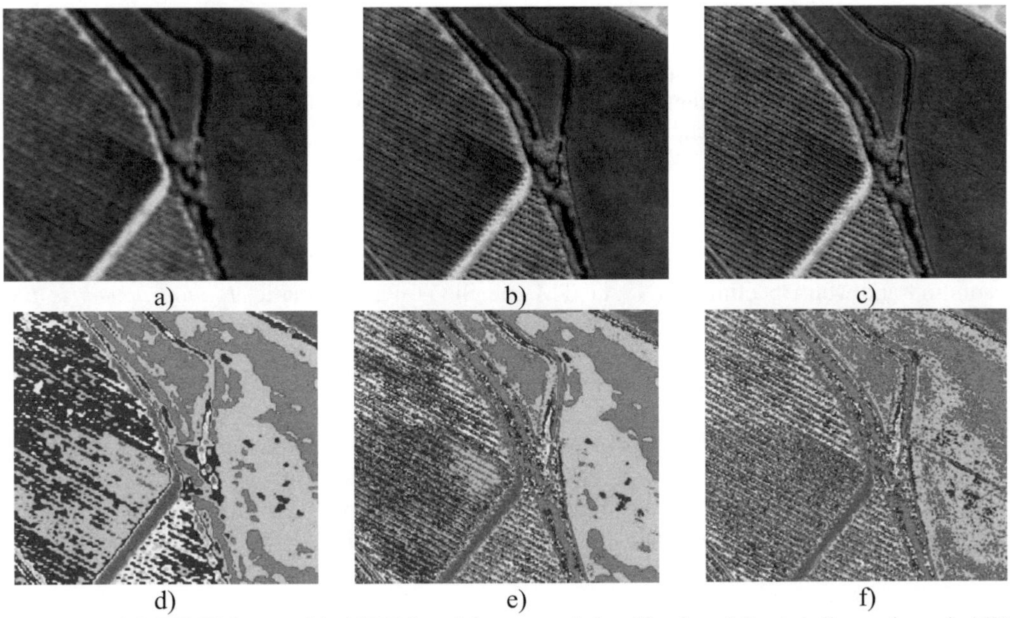


Figure 5: a) MULTI image, b) AWT fused image weighted by local fractal dimension, c) AWT fused image. Figure 5: d), e) and f) are classified image from a), b) and c), by means k-means algorithm.

## 4 CONCLUSIONS

A new methodology to fuse multispectral and panchromatic images has been proposed. The main feature of the new methodology is the local weighting of the information injected from the panchromatic image in each of the spectral bands of the multispectral image. For that the information provided for the local fractal dimension maps of the source images computing by the box-counting algorithm have been used.

From the experimental results can be concluded that the local fractal dimension, calculated by means the box-counting algorithm, is an appropriate tool to differentiate the land cover characteristics, in order to obtain a fused image without blow-up information.

Since it has been proved that the fused results depend on land covers, it is required the use of local weighting maps, consequently the quality of the fused images should be evaluated locally. It has been proved that the local quality of the fused image by the proposed fusion methodology is higher in the whole image than for the fusion without weighting. This improvement is due that this method preserves the high frequency information of the panchromatic image in areas with a high detail and conserved the low frequency information of the multispectral image in homogeneous areas. That is especially interesting for some remote sensing applications like classification. Thus it has been showed that the AWT fused image, weighted by local fractal dimension, provides the same classification results than the multispectral image in the homogeneous zone, and simultaneously it provides high discrimination performances in areas with high detail.

## ACKNOWLEDGEMENTS

The work presented in this paper has been jointly supported by the Chile Research Council (FONDECYT 11060056) and Technical University of Madrid, Spain (AL08-P(I+D)-19) TEC2007-60607/TCM (Ministerio de Educación y Ciencia, Spain).

## REFERENCES

Chavez, P.S., Sides, S.C., Anderson, J.A. 1991. Comparison of the Three Methods to Merge Multiresolution and Multispectral: LANSAT TM and SPOT Panchromatic, *Photogrammetric Engineering and Remote Sensing*, vol. 57: 295-303.

Dutilleux, P. 1987. An implementation of the algorithm à trous to compute the wavelet transform. *Compt-rendus du congrès ondelettes et méthodes temps-fréquence et espace des phases*, J. Combes, A. Grossmann and Ph. Tchanitchian (eds.), Marseille, Springer-Verlag, 298-304.

González-Audícana M., Otazu X., Fors O. y Seco A. 2005. Comparison between the Mallat's and the 'à trous' discrete wavelet transform based algorithms for the fusion of multispectral and panchromatic images. *Internacional Journal of Remote Sensing*, vol.26: 597-616.

Hallada, W.A. and Cox, S. 1983. Image Sharpening for Mixed Spatial and Spectral Resolution Satellite Systems. *Proceedings of the 17th International Symposium on Remote Sensing of Environment.* 1023-1032.

Huang, H., Bingfang, W. and Jinlong, F. 2003.Analysis to the relationship of classification accuracy, segmentation scale, image resolution. *IEEE Geoscience and Remote Sensing Symposium, IGARSS '03.Proceedings*, vol. 6: 3671-3673.

Lillo-Saavedra and M., Gonzalo. 2006. Spectral or spatial quality for fused satellite imagery? A trade-off solution using wavelet à trous algorithm. *International Journal of Remote Sensing*, vol. 27, 7: 1453-1464.

Pentland, A. 1984. Fractal Based Description ofNatural Scenes. *IEEE Trans. Pattern Analysis Mach. Intell. PAMI-6 (6)*, vol. 66: 1-674.

Ranchin, T. and Wald, L. 2000. Fusion of High Spatial and Spectral Resolution Image: The ARSIS Concept and its Implementation. *Photogrammetric Engineering and Remote Sensing*, vol. 66: 49-61.

Zhou, W. 2002. A universal image quality index. *IEEE Signal Processing Letter,* vol. 9, 3: 81-84

*Remote Sensing for a Changing Europe*
*D. Maktav (Ed.)*
*IOS Press, 2009*


*doi:10.3233/978-1-58603-986-8-239*

# Application of remote sensing to heritage conservation: a case study in central Portugal

L. M. S. Gonçalves
*Polytechnic Institute of Leiria, School of Technology and Management, Portugal*

C. C. Fonte
*Institute for Systems and Computers Engineering at Coimbra, Portugal*

*Department of Mathematics, University of Coimbra, P - Coimbra, Portugal*

E. N. B. S. Júlio
*ISISE, Civil Engineering Department, University of Coimbra, Portugal*

M. Caetano
*Portuguese Geographic Institute (IGP), Remote Sensing Unit (RSU), Lisboa, Portugal*

*CEGI, Instituto Superior de Estatística e Gestão de Informação, ISEGI, Universidade Nova de Lisboa, 1070-312 Lisboa, Portugal*

Keywords: soft classification, hybrid classification, measures of uncertainty, building roofs pathologies, heritage conservation

ABSTRACT: The aim of this work is to evaluate the applicability of Very High Spatial Resolution (VHSR) images to heritage conservation, by using multi-spectral images with different resolutions to extract information concerning the state of conservation of building roofs. To this end, a hybrid classification method was developed that incorporates uncertainty information in the automatic classification of VHSR images. The developed classification methodology includes the following steps: 1) a soft pixel-based classification; 2) computation of the classification uncertainty; and 3) buildings classification based on decision rules which include the results of the soft pixel-based classification and its uncertainty.
The proposed methodology was tested on the classification of the historical downtown centre of Coimbra, in Portugal, and validated with the results obtained from an independent pathology survey performed by the University of Coimbra. The test was carried out with a multi-spectral aerial image and a QuickBird satellite image with spatial resolutions of respectively 0.5 m and 2.4 m.

## 1 INTRODUCTION

One of the major tasks to define strategies for intervention in historical urban spaces, such as cities, blocks or buildings, is the survey of what exists. Preliminary examination plays a crucial role concerning the planning of construction and restoration tasks. However, since traditional techniques consist in manually performing pathology mapping, which is laborious and time consuming, and taken into account that some parts of the buildings are often inaccessible, this examination process can become extremely difficult. Therefore, it is desirable to explore and evaluate new methods and other possible sources of information capable of providing reliable data on existing anomalies, like corrosion and natural covers (*e.g.* moss), and also capable of enabling information concerning inaccessible parts of the buildings' roofs.

The continuous improvement of the spatial resolution of remote sensing sensors and the availability of Very High Spatial Resolution (VHSR) images allow the identification of smaller objects and, therefore, more fine features can be delineated, which lead to a broad number of new applications. This study intends to evaluate the applicability of VHSR multispectral satellite images to detect anomalies on roofs, to identify their state of conservation and to determine if increasing spatial resolution results in improvements in the identification and extraction of the information concerning roof pathology.

There are several available techniques for the automatic extraction of information from VHSR images. They can be performed on a pixel basis, object basis (*e.g.* Gonçalves and Caetano, 2004; Blaschke *et al.* 2004) or a combination of both (*e.g.* Wang *et. al.*, 2004; Plantier and Caetano, 2007), also called per-field classification when the objects result from digital vector data and Geographic Information Systems (GIS) (Aplin *et al.,* 2000). In this study a hybrid classification method that combines pixel and object classification and incorporates uncertainty information in the soft automatic classification is developed, to obtain information on buildings roof pathology of the historical downtown centre of Coimbra, Portugal. This approach has similarities with per-field classification since the objects used in the classification process were obtained from a vector cartographic map of the region. The results are then compared with those obtained with a traditional pathology survey performed by the University of Coimbra, and conclusions are drawn.

## 2 DATA SET AND CASE-STUDY AREA

The case-study area is the historical downtown centre of Coimbra with an area of approximately 15 ha (Figure 1 a)). The ancient downtown spreads along the flat riverside and is occupied mainly by old buildings, most of the latter from the 16th century, innumerable alleyways and small squares.

The data set used in the study includes two multiespectral images and a vector cartographic map at the 1:1000 scale (Figure 1). The images are a multispectral image obtained by the QuickBird sensor in December 2004, with a spatial resolution of 2.4 m in mode (XS), and multispectral aerial images with a resolution of 0.5 m. The pixels are recorded in 16 bits in the QuickBird image whereas in the aerial image only 8 bits are used. The study was performed using four multispectral bands. A vegetation index was also calculated for each data image, namely the Normalized Difference Vegetation Index (NDVI) which was used as additional band information.

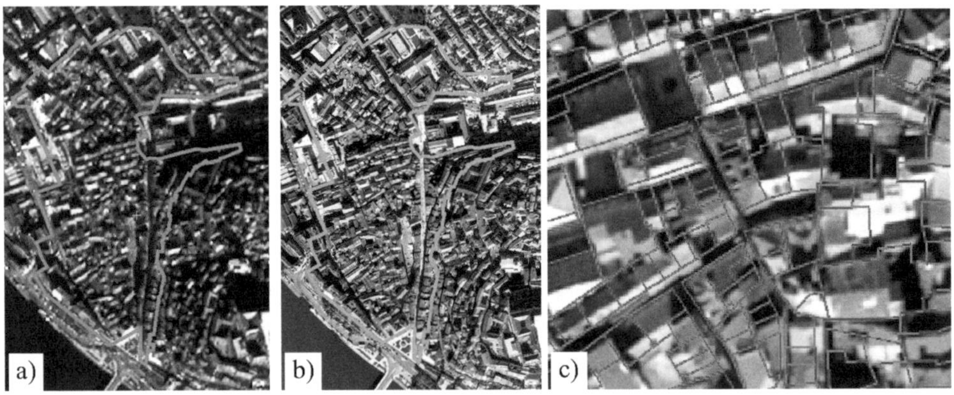


Figure 1 – Study area. a) QuickBird image (RGB 321), b) aerial image (RGB 321), c) overlay of the vector buildings map and the aerial image (RGB321).

## 3 METHODOLOGY

Before the definition and implementation of the methodology, an analysis of both images was made to evaluate if it was possible to identify, just through visualization, the main materials existing in the case-study area, the presence of anomalies in the buildings roofs or, in this case, the different kind of anomalies. This analysis was extremely important to identify the kind of information that would be possible to extract and the definition of the best extraction technique.

The main roof materials in the study area are: ceramic tiles; fibrocement corrugated sheets; and steel panels. The ceramic tile is the most representative material, used in 85% of roofs. For this reason, the present study focuses only in the identification of ceramic tile roof pathologies.

From the performed previous analysis it was possible to conclude that, even though the images have very high spatial resolution, these only enable the identification of the roofs with anomalies and do not allow the identification of the different types. Since the survey made by the University of Coimbra using traditional methods produced a classification of the buildings conservation level, the authors also worked at the building level, to allow the comparison of both methods. Furthermore, classifications such as 'Building Roofs with Pathology', or, for example, 'Building Roofs With 25% of Anomalies' can not be made only with pixel information. The identification of these classes requires not only the classification of the elementary entities (pixels in this case), but also the analysis of their arrangement. For this reason, the classification methodology includes a preliminary classification of the images, constructing a Surface Elements Materials Map (SEMM) and a Surface Elements Pathology Map (SEPM) based on a soft probabilistic pixel classification.

The subsequent step was the identification of the Building Units (BU), which were obtained from the vector cartographic map, available at 1:1000 scale. This information was converted to a raster format, generating a raster buildings map, and it was used as a mask, so that only the parts of the images inside the building were processed. At this stage the BU becomes the basic units rather than the individual pixels.

To analyze if the increase of the spatial resolution can improve the identification and extraction of the required information, the proposed method was applied to both images. The proposed methodology (see Figure 2) includes the following steps: 1) test the classes spectral separability; 2) soft pixel-based classification to obtain the SEMM and the SEPM from both images; 3) evaluation and comparison of the classification of both images; 4) evaluation of the uncertainty of the classification; 5) pathology classification of buildings, with ceramic tile material, based on decision rules using the SEMM and the SEPM that presented better accuracy; and 6) evaluation of the classification accuracy.

### 3.1 *Sampling Design*

In this study five sampling protocols were established, two for training respectively the SEP and the SEM classification, two to evaluate these classifications and another to evaluate the final building map classification. The training dataset consisted of a semi-random selection of sites. For each class, fifteen building polygons were considered and a stratified random selection of 100 samples inside the chosen buildings was performed. To evaluate the soft pixel classification accuracy a stratified random sampling of 100 pixels per class was selected in another fifteen building polygons for each class. For the accuracy assessment of the final buildings classification map a stratified random selection of 100 samples was performed considering all buildings. The accuracy assessment was made with an error matrix. At this stage, the survey performed by the University of Coimbra regarding all buildings was used as reference information to identify the ceramic building roofs, with anomalies and not deteriorated, in the study area. For the accuracy evaluation, only the regions inside the BU were considered. The same sample protocols were used for both images. Since there are some discrepancies between both images, when the position of the aerial image surface element does not match the same surface element in the QuickBird image, it was necessary to move the pixel sample to the closest surface element.

### 3.2 *Classification*

The classification process included several steps. First, the identification and mapping of the SEP and the SEM were performed using a soft Maximum Likelihood Classifier (MLC) based on Bayesian modelling. The MLC is the most widely used image classifier, but it has been mainly used in its crisp version. However, the output can be in the form of posterior class probabilities providing a soft classification (*e.g.* Wang, 1990; Foody, 1996; Foody, 2004). Unlike traditional hard classifiers, the output is not a single classified map, but rather a set of images (one per class) that expresses the probability that each pixel belongs to the class in question. This extra data also provided additional information at the pixel level which allowed the assessment of the classification uncertainty. The surface elements classes used in this study are, for the SEMM, "Ceramic Roof Tile" (C), "Fibrocement Corrugated Roof Sheet" (F) and "Steel Roof Panel" (S); and for the SEPM "Ceramic Roof Tile Anomalies" (C-A) and "Ceramic Roof Tile Not Deteriorated" (C-ND).

Secondly, a comparison of the classes spectral separability at the pixel level was performed for both images, using the Bhattacharya Distance (B-Distance) and a dimensional scatter plot.

The uncertainty of the probabilistic fuzzy classification was evaluated using an indicator of the classification uncertainty (CU), available in the commercial software IDRISIS, given by:

$$CU = 1 - \frac{\max\limits_{i=1,\dots,n}(p_i) - \frac{\sum_{i=1}^{n} p_i}{n}}{1 - \frac{1}{n}} \tag{1}$$

where $p_i$ (i=1,…,n) are the probabilities associated with the several classes and n is the number of classes under consideration. This indicator assumes values in the interval [0,1] and only depends on: the maximum probability; the sum of all probabilities assigned to the class; and the total number of classes. CU evaluates up to which point the classification is dispersed over more than one class and the degree of compatibility with the most probable class, providing information regarding the classifier difficulty in assigning only one class to each pixel.

The Building Pathology Map (BPM) was built through the combination of the SEPM, the SEMM, their uncertainty information and the buildings polygons converted to the raster format and used as a mask. The next step corresponds to the development of rules that incorporate the information on the probabilities assigned to the several classes at each pixel in the previous soft pixel based classifications and the degree of uncertainty associated with these assignments. The rules construction requires a preliminary analysis of the probabilities assigned to the SEP and SEM classes and their uncertainty, in order to choose the appropriate thresholds. The transformation of a SEPM and a SEMM into a BPM is similar to a decision tree. The BU classes used in this study are: "Buildings with Ceramic Roof Tile With Anomalies"(B-C-A), varying from "State of Conservation 1" (SC1) to "State of Conservation 4" (SC4); and "Buildings With Ceramic Roof Tile Not Deteriorated" (B-C-ND), corresponding to "State of Conservation 5" (SC5). Figure 3 shows the BU classes classification workflow and Table 1 shows the used rules.

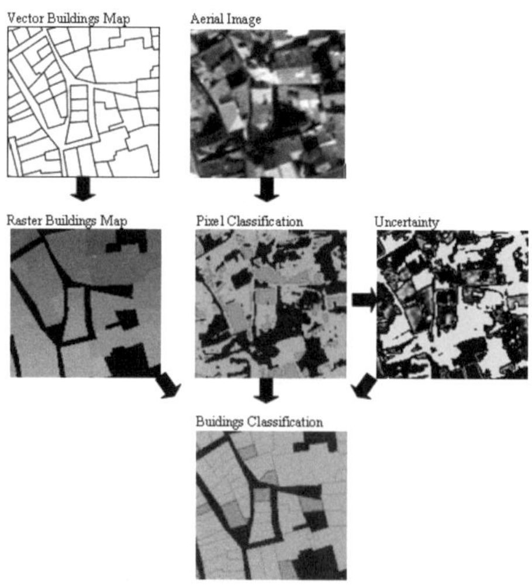


Figure 2. Flowchart of the methodology to grate uncertainty in the aerial image classification.

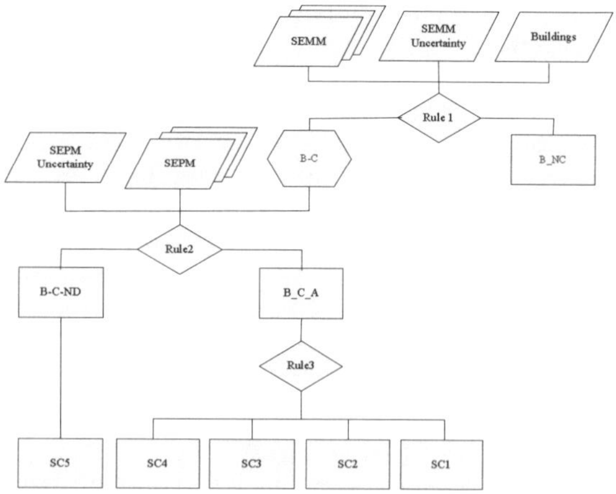


Figure 3. Building Unit Classes classification workflow.

The aim of rule 1 is to make a distinction between 'Buildings With Ceramic Roof Tile' (B-C) and 'Buildings With No Ceramic Roof Tile' (B-NC). Rule 2 classifies the 'Buildings With Ceramic Roof Tile' into "Buildings With Ceramic Roof Tile With Anomalies" and "Buildings With Ceramic Roof Tile Not Deteriorated", corresponding to "State of Conservation 5". Rule 3 assigns the "Buildings With Ceramic Roof Tile With Anomalies" (B-C-A) to one of four possible classes: "State of Conservation 1", "State of Conservation 2", "State of Conservation 3" and "State of Conservation 4"

Table 1. Classification Rules

| Rules | Test | Class if true |
|---|---|---|
| Rule 1 | At least 50% of the pixels of the SEMM, located inside the BU, are classified as "Ceramic Roof Tile" with a probability higher than 0.75 and uncertainty less than 0.25 | B-C |
| | Less than 50% of the pixels of the SEMM, located inside the BU, are classified as "Ceramic Roof Tile" with a probability higher than 0.75 and uncertainty less than 0.25 | B-NC |
| Rule 2 | The mode of the pixels of the SEPM inside the BU classified as "Buildings With Ceramic Roof Tile", is " Ceramic Roof Tile With Anomalies" with a probability higher than 0.75 and uncertainty less than 0.25 | B-C-A |
| | The mode of the pixels of the SEPM inside the BU classified as "Buildings With Ceramic Roof Tile", is "Ceramic Roof Tile Not Deteriorated" with a probability higher than 0.75 and uncertainty less than 0.25 | B-C-ND |
| Rule 3 | Buildings classified as "Buildings with Ceramic Roof Tile With Anomalies" have more than 75% of pixels classified as "Ceramic Roof Tile With Anomalies" with a probability higher than 0.75 and uncertainty lower than 0.25 | SC1 |
| | Buildings classified as "Buildings with Ceramic Roof Tile With Anomalies" have between 50% and 75% of pixels classified as "Ceramic Roof Tile with Anomalies" with a probability higher than 0.75 and uncertainty lower than 0.25 | SC2 |
| | Buildings classified as "Buildings with Ceramic Roof Tile With Anomalies" have between 25% and 50% of pixels classified as "Ceramic Roof Tile With Anomalies" with a probability higher than 0.75 and uncertainty lower than 0.25 | SC3 |
| | Buildings classified as "Buildings with Ceramic Roof Tile With Anomalies" have between 0% and 25% of pixels classified as "Ceramic Roof Tile With Anomalies" with a probability higher than 0.75 and uncertainty lower than 0.25 | SC4 |

## 4 RESULTS AND DISCUSSION

Figure 4 presents the spectral separability of the SEMM and SEMP classes in the aerial image and the QuickBird image. The comparison of the scatter plots shows that the overlap of the classes spectral signatures decreases with the higher spatial image resolution. It can be observed that the C, F and S classes present a total overlap of their ellipses in the QuickBird image while in the aerial image the major confusion is between F and S. The results given by the B-distance reinforce the results shown by the scatter plot. The average separability measure of the SEMM classes with the aerial image is 1.8 and with the QuickBird image is 1.4. For the SEPM the average separability between Ceramic Tile with and without pathology with the aerial image is 1.9 and with the QuickBird image is 1.7.

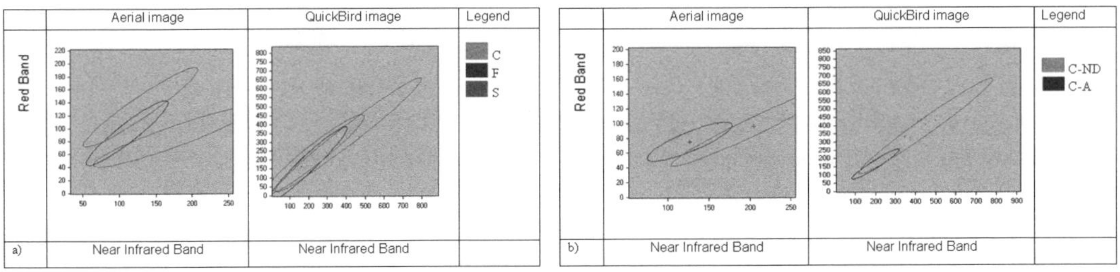


Figure 4. a) Spectral separability of the Surface Elements Materials classes: Ceramic Roof Tile (C); Fibrocement Corrugated Roof Sheet" (F) and "Steel Roof Panel" (S). b) Spectral separability of the Ceramic Roof Tile with Anomalies (C-A) and Not Deteriorated (C-ND).

The Global Accuracy (GA) of the SEMM and SEPM classifications with aerial image was 82% and 93% respectively, and the GA obtained with the QuicKbird image was 79% and 88%, respectively. The accuracy of results obtained with the aerial image was slightly better than the one obtained with the QuickBird image, which proves that, for this kind of application, an increase in the spatial resolution results in an improvement in the identification and extraction of this type of information. The User's and Producer's accuracy of the SEMM and SEMP are presented respectively in Figure 5.

The User´s and Producer's accuracies values per class show that Ceramic Roof Tile (C) class was very well identified. Significant confusion was observed between Fibrocement Corrugated Roof Sheet (F) and Steel Roof Panel (S). This confusion was due to the proximity of the corresponding spectral signatures and is higher in the QuickBird image than in the aerial image, confirming the results of the spectral analysis.

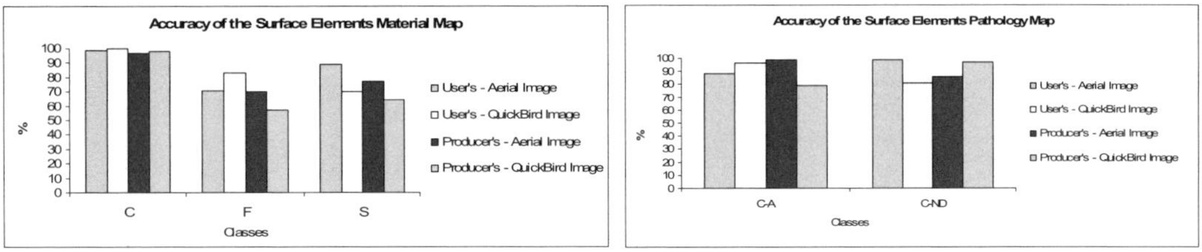


Figure 5. User's and Producer's accuracy of the Surface Elements Material Map (SEMM) and the Surface Pathology Map (SEPM) produced with the QuickBird and aerial images.

The classification of Ceramic Roof Tile With Anomalies (C-A) with the aerial image presents higher values for the Producer's accuracy than for the User's accuracy while the opposite occurs for the Ceramic Roof Tile Not Deteriorated (C-ND). However the classification results for the QuickBird image presented an opposite behaviour, that is, the C-A presented more commission error than C-ND. This different behaviour may possibly be due to the fact that the QuickBird image presents more shadows than the aerial image (Figure 1), since it was captured in December, and the spectral signatures of the C-A and shadows are significantly close.

Figure 6 shows the average uncertainty per class of the SEMM and SEPM respectively. The comparison of these results with those of Figure 5 shows that the results are consistent. For the SEMM, the C class, for instance, presents the lowest value of uncertainty and the highest value of accuracy. The other two classes in the SEMM show higher uncertainty and lower accuracy.

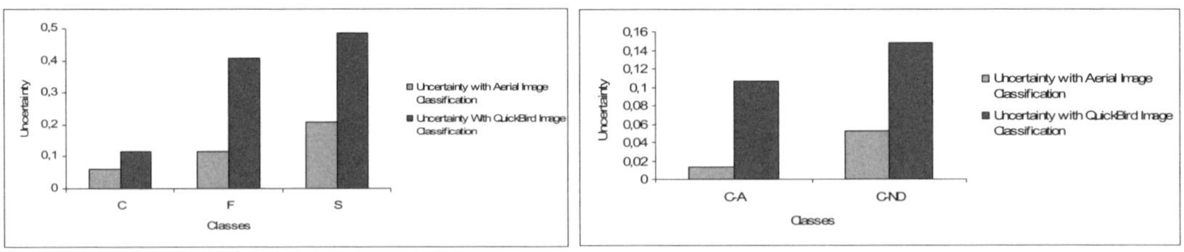


Figure 6. Average Classification Uncertainty per class of the Surface Elements Material Map (SEMM) and the Surface Elements Pathology Map (SEPM) obtained both images.

Since the results of pixel classification obtained with the aerial image were slightly better than the ones obtained with the QuickBird image, to perform the BPM the authors used the SEMM and the SEPM obtained with the aerial image classification. The global classification accuracy of the BPM was 74%. The values of the User's and Producer's accuracy are shown in Figure 7.

This hybrid method of classification proved that the uncertainty information allowed the identification of the misclassified surface elements, avoiding their use in the construction of BPM, and revealed to be adequate for the transformation of surface elements into building classification units.

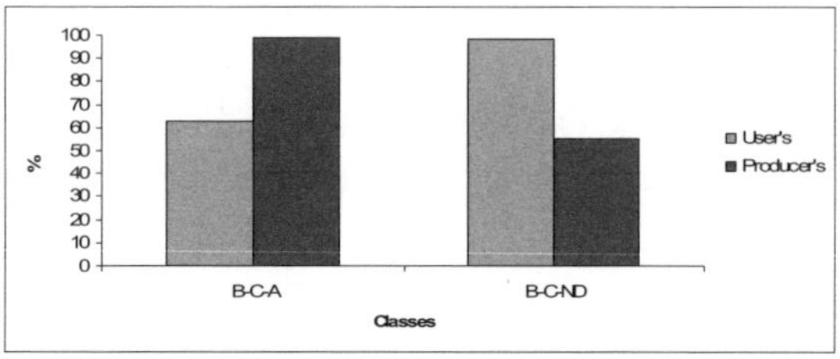


Figure 7. The User's and Producer's accuracy of the Building Pathology Map (BPM) obtained with the hybrid approach applied to the aerial image.

The comparison between the BPM and the map made by the University of Coimbra with the traditional methodology showed that this new approach identified 66% of the buildings to have ceramic tile roofs, whereas the traditional survey identified 85%. In addition, the previous survey showed that 8% of the ceramic tile roofs did not have anomalies, while, with this study, this number increased to 12%. This difference is mainly due: to the shadows; to the fact that the images and the vector cartographic map do not match perfectly, as is shown in Figure1; and mainly because the survey of the University of Coimbra was made between 2003 and 2004 and the aerial image used digital photos obtained in 2005 and 2006. Between these two periods several roofs were repaired or renewed, and therefore presented no deterioration in the aerial image being, for this reason, well identified with the new classification approach, which explains the increase of roofs with no anomalies (Figure 8).

Figure 8. a) Extract of the aerial image (RGB321) overlaid with the buildings with ceramic tile roof not deteriorated classified by the University of Coimbra using traditional methods. b) Extract of the aerial image (RGB321) overlaid with the buildings with ceramic tile roof not deteriorated classified by the new approach. The circles mark some of the buildings which roofs were repaired or replaced by new ones and were correctly identified with the new classification approach.

## 5 CONCLUSIONS

The results obtained show that the applicability of VHSR images to heritage conservation is very promising. The use of hybrid pixel-object classification integrating the surface elements classification uncertainty proved to be valuable in the classification process and therefore this approach should be further studied. From a methodological point of view, the hybrid approach also proved to be adequate for the transformation of surface elements into Bulding Pathology Map with a format well suited to be integrated in a GIS.

## ACKNOWLEDGEMENTS

The authors thank the City Council of Coimbra for making the QuickBird image available for this study and the University of Coimbra for providing data related to the state of conservation of buildings.

## REFERENCES

Aplin, P., P. Atkinson, and P. Curran, 1999. Per-field classification of land use using the forthcoming very fine spatial resolution stellite sensors: problems and potential solutions. Advances in Remote Sensing and GIS Analysis. Edited by P.Atkinson, and N. Tate (Chichester: John Wiley and Sons): 219-239.

Blaschke T., C. Burnett, and A. Pekkarinen, 2004. Image segmentation methods for object-based analysis and classification. Edited by S. M. de Jong,. and F.D. van der Meer. Remote Sensing Image Analysis, pp. 211-234.

Donnay, J.-P., M. Barnsley, e P. Longley, 2001. Remote sensing and urban analysis. Remote Sensing and Urban Analysis. Edited by J.- P. Donnay, M. Barnsley, P. Longley (New York: Taylor & Francis): 3-18.

Foody, G. M., 1996, Approaches for the production and evaluation of fuzzy land cover classifications from remotely-sensed data. International Journal of Remote Sensing, 17, 1317-1340.

Foody, G.M..2004. Sub-pixel methods in remote sensing. In, de Jong, S.M. and van der Meer, F.D. (Eds.) Remote sensing image analysis: including the spatial domain. Dordrecht, The Netherlands, Kluwer, 37-49.

Gonçalves, L., Caetano, M., 2004, Classificação das imagens do satélite IKONOS utilizando uma abordagem orientada por objectos, Actas da Conferência de Cartografia e Geodesia, (ed. L. Bastos e J. Matos) Lidel, Lisboa, 287-298.

Plantier T. and M. Caetano, 2007. Mapas do Coberto Florestal: Abordagem Combinada Pixel/Objecto. Acta da V Conferência Nacional de Cartografia e Geodesia, pp. 157-166.

Wang L., W.P. Sousa and P. Gong, 2004. Integration of object-based and pixel-based classification for mapping mangroves with IKONOS imagery. International Journal of Remote Sensing, V. 20, No. 24, pp. 5655-5668.

Wang, F., 1990, Improving remote sensing image analysis though fuzzy information representation. Photogrammetric Engineering and Remote Sensing, 56, 1163-1169.

*Remote Sensing for a Changing Europe*
*D. Maktav (Ed.)*
*IOS Press, 2009*


*doi:10.3233/978-1-58603-986-8-247*

# Assessment of crop growth conditions for agri–environment ecosystem for modern landscape management

K. Dabrowska-Zielinska, M. Budzynska, I. Malek , J. Bojanowski, Z. Bochenek, St. Lewinski
*Institute of Geodesy and Cartography, Remote Sensing Centre, Modzelewskiego 27, 02-679 Warsaw, Poland*

Keywords: surface reflectance, surface temperature, backscattering coefficient, heat flux, evapotranspiration, soil moisture, LAI

ABSTRACT: The area of the research is a part of the Landscape Park and belongs to agri–environment ecosystem of South-West Poland. The study on vegetation growth conditions has been carried out applying optical data from TERRA/ASTER, TERRA/MODIS, and ENVISAT/MERIS as well as microwave data from ENVISAT/ASAR. *In situ* data collected at the time of satellite observations were as follows: soil moisture, leaf area index, biomass, vegetation moisture, crop height, types of crop and its actual vegetation phenological stage. The supervised maximum likelihood classification method was applied to multispectral optical and multitemporal microwave satellite data to distinguish different crop type. From optical satellite data different vegetation and soil moisture indices have been calculated on the basis of surface reflectance and surface temperature supported with meteorological data. Assessment of heat fluxes applying surface temperature calculated from TERRA/ASTER and TERRA/MODIS images was very important for water balance assessment. From microwave satellite data, registered in various polarisations and incidence angles, the backscattering coefficients have been derived and related to soil moisture. It was presented that the classification obtained from ENVISAT ASAR VV HH IS4 and IS6 and assessment of soil moisture yield into good results close to these obtained from optical data, what is very important in regard to often cloud cover. It was very important to get evapotranspiration from optical data. The results should be implemented into management of agri–environment ecosystem. ENVISAT images have been obtained from ESA for CAT-1 1427 project, TERRA images for the national project 4T12E02630.

## 1 INTRODUCTION

Proper management of agri–environment ecosystem requires accurate and rapid obtained data about crop distribution and it's growth conditions. For this goal remote sensing methods play a significant role, because of spatial and temporal coverage capability. Optical and microwave satellite data provide valuable information for a good discrimination between crop types as well as for the monitoring of crop growth conditions for proper agriculture management. This paper demonstrates methods of classifying agricultural crops and monitoring agri–environment ecosystem using optical and microwave satellite images acquired for spring 2007. This time of observation is crucial for assessing the soil water retention, calculating the soil water capacity in spring, as it is important time due to soil water storage in the start of vegetation growth. The present results show that in cloudy conditions radar data may prove to be a potential tool in the measuring and monitoring soil and vegetation growth conditions. Developed algorithms were used in preparation of maps of soil -

crop biophysical parameters (LAI, soil moisture, evapotranspiration) that give the crucial information for monitoring of vegetation and it is very important task for modern landscape management.

## 2 STUDY AREA

The study area is located in the South-West part of Poland (Longitude E: 16° to 17°, Latitude E: 52° to 52°10′) in Wielkopolska region. This site has been chosen because of the presence of big agricultural plots and various crop types. It is flat agricultural areas with the mean elevation of 85 m above see. The climate is between climatic zones transient oceanic and continental. The prevailing soil types are sandy loam on arable land (*Ortio Podzols* according to FAO classification) and moorsh on grasslands (*Eutric Histosols*), with stable soil texture condition. The dominating crops are: winter and spring wheat (*Triticum aestivum ssp. vulgare*), winter rye (*Secale cereale*), winter and spring triticale (*Triticale*), winter and spring barley (*Horoleum sativum*), winter rape (*Brassica napus*), corn (*Zea mays*) and sugar beet (*Beta vulgaris saccharifera*).

## 3 MATERIALS AND METHODS

The paper focused on the spring season of the year 2007, when spring cereals were in the emergence stage, winter cereals in the jointing stage, rape in the flowering stage, alfalfa in the full biomass development, and corn and sugar beet in germination (bare soil) or emergence stage. Table 1 presents dates of ground and satellite observations. The measurements of various soil and vegetation parameters were carried out at the selected plots for each of the different crops simultaneously to the satellites overpasses. For each plot, several measurements were taken and then averaged. The following parameters have been measured at the test area:

- soil moisture (SM), [% vol.] to the depth of 0–15 cm using TDR method,
- wet and dry biomass weights,
- height of the crops,
- development stage,
- Leaf Area Index (LAI), [$m^2/m^2$] measured with the Plant Canopy Analyzer 2000,
- meteorological data using portable station.

The following satellite data have been applied: optical data from TERRA/ASTER (pixel size 15, 30, and 90 m-thermal), TERRA/MODIS (pixel size 250, 500, and 1000 m-thermal), and ENVISAT/MERIS (pixel size 300 m after resampling) and microwave data from ENVISAT/ASAR (pixel size 12.5 m) from descending orbits, (tab. 1). All optical satellite images were radiometrically (atmospheric correction) corrected. Microwave ENVISAT/ASAR images were processed using BEST programme. All images were geometrically corrected to the ellipsoid WGS-84 (UTM zone N 33) using ground control points and nearest neighbour resampling. All resamplings and classifications were performed using ERDAS system.

Table 1. Dates of ground truth and satellite observations

| GROUND Observation | ASTER | ENVISAT ASAR | MERIS | MODIS |
|---|---|---|---|---|
| 15.04-23.04.2007 | 23.04.2007 | 15.04.2007 IS6 HH VV<br>07.05.2007 IS4 HH VV | 26.04.2007 | 23.04.2007 |

Optical satellite images have been applied for crop classification and calculation of vegetation indices. Surface temperature (corrected for atmospheric influence) derived from thermal channels of ASTER and MODIS images in conjunction with meteorological data (air temperature, wind speed, net radiation) has been used for calculation of latent heat flux as a residual of the energy budget equation:

$$LE = RN - H - G \quad (1)$$

where: LE = the rate of loss of latent heat by evapotranspiration, $W/m^2$; RN = the rate of net gain of heat from radiation measured at the station, $W/m^2$; H = the rate of sensible heat loss by convection, $W/m^2$; and G = the rate of heat loss into the ground (measured), $W/m^2$.
The values for the rate of sensible heat loss by convection (H) were calculated in line with the formula:

$$H = (pC_p(T_s\text{-}T_a))/r_a \quad (2)$$

where: p = air density, $kg/m^3$; $C_p$ = specific heat of air, J/kg/K; $T_s$ = surface temperature from ATER or MODIS corrected due to atmospheric water vapor, °C; $T_a$ = air temperature measured at the study area, °C; and $r_a$ = air resistance for heat transfer, s/m.
The values of air resistance ($r_a$) were in turn calculated using:

$$r_a = [\ln((z\text{-}d)/z_o)]^2/(K^2u) \quad (3)$$

where: z = the level of measurements above the ground, m; d = the zero-plane displacement level, m; $z_o$ = the roughness length, m; K = Von Karman's constant equal to 0.41; and u = the wind speed, m/s.
The values for roughness length ($z_o$) and zero-plane displacement level (d) were calculated as the function of vegetation height (h) after Monteith (Monteith 1973): $z_o = 0.13h$ and $d = 0.66h$. Air resistance was corrected for air stability in the manner of Jackson (Jackson *et. al,* 1987) and Kustas (Kustas *et. al.* 1989). Daily evapotranspiration ($E_{daily}$) has been calculated using instantaneous values of LE and instantaneous values of RN and daily values of RN as similar approach to Jackson (Jackson et al. 1983, Dabrowska–Zielinska 1995).

As it is important to find which part of incoming radiation has been used for sensible heat and which part for latent heat flux, therefore the ratio of sensible to latent heat flux seems to be good indicator of soil moisture. Soil moisture was calculated from the relationship of soil moisture to (H/LE) index. Leaf Area Index (LAI) represents the growth condition of the plant. It was found that LAI can be calculated from vegetation index NDVI and soil moisture index H/LE (Dabrowska–Zielinska et al. 2007). Two microwave satellite images have been applied as multitemporal image for crop classification as well as for soil moisture assessment. Soil moisture has been estimated from the relationship with radar backscatter derived separately for each class presenting similar roughness conditions and similar attenuation of the radar signal by vegetation.

## 4 RESULTS AND DISCUSSION

### 4.1 *Crop classification*

Crop classification is essential for the monitoring of soil moisture conditions for specific crops or bare soil. Crop classification has been performed for the layer of arable land derived from CORINE Land Cover map. Also, grassland area has been taken from this map. The supervised maximum likelihood classification method was applied to multispectral optical and multitemporal microwave satellite data to distinguish different crop type. For each of the optical images and one multitemporal ASAR image (composition of VV HH IS4 and IS6) the sample training dataset has been generated. The validation has been performed using ground truth observations and the accuracy of the four classifications has been evaluated. From the classification of ASTER image four crop types have been distinguished: spring cereals, winter cereals, rape, alfalfa, and bare soil (Fig. 1, left). From the classification of MERIS image three crop type have been distinguished: spring cereals, winter cereals, rape, and bare soil (Fig. 1, right).

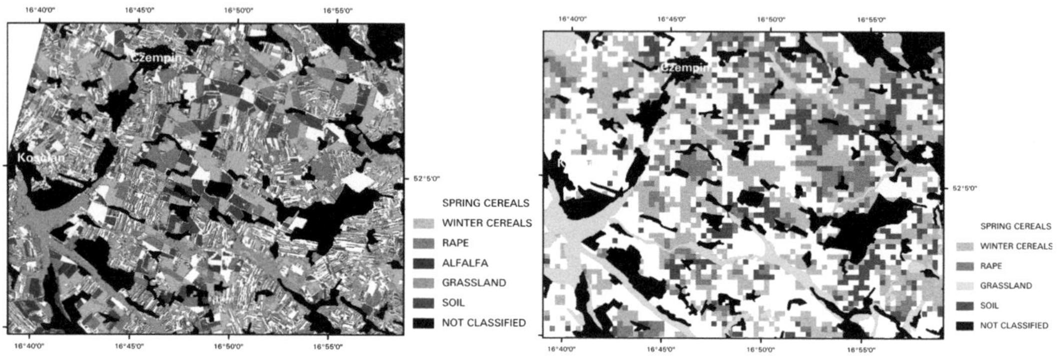


Figure 1. Crop classification based on ASTER (left) and MERIS (right) data.

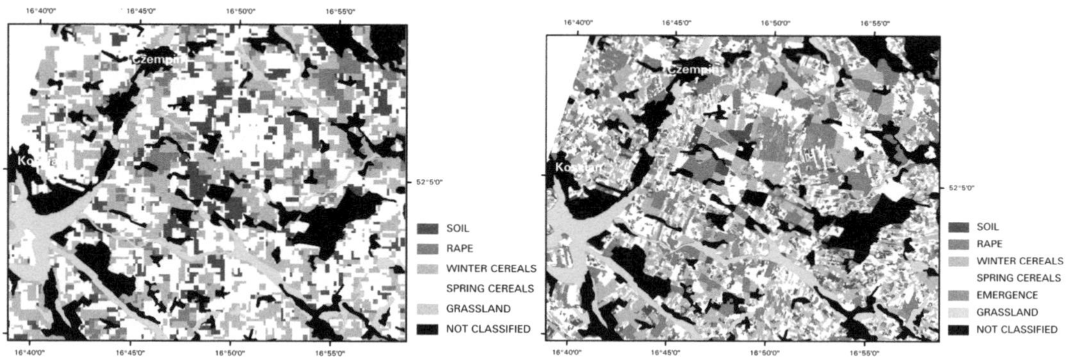


Figure 2. Crop classification based on MODIS (left) and multitemporal ASAR (right) data.

From the classification of MODIS image also three crop classes have been distinguished: spring cereals, winter cereals and rape, and bare soil (Fig. 2, left). From the classification of multitemporal ENVISAT ASAR data (acquired on 15.04.2007 and 07.05.2007) the following crops have been distinguished: spring cereals, winter cereals, rape, crops in emergence stage, and bare soil (Fig. 2, right). The classification results of these four satellite images show that ASTER and multitemporal ASAR images give the best results in distinguishing crop types. Due to lower spatial resolution of MERIS and MODIS images (300 and 250 m, appropriately) the classification of crop type was poorer - only three crop type classes were distinguished. The overall accuracy calculated from error matrix was: for ASTER image 95%, for ASAR multitemporal image 82%, for MERIS image 72%, and for MODIS image 76%.

## 4.2 *Evapotranspiration*

Latent heat flux (LE) has been calculated using equation (1). Daily values of evapotranspiration ($E_{daily}$) were calculated for the days of satellite acquisitions. The figure 3 presents the maps of LE calculated using the surface temperature from ASTER (left) and MODIS (right) data and meteorological parameters.

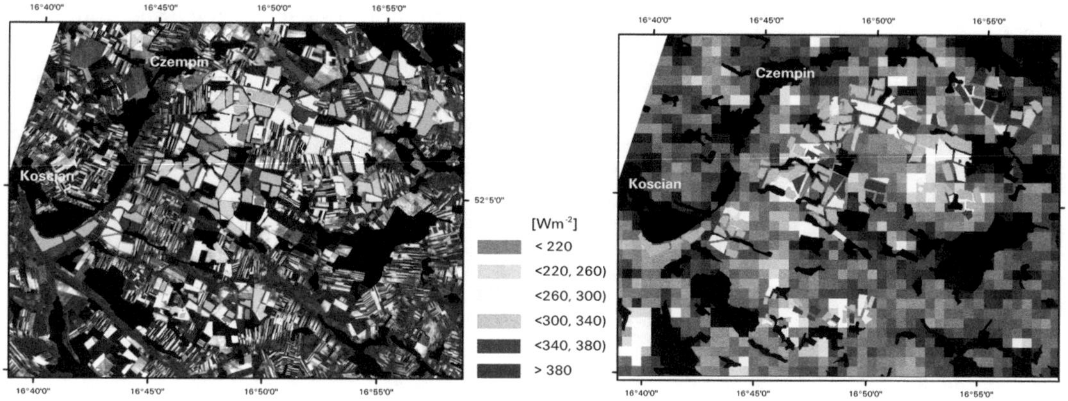


Figure 3. Map of LE generated based on ASTER (left) and MODIS (right) thermal data.

The figure 4 presents the distribution of daily evapotranspiration ($E_{daily}$) assessed from ASTER (left) and MODIS (right) data. Daily values of evapotranspiration were in most occasions between 2,5-3.0 mm per day. In the Northern part of the area the daily values of evapotranspiration was higher than 3 mm. It was noticed that the area with higher evapotranspiration from 3-3,5 mm/day predominates at the test site. The MODIS thermal pixel is 1km$^2$ and the temperature is averaged from the larger area than the temperature obtained from ASTER where resolution is 90 m. Within 1 km$^2$ pixel there is bare soil and different crops.

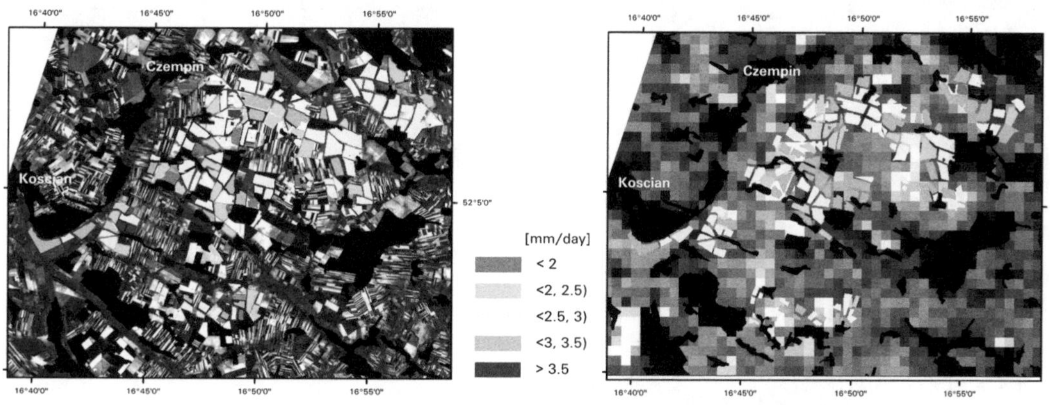


Figure 4. Map of $E_{daily}$ generated based on ASTER (left) and MODIS (right) thermal data.

## 4.3 *Soil moisture*

Soil moisture has been assessed from the ratio of sensible heat to latent heat (H/LE) obtained using optical data and meteorological parameters. High values of soil moisture related to low values of the index because in wet conditions most of the incoming energy is consumed by the latent heat flux - the surface evaporates and its temperature decreases. The difference between surface and air temperature is low and sensible heat flux is low. The figure 5 presents the relationships of soil moisture and H/LE index calculated based on ASTER (left) and MODIS (right) thermal data. The developed equations have been used for the generation of soil moisture maps (Fig. 6). In April during early spring the soil moisture is high. The higher soil moisture values (20-25%) were obtained using the H/LE index where the surface temperature was obtained from ASTER data (Fig. 6 left). Lower values of soil moisture were noticed on the map based on MODIS data (generated using Fig.

6 right) what was the consequence that the surface temperature recorded by MODIS is from 1km$^2$. The MODIS thermal pixel most often contained the plots of rape with lower soil moisture, which dominated within the pixel (Fig. 7).

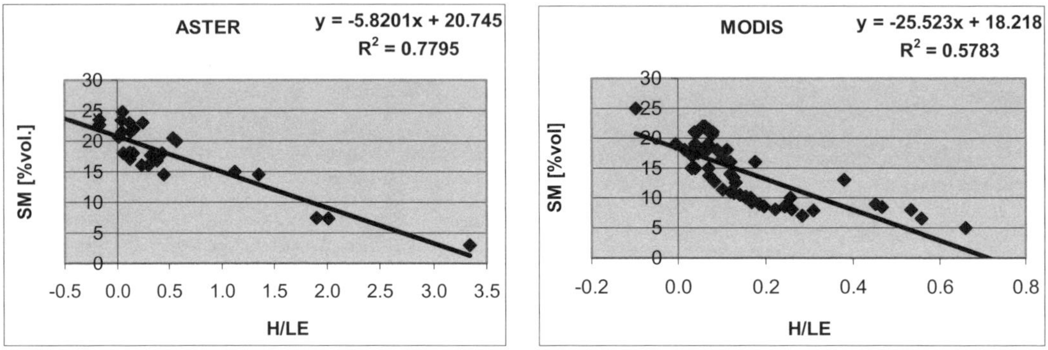


Figure 5. Relationship between soil moisture (SM) observed and H/LE calculated based on surface temperature derived from ASTER (left) and from MODIS (right) data.

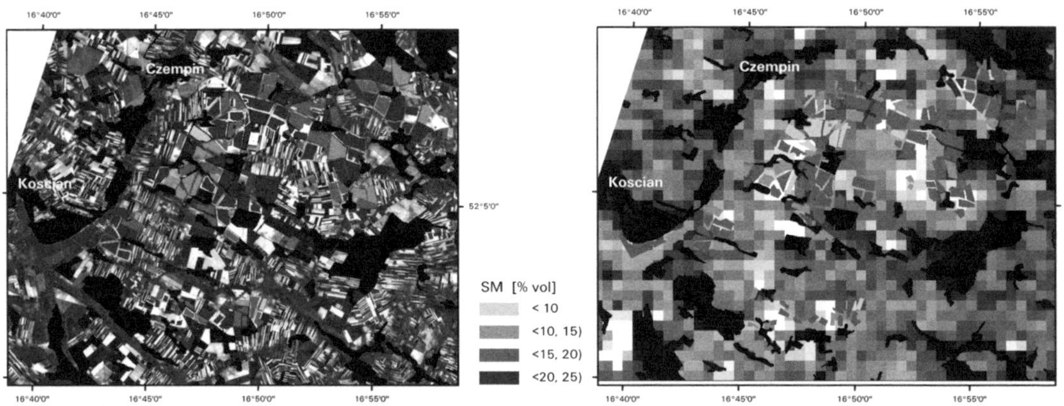


Figure 6. Map of soil moisture generated based on ASTER (left) and MODIS (right) thermal data.

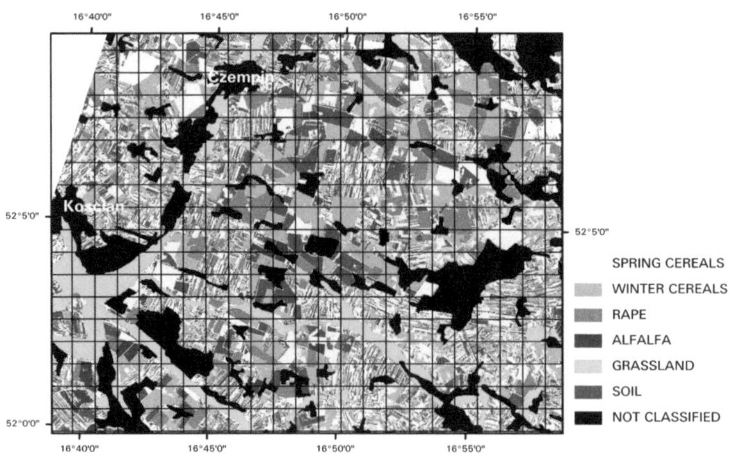


Figure 7. Crop classification based on ASTER data overlaid MODIS thermal pixels.

The other method of soil moisture calculation, presented in this study, was to apply microwave data from ENVISAT ASAR IS6 VV. For each of the agricultural class presenting similar roughness conditions and similar attenuation of the radar signal by vegetation there was performed the relationship between backscattering coefficient and soil moisture. The figure 8 presents the distribution of soil moisture values estimated using developed algorithms. In general, the values were high i.e. 20-25 % for crop area and over 50% for grassland.

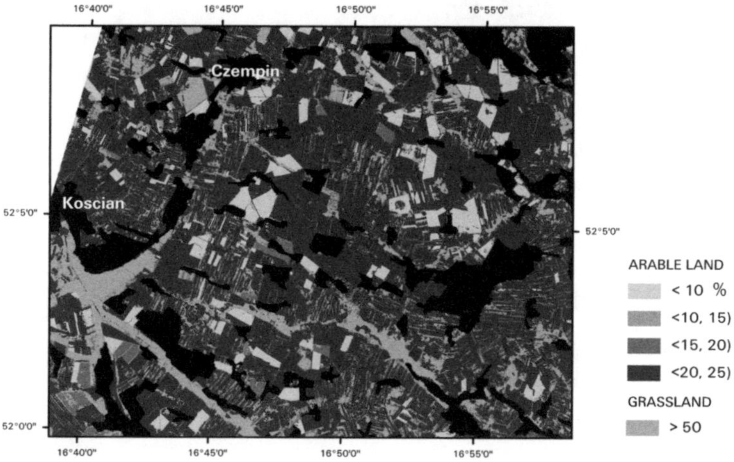


Figure 8. Map of soil moisture generated from ENVISAT ASAR IS6 VV image acquired on 15.04.2007.

## 4.4 *LAI*

The LAI index describes the geometry and density of a crop. LAI changes during the development of crops from low at the beginning of growth to a maximum at their full biomass stage (heading stage for cereals). LAI has been calculated from NDVI index and soil moisture index H/LE derived from ASTER data for particular crops distinguished from the same image. The figure 9 presents equations developed for winter cereals (left, $R^2$=0.85) and rape (right, $R^2$=0.86) as an example. The similar statistical analysis has been applied for the remaining classes (spring wheat, $R^2$=0.95; grass-lands and alfalfa, $R^2$=0.96) distinguished from ASTER data.

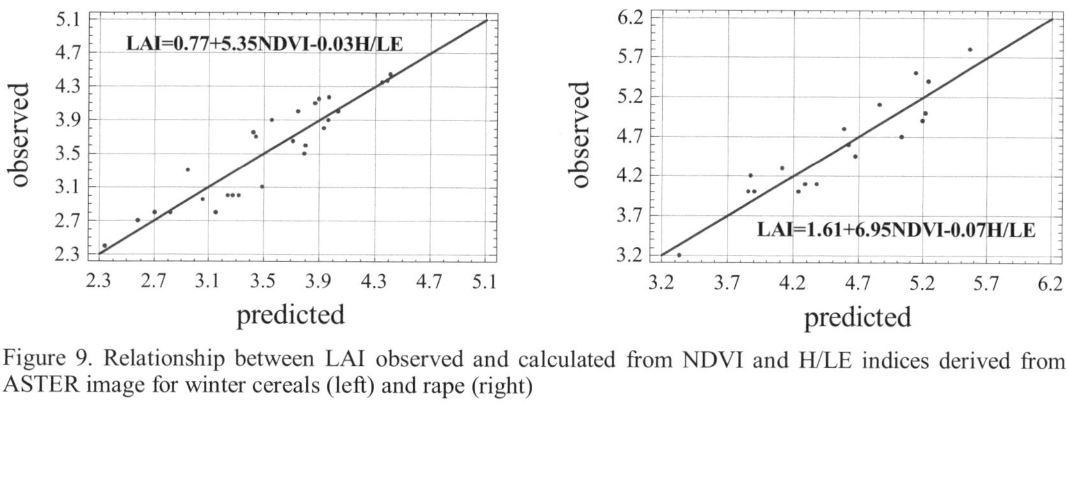


Figure 9. Relationship between LAI observed and calculated from NDVI and H/LE indices derived from ASTER image for winter cereals (left) and rape (right)

The figure 10 presents distribution of LAI calculated using NDVI assessed from ASTER data and H/LE calculated using surface temperature from ASTER and meteorological data, according to (1) and (2). The higher LAI was noticed for rape plots and grassland. The lowest for spring wheat and some winter wheat plots.

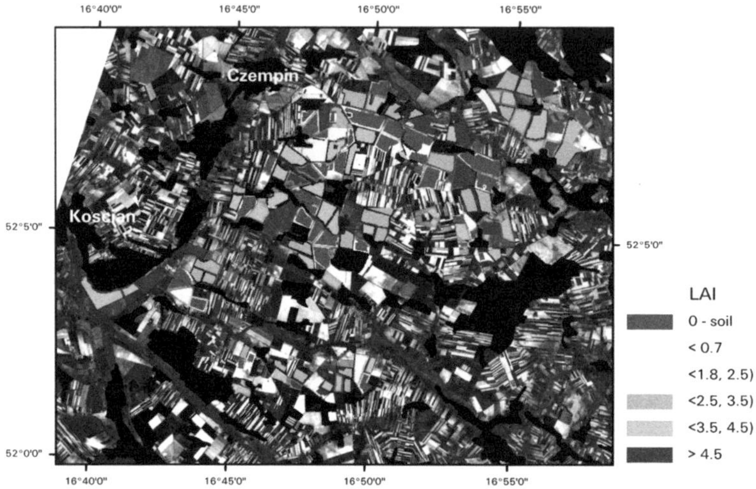


Figure 10. Map of LAI generated based on ASTER data

## 5 CONCLUSIONS

Crop growth conditions focused on biomass expressed by LAI, evapotranspiration, and soil moisture can be monitored using optical and microwave satellite data. The first step for building the monitoring system for agri-environment is crop classification. The next step is to assess various soil-crop biophysical parameters like evapotranspiration, soil moisture, LAI for the distinguished classes. Application of multispectral TERRA/ASTER, TERRA/MODIS, ENVISAT/MERIS, and multitemporal ENVISAT/ASAR data proved the possibilities of using this information as the regular input to the system of modern landscape management. Due to often clouds cover the presented results based on microwave data were considered as great possibilities for the measuring soil-water retention and monitoring crop growth conditions. Assessment evapotranspiration and soil moisture based on ASTER and MODIS thermal data gives essential information as the input to water balance of the agri-environment ecosystem. The described results show that proposed method of the assessment of crop growth conditions should be implemented into management of agri–environment ecosystem.

## REFERENCES

Dabrowska–Zielińska K. 1995. Szacowanie ewapotranspiracji, wilgotności gleb i masy zielonej łąk na podstawie zdjęć satelitarnych NOAA. *Prace Geograficzne* nr 165.

Jackson, R. D., Hatfield J. L., Reginato, R. J., Idso S. B., and Pinter P. J., 1983. Estimation of daily evapotranspiration from one time of day measurements. *Agric. Water Management* 7, pp. 351 – 362.

Jackson, R. D., Moran, M. S., Gay, L. W., Raymond, L. H., 1987. Evaluating evapotranspiration from field crops using airborne radiometry and ground based meteorological data. *Irrigation Science* 8 (2), pp. 81-90.

Kustas, W. P., Choudhury, B. J., Moran, M. S., 1989. Determination of sensible heat flux over sparse canopy using thermal infrared data. *Agriculture, Forest, Meteorology* 44, pp. 197-216.

Monteith, J.L., 1973. *Principles of environmental physics*. Arnold E. (Eds.), London.

*Remote Sensing for a Changing Europe*
*D. Maktav (Ed.)*
*IOS Press, 2009*


*doi:10.3233/978-1-58603-986-8-255*

# Monitoring of maize damage caused by western corn rootworm by remote sensing

G. Nádor & D. Fényes
*Institute of Geodesy, Cartography and Remote Sensing, Budapest, Hungary*

Gy. Surek
*MLOG Ltd., Budapest, Hungary*

L. Vasas
*Agricultural Office of County Békés, Plant Protection and Soil Conservation Directorate, Békéscsaba, Hungary*

Keywords: western corn rootworm, Diabrotica virgifera, agriculture, plant protection, remote sensing, polarimetric radar satellite data, ALOS PALSAR

ABSTRACT: As one of the consequences of global warming, the gradual dispersion of western corn rootworm (WCR) is becoming a serious maize pest in Europe and all over the world. The WCR was introduced to Europe from the USA. First it was detected in Europe near Beograd in 1992. The WCR has spread from its initial infestation point to a range of several hundred kilometres, affecting many countries in the region including Hungary first in 1995. The structure of a healthy, WCR free maize field shows straight rows in a clear order and upstanding maize stalks. WCR infection can cause wilted broken corn stalks randomly lying on the ground. That means the damage itself causes physical and visible disorder in the maize field. Our goal is to assess and identify the disorders and structural changes caused by WCR using polarimetic radar images (ALOS PALSAR). The initial results of our assessment will be presented in this paper. This project has been implemented in the framework of the tender called by the Ministry of Environment and Water and with the support of the Hungarian Space Office. To develop and extend the method is worthwhile as the WCR monitoring system is expeditious, objective and economical. The development of the system will create the possibility to monitor the spread of WCR on larger areas and to forecast the following year's damage. The amount of information can be used efficiently in plant protection; moreover can be useful for farmers, pesticide producers, state authorities and research institutes.

## 1 INTRODUCTION

The Institute of Geodesy, Cartography and Remote Sensing (FÖMI) implemented the present experimental project called "Identification of the Western Corn Rootworm damage by integrated utilization of optical and radar images" in the framework of the tender called by the Ministry of Environment and Water and with the support of the Hungarian Space Office.

According to our experience, to apply remote sensing in a proper way, the key is to have the most possible detailed and versatile description based on ground assessment of the differences between the healthy and the WCR damaged fields.

The healthy maize field is standard, life cycle of the maize plant suits the yearly average and it is free from WCR damage. The damaged maize parcel is destroyed by WCR larvae.

The structure of a healthy, WCR free maize field represents straight rows in a clear order and upstanding corn stalks. Most of the damage in corn is caused by WCR larval feeding. WCR larvae feed on and destroy maize roots in different scale. The serious WCR larval damage like it was in Hungary, 2007 can cause wilted broken corn stalks. In injured corn fields, WCR larval damage results in disordered rows and corn stalks lying on top of each other randomly on the ground. Rootworm damage also can result in extensive "gooseneck" lodging and harvest difficulties. The damage usually appears in spots and causes chaos, disarray and disorder in corn fields.

It is appropriate to describe the degree of the root injury by using the Iowa corn root rating scale. It is a 1-6 root damage scale depending on the root injury intensity. The important point of the remote sensing analysis is that ratings from and over 4 indicate “gooseneck” prone lodging corn stalks. Although sometimes root rating 3 combined with extreme weather condition can also show the same result.

## 2 METHODS AND DATA

Our goal with the present project is to assess and identify the disorder and structural changes caused by WCR larvae using optical (IRS-P6 AWiFS) and polarimetic radar (ALOS PALSAR) satellite images.

It is important to clarify some details about the reference database and its establishment. The Central Agricultural Office of Békés County (CAO) provided the GPS recording of the WCR damaged field, the related data and the underlying basic ground assessment. We represented the received data from CAO into GIS to make them more visible and create an easier access in the future assessment. In Békés County the tested maize parcels were selected based on surveyed monoculture maize fields by CAO as well as randomly on the source of public announcements. We determine and classify the WCR larval damage degree by 1-6 Iowa corn root rating scale created by Hills-Peters. The injury of test maize fields reached different degrees. It was between 1-5,3 rating scale. The damage percentage of test fields was 0-85%.

We used the reference ground assessment reports from the year 2007 recorded in Békés County, Hungary in cooperation with the local agency of CAO. We attempted to identify rootworm damage among the 26 selected test parcels by remote sensing. Table 1 shows the main features of the test parcels. Figure 1 gives information about the spatial location of the selected parcels.

We used 3 different individual features (Mono-maize feature, Optical feature, Radar feature) derived from remote sensing data to identify larval damage. These are as follows:

- *The result of monoculture maize field assessment (Mono-maize feature)*
  Monoculture maize production increases the risk of WCR infection. Satellite data time series 1997-2006 (during 10 years, before 2007) are available in FÖMI which we analysed and determined the frequency of maize production in the tested parcels. This became the value of mono-maize feature. If this value is higher than the limit value it means that the parcel is probably damaged by WCR otherwise we assume the parcel is not damaged. If the value equals the limit value it is an ambiguous case. In this special case it is impossible to determine whether the parcel is damaged or not. The limit value was created by crosschecking the reference data and the monoculture maize map.

- *Vegetation index derived from optical satellite data decreases if there is larval damage (Optical feature)*
  The WCR larval damage results in stress in corn plants. It leads to the vegetation index decreasing. If the vegetation index curve of the cornfield is lower in a certain period than a given limit value it means that the parcel is probably damaged by WCR otherwise it is not. The optical feature measures the relative difference rate of the limit value. If its value is negative there is WCR damage, if it is positive there is no WCR damage. The ambiguous case is that if the relative difference is in +/- 2 percentages. In this case based on this

feature it is impossible to determine whether the parcel is damaged or not. The limit value was created by crosschecking the reference data and optical data time series.

- *Identification of the disorders and structural changes caused by WCR using polarimetic radar images (Radar feature)*
  According to the theories and the bibliography issued about the radar polarimetry (Pierce at.al. (1994), Cloude & Pottier (1997), Ferrazzoli at.al, (1999), Skriver at.al (1999), Schou & Skriver,(2001)) we should recognize the disorder caused by WCR damage through the change of polarimetric state of radar wavelength which is similar to the structural changes in corn field. Considering the above mentioned theory we decided to apply L-band (wavelength: 23 cm) polarimetric radar images in our remote sensing methodology. We derived the entropy value as described in Refregier & Morio (2006) by analysing polarimetric radar image. This value describes the disorder of the parcel. If this value is higher than a limit value it means that the parcel is probably damaged by WCR otherwise it is not. The radar feature measures its relative difference rate of the limit value. If its value is positive there is WCR damage, if it is negative we assume there is no WCR damage. Ambiguous cases are those in which the difference is in +/- 2 percentages. In such cases based on this feature it is impossible to decide whether the parcel is damaged or not. The limit value was created by crosschecking the reference data and entropy map derived from radar image.

We defined the above detailed three distinct features regarding the tested corn fields to identify WCR damage. If the features shown contradictory damage findings of the individual test parcels we classified those parcels into categories in which the image included the most pixels. For example if two features give the same result that a parcel is WCR damaged and one feature shows it is not damaged we classify the parcel as damaged. We decided the final category based on the majority.

We should point out some details in connection with the assessment. Separately these features can not identify the WCR damage unmistakably. The monoculture maize production increases the possibility of WCR infection. If the plant protection is appropriate the damage can not develop. The most effective measure to control the WCR is crop rotation. Vegetation index decrease can be caused by any other stress (e.g.: drought, heavy weed infection, wrong pesticide use, etc.). Furthermore usually lower larval damage does not always result in so heavy visible and physical change (disorder) in the fields that the damage can be identified by polarization change of the backscattered radar signs. The integrated assessment of the 3 features can give more accurate WCR damage identification than to assess features separately one-by-one.

Table 1. Classification of the tested reference cornfields based on WCR damage, Békés County 2007

| IOWA root rating scale | No. of fields | Area (ha) | Area percentage (%) | Average area of the fields (ha) |
|---|---|---|---|---|
| 1 – no damage (control) | 11 | 398,6 | 75,7 | 36,2 |
| 2-3 | 2 | 12,7 | 2,4 | 6,3 |
| 3-4 | 4 | 79,0 | 15,0 | 19,8 |
| 4-5 | 6 | 32,0 | 6,1 | 5,3 |
| >=5 | 3 | 3,9 | 0,7 | 1,3 |
| **Total** | **26** | **526,3** | **100,0** | **20,2** |

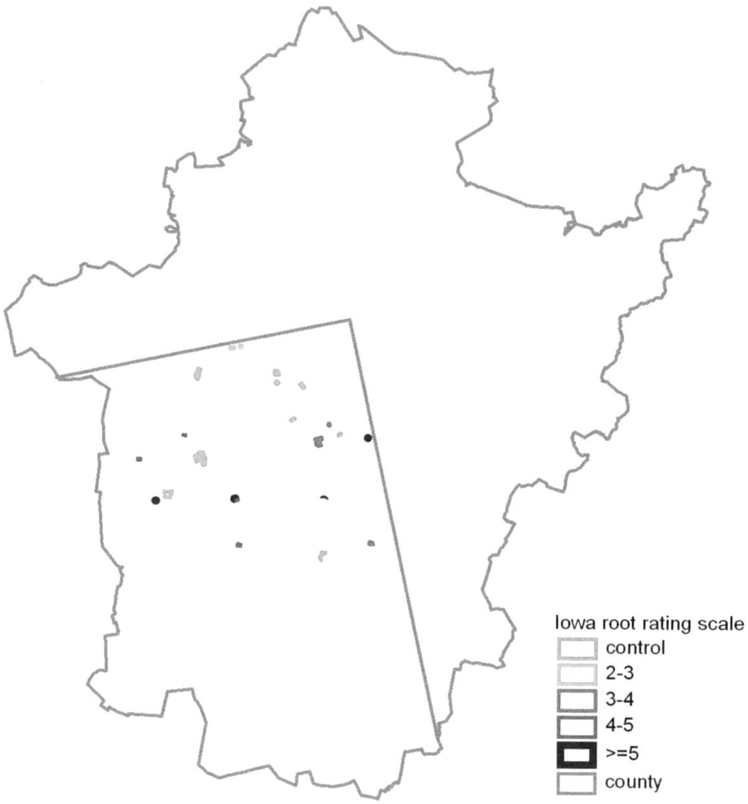


Figure 1. Spatial position of WCR larva damage and control reference data, Békés County 2007 (Source: Agricultural Office of County Békés, Plant Protection and Soil Conservation Directorate) (control, WCR free parcels are green)

Table 2. Features of optical data, 2007 (a) and radar image (b) in connection with the remote sensing assessment of WCR larval damage

| Acquisition date | Type | Spatial resolution (ha) | Spectral resolution |
|---|---|---|---|
| 2007.04.22 | IRS-P6 AWiFS | 0,3 | 4 bands in the visible-near infrared range |
| 2007.05.02 | | | |
| 2007.05.11 | | | |
| 2007.05.21 | | | |
| 2007.06.09 | | | |
| 2007.07.09 | | | |
| 2007.07.23 | | | |
| 2007.07.27 | | | |

a.

| Acquisition date | Type | Polarization | Wavelength (cm) | Spatial resolution (ha) |
|---|---|---|---|---|
| 2007.07.22 | ALOS PALSAR | dual (HH+HV) | L-band: 23 | 0,1 |

b.

## 3 RESULTS

We determined all the remote sensing features identifying WCR damage for all the tested cornfields. We created the remote sensing features (Optical feature, Radar feature) from optical and radar data described in Table 2. We created the monoculture feature (Mono-maize feature) by using about 50 high resolution optical images (1997-2006). Notice that Table 2 does not include the data of those images for the reasons of size.

Figure 2 introduces a WCR damaged and a control cornfield. The differences of the optical (2.a.) and radar (2.b.) satellite images are well recognizable. The Landsat TM image taken 24/07/2007 (2.a.) and the entropy map derived from ALOS PALSAR radar satellite image taken 22/07/2007 (2.b.) demonstrate a WCR damaged (delimited by bourdon line) and a control (delimited by yellow line) maize field. A ground assessment of the presented WCR damaged maize field was taken on 06/07/2007. According to this assessment the corn plants were completely lodged and broken corn stalks were randomly lying on the ground, the Iowa rating shown 4,7 root damage in 55% of the tested cornfield. The WCR damage maize field has bluish spots in the optical image (2.a.). The healthy, WCR free maize field does not have the mentioned spots, it is more homogeneous. The measured entropy of the root damaged maize field (light tone) is different than the control (dark tone) field in the entropy map based on radar data (2.b.).

We created the vegetation index curves of the reference test fields based on optical satellite images. The optical feature is derived from vegetation index curves. Based on a radar image we determined the radar feature which presents the entropy of the tested field. Finally, based on optical satellite data time series (1997-2006) we defined the mono-maize feature for all the tested fields which represents the frequency of maize production in 1997-2006.

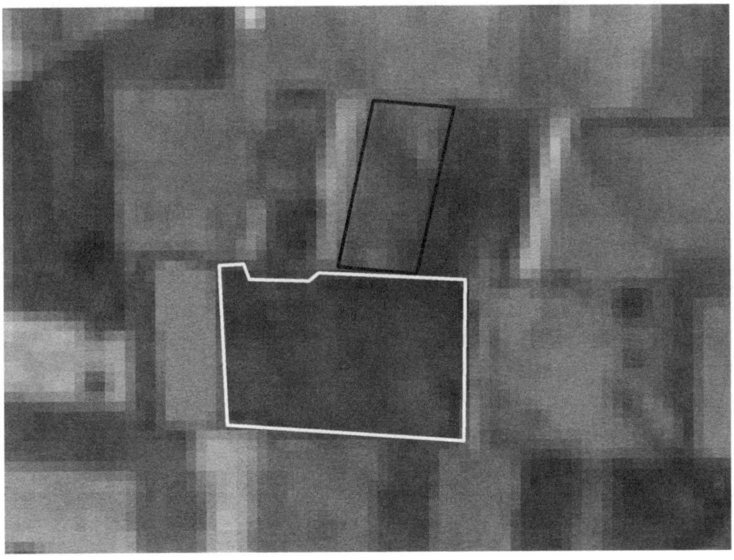

a.

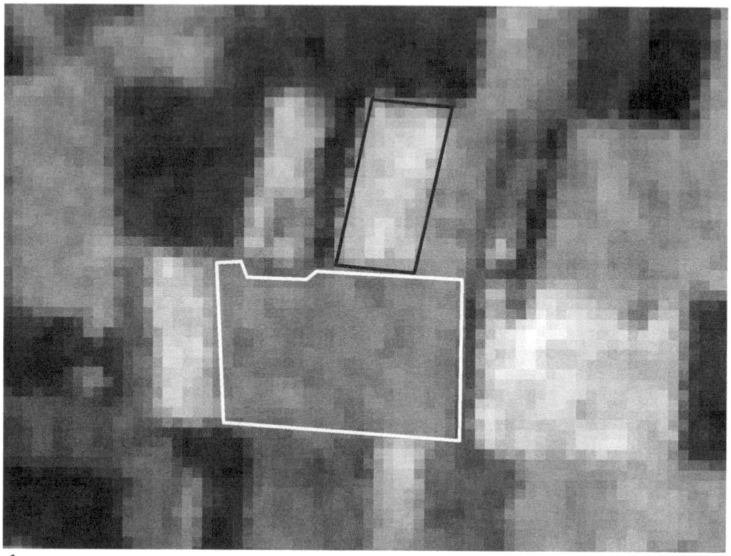

b.

Figure 2. Optical (a) and radar (b) images

The evaluation of the results clearly shows that there was not any tested field which all three features classified incorrectly. At the same time based on all of three features we could classify the WCR damage of 6 fields correctly in harmony with the ground assessment. Five fields were classified incorrectly. The reasons of incorrect classifications in two of the five cases were the small coverage of medium damage (it affected 30-40% of the field) and the 16 day difference between the acquisition date of radar image and the ground assessment. During 16 days the two fields might regenerate (in these fields the ground assessment recorded, "gooseneck" lodging and broken cornstalks tried to be upstanding and regenerated). Probably for the above mentioned reasons the damage stress and entropy of these two maize fields was not observable.

Table 3 shows the accuracy of the methodology based on the integrated and one-by-one evaluation of the remote sensing features. Based on the results of Table 3 the accuracy of damage identification was 61-70 % with the one-by-one evaluation of the remote sensing features. The monoculture feature gave the least accurate results. The other two features presented about the same degree of accuracy, did not show significant differences. It is an important fact that we could achieve as accurate results with one 2-band polarimetric radar image as it is possible with 8 of 4-band optical images. It clearly proves that considering the theory we made a good decision about using radar data to identify WCR damage. The important outcome of our mission is that polarimetric radar data can add a competitive advantage in remote sensing vegetation research and detecting structural changes.

The accuracy of damage identification is higher than 80% with the integrated evaluation of the 3 features. This result is 10 % better than with the one-by-one evaluation.

Table 3. Quantitative results of accuracy assessment achieved in tested maize fields. It includes the integrated and the one-by-one evaluation of the remote sensing features focusing on WCR damages.

| Categories of accuracy | Mono-maize feature | | Optical feature | | Radar feature | | Altogether | |
|---|---|---|---|---|---|---|---|---|
| | No. of fields | Accuracy (%) | No. of fields | Accuracy (%) | No. of fields | Accuracy (%) | No. of fields | Accuracy (%) |
| Correct | 16 | 61,5 | 18 | 69,2 | 18 | 69,2 | 21 | 80,8 |
| Ambiguous | 1 | 3,8 | 3 | 11,5 | 2 | 7,7 | 0 | 0,0 |
| Incorrect | 9 | 34,6 | 5 | 19,2 | 6 | 23,1 | 5 | 19,2 |
| **Total** | **26** | **100,0** | **26** | **100,0** | **26** | **100,0** | **26** | **100,0** |

## 4 CONCLUSIONS

To sum up the results, the retroactive ground assessments verified the goal of our project. We vividly demonstrated a methodology which clearly identifies WCR larval damage efficiently by remote sensing. We are eager to develop and fine-tune the methodology and regional assessment in 2008.

Based on the achieved objectives of the project there is potential in the integrated analysis of optical and radar images to assess and identify disorders and structural changes caused by WCR larvae.

WCR is quarantine pest in the European Union. It is necessary to detect WCR damage in order to comply with the statutory management requirements and good agricultural and environmental conditions which are the basic requirements to receive agricultural subsidies.

The outcome of this project can be used directly to create a nationwide and regional maize damage risk map. This project is unique in the European Union. Based on the initial results, introduction of polarimetric radar technique significantly increased the accuracy. Further development of this technology and the use of additional radar images will create the possibility to accomplish a more accurate damage identification system. With the on-going development, this project can effectively contribute to WCR identification, spread-monitoring and control in Hungary as well as in WCR affected EU Member States.

## ACKNOWLEDGEMENTS

The project was carried out with the support of the Hungarian Space Office and Ministry of Environment and Water. The ALOS PALSAR data were provided by ESA (ESA EO CAT-1 5162) and FÖMI. The reference data were collected by the experts of the Central Agricultural Office of Békés County. It was translated by Diána Fényes.

## REFERENCES

Cloude J. & Pottier E. 1997: *An entropy based classification scheme for land applications of polarimetric SAR IEEE* Trans. Geosci. Remote Sensing, vol. 35, pp. 68–78, Jan. 1997.

Ferrazzoli, P., Guerriero, L. & Schiavon, G. 1999. *Experimental and model investigation on radar classification capability*, IEEE Trans. Geosci. Rem. Sens., 37, pp. 960-968, 1999.

Pierce, L.E., Ulaby, F.T., Sarabandi, K. & Dobson, M.C. 1994. *Knowledge-based classification of polarimet-*

*ric SAR images*, IEEE Trans. Geosci. Rem. Sens., 32, pp. 1081-1086, 1994.
Refregier, P. & Morio J. 2006: *Shannon entropy of partially polarized and partially coherent light with Gaussian fluctuations*, JOSA A, Vol. 23, Issue 12, pp. 3036-3044, December 2006
Schou, J. & Skriver, H. 2001. *Restoration of Polarimetric SAR Images using Simulated Annealing*, IEEE Trans. Geosci.And Remote Sensing, 39, pp. 2005-2016, 2001.
Skriver, H., Svendsen. M. & Thomsen, A.G. 1999. *Multitemporal C- and L-band polarimetric signatures of crops*, IEEE Trans. Geosci. Rem. Sens., 37, pp. 2413-2429, 1999.

*Remote Sensing for a Changing Europe*
*D. Maktav (Ed.)*
*IOS Press, 2009*


*doi:10.3233/978-1-58603-986-8-263*

# 3D models of trees for the Discrete Anisotropic Radiative Transfer model

L. Halounová
*CTU in Prague, Faculty of Civil Engineering, Prague 6, Czech Republic*

T. Dolanský
*UJEP, Faculty of the Environment, Ústí nad Labem, Czech Republic*

Keywords: DART model, 3D model of trees, geodetical measurement, lidar measurement

ABSTRACT: DART model (Discrete Anisotropic Radiative Transfer). is a model of the earth surface layer simulating its radiative transfer. The model simulates 3-dimensional reflectance of surface layers. The model needs the detailed information of the layer in the form of cubes determining individual surface elements. The layer comprising forest areas consist of atmosphere and trees. The paper presents two methods for 3-dimensional models of trees. The first one is the geodetical one performing ground measurement of individual trees by total stations and the second one is the lidar measurement.

Terrestrial lidar for studying tree characteristics was used in two study areas. Point cloud of lidar measurement produces accurate tree canopy information; height distribution of lidar data is related to the vertical and horizontal structure of the tree canopy. The National Park Czech Switzerland in the northwest part of Czech Republic was the first area where the measurement was performed. Terrestrial lidar data are acquired for tree characteristic determination in a compact forest of Fagus sylvatica L. and solitary standing Quercus robur L. In the second area in the Beskydy Mountains, terrestrial scanner was used for scanning of Picea abies L. Statistical values as height and density of trees, branches structure and spatial distribution of foliage.were derived from these data.

## 1 INTRODUCTION

Discrete Anisotropic Radiative Transfer model simulates directional reflectivity of the 3-D model of earth surface layers. The model is formed by voxels with defined type classifications determining structural and radiative characteristics of individual volume units. Three-dimensional tree models, resp. vegetation models seem to be the most complicated due to variety of kinds and shapes of individual trees.

Physically based models of radiative transfer simulating directional reflectance are used for quantitative estimation of biophysical and biochemical parameters from hyperspectral or multispectral remote sensing data. The parameters – as chlorophyll content or Leaf Area Index (LAI) are then determined for large areas from the remote sensing data (Malenovsky 2006).

The project focused on tree volume characteristics was processed in co-operation with the Institute systematic biology and ecology of the Czech Academy of Science (ISBC of CA) The institute uses FieldMap system which is a computer based tool for *in situ* data collection being applied for a forest inventory work, monitoring, landscape engineering in GIS and remote sensing. The system is joined with GIS creating directly GIS layers.

Fig. 1 shows discretisation of *Picea abies* made for the real tree (see Fig. 1) in Sumava region. The example was processed by ISBC of CA using FieldMap. The tree is formed by the tree trunk with changing diameter, individual branches of the first order in horizontal and vertical directions, spatial distribution of the leaf biomass, and atmospheric volume standing in a clear-cut area. The 3D model was created from the 2D model by rotation. However, its positional accuracy is not sufficient and therefore new approaches were tested. Two independent methods are presented below – a geodetical one and lidar one.

Fig. 1. *Picea abies* – photograph (a) FieldMap model (b) made in 2D and rotated to form 3D

## 2 . METHODOLOGY

### 2.1 *Geodetical measuremenst*

Forty sample tree types were selected for testing by geodetical tools for Pinus silvestris and Quercus robur in the forest area alongside D11 highway around 40 km eastwards from Prague (near Velký Osek). The tree age was 40 years.

The following list shows individual tree parameters where only branches of the first are measured:

- $h_c$ tree height
- $h_m$ trunk height up to the dead crown
- $h_z$ trunk height up to the live crown
- $d_k$ crown height
- $d_m$ dead crown height
- $v_k$ live crown height
- $p_k$ crown horizontal projection
- $d_v$ branch length
- $h_v$ branch height
- $z_v$ branch zenith
- $a_v$ branch azimuth
- r trunk diameter (in 1,3 m height).

All height measurements were performed according to fig. 2. Lengths, heights and zeniths of branches were calculated from coordinates of their starting and ending points. The geodetical measurements were performed in the forest stand by the Trimble 5000 total station data were processed in Matlab.

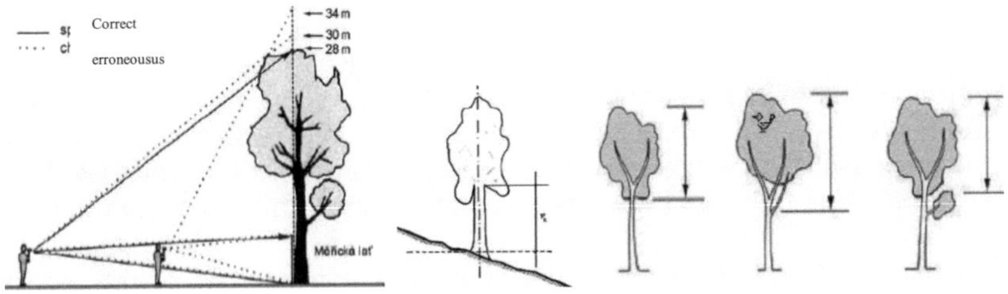


tree height live crown height live crown height

Fig. 2 Geometrical parameters of trees

## 2.2 *Laser scanning*

The scanning of *Picea abies* at the first location in the Moravian and Silesian Beskydy Mountains (Bílý Kříž - 18.54 °E, 49.50 °N, 936 m above sea level) was carried out from two stations only on the basis of the first reflection. The average age of the trees is estimated around 120 years, the average height is approximately 40 m. A total of 20 trees were studied and processed, one by one, at this site.

The stations were approximately 60 m from the border of the stand. The scanning density was set at 25 mm for the 60 m distance. With a similar value of the width of a beam of rays, the set density ensures continuous surveying of the area. The type of scanner used allows scanning the area in the full range (360°), namely to a distance of up to 1,000 m. The wavelength of a laser beam is 1,500 nm, thus it also allows differentiating needles from the trunk and branches on the basis of reflectivity.

The second site was selected at a place of freshly felled bark beetle trap trees in a continuous dense stand with an average age of the trees of approximately 75 years. The average height of the trees on this site is lower and amounts to 25–30 m. Another 9 trees were processed at this site.

The second site was surveyed from a single station approximately 40 m from the trees, and here the density of points was also set at 25 mm at the distance of the trees (40 m).

Control points (6 points at the first location, 4 points at the second) were surveyed by a GPS method. These control points served for the transformation of the entire point cloud into a UTM 34N geodetic coordinate system on one hand and for connecting both clouds at the first location on the other hand. The accuracy of transformation into the UTM coordinate system was 0.11 m at the first location and 0.70 m at the second. The deterioration in the accuracy is due to high screening of the sphere by surrounding high-grown trees.

Initial processing consisted partly of the connection of individual scans and their transformation into the UTM 34N coordinate system and partly of the extraction of individual selected trees from the entire point cloud. Points indicating wood mass (branches and trunk) in individual trees were further differentiated from points reflected from needles based on an analysis of the intensity of reflection. The result comprises ASCII files with a list of points belonging to individual trees (a separate file for each tree) with point coordinates and the intensity of reflected radiation. These files serve for further analyses, which include the basic determination of the tree height, the height of the live and dead part of the crown, the length of overfall of needles, number of branches, zenith angle of the emergence of first order branches, crown radius and others. These parameters were determined manually on selected trees by means of PolyWorks software by ISBE specialists. Furthermore, data on the spatial arrangement of tree mass in space are of importance for a DART model, and these parameters have already been ascertained by methods of the statistical analysis of a point cloud in GIS systems, which allow solving a number of spatial tasks in an effective manner. Small mutual shielding is a prerequisite for processing a point cloud. Such shielding manifests itself negatively by a decrease in points when passing through the tree crown. The tree being scanned then contains much fewer points on the reverse side of the crown from the scanner than on the near side. The symmetry of the crown is another prerequisite. The last prerequisite is the fact that the tree trunk grows vertically and it does not branch into more trunks. Subject to these three prerequisites being accepted, we can divide the tree crown vertically into two halves and work in analyses only with the near half while the other half is axially symmetric. These are simplifications that result in real values being slightly distorted while the analyses being conducted are simultaneously markedly simplified.

The determination of the near half and the centre of the trunk were made for the analyses in a TerraScan program, as well as in a specialized upgrade for processing the data of laser scanning in a Bentley MicroStation environment. The vertical section plane was determined at the first location as a parallel with the connecting line of both stations going through the centre of the trunk. The centre of the trunk was determined for this purpose in the mean height of the live crown. On the other site, the vertical section plane was led perpendicularly to the direction of scanning. Individual points from the near half of the tree and classified as leaf mass according to the intensity of reflected radiation and the centre of the trunk were exported into a personal geodatabase, with which the ArcGIS system continued to work.

The analysis in the ArcGIS consisted partly of the determination of a series of statistical values, whose calculation was modelled by means of ModelBuilder and partly of the generation of a voxel model of the tree, i.e. a spatial model of the distribution of leaf mass inside the crown (the voxel edge was set at 0.1 m). The voxel model was further supplemented with a spatial model consisting of layers with a height of 1 m, which were divided into sectors of 20° each and further into individual annular rings with a thickness of 0.2 m. The radius of a cylinder, which contains 95% of all the points of the cloud, was calculated for the whole tree as a basic characteristic of the crown diameter. Further parameters for the voxel model include, for example:

- total number of non-empty voxels and the average number of points in such voxels including a standard deviation
- average and maximum distance of a voxel from the centre of the tree

Among others, the following parameters were further calculated for the sectors and layers:

- average number of points in individual elementary areas
- maximum distance for each sector and each layer.

Tab. 2: Basic parameters for tree No. L_2_9 from the second site

| Parameter | Value |
|---|---|
| Total number of non-empty voxels | 20,402 |
| Average number of points per voxel | 2.2 |
| Standard deviation (m) | 1.5 |
| Crown radius (m) | 3.96 |

Two sites were processed near the territory of the national park – Vysoká Lípa and Růžový vrch. The Vysoká Lípa site (50°51'N, 14°21'E) is located in the north-western part of the Czech Republic in the vicinity of the Czech Saxonian Switzerland National Park, which is known for its sandstone rock cities. A high-grown solitary sessile oak (*Quercus petraea*), which is declared as a protected tree, was studied here. The second site was a slope of nearby Růžový vrch (50°50'N, 14°19'E) with a stand of European beech (*Fagus silvatica L.*) with partially interspersed sycamore maple (*Acer pseudoplatanus L.*).

Scanning took place partly to ascertain the spatial characteristics of forest stands but also to check the accuracy of a digital model of the terrain created on the basis of aerial laser scanning. Therefore, an HDS 3000 system by the Leica Company was selected. It allows accurate scanning by a very narrow beam of rays with a trace size of a mere 6 mm. For scanning it utilizes green laser light. The range of the scanner is up to approximately 300 m, which is fully sufficient for application at both locations, as scanning at a distance over 100 m is not expected. For the set purposes it is an ideal system that allows verifying data acquired by aerial scanning.

During April 2007, both sites were scanned with this ground scanner even before the beginning of the growing season. Scanning on the first site was performed from four stations evenly distributed along the perimeter at maximum distance of 20 m from the tree trunk, which ensured homogeneous scanning of the whole tree with a minimum of covered parts of branches. The scanning density is, considering the spatial structure of the crown, not uniform and corresponds to a density of 5 cm/50 m. Standard reflective control targets, by the survey of which the whole model was also spatially positioned into a JTSK coordinate system, were used for the connection of the clouds from individual stations. Surveying of the targets took place by means of a total station with connection to two near trigonometric points.

The second stage of scanning took place in June of the same year, with full foliage. In this way, data were acquired for an analysis of the leaf mass and analyses of the wood mass.

Points lying on the ground and points lying on a separate tree were filtered out of the united point cloud. Individual tree parameters, such as tree height, crown width and height of the emergence of first branches, spatial orientation of branches, density of branches and leaves in a horizontal and vertical direction are analysed from the thus prepared point cloud. Evident in figure 3, which shows a horizontal section (with a thickness of 1 m) through the tree in a leafless and leafy state, is the difference in the volume of mass. In the graphs that display a histogram of the density of points dependent on the distance from the tree trunk, it is possible to notice the internal distribution of mass inside the crown.

## 3 RESULTS

The geodetical measurements results consisted of all parameters except for distinguishing live and dead crown for *Quercus robur*. The measurements were performed early in April to avoid period with leaves covering branches and creating geodetically non-measurable object having endless movement. The resulting data were in the form of ASCII coordinates of individual tree characteristics (listed on the second page of the paper).

Figure 3 shows voxel model of *Picea abies* already in the voxel form where individual colors describe voxel number of points.

Fig 3. Voxel model of *Picea abies* (left), photographs and results of *Quercus petraea* scanning (position, crown in a horizontal level, tree photo, result of the scanning)

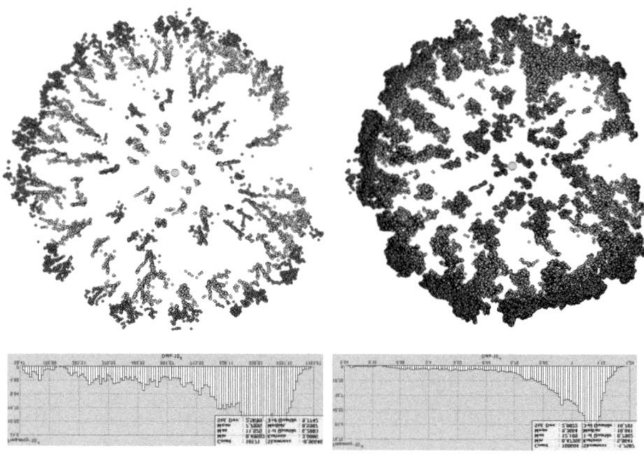

Fig. 4. Comparison of change in the volume of *Quercus petraea* leaf mass between scanning in April and in June on a horizontal section with a thickness of 1m. The graphs display the density of mass dependent upon the distance from the trunk

## 4 CONCLUSIONS

Geodetical measurement can be performed even in a forest stand; only horizontal crown projection can be performed only for solitary standing trees. Branch zenith and azimuth represent laborious measurement. Oblique distances must be measured and the accuracy can be worsened by erroneous laser beam reflection caused by irregular shapes of branches. Their length is in fact a horizontal projection instead of the real curved length. And furthermore, the measured end of branch does not represent the real end. The method brings only a simplified tree model data.

It is possible to acquire quite a number of statistical data from a point cloud in a very efficient manner. Calculations can be made entirely automatically for an absolutely random number of trees. It is necessary to prepare source data in advance by a simple separation of points required for analysis – based on spatial and spectral characteristics. With a suitable density of points, it is possible to determine other parameters necessary for the DART model from the point cloud such as the azimuth and zenith of the first order branches, the thickness of the trunk at a random height, the total tree height, height of the emergence of the live and dead crown. These parameters were measured manually by ISBE workers by means of suitable software, in particular, for the speed and uniqueness of measurement. Nevertheless, even these data could be acquired by a suitable analysis of narrow layers where the branches are displayed by a unique circular cluster of points, and thus they can be monitored safely.

However, the geodetical measurement can be performed also in a forest stand unlike the laser scanning; the measurement station is within meters distance unlike in laser scanning case where it is in tens of meters.

## ACKNOWLEDGEMENT

The presented results were performed in the framework of 2 projects; the first one is financially supported by the MŠMT grant No. MSM 6840770006 of the Czech Republic and the second one was processed with the support of ESA/PECS projects, Project No. 98029 “Spectral-spatial scaling from leaf to canopy level using spectro-directional approaches in support of the GMES Sentinel 2: 'Superspectral' mission” and the Czech Science Foundation 205/07/P331.

## REFERENCES

Hanuš, J., Malenovský, Z. 2006. Project of aerial campaign for hyperspectral data collection alongside the D11 highway Prague – Hradec Králové. Institute of systems biology and ecology, Academy of Sciences of the Czech republic

Dolanský, T. 2004: Lidar and laser scanning. Acta Universitatis Purkynianae n. 99. Ústí nad Labem, UJEP FŽP, 2004, p.100. ISBN 80-7044-575-0

*Remote Sensing for a Changing Europe*
*D. Maktav (Ed.)*
*IOS Press, 2009*

*doi:10.3233/978-1-58603-986-8-270*

# Analysis of forest health at landscape level through very high resolution satellite images

F. Giannetti
*I.P.L.A. S.p.A. – C.so Casale 476-10032, Torino, Italy – Phone: +39-011 8998933, Fax: +39-011 8989333; e-mail: giannetti@ipla.org*

A. Grignetti
*CRA – Unità di Ricerca per le Produzioni Legnose fuori Foresta – Strada per Frassineto, 35, 15033 Casale Monferrato (AL) – Phone:+ 39-0142 330925, fax. +39-0142 55580; e-mail: grignetti@populus.it*

Keywords: trees mortality, object oriented classification, forest health

ABSTRACT: Analysis of satellite images, combined with Geographic Information System (GIS) data and field surveys, proved to be an effective tools for studying forest health conditions. Most of the past works examined conifer trees but structure and health of hardwood trees were also analyzed. In recent years the traditional use of moderate to medium-resolution imagery (10–30 m) were integrated by the availability of very high resolution satellite data which allows monitoring individual trees.

In the present work a study area (Trino forest protected areas) in north-western Italy was selected in the framework of a project devoted to the analysis of the extensive phenomena of forest declining recently registered in some Oak-hornbeam stands. Following a sequence of very dry summer seasons a large number of dead trees were found in these stands, representing some of the most significant forest remnants in the western part of Po Plain.

In order to monitor this phenomena an Ikonos image was acquired and processed with the aim of classifying the dead trees and possibly identifying areas in which trees are subject to drying phenomena. Image enhancement methods based on texture measures and vegetation indexes were applied with the aim of evaluating vegetation vigour and defining health classes. The same procedure was applied to homogeneous groups of trees resulting from the application of an object-based approach integrating segmentation algorithms and semi-automatic classifiers. Information about the distribution of dead trees and declining forest areas extracted from satellite images were then correlated to field collected data on crown condition and soil moisture.

Remotely sensed data in combination with field surveys revealed very useful for a landscape-scale pattern analysis of trees mortality considering the infrared sensitivity in registering canopy changes related to health condition. The identification of the dead trees allowed to better understand the spread and impact of forest decline in the examined areas as a function of the interaction between spatial heterogeneity of the environment, landscape structure and specific characters at site level.

## 1 INTRODUCTION

Climate-induced forest declining and mass mortalities are recent phenomena that can be correlated with a number of interacting environmental stress factors such as general temperatures increasing, recurrent dry seasons, disease spread due to pathogens affecting weakened trees, wildfires and so on. In recent years many oak-hornbeam stands of the plain areas in north-western Italy suffered a significant decline with a considerable increase of oak trees mortality that need to be monitored. The Piemonte Region administration through its forest department commissioned IPLA a research for investigating the dynamic of oak mortality and understanding the main causes.

In this context remote sensing together with GIS technologies are valuable tools considering the capacity of analyzing and monitoring the spatial components of tree mortality pattern. In fact analysis of forest pathology at landscape level through remote sensing data is an emerging field that seek to understand the distribution of dead trees as related to the spatial heterogeneity of the environment and to the landscape structure.

Medium to large spatial resolution imagery has been used in the past for deriving information about forest structure and health (Royle and Lathorp, 1997, Muckoney and Haack, 1994), but only the recent availability of very high resolution imagery allows mapping and monitoring forest at individual tree crowns level. Some recent researches made use of high resolution satellite data for studying individual tree biometric parameters in tropical forest (Clark et al., 2004) and airborne digital data were recently used for mapping tree mortality of coastal forests of central California with particular reference to oaks stands attached by a new pathogen (Guo et al., 2007). In this framework various techniques of individual trees recognition were adopted with the use of both hybrid supervised – unsupervised classification methods (Wayman et al., 2001; Kelly et al., 2004) and object-based approach including preliminary segmentation and knowledge based classification of the resulting objects.

This kind of approach performed well with high resolution satellite images in which the raise in spatial resolution negatively affects the accuracy of pixel-based classifiers due to increased within class spectral variability. Preliminary definition of homogeneous areas (polygons or segments) delineated with reference to textural and/or spectral properties gives the advantage of incorporating geometric properties and spatial topology into the classification process.

Our research goal is to detect gaps in the forest structure related to dead trees presence through very high resolution satellite data. Some discontinuities in the forest canopy are actually related to trees die standing, snap off, or blown to the ground. Declining crowns before dying pass through a remarkable reduction of foliar coverage often associated with production of epicornic sprouts. This process resulted in a significant reduction of infrared reflectance and, after the trees death, in the formation of dark "holes" in the image, corresponding to forest canopy gaps that are rapidly colonized by invasive species such as black locust (*Robinia pseudoacacia*) and hazelnut (*Corylus avellana*). Gaps detection was carried out through the application of both pixel-based and object-based classification approaches. The results of the two methods were compared and correlated with surveyed data about the dead trees occurrence.

The present paper reported mainly the methodological approach and some preliminary results that were obtained through the analysis of a single-date image while monitoring the oaks mortality phenomena through time is the objective of next project developments.

## 2 STUDY AREA AND AVAILABLE DATA

This study focused on the analysis of the last residual plain forest area in the eastern part of the Piemonte region named "Bosco delle Sorti della Partecipanza" of Trino Vercellese. The conservation of this forest patches (650 he) in the middle of an highly specialized agricultural areas (Predominant rice crops) is mainly due to a community land management system lasting from the medieval period.

This area is characterized by the presence of east-west elongated hills presently located about 3 kilometres from the Po river and composed by a series of old terraces. The higher surfaces are covered by very old soils (Alfisols) with a high clay content while intermediate terrace levels show undulated surfaces formed by sandy-gravel alluvial deposits with hydromorphic soils.

The forest cover is essentially composed of an oak-hornbeam (*Quercus robur*) association with presence of black locust, red oaks (*Quercus rubra*) and alder (*Alnus glutinosa*) stands; the forested surface has been divided in parcels subject to coppice management according to a forest plan.

An Ikonos image acquired the $14^{th}$ of July 2007 on the Trino area was used for developing the study. Preliminary processing, including orthorectification and pansharpening, were applied to the raw panchromatic band (1 meter resolution) and multispectral bands (blue, green, red, near infrared; 4 meters resolution). Geometric rectification was done using a rational polynomial coefficients model with less than 2 meters of residual RMSE error on independent control points.

Vector data were also used for developing the research:

- Forest map based on a forest type classification scheme.
- Survey data of dead trees and of different kind of gaps. These data were collected as ground points through a GPS device. A differential correction phase was applied to the collected points and, in consideration of local signal quality degradation due to tree canopy obstruction and multipath, a maximum positional error of 2 meters was obtained.

## 3 METHODS

The following image enhancements techniques were preliminary applied to the image bands.

- High pass filter (3 × 3 kernel) on the near infrared band. The filter was applied in order to eliminate low frequency phenomena and to emphasize edges between land cover classes, particularly those related to presence and abundance of healthy vegetation.
- Texture filter (3 × 3 kernel) based on the co-occurrence matrix that measures the relative frequencies with which pixel values occur in two neighbouring processing windows separated by a specified distance and direction. It shows the number of occurrences of the relationship between a pixel and its specified neighbour.
- Calculation of derivative bands such as NDVI (Normalized Difference Vegetation Index), simple bands ratios (green/infrared and red/infrared bands) and principal component bands.

After image processing a set of bands to be used for the classification procedures was selected basing on the application of a separability test; the most useful data set comprehends the four raw bands, texture co-occurence measures (Data range and mean) and green/infrared bands ratio.
A preliminary field campaign was carried out moving along east-west transects in the southern part of the Trino forest and allowed surveying 44 points with presence of dead trees and 5 gaps colonized by hazelnut and black locust.

### 3.1 *Pixel-based classification*

Samples of the main forest types and forest gaps were first collected and a Jeffries-Matusita test was applied in order to compute the spectral separability between selected samples pairs for the given input file.

The sampled classes are the following:

- Oak-hornbeam stands that is the dominant forest association (Mixed oak forest).
- Black locust stands that, mixed with other species such as hazelnut and other shrubs, are invasive species always present in the understory (Mixed forest with black locust).
- Small-scale forest gaps normally characterized by an abrupt decrease of the reflectance in the visible and infrared domains (Forest gaps).
- Red oaks stands (Mixed forest with red oak).

Jeffrey-Matusita test values ranges from 0 to 2.0 and indicate how well the selected pairs are statistically separate; values greater than 1.7 indicate that pairs have adequate separability to be discriminated by classification algorithm (Richards, 1999).

All the classes resulted quite separable with the exception of mixed forest with red oak, exotic species concentrated only in some part of the study area that resulted close to the black locust stands from the spectral point of view.

A maximum likelihood supervised classification was applied to the selected band set for the three classes mixed oak forest, Mixed forest with black locust, and forest gaps.

Table 1. Separability measures resulting from the application of Jeffries-Matusita test to the sampled classes

| | Mixed forest with red oak | Mixed forest with black locust | Mixed oak forest | Forest gaps |
|---|---|---|---|---|
| Mixed forest with red oak | | 1.450 | 1.659 | 1.997 |
| Mixed forest with black locust | 1.450 | | 1.795 | 1.993 |
| Mixed oak forest | 1.659 | 1.795 | | 1.850 |
| Forest gaps | 1.997 | 1.993 | 1.850 | |

Figure 1. Maximum likelihood classification result on the study area.

## 3.2 *Object-based classification*

The multi-resolution segmentation procedure, implemented in the Definiens®Developer software, was applied on a part of the image (1040 × 649 pixels) corresponding to the Ramezzana area, heavily affected by oaks mortality. This procedure is based on a region growing tecnique which starts from a single pixel, compares it with its neighbours and carries on clustering pixels pair-wise in order to minimize the resulting heterogeneity. Clusters are then compared with their neighbours and progressively merged into larger segments until a user-defined homogeneity threshold (scale factor) is reached (Baatz and Schäpe 2000). The homogeneity threshold is composed of criteria such as pixel values (colour) and polygon shape properties. The later can further be divided into two weighted elements: segment border smoothness and segment compactness.

A scale factor of 25 with colour/shape weighted 0,7/0,3 and smoothness/compactness 0,5/0,5 were used as these segmentation parameters allowed to delineate even the smaller forest gaps.

Delineated segments were then classified into three classes (mixed oak forest, mixed forest with black locust and forest gaps) following specific fuzzy rules based on thresholds (Table 1) defined

Table 2. Thresholds applied to image bands as object-based classification rules

| | Brightness | Mean B1 | Mean B2 | Mean B3 | MeanB4 | NDVI |
|---|---|---|---|---|---|---|
| Mixed oak forest | 262-350 | | | | 551-750 | |
| Mixed forest with black locust | 352-442 | 258-280 | 257-303 | 154-193 | 751-987 | |
| Forest gaps | 208-261 | 224-247 | 195-231 | 94-133 | 411-566 | 0.55-0.69 |

Table 3. Comparison among the number of polygons classified as gaps and the number of polygons associated to dead trees presence in the Ramezzana area according to the applied classification methods

| | Classification methods | |
|---|---|---|
| | Maximum likelihood | Object oriented |
| Number of polygons classified as gaps (Area in hectares) | 1110<br>(3,80 ha) | 592<br>(6,60 ha) |
| Number of gaps polygons associated to dead trees presence (Area in hectares) | 28<br>(0,3 ha) | 32<br>(0,42 ha) |

for brightness, NDVI and mean spectral reflectance values of the 4 satellite bands. The brightness band is formed through the composition of the whole data set (raw spectral bands, texture bands, NDVI and high pass band).

### 3.3 *Geometric properties extraction*

After classification forest gaps obtained with the two classification approaches were exported as polygons in order to be analyzed in a GIS environment.

In order to select forest gaps that can be associated with the surveys a buffer of 4 meters was first created around the 44 GPS points for taking into account positional uncertainty due to image rectification and GPS residual errors. Then polygons intersected by those buffers and labelled as gaps according to the two classifications approaches were associated to the collected information about presence, number and characters of dead trees.

Ecologically meaningful geometric parameters, such as shape index, compactness, length/width ratio, were also calculated and assigned to the polygons. In particular the shape index gives a measure of the fractal dimension of an object and is defined as $P/4\sqrt{A}$ where P and A are the perimeter and the area of an object, respectively. Compactness of an image object is calculated by the product of the length ($l_v$) and the width ($w_v$) and divided by the number of its pixels ($\#P_v$) and Length/Width ratio index is based on a complex algorithm of the eigenvalues of the covariance matrix (Baatz et al., 2000).

## 4 RESULTS

The Table 2 reported the general number of gaps polygons resulting from the two classification approaches for the Ramezzana area and the respective number of polygons that could be associated to the surveyed data about dead trees presence.

In order to understand if the gaps with trees mortality delineated through the two approaches showed significant differences from the spectral and morphological point of view mean, standard deviation, minimum and maximum values of the geometric indexes (Shape index, length/width ratio, compactness) and of NDVI band were also computed and compared (Table 3).

## 5 DISCUSSION AND CONCLUSIONS

The application of proposed method seems to be encouraging for the purpose of identifying and delineating forest gaps related to trees mortality occurrence. In fact, with reference to field survey

Table 4. Statistical parameters of geometric properties and NDVI calculated for the gaps polygons with dead trees classified with the object and pixel based approach

| | *Shape index* | | *Length/Width* | | *Compactness* | | *NDVI* | |
|---|---|---|---|---|---|---|---|---|
| | MaxLike | Object | MaxLike | Object | MaxLike | Object | MaxLike | Object |
| Mean | 1.617 | 1.828 | 1.633 | 1.752 | 1.674 | 1.976 | 0.6367 | 0.642 |
| Minimum | 1.000 | 1.179 | 1.000 | 1.000 | 1.000 | 1.250 | 0.5853 | 0.581 |
| Maximum | 2.538 | 3.322 | 2.774 | 3.357 | 2.489 | 3.372 | 0.7178 | 0.706 |
| Standard deviation | 0.421 | 0.492 | 0.514 | 0.596 | 0.372 | 0.503 | 0.0306 | 0.031 |

data, image analysis allows to identify 32 out of 44 (72%) gaps with dead trees through the application of object-oriented approach and 28 (63%) through the pixel-based classifier.

The object-oriented classification approach resulted more suitable respect to the conventional maximum likelihood pixel classifier. Less gaps polygons were actually produced in the segmentation phase reducing the salt-and pepper effect of the pixel-based classification and creating more compact and homogeneous objects. In turn more of those objects (32 compared to the 29 classified by the pixel based approach) could be associated to trees mortality phenomena according to field surveys. This is mainly due to the tendency of the pixel-based algorithm of including in the gaps class big trees shadows and other small discontinuities in the forest canopy.

According to the analysis of the geometric parameters and NDVI reported in Table 4 the gaps polygons classified by the two methods are very similar from the geometric and spectral point of view.

In the prosecution of the research more field data are certainly needed in order to verify the following hypothesis:

- Canopy gaps associated to trees mortality are more regularly shaped than others forest openings and bare areas as the shapes of dead oak crowns are relatively compact.
- There is a possible correlation between the position of those gaps and site-level soil characters with particular reference to micro-morphology of the plateau surface.

## REFERENCES


Baatz, M., Schäpe, A., 2000. Multiresolution segmentation – an optimization approach for high quality multi-scale image segmentation. In *Angewandte Geographische Informationsverarbeitung*. Wichmann-Verlag, Heidelberg: 12–23.

Clark, D.B., Castro, C.S., Alvarado, L.D.A., and J.M. Read, 2004. "Quantifying Mortality of Tropical Rain Forest Trees Using High-Spatial-Resolution Satellite Data," *Ecological Letters*, 7: 52–59.

Guo Q., Kelly M., Gong P., D. Liu, 2007. An object-based classification approach in mapping tree mortality using high spatial resolution imagery. GIScience & Remote Sensing 444 N. 1: 24–47.

Kelly, M., Shaari, D., Guo, Q. H., and D. S. Liu, 2004. "A Comparison of Standard and Hybrid Classifier Methods for Mapping Hardwood Mortality in Areas Affected by 'Sudden Oak Death'," *Photogrammetric Engineering & Remote Sensing,* 70:1229–1239.

Liu, D., Kelly, M., Gong, P., Guo, Q., 2007. Characterizing Spatial-temporal Tree Mortality Patterns Associated with A New Forest Disease. Forest Ecology and Management. 253: 220–231.

Muchoney, D.M., and Haack B.N., 1994. Change detection for monitoring forest defoliation. Photogrammetric Engineering and Remote Sensing 60: 1243–1251.

Richards A., 1999, Remote Sensing Digital Image Analysis, Springer-Verlag, Berlin, p. 240.

Royle D.D. and Lathrop R.G., 1997. Monitoring hemlock forest health in New Jersey using Landsat TM data and change detection techniques. Forest Science 43 (3): 327–335.

*Remote Sensing for a Changing Europe*
*D. Maktav (Ed.)*
*IOS Press, 2009*

*doi:10.3233/978-1-58603-986-8-276*

# Multi-temporal spectral unmixing to characterise urban change in the Greater Dublin area

T. Van de Voorde, L. Demarchi & F. Canters
*Cartography and GIS Research Group, Department of Geography, Vrije Universiteit Brussel, Brussels, Belgium*

*Email: tvdvoord@vub.ac.be*

Keywords: urban remote sensing, spectral mixture analysis, change detection

ABSTRACT: Urban growth models are useful tools to assess the impact of alternative public policy scenarios. The calibration of such models requires historical land-use data, which is not always available at frequent time intervals. In this research, information on spatio-temporal change of land-cover gradients is extracted from two multidate Landsat images by linear spectral unmixing. For this purpose, the two images were subjected to a relative radiometrical calibration in order to reduce the impact of sensor drift and differences in illumination and atmospheric conditions. The resulting gradient information can be used to derive spatial metrics, which in turn characterise urban form and morphology and may serve as a proxy to land-use data. Because the sub-pixel proportions do not contain implicit information as to whether a particular pixel belongs to the urban fabric, an urban mask was required to make that distinction. This mask was created by first applying an unsupervised classification based on Kohonen self-organising maps, and subsequently enhanced by applying ad-hoc knowledge-based post-classification rules. The results of this study demonstrate that sub-pixel change maps are useful to identify urban growth patterns. In combination with the urban mask and unsupervised classification, sub-pixel gradients will be used in future research to investigate if they provide useful information on urban structure, and can maybe even be used to infer certain types of land use.

## 1 INTRODUCTION

The world is urbanising at an increasing pace. Today, more than half of the earth's population lives in cities and the number of urban residents is expected to increase to over 5 billion by 2030 (UN-HABITAT 2006). Despite the many advantages city life offers, uncontrolled urban growth affects both the human and natural environment and calls for effective urban management strategies.

Developing such strategies and monitoring policy effects requires reliable and sufficiently detailed information on the urban environment and its dynamics, including an understanding of urban change processes. For this purpose, analysing changes in urban land use is of great consequence and is facilitated by computer-based models that predict urban development patterns and determine the future impacts of public policy choices. Current and historical land-use maps are required for calibrating urban growth models and given the large effort it takes to create such maps, they are often not available at short intervals. In the European MOLAND project, for instance, urban land use is mapped for four periods: the early 1950s, the late 1960s, the 1980s and the late 1990s (MOLAND 2008). Data from earth observation satellites may provide intermittent information on urban development and could in that way improve the calibration of MOLAND'S growth model. Land use is, however, tied to socio-economic activities. It can therefore not be directly inferred from spectral information as opposed to land cover, which refers to the physical properties of the earth's surface. Previous studies have nevertheless demonstrated a relationship between the spatial structure of the built-up environment and its functional characteristics (Barr & Barnsley 1997). A rather novel approach in this research area is to describe urban form and structure by means of spatial metrics. Spatial metrics describe various properties of the spatial heterogeneity and configuration of land cover in a given area. They have recently shown considerable potential for structural analysis of urban environments (Herold et al. 2005). Spatial metrics derived from satellite imagery might therefore be the key to complement existing land-use maps and improve the calibration and validation of urban growth models.

Despite the currently available high resolution satellite images, which provide increasingly detailed information about urban surface materials, most of the available historic archive imagery consists of medium resolution data such as from the Landsat or SPOT programmes. Notwithstanding the advantages of medium resolution data in terms of cost, the extensive historic archives and large mapping extents, their relatively low spatial resolution may lead to low mapping accuracies because the sensor's instantaneous field of view (IFOV) often contains different types of land cover, especially in urban areas. Traditional classification algorithms that derive land-cover maps from digital images assign pixels individually to a single class, and will run into difficulties when dealing with such mixed pixels. Spectral mixture analysis addresses this problem by unmixing (deconvolving) each pixel spectrum into fractional abundances of its surface constituents or endmember spectra (van der Meer 1999). It might therefore be a useful technique to extract gradient-based spatial metrics to characterise urban structure and urban change.

In this research, we extracted urban gradient information for the city of Dublin by producing sub-pixel proportions with linear spectral mixture analysis for two dates. Because of the spectral confusion between urban and some none-urban surface types such as bare soil in rural areas, we developed an urban mask to constrain the unmixing to the urban fabric. This mask was created with an unsupervised classifier based on Kohonen self-organising maps (Kohonen 2001) and enhanced with knowledge-based post-classification rules. The produced sub-pixel proportions for each date will be used in the MAMUD project (http://www.mamud.be) together with the unsupervised classification in order to infer information on urban structure and context and, ideally, on land use.

## 2 STUDY AREA

The city of Dublin was chosen as study area for this research. Dublin is the political, social, economical and cultural capital of Ireland, and is home to over 40% of the country's population. As Ireland's most prominent urban centre, the significant demographic and economic changes that have been taking place in the country since the mid 1980's are especially apparent in the city (Kitchen 2002). Dublin experienced rapid urban expansion in the 1980's and 1990's, fuelled by the building of new roads that drove residential and commercial development rapidly outward into the urban fringe. While the Greater Dublin area as a whole experienced only a moderate population growth of 3.6% between 1986 and 1996, population in the urban periphery increased more rapidly with as much as 9.6% in South Dublin and 21.1% in Fingal, to the north. This has resulted in a hollowing of the central city and a simultaneous growth and movement into Dublin's low density, car-oriented and seemingly unplanned periphery (Bannon 1999).

## 3 DATA AND PREPROCESSING

Two Landsat image datasets (path 206, row 23) were used to extract sub-pixel gradient information for Dublin: a TM image of 13$^{th}$ June 1988 and an ETM+ image of 24$^{th}$ May 2001. While the 1988 image was nearly cloud free, the 2001 image was partly covered by clouds to the east, mainly above the sea. A cloud mask was used to minimise the impact on the formation of spectral clusters by the unsupervised classification algorithm. The images were geometrically co-registered to the Irish Grid projection system by a first-order polynomial transformation. The RMS error on an independent set of control points was 27.61 m for the 1988 image and 27.62m for the 2001 image, which implies that on the average the geometric shift is less than a 30m Landsat pixel. The raw digital numbers of both images were converted to exoatmospheric reflectance according to the formulas and calibration parameters presented by The Landsat 7 Users Handbook (Irish 2007). While this conversion removes predictable effects caused by differences in solar irradiance and solar angle, it does not take into account the influence of atmospheric condition and sensor drift on the measured radiances. To quantify changes in surface reflectance between the two acquisition dates with spectral mixture analysis, the impact of temporal spectral variability that is not caused by changes in surface reflectance should therefore be minimised. Because no atmospheric data or field measurements of ground reflectance were available, a relative reflectance calibration based on the identification of pseudo-invariant features (Schott 1988) was applied. By visually comparing the 1988 and the 2001 images, nine sites were selected for which the surface reflectance was not expected to have changed in between the two dates. Three were chosen to represent high albedo surfaces: highly reflective roofs of commercial or industrial buildings in the port area. Three low albedo sites were selected on the Liffy River in central Dublin, and three sites with asphalt were selected on the airport runway. Because at-sensor radiances vary linearly with ground reflectances for visible and short wave infrared wavelengths (Conel 1990) and because this relationship can be extended to multidate images (Caselles & Garcia 1989), the pseudoinvariant features of the 2001 image can be linearly transformed to give them the same apparent reflectance as in the 1988 image (Hall et al. 1991). The estimated linear function between the pseudoinvariant sites of the two images can then be applied on the entire 2001 image to reduce temporal spectral variability caused by the combined impact of differences in illumination, sensor drift and atmosphere.

## 4 METHODS

### 4.1 *Linear spectral mixture analysis*

Linear spectral mixture analysis (LSMA) is a common approach to sub-pixel classification whereby a pixel's observed reflectance is modelled as a linear combination of spectrally pure "endmember" reflectances. Each endmember contributes proportionally to the overall spectral response according to its relative abundance within the sensor's instantaneous field of view (IFOV). To estimate the fractional cover of each endmember within a given pixel, the following equation has to be solved for all image bands simultaneously, using a least squares approach:

$$R_b = \sum_{i=1}^{n} f_i r_{i,b} + e_b \quad (1)$$

where $R_b$ is the reflectance of the pixel for band $b$, $f_i$ is the proportion of endmember $i$ within the pixel, $r_{i,b}$ is the reflectance of endmember $i$ for band $b$, $n$ is the number of endmembers and $e_b$ the error of fit for band $b$ (van der Meer et al. 1999). Inverting this system of mixing equations to retrieve endmember fractions that best fit the observed mixed reflectances implies determining the optimal location of endmembers in feature space.

LSMA has recently received quite some attention in studies that aim to characterise urban environments (e.g. Rashed et al. 2005, Small 2003). For this purpose, the VIS model is a useful conceptualisation of the urban environment because it allows representing any urban area by three physical components: vegetation (V), impervious surfaces (I) and soil (S), in addition to water (Ridd 1995). If these components could be unambiguously represented as endmembers in feature space, fractions derived from unmixing an urban area would allow to position urban pixels in the VIS triangle. This in turn would make it possible to analyze urban morphology, form and function starting from medium resolution satellite imagery. However, not all pure vegetation, impervious surfaces or bare soil pixels occupy extreme positions in feature space and can, as such, not be directly used as endmembers for unmixing. Instead, the apexes of the typical triangular shaped feature space correspond to true biophysical endmembers representing high albedo substrate (S), bright vegetation (V) and dark surfaces (D) (Small 2004). Any pixel falling inside the convex hull circumscribing the apexes can be considered as a mixture of these three components, and not of V-I-S. One reason for this is that endmembers are spectrally variable because of brightness differences (Wu 2004). In mixture space, pure vegetation pixels are mostly located on the vegetation – dark axis, indicating binary mixing between these two endmembers. Darker vegetation types such as trees are located closer to the dark endmember, while brighter vegetation types such as grass or crops are typically found closer to the vegetation endmember. Binary mixing on the "grey axis" between the dark and substrate endmembers represents different types of urban surfaces, e.g. asphalt versus concrete, while binary mixing on the vegetation – high albedo substrate axis is extremely rare (Small 2004). This complicates the direct use of the VIS ternary as an appropriate model for unmixing. Although bare soil was indeed present as a separate endmember in some studies carried out on other areas (e.g. Phinn et al. 2002), man-made impervious surfaces and exposed soils may indeed be spectrally very similar, depending on the soil type and characteristics on broadband image data. For instance, Van de Voorde *et al.* (2007) reported high levels of spectral confusion between exposed soils near Brussels and red-clay roof tiles, very common in the city. This will further complicate the unmixing process and will lead to some degree of confusion between these two land-cover types if they are chosen to represent endmembers of a VIS unmixing model.

Because bare soil could not be used as an endmember in the Dublin study area, the SVD unmixing model was applied. The substrate (S) and dark (D) endmembers for each image were chosen to be identical to the bright and dark pseudo-invariant features that were used for the radiometric calibration. The vegetation endmembers were selected from the extreme pixels on the vegetation axis in a feature space visualisation of each image by means of high-order principal components.

## 4.2 *Creating urban masks with self-organising maps and knowledge-based post-classification*

In order to characterise urban structure with spatial metrics based on sub-pixel proportions of SVD endmembers, it is necessary to distinguish urban from non-urban fabric. An urban mask is required because pixels consisting of rural bare soil or urban surface types may exhibit a similar substrate-dark mixture, and will therefore also share the same metric signature. To create this mask, we applied a non-parametric unsupervised classification approach and enhanced the resulting thematic map with knowledge-based post-classification rules.

The unsupervised classification approach was based on Kohonen self-organising maps (SOM) (Kohonen 2001). A SOM is a type of artificial neural network that was originally developed to visualise topologies and hierarchical structures of multi-dimensional data by transforming the input space into an ordered two dimensional map. The SOM architecture consists of two network layers: an input layer, which is fully connected to a typically two dimensional array of nodes called Kohonen layer or codebook vector map. The SOM is trained by passing an input vector (i.e. a pixel's spectral values) to the network, and by choosing a winning node based on its distance from that input vector. Then, the weights of the winning node and its neighbours are adjusted in order to reduce the node's distance to the input vector. After each image pixel or a representative set of image pixels is passed to the SOM during training, the built model can be applied to any part of the image and even to other images when atmospheric or other calibration constraints are taken into account. Because the trained SOM network assigns each pixel to a particular node in the codebook vector, each such node can be considered to represent a certain information category. This is similar to other unsupervised classification approaches, except that nodes or classes that are closer to each other on the codebook vector are also more spectrally similar. In this research, we applied a SOM with a 3 by 5 Kohonen layer, which divides the image into 15 spectral classes.

Although Kohonen SOM is a rather advanced approach to unsupervised classification, spectral confusion between certain surface types is still likely to occur and will cause errors in the urban mask. To enhance the classification output with respect to the intended purpose, i.e. distinguishing urban from non-urban areas, a rule-based post-classification approach was adopted (Van de Voorde et al. 2007). This method uses post-classification rules that operate on clumps or groups of adjacent pixels of the same class. If certain user defined criteria are met, a clump's original class label (O) is changed to a target class (T). For the purpose of developing an urban mask in this paper, the target class is always a neighbour's class label. Each rule uses two criteria: area (A) and adjacency (Ad). The area criterion constrains a rule's operation to clumps of a certain maximum size, expressed as number of pixels within the group. Adjacency constrains a rule to pixel groups that are next to clumps of a given type. The value of the adjacency criterion represents the fraction of the clump's border that is shared with the target class. For example: a rule might be developed to assign small bare soil groups (O = bare soil, A < 20 pixels) that share more than half of their border with urban clumps (Ad > 0.5) to the urban class (T). The values of O, T, Ad and A are determined from a visual inspection of the unsupervised classification, keeping in mind the intended use of the post-classification. For the purpose of developing an urban mask, it is especially important that confusion between rural bare soil and urban fabric is resolved, and that each class from the unsupervised classification is unambiguously assigned to either urban or non urban. In addition to the knowledge-based rules, all single pixel groups were removed from the map before and after post-classification to reduce the amount of noise. This was done by applying a majority filter on a 3 by 3 window centred on groups that consisted of only one pixel.

## 5 RESULTS AND DISCUSSION

The 15 classes of the land-cover maps produced by the SOM classifier were recoded into 6 meaningful information classes (figures 1a and 1c). Both the 1988 and the 2001 maps show a high level of spectral confusion between urban and bare soil, which makes it impossible to directly use the classification output as an urban mask. In the 2001 classification, clearly even more confusion occurs between urban and bare soil surface types compared to the 1988 map. This is partly caused by the presence of many construction sites, but also by a higher level of spectral confusion in the 2001 SOM. To enhance the unsupervised classification output, three knowledge-based post-classification rules (table 1) were developed and applied in combination with a filter to remove individual, isolated pixels.

Table 1. Overview of the knowledge-based post-classification rules and their parameters

| Rule nr. | Original class (O) | Target class (C) | Area (A) | Adjacency (Ad) |
|---|---|---|---|---|
| 1 | Bare Soil | Urban | Unlimited | 0.75 |
| 2 | Peat bog / mixed urban-vegetation | Urban | < 200 | Unlimited |
| 3 | Shadow | Urban | < 1000 | 0.75 |

The first rule was intended to re-assign erroneous bare soil clumps within the urban fabric to neighbouring urban type clumps. The adjacency threshold was set to a relatively high 0.75, meaning that bare soil clumps that shared at least 75% of their border with an urban clump had their class changed to urban. Because actual fallow fields do not share a large part of their border with urban surfaces, setting an area threshold was not necessary. This rule effectively cleaned up most of the bare soil/urban confusion in the city centre for both time steps. In the 1988 map, part of the ring road remains classified as bare soil probably because it was still under construction at the time.

The higher degree of confusion in the 2001 classification produced some larger misclassified areas near the western part of the city, which the post-classification was not able to resolve fully. A few large misclassified clumps were therefore manually corrected after visual inspection. A second post-classification rule was developed to operate on parts of low density urban areas that were classified into the same class as *peat and bog*, a type of vegetation cover that is common in the Wicklow Mountains (coloured yellow on figures 1a and 1c). Because all misclassified clumps of this type within the city are relatively small compared to actual *peat and bog* regions, an area threshold was sufficient to improve the classification. A third and final rule was applied to remove small groups of shadow pixels in the city. To avoid confusion with shadows near vegetation, an adjacency threshold was used together with the area threshold. After post-classification, the two land-cover maps clearly improved (figures 1b and 1d), and could be used to derive an urban mask.

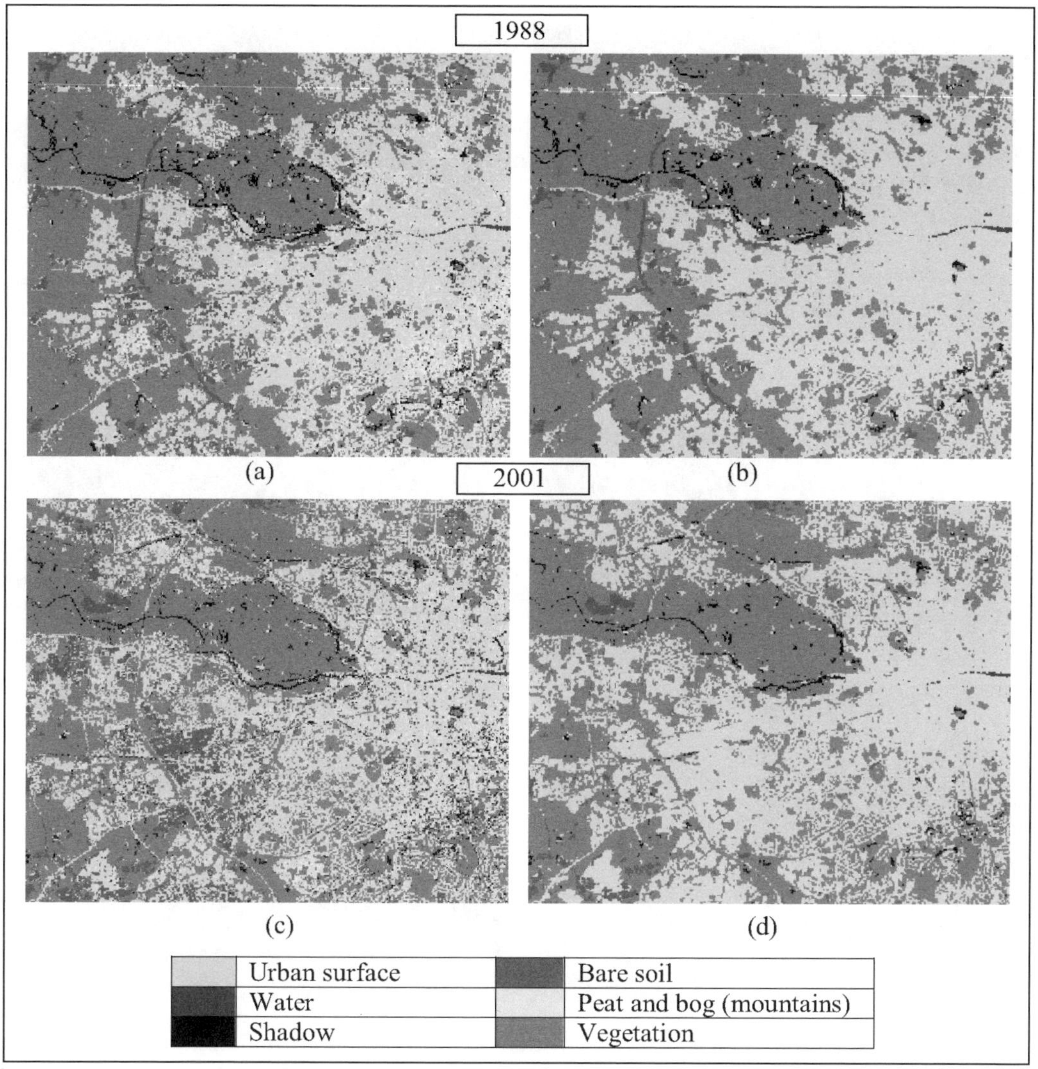


Figure 1. Land-cover classifications for 1988 and 2001 created by self-organising map (left) and enhancement by knowledge-based post-classification (right)

The proportion maps for the three end-members (figure 2) prove useful to visually detect changes in the urban fabric of Dublin. Especially the high albedo substrate and vegetation maps indicate urban expansion to the west of the city centre. The substrate map also shows the expansion of an industrial or commercial area to the southwest, which has a bright appearance because most structures in that area have roofs made of metal or other highly reflective materials. The ring road under construction is also clearly present in the 1988 substrate map, while the asphalt of the finished ring road is no longer detectable in the 2001 substrate map. Interpretation near the city's outer edge is, on the other hand, complicated by the presence of fallow land in 2001 that was covered by crops in 1988. This demonstrates the necessity of using an urban mask. The dark endmember proportion maps appear to be somewhat less useful for visual change detection, but may prove useful for quantitative structural gradient analysis because the amount of shadow, for instance, may provide information on the size and type of urban structures.

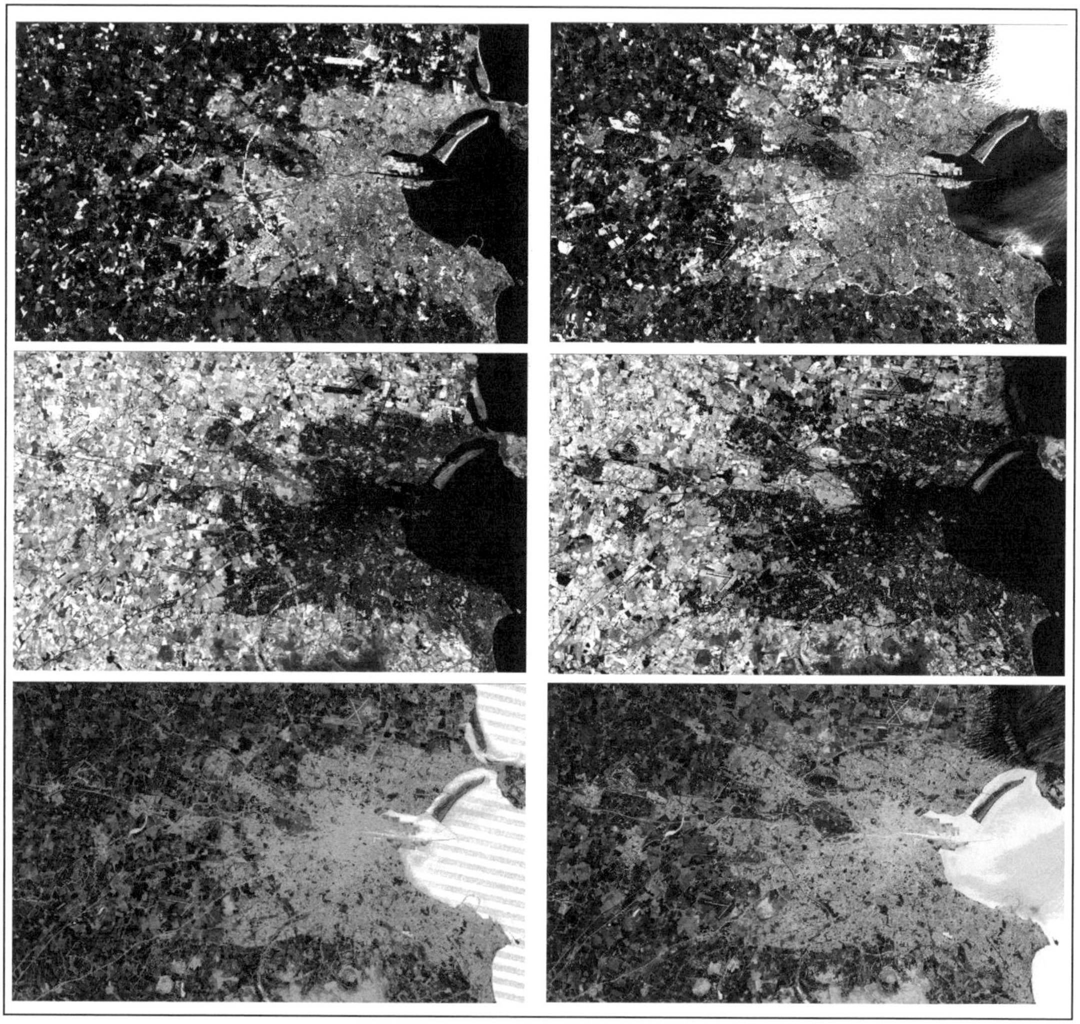

Figure 2. Sub-pixel proportion maps for substrate (top), vegetation (middle) and dark materials (bottom), obtained by unmixing the Landsat images of 1988 (left) and 2001 (right)

## 6. CONCLUSIONS

The objective of this research was to extract urban gradient information by applying linear spectral unmixing on a set of two multitemporal Landsat images. Spectral confusion between urban surfaces and non-urban land-cover types made it impossible to use VIS endmembers, and required the development of a mask to restrict the unmixing to urban areas. This mask was created with a SOM-based unsupervised classification and was then successfully enhanced with knowledge-based classification rules. The outcome of this study will be used to develop spatial metrics to characterise urban spatial form, which will aid the calibration of urban land-use models.

## ACKNOWLEDGEMENTS

The research presented in this paper is funded by the Belgian Science Policy Office in the frame of the STEREO II programme - project SR/00/105.

## REFERENCES

Bannon, M. J. 1999. The Greater Dublin Region: Planning for its transformation and development. In: Killen, J. & MacLaran, A. (eds.) *Dublin: Contemporary trends and issues for the twenty-first century*. Geographical Society of Ireland, Special Publication 11, Dublin: 1-19.

Barr, S. & Barnsley, M. 1997. A region-based, graph-oriented data model for the inference of second order information from remotely-sensed images. *International Journal of Geographical Information Science* 11: 555-576.

Caselles, V. & Garcia, M.J.L. 1989. An alternative simple approach to estimate atmospheric correction in multitemporal studies. *International Journal of Remote Sensing* 11: 783– 828.

Conel, J.E. 1990. Determination of surface reflectance and estimates of atmospheric optical depth and single scattering albedo from Landsat Thematic Mapper data. *International Journal of Remote Sensing* 11: 783–828.

Hall, F.G., Strebel, D.E., Nickeson, J.E., & Goetz, S.J. 1991. Radiometric rectification: toward a common radiometric response among multidate multisensor images. *Remote Sensing of Environment* 35: 11– 27.

Herold, M., Couclelis, H. & Clarke, K.C. 2005. The role of spatial metrics in the analysis and modelling of urban land use change. *Computers, Environment and Urban Systems* 29: 369-399.

Kitchen, P. 2002. Identifying changes of urban social change in Dublin – 1986 to 1996. *Irish Geography* 35:156-174.

Irish, R.R. 2007. *Landsat 7 science data users handbook.* NASA: Greenbelt, MD.

Kohonen, T. 2001. *Self-Organzing Maps. Series in Information Sciences Vol. 30*, 3rd extended edition. Berling: Springer.

MOLAND - Monitoring Land Use/Cover Dynamics. *MOLAND project website* URL: http://www.moland.jrc.it. Date last accessed: 12 March 2008.

Phinn, S., Stanford, M., Scarth, P, Murray, A.T. & Shyy, P.T. 2002. Monitoring the composition of urban environments based on the vegetation-impervious-soil (VIS) model by subpixel analysis techniques. *International Journal of Remote Sensing 23*: 4131 – 4153.

Rashed, T., Weeks, J.R., Stow, D. & Fugate, D. 2005. Measuring temporal composition of urban morphology through spectral mixture analysis: towards a soft approach of change analysis in crowded cities. *International Journal of Remote Sensing* 26: 699 – 718.

Ridd, M.K. 1995. Exploring a V–I–S (Vegetation–Impervious Surface–Soil) model for urban ecosystem analysis through remote sensing - Comparative anatomy for cities. *International Journal of Remote Sensing* 16: 2165 – 2185.

Schott, J. R., Salvaggio, C. & Volchok, W. J. 1988. Radiometric scene normalization using pseudoinvariant features. *Remote Sensing of Environment* 26: 1 – 16.

Small, C. 2003. High spatial resolution spectral mixture analysis of urban reflectance. *Remote Sensing of Environment 88*: 170 – 186.

Small, C. 2004. The Landsat ETM+ spectral mixing space. *Remote Sensing of Environment* 93: 1-17.

UN-HABITAT 2006. *The State of the World's Cities Report.* London: Earthscan.

Van de Voorde, T., De Genst, W. & Canters, F. 2007. Improving pixel-based VHR land-cover classifications of urban areas with post-classification techniques, *Photogrammetric Engineering and Remote Sensing 73*: 1017 – 1027.

van der Meer, F. 1999. Image classification through spectral unmixing. In: Stein, A., van der Meer, F. & Gorte, B. (eds.). *Spatial Statistics for Remote Sensing;* Kluwer Academic Publishers: Dordrecht, The Netherlands: 185 – 193.

Wu, C. 2004. Normalized spectral mixture analysis for monitoring urban composition using ETM+ imagery. *Remote Sensing of Environment 93*: 480 – 492.

*Remote Sensing for a Changing Europe*
*D. Maktav (Ed.)*
*IOS Press, 2009*

*doi:10.3233/978-1-58603-986-8-284*

# Selection criteria of training set for optimal land cover discrimination with a view to automatic segmentation

Ernesto Marcheggiani, Andrea Galli
*Università Politecnica delle Marche, DiSASC, 60131 Ancona, Italy*

Annamaria Bernardini, Eva S. Malinverni
*Università Politecnica delle Marche, DARDUS, 60131 Ancona, Italy*

Primo Zingaretti
*Università Politecnica delle Marche, DIIGA, 60131 Ancona, Italy*

Keywords: multispectral automatic classification, GIS ancillary dataset, interoperability

ABSTRACT: This work is part of a wider project whose general objective is to develop an automatic classification methodology, congruent with the CORINE land-cover thematic legend, by high resolution multi-spectral IKONOS images datasets. The imagery were provided by Regione Marche institution thanks to a research agreement signed together with the Technical University of Marche jointed departments (DARDUS, DIIGA and DiSASC).
According to a hierarchical approach, different phases exist for image classification. At the lowest level classification was merely based on specific pixels spectral value, while moving toward highest levels, segmentation methodology can be used according to features spatial pattern to perform better results.
The present paper aims to deepen the first of the above mentioned hierarchical levels. The methodological approach focus on the optimal selection criteria, in order to define the best setting (spectral bands, Ground Truth, etc.) for the training stage. With this intent different supervised classification algorithms have been tested. In particular, the training stage was carried out tacking advantage of a dedicated-GIS platform implementation. Each Ground Truth sample was collected by means of specific campaign and/or pan-sharpened IKONOS dataset visual interpretation. The depicted classes were grouped in different levels of increasing detail. To be more confident, the CORINE standard legend has been modified according to current study case specificity, in order to obtain an optimal distribution of the samples in the training set. This also allowed to improve the training stage setting by excluding clusters whose spectral values largely ranging far from corresponding average class values.

## 1 INTRODUCTION AND OBJECTIVES

Nowadays the needs for a more updated land use/coverage database, at proper scales of representation (i.e. 1:10'000), represents a urgent issue for the public bodies in charge of Marche Region land administration. On one side, the 1:10'000 scale of representation is the cartographic reference base according to which both local and regional planes (respectively, Piano Regolatore Generale, PRG and Piano Territoriale di Coordinamento, PTC) are conceived and implemented. Moreover, database represents the starting point to upper level land management plans updating, such as the Regional Environmental and Landscape Plan (Piano Paesistico Ambientale Regionale, PPAR). Secondly, to answer to several recent environmental, agricultural and landscape European Union's directives, the Regional Informative Systems should strengthen their structures and provide new tools devoted to landscape resources transformation monitoring. Last but not least, current Marche Region brand land use thematic cartography is obsolete; about 25 years old. According to the above considerations, this paper is part of a wider project whose general objective is an automatic classi-

fication methodology, at the 1:10'000 scale, congruent with the CORINE land-cover thematic legend, starting from monotemporal high resolution multi-spectral IKONOS imagery. The images were provided by Regione Marche authorities thanks to a research agreement, signed together with the Technical University of Marche departments (DARDUS, DIIGA and DiSASC). In order to provide a permanent monitoring service of Marche region territory, the focus is on the possibility to take advantage of relative "low cost–but low quality" datasets availability to improve the use of modern Remote Sensing (RS) techniques in Public Bodies sectors at regional level.
The authors are conscious that so far most expensive imagery (even thought better) are not budgeted for the poor state of regional administration cash; but fostering interoperability among several public or private bodies' databases, it is possible to take advantage and recycle a lot of strategic information with which also a monotemporal –low cost- satellite image can be turned up as useful. Consequently the main goal of this work has been the development of a methodology according to whom a suitable Ground Truth (GT) set, congruent with the IKONOS images, has been built integrating already existing databases, such as remote sensing image collections, vector and punctual Geographic Information System (GIS) database, etc., owned by different bodies of public administration at regional level. Besides a suitable Ground Segment -thus a trustable base of knowledge of the real state of coverage of a defined territory in a defined time- availability since form early stages of a RS project represents the main key issues, affecting all consequent steps and final result.

| Information Class | User Code | CLC Code | CLC Class |
|---|---|---|---|
| brick coverage Building | 1101 | 1.1 | Urban Fabric |
| Reflective mat. (concrete, gravel, metal) | 1102 | 1.2.1 | Industrial, commercial and transport |
| Asphalt road, railway, bitum. hot-mix | 1103 | 1.2.2 | Road and rail network and ass. land |
| Wheat | 1201 | 2.1.1 | Non irrigated arable land |
| Vineyard | 1202 | 2.2.1 | Vineyards |
| Olive grow | 1203 | 2.2.3 | Olive growes |
| Fruit tree | 1204 | 2.2.2 | Fruit trees |
| Medicago Sativa (Alfa-Alfa) | 1205 | 2.1.1 | Non irrigated arable land |
| Barley | 1207 | 2.1.1 | Non irrigated arable land |
| Meadow | 1208 | 2.1.1 | Non irrigated arable land |
| Sugar Beet | 1209 | 2.1.1 | Non irrigated arable land |
| Pasture | 1210 | 2.3.1 | Pastures |
| Wet bare soil | 1211 | 2.1 | Arable land |
| Dry bare soil | 1212 | 2.1 | Arable land |
| Coniferous reforestation | 1301 | 3.1.2 | Coniferous forest |
| Orno-Quercetum ilicis (xerophilous) | 1302 | 3.1.1 | Broad-leaved forest |
| Gariga | 1303 | 3.2.4 | Transitional wood and shrub |
| Quecus cerrus | 1304 | 3.1.1 | Broad-leaved forest |
| Riparian buffer strip | 1305 | 3.1.1 | Broad-leaved forest |
| Fraxinus ornus and Ostrya carpinifolia | 1306 | 3.1.1 | Broad-leaved forest |
| Quecus pubescens | 1307 | 3.1.1 | Broad-leaved forest |
| Water course | 1401 | 5.1.1 | Water courses |
| Water body | 1402 | 5.1.2 | Water bodies |
| Sea | 1403 | 5.2.3 | Sea and ocean |

Table 1 – Concordances among case study information classes and the CLC legend

Unfortunately different source large databases semi-automated integration needs a level of interoperability that notwithstanding the Inspire Directive (COM 516), so far in Italy, have to be considered as inadequate. Several limitations impede or hamper data access and exchange. Such sensitive limitations both from knowledge management technologies point of view, as well as flexibility and interoperability relationship, could partially be overcome by means of innovative approaches such as the Geospatial Semantic Web (Tummarello G., Marcheggiani E., 2007). This last one represent a very novel approach that proposes geographical information advanced treatment technologies implemented through semantic and specific ontology rules to overcome the rigidities of the traditional relational geographical database.

## 2 METHODOLOGY

In the framework of the above mentioned agreement, researchers from DARDUS, SASC and DIIGA departments have tested two main steps of the whole process: on one side, the on ground operations, such as GTs acquiring, according to the legend classes previously defined, and on the other, IKONOS dataset processing (from ortho-rectification and geometrical distortion correction to unsupervised and supervised classifications). In particular, the case study refers to an area belonging to the Ancona Province in Italy; the IKONOS image covers an extension of approximately 150 km$^2$, comprising urban and rural landscapes and natural environment, among which the Conero mountain Natural Park have to be mentioned.

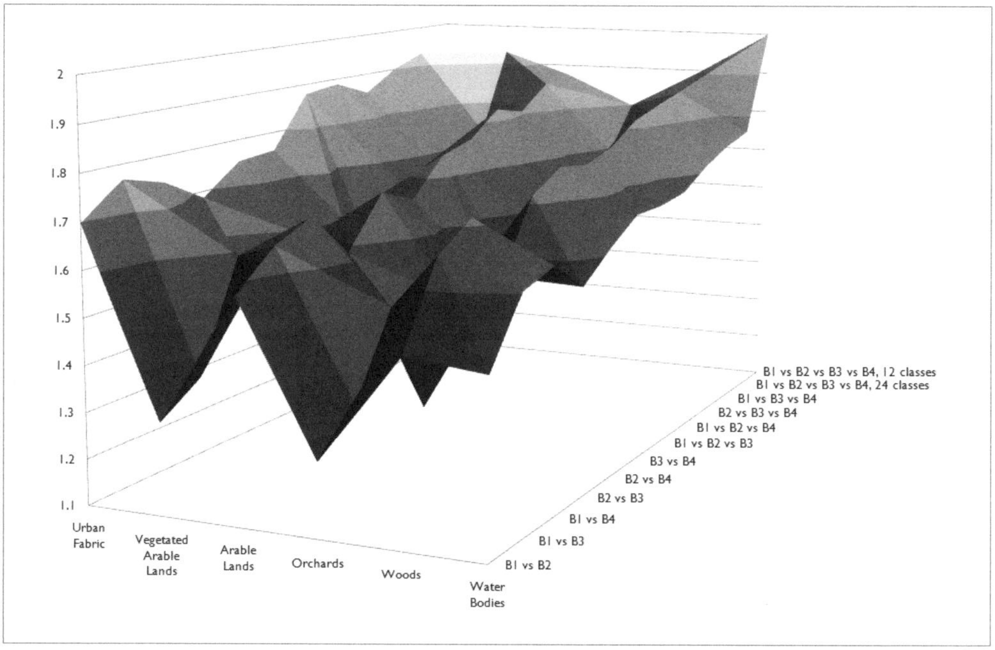


Figure 2 - Optimal bands set combination basing on average JM separbility values (vertical axes in figure, ranging from 1.1 to 2). Classes have been grouped together according to 2nd level of CLC thematic legend (left hand bottom axes). The assessment has been iteratively repeated for each possible combinations of spectral bands (B1 = Red, B2 = Green, B3 = Blue; B4 = NearIR). Besides, optimal band combinations found to be that in which the four available spectral bands were used. For that combination, JM assessment was also performet by means of the restricted set of 12 GTs layers.

A pixels approach has been followed to IKONOS dataset classification, according to both unsupervised and supervised methods, by means of traditional algorithms implemented on commercial software ENVI®. Furthermore object oriented approaches are scheduled to be tested. So far, present paper refers to the supervised classification procedure results. The supervised classification represents the most classical and widely used methodology; easy to use and friendly to approach

also for all those practitioners and civil servants, often barley skilled in RS techniques, who are in charge in the different Region Marche land management and planning departments.
In this framework, the reuse of preexistent images and GIS database collections represents the attempt to valorise investments already set up (such as, the ANCONA PACO Interreg III project) by the different entities and services of the Marche Region bodies, in order to produce low cost thematic cartographies. In spite of being scantily updated and scarcely interoperable, most of the acquired database collections can be useful to implement a GIS dedicated both to information classes thematic legend definition and to training set extraction. These last two are fundamental key issues affecting the whole methodology results.

## 2.1 *GTs selection crtiteria (ancillary database)*

In lack of a GTs set contemporaneously acquired with the IKONOS image (Curran, P. J. and Williamson, H. D., 1985), the methodology to built a reliable set is based on an integrated database retrieved from the different Information System (IS) owned by Marche Region, most of those WebGIS available. The main reference for GTs legend definition is the Corine Land Cover (CLC) European project, by the European Environmental Agency (EEA).
A classical visual interpretation procedure has been used to assess the GTs relative to the 1st CLC classes (Artificial Surface), displaying in GIS environment a view of the IKONOS set after its orthrectification (see, Pre-processing image orthorectification). Whereas, to asses both agricultural and wooded areas GTs, the method has consisted in collecting and consequently cross-querying relevant ancillary database (such as, GIS database), considering among all database that were available on the different Marche Region departmental Information Systems (IS), only those sufficiently up to date with IKONOS images (acquisition date May 7, 2006).
In particular agricultural GTs have been assessed overlying cadastral information layer and farmers Common Agricultural Policy (CAP) yearly financing database, available on National Agricultural Information System (SIAN). In this way, the geographical position (where) of a specific cadastral parcel is crossed with a trustable source of information about which crops (what) are really cultivated, during a specific period of time. Through this first screening 20’259 particles have been identified (covering 12’025 ha of Usable Agricultural Area, UAA), classed according to 75 different types of land uses. which were restricted according to the crop phenological state to 16 classes. Of these last ones, only 9 classes (namely, Wheat, Vineyard, Olive grow, Fruit tree, Medicago Sativa (Alfa-Alfa), Barley, Meadow, Sugar Beet, Pasture) were present in the IKONOS acquired imagery. The remaining classes were not present or were at a too early phenological stage (emergence) and consequently clustered as bare soil.
In a similar way the definition of the GTs relative to wooded lands and seminatural areas, was carried out by overlaying and querying the IKONOS imagery by means of Forestry Regional Information System (SIFR), provided thanks to a project aiming to meet the demands of an effective tool to support regional forest planning, management and monitoring. The project was implemented through systematic sampling of several test areas. The density of sampling ranging from 50 to 100 hectares. Thanks to this inventory, at the scale of 1:25’000, 2’406 sample areas were located and identified, and the regional surface has been divided into 17 different forestry departments, comprises into 13 mountain communities and 4 individual provinces. In the acquired IKONOS imagery 23 different forestry coverage were present, encased into 7 different categories: Coniferous reforestation, Orno-Quercetum ilicis (xerophilous), Gariga, Quecus cerrus, Riparian buffer strip, Fraxinus ornus and Ostrya carpinifolia, Quecus pubescens.

# 3 IMAGE PROCESSING AND RESULTS

The image processing from the geometrical and spectral correction until to the classification consists of the following phases.

## 3.1 *Pre-processing image orthorectification*

IKONOS Multispectral imagery, was acquired on July 2006, with a 29 degrees solar zenith angle, ad issued both in panchromatic mode (0.45-0.90 μm) with 1 meter ground resolution and in multispectral mode with 4 meter ground resolution. The data are available in 11-bit radiometric resolution. They are delivered georeferenced according to the UTM projection and WGS84 Datum. The whole image dataset was orthorectified using the third order rational function model implemented in the software PCI - Orthoengine. The 15 Ground Control Points (GCPs) were selected on the Technical Regional Map at the scale 1:10.000 and identified on the panchromatic image. Using also the DTM information we obtained an RMS error below 1 pixel (1 meter ground resolution). The radiometric interpolation, applied by means of nearest neighbour resampling method, preserves the original image values.

| | 1101 | 1102 | 1103 | 1201 | 1202 | 1203 | 1204 | 1205 | 1207 | 1208 | 1209 | 1210 | 1211 | 1212 | 1301 | 1302 | 1303 | 1304 | 1305 | 1306 | 1307 | 1401 | 1402 | 1403 |
|---|---|---|---|---|---|---|---|---|---|---|---|---|---|---|---|---|---|---|---|---|---|---|---|---|
| 1101 | | | | | | | | | | | | | | | | | | | | | | | | |
| 1102 | 1.80 | | | | | | | | | | | | | | | | | | | | | | | |
| 1103 | 1.59 | 1.41 | | | | | | | | | | | | | | | | | | | | | | |
| 1201 | 2.00 | 2.00 | 2.00 | | | | | | | | | | | | | | | | | | | | | |
| 1202 | 1.83 | 1.88 | 1.89 | 1.72 | | | | | | | | | | | | | | | | | | | | |
| 1203 | 1.93 | 1.97 | 1.96 | 1.11 | 1.10 | | | | | | | | | | | | | | | | | | | |
| 1204 | 1.99 | 1.99 | 1.98 | 1.54 | 1.47 | 1.11 | | | | | | | | | | | | | | | | | | |
| 1205 | 2.00 | 2.00 | 2.00 | 0.89 | 1.70 | 1.29 | 1.39 | | | | | | | | | | | | | | | | | |
| 1207 | 2.00 | 2.00 | 2.00 | 1.88 | 1.93 | 1.94 | 1.86 | 0.99 | | | | | | | | | | | | | | | | |
| 1208 | 2.00 | 2.00 | 2.00 | 0.91 | 1.85 | 1.39 | 1.66 | 0.89 | 1.61 | | | | | | | | | | | | | | | |
| 1209 | 1.97 | 1.96 | 1.96 | 1.95 | 0.69 | 1.50 | 1.81 | 1.94 | 1.99 | 1.98 | | | | | | | | | | | | | | |
| 1210 | 2.00 | 2.00 | 2.00 | 1.53 | 1.73 | 1.69 | 1.42 | 1.52 | 1.92 | 1.72 | 1.93 | | | | | | | | | | | | | |
| 1211 | 1.49 | 1.83 | 1.81 | 2.00 | 1.67 | 1.69 | 2.00 | 2.00 | 2.00 | 2.00 | 1.92 | 1.96 | | | | | | | | | | | | |
| 1212 | 1.89 | 1.88 | 1.99 | 2.00 | 1.35 | 1.89 | 2.00 | 2.00 | 2.00 | 2.00 | 1.51 | 2.00 | 1.99 | | | | | | | | | | | |
| 1301 | 1.99 | 1.99 | 1.99 | 2.00 | 1.89 | 2.00 | 1.99 | 2.00 | 2.00 | 2.00 | 1.98 | 2.00 | 1.31 | 1.99 | | | | | | | | | | |
| 1302 | 2.00 | 2.00 | 2.00 | 1.74 | 1.68 | 1.84 | 1.79 | 1.91 | 1.99 | 1.85 | 1.96 | 2.00 | 1.53 | 2.00 | 2.00 | | | | | | | | | |
| 1303 | 1.92 | 1.96 | 1.82 | 1.99 | 1.82 | 1.91 | 1.38 | 1.98 | 2.00 | 1.89 | 1.94 | 2.00 | 1.72 | 2.00 | 2.00 | 2.00 | | | | | | | | |
| 1304 | 2.00 | 2.00 | 2.00 | 1.68 | 1.79 | 1.90 | 1.98 | 1.92 | 1.98 | 1.84 | 1.98 | 1.67 | 2.00 | 2.00 | 0.64 | 1.38 | 1.49 | | | | | | | |
| 1305 | 1.98 | 1.99 | 1.98 | 0.77 | 1.53 | 1.06 | 0.97 | 0.68 | 1.39 | 0.82 | 1.87 | 1.99 | 2.00 | 2.00 | 2.00 | 1.81 | 2.00 | 0.70 | | | | | | |
| 1306 | 1.99 | 1.99 | 1.99 | 1.31 | 1.62 | 1.58 | 1.38 | 1.48 | 1.87 | 1.35 | 1.93 | 1.41 | 2.00 | 2.00 | 1.67 | 1.60 | 2.00 | 0.48 | 1.95 | | | | | |
| 1307 | 2.00 | 2.00 | 1.99 | 1.56 | 1.66 | 1.78 | 1.75 | 1.74 | 1.97 | 1.70 | 1.95 | 2.00 | 1.49 | 2.00 | 2.00 | 1.97 | 2.00 | 2.00 | 0.75 | 2.00 | | | | |
| 1401 | 2.00 | 2.00 | 2.00 | 2.00 | 2.00 | 2.00 | 2.00 | 2.00 | 2.00 | 2.00 | 2.00 | 1.91 | 2.00 | 2.00 | 2.00 | 1.70 | 2.00 | 2.00 | 1.49 | 1.62 | 1.41 | | | |
| 1402 | 1.93 | 1.98 | 1.97 | 2.00 | 2.00 | 2.00 | 1.96 | 2.00 | 2.00 | 2.00 | 2.00 | 2.00 | 2.00 | 2.00 | 0.51 | 1.40 | 2.00 | 2.00 | 1.95 | 1.16 | 2.00 | 2.00 | | |
| 1403 | 1.99 | 2.00 | 2.00 | 2.00 | 2.00 | 2.00 | 2.00 | 2.00 | 2.00 | 2.00 | 2.00 | 2.00 | 1.94 | 2.00 | 2.00 | 2.00 | 2.00 | 1.78 | 2.00 | 0.82 | 1.96 | 2.00 | 2.00 | |

Figure 3 - JM separbility class pairs by means of 24 classes GTs set, applied to the 11bit IKONOS imagery, considering the whole set of 4 available bands (namely, Red, Green, Blue, Near Infrared).

## 3.2 *Spectral separability assessment*

In order to determine both the optimal spectral bands combination and less redundant GTs set, to be used during supervised classification step, a separability assessment was performed by means of Jeffries-Matusita (JM) indicator, whose algorithm is implemented on ENVI® software. The JM measurement is based on Bhattacharrya distance and allows to indicating how well a selected spectral class pair is statistically separate (class separability). In particular, on ENVI®, JM values ranging from 0 to 2. Values greater than 1.9 indicate that the pair has a good separability, for insights, see referred international literature (Richards J.A., 1999).The separability assessment was conducted on the 11bit IKONOS imagery, using first the available whole GTs set (24 levels). The separability assessment has been iteratively repeated for each possible combinations of spectral bands. In the present case study, the optimal combination among spectral bands results to be that in which the four available spectral bands were used (see fig. 2).

Note that using the whole set of GTs found to be redundant considering that many combinations are no longer separable showing JM values often less than 1.5 (see Fig. 3). Consequently, by choosing 1.8 as JM threshold value, all classes with separability values less than that limit have to be considered as redundant and eliminated (namely, 1201, 1202, 1203, 1204, 1205, 1208, 1210, 1301, 1302, 1306, 1307, 1402, 1403); assuming samples spectral signature distribution as normal in images spectral space. Afterward, separability assessment procedure was repeated using the GTs set restricted to 12 classes. Further to that, results show JM values all greater than 1.8 for all classes (see Figure 2). The GTs set restricted to 12 classes has been used for subsequent phases of supervised classification.

| | JM | Com ML | Com SAM | Om ML | Om SAM | PxlClsd ML | PxlClsd SAM | PAc ML | PAc SAM | UAc ML | UAc SAM | Clst ML | Clst SAM |
|---|---|---|---|---|---|---|---|---|---|---|---|---|---|
| 1101 | 1.88 | 94.84 | 26.61 | 22.16 | 56.76 | 324435 | 215954 | 77.84 | 43.24 | 5.16 | 73.39 | A | A |
| 1102 | 1.89 | 10.63 | 55.42 | 22.13 | 34.68 | 3676196 | 197596 | 77.87 | 65.32 | 89.37 | 44.58 | C | A |
| 1103 | 1.87 | 35.18 | 85.36 | 18.72 | 28.84 | 420010 | 301857 | 81.28 | 71.16 | 64.82 | 14.64 | A | A |
| 1207 | 1.94 | 8.95 | 49.64 | 14.49 | 32.03 | 831483 | 2071521 | 85.51 | 67.97 | 91.05 | 50.36 | B | C |
| 1209 | 1.92 | 19.91 | 22.42 | 19.49 | 31.76 | 1739145 | 1357361 | 80.51 | 68.24 | 80.09 | 77.58 | D | B |
| 1211 | 1.86 | 19.14 | 30.99 | 30.37 | 54.69 | 751163 | 744614 | 69.63 | 45.31 | 80.86 | 69.01 | B | A |
| 1212 | 1.89 | 8.51 | 17.29 | 22.94 | 45.36 | 1176701 | 1305098 | 77.06 | 54.64 | 91.49 | 82.71 | B | B |
| 1303 | 1.90 | 24.42 | 33.71 | 12.53 | 27.03 | 423414 | 546046 | 87.47 | 72.97 | 75.58 | 66.29 | A | A |
| 1304 | 1.94 | 0.72 | 8.03 | 8.29 | 33.84 | 817597 | 482990 | 91.71 | 66.16 | 99.28 | 91.97 | B | A |
| 1305 | 1.85 | 34.64 | 63.41 | 18.82 | 51.71 | 3542255 | 2670579 | 81.18 | 48.29 | 65.36 | 36.59 | C | D |
| 1401 | 2.00 | 0.00 | 2.06 | 67.81 | 34.93 | 1918 | 8445 | 32.19 | 65.07 | 100.00 | 97.94 | A | A |
| 1403 | 2.00 | 0.00 | 0.00 | 0.56 | 0.02 | 1120483 | 1153126 | 99.44 | 99.98 | 100.00 | 100.00 | B | B |

Tab 2 –Cluster Analysis results considering both Jeffries Matusita separability and supervised classifications, using restricted 12 layers Ground Truth set. For both algorithms either Commission (Com) and omission errors (Om) and Producer (PAc) and User (UAc) accuracy, together with the number of classified pixels (PxlClsd) and specific membership cluster (Clst), are for all layers provided.

| Centroid | Average JM | Average Com | Average Om | Average PxlClsd | Average PAc | Average UAc | Number of Clst |
|---|---|---|---|---|---|---|---|
| Cluster A | 1.91 | 38.61 | 30.31 | 292444.25 | 69.70 | 61.39 | 4 |
| Cluster B | 1.93 | 7.46 | 15.33 | 939485.40 | 84.67 | 92.54 | 5 |
| Cluster C | 1.87 | 22.64 | 20.48 | 3609225.50 | 79.53 | 77.37 | 2 |
| Cluster D | 1.92 | 19.91 | 19.49 | 1739145.00 | 80.51 | 80.09 | 1 |

| Centroid | Average JM | Average Com | Average Om | Average PxlClsd | Average PAc | Average UAc | Number of Clst |
|---|---|---|---|---|---|---|---|
| Cluster A | 1.91 | 34.60 | 38.68 | 356786.00 | 61.32 | 65.40 | 7 |
| Cluster B | 1.93 | 13.24 | 25.71 | 1271861.67 | 74.29 | 86.76 | 3 |
| Cluster C | 1.94 | 49.64 | 32.03 | 2071521.00 | 67.97 | 50.36 | 1 |

Tab 3 - Cluster Analisys according to both Maximum Likelihood (up) and Spectral Angle Mapper (down) applied by means of the restricted set of 12 layers.

## 4 CLASSIFICATION

### 4.1 *Supervised classifications using the whole 24 GTs classes*

In order to benchmark classifications, a 24 classes supervised classification was previously performed by means of two of most classic sound classifier algorithms: Maximum Likelihood (ML) and Spectral Angle Mapper (SAM). In both cases ENVI® default standard parameters input setup has been used. The first classifier (ML) obtains an Overall Accuracy of 86.08% (Kappa Coefficient equal to 0.7247) and no unclassified pixels, while the second (SAM) tends to Overall Accuracy value of 81.94% (Kappa Coefficient equal to 0.6419) and 25.85% of unclassified pixels. All the values obtained to each information classes, (JM separability, ML and SAM 24 GTs supervised classifications classes), have been used to a Cluster Analysis evaluation in order to compare the performances of the two classification algorithms (SAM and ML). For clustering a Average Euclidean Hierarchical Model has been used, imposing as threshold, 10 as the maximum number of clusters and gradually reducing that limit until to the numbers of singularity (cluster representing just one entity) lower than 20%. Cluster analysis results in case of ML shows ambiguous mixed clusters, where wooden, agricultural and urban land uses are mixed together. On the contrary, clustering according to SAM has spawned more homogeneous clusters.

### 4.2 *Supervised classifications using the 12 GTs classes restricted set*

The above mentioned ML and SAM supervised classifications (Lillesand, T. M. and Kiefer, R. W., 1994, Kruse F.A *et Al.* 1993) has been repeated, using the restricted set of 12 GTs classes. Through ML classifier a 94.13% Overall Accuracy (Kappa Coefficient equal to 0.8550), and no unclassified pixels, have been obtained. While SAM has produced a 89.60% Overall Accuracy (Kappa Coefficient equal to 0.7422) and 25.43% of unclassified pixels.
In particular, referring to Table 2, ML classifier turn out a first group of classes (1304, 1403) showing a low Commission and Omission errors (closer to 0%) and high Producer and User accuracies (higher than 90%), while a second group (1103, 1207, 1209, 1303, 1305) shows high accuracies values (around 80%) and relatively low errors (lower than 35%). Moreover, a third group (1211 and 1212) shows either errors (lower than 30%) and accuracies lower than two previous groups. Finally, the 1401 class differs from others, especially for a very low Producer Accuracy.
Some considerations need to be given both on errors (Omission and Commission) and individual classes accuracy (Producer and User). As in common knowledge (Congalton, R.G., 1991, Wickham, J.D., *et Al.*, 2004, Stehman, S.V., *et Al.*, 2003), Ommission errors occurs when specific classes pixel assignment hypothesis, according to GTs set, is rejected by the classifier. While Commission errors are due in case of false hypothesis acceptance, by the classifier, assigning a pixel to an untrue class. Omission errors are evaluated taking into account the confusion matrix rows, excluding those cells along the matrix diagonal, these last ones are accounted to assess the Overall Accuracy value. Consequently, Omission error involves an underestimation of information class while the Commission error an overestimation. The SAM 12 GTs classes supervised classification (see Table 3), produce a first group (1403) with low errors (closer to 0%) and high accuracies (closer to 100%), a second group (1209, 1303, 1304) which presents either relatively low errors (less than 35%) and accuracies (around 70%). Groups (1102 and 1103) with intermediate values. The JM and supervised classifications, in case of 12 GTs classes restricted set, values has been used to repeat the Cluster Analysis (see Table 3). In this case, excluding the abnormal behavior of 1401 and 1403 classes, the Cluster Analysis trends seems to indicate that SAM alghoritms generate more homogeneous clusters. This induce to argue that in the case study contest, SAM act better performances.

## CONCLUSION

The presented work proves that, in lack of investments and resources for the implementation of a specific regional service to permanently monitoring the state of land use, and its dynamics of transformation with the times, regional public utilities can profitably retrieve and enhance, heterogeneous information residing in its archives. Through the integration of both classic RS techniques approach (even thought by means of low cost imagery) and preexistent information, shuffled to selecting a trustable and representative set of GTs, it is possible to produce a useful land cover database, flexible and adaptable to different Marche Region public utilities needs. In fact, knowledge arising from updated source of geo-information is the necessary condition to perform detailed survey (such as using GIS operators) on the dynamics taking place in the whole territory.
The case study has also shown how the implementation of such a system, especially at a detailed scale of representation (such as 1:10'000), is not easy in Marche Region, given several current operational limits characterizing most of Italian public bodies. Structural inadequacy prevents many public utilities to fully take advantage from the potentiality offered today by remote sensing and GIS techniques. This is particularly due to the lack of investment in terms of both human and eco-

nomic resources. The depicted situation found to be seriously affecting most of the services that regional land governing public bodies compulsorily must provide both to citizens and institutions.
As is common knowledge, generally there is not a unique relationship between information classes (namely, the selected ground truths) and their spectral behavior, especially when using, as in case study, mono-temporal imagery. Therefore, to improve performances and overcame major technical limitations a separability assessment of the selected ground truth samples has been performed. The class separability performed by means of classical Jeffries-Matusita distance, enable to define both the optimal bands set and to eliminate GTs classes which generate more confusion than real information. In case study the number of GTs layers was restricted from 24 to 12. The final stage of the work, devoted to IKONOS imagery supervised classification by means of two of the most classical and widely used algorithms (ML and SAM), clearly shows significant benefits, in terms of accuracies, after JM based GTs set restriction. In addition, a statistical analysis shows that SAM supervised classification tends to provide results more consistent with the on ground reality. So far, experimental tests to the scheduled object oriented approaches are in progress; obtained results in further papers.

## REFERENCES

Ancona PACO, INTERREG IIIA (http://anconapaco.regione.marche.it/)

Civico, D.L., J. D. Hurd, E.H. Wilson, C.L. Arnold, and M.P. Prisloe, 2002. Quantifying and Describing Urbanizing Landscapes in the Northeast United States, Photogrammetric Engineering and Remote Sensing, 68: 1083-1090.

Congalton, R.G., 1991, A review of assessing the accuracy of classifications of remotely sensed data. Remote Sensing of Environment, 37, 35-46

Curran, P. J. and Williamson, H. D., 1985. The accuracy of ground data used in remote-sensing investigations- International Journal of Remote Sensing, 6 (10), 1637-1651.

Homer, C., J. Dewitz, J. Fry, M. Coan, N. Hossain, C. Larson, N. Herold, A. McKerrow, J.N. VanDriel, and J. Wickham, 2007. 'Completion of the 2001 National Land Cover Database for the Conterminous United States', Photogrammetric Engineering and Remote Sensing, 73: 337-341.

Irwin, E.G., N.E. Bockstael and H.J. Cho, 2006. 'Measuring Modeling Urban Sprawl: Data, Scale and Spatial Dependencies', Proceedings of the 53rd Annual North American Regional Science Association, Toronto, Canada, November 16-18, 2006.

Kruse F.A., Lefkoff A.B., Dietz J.B. (1993) Expert System-based mineral mapping in northern Death Valley, California/Nevada using the Airborne Visible/Infrared Imaging Spectrometer, Remote Sensing of the Environment, 44:309–336

Kruse F.A., Lefkoff A.B., Boardman J.W., Heiderbrecht K.B., Shapiro A. T., Barloon P.J., *et Al.* (1993) The spectral image processing system (SIPS), Interactive visualization and analysis of imaging data, Remote Sensing of the Environment, 44: 145-163

Lillesand, T. M. and Kiefer, R. W., 1994. Remote Sensing and Image Interpretation. John Wiley and Sons, Inc., New York, p. 468

Richards, J. A. (1995), Remote Sensing Digital Image Analysis. An Introduction, 2nd ed., Springer-Verlag, Berlin.

Stehman, S.V., J.D. Wickham, J.H. Smith and L. Yang, 2003. 'Thematic Accuracy of the 1992 National Land-Cover Data for the Eastern United States: Statistical Methodology and Regional Results', Remote Sensing of Environment, 86: 500-516.

Swain, P. H. (1978), Fundamentals of pattern recognition in remote sensing. In Remote Sensing: The Quantitative Approach (P. H. Swain and S. M. Davis, Eds.), McGraw-Hill, New York, pp. 136–187.

Tummarello G., Morbidoni C., Nucci M., Marcheggiani E. "Interacting Geo Semantic Web Communities with the DBin platform: use cases and perspectives" in Scharl, A. and Tochtermann, K., Eds. (2007). The Geospatial Web - How Geobrowsers, Social Software and the Web 2.0 are Shaping the Network, 2007, London: Springer, pp 171-176, ISBN 1-84628-8

Wickham, J.D., S.V. Stehman, J.H. Smith and L. Yang, 2004. 'Thematic Accuracy of the 1992 National Land-Cover Data for the Western United States', Remote Sensing of Environment, 91: 452-468.

*Remote Sensing for a Changing Europe*
*D. Maktav (Ed.)*
*IOS Press, 2009*

*doi:10.3233/978-1-58603-986-8-292*

# Estimation of the troposphere phase delay in interferometric SAR using MERIS data: experiments over the Campania area

G. Fornaro & F. Serafino
*Institute for Electromagnetic Sensing of the Environment, National Research Council, Naples, Italy*
L. Paglia
*Remote Sensing Laboratory for Environmental Hazard Monitoring, University of Salerno, Italy*

Keywords: *synthetic aperture radar, atmosphere*

ABSTRACT: in modern multipass DInSAR processing techniques one of the main point regards the mitigation of the atmospheric contribution. To this end we investigate the potential use of data acquired by multispectral sensors. In particular we refer to the instruments onboard the Envisat sensor and specifically to the ASAR and MERIS sensors. Such sensors have the advantage to simultaneously acquire data along the same orbit, thus imaging the scene under same conditions. In particular we carried out an experiment aimed at using the MERIS Water Vapour (W) product to estimate Atmospheric Phase Delay (APD) and to calibrate the measurements of the ASAR sensor. The W product measures the integrated amount of atmospheric water vapour by using the water absorption and non-absorption bands. The work describes preliminary results achieved by processing a dozen ASAR acquisition and MERIS data over the Campania region in Italy with descending passes. We have compared the APD estimated by MERIS data with the APD signal extracted by the multipass DInSAR analysis on long term coherent points. The final aim is to exploit the MERIS water vapour for compensating the APD signals in a long series of acquisition used in the multipass DInSAR techniques to achieve higher accuracy and/or extend the applicability of the technique to emergency cases.

## 1 INTRODUCTION

Multipass SAR Differential Interferometry is a powerful technique that allows monitoring ground deformations with high accuracy and represents an instrument of support for quantitative analysis in evaluating the environmental risks associated, for instance, with volcanic or seismic activity, landslides, etc. The availability of large data archives and their easy accessibility permits, thanks to the use of advanced techniques of SAR data processing (eg, Permanent Scatterers, Small BAseline Subset, Enhanced Spatial Differences Methods , etc.), to generate long term deformations time series. However DInSAR techniques are limited by many factors that cause spurious phase variations, hence affecting the quality of estimated results. Beside the presence of the wanted ground deformation signal, interferograms are affected also by the presence of several noise sources originating from digital elevation model (DEM) errors, baseline and temporal decorrelation and by correlated noise such as atmosphere effects (Hanssen 2001). The latter represents the mainly source of errors in the applications of DInSAR techniques. The atmospheric phase delay is mainly due to the presence of water vapor concentration that causes a refractive index variation and, as a consequence, a variation of the electromagnetic radiation propagation delay through the troposphere. This phenomenon has limited for a long time the use of the DInSAR techniques based on the analysis of individual pairs of differential interferograms. Modern techniques based on the jointly

use of multi-pass SAR data (Stacking) (Fornaro et al. 2007) (Berardino et al. 2002) (Ferretti et al. 2000), allow overcoming these problems by filtering the atmospheric noise contribution based on the space-time statistical characterization of the atmospheric delay.
A completely alternative methodology for the purification of atmospheric contributions in radar interferograms (Puysségur et al. 2007) (Li et al. 2006) is based on the availability of independent measures of the atmospheric signal, such as those obtained by the multispectral sensors. These sensors, measuring radiation in the near-IR channels, allow obtaining, as a product, a rather accurate estimate of the Water Vapor (W) amount. This product, after proper scaling, is used, as external information, in the Stacking process in order to reduce the atmospheric phase delay of the data and improve the estimation of ground deformation velocity and topography errors.
In this work we analyze the possibility to correct the atmospheric phase patterns of SAR interferometric pairs by using the W distribution obtained from Multispectral Medium Resolution Imaging Spectrometer Instrument (MERIS) data. To validate this technique ASAR (Advanced Synthetic Aperture Radar) images and MERIS data, simultaneously acquired by sensors mounted on the ENVISAT platform, have been used.

## 2 MERIS AND WATER VAPOR PRODUCT (W)

MERIS is an Imaging multi-spectral radiometers (vis/IR). It is a passive sensor with fifteen spectral bands, each of which has a programmable width and a programmable location in the 390 nm to 1040 nm spectral range. The MERIS sensor has two channels for the water vapour measurements, located at 890 nm (channel 14) and 900 nm (channel 15) both with 10 nm bandwidth.
The column amount of water vapour (W) may be obtained by using the algorithm proposed by J.Fischer and R.Bennartz (Fischer et al. 1997). They distinguished three different cases in water vapour content evaluating: above cloud-free land surfaces, above the top of clouds, and above water. In this work we consider only the column amount of water vapour related to land pixels in cloud-free areas, which can be selected using the cloud mask provided with MERIS images. In this case it is given by:

$$W = k_0 + k_1 \log\left(\frac{L_{Ch15}}{L_{Ch14}}\right) + k_2 \log^2\left(\frac{L_{Ch15}}{L_{Ch14}}\right) \tag{1}$$

Where $L_{Ch14}$ and $L_{Ch15}$ are the radiances measured in the MERIS channels 14 and 15, and $k_0$, $k_1$ and $k_3$ are regression constants. This product is directly provided by ESA in the MER_FR_2P (MERIS Full Resolution Geophysical Product), with 300 m of ground resolution, and in MER_RR_2P (MERIS Reduced Resolution Geophysical Product), with a spatial resolution of 1.2 km. In this work we used only the reduced resolution product. In the next session it is show the relationship between this term and the SAR phase delay.

## 3 SAR PHASE SIGNAL AND ATMOSPHERIC DELAY

Synthetic Aperture Radar (SAR) is an active microwave remote sensing system, usually mounted onboard satellite or airborne, that exploits the coherent nature of the employed radiation to achieve significant discrimination capabilities in the azimuth (along track) and range (cross track) direction. The presence of the atmosphere translates to a slant phase delay with respect to the propagation velocity in vacuum (Hanssen 2001). This is mainly due to the velocity variations along the path between the antenna and the ground. The electromagnetic properties concerning radio waves propagation are governed by the refractive index $n(x,y,z,t)$. In nominal conditions the refractive is approximately equal to one. In standard atmosphere the deviation from unitary refractive index is small; hence the refractive index is usually scaled and shifted as follows: $N=(n-1)\times 10^{6}$. The two-way SAR atmospheric phase delay is directly related to the refractivity index as:

$$\varphi = \frac{4\pi}{\lambda}\left(2\times 10^{-6}\int_{0}^{H}\frac{N}{\cos\theta_{inc}}dh\right) \tag{2}$$

Where $H$ is the height of the satellite and $\theta_{inc}(h)$ is the incidence angle $\lambda$ is the wavelength of the transmitted radiation. For common radar frequencies the refractivity index is:

$$N=(n-1)\times 10^{6}=k_1\frac{P}{T}+k_2\frac{e}{T}+k_3\frac{e}{T^2}+k_4W_{cloud}+k_5\frac{n_e}{f^2} \tag{3}$$

where $n$ is the refractive index of air at radio frequencies, $P$ is the total pressure (mbar), $e$ is the water vapour partial pressure (mbar), $T$ is the temperature (K), $W_{cloud}$ is the cloud water content (g $m^{-3}$), $f$ is the radar frequency, $n_e$ is the electronic density of the ionosphere. Substituting eq. (3) in eq. (2) we have:

$$\varphi = \frac{4\pi}{\lambda}\left[\frac{2\times 10^{-6}}{\cos\theta_{inc}}\left(\int_{0}^{H}k_1\frac{P}{T}+\frac{RW}{10M_w}\left(k_2+\frac{k_3}{T_m}\right)dh+\int_{0}^{H}k_4W_{cloud}dh+\int_{0}^{H}k_5\frac{n_e}{f^2}dh\right)\right] \tag{4}$$

were $W$ (g $cm^{-2}$) is the water vapour content given by:

$$W=10\frac{M_w}{R}\int_{h}^{H}\frac{e}{T}dh \tag{5}$$

where $M_w$ is the molar mass of water ($18.0152 \times 10^{-3}$ kg $mol^{-1}$), $R$ the perfect gas constant (8.314 $J.K^{-1}$ $mol^{-1}$), $T$ is the temperature (K).

The dominant phase contribution in eq. (4) is given by the water vapour content, for a preliminary analysis the first integral of the equation, called dry phase delay, may be neglected, for the same reason we can delete the term depending on $k_3$.

The $W_{cloud}$ term associated to the areas covered by clouds, is neglected by applying a mask related to the cloudy pixels, provided by each MERIS product. Since these masks are not sufficiently accurate we generate other masks to reduce, as well as possible, the presence of clouds in the MERIS product.

The last integral of eq. (4) may be considered constant within the scene and then it may be neglected for interferometric applications.

Under this conditions the atmospheric phase delay is given by:

$$\varphi = \frac{4\pi}{\lambda}\left(\frac{2\times 10^{-6}}{\cos\theta_{inc}}\int_{0}^{H}\frac{RW}{10M_w}k_2dh\right) \tag{6}$$

The water vapour content W is one of the products derived from the multispectral sensors, see eq. (1).

## 4 DATA PROCESSING

The ASAR data are in RADAR coordinates: azimuth (along track) and range (cross track) direction while the MERIS ones are in maps coordinates. To compare the MERIS and ASAR images to each other it is needed that both data are in the same coordinates system. To this end we move MERIS images to the RADAR coordinates by applying the following steps.
First of all we select the MERIS images, common to the ASAR ones, with a low clouds coverage in order to minimize errors due to the overestimation of the MERIS water vapour measures over clouds. The ASAR stack refers to eleven images associated to the free-clouds MERIS ones, see table I, related to the descending passes of ENVISAT satellite over the region of Naples corresponding to an area of about 100x100 Km, track 036, frame 2781, in the period from 1992 to 2006. The total baseline span is of 1890 metres and the Doppler Centroid excursion is 250 Hz.

| IMAGES | DATE | ORTOGONAL BASELINE | DOPPLER CENTROID |
|---|---|---|---|
| 1 | 10072003 | -1077.87 | 198.887 |
| 2 | 14082003 | -1012.35 | 157.617 |
| 3 | 05062003 | -928.567 | 245.885 |
| 4 | 31032005 | -917.827 | 159.836 |
| 5 | 25052006 | -630.065 | 197.919 |
| 6 | 24062004 | -384.910 | 175.973 |
| 7 | 27112003 | -213.853 | 121.309 |
| 8 | 20052004 | 0.00000 | 0.00000 |
| 9 | 07102004 | 12.6046 | 190.294 |
| 10 | 27102005 | 346.519 | 203.324 |
| 11 | 14072005 | 671.750 | 191.867 |

Table I. List of ENVISAT ASAR and MERIS images.

The MERIS free-clouds images corresponding to the ASAR stack are seventeen and are indicated in table I.
After the MERIS image selection, each image was cropped and resampled to the ASAR spatial resolution of 100 metres. The results have been georeferred in UTM-WGS84 cartographic coordinates system. The MERIS images were subsequently reported in RADAR geometry, transforming the cartographic coordinates in azimuth and range by mean of an polynomial interpolation.
The atmospheric ASAR images resulting from a stacking process are expressed in radians, whereas, the MERIS total water vapour column content is expressed in g/cm$^2$. To obtain the phase delay corresponding to the MERIS total water vapour column content, we used eq. 6. The estimated ASAR atmospheric images, may contain plains due to the errors of the unwrapping procedure, to the orbital inaccuracy or to the linear variation of the atmospheric pattern. To distinguish the plains associated to atmospheric signal and to estimate the possible residual offsets between SAR ASAR images and MERIS ones we used the Least Squares (LS) method.
At this point it is possible to compare the phase delay obtained by the MERIS data with the phase delay associated to the atmospheric signal estimated by the ASAR Stacking procedure.
In fig. 1 two atmospheric pattern delay difference, related to the acquisition pairs: 05.06.2003 and 14.08.2003 estimated by Stacking technique are shown. In figure II are shown the corresponding

atmospheric delay obtained by MERIS. All pictures of fig. 1 and fig. 2 have been masked to take into account free-clouds areas and spatial-temporal coherence associated to the Stacking procedure.

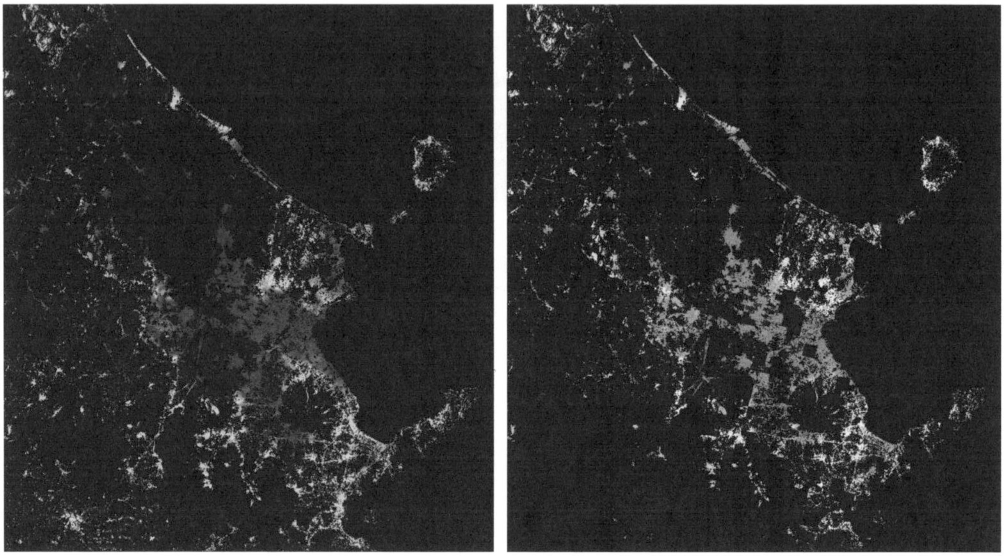

Figure 1. Atmospheric SAR phase delay difference associated to pairs: 14082003 – 05062003 and 14082003 – 31032005, colours are saturated in [-10 rad, 10 rad] .

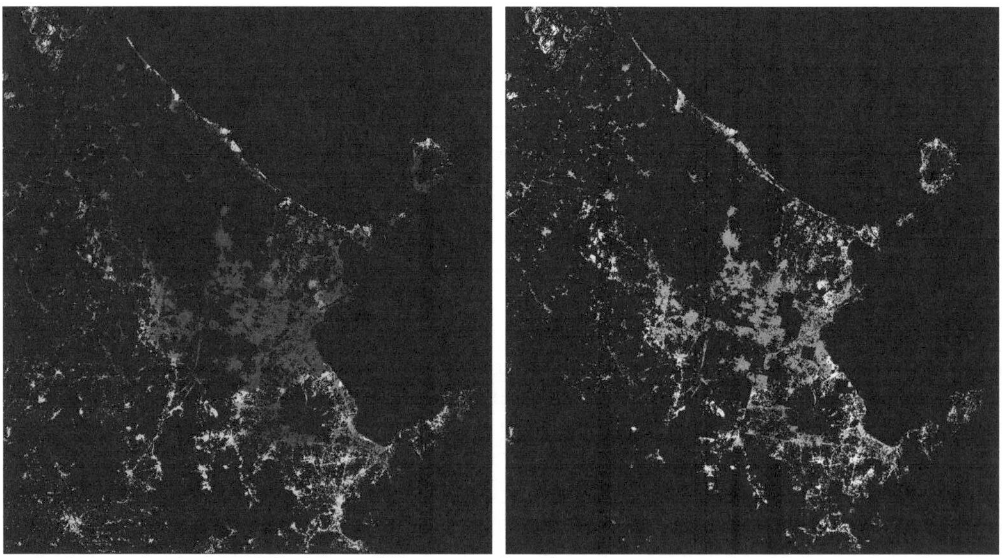

Figure 2. Atmospheric MERIS delay difference related to the same pairs of figure I, colours are saturated in [-10 rad, 10 rad] .

Figure 3 shows the residual atmospheric pattern after the removal of MERIS atmospheric contribution from the ASAR images. The differences, marked by circles in fig.3 are mainly due to errors in the unwrapping procedure and/or to the the lower resolution of the MERIS data.

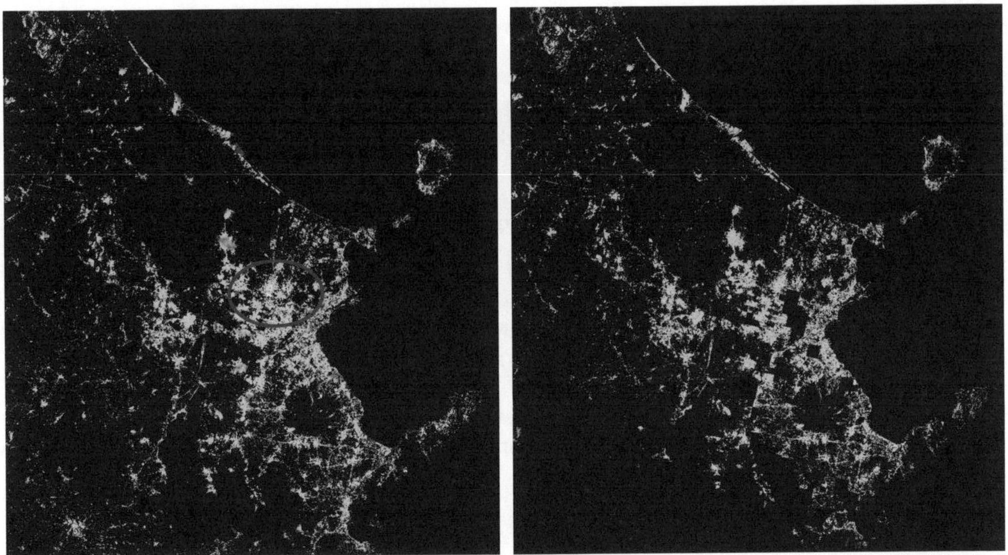

Figure 3. Atmospheric SAR phase delay images, of figure 1, compensated by using MERIS data of figure 2, colours are saturated in [-10 rad, 10 rad] .

## 5 CONCLUSIONS

In this work we have demonstrated that SAR atmospheric phase delay may be compensated by exploiting the measures of the water vapour content derived from multispectral sensors. This method can be used in the DInSAR multipass techniques to improve the estimation of ground deformations and can play an important role in crisis situations (earthquakes, volcanic eruptions, etc.) where only a few SAR data are available and then the atmospheric phase delay can lead to inaccuracies in estimates of deformation. The main limitation of this technique is due to the frequent presence in the scene of clouds, where the water vapour content is not reliable.

## ACKNOWLEDGMENT

We would like to thank Paolo Berardino for his support about the use of the Cartographic projection systems.

## REFERENCES

Berardino, P., Fornaro, G., Lanari, R., Sansosti, E. A new Algorithm for Surface Deformation Monitoring based on Small BaselineDifferential SAR Interferograms. *IEEE Trans. Geosci. Remote Sens.*, 40, pp.2375 2383, 2002.

Ferretti, A., Prati, C., Rocca, F. Nonlinear subsidence rate estimation using the permanent scatterers in differential SAR interferometry. *IEEE Trans. Geosci. Remote Sens.*, 38 (5), pp.2202 2212, 2000.

Fischer, J., Bennartz, R. 1997 *ATBD Retrieval of Total Water Vapour Content*. Doc. No: PO-TN-MEL-GS-0005

Fornaro, G., Pauciullo, A., Serafino, F. Deformation Monitoring over large areas with Multipass Differential SAR Interferometry: a new approach based on the use of Spatial Differences, *Int. Journal of Remote Sens.*, (Feb. 2007), accepted (Oct. 2007).

Hanssen, R. 2001 Radar Interferometry - Data Interpretation and Analysis. *Kluwer Academic Publishers*, vol. 2.

Li, Z., Fielding, E. J., Cross, P., Muller, J. 2006 Interferometric Synthetic Aperture Radar Atmospheric Correction: Medium Resolution Imaging Spectrometer And Advanced Synthetic Aperture Radar Integration. *Geophysical Research Letters*, Vol. 33, L06816, doi:10.1029/2005gl025299, 2006

Puysségur, B., Michel, R., Avouac, J.P. 2007 Tropospheric phase delay in interferometric synthetic aperture radar estimated from meteorological model and multispectral imagery. *Journal of Geophysical Research*, vol. 112, b05419, doi:10.1029/2006jb004352, 2007

*Remote Sensing for a Changing Europe*
*D. Maktav (Ed.)*
*IOS Press, 2009*

*doi:10.3233/978-1-58603-986-8-299*



# Analysis of the provincial structure of Sariyer / Istanbul using Remote Sensing and GIS

H.G. Coskun, U. Alganci & G. Usta
*Istanbul Technical University, Remote Sensing Division*

H. Celik
*3rd Army Corps*

Keywords: Remote sensing, Satellite Images, Digital Image Processing Techniques, Provincial Structure, Geographical Information System, Sariyer Province.

ABSTRACT: Basis of the economical development of the countries depends on the use of the natural resources in a most sustainable and productive manner. While developed countries are able to use their resources in this way; developing countries, in general, are still in lack of the sufficient knowledge and monitoring technologies about the quality and quantity of their natural resources. So, as a result of insufficient mapping of the natural resources; complete and well-established data sets about the land and water resources, distribution of cultivated lands and forests, grasslands and mines are not available in developing and under-developed countries.
In this study, analysis of the provincial structure of Sarıyer/Istanbul is aimed by using "Remote Sensing" and "Image Processing" techniques to the digital images. The satellite images used usually belong to the different years and of different resolutions. During present study 1998 dated IRS, 2001 and 2005 dated LANDSAT raster satellite images and the vector data of wards, parcels, fields from Sarıyer Province are used. The images, relevant data and software were all provided by Istanbul Technical University (I.T.U.) Civil Engineering Faculty, Department of Remote Sensing. By making use of these materials, initially, provincial boundaries have been established and then this analysis is integrated with a Geographical Information System (GIS). Several data layers have been constituted and finally a difference analysis among different years has been performed. As a result, important statistical results have been obtained, which will illuminate the future development and landcover-landuse studies in the Sarıyer Province.

## 1 INTRODUCTION

Remote Sensing and Geographic Information System are the methods that developed due to the scientific and technologic improvement in the $20^{th}$ century. Importance of these methods occurs with determination of the human effects to the water, air and the soil (Burrough 1998). Besides that, defining the natural and cultural sources of the earth and monitoring the changes in these sources continuously and rapidly make RS and GIS indispensable. With the uses of these technologies, national and international, all kind of data and knowledge analysis are possible and sustainable development plans can be produced in social, economical and even political arena.

Classification results of the satellite imagery make it possible to determine the land use characteristics in numerical values. Usage of multitemporal images in this process also gives the opportunity of detecting the changes of land use (Coskun et al., 2006, Yıldırım at al., 1998 and Yıldırım at al., 2001). At this point, an integrated GIS model will help to analyze the affect of change in one land use class to another. Moreover, vector data of urban areas can be used in combination with RS data in order to detect if the increase in urban areas are planned or unplanned.

The main purpose of this paper is to analyse the provincial structure and land-use distribution of Sariyer, located in Istanbul, by using RS and GIS techniques with multi-temporal satellite data that have different resolutions (Mermer, 2004; Yomralıoğlu and Reis, 2003). For this aim, satellite data and vector data of the districts of Sariyer have been handled together and achieved natural and cultural inventory.

## 2 STUDY AREA

Sariyer district of Istanbul is located on the European side of Istanbul with a coverage of 151 $km^2$ area. It is surrounded by Black Sea on the north side, Eyup on the west side, Sisli and Besiktas on the south side and the Bosphorus on the east side.

Sarıyer, has lots of precious historical trace. Especially Belgrat Forest in Bahçeköy has an old aqueduct which is constructed by Architecture Sinan who was lived Ottoman Empire. There are seven hanging dams in Belgrat Forest that have provided water for Istanbul. This aqueduct is interesting for tourists so summer population of Bahcekoy Willage arrise up to 10000. The first arboretum of Turkey, Atatürk Arboretum, is located in Belgrat Forest. Also, there is one hunt-protect-produce field and one fish breeding station (www.istanbul.gov.tr).

In relation to the rapid urbanization in Istanbul in the last two decades, the population of Sariyer reached 276.407 according to (http://tuikapp.tuik.gov.tr/adnksdagitimapp/adnks.zul, 2007).

## 3 GROUND TRUTH DATA

The vector data such as Sariyer region's vector data, Sariyer Zone 1 (North Zone) parcel vector data, Sariyer Zone 2 (Middle Zone ) parcel vector data, Sariyer Zone 3 (South Zone) parcel vector data and Sariyer fields vector data from Sariyer Province helped to understand the structure of the town.

Some ground truth measurements were taken from orthophotos, aerial photographs, 1:5000 scaled digital topographic maps, fieldworks and various thematic maps, in order to use both in geometric correction and classification process.

## 4 METODOLOGY

### 4.1 *Image Processing*

In this study, satellite data set of IRS and LANDSAT-5 TM dated 1998, 2001, and 2005 respectively, were used for land use – land cover analysis. The spatial resolutions for IRS and LANDSAT-5 TM are 23m and 30 m.

The digital satellite data sets were transformed into the UTM coordinate system using the orthophotos, 1:5000 scaled digital topographic maps,.Resulting root mean error (RMSE) of coordinate transformation is smaller than 0.5 pixels.

### 4.2 *Classification*

#### 4.2.1 *Land cover / use analysis*

1998 dated IRS, 2001 and 2005 dated LANDSAT images were subjected to a classification procedure with 150 clusters, using the supervised Maximum Likelihood Classification algorithm. Finally, 150 clusters reduced to 8 final classes by recode operation. During the accuracy assessment, 100 random pixels were chosen for each classification up to three different years and these pixels were compared with the results of the fieldwork and ground truth data. Accuracy assessment results for all classes at IRS and LANDSAT-5 TM satellite data dated 1998, 2001 and 2005 are 82%, 74% and 80%, respectively. Figure 1 (a, b, c) shows the result images of the land cover and land-use classification for years of 1998, 2001 and 2005 in Sariyer.

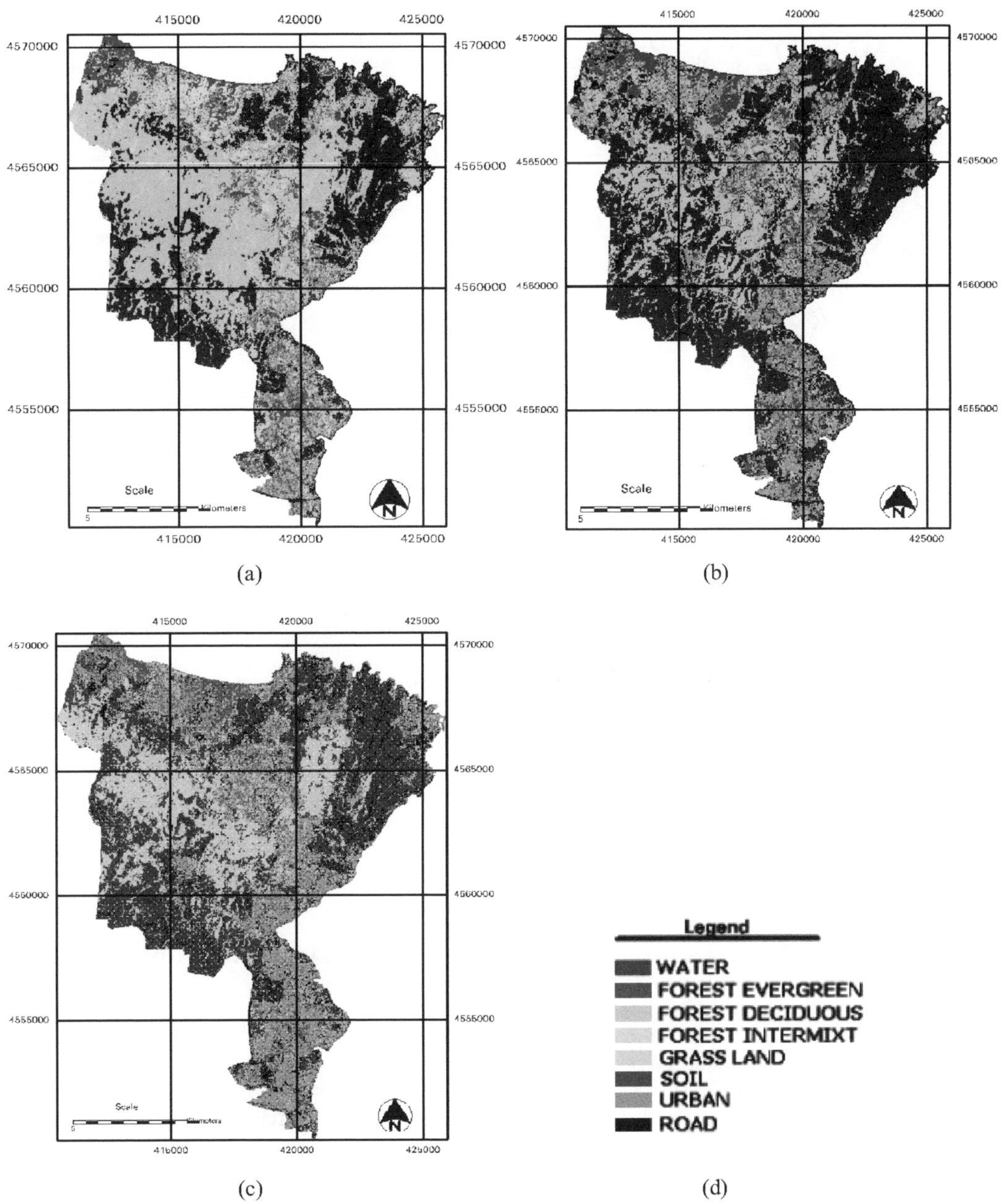


Figure 1. Classified images of Sariyer (a) IRS dated 1998, (b) LANDSAT5 TM dated 2001, (c) LANDSAT-5 TM dated 2005 image, (d) Legend of land-use/land-cover classes

## 4.2.2 *GIS Modeling Using Raster-Vector Integration*

IRS dated 1998, LANDSAT dated 2001 and 2005 used forming the data layers for understanding the structure of the district. In figure 2, Sariyer wards vector data has overlayed with the classified IRS image.This figure shows the visual interpretation of settlements with names, geographic boundries, areas and locations of 23 district and 9 village.

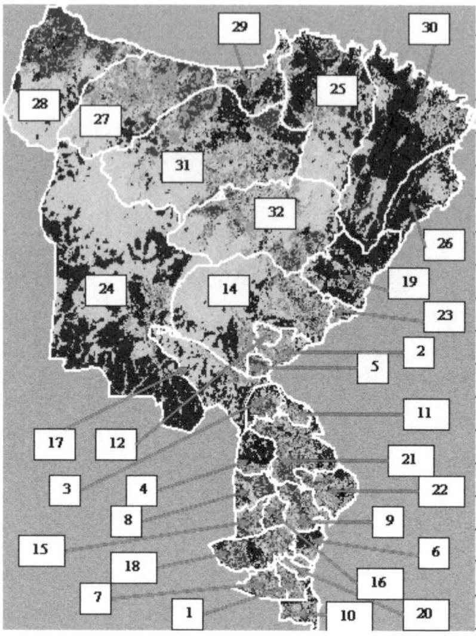


| | TOWN/VILLAGE NAME | AREA (hectare) | | | AREA (hectare) |
|---|---|---|---|---|---|
| 1 | Baltalimanı | 79,27 | 17 | PTT Evleri | 443,73 |
| 2 | Büyükdere | 117,59 | 18 | Reşit Paşa | 278,82 |
| 3 | Cumhuriyet | 114,60 | 19 | Rumelikavağı | 449,29 |
| 4 | Çamlıtepe | 142,48 | 20 | Rumelihisarı | 85,16 |
| 5 | Çayırbaşı | 47,93 | 21 | Tarabya | 370,96 |
| 6 | Emirgan | 125,69 | 22 | Yeniköy | 202,29 |
| 7 | Fatih | 107,17 | 23 | Yenimahalle | 57,74 |
| 8 | Ferahevler | 128,27 | 24 | Bahçeköy | 2847,07 |
| 9 | İstinye | 178,94 | 25 | Demirciköy | 1095,01 |
| 10 | Kazım Karabekir Paşa | 106,56 | 26 | Garipçe | 391,25 |
| 11 | Kireçburnu | 62,26 | 27 | Gümüşdere | 940,92 |
| 12 | Kocataş | 45,48 | 28 | Kısırkaya | 997,67 |
| 13 | Maden | 70,18 | 29 | Kumköy (Kilyos) | 301,31 |
| 14 | Merkez | 1221,31 | 30 | Rumelifeneri | 1294,27 |
| 15 | Pınar | 89,48 | 31 | Uskumruköy | 1680,17 |
| 16 | Poligon | 67,42 | 32 | Zekeriyaköy | 1162,98 |

Figure 2. Sariyer wards vector data with classified IRS image

When the Zone 1 parcel data is overlayed over classified image, it can be seen that urban class with pink colour exceeds the boundries of vector data.This analaysis shows that, urbanization in Sariyer increases with a damage on the forest area (Figure 3a). The amount of damaged forest area can be calculated with this analyses in hectare units.

For Zone 2 it can be seen that, urban class in classified image exceeds the boundry of vector data. When figure 3a and 3b analysed together, it can be said that urbanization increase and forest damage are in less amounts in the middle part of Sariyer compared to north part of the district. This result can visually interpreted in the Figure 3b which the parcel vector data and urban classes are mostly overlaped.

In the south part of Sariyer, urbanization is intensive because of the zone's nearness to the city center. However, when the zone three parcel data is overlayed with classified image, it can be seen that urban class and vector data are mostly in coherence. Beside that, on the west side of this zone there are unplanned buildings. In the third zone, on the east part, some forest area enrolled as parcellation area (Figure 3c).

Finally wards vector data is analysed together with classified image.While vector data and buldings are overlapping in the middle and south part, in the north side there are differences between the urbanization and vector data (Figure 3d).

For change detection analysis, IRS satellite data of 1998 and LANDSAT 5 TM satellite data of 2005 used and changes in the land cover types such as forestery, urbanization, soil, grassland, roads and water computed (Table 1). In Sariyer, from 1998 to 2005, while there is a decrease at forest area and grassland, urban area has increased. This process shows that, the main reason of the forest destruction is the urbanization. In addition, increasing of the soil area is the first stage of transformation from forest to urban so the increase on the soil class can be explained as a continious increase in urbanization. Also,an increase for the road class is the another proof of urbanization.

Table 1. The analysis of the landcover/landuse distrubition for Sariyer for the years 1998 and 2005

| | IRS(1998) | | LANDSAT (2005) | | Difference | |
|---|---|---|---|---|---|---|
| | Area (ha) | Percent(%) | Area (ha) | Percent (%) | Area (ha) | Percent (%) |
| Total Forest | 9793,55 | 64,08 | 9365,57 | 61,26 | -427,98 | 2,82 |
| Urban | 1950,66 | 12,76 | 2505,24 | 16,38 | 554,58 | 3,62 |
| Soil | 1100,98 | 7,20 | 1328,76 | 8,69 | 227,78 | 1,49 |
| Grassland | 854,44 | 5,59 | 412,11 | 2,69 | -442,33 | 2,90 |
| Roads | 1492,95 | 9,76 | 1563,66 | 10,22 | 70,71 | 0.46 |
| Water | 89,53 | 0,58 | 110,70 | 0,72 | 21,16 | 0.14 |

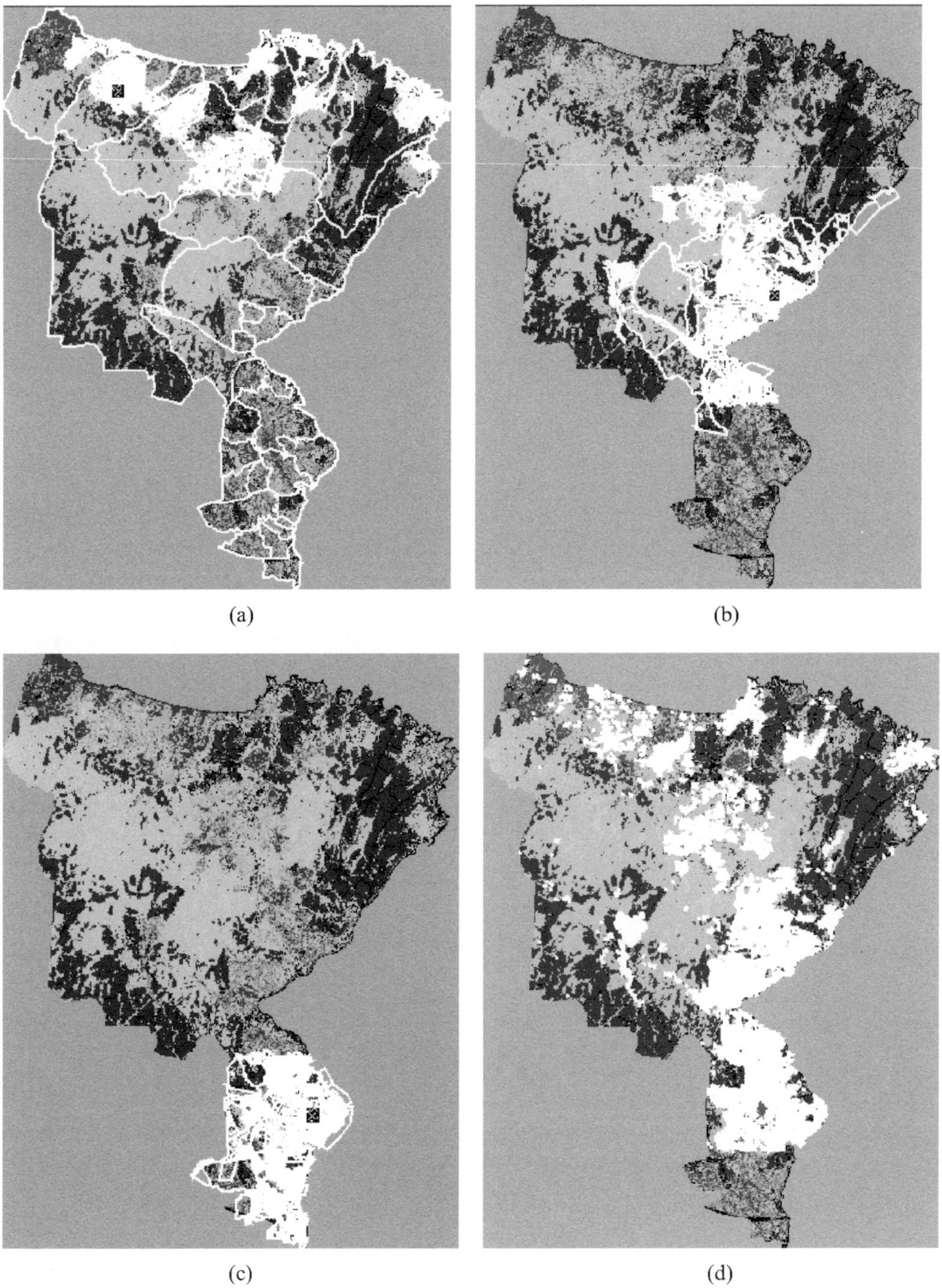


Figure 3. Classified IRS images of Sariyer dated 1998 with (a) Zone 1 parcel data (districts and villages numbered 26-32 on the north) (b) Zone 2 parcel data (districts numbered 3,11-32 in the middle) (c) Zone 3 parcel data (districts numbered 1-11 except 3 on the south) (d) wards vector data with classified image

## 5 RESULTS AND DISCUSSION

Sariyer is the one of the most important district in Istanbul with 23 towns and 9 villages.Urbanization is increasing fast and uncontrolled with destructing the natural beauty of the district.To prevent the illegal housing and the determine the old effects of urbanization, RS and GIS techniques are used.

In this study geometric correction, image enhancement and classification have done for IRS and LANDSAT-5 TM dated 1998, 2001, and 2005 respectively. Then a change analysis is done and numerical results are obtained in hectare units. These analysis shows that, the change in forest area is in relation with the changes in urban, road and soil. While buildings and roads are increasing, forest and soil area are decreasing.

As the vector data, Sariyer wards vector data, Zone 1(North Zone) parcel data , Zone 2(Middle Zone ) parcel data, Zone 3 (South Zone) parcel data, Sariyer fields vector data used together with IRS and LANDSAT-5 TM dated 1998, 2001, and 2005.

In the 21th century, importance of the Remote Sensing is undeniable. With the usage of the vector data and images illegal housing and occupied forest area is determined. Consequently, incoming new buildings should be kept under the control.

## REFERENCES

Burrough, P.A., 1998. Principles of Geographical Information Systems for Land Resources Assessment. Oxford Universty, Press 2.ed.

Coskun H. G. Et al, 2006. Monitoring of Protected Bands of Terkos Drinking Water Reservoir of Metropolitan Istanbul Near the Black Sea Coast Using Sattelite Data,International Journal of Applied Earth Observation and Geoinformation 8 (2006) 49-60.

Mermer, A., 2004. Uydu Görüntüleri Kullanarak Gaziantep ilindeki Tarımsal Alanların Belirlenmesi Projesi. ISPRS Kongresi, İstanbul.

Yıldırım, H., Özel, M.E., et al., 1998. Yeşilırmak Watersbed Development Project (YWDP) Geographical Information Systems Infrastructure: Land Cover/Land Use Mapping via Recent Satellite Imagery. International Symposium on Arid Region Soils(YISARS), Menemen, İzmir.

Yıldırım, H., Özel, M.E., et al., 2001. Satellite Monitoring of Land-Cover/Land Use Change in Gebze-Kocaeli, Over 15 Years and Its Impact on Environment. Turkish Journal of Agriculture and Porestry, 3.

Yomralıoğlu, T., Reis, S., 2003. Landsat ETM+ Uydu Görüntüsü ile Trabzon İli Arazi Örtüsünün Belirlenmesi. Doğu Karadeniz Bölgesinde Kırsal Alanda Ulaşım, Yerleşim Sorunları ve Çözümleri Sempozyumu, KTÜ, Trabzon.

(www.istanbul.gov.tr).

(http://tuikapp.tuik.gov.tr/adnksdagitimapp/adnks.zul)

*Remote Sensing for a Changing Europe*
*D. Maktav (Ed.)*
*IOS Press, 2009*


*doi:10.3233/978-1-58603-986-8-305*

# Hydrological modeling of ungauged basins using Remote Sensing and GIS: a case study of Solaklı Watershed in Turkey

U. Alganci, H. G. Coskun & G. Usta
*Istanbul Technical University, Remote Sensing Department, Istanbul Turkey*

N. Agiralioglu, , H. K. Cigizoglu, H. Aksoy, L. Yilmaz & E. Eris
*Istanbul Technical University, Hydraulics Department, Istanbul, Turkey*

M. Dagci
*Koklu Engineering*

Keywords: remote sensing, surface runoff, ungauged basin, rainfall-runoff model, digital elevation model, geographic information system.

ABSTRACT: In Turkey, the population growth and industrial development have caused a scarcity in energy. Turkey has a great hydroelectric energy potential. However, this potential is evaluated only to a certain extent. The major step in the determination of hydroelectric potential is to estimate the river runoff, which is one of the most important hydrological variables used in most of the water resources applications. Reliable estimation of runoff from land surface into streams and rivers is difficult and time consuming for ungauged basins. However, Remote Sensing (RS) and Geographic Information System (GIS) technologies can augment to a great extent the conventional methods used in rainfall-runoff studies. These techniques can be used to estimate the spatial variation of the hydrological parameters, which are useful as input to the rainfall-runoff models.

The main objective of this study was to model the rainfall-runoff process in a selected ungauged basin for the purpose of calculation of water potential of Solaklı Watershed in Eastern Black Sea Region, Turkey. This model simulation was carried out using a hydrological modeling system assisted by GIS. The merging of LANDSAT 7 ETM multisensor satellite images that is established using an image processing software shows an effective means of extracting information from different source of sensors, reaching 15 m spatial resolution. The classification of satellite data set was carried out using maximum likelihood classification method to obtain land cover/land use (LULC) information of the basin. GIS analysis is then used to combine inventory field data collected on the basin by ARC-GIS. Additionally, Digital Elevation Model (DEM) of Solaklı Watershed derived from 1:25000 scaled digital topographic maps with 10 m spatial resolution. DEM is used for determining flow accumulation and direction data. A synthetic stream network is then obtained from this model and area, perimeter, minimum and maximum elevation values are calculated for the sub basins of Solaklı Watershed.

## 1 INTRODUCTION

Solaklı Basin, which is located in the East Black Sea Region, has a great hydrological potential with its rainy climatic characteristics and rough topographic structure with huge elevation differences. Unfortunately, the basin is poorly gauged and reliable runoff estimations have never been done before. For the determination of hydroelectric potential in a basin, runoff and gradient height information is needed.

Runoff is one of the most important hydrological variables used in water resources studies. Reliable prediction of direct runoff for ungauged basins is difficult and time consuming. Conventional models for predicting stream discharge require considerable hydrological and meteorological data. Remote Sensing (RS) and Geographic Information Systems (GIS), in combination with appropriate rainfall runoff models, provide ideal tools for the estimation of direct runoff volume, peak discharge and hydrographs (Miloradov and Marjanovic, 1991; Demayo and Steel, 1996; Bellal et al., 1996). The role of remote sensing in runoff estimation is generally to provide a source of input parameters for the models. Satellite data can provide thematic information on land use, soil, vegetation, drainage, etc., and topographic parameters (area, elevation and slope) which, combined with conventionally measured climatic parameters (precipitation, temperature, etc.), constitute the necessary input data for the rainfall-runoff models. Today, Digital Elevation Models (DEM) and land use – land cover (LULC) maps, produced from high resolution satellite imagery, provide the stated information in a quick, accurate and reliable way compared to classical terrestrial survey methods. The information, extracted from remote sensing and other sources, can be stored as a georeferenced database in a GIS (Dudhani et al., 2006; Hammouri and El-Naqa, 2007).

Digital topographic maps are used for producing DEMs that are needed in determination of elevation, slope, and aspect information of the basin. In addition, sub basin areas and synthetic drainage network can be derived with processing the DEM data in a specific software module (Kumar and Singhal, 1999). Land cover characteristics, which are another important parameter in runoff modeling, can be determined by applying image classification to multispectral satellite data. All this information combined with hydro meteorological information can be analyzed in GIS environment in order to make a runoff computation (Cheng et al., 2006).

In this study, three important parameters are determined for the Solaklı Basin and a sub basin named Ogene - Alcakkopru via RS and GIS. These parameters are elevation, area and LULC information accordingly.

## 2 STUDY AREA

Solaklı Basin is located in Trabzon in the Eastern Black Sea Region with a coverage of 767, 25 $km^2$. Basins location is between 41o North latitude and 39o 43' East longitude. Forest area covers % 38; grassland covers %24 of total basin area. In the south part there are Haldizen and Soganli Mountains. The steepest area of the basin is Demirkapı Hill with 3376 m elevation. The main river of the basin is same named Solaklı River. Heavy rainfall especially in spring season causes floods and landslides in the bed of river and annual average rainfall is 830 mm /$m^2$ for the basin (Fig. 1).

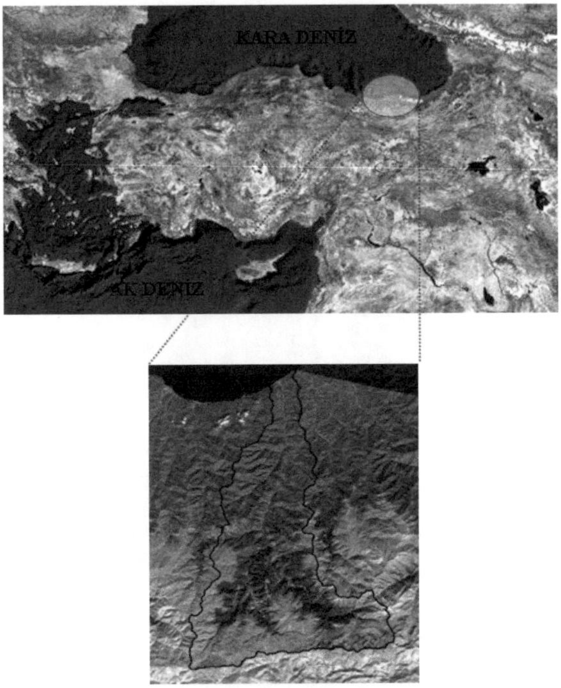


Figure 1. Multispectral LANDSAT 7 ETM image of Solaklı Basin with 15 m geometric resolution dated 10.07.2000 (Band 5/4/3).

## 3 METHODOLOGY

### 3.1 *Determination of Topographic Parameters*

The first step to construct a basin model for Solakli basin is preparing a DEM. For this purpose, 1: 25000 scaled digital topographic maps are used. A Triangulated Irregular Network (TIN) model is created in the GIS software and a grid file of DEM is constituted with 10 meter spatial resolution (Fig. 2a).

Such DEMs include pits or ponds that should be removed before being used in hydrological modeling (Ashe, 2003). These are cells where water would accumulate when drainage patterns are being extracted. Pits are a sign of errors in the DEM arising from interpolation. These pits were removed by an algorithm known as sink filling. After filling the DEM sinks, a flow direction map was computed by calculating the steepest slope and by encoding into each cell the eight possible flow directions towards the surrounding cells. Flow direction is then used to generate the flow accumulation map. The flow accumulation, generated by addressing each cell of the DEM, counts how many upstream cells contribute to flow through the given cell. (Fig. 3a,b) Flow direction and accumulation maps are then used to delineate the stream network.

Sub basins were determined according to the runoff stations and vector data of these sub basins are constituted. The grid structured DEM and this vector data are analyzed together in order to obtain area, perimeter and average elevation information of the sub basins (Fig. 2b). Also two important physical parameters affecting the surface runoff; slope and aspect information are produced from DEM data (Fig. 4a,b).

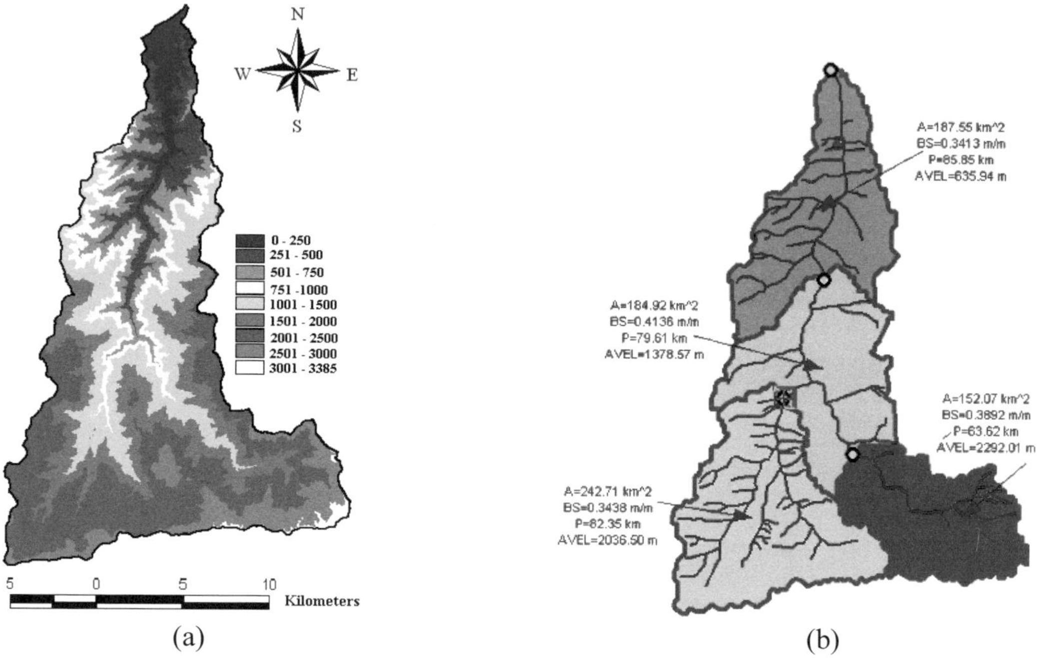


Figure 2. a) Digital elevation model of the basin, b) sub basins and drainage network in vector format and topographic information determined.

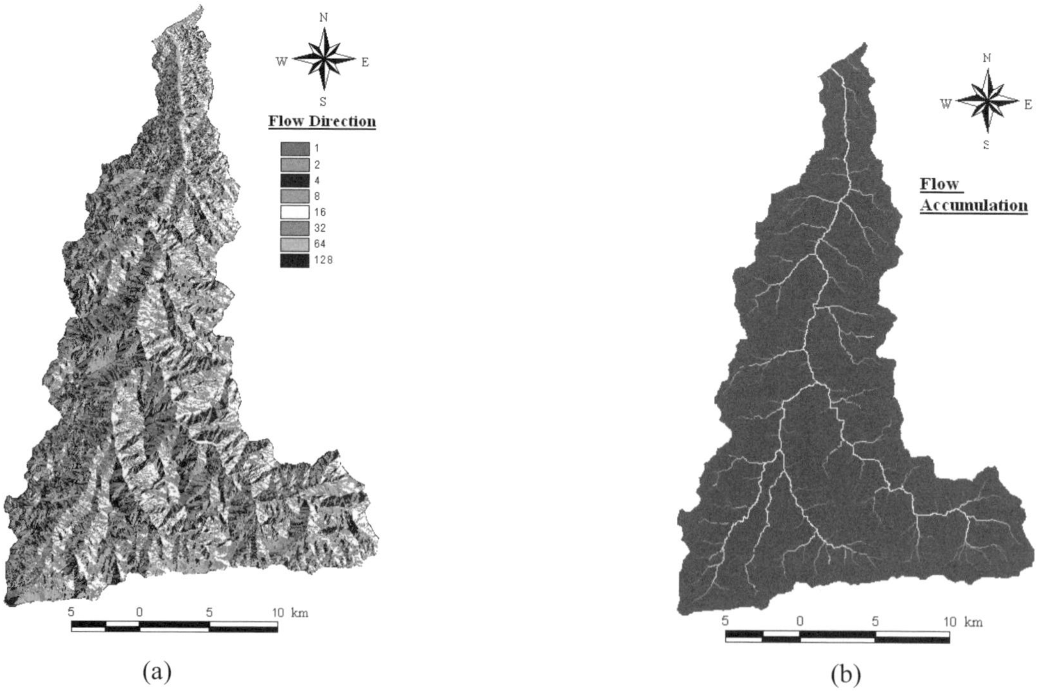


Figure 3. a) flow direction, b) flow accumulation map of the Solaklı basin

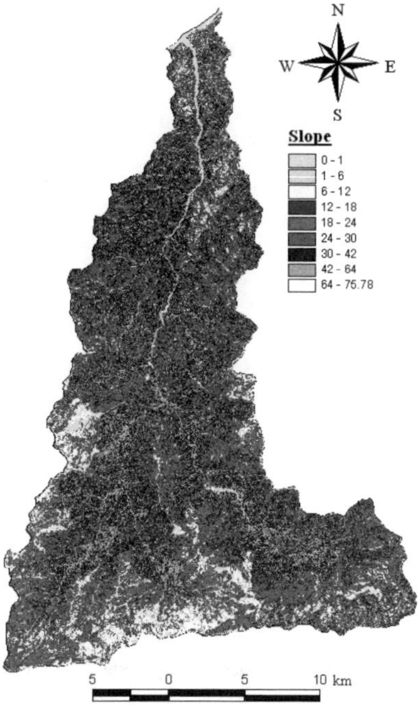


(a)

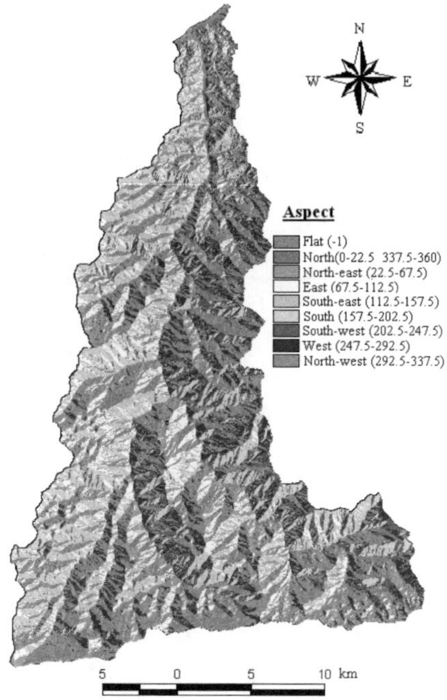


(b)

Figure 4. a) slope map b) aspect map of the Solaklı Basin

## 3.2 *Land Use – Land Cover Analysis*

In hydrology researches, LULC is one of the important parameters that affecting the surface runoff. Satellite imagery is a convenient data source for the determination of LULC in a quick and reliable way. In this study, 2000 dated LANDSAT 7 ETM MS-PAN data set is used for LULC analysis. Dataset is firstly rectified to UTM coordinate system using the ground control points measured with GPS. The RMSE of affine transformation was ± 0.2 pixel. Then, the rectified images are merged using HIS transformation algorithm. After this operation, a three band image with 15 m spatial resolution is obtained.

Rectified and merged image is subjected to maximum likelihood unsupervised classification process. Obtained 80 clusters are reduced to 11 final classes that represent the LULC structure of the basin. Overall classification accuracy was found %82 with a kappa result of 0.7497 for randomly selected 100 pixels.

# 4 OGENE – ALCAKKOPRU SUB BASIN SAMPLE

This sub basin is located in the south-west part of the Solaklı Basin. According to the grid –vector analyses, it has coverage with 242.71 $km^2$ area and 82.35 km perimeter. With its average elevation of 2036.50 m, it is the second steepest sub basin of the study area. The results of classification states that LULC characteristics of this sub basin consists of four basic classes: deciduous forests with 8363.16 ha, grassland with 7511.79 ha, bare land with 6607.32 ha and evergreen forests with 1842.61 ha (Fig.5).

In the other step, longitudinal section of the drainage for this sub basin is constituted as necessary information needed in determination of gradient height (Fig. 6). This height information is used with runoff information in the process of hydroelectric potential.

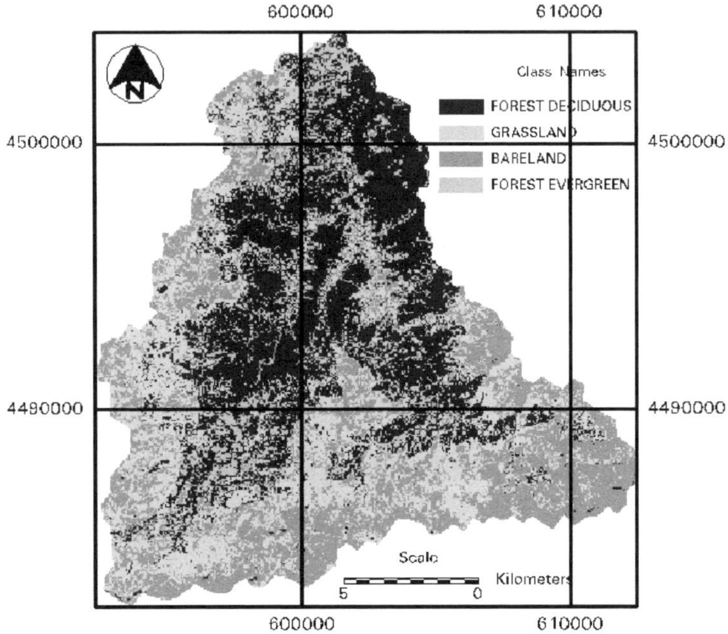


Figure 5. Classification result of multispectral LANDSAT 7 ETM image belongs to Ogene River basin area Solaklı Basin 10.07.2000 (Band 5/4/3).

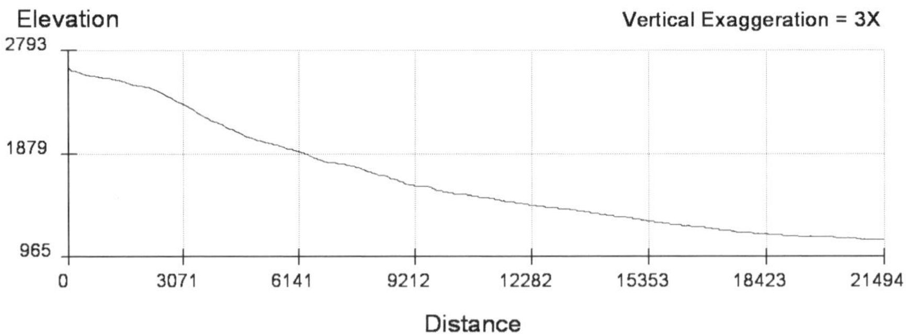


Figure 6. Longitudinal section of Ogene sub basins' drainage

## 5 FLIGHT SIMULATION

A 3D synthetic flight simulation model is constructed in order to analyse the topographic structure of the basin visually. LANDSAT 7 ETM satellite image is overlayed with 10m resolution DEM and 3D rendering is done accordingly in th image processing software. Figure 7 represents an ilustration of this simulation for Solaklı Basin

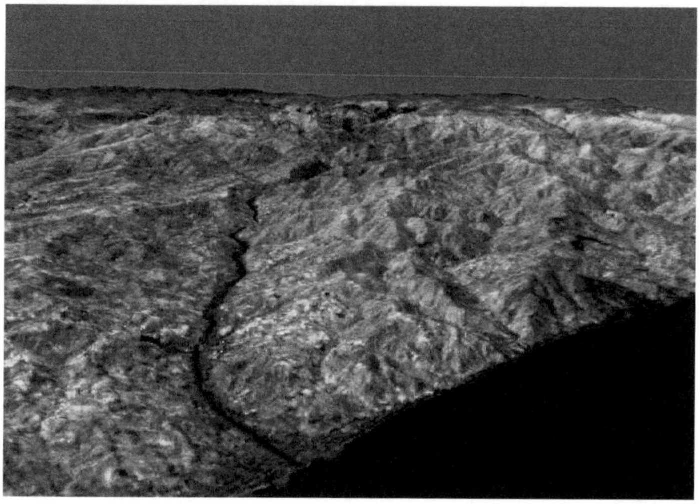

Figure 7. 3D North side perspective view from the flight simulation for Solaklı Basin

## 6 RESULTS

In hydrology researches, computing the runoff is one of the most important step. Up to last decade , conventional models were being used for estimation off the discharges that need a huge amount of data. In ungauged or poorly gauged basins, neural networks or fuzzy logic techniques were used to estimate the runoff directly from long term rainfall information due to the lack of physical parameters of the basin. For now, RS and GIS techniques make it possible to determine physical parameters of the basin. With a hydraulic model constructed with mentioned parameters in a GIS environment, reliable, quick and accurate runoff estimations can be performed.

## ACKNOWLEDGEMENT

This study is prepared under the project named "Determination of Technical Hydroelectric Potential in ungauged basins" which is founded by TUBİTAK

## REFERENCES

Ashe, R., 2003, Invesigating runoff source area for an irrigation drain system using ArcHydro tools, in Proceedings of the Twenty-Third Annual ESRI User Conference, San Diego.

Bellal, M., Sillen, X., Zeck, Y., 1996, Coupling GIS with a distributed hydrological model for studying the effect of various urban planning options on rainfall-runoff relationship in urbanised basins, in Kovar, K., Nachtnebel, H.P. (eds.), Application of Geographic Information Systems in Hydrology and Water Resources Management: International Association of Hydrological Sciences, Series of Proceedings and Reports, 235, 99-106.

Cheng, Q., Ko, C., Yuan, Y., Ge, Y., Zhang, S., "GIS Modeling for predicting river runoff volume in ungauged drainages in Greater Toronto Area, Canada", Computers and Geosciences, 2006, 32:1108-1119.

Demayo, A., Steel, A., 1996, Data handling and presentation, in Chapman, D. (ed), Water Quality Assessments, A Guide to the Use of Biota, Sediments and Water in Environmental Monitoring: London, United

Nations Educational, Scientific and Cultural Organization, World Health Oranization , United Nations Environment Programme, 2nd edition, Chapter 10, 511-612.
Dudhani, S., Sinha, A. K., Inamdar, S. S., "Assessment of small hydropower potential using remote sensing data for sustainable development in India", 2006, Energy Policy, 34:3195-3205.
Kumar, A., Singhal, M. K., "Hydropower assessment for small ungauged catchments in Himalayan region using GIS techniques", 1999, Workshop on GIS Applications in Power Sector, Map India.
Miloradov, M., Marjanovic, P., 1991, Geographic information system in environmentally sound river basin development, in 3rd Rhine-Danube Workshop, Proceedings, 7-8 October: Delft, the Netherlands, Technische Universiteit Delft.
Nezar Hammouri and Ali El-Naqa 'Hydrological modeling of ungauged wadis in arid environments using GIS: a case study of Wadi Madoneh in Jordan', Revista Mexicana de Ciencias Geológicas, v. 24, núm. 2, 2007, p. 185-196

*Remote Sensing for a Changing Europe*
*D. Maktav (Ed.)*
*IOS Press, 2009*

*doi:10.3233/978-1-58603-986-8-313*



# Thermal remote sensing imagery in permafrost studies

R. Van De Kerchove, R. Goossens, A. De Wulf
*Department of Geography, Ghent University, Ghent, Belgium*

J. Bourgeois, W. Gheyle
*Department of Archaeology, Ghent University, Ghent, Belgium*

Keywords: Permafrost, modelling, Altai, MODIS, LST

ABSTRACT: Mountain permafrost modelling in remote, continental mountain ranges (e.g. Russian Altai Mountains) holds several difficulties due to the limitations these environments pose. The lack of meteorological input data and impossibilities for BTS-validations (Bottom Temperatures of winter Snow cover) makes conventional modelling strategies inapplicable. Statistical methods, however, based on correlation coefficients between different parameters, offers good alternative but requires lots of observations to be significant. As a solution, spatially covering land surface temperature (LST) values might be used as a proxy replacing the interpolated air and near ground surface temperatures. This article proposes 2 strategies, one statistical and one adapted TTOP (temperature at the top of permafrost), based on remote sensing data and ground measurements. Although these methods seem promising, they require a detailed understanding of the relation between LST and the air and near ground surface temperature. Therefore, before installing field equipment, we compared filtered MODIS LST time-series with corresponding ground temperature measurements recorded by Sergei Marchenko (Geophysical institute of Alaska, Fairbanks) in the Ulandryk Valley. Despite the cloudy conditions of this test site, a good correlation is showed between both time-series.

## 1 INTRODUCTION

Future climate models indicate a pronounced warming throughout the 21$^{st}$ century. (McCarthy *et al.*, 2001) Therefore a significant global retreat of glaciers and degradation of permafrost with a corresponding altitudinal shift in alpine landscape zones processes is expected. (Marchenko, 2007 after Haeberli & Burn, 2002) Besides this global warming, circulation models suggest that the increase in summer diurnal temperatures over Central Asia is likely to be higher relative to other regions. (McCarthy *et al.*, 2001) As the higher parts of the Altai Mountains are situated in areas dominated by continuous and discontinuous permafrost, they are extremely vulnerable towards such a possible temperature increase.

Many archaeological monuments (burial mounds) are situated within the permafrost area of the Russian Altai Mountains and some of them have a frozen core. (Marchenko, 2007) These so called 'kurgans' are burial mounds which belong to the Scythian Culture. These pastoral nomads lived across the whole Eurasian Steppe Belt during the 1$^{st}$ millennium BC. (Bourgeois and Gheyle, 2007) In particular those tombs located at the lower fringe of the permafrost area, are likely to defrost in several decades. (Bourgeois and Gheyle, 2007) Therefore, extensive permafrost research regarding the thermal state of the frozen tombs and the spatial distribution of the mountain permafrost is necessary to forecast which of the tombs are endangered by thawing.

Multi-approach periglacial research has been carried out in the Russian Altai Mountains. However, no large scale permafrost distribution map has been presented yet. (Fukui *et al.*, 2007) investigated the lower limits of mountain permafrost in the Akkol Valley, by means of indicator features such as rock glaciers, pingo's and ice wedge polygons. The distribution of these features suggests that the altitudinal range of the sporadic-patchy and discontinuous-continuous permafrost zones are 1800-2000m ASL and above 2000m, respectively. Dr. Sergei Marchenko (Geophysical institute of Alaska, Fairbanks) examined the thermal state of diverse kurgans and the thermal regime of diverse altitudinal landscape settings in the Ulandryk Valley, as part of the UNESCO-project *'The Frozen Tombs of the Altai'*. (Marchenko, 2007) The geothermal observations and modelling he carried out, indicate that favorable conditions for permafrost occurrence and preservation exist in coarse blocky material where mean annual temperatures are typically 3-5°C colder than the mean annual air temperature (MAAT). (Marchenko, 2008)

Due to the vast and remote character of the Russian Altai Mountains, conventional permafrost modelling approaches cannot be applied. Process-oriented energy balance models (e.g. Stocker-Mittaz *et al.*, 2002) namely require lots of energy input data, while traditional statistical 'rules of thumb' variants depend on BTS-measurements under a substantial snow cover in winter months (Brenning *et al.*, 2005; Julian and Chueca, 2007) As the Altai Mountains are characterized by a summer precipitation regime and powerful winds clear the valley floor at regularly time, no such cover is able to form in winter months. The originally developed regional TTOP-model (temperature at the top of the permafrost) (Smith and Riseborough, 1996) and its alpine adaptation (Juliussen and Humlum, 2007) however, offers good modelling alternative. This model uses accumulated degree-days (thawing and freezing) and n-factors (both summer and winter) to model the mean annual ground-surface temperature (MAGST). Finally by using the ratio between thawed thermal conductivity and frozen thermal conductivity the TTOP can be modelled. (Juliussen and Humlum, 2007). Despite its simplified form, this model still needs a parameterization of both summer and winter n-factors, based on respectively potential incoming solar radiation and snow depth. As this last parameter is impossible to obtain, this traditional method can't directly be used in the Russian Altai context.

Therefore, two types of adapted conventional methods, based on remotely sensed obtained LST values, might be used to model the spatial distribution and thermal properties of permafrost in the Altai Mountains. These methods are based on the relation between LST values on the one hand and air/ground temperatures on the other. As showed in previous studies, LST-values might improve lapse rate based interpolation algorithms, to obtain a spatially covering air temperature grid (Colombi *et al.*, 2007; Pridhoko and Goward, 1997). This might especially enhance air temperature modelling in the Altai, as mountain meteo stations are limited in the region. Correlation between LST and near surface temperatures however is only seldom described. Therefore, an extensive field campaign will install about 100 thermochron DS1922L-iButtons, together with 19 HOBO U23-series temperature data-logger to monitor year round near surface temperatures. These measurements will be used to understand the relation between LST and ground temperatures with respect to different environmental conditions. (topography, lithology, snow cover, vegetation). To test this relationship, MODIS LST-time series are in this article compared with corresponding ground temperature time-series from the Ulandryk valley, kindly provided by Sergei Marchenko.

## 1.1 *Study Area*

The Altai Mountains are situated in Central Asia at the meeting point of four countries: China, Kazakhstan, Mongolia and the Russian Federation. (Figure 1) Traditionally, this mountain range is subdivided into the Russian, Mongolian and Gobi Altai. (Fukui *et al.*, 2007) The Russian Altai Mountains extend between approximately 48 an 53°N latitude and 83 and 92°E longitude.

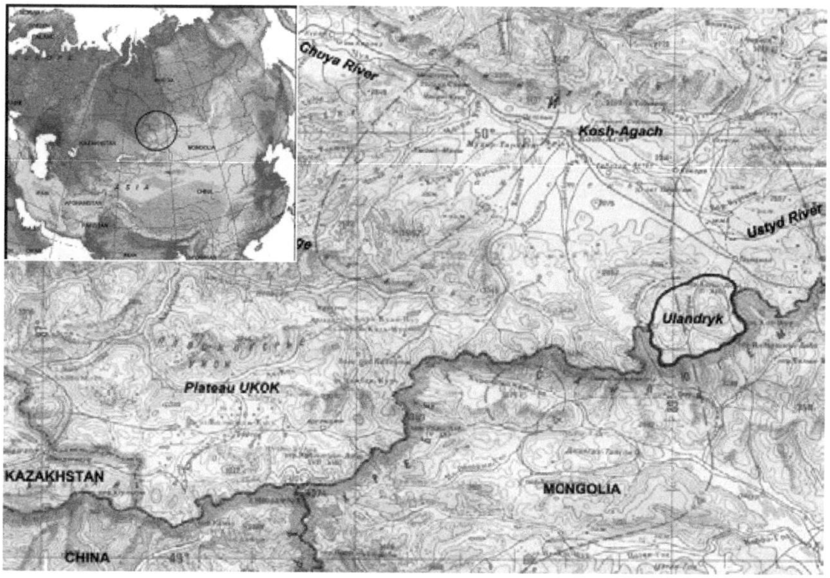


Figure 1. Location of the Ulandryk Valley in the Altai Republic (after Marchenko, 2007)

This article focuses on the Ulandryk Valley depression which is situated at the border between Russia and Mongolia (Figure 1) Climate of the valley is dry and distinctly continental. (Marchenko, 2007). Mean annual air temperature (MAAT) was -4.6°C during the 1951-1998 period, according to the Ulandryk weather station. The amount of annual precipitation is 184 mm. It is an area of continuous and discontinuous permafrost. (Marchenko, 2007)

## 2 METHODS

### 2.1 *MODIS LST time series*

MODIS TERRA and AQUA, daily LST (MOD11A1 and MYD11A1)) and 8-day composite grid data (MOD11A2 and MYD11A2)) were acquired between January 2006 and December 2007 from the NASA Earth Observing System (EOS) data gateway. Details documenting the MODIS products can be found at NASA's MODIS website. By selecting those daily and composite pixel values that correspond to the different test sites, site-specific LST time-series can be derived.

These series contain however, many missing and poor quality values due to the substantial amount of overcast days in the study area. Therefore time-series filtering and interpolation is an essential step in this study.

#### 2.1.1 *LST time-series filtering: CLEAN algorithm*

Following the method described in (Lunetta *et al.*, 2006) for filtering MODIS NDVI profiles, the LST data set is first filtered for quality assurance based on MODIS quality bit flags QC_day and QC_night. Only these pixels with good quality were retained

The filtered data is then transformed into its frequency domain using a discrete Fourier transformation and 'cleaned' with a nonlinear deconvolution algorithm (CLEAN) described by (Roberts *et al.*, 1987). This procedure utilises knowledge of the sampling function to perform an iterative deconvolution of the spectral window in the frequency domain. Next to this ability to remove noise out of the signal spectrum, (Roberts *et al.*, 1987) showed the robustness of their CLEAN procedure in the analysis of unevenly sampled data points and data sets containing missing values.

A disadvantage of this CLEAN technique is that there is no simple calculation to determine the statistical significance of the peaks in the observed frequency spectrum. Therefore, a technique based upon Monte Carlo simulations, is used for calculating these significance levels. (Heslop and Dekkers, 2002) This Monte Carlo technique uses tree different types of simulation involving data stripping and the addition of red and white noise to the input time series. In this study we used the data stripping method.

### 2.1.2 *LST time-series filtering: Savitzky Golay method*

As a second method for filtering and interpolating MODIS LST time-series we used the Timesat 2.3 program, designed by Per Jönsson and Lars Eklundh. (Jönsson and Eklundh, 2002, 2004) This application encompasses an adaptive Savitzky-Golay filter, and takes the quality of the different LST values into account, by applying a weight to the different data points.

This filter uses a moving window to replace each data point by the value of a fitted polynomial at the same position. Choosing a small window size and only 1 fitting step, enables small changes detection. (Figure 2)

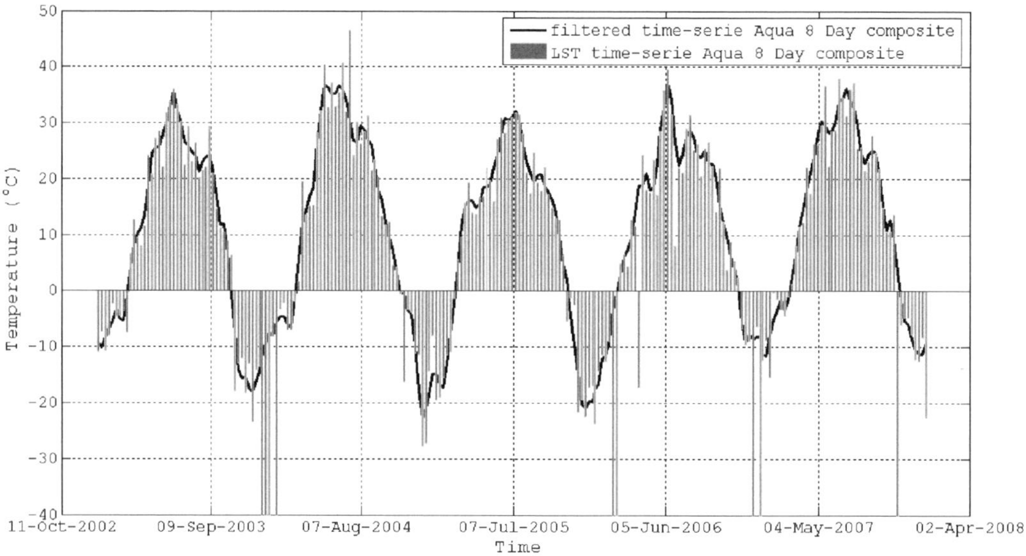


Figure 2. Savitzky-Golay filtered time-serie of an Aqua 8-day composite dataset. The original data values are shown for comparison

## 2.2 *Near ground temperature time-series*

During the 2006 summer, Dr. Sergei Marchenko installed several temperature data loggers in the Ulandryk River Valley for the round-year ground temperature monitoring at the different altitudes and landscape settings. (Marchenko, 2007). For this study we selected only "Site 1" located at 49.60692N and 89.06966E (2300m ASL) equipped with 2 HOBO U22 Water Temp Pro v2 data loggers, respectively installed at surface and at 0.7m depth. Temperature was recorded from 8 July 2006 until 17 July 2007 at a 2 hour interval. (Figure 3)

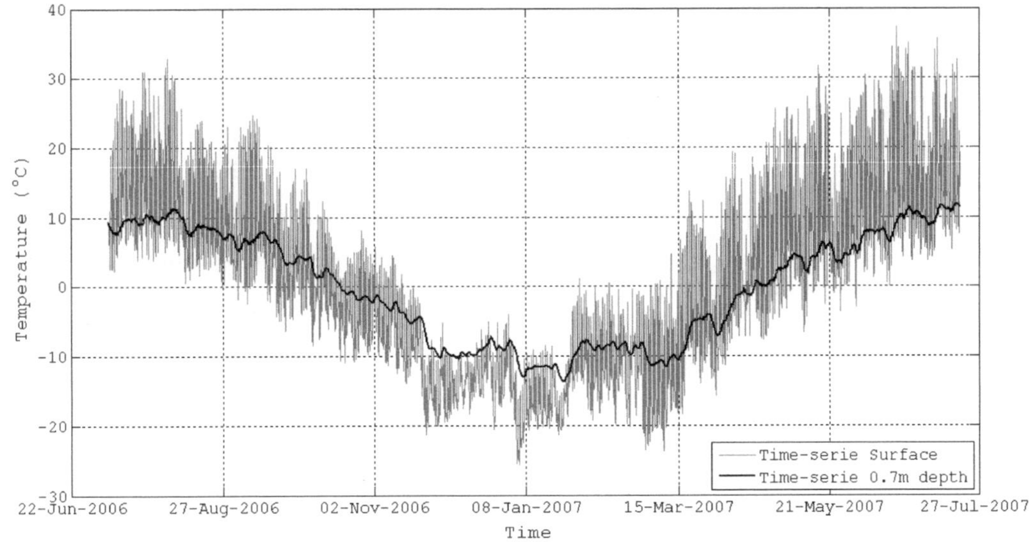


Figure 3. Year-round time-series of 'Site 1' recorded by 2 HOBO U22 Water Temp Pro V2. (Marchenko, 2008)

## 3 RESULTS

### 3.1 *CLEAN algorithm*

Although this method seemed very promising for interpolating missing values, and removing the noise out of the LST time-series, results showed the impotence of the CLEAN algorithm (Figure 4)

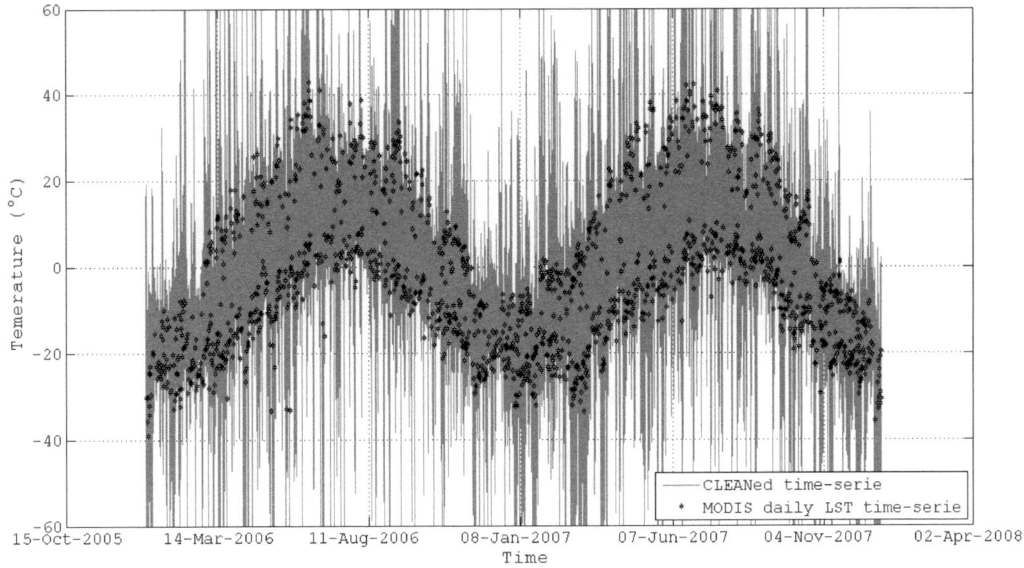


Figure 4. CLEAned MODIS daily LST time-serie. The original data points are showed for comparison

Due to lack of trend, questionable data quality and lots of missing values, no significant peaks are distinguished in the frequency spectrum. This results in a meaningless Fourier transform. Monte Carlo simulation shows a 0.55 significance of the highest peaks in the frequency spectrum. By only retaining those frequencies with these significances, results in the same meaningless time-serie.

## 3.2 *Savitzky-Golay method*

The best results with this algorithm were obtained by first filtering the 4 different time-series (Aqua Day, Aqua Night, Terra Day and Terra Night) separately by using the Timesat-program. Next, these filtered datasets were alternated conforming their time of recording. Results are showed in Figure 5

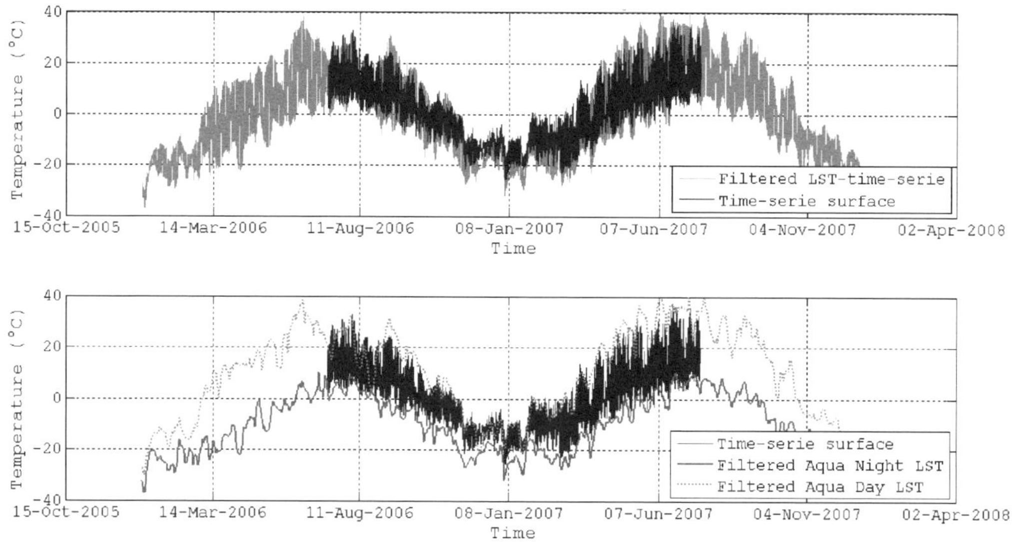


Figure 5: On top: Savitzky-Golay-filtered alternated daily MODIS LST time-serie compared with corresponding ground temperatures. Bottom: Separated filtered AQUA LST time-series compared with the corresponding ground temperatures.

Comparing these filtered-time-series with the dataset obtained by ground measurements indicate a good correlation. MODIS filtered daily LST values follow closely the small trends visible in the ground temperatures. However at some points (Figure 5, bottom), the filtered line crosses the surface temperature line, which might be caused by the significant amount of overcast days in these periods, and the resulting no data values.

# 4 CONCLUSIONS

A first test for comparing LST with near surface temperatures was carried out to test the potential of this method. As these LST values will be used as proxy for near surface temperatures in 2 different permafrost models, a good understanding of this relationship is essential. However, to implement these values in a statistical permafrost model on the one hand, and an adapted TTOP on the other, further research is needed which we will carry out next summer in the Altai Mountains.

Two methods for filtering MODIS LST time-series were tested. The CLEAN algorithm wasn't successful, due to missing values, missing trends, and questionable data. The Savitzky-Golay algorithm however, showed good potential for interpolating missing values, and taking the quality of the data into account. Although, this method had some difficulties regarding large periods with no data values, it might be used with some adaptations for further applications.

It is important to mention, that the use of these MODIS LST values will be limited to broad intermountainous valleys and quite homogenous areas, to restrict spectral mixing. However in further studies, we will compare ASTER data, to test correlation at a larger scale

## ACKNOWLEDGEMENTS

This study is funded by the "Institute for the Promotion of Innovation through Science and Technology in Flanders (IWT-Vlaanderen)". We thank Dr. Sergei Marchenko for the usage of his data and Dr. Stephan Gruber for the helpful discussion and invaluable suggestions.

## REFERENCES

Bourgeois, J. and Gheyle, W. 2007: The Frozen Tombs of the Altai Mountains, Phase 1: 2005-2006.

Brenning, A., Gruber, S. and Hoelzle, M. 2005: Sampling and Statistical Analyses of BTS Measurements. *Permafrost and Periglacial Processes* 16, 383-393.

Colombi, A., De Michele, C., Pepe, M. and Rampini, A. 2007: Estimation of daily mean air temperature from MODIS LST in alpine areas. *EARSeL eProceedings* 6, 38-45.

Fukui, K., Fujii, Y., Mikhailov, N., Ostanin, O. and Iwahana, G. 2007: The Lower Limit of Mountain Permafrost in the Russian Altai Mountains. *Permafrost and Periglacial Processes* 18, 129-136.

Heslop, D. and Dekkers, M.J. 2002: Spectral analysis of unevenly spaced climatic time series using CLEAN: signal recovery and derivation of signifance levels using a Monte Carlo simulation. *Physics of the Earth and Planetary Interiors* 130, 103-116.

Jönsson, P. and Eklundh, L. 2002: Seasonality extraction by function fitting to time-series of satellite sensor data. *IEEE transactions on geoscience and remote sensing* 40, 1824-1832.

Jönsson, P. and Eklundh, L. 2004: TIMESAT - a program for analysing time-series of satellite sensor data. *Computers and Geosciences* 30, 833-845.

Julian, A. and Chueca, J. 2007: Permafrost Distribution from BTS Measurements (Sierra de Telera, Central Pyrenees, Spain): Assessing the Importance of Solar Radiation in a Mid-elevation shaded Mountainous Area. *Permafrost and Periglacial Processes* 18, 137-149.

Juliussen, H. and Humlum, O. 2007: Towards a TTOP ground temperature model for mountainous terrain in Central-Eastern Norway. *Permafrost and Periglacial Processes* 18, 161-184.

Lunetta, R.S., Knight, J.F., Ediriwickrema, J., Lyon, J.G. and Worthy, L.D. 2006: Land-cover change detection using multi-temporal MODIS NDVI data. *Remote Sensing of Environment* 105, 142-154.

Marchenko, S. 2007: On Preliminary Permafrost Investigation in the Ulandryk Valley, Altai Mountains, within the framework of the UNESCO project "The Frozen Tombs of the Altai Mountains" in the summer of 2006. In Bourgeois, J. and Gheyle, W., editors, *UNESCO report: The Frozen Tombs of the Altai Mountains, Phase 1: 2005-2006*, 63-66.

Marchenko, S. 2008: Climate Change and its Impact on the Frozen Tombs of the Altai Mountains. In Han, J. and Tresilian, D., editors, *Preservation of the Frozen Tombs of the Altai Mountains,* Paris: UNESCO.

McCarthy, J.J., Canzani, O.F., Leary, N.A., Dokken, D.J. and White, K.S. 2001: *Climate Change 2001: Impacts, Adaptation, and Vulnerability Report of IPCC Working Group II*: Cambridge University Press.

Pridhoko, L. and Goward, N. 1997: Estimation of air temperature from remotely sensed surface observations. *Remote Sensing of Environment* 60, 335-346.

Roberts, D.H., Lehar, J. and Dreher, J.W. 1987: Time series analysis with CLEAN. I. Derivation of a spectrum. *The astronomical journal* 93, 968-989.
Smith, M.W. and Riseborough, D.W. 1996: Permafrost monitorring and detection of climate change. *Permafrost and Periglacial Processes* 7, 301-309.
Stocker-Mittaz, C., Hoelzle, M. and Haeberli, W. 2002: Modelling alpine permafrost distribution based on energy-balance data: a first step. *Permafrost and Periglacial Processes* 13, 271-282.

*Remote Sensing for a Changing Europe*
*D. Maktav (Ed.)*
*IOS Press, 2009*


*doi:10.3233/978-1-58603-986-8-321*

# First results of man-induced terrain changes in Karvina region

M. Hanzlova, P. Rapant
*Institute of geoinformatics, VŠB-Technical University of Ostrava, Ostrava, Czech Republic*

D. Böhmova, E. Jirankova
*Institut of geological ingeneering, VŠB-TU Ostrava, 17. listopadu 15, 70833 Ostrava-Poruba, Czech Republic*

L. Halounová, I. Hlaváčová, B. Knechtlová
*Department of Mapping and Cartography, Czech Technical University in Prague, Prague, Czech Republic*

P. Blaha, R. Duras, J. Fousek
*GEOtest Brno, a.s. (Ostrava), Ostrava, Czech Republic*

Keywords: man-induced effects, radar interferometry, land subsidence

ABSTRACT: Geological processes such as landslide and land subsidence cause extensive problems to humans. The sources of these processes are mainly effected by human activities, in our case mining activities. Since the eastern part of the Moravian-Silesian Region has been known mainly for the mining industry, the mining subsidence led inevitably to problems such as land subsidence and slope instabilities. Along with mining subsidence there were also numerous slope instabilities (landslides) identified within the Moravian-Silesian Region. These geological processes cause damage to buildings and other structures (e.g. communications), cause financial losses and jeopardize human lives. Areas where mining subsidence occurs are densely urbanised and industrialised locations, Silesian part of Ostrava, Karvina and its surrounding. Some locations have been under "ground surveillance" since 1993 with 6-month period. Nevertheless, there is need for remote sensing methods to support and to supplement field measurement. Comprehensive information such as these can be valuable and helpful in land use planning and other decision support systems.

Two areas were chosen as test areas – 1] observed subsidence area in Holkovice (Stonava, Zolkov, subsidence reached 2 m in 6 years), where mining activities started in 1996 and finished in 2002 (new mining activities started on the north of the area in 2002), 2] observed subsidence effect (vertical movement spans from +51 mm to –187 mm in 10 months) on communication in Orlova and Doubrava where measurement has begun together with the mining activity progress in 2002.

Used methodology is based on DInSAR (Differential Interferometric SAR). The terrain change time line that is subjected to be investigated follows processing of 1 upto 10 radar images per year due to image availability (time extend spans from 1996 to 2007). First data which were needed to analyze cover year 1999 and year 2007 because of field measurements organized in the areas at the time.

## 1 LAND SUBSIDENCE SITUATION IN THE INVESTIGATED REGION

The subsidence areas within Moravian-Silesian Region are mostly densely urbanized and industrialized locations (See Fig. 1) – Silesian part of Ostrava and Karvina surrounding. Mentioned areas are affected by long walling with caving mining methods, thus stresses increase around the work-out space. Periodical ground measurement has been carried out in particular areas since 1993. In some areas the subsidence reached 2 m in 6 years, there are also observed events such as vertical movements within interval of +51 mm to –187 mm in 10 months.

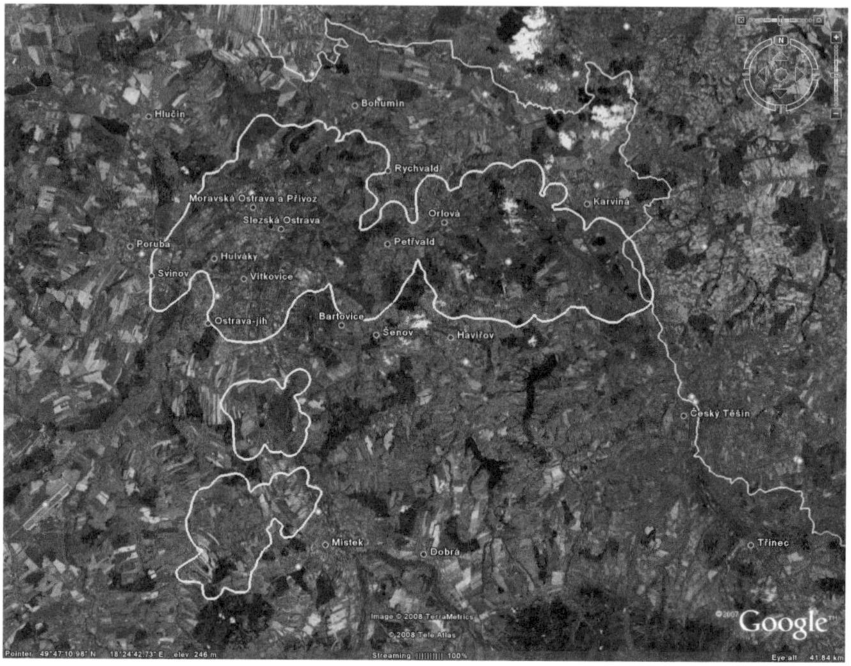


Figure 1. Areas where mining subsidence occurs (turquoise colour).

## 2 USED METHODS

For the first results the DInSAR (Differential InSAR) method was used. This method uses SLC (Single-Look Complex) radar data to detect vertical topographic changes over large areas. SLC data consist from real and imaginary values (amplitude and phase). InSAR uses the phase difference between two radar data sets.

Nowadays there are many studies which have been carried out using SAR interferometry for applications such as subsidence caused by mining activities. The geographically and thematically closest study to our investigation is situation in Upper Silesia. Repeated ERS-1 and ERS-2 SAR observations of selected test sites in the Upper Silesia (Katowice area) have been used to process interferograms which present a pattern of elliptical fringes – the indicators of the phase shift of radar signals caused by the mining subsidence (Perski, Jura 1999 in Perski, 2000).

DInSAR method uses repeat pass interferometry that is used for deformation detection, in our case we used three-pass interferometry which resulted in deformation influeced by atmospheric noise. Two-pass interferometry is beening carried out at this moment using SRTM data as DEM for topography elimination from the interferogram (surface changes – subsidence).

As for geodetic measurement we collected several geodetic datasets from an area that covers monitored subsidence in Holkovice (Stonava, Zolkov). LIDAR data for the same area were obtained in 2007.

## 3 FINDINGS

### 3.1 *Data selection*

ERS data capability for interferometry can be reduced by unknown orbital parameters, weather conditions and land cover such as vegetation (thus seasonal effects can influence the results).

| Date | Orbit | Baseline [m] | ΔDays | f_DC [Hz] |
|---|---|---|---|---|
| 4-Jan-1999 | 19386 | -1368 | -140 | 208,256 |
| 8-Feb-1999 | 19887 | 18 | -105 | 304,106 |
| 15-Mar-1999 | 20388 | -295 | -70 | 270,675 |
| 19-Apr-1999 | 20889 | 257 | -35 | 219,108 |
| 24-May-1999 | 21390 | 0 | 0 | 262,424 |
| 28-Jun-1999 | 21891 | 144 | 35 | 265,842 |
| 2-Aug-1999 | 22392 | 20 | 70 | 221,836 |
| 6-Sep-1999 | 22893 | -736 | 105 | 181,316 |
| 11-Oct-1999 | 23394 | -702 | 140 | 198,932 |
| 15-Nov-1999 | 23895 | 61 | 175 | 194,567 |
| 20-Dec-1999 | 24396 | -342 | 210 | 248,163 |

Figure 2. ERS-2 datasets overview (year 1999).

| f_DC [Hz] | 208,256 | 304,106 | 270,675 | 219,108 | 262,424 | 265,842 | 221,836 | 181,316 | 198,932 | 194,567 | 248,163 |
|---|---|---|---|---|---|---|---|---|---|---|---|
| 208,256 | 0 | 95,85 | 62,419 | 10,852 | 54,168 | 57,586 | 13,58 | -26,94 | -9,324 | -13,689 | 39,907 |
| 304,106 | -95,85 | 0 | -33,431 | -84,998 | -41,682 | -38,264 | -82,27 | -122,79 | -105,174 | -109,539 | -55,943 |
| 270,675 | -62,419 | 33,431 | 0 | -51,567 | -8,251 | -4,833 | -48,839 | -89,359 | -71,743 | -76,108 | -22,512 |
| 219,108 | -10,852 | 84,998 | 51,567 | 0 | 43,316 | 46,734 | 2,728 | -37,792 | -20,176 | -24,541 | 29,055 |
| 262,424 | -54,168 | 41,682 | 8,251 | -43,316 | 0 | 3,418 | -40,588 | -81,108 | -63,492 | -67,857 | -14,261 |
| 265,842 | -57,586 | 38,264 | 4,833 | -46,734 | -3,418 | 0 | -44,006 | -84,526 | -66,91 | -71,275 | -17,679 |
| 221,836 | -13,58 | 82,27 | 48,839 | -2,728 | 40,588 | 44,006 | 0 | -40,52 | -22,904 | -27,269 | 26,327 |
| 181,316 | 26,94 | 122,79 | 89,359 | 37,792 | 81,108 | 84,526 | 40,52 | 0 | 17,616 | 13,251 | 66,847 |
| 198,932 | 9,324 | 105,174 | 71,743 | 20,176 | 63,492 | 66,91 | 22,904 | -17,616 | 0 | -4,365 | 49,231 |
| 194,567 | 13,689 | 109,539 | 76,108 | 24,541 | 67,857 | 71,275 | 27,269 | -13,251 | 4,365 | 0 | 53,596 |
| 248,163 | -39,907 | 55,943 | 22,512 | -29,055 | 14,261 | 17,679 | -26,327 | -66,847 | -49,231 | -53,596 | 0 |

Figure 3. Doppler centroid differences (1999 datasets).

Our datasets spans from year 1995 up to 2008 (ERS-2 and ERS-1 SLC data). Knowing that after the year 2001 there has been problems with gyroscope onboard ESR-2, were selected data sets separately for year 1999 and for years 2005–2008 (searching for some available pair within the data set stack).

There are 11 datasets available for year 1999 (track 494, frame 2601). According to baseline analyses, the ERS-2 datasets were selected as follows:

- 24 MAY 1999, orbit 21390, track 494, frame 2601
- 28 JUNE 1999, orbit 21390, track 494, frame 2601
- 2 AUGUST 1999, orbit 21390, track 494, frame 2601

As a topographical reference data from May and June 1999 (topopair) were selected. Deformation is dated to August 1999 (70 days).

Datasets for year 2005 up to 2008 were selected as interferometric deformation pairs based on Doppler centroid difference and perpendicular baseline value. Figure 5 shows selected interferometric pairs. Pairs were selected among 30 datatsets, which pair combination resulted into 69 possible pairs according to Doppler centroid difference. Offset vectors calculations ended up with 7 pairs that are possible to coregistrate.

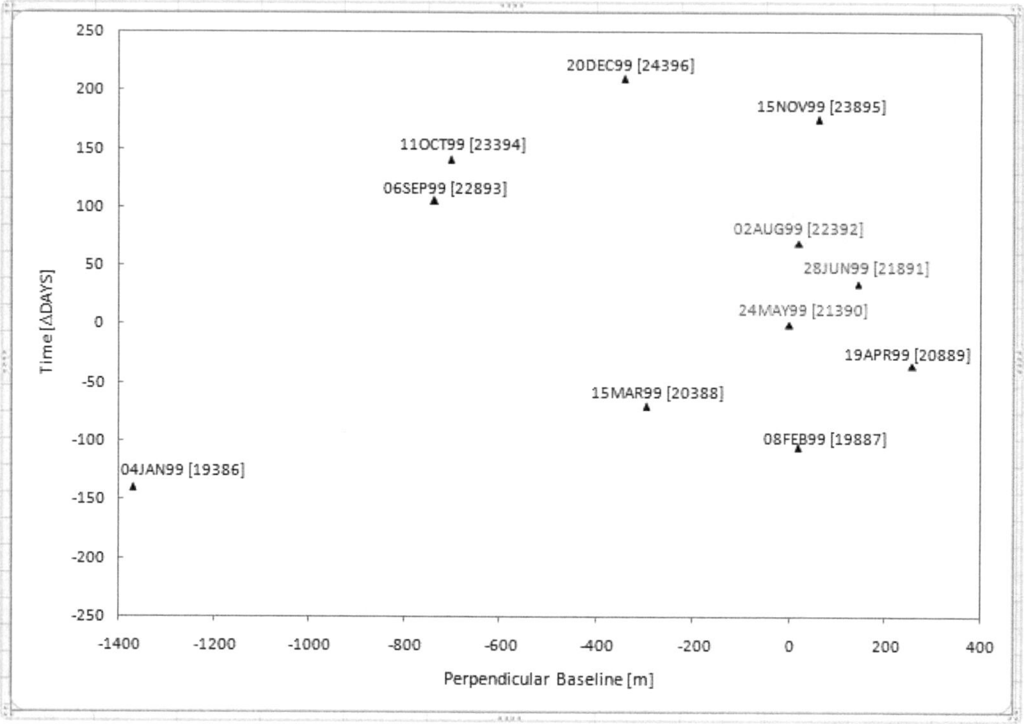


Figure 4. Perpendicular baseline analysis, MAY was used as reference dataset.

| | Pairs | | Δf_DC | ⊥baseline | Δdays |
|---|---|---|---|---|---|
| BC | 28MAR05 | 02MAY05 | 34,46 | 492 | **35** |
| OY | 31JUL06 | 20AUG07 | 230,497 | 226 | **385** |
| HZ4 | 28NOV05 | 03DEC07 | 130,25 | no info [DESCW] | |
| DQ | 06JUN05 | 09OCT06 | -73,244 | 216 | **490** |
| EH | 11JUL05 | 28NOV05 | 138,063 | -751 | **140** |
| FK | 15AUG05 | 13MAR06 | 371,65 | -427 | **210** |
| RZ2 | 13NOV06 | 24SEP07 | 260,33 | no info [DESCW] | |

Figure 5. Selected pairs for dataset stack (years 2005–2008).

## 3.2 *First results*

First results (three-pass interferometry) showed major influence of atmosphere around Ostrava and Karvina cities area (see Fig. 6). The atmospheric affect is cause probably by used topographic pair and will be eliminated by using two-pass interferometry method (topography elimination by SRTM DEM data).

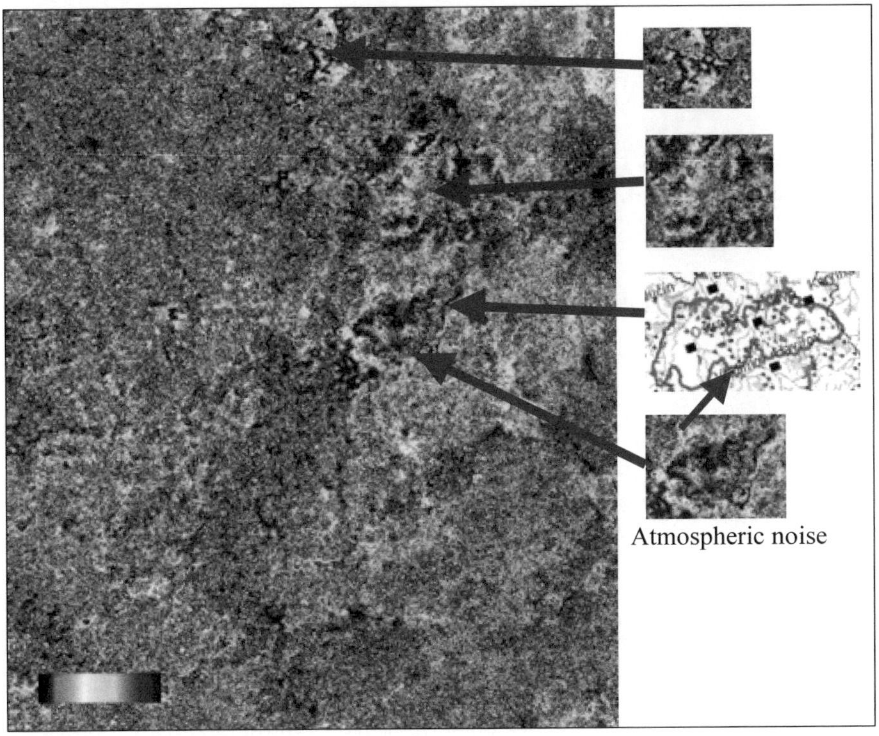


Figure 6.

## 4 CONCLUSIONS

Even first results proves occurrence of ground displacements in the areas where subsidence is monitored. Further processed will emphasize on clearing interferograms from topography and their subsequent interpretation. Terrain changes extracted from interferograms will be compared with geodetic measurement datasets.


## ACKNOWLEDGEMENTS

The investigation is done within project "Radar Interferometry within geological risks assessment" (client: Ministry of Environment, Czech Republic) and EOPI/ESA/ESRIN project "Surveillace of Man-induced Terrain Changes Using EO Data" (C1P.4578).

Outcomes contribute to project FLOREON (FLOods REcognition On the Net) that is focused on DSS development with aim to reduce severe influences of natural phenomena and effects of man-induced activities.


## REFERENCES


Perski Z., Jura D., 1999. ERS SAR Interferometry for Land Subsidence Detection in Coal Mining Areas. Earth Observation Quarterly No. 63, pp. 25–29. in Perski, Z. (2000) The Interpretation of ERS-1 and ERS-2 InSAR Data For The Mining Subsidence Monitoring In Upper Silesian Coal Basin, poland. IAPRS Vol. XXXIII,Amsterdam, 2000, 5p.

*Remote Sensing for a Changing Europe*
*D. Maktav (Ed.)*
*IOS Press, 2009*

*doi:10.3233/978-1-58603-986-8-326*

# The causes of the floods on the streams flowing to Bosphorus and the precautions needed: Kozdere case study

H. E. Celik, A. Aydin, I. Esin
*Istanbul University Faculty of Forestry,Istanbul,Turkey*

H. G. Coskun, U. Alganci, G. Usta
*Istanbul Technical University Remote Sensing Department, Istanbul, Turkey*

H. K. Cigizoglu
*Istanbul Technical University,Hydraulics Department, Istanbul, Turkey*

Keywords: : flood control, remote sensing, satellite images, digital elevation model, classification, land use, time variant analysis

ABSTRACT: Istanbul faced several floods in different districts after the heavy rainfall dated 11 August 2004 which affected the whole Marmara Region. These floods caused 3 deaths and huge economical losses. One of these floods occurred on the Kozdere River flowing into the middle Bosporus in Istanbul. The aid amount to suffering losses made by different governmental and non governmental bodies is around 350.000 $. Climatic and physiographic factors are the main factors of floods. Rainfall type, rainfall intensity etc. are some of the climatic factors. Basin shape, elevation information etc. and land use changes are some of the physiographic factors. According to land use classification, lands classified as forest area must not be used as settlement or agriculture. This kind of miss land uses encourages flood formation. In this connection, it should be investigated whether forest area is used as settlement or agriculture or are there any land use changes against forest area.

In this study, in the context of the reasons of the Kozdere flood, land use and land use changes are investigated. Detailed elevation of the basin is extracted from the Digital Elevation Model (DEM) that is produced from 1/5000 scaled orthophotos. Remotely sensed data is used for the determination of land use characteristics and time variant land use analysis. 1992 dated Landsat 5 TM - SPOT PAN and 2005 dated Quickbird MS - PAN Datasets are geometrically corrected and subjected to merge operation. These datasets are then classified using maximum likelihood supervised classification algorithm.

## 1 INTRODUCTION

Floods occurred on several locations of Istanbul on 11 August 2004 following the intense rainfall event. Floods caused 3 human losses and huge damage. One of the floods was on the Kozdere basin in Sariyer region. The base floors of PTT houses whith 5000 residents were flooded and approximately 300 apartments were damaged severely. The official aid provided to 200 families by the official institutions is around 500.000 YTL (350.000 $).

The rainfall event which resulted in the flood started at 00:30 on the night of 11 august 2004. The meteorology station in Bahçeköy reported 91.6mm rainfall between the midnight and 7 am and 76.4mm rainfall till 14:00 afterwards. So the total rainfall was 168mm and was interrupted afterwards.

The flood term covers the huge water quantity which containing intense sediment (both suspended load and bed load) and coming suddenly from the tributaries of the river. The floods coming from the tributaries reach the main river and the water level increases through the valley spreading on the flood zone and the neighboring area. The increase on the flow is due to the climatic and physiographic factors. The climatic factors are the precipitation characteristics, the interception, the evaporation and transpiration whereas the physiographic factors represent slope, elevation, soil type, drainage, land use and bed characteristics (Gorecelioglu, 2003). The human does not have influence on the majority of the climatic and physiographic factors. But the human can destroy the forest on the basin for agriculture, settlement or industry changing the land use on the basin.

The flow measurements show that a productive forest can keep 50-80% of the rainfall on the upstream. Some of the rainfall returns to the atmosphere and some is kept by the soil. The humus layer on the soil increases the seepage within the soil. In our country it is seen that the flood occurrence on some basins is prevented when the flood and erosion control measures are taken and when the basin is implented with forest.

On Europe or regional scale the forestry (due to the excessive precipitation) has a relative effect on the high flows (Robinson et al., 2003). When the rainfall is intense and long, the water storage capacity of the soil is exceeded and the rainfall is transformed to surface flow and high flow. The high flow has tendency to overflow the normal bed cross-section and spread on the flood zone. If the flood zone or even the normal river bed is narrowed due to the diferent effects then floods can occur. On the urban areas the upstream zone is also covered with settlement frequently causing the rainfall to be transformed to surface flow directly increasing the flow and floods occur even after low precipitation events.

In this study, the causes for Kozdere flood will be investigated using remote sensing technique. The change on the land use and land cover will be interpreted using two satellite images from two different times.

## 2 STUDY AREA

The investigated area is in the Sarıyer Region of Istanbul between northern latitudes 41°08′ 30″-41°11′ 30″ and eastern longitudes 28° 59′30″ – 29° 02′ 00″. It is the drainage area of Kozdere River which flows to Bosporus (Fig. 1). The yearly mean precipitation measured in Bahçeköy meteorology station is 850.55mm. The highest mean monthly precipitation is measured in December (129.2 mm), November (108,3 mm) and January (101,1 mm). The precipitation does not have a homogeneous distribution within the year. 65.7% of the precipitation falls in autumn and winter months. The yearly mean temperature is 12.8°C. The yearly mean relative humidity is 79%. The highest relative humidity is in December and January whereas the lowest values are observed in April and June.

According to the Thorntwaite method it is determined that Kozdere basin has a climate which is semi humid, mesothermal and water deficient in summer season. The plant cover on the basin is 95% leafy and 5% with needle. 90% of the forests is productive and the remaining 10% not. The ratios of the closed forest area are 71-100% for 59%, 41-70% for 20%, 11-40% for 12%, 11-40% for 9% (destroyed) ((Anonymous 2002, Anonymous 2003a, Anonymous 2003b, Anonymous 2007). As provided by the management plan, Kozdere Basin consists of forest (71.7%), open field (0.8%), settlement (26,8%), and nursery (0,7%) (Figure 2). According to the Istanbul Ability Class Map (KHGM 1987); 1.3%, 77.6%, and 21.1% of Kozdere basin consists of IV, VII and settlement type class fields, respectively (Fig. 2).

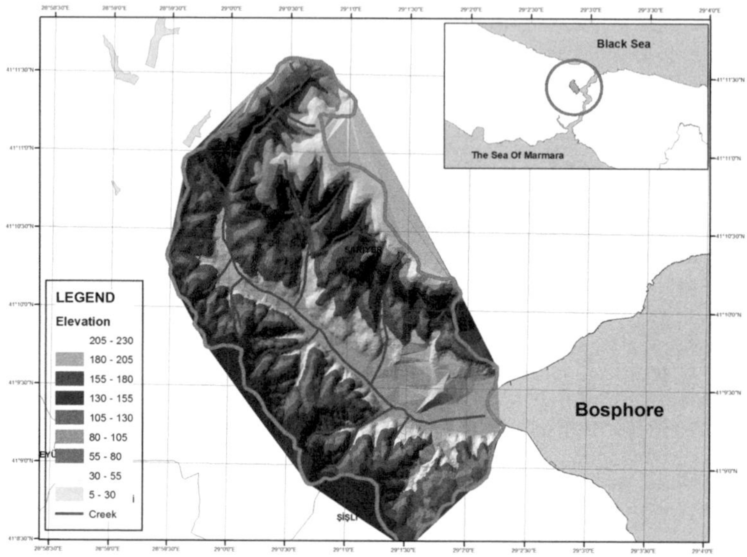


*Figure 1: Drainage Basin of Kozdere River.*

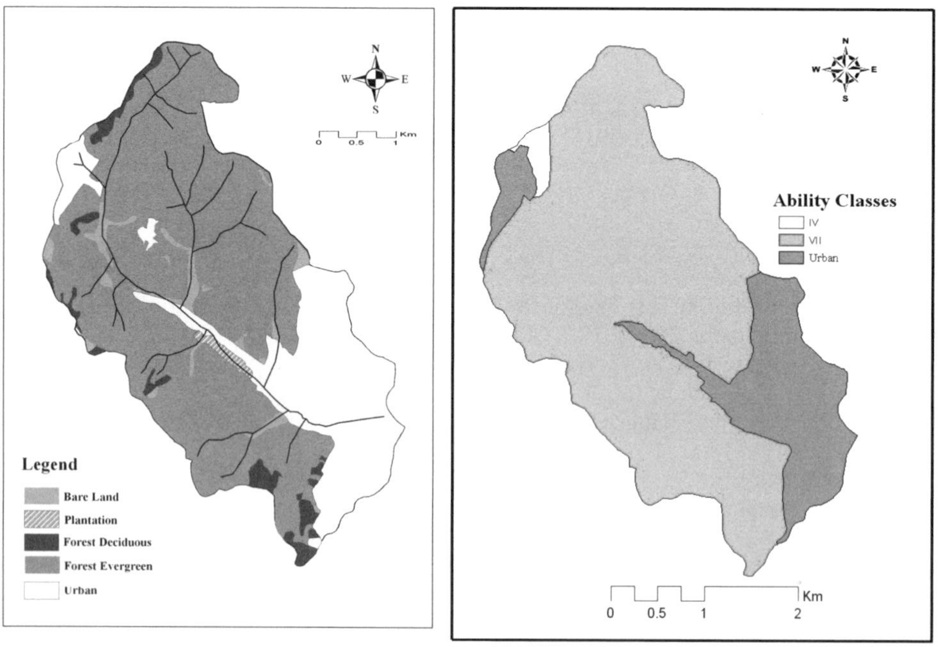


(a) (b)

Figure 2. Management plan (a) and Ability Class Map (b) of Kozdere basin.

## 3 METHODOLGY

### 3.1 *Image Processing*

In this study, multispectral digital satellite data set of Landsat 5 TM dated 12 July 1992, SPOT-PAN dated 13 June 1993, and Quickbird MS – PAN dated 18 May 2005 were used for the Kozdere Basin land-use and land cover investigation. The spatial resolutions for Landsat 5 TM is 30 m, for SPOT-PAN is 10 m, for Qicckbird MS and PAN are 2.8 m and 0.6 m.

First, the digital satellite data set was transformed into the UTM coordinate system using the 1: 5000 digital topographic maps in order to achieve the necessary geometric registration. Taking 50 ground control points from the maps, the images are registered and then geometrically corrected before applying image merging and classification. From a test of the registration accuracy on the test points the resulting root mean error (RMSE) amounted to ±0.1 pixels (Figure 2). A coordinate transformation was applied using a polynomial transformation equation and resampling is done using the cubic convolution algorithm. The HIS transformation method was used to obtain merged, multi-sensor images. This method includes color transformations that result in a new image product, in which distinct spectral patterns are revealed. High-resolution PAN digital satellite data is combined with multi-spectral data (Coskun et al., 2006).

### 3.2 *Land use- land cover classification*

The rectified and merged images were subjected to a classification procedure with five classes, using the supervised Maximum Likelihood Classification algorithm. These five classes were verified using pattern recognition, referring to the ground truth data and fieldwork. The classification results were evaluated in an accuracy assessment. In this accuracy assessment, for each classification up to two different years, 100 random pixels were chosen and these pixels were compared with the results of the fieldwork. Total accuracy results are 80.00% and 91.00% for the years 1992 and 2005, respectively. Figure 3 shows the result images of the visual differences of the urbanized area between years of 1992 and 2005 in Kozdere Basin.

The classification results were evaluated as area of land use / land cover in hectares and differences according to years are shown in Table 1. Figure 3 illustrates the graphical presentation of urban analyses.

Land use and cover change associated with human activities may change the hydrological processes and increase flood risks. The results of land use and land cover analysis shows that urbanization increased and this increment shows a characteristic of being dense nearby the bed of Kozdere stream. Also, roads are increased in the basin according to the urban increment.

Table 1: Classification results of images dated 2005 and 1992.

| Class name | 2005 (Area) | % | 1992 (Area) | % |
|---|---|---|---|---|
| Deciduous | 969.84 | 80.60 | 934.41 | 77.65 |
| Coniferous | 64.15 | 5.33 | 47.60 | 3.96 |
| Open Land | 62.15 | 5.17 | 167.23 | 13.90 |
| Road | 28.91 | 2.40 | 10.00 | 0.83 |
| Settlement | 78.00 | 6.49 | 44.11 | 3.67 |
| | 1203.35 | 100.00 | 1203.35 | 100.00 |

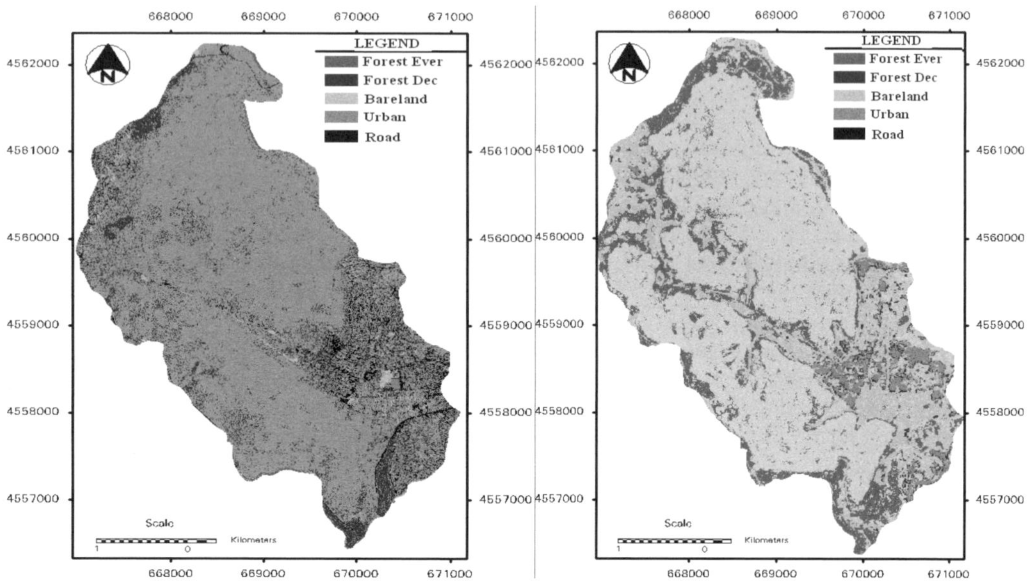


(a) (b)

Figure 3. Land use in Kozdere basin according to satellite images. a) Quickbird dated 2005; b) Landsat-Spot dated 1992

## 3.3 *Determination of physiographic factors*

Physiographic factors are one of the two main factors that affects flood. A Digital Elevation Model (DEM) is produced from 1:5000 scaled digital topographic maps in order to determine the physiographic parameters such as boundaries, area, elevation information and drainage of the Kozdere Basin (Fig. 4). This procedure involves filling the sinks of DEM, flow accumulation and flow direction map creation, synthetic drainage network determination and watershed delineation.

When the DEM is analyzed together with the high resolution Quickbird image, it can be seen that a considerable amount of the settlements in the south part of the basin are located in the accumulated sediment zone.

Figure 4. A 3D looking to the bed of Kozdere basin which is created by image drape of Quickbird satellite image and DEM

## 4 DISCUSSION AND CONCLUSION

The present land use distribution is obtained with the help of the forest management plan and classified satellite images. The correctness of the classification according to the satellite images is determined as 80% for year 1992 and 91% for year 2005.

The classification results of the Quickbird sattelite images with spatial resolution 0.6m dated 2005 demonstrate that Kozdere Basin is covered by the forest (85.9%), the open land (5.1%), and the settlement (6.4%). According to the Landsat-Spot image with 10 m spatial resolution in 1992, however, the basin is covered by the forest (81.6%), the open land (13.9%), and the settlement (3.6%) (Figure 3, Table 1). In contrast, the management plan shows that Kozdere basin has a land distribution consisting of 71.7% forest, 0.8% open land, 26.8% settlement, 0.7%.

The higher forest land area value on the sattelite images compared to the management plan is mainly due to the inclusion of the wooded areas outside the forest boundaries. On the sattelite image, also the wooded areas in the settlement zone are classified as forest. The forest part on the basin is nearly 3 times of the mean forest area (27%) on Turkey and 7 times of the productive forest area (13%) on Turkey. As a conclusion we can state that the occurred flood can not be attributed to the insufficient forest area on the basin.

The land ability classes and the present land use are compared with GIS and it is seen that nearly the entire forest area (VII class) of the basin is kept as forest (Figure 3). 90% of the forest is productive and the remaining 10% not. Since the basin is covered with productive and quite intense forest land it can be concluded that there is no negative effect on the upper basin due to mis-land use. The occurrence of the flood, on the other hand, can be explained with the hydraulic conditions on the river cros-sections which are quite close to the present settlement location. The settlement area is so close to the river cross-section preventing the use of the neighbouring are as flood protection zone.This part of the settlement area also prevents the enlargement of the critical river cross-sections. Since the present river cross-section can not transport the flood discharge the water level increases dramatically flooding the neighbouring settlement zone causing human losses and damage.

## REFERENCES

Anonymous 2007: İÜ Orman Fakültesi Eğitim ve Araştırma Ormanı Amenajman Planı, İstanbul.

Anonymous 2002: Sarıyer Orman İşletme Şefliği Tekel Ezba Özel Ormanı Amenajman Planı, İstanbul.

Anonymous 2003b: Sarıyer Orman İşletme Şefliği Amenajman Planı, OGM, İstanbul.

Anonymous 2003a: Bentler Orman İşletme Şefliği Amenajman Planı, OGM, İstanbul.

Coskun H. G. Et al, 2006. Monitoring of Protected Bands of Terkos Drinking Water Reservoir of Metropolitan Istanbul Near the Black Sea Coast Using Sattelite Data,International Journal of Applied Earth Observation and Geoinformation 8 (2006) 49-60.

GÖRCELİOĞLU, E. 2003: Torrent and Avalanche Control, Istanbul University Faculty of Forestry Publication Number 473, İstanbul, Turkey.

KHGM 1987: Land Capability of Istanbul Province, General Directorate of Rural Services, Report number 09, Ankara, Turkey.

ROBINSON, M., A.-L. COGNARD-PLANCQ, C. COSANDEY, J. DAVID, P. DURAND, H.-W. FUHRER, R. HALL, M.O. HENDRIQUES, V. MARC, R. McCARTHY, M. McDONNELL, C. MARTIN, T. NISBET, P. O'DEA, M. RODGERS, A. ZOLLNER, 2003: Studies of The Impact of Forests on Peak Flows and Baseflows: A European Perspective, Journal of Forest Ecology and Management, 186.

*Remote Sensing for a Changing Europe*
*D. Maktav (Ed.)*
*IOS Press, 2009*

*doi:10.3233/978-1-58603-986-8-332*

# Drought monitoring with the integrated utilization of satellite images in Hungary

Zs. Suba, G. Nádor, G. Csornai, I. László & Cs. Wirnhardt
*Institute of Geodesy, Cartography and Remote Sensing (FÖMI), Budapest, Hungary*

Keywords: drought monitoring, crop development assessment, remote sensing, integrated usage of satellite images, ENVISAT MERIS

ABSTRACT: The Hungarian Agricultural Remote Sensing Program (HARSP) led to the operational remote sensing based National Crop Monitoring and Production Forecast Program (CROPMON, 1997-2003) with the support of the Ministry of Agriculture and Regional Development. The Remote Sensing Center of Institute of Geodesy, Cartography and Remote Sensing, (FÖMI RSC) provided nation-wide and local crop production forecasts exclusively on remote sensing basis. The technology basis of CROPMON served as the background of operational remote sensing based drought monitoring and impact assessment.

The ESA–FÖMI Prodex-ENVISAT R&D project (2000-2004) aimed at the rapid regional monitoring of flood, waterlog and drought affected areas. The achieved results demonstrated and confirmed that a remote sensing based rapid, regional disaster monitoring model could be developed and applied effectively with the integration of available medium and low resolution optical satellite data together with the involvement of new generation ENVISAT sensors. As the follow-up, the ESA-FÖMI PECS-ENVISAT project (2004-2007) aimed at the improvement of the previous R+D activities. The tasks focus on the utilization of ENVISAT sensors by assessing a longer monitoring period and a larger ENVISAT dataset.

With the climatic change it can be expected that drought will occur more and more frequently in the future. For the effective prevention and the compensation of loss caused by drought, the remote sensing based characterization of crop development is very important on parcel, settlement and regional level. The paper summarizes the drought monitoring activities carried out by FÖMI RSC since 2000.

## 1 INTRODUCTION

More than 300 man-years R&D investment during the Hungarian Agricultural Remote Sensing Project (HARSP 1980-) led to the operational satellite image based National Crop Monitoring and Production Forecast Program (CROPMON 1997-2003) at the Remote Sensing Centre of the Institute of Geodesy, Cartography and Remote Sensing (FÖMI RSC); see (Csornai 1999). The operational CROPMON program provided county and country level crop production forecast to the Ministry of Agriculture and Rural Development (MARD) by a strict calendar through the growing period. This data provision was based on remote sensing, measuring the areas and expected yields of the eight main crops.

The methodology of CROPMON was later applied from drought monitoring. Its main component was the processing and the comparative analysis of low- and medium resolution operational satellite data (NOAA AVHRR, SPOT VEGETATION, IRS-1C/1D WiFS) using large data set

(1998-2003) and a regional test area. Further extension of the already developed methodologies was proposed by increasing the availability and the quality (temporal, spatial or spectral) of satellite data.

Making use of the ESA Scientific Experiment Development Programme (Prodex), the ESA–FÖMI Prodex-ENVISAT R&D project was going on between 2000 and 2004. The main objectives of the project were to extend the already developed methodologies to the rapid, regional monitoring of flood/waterlog and drought affected areas using currently available low- and medium resolution operational satellite data with the involvement the of data acquired by the new ENVISAT satellite (MERIS, ASAR). The results were presented in (Suba *et al.* 2004), (Csornai *et al.* 2003), (Csornai *et al.* 2005).

As the follow-up, the ESA-FÖMI PECS-ENVISAT project (Programme for European Co-operating States; 2004-2007) aimed at the improvement of the previous R&D activities. The tasks focus on the utilization of ENVISAT sensors by assessing a longer monitoring period and a larger ENVISAT dataset. The results of the first two years (of phases) of the project were presented in (Csornai *et al.* 2007). This paper summarizes the results of these projects achieved between 2000 and 2007 in the frame of these two projects, focusing on the latter one.

## 2 DROUGHT MONITORING AT REGIONAL AND COUNTRY LEVEL IN THE FRAMEWORK OF PRODEX PROJECT (2000-2004)

Drought maps have been produced for the whole area of Hungary from NOAA AVHRR, IRS WIFS and SPOT VEGETATION satellite images for the years 2000 and 2003, when severe drought hit Hungary. To assess the severity of drought, the NDVI (Normalized Difference Vegetation Index) of the year in subject has been compared to the mean NDVI value of several years. These yearly values were calculated using the maximum NDVI values of several images over the period to be examined (usually a decade, i.e. 10 days), instead of individual images. The areas affected by drought can be clearly identified on the eastern part of the country in 2000 from the NDVI difference maps derived from SPOT VEGETATION and IRS WIFS data (see Fig. 1).

To provide the stability and the continuity of drought monitoring over a several years' period, images from different satellite sensors can be used. However, to use them in a coherent model, the quantities used in the estimation –in our case, vegetation indices– have to be calibrated to each other. Drought maps from different data sources have been compared and their relationships have been examined. In this way drought mapping can be done by using one satellite data source for the current year's monitoring and comparing that result to the previous years' data from other satellites.

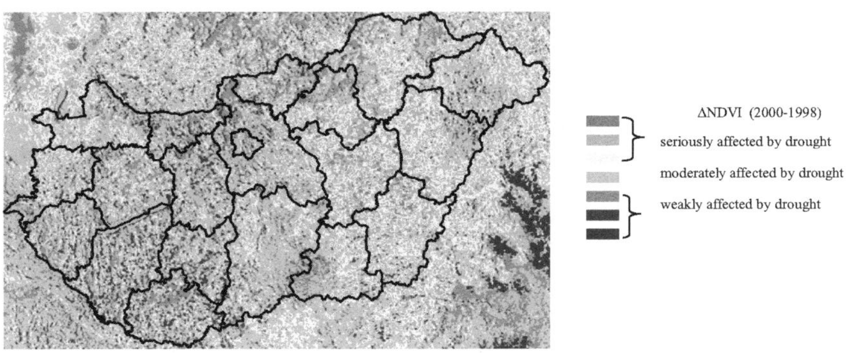


Figure 1. NDVI difference (June 2000 – June 1998) map derived from SPOT VEGETATION images, showing the drought affected areas in the Great Hungarian Plain.

## 3 DROUGHT MONITORING AT COUNTRY AND REGIONAL LEVEL IN THE FRAME OF PECS PROJECT (2004-2007)

The results of Prodex project were very promising in applying MERIS data for regional drought and crop development monitoring, despite the short processing time (first quarter of 2004) and available ENVISAT data (only for the May-June 2003 monitoring period). The monitoring activities started in Prodex have been continued in the first phase of PECS. In 2003, severe drought hit Hungary. Its monitoring was carried out with several methods in the first and second phase of PECS project, using different image data sets (IRS WiFS, ENVISAT MERIS) and reference periods (1997-1999, 2003-2006). From 2004 to 2006, no large area drought hit the country. The crop development in 2004 has been monitored with several methods, similarly to the drought monitoring in 2003.

The model applied in the first phase of PECS project was based on the integrated utilization of ENVISAT MERIS and IRS WiFS data using multitemporal and multiannual satellite data set (MERIS 2003-2004, WiFS 1997-1999, 2003-2004) and an intercalibration approach between WiFS and MERIS. We have generated country-wide MERIS-based drought maps for the May-June period of 2003 (later extended until August). As no large area disaster (flood/waterlog or drought) hit the country between 2004 and 2005, the first phase of PECS project activities covered mainly the methodological testing of ENVISAT MERIS data for retrospective drought monitoring. More emphasis was put on the improvement of processing and evaluation of available ENVISAT MERIS data set (2003-2004).

Regarding the year 2003 drought, the focus of its monitoring was put on three time intervals: the middle of May, the beginning of June and the middle of August. Improved calibration procedure was performed between MERIS and WiFS images using an extended data set. This means, pairs of data values belonging to the same location were taken from the images of different years, correspondence – an appropriate analytical function – was sought between the value pairs. The function describing the correspondence was used to transform further images (meaning also derived quantities, for example vegetation indices) of one kind to simulate another kind of image (Fig. 2).

Maps showing the area distribution of drought were derived both from MERIS MGVI (Gobron *et al.* 1999) and from WiFS NDVI. The MGVI values were transformed to "simulated" NDVI values. For the creation of drought maps, the current (year 2003) NDVI (actual or simulated) values were compared to the average of appropriately selected reference years (1997-1999)[1]. In order to minimize loss of information caused by cloudiness, the pixelwise maximum of several (2-3) images belonging to the same time period were taken into account. To measure the severity of drought, the difference between the vegetation index of the current year and the average vegetation index of the reference years was calculated. The results confirmed the twofold nature (intensive hit in spring and summer) of the drought event of 2003 in Hungary. Detailed quantitative analysis of the obtained drought maps was conducted together with the comparison of different years (2003-2004) based on MERIS data. The development of drought is clearly identified: it hit the whole country in August 2003, but more severely the south-eastern part of the country, while crop development was much more favourable in 2004 than in 2003 (Fig. 3 a-b).

In the second phase of PECS project ENVISAT MERIS time series of years 2003 to 2006 were also available. Therefore the average crop development of these years could be assessed and it could be used to develop drought and crop development maps solely on the basis of MERIS images (Fig. 4). This means that the calibration was not necessary, the newly derived maps were based exclusively on the MGVI values, appropriately reflecting the crop development. Years 2003-2006 could be used as reference period, the positive or negative difference of current year data from data in this reference period was used to generate the maps. Besides, regarding years 2003 and 2004, comparative analysis has been carried out between the drought and crop development maps generated in the first phase (calculated from IRS WiFS or calibrated ENVISAT MERIS images) and the

[1] Data from 2000 were excluded from WiFS data set in this new procedure in contrast with the previous one followed in Prodex because of the very extreme climate (intensive drought) of that year.

ones created in the second phase, based exclusively on ENVISAT MERIS data. To assess the adequacy of derived drought and crop development maps, remote sensing-based maps have been compared to statistical yield data, beside the comparative analysis between the maps created with different methods (image type, reference period).

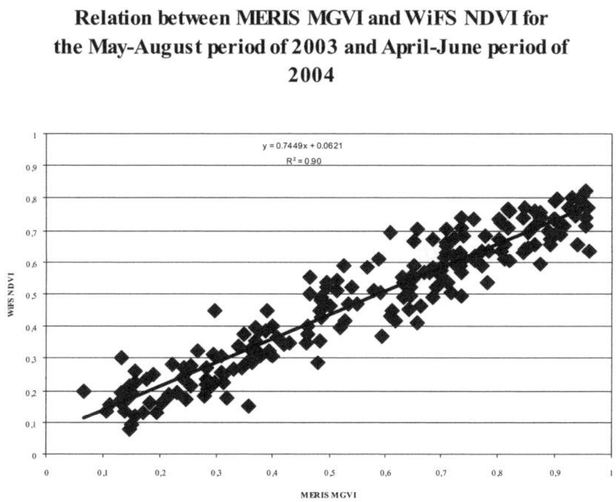


Figure 2. The extension of drought monitoring model using improved calibration procedure on an extended MERIS-WiFS data set (2003-2004). The relation is linear, the correlation is good ($R^2$=0.90)

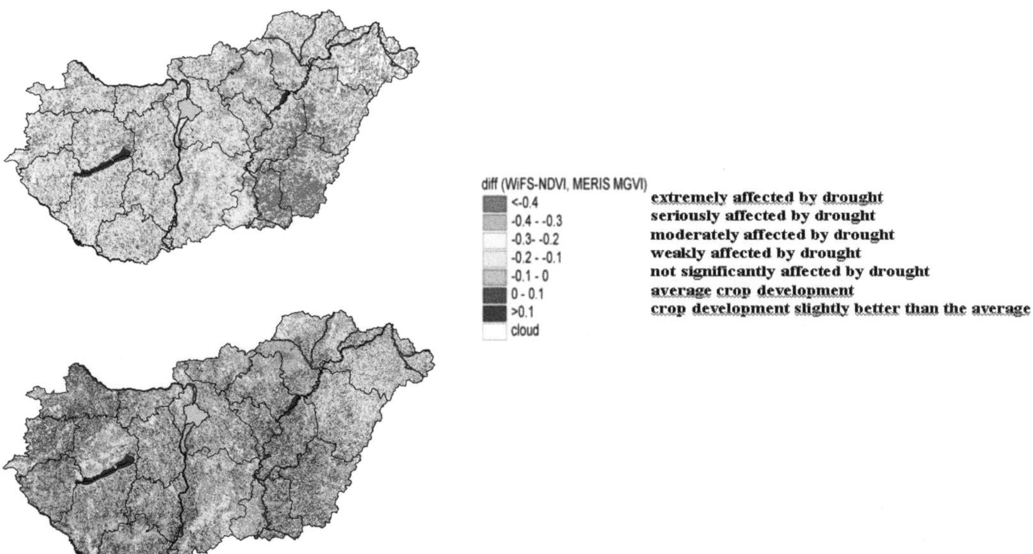


Figure 3. Drought monitoring for Hungary in 2003 (top) and 2004 (bottom) using calibrated MERIS data (2003-2004) and WiFS data (1997-1999)

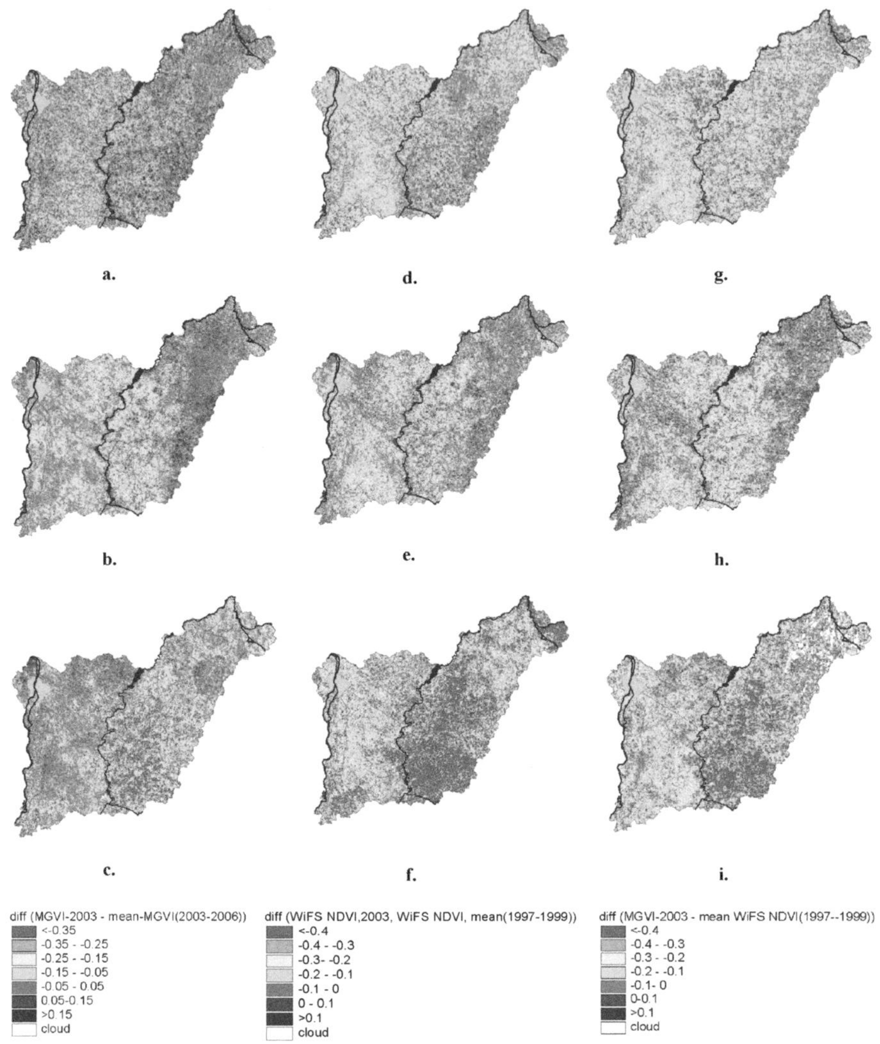


Figure 4. Drought monitoring for the Great Plain region of Hungary (7 counties) in 2003 using ENVISAT MERIS (a., b., c.), IRS WiFS (d., e., f.) and calibrated ENVISAT MERIS-IRS WiFS satellite data (g., h., i.); middle of May (a., d., g.), beginning of June (b., e., h.) and August (c., f., i.)

The maps generated in the first two phases of PECS project have undergone some checks to assess their validity. In the majority of cases we experienced good matching among the maps created with different methods. Besides, maps generated from satellite images are in good accordance with statistical yield data in most cases. However, some specific differences were found, depending on the year. In August of 2003, in the map derived from IRS WiFS images drought is shown much more serious than in the other two maps. The potential reason for this is that the majority of parcels are rather small in this area, and MERIS images – with about three times lower resolution than that of WiFS – less reflect the extremities in the reflectance of ground cover, the differences are more averaged, filtered.

In the third phase of PECS project the ENVISAT MERIS based drought model was extended to the whole country in 2007. Extreme weather conditions led to the acceleration of crop development in the spring and resulted in serious drought condition by the end of summer, which affected large part of the country. Drought monitoring tasks aimed at the countrywide monitoring of spatial and temporal development of drought in four periods throughout the year (May, June, July and August) and detecting also the effect of drought in crop development at parcel level using MERIS data. The derived drought maps (Fig. 5) reflected well the differences in crop developments as the drought progressed – the central part of Great Plain was affected the most seriously in July and August. The latter was justified by the deviation (yield loss) and strong spatial differences of the observed yields of the main crops (wheat and maize) at the level of the counties. The source of observed yield were data of Central Statistical Office of Hungary (CSOH).

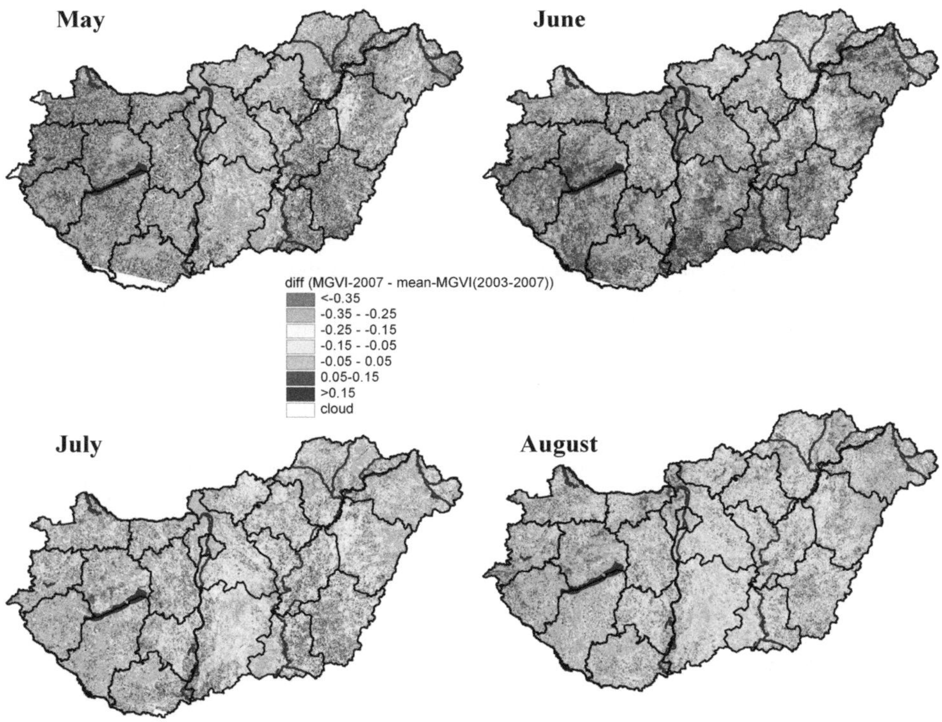


Figure 5. Drought maps derived from ENVISAT MERIS satellite data in 2007

## 4 CONCLUSIONS

When large extent natural disasters – like drought – happen regularly, there is a need for a complex model for monitoring them at regional level to provide quick, objective and homogeneous information to the various end users. The spatial extent, intensity and temporal development of these disastrous events can be rapidly and objectively assessed by remote sensing techniques. The achieved results demonstrated and confirmed that an effective, rapid regional drought/crop monitoring model can be developed and applied, which is based on remote sensing. The integration of available medium and low resolution optical satellite data (NOAA AVHRR, SPOT VEGETATION, IRS-1C/1D WiFS) of different available sensors together with the involvement of new generation sensors (ENVISAT MERIS) serve as an efficient basis to this model. Low price, large area coverage, frequent availability (within 1-3 days) and quick processing capability could compensate for their low or medium spatial (120 ha and 3.6 ha) and spectral (2-5 wide spectral band) resolution.

To assess the adequacy of the drought and crop development maps derived in the PECS project, comparative analysis has been carried out among the maps created with different methods (image type, reference period). Besides, remote sensing-based maps have been compared to statistical yield data. The conclusion of comparisons is that the kinds of satellite images used in this project appropriately reflect the crop development and show the presence of drought in the majority of cases. However, the result and accuracy depend on some factors:

- spatial resolution (finer pixel size allows more detailed distinction between drought categories, that is, higher spatial accuracy yields higher thematic resolution),
- temporal resolution (the usage of more images for a given period helps better in eliminating the effect of clouds),
- the applied vegetation index (the MERIS MGVI has proven to follow better the progress of crop development) and
- intercalibration (converting values between vegetation indices may cause some loss of information).

Nevertheless, with the proper choice of satellite images used and reference period, and taking into account the limitations of the selected kind of data source, appropriate drought or crop development map can be obtained at country and regional level.

The presented examples show that this approach and the involvement of the new generation sensors (particularly, MERIS) on board the ENVISAT satellite increase the data availability and the effectiveness of remote sensing based drought/crop monitoring model. The ENVISAT MERIS sensor's capabilities in regional drought and crop monitoring have been described in this paper. These show that ENVISAT MERIS data can be inserted the multi-source satellite dataset and can be effectively used together with other low and medium resolution data in operative applications.

High, medium and low resolution images provide fast, objective and accurate information to the monitoring of spatial and temporal progress of drought. They can serve as a basis to drought loss compensation measures and to the consideration of such claims. Due to the climate change, the occurrence of drought is expected to get more frequent. The usage of remote sensing is very important in the assessment of possibilities of cropping technologies so that protection and loss compensation can be effectively carried out. This is supported by remote sensing based objective, quantitative crop development monitoring at parcel, settlement, region and country level. The results of ENVISAT MERIS based drought and crop development monitoring provide a reliable basis to the drought impact assessment at regional and country level.

## ACKNOWLEDGEMENTS

The ESA-FÖMI Prodex-ENVISAT R&D (ESA Contract no. 14525/00/NL/SFe(IC) project is carried out from ESA Prodex (Scientific Experiment Development Programme) fund and FÖMI's own resources through the use of CROPMON program supported by MARD (Ministry of Agriculture and Rural Development). The ERS and ENVISAT datasets had been provided by ESA (ESA EO CAT-1 1139).

The FÖMI-ESA PECS project (2004-2007) (PECS Experiment Arrangement, ESA contract No. 98016) was carried out from ESA PECS fund and FÖMI's own resources. The ENVISAT and other datasets had been provided by ESA (ESA EO CAT-1 3504) and FÖMI.

## REFERENCES

Csornai, G. 1999. Operational crop monitoring by remote sensing in Hungary. G.J.A. Niewenhuis, R.A. Vaughan & M.Molenaar (eds), *Operational Remote Sensing for Sustainable Development,* Rotterdam, Balkema, 89-95..

Suba, Zs. Csornai, G., Wirnhardt, Cs., Nádor, G., Tikász, L., László, I., Csekő, Á., Martinovich., L. 2004. Application development for the utilisation of ENVISAT data in remote sensing based regional flood/waterlog and drought monitoring. *Remote Sensing in Transition – Proceedings of the 23rd Symposium of EARSeL,* Millpress, Rotterdam, Netherlands

Gobron, N., Pinty, B., Verstraete, M. , Govaerts, Y. 1999. The MERIS Global Vegetation Index (MGVI): description and preliminary application. *International Journal of Remote Sensing,* 1999, Vol. 20. No. 9. 1917-1927

Csornai, G., Suba, Zs., Wirnhardt, Cs., Nádor, G., Tikász, L., László, I., Csekő, Á., Martinovich, L. 2003. Application development for the utilisation of ENVISAT data, *ESA MERIS Workshop,* 11-14 November 2003, Frascati

Csornai, G., Suba, Zs., Nádor, G., Csekő, Á., Wirnhardt, Cs., Tikász, L., Martinovich, L. 2004. Evaluation of a remote sensing ba-sed regional flood/waterlog and drought monitoring model utilizing multi-source satellite data set including ENVISAT data. *Proceedings of the 2004 Envisat & ERS Symposium,* ESA, 2005.

Csornai, G., Suba, Zs., Nádor, G., László, I., Wirnhardt, Cs. 2007. Disaster monitoring with the integrated utilization of ENVISAT and other satellite data sets in the 2004-2006 period in Hungary. *Proceedings of ENVISAT Symposium 2007,* ESA.

*Remote Sensing for a Changing Europe*
*D. Maktav (Ed.)*
*IOS Press, 2009*

*doi:10.3233/978-1-58603-986-8-340*

# Quantification of bio-physical intertidal sediment properties using hyperspectral measurements

S.Adam, J. Monbaliu & E.A. Toorman
*Dept. of Civil Engineering, Katholieke Universiteit Leuven, Belgium*

Keywords: intertidal sediments, bio-physical properties, imaging spectroscopy, absorption, HyMAP

ABSTRACT:
In this work, the use of hyperspectral remote sensing for the quantification of bio-physical characteristics of the surface intertidal sediments is explored.
A two-step approach is followed. Firstly, algorithms based on absorption features for the quantification of bio-physical sediment properties are developed using laboratory measurements of known sediment mixture compositions. Secondly, the appropriateness of the method in natural conditions is explored by i) applying the algorithms on in situ spectra, and ii) applying the methods on these spectra resampled to a lower spectral resolution, namely the resolution of HyMAP, a typical airborne hyperspectral sensor.
In laboratory conditions, high correlation coefficients ($R^2 > 90\%$) are found between absorption features and clay content or moisture content lower than 20%.
In natural conditions, the relative moisture content lower than 20% is predicted with an error of ±3.5%. Clay content cannot be estimated from our field dataset, which can be caused by the limited amounts of clay or by noise in the spectral region of the absorption. High correlations (>75%) are obtained between the absorption feature of chlorophyll *a* and chlorophyll *a* content using in situ spectral measurements.
Interestingly, resampling the very high spectral resolution measurements to the spectral resolution of a HyMAP sensor influences the results to a very limited extent, leading to the conclusion that absorption features obtained by hyperspectral remote sensing might be suitable for quantifying bio-physical intertidal sediment properties. However, the appropriate spectral resolution for a particular application should be investigated.

## 1 INTRODUCTION

Mudflats are important for coastal zone ecosystems by providing wildlife habitats and by acting as natural sea defences that serve to dissipate tidal and wave energy. The primary production on intertidal flats can be considerable and is in our regions mainly accomplished by microphytobenthos (MPB), which typically consists of unicellular eukaryotic algae and cyanobacteria that grow within the upper millimeters of illuminated sediment (MacIntyre *et al.* 1996). The erodibility of cohesive sediments on these mudflats is dependent on physical sediment properties such as grain size distribution, density and moisture content, and on the biota living on the sediment which can have a stabilizing or destabilizing influence (Paterson 1997).

Remote sensing offers a means to acquire the necessary data to study these large mudflats.

Since the bio-physical sediment properties investigated in this study, namely clay content, relative moisture content (RMC) and chlorophyll *a* content, have specific absorption features within the 350-2500nm region of the spectrum, these features are investigated in this work.

Absorption features of water at around 970nm and 1200nm have been used for the quantification of relative water content in vegetation (Peñuelas *et al.* 1993; Pu *et al.* 2003). Clay absorption features are successfully used for the identification of clay minerals in geological applications (Hunt 1977; Van der Meer & Bakker 1988), and a recent study (Lagacherie *et al.* 2008) showed the capability of the absorption feature at 2206nm for clay content prediction both in laboratory and in the field for relatively dry soil samples. Microphytobenthos on mudflats have been studied using hyperspectral remote sensing and the absorption of chlorophyll *a* at around 673nm (Carrère 2003; Méléder *et al.* 2003).

The objective of the presented work is the development of algorithms for clay, chlorophyll *a* and moisture content prediction using absorption features, and the assessment of the accuracy of these algorithms on field spectra and on field spectra with the spectral resolution of HyMAP.

The research is performed in the frame of the CISS project (Bio-physical Characterization of Indicators Sediment Stability using Remote Sensing), financed by the Research Foundation – Flanders.

## 2 METHODOLOGY

### 2.1 *Measurements*

Hyperspectral laboratory and field measurements are acquired with the Analytical Spectral Device (ASD), which records the reflectance from 350nm till 2500nm, i.e. in the visible (VIS), near infra-red (NIR) and short wave infra-red (SWIR) region of the spectrum. The spectral resolution is 3nm from 350 till 1000nm and 10nm from 1000 till 2500nm.

#### 2.1.1 *In the laboratory*

The height of measuring of the ASD is 50cm, and the instrument is nadir looking. The field of view is set to 1 degree. The experimental setup is described by Biliouris *et al.* (2005). Calibration is performed every 30 minutes using a Spectralon panel (Biliouris *et al.* 2005).

Sediment is collected at the IJzermonding, a small intertidal flat located at the outlet of the IJzer river at the Belgian coast. Fine sand (125-250µm) is the most abundant and therefore the most often used in the experiments. Water and a clay fraction without organic matter bought in a shop are also used to make the sediment mixtures in small black pots with minimal reflectance (<0.05%).

Two sets of spectral measurements are carried out (Table 1).

- With varying grain size distribution of sand/clay mixtures (n=193).
- With varying moisture content of sand/clay mixtures (n=972) by drying saturated sediment mixtures on a warm plate to ensure homogeneous drying. Relative moisture content (%) is calculated as

$$RMC(\%) = \frac{WS - DS}{WS} * 100$$ with WS and DS are the weights of the wet and dry sample.

Table 1: Overview of the laboratory measurements. The third column giving the number of samples with RMC smaller than 20% is a subset of the second column.

| **Clay content** (0 - 64% by weight) | | **Moisture content** (0 - 36% RMC) | | **Moisture content** RMC < 20% | |
|---|---|---|---|---|---|
| fine sand + clay | n=98 | fine sand + water | n=535 | fine sand + water | n=278 |
| medium sand + clay | n=95 | fine sand + clay + water | n=437 | fine sand + clay + water | n=305 |
| *Total* | *n=193* | *total* | *n=972* | *Total* | *n=583* |

As long as the RMC is higher than ±20%, the sediment is saturated and a water film exists on the sediment surface. Once the RMC is lower than 20%, the water film disappears. The threshold is dependent on the sediment mixture.

### 2.1.2 *In the field*

In situ spectral measurements from field campaigns at the IJzermonding in 2005 and at the Molenplaat and IJzermonding in 2007 are used. The Molenplaat is an intertidal sandbank in the Westerschelde estuary, the Netherlands, described in e.g. Deronde *et al.* (2006). The height of the ASD sensor is 0.7m and the field of view 25degree, leading to a diameter of ±31cm. Calibration using the Spectralon panel is performed every 30 minutes under stable illumination conditions. If clouds are present, calibration is performed more frequently.

Surface sediment samples are collected for pigment, grain size and moisture content analysis with a 2.5cm diameter contact corer (Ford & Honeywill 2002), which freezes a layer of 2mm including all photosynthetically-active algal cells, as well as the bulk of sediment chlorophyll (Forster & Kromkamp 2004). Samples for pigment analysis are freeze-dried and extracted using 90% acetone. The extracts are then analysed using the HPLC-method by Wright *et al.* (1991) and quantified using a calibration with commercial standards. The sediment water content is determined by calculating the weight difference after 12h drying at 105°C. Grain size is determined using a Coulter Counter (Coulter LS Particle Size Analyser; Beckman Coulter).

## 2.2 *Absorption*

The degree of absorption by clay, water and chlorophyll *a* can be characterized by i) the ratio between minimal reflectance in the absorption feature and reflectance outside the absorption feature: $R_b/R_c$ ($R_b$ =reflectance at maximum absorption, $R_c$ =reflectance out of absorption feature), ii) the scaled band depth after continuum removal, and iii) the scaled band area of the absorption feature after continuum removal (figure 1). Continuum removal is a normalization technique that allows to compare absorption features from a common baseline. This continuum consists of straight-line segments that connect local spectra maxima. It is then removed by taking the ratio between the spectrum and the continuum (Clark and Roush, 1984). Before the calculation of the continuum, the spectral region of interest is isolated and the continuum is calculated between the two local maxima. It is observed that the first maximum can shift to lower wavelengths for deeper absorption features. However, the position of the second maximum can be assumed constant.

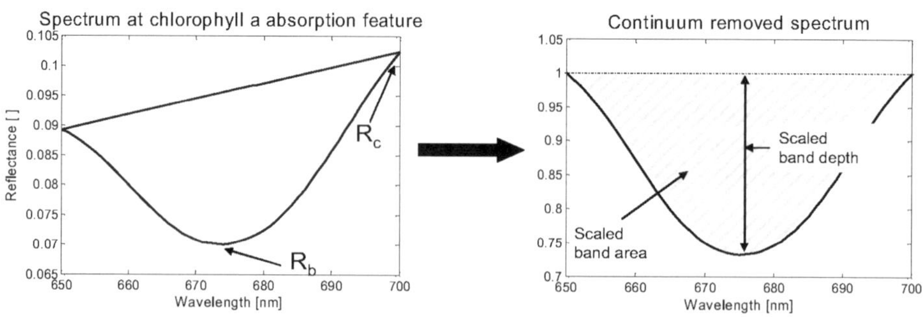


Figure 1: Measures that quantify absorption features. $R_b$ is the reflectance at maximum absorption, $R_c$ is the reflectance out of the absorption feature. Example for the chlorophyll *a* absorption feature.

The relation of the bio-physical sediment parameters to the measures of the absorption feature is investigated using regression analysis. A regression model, which can be linear or non-linear, is chosen by minimizing the mean squared error of the residuals. The model parameters are estimated using the laboratory dataset. The goodness of fit and the usefulness of the absorption features to assess the clay, moisture and chlorophyll *a* content are expressed by the R-squared value. The developed models are then applied on the in situ datasets.

### 2.3 *Spectral resampling*

In order to assess the influence of lower spectral resolutions on the use of absorption features for quantification of sediment properties, the field spectra are resampled using a Gaussian model based on the full width at half maximum (FWHM) and band center properties of the HyMAP sensor.

## 3 RESULTS

### 3.1 *Laboratory reflectance measurements*

The spectrum of fine sand with clay shows absorption features at around 1420nm, 1950nm and 2204nm. The absorption feature at 2204nm is caused by Al-OH present in the clay mineral (Yang *et al.* 2000) and will be considered. The spectrum of moist fine sand shows clear water absorption features at 1450nm and 1950nm. If the sediment is saturated, light is absorbed at around 970nm and 1190nm. However, these water absorption features disappear when the sediment is drier. The water absorption around 1450nm will be used.

The relation of clay content to the measures of the clay absorption feature at 2204nm in dry sediment mixtures appears to be 4$^{th}$ order polynomials (figure 2) and independent on the grain size of the sand fraction, with R-squared values equal to 98%.

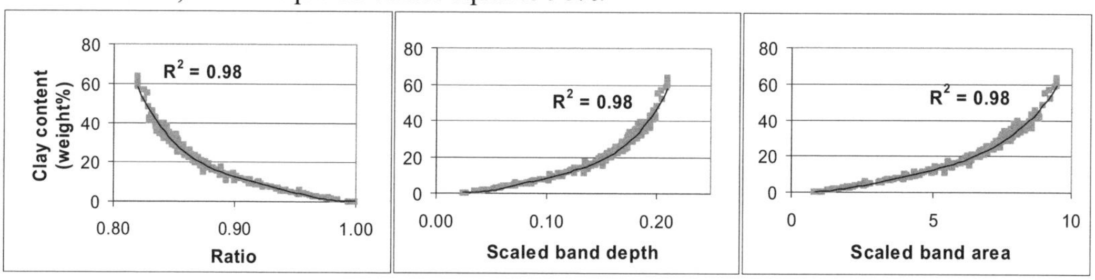


Figure 2: Clay content versus measures of the clay absorption feature for mixtures of fine sand with clay and medium sand with clay, n=193.

For moisture content, a 2$^{nd}$ order polynomial fit gives R-squared values of 89% between measures of absorption feature and RMC lower than 20%. R-squared values of 58% are found between the scaled band area and RMC higher than 20%, while for the ratio and the scaled band depth the R-squared values for RMC larger than 20% are smaller (figure 3).

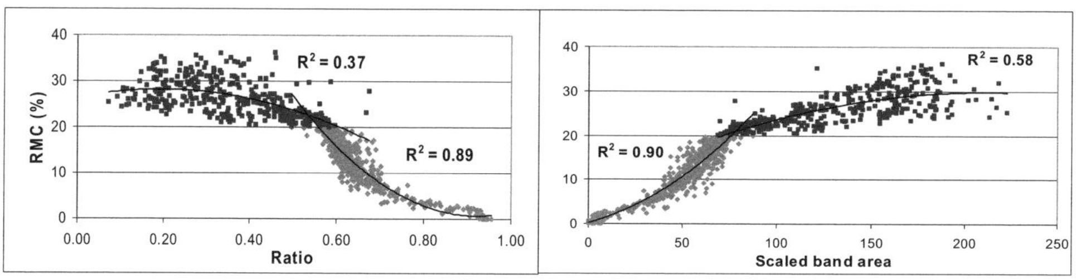


Figure 3: Relative moisture content versus measures of the water absorption feature at 1450nm for mixtures of fine sand with water and for mixtures of fine sand, clay and water, n=972.

The clay absorption feature becomes less apparent when the RMC increases, probably due to hindering of absorption by water. Therefore, a model using two independent variables, namely the scaled band area of the clay absorption feature at 2204nm (area_clay) and the scaled band area of the water absorption feature at 1450nm (area_water) with a R-squared value of 87%, is developed:

$$Claycontent(\%) = b_0 + b_1 * area\_clay + b_2 * area\_water$$

## 3.2 *Field reflectance measurements*

### 3.2.1 *Clay content*

The clay absorption feature is not visible in the field spectra, so it cannot be used for clay prediction.

### 3.2.2 *Relative moisture content*

During the field campaign at the IJzermonding in 2005, 69 sites are spectrally characterized and sampled for moisture content analysis. As shown in figure 4, the model developed in the laboratory is suited for predicting the RMC from spectral measurements as long as the RMC is lower than 20% (n=11). For this range, the root mean square error of prediction is equal to 4.5%, 3.6% and 3.3% using the ratio, scaled band depth and scaled band area respectively.

For the measurements with RMC higher than 20% (n=58), the model using the scaled band area from the laboratory measurements is used and the results are shown in figure 4. The error of prediction is 14.4%RMC, while the mean value of the measured RMC is 40.2%. The model for the ratio and scaled band depth is not used, due to the low correlations obtained in the laboratory.

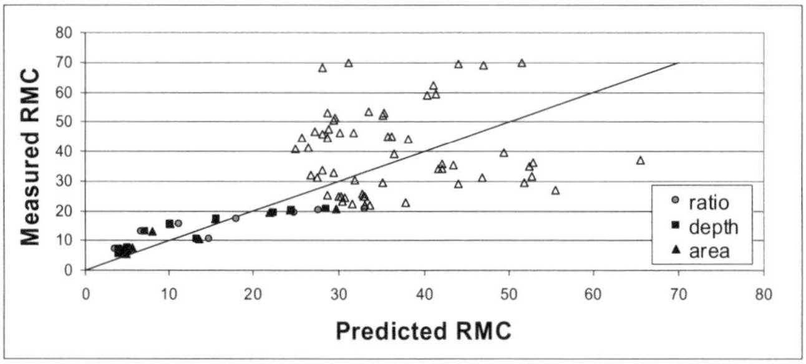


Figure 4: Measured and predicted RMC values for field measurements at the IJzermonding (2005). The predicted RMC values are obtained using the model developed in the laboratory. The filled symbols are for data with RMC lower than 20%.

### 3.2.3 *Chlorophyll a content (chl a)*

For chlorophyll *a*, the datasets of the field campaigns at the Molenplaat and IJzermonding in 2007 are used, with a total number of samples equal to 66.

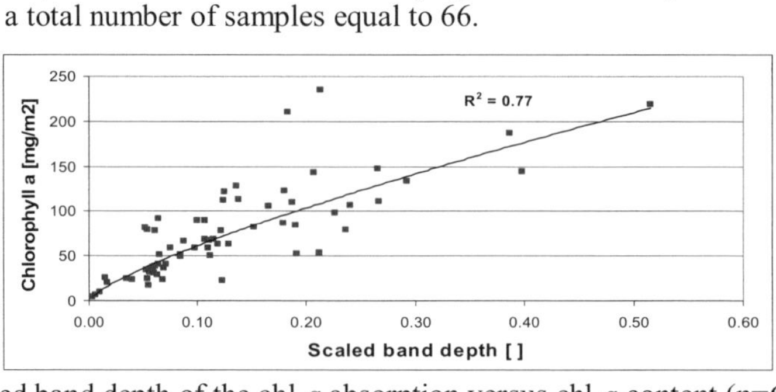


Figure 5: Scaled band depth of the chl *a* absorption versus chl *a* content (n=66)

## 3.3 *Spectral resampling*

The spectral resolution of the HyMAP sensor is 15-16nm in the visible region and in the short wave infrared 1 region, corresponding to the regions of the absorption dip of chlorophyll *a* and water respectively.

The root mean square error of prediction for RMC lower than 20% using the resampled spectra and the model developed in laboratory conditions are 4.4%, 3.7% and 3.6% for the ratio, scaled band depth and scaled band area respectively. For RMC higher than 20%, the prediction error is 14.4% using the scaled band area.

For chlorophyll *a*, the model obtained from the in situ original field spectra is applied on the measures of the absorption feature obtained from the resampled data. The root mean square error of prediction using the resampled spectra are 21.8, 20.3 and 27.3mg/m$^2$ for the ratio, scaled band depth and scaled band area respectively, whereas the average of the measured chlorophyll *a* values is 74.7mg/m$^2$.

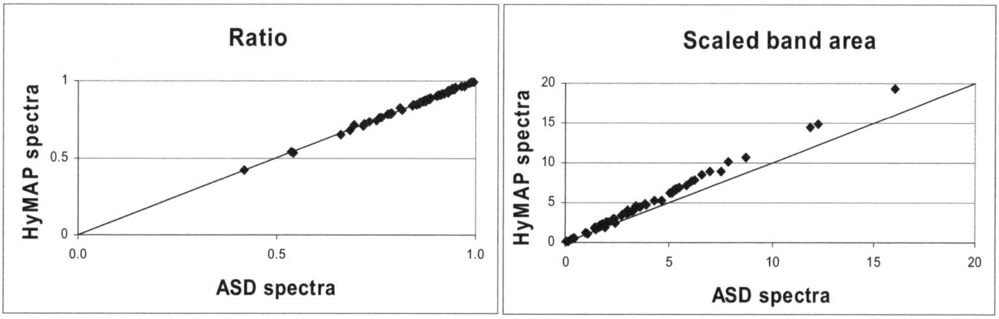


Figure 6: Comparison of measures of the absorption feature of chlorophyll *a* between the ASD spectra and the spectra resampled to HyMAP resolution.

As can be seen in figure 6, the difference in the values of the ratio between the spectra with ASD resolution and HyMAP resolution is negligible. For the scaled band area, there is a small offset for the higher chlorophyll *a* contents, which can be due to a deteriorated representation of the absorption dip after resampling. This means that the potential of HyMAP imagery for prediction of the chlorophyll *a* content is higher if a model can be calibrated based on the HyMAP spectra and field data.

# 4 CONCLUSION

## 4.1 *Clay content prediction*

In laboratory conditions, the hyperspectral signal in the short wave infrared (SWIR) and the absorption feature at 2204nm can be used to assess the clay in the surface layer of sand/clay mixtures. The R-squared values between absorption features and clay content are very high, as long as the sediment mixtures are relatively dry.

The clay absorption feature is not visible in our field measurements. Possible explanations are i) the low values of clay content, ii) the high moisture content of intertidal sediments, and iii) the high level of noise in the SWIR due to the lower energy amount and spectral resolution in the SWIR. Although Lagacherie *et al.* 2008 demonstrated the use of the clay absorption feature for clay content prediction in the field for relatively dry soils, the presented work brings about that the clay absorption feature cannot be used for clay prediction on intertidal mudflats.

### 4.2 *Relative moisture content prediction*

Relative moisture content is accurately predicted in laboratory conditions as long as the RMC is lower than 20%. This corresponds to the point of the appearance of a surface water film or the point of saturation. If the RMC is higher than 20%, the scaled band area performs better than the ratio and the scaled band depth, which can be explained by the fact that the scaled band area takes into account the widening of the absorption feature observed once the RMC is higher than 20%.

The relation obtained in the laboratory is successfully used on field measurements and on the resampled field measurements for RMC lower than 20%. This means that RMC lower than 20% can be estimated from HyMAP imagery.

### 4.3 *Chlorophyll a content*

Unfortunately, laboratory measurements of microphytobenthos are not available at the moment of writing this paper. However, high correlations are obtained between the measures of the absorption feature at 673nm from field spectra and chlorophyll *a* content, which is the main pigment of microphytobenthos.

After resampling these spectra to the HyMAP spectral resolution, the same model can be used to predict the chlorophyll *a* content from the resampled spectra with acceptable root mean square errors.

It is concluded that bio-physical sediment properties can be quantified using absorption features within certain limits. The three measures of absorption perform equally well, although the scaled band area should be used if the absorption feature becomes wider for higher quantities.

It is shown that it is feasible to extrapolate models developed in the laboratory to field measurements. However, this needs more investigation, especially for chlorophyll *a*, the most abundant pigment in microphytobenthos. For relative moisture contents lower than 20%, the model developed in the latoratory is suitable for field measurements. For higher moisture contents, several approaches can be investigated, e.g. using another water absorption feature or the overall spectrum.

However, we believe that a model calibrated using the field spectra or HyMAP images will yield better correlations and lower prediction errors.

Finally, the spectral resolution of the HyMAP sensor is appropriate for quantifying relative moisture content lower than 20% and chlorophyll *a* content.

## ACKNOWLEDGEMENTS

The CISS-project is funded by the Research Foundation-Flanders (FWO Vlaanderen) under contract no. G0480.05. The third author's position as research associate is partially financed by the KULeuven Special Research Fund.

The field campaigns are supported by the Belgian Science Policy in the framework of the STEREO program – project 043, 072 and 109.

The authors are grateful to Annelies De Backer and Dr. Steven Degraer from the Marine Biology Section of the University Gent, to Prof Koen Sabbe and Renaat Dasseville from the Laboratory for Protistology and Aquatic Ecology of the University Gent, to Dr. Daphne van der Wal from the Netherlands Institute of Ecology (NIOO-CEME), and to Dr. Rodney Forster from the Centre for Environment, Fisheries and Aquaculture Science (Cefas) for their help during the field campaigns, the loan of equipment and the sediment analysis. The comments of Dr. Aaike De Wever of the Laboratory for Protistology and Aquatic Ecology of the University Gent were highly appreciated and helpful.

## REFERENCES


Biliouris, D., vom Berge, K., Van Aardt, J., Muys, B. & Coppin, P. 2005. Hyperspectral bidirectional reflectance measurements of *Fagus sylvatica* leaves. *Proc. 4th EARSeL Workshop in Imaging Spectroscopy. New quality in environmental studies*. Zagajewski B., Sobczak M., Wrzesien M. (Eds)

Carrère, V. 2003. Mapping microphytobenthos in the intertidal zone of Northern France using high spectral resolution field and airborne data. *Proc. 3rd EARSeL Workshop on Imaging Spectroscopy*, Herrsching, 13-16 May 2003: 395-409

Clark, R.N., Roush, T. 1984. Reflectance spectroscopy: quantitative analysis techniques for remote sensing applications. *Journal of Geophysical Research* 89: 6329-6340

Deronde B., Kempeneers, P., Forster, R.M. 2006. Imaging spectroscopy as a tool to study sediment characteristics on a tidal sandbank in the Westerschelde. *Estuarine, Coastal and Shelf Science* 69: 580-590

Ford, R.B., Honeywill, C. 2002. Grazing on intertidal microphytobenthos by macrofauna: is pheophorbide a a useful marker? *Marine Ecology-Progress Series* 229: 33-42

Forster, R.M., Kromkamp, J.C. 2004. Modelling the effects of chlorophyll fluorescence from subsurface layers on photosynthetic efficiency measurements in microphytobenthic algae. *Marine Ecology Progress Series* 284: 9-22

Hunt, G.R. 1977. Spectral signatures of particulate minerals in the visible and near-infrared. *Journal of Geophysical Research* 95: 12653-12680

Lagacherie, P., Baret, F., Feret, J.-B., Netto, J.M., Robbez-Masson, J.M. 2008. Estimation of soil clay and calcium carbonate using laboratory, field and airborne hyperspectral measurements. *Remote Sensing of Environment* 112: 825-835

MacIntyre, H.L., Geider, R.J., Miller, D.C. 1996. Microphytobenthos: the ecological role of the "Secret Garden" of unvegetated, shallow-water marine habitats. I. Distribution, abundance and primary production. *Estuaries* 19(2A): 186-201

Méléder, V., Barillé, L., Launeau, P., Carrère, V., Rincé, Y. 2003. Spectrometric constraint in analysis of benthic diatom biomass using monospecific cultures. *Remote Sensing of Environment* 88: 386-400

Paterson D M. 1997. Biological mediation of sediment erodibility: Ecology and physical dynamics. In N. Burt, R. Parker & J. Watts (Eds), *Cohesive sediments* (pp. 215-229). Wiley

Peñuelas, J., Filella, I., Biel, C., Serrano, L., Savé, R. 1993. The reflectance at the 950-970nm region as an indicator of plant water status. *International Journal of Remote Sensing* 14(10): 1887-1905

Pu, R., Ge, S., Kelly, N.M., Gong, P. 2003. Spectral absorption features as indicators of water status in coast live oak (*Quercus agrifolia*) leaves. *International Journal of Remote Sensing* 24(9): 1799-1810

Van der Meer, F., Bakker, W. 1998. Validated surface mineralogy from high-spectral resolution remote sensing: a review and a novel approach applied to gold exploration using AVIRIS data. *Terra Nova* 10: 112-118

Wright, S.W., Jeffrey, S.W., Mantoura, R.F.C. Llewellyn, C.A., Bjornland, T., Repeat, D. & Welshmeyer, N. 1991. Improved HPLC method for the analysis of chlorophylls and carotenoids from marine phytoplankton. *Marine Ecology Progress Series 77*: 183-196

Yang, K., Huntington, J.F., Browne, P.R.L., Ma, C. 2000. An infrared spectral reflectance study of hydrothermal alteration minerals from the Te Mihi sector of the Wairakei geothermal system, New Zealand. *Geothermics* 29: 377-392

*Remote Sensing for a Changing Europe*
*D. Maktav (Ed.)*
*IOS Press, 2009*

*doi:10.3233/978-1-58603-986-8-348*

# Assessment of unsupervised classification techniques for intertidal sediments

E. Ibrahim, S. Adam, J. Monbaliu
*Dept. of Civil Engineering, Katholieke Universiteit Leuven, Belgium*

Keywords: intertidal sediments, unsupervised classification, imaging spectroscopy, clustering, artificial data

ABSTRACT:
The aim of this study is to explore three techniques for unsupervised classification of airborne hyperspectral imagery of intertidal flats. The unsupervised classification techniques considered are k-means (hard clustering), the Gustafson-Kessel algorithm (Fuzzy clustering), and the mixture of Gaussians model (probabilistic clustering). The behavior and suitability of these techniques is analyzed for sediment classification. Artificial data sets based on real airborne and field spectra are used for this purpose. The sensitivity of the techniques is investigated on two spectral aspects: the effect of within class (intra-class) variability and the effect of spectral dimensionality using feature selection. This sensitivity is expressed as classification accuracy in terms of the Kappa statistic (κ) that indicates how better the classification is than chance agreement. The results show that the three techniques are suitable for sediment classification. When there is no feature selection involved, the mixture of Gaussians results in the best classification results. When feature selection is considered, sediment classification accuracy increases for all three techniques applied on the artificial imagery.

## 1 INTRODUCTION

An intertidal zone in a marine aquatic environment is the area of the foreshore and seabed that is exposed to air at low tide and submerged at high tide generating important geochemical processes (Silva et al., 2005). To describe these processes, extensive field knowledge is required. Since accurate data collection is often costly, inefficient, or unattainable, remote sensing can be a fine and resourceful alternative. Specifically, unsupervised classification techniques are of high interest where an image is classified on the basis of its reflectance values without taking field measurements into account. Field knowledge plays a role only in the analysis and identification of the classified groups. In practice, there is globally no "absolute best" unsupervised classification methodology, where the reliability of a result is based on the aim of the classification and its use (Everitt, 2001). Therefore, various unsupervised classification techniques are explored and assessed in this study for their performance in sediment classification. The techniques selected are K-means, Gustafson-Kessel algorithm (GK), and the mixture of Gaussians model (MG). In order to be in total knowledge and control of the test records, synthetic data sets are utilized as real imagery can include various uncertainties.

# 2 METHODOLOGY

## 2.1 *An introduction to clustering*

Clustering is a means of unsupervised classification defined as finding a structure in a collection of unlabeled data. It partitions an $N \times n$ data set $X$ into $c$ clusters (Lillesand and Kiefer, 2000) where $X$ is the studied image, $N$ is the number of pixels, and $n$ is the spectral dimensionality. Each of the $N$ pixels is characterized by a spectrum of the $n$ frequency bands and is represented by a row vector (Balasko et al., 2004). Each type of clustering entails its limitations and specialties based on the concept of dissimilarity or *distance measures* (Everitt, 2001). Two popular distance norms are utilized in this study: the Euclidean and the Mahalanobis distance measures. When these distance norms are used for finding clusters as fixed or non-adaptive norms, limitations arise as the techniques then impose defined geometrical structures. Yet, when an adaptive norm is used for each cluster, more freedom in cluster features such as cluster shape, size, and orientation becomes possible (Kim et al, 2005).

### 2.1.1 *The clustering techniques*

The investigated techniques are of three clustering types: hard, fuzzy, and probabilistic. From hard clustering techniques, *k-means* is chosen where a pixel is allocated to a cluster minimizing the within cluster sum of squares using a non-adaptive Euclidean distance measure (Everitt, 2001, Balasko et al., 2004). In fuzzy techniques, each pixel can belong to more than one cluster yet in a different degree of belonging. The *Gustafson-Kessel algorithm (GK)* applies an adaptive distance norm of the Mahalanobis distance measure (Balzano and Del Sorbo, 2007, Gustafson and Kessel, 1979). The hard and fuzzy clustering are based on the "Fuzzy Clustering and Data Analysis Toolbox" by Balasko B, Abonyi J., and Feil B. (2004).

As for probabilistic clustering, the *mixture of Gaussians (MG)* model is used. It is a model-based probabilistic approach that constitutes of models describing each cluster. Clusters are considered as various Gaussian distributions according to their covariance structure (Beaven et al., 2000, Banfield et al., 1993). For this study, the popular expectation-maximization algorithm (EM) optimizes the fit between the data and the models (Dempster et al., 1977), and the Bayesian Information Criterion (BIC) selects the most suitable model describing the data (Fraley and Raftery, 1998). MG was performed by means of the Mixture Modeling software, MIXMOD (Biernacki et al, 2006).

### 2.1.2 *Accuracy assessment*

An artificial image is a combination of various artificial *classes* built on the basis of specific sediment *types* acquired from imagery and field spectra. The clustering techniques applied to the artificial data sets result in *clusters*. The comparison between *classes* and *clusters* indicates the accuracy of the classification. Confusion matrices are used to assess this accuracy as they compare the relationship between the artificial data as the reference data and the "corresponding" results of the unsupervised classification techniques (Lillesand and Kiefer, 2000). Yet with unsupervised classification, a resulting cluster is not automatically labeled nor identified as corresponding to a specific class. So, a class is investigated with respect to all clusters, and the cluster containing most of the pixels closest to the mean of that class is considered as its corresponding cluster. Based on the confusion matrix, the accuracy is then expressed in terms of the kappa statistic ($\kappa$) where the difference between the clustering accuracy and the chance agreement between the classes and the clusters is calculated (Lillesand and Kiefer, 2000). It results in a value between 0 and 1 for each classification, where 0 indicates that the clustering is no better than grouping the data by chance:

$$\kappa = N\sum_{i=1}^{c} y_{ii'} - \sum_{i=1}^{c} (y_i . y_{i'}) / N^2 - \sum_{i=1}^{c} (y_i . y_{i'}) \quad (1)$$

Where

$c$ is the number of classes or clusters, $1 \le i \le c$.
$N$ is the total number of pixels in the artificial image classified
$i'$ refers to the cluster corresponding to class $i$
$y_{ii'}$ is the number of observations in row $i$ and column $i'$ in the confusion matrix
$y_i$ is the number of observations in row $i$ in the confusion matrix
$y_{i'}$ is the number of observations in column $i'$ in the confusion matrix

## 2.2 *Building artificial data*

### 2.2.1 *Data available*

To build the synthetic images, data extracted from a hyperspectral image is taken into consideration. This image acquired by Airborne Hyperspectral Sensor (AHS) on the $17^{th}$ of June 2005 covers the IJzermonding estuary located on the Belgian coast. Its spatial resolution is 3.4 meters and 19 non-corrupted bands are available covering the visible part of the spectrum VIS (10 bands, 0.455-0.746μm, 30nm wide), the Near Infra-red NIR (8 bands, 0.774-1.004μm, 30nm wide), and the short wave infra-red SWIR-1 (1 band, 1.622μm, 200nm wide). Furthermore, field sampling was carried out at a timeframe close the image acquisition. The resulting samples were analyzed for relative moisture content (RMC), mud content (MC), and chlorophyll *a* content (chl *a*). The sediments are considered "wet" when RMC> 30%, high chl *a* when the chl *a* content > 40 mg/m$^2$, and either muddy or sandy based on the clay and silt content of a threshold equal to 30% of particles (Deronde et al., 2006). The spectral measurements were carried out by means of an Analytical Spectral Device (ASD) that records reflectance covering the spectrum from the VIS to SWIR with a resolution varying from 3nm to 10nm.

### 2.2.2 *Sediment types*

To build the synthetic data, three major sediment *types* identified at the study area are selected. Representative spectra of those *types* are extracted from the AHS image and the basic statistics of each of the types is calculated such that for $type_i$, mean $\mu_i$ and standard deviation $\sigma_i$ are calculated per spectral band ($\mu_{ib}$ and $\sigma_{ib}$). *Type1* is wet, clayey, and with high chl *a* content; *type2* is wet, clayey, yet with low chl *a* content. *Type3* is dry, sandy, and with low chl *a* content.

### 2.2.3 *Sediment properties in artificial data*

Synthetic data sets are constructed using all the $n$ bands of the AHS image. An artificial image is a combination of a number of *classes* $X_i$. Each class is of $N_i$ pixels and built on the basis of the statistics of each *type*; a mean spectrum, $\mu_{ib}$, and a standard deviation in terms of $\sigma_{ib}$. To relate the artificial data to the reality of spectral distribution in the field, three properties, RMC, chl *a* content, and mud content, that distinguish the three sediments types play a role in building the data. The first step in building theses images consists of applying the effect of moisture content to the different types. This is explained in 2.2.4. In the second step, the effect of chl *a* content is applied as in 2.2.5 to the spectra of *type1* and *type2* generated in the first step. These types show variability in chl *a* content where a chl *a* absorption feature appears in the red region at around 673nm.

The effect of clay content is not included directly as the clay content absorption features are not visible with field or airborne spectra (Adam et al., 2008). However, correlations between mud content and chl *a* content have been demonstrated (Van Engeland, 2008). Therefore, by considering the effect of chl *a* and RMC in the data, the effect of clay content is considered indirectly.

### 2.2.4 *Effect of moisture content*

For changes in RMC, an increase or decrease of the whole spectrum is considered (Weidong et al., 2001; Muller & Décamps, 2000). Reflectance normally decreases as water content increases. Yet, at a value of RMC referred to as "the cut-off thickness" that is soil dependent, reflectance increases (Neema et. al, 1987). Assuming at each spectral band of a $type_i$ the spectra are normally distributed, a spectrum of $class_i$ represented by the $N_i$ *by n* matrix $X_i$ is generated by $\mu_i$ and a standard deviation represented in terms of $\sigma_i$. So, a spectrum *k* in $class_i$ is calculated as follows:

$$\mathbf{X_{ik}} = \boldsymbol{\mu}_\mathbf{i} + r.\boldsymbol{\sigma}_\mathbf{i}.s \qquad (2)$$

Where

$1 \le i \le c'$, $c'$ is the number of classes
$1 \le k \le N_i$
$\mu_i$ is a matrix containing the mean values per band for $type_i$
$\sigma_i$ is a matrix of standard deviations per band of $type_i$
*s* is a factor multiplied by $\sigma_i$ and determines the standard deviation of a class ($s > 0$)
$N_i$ is the number data points in $class_i$
*r* is a random value following a normal distribution: *mean=0* and *standard deviation=1*

In addition, random noise factors are added to each band in the order of ±1% of the reflectance values.

### 2.2.5 *Effect of chl a content*

In the field samples acquired from the IJzermonding, RMC and chl *a* show a significant correlation with an $r^2 = 0.56$. Due to the cut-off thickness in spectra affected by RMC, the positive correlation between moisture content and chl *a* can be demonstrated in the artificial data by allowing the presence of a random chl *a* dip. This is introduced to the spectra generated in 2.2.4 for *type1* and *type2*. The ratio between the minimal reflectance in the absorption feature and the reflectance outside the absorption feature has a positive correlation with chl *a* content (Adam et al., 2008). With airborne spectra, Pearson's correlation resulted in $r^2 = 0.57$ between chl *a* content and the ratio between the AHS bands 9 and 8 of 718nm and 689nm respectively.

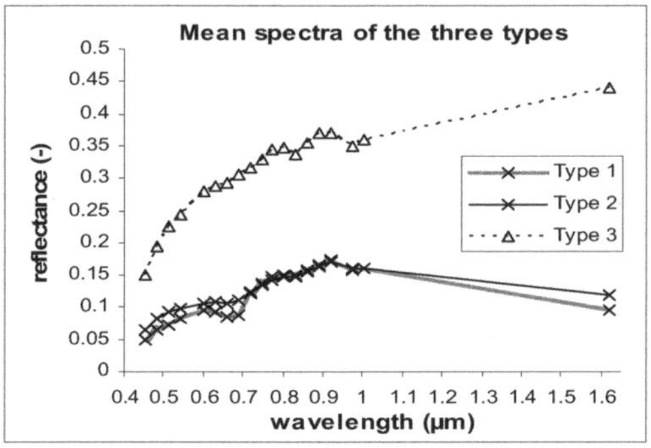


Figure 1: Mean spectra $\mu_i$ of each *type*

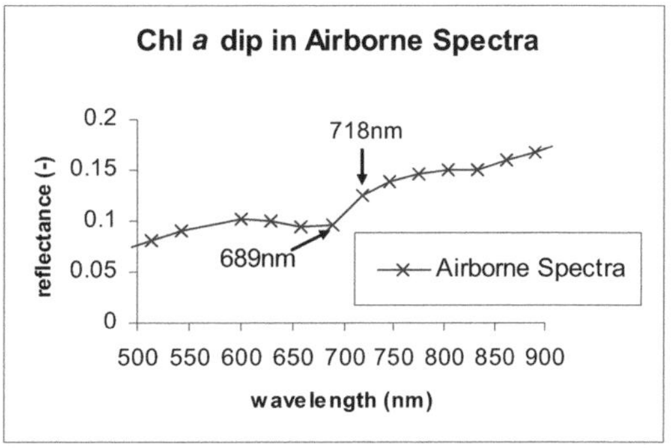


Figure 2: Chl *a* dip in AHS spectra

Considering that the chl *a* dip is bounded by band 5 and band 9 of 0.601µm and 0.718µm respectively, the effect of chl *a* is first introduced at band 8 of 0.689µm in the following manner:

$$X_{ik8'} = min_{i8}.\, s + r'.min_{i8}.\, s \qquad (3)$$

Where

$X_{ik8'}$ is the new value at band 8

$min_{i8}$ is the minimum reflectance value of $band_8$ in $type_i$

$r'$ is a random value / $0< r' \leq$ m, where m varies fulfilling the condition: $X_{ik8'} < X_{ik5}$ as band 5 notes the start of the dip

Finally, bands 6 and band 7 of 0.689 and 0.630μm respectively are altered according to the new values of band 8.

## 3 RESULTS

### 3.1 *Clustering and inter/intra class variability*

The dependence of the clustering techniques on spectral variance in the data is analyzed while spatial properties are preserved. So, the classes are set to have the same number of pixels $N_i = 500$. First, the artificial data is built containing the three *types* and of various standard deviations represented by *s*. With *c=c'=3*, *figure 3(a)* shows the results obtained by the three techniques. They all resulted in acceptable $\kappa$ values through the range of intra-class variability. Yet, referring to the confusion matrices, a good classification by the three techniques for *type3* is noticed while *type1* and *type2* are more difficult to discriminate. Therefore, the second step is to reduce the inter-class variability. So, the images are built using 2 *types* at a time, *c=c'=2*. The clustering results show a better performance in distinguishing *type1* and *type2* by k-means; MG still resulted in $\kappa$ =1, while GK resulted in a poor classification of the spectra *(figure 3)*. As for distinguishing *type3* from the other two types, all methods lead to high values of kappa. Therefore, in real imagery, since *type3* can be detected easily, one could mask it out of the image and then apply the clustering techniques on the other two types for better classification accuracy.

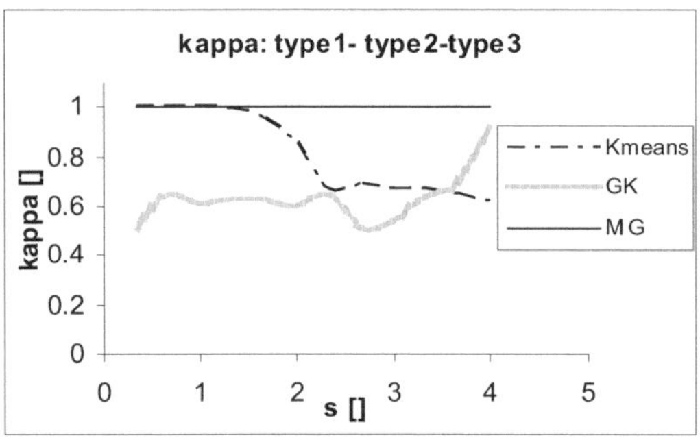


(a)

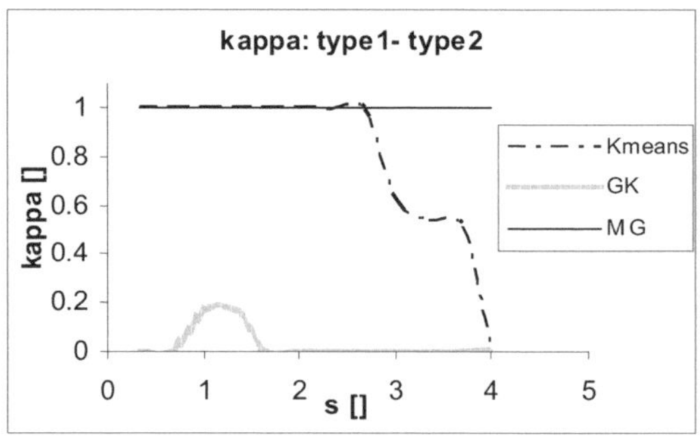


(b)

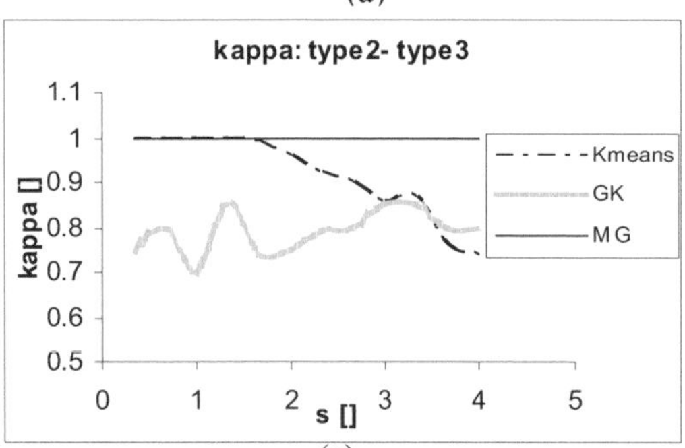


(c)

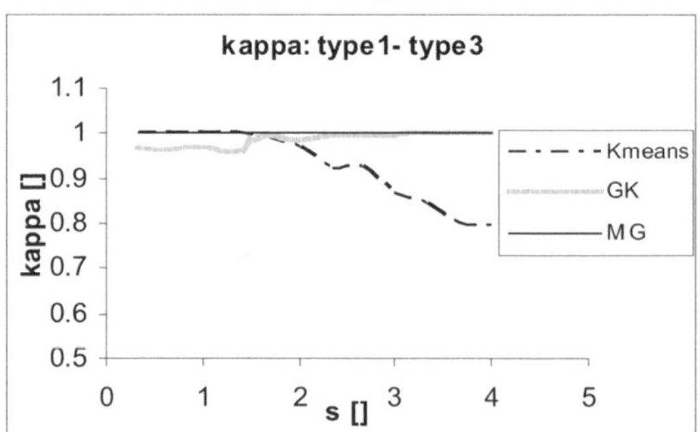


(d)

Figure 3: Accuracy in terms of Kappa with respect to standard deviation of (a) the three types (b) *type1* and *type2* (c) *type2* and *type3* (d) *type1* and *type3*

## 3.2 *Feature selection*

### 3.2.1 *Based on supervised classification*

According to Deronde et al. (2006), spectral dimensionality can be decreased resulting in not only computational efficiency, but also an increase in classification accuracy. A band selection by the sequential floating forward selection algorithm (SFFS) based on the methodology of Deronde et al (2006) is carried out for the AHS image to test the clustering techniques. The bands selected for RMC are 455nm, 513nm, 918nm, and 1622nm, for mud content 689nm, 833nm, 1004nm, and for Chl *a* content 718nm, 689nm, 774nm.

After building the artificial data, these bands are selected and a comparison is carried out between the results obtained by using these few bands and all the *n* bands (*figure 4)*. Using bands chosen for any of the three properties, similar or better results to using the *n* bands are obtained for k-means, better results for GK, and the same result of $\kappa$=1 for MG. Therefore, using only three or four bands, the computational efficiency increases, and the results are generally better than using all the 19 bands of the AHS image.

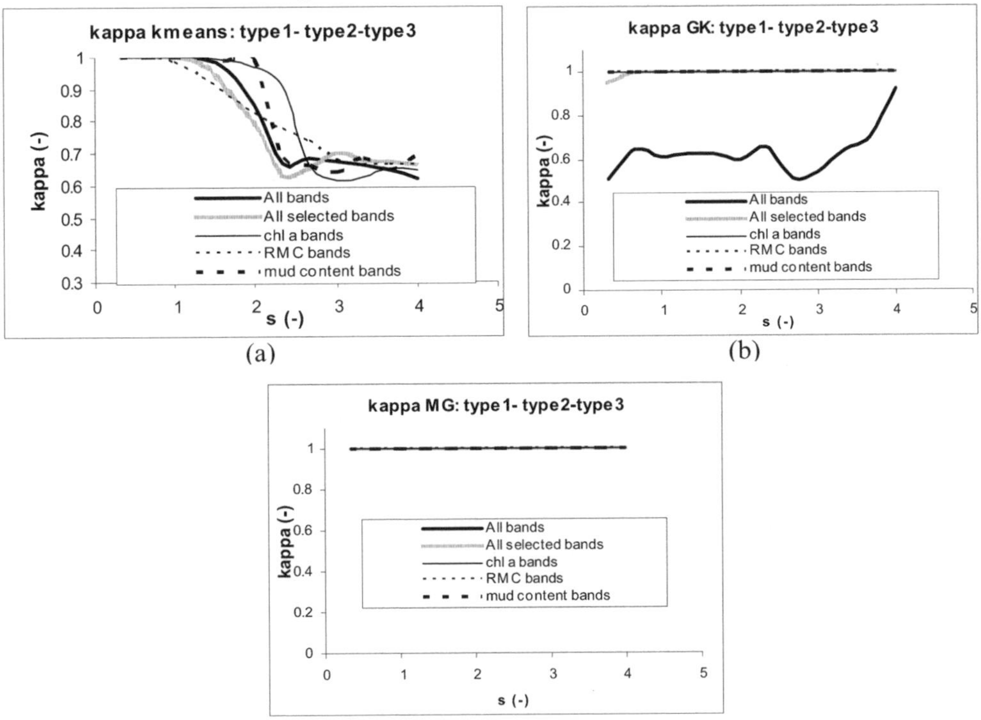


(a) (b)

(c)

Figure 4: the effect of feature selection on kappa accuracy for (a) k-means (b) GK (c) MG

### 3.2.2 *Based on random band selection*

Hyperspectral data within adjacent bands are usually highly correlated. As a random attempt to reduce these correlations, 4 bands are selected from the blue, green, red, and NIR parts of the spectrum repetitively. With a range of standard deviations considered, $0 < s \leq 4$, the three techniques result in high classification accuracy. MG and GK resulted in kappa values of minimum 0.95. The k-means results varied between $0.6< \kappa< 0.8$ for higher and lower standard deviations respectively.

## 4 CONCLUSIONS

Unsupervised classification is essential for intertidal sediment characterization due to the many difficulties incorporated in field work on such regions. Three techniques are investigated in this work, namely k-means, the Gustafson-Kessel algorithm (GK), and the mixture of Gaussians model (MG). Artificial data based on real imagery and sediment properties are built to test the data with varying intra-class variability and spectral resolutions.

From this study, it can be concluded that the unsupervised techniques are capable of discriminating sediment types successfully. With varying inta-class variability, the three techniques lead to acceptable results with superiority to MG.

Furthermore, it is revealed that the techniques are dependent on inter-class variability represented by the occurrence of different sediment types in an image. Therefore, a good approach can be to cluster first the easily distinguished types and mask them out from the image. This can lead to an increase in the classification accuracy of the remaining types.

Feature selection also increases the classification accuracy. By referring to the work of Deronde et al., 2006, the extracted bands from the AHS image lead to generally better classification accuracy for the three techniques. GK and MG show a perfect classification of the three types on a range of intra-class variability $0 < s \leq 4$.

Furthermore, if four bands are selected randomly, one from each of the blue, green, red, and NIR regions of the spectrum, the classification accuracy is superior to using all the 19 bands. MG and GK result in very high kappa values ($\kappa > 0.95$) for any combination of the four bands on the whole range of standard deviations. The accuracy of k-means also increases resulting in $\kappa > 0.6$.

In conclusion, all three techniques are able to distinguish the three sediment types considered for this study. Yet, if there is no feature selection involved, MG can be considered the most robust. If there is feature selection involved, the three techniques lead to good classification accuracies, yet with superiority to MG and GK.

## 5 FURTHER WORK

Further study is being done to interpret the results obtained using artificial imagery for the real imagery. There are three major points addressed. First, the same class sizes are considered for the artificial images. Therefore, the effect of having various class sizes in an image on classification accuracy is investigated. Furthermore, the spatial distribution of the spectra and the role of neighboring pixels are examined. Although these clustering techniques do not take spatial information into account, the spatial distribution can affect the choice of suitable seed pixels initiating the different algorithms. Finally, the issue of the number of clusters to be retrieved from an image is considered. In a real image, the number of clusters in the image is not known from beforehand as it is in this study. Therefore, this aspect is investigated in order to be able to apply the clustering techniques successfully on real imagery.

## ACKNOWLEDGEMENTS

The research presented in this paper is funded by the Belgian Science Policy Office in the frame of the STEREO II programme – project SR/00/109 - ALGASED (*remote sensing for characterization of intertidal sediments and microphytobenthic algae*)

For this reseach, there was also financial contribution from the CISS-project funded by the Research Foundation-Flanders (FWO Vlaanderen) under contract no. G0480.05.

The field campaigns are supported by the Belgian Science Policy in the framework of the STEREO program – project 043, 072 and 109.

The authors are grateful to the ALGASED team members

Finally, the authors express their gratitude to the contribution of Ms. Sindy Sterckx (Flemish institute for technological research - VITO). The work of Ms. Sterckx work was in the framework of the SEDOPTICS project funded by the Belgian Science Policy Office in the frame of the STEREO programme – project SR/00/72

## REFERENCES


Adam, S., Monbaliu, J., Toorman, E. 2008. Quantification of bio-physical intertidal sediment properties using hyperspectral laboratory and in situ measurements. *Presented at the 28th EARSeL Symposium and Workshops, Remote Sensing for a Changing Europe*, 2-7 June, 2008, Istanbul.

Balasko, Abonyi, B., J. and B. Feil. 2004, Fuzzy Clustering and Data Analysis Toolbox: For Use with Mat lab, University of Veszprem, Hungary

Balzano W. and Del Sorbo M.R. 2007. Genomic comparison using data mining techniques based on a possibilistic fuzzy sets model. *Biosystems* 88, Issue 3: 343-349.

Banfield, J and Raftery. 1993. A. Model-based Gaussian and non-Gaussian clustering, *Biometrics* 49: 803–821

Beaven, S.G., Stein, D., Hoff, L.E. 2000. Comparison of Gaussian mixture and linear mixture models for classification of hyperspectral data. *Proceedings of the IEEE Geoscience and Remote Sensing Society. Honolulu, Hawaii, USA*: 1597-1599

Biernacki, C., Celeux, G., Govaert, G., Langrognet, F. 2006. Model-based cluster and discriminant analysis with the MIXMOD software. *Computational Statistics & Data Analysis* 51: 587-600

Dempster, AP, Laird N. M. and Rubin D. B. 1977. Maximum likelihood from incomplete data via the EM algorithm. *J. Roy. Statist. Soc.* 39: 1–38

Deronde, P. Kempeneers and R.M. Forster. 2006. Imaging Spectroscopy as a Tool to Study Sediment Characteristics on a Tidal Sandbank in the Westerschelde. *Estuarine, Coastal and Shelf Science* 69: 580-590

Everitt, B. S., 2001, Cluster Analysis, Edward Arnold, London.

Gustafson D. E. and W.C. Kessel, 1979, Fuzzy clustering with fuzzy covariance matrix. *In Proceedings of the IEEE CDC, San Diego*: 761-766.

Kim D., Kwang H. Lee, and Doheon Lee, 2005, Detecting clusters of different geometrical shapes in microarray gene expression data. *Bioinformatics Advance Access, Bioinformatics* 21: 1927-1934

Lillesand, T.M. and Kiefer, R.W. 2000. Remote sensing and image interpretation. John Wiley & Sons, Inc., New York, NY, USA.

Muller E. and Décamps H. 2000. Modeling soil moisture-reflectance. *Remote sensing of environment* 76: 173-180

Neema, D., Shah, A, & Patel, A. N. 1987. A statistical model for light reflection and penetration through sand. *International Journal of Remote sensing* 8: 1209-1217

Silva J., Santos R., Calleja M., and Duarte C. 2005. Submerged versus air-exposed intertidal macrophyte productivity: from physiological to community-level assessments. *Journal of Experimental Marine Biology and Ecology* 317, Issue 1: 87-95

Van Engeland, T. 2005. Using field data and hypsespectral remote sensing to model microalgae distribution and primary production on an intertidal mudflat. Unpublished MSc thesis. Universiteit Gent, Belgium.

Weidong, L., Baret F., Xingfa G., Qingxi T., Lanfen Z., Bing Z. 2002. Relating soil surface moisture to reflectance. *Remote sensing of environment* 81: 238-246

*Remote Sensing for a Changing Europe*
*D. Maktav (Ed.)*
*IOS Press, 2009*

*doi:10.3233/978-1-58603-986-8-356*

# Land cover classification in Portugal with multitemporal AWiFS images: a comparative study

H. Costa, H. Carrão & M. Caetano
*Portuguese Geographic Institute (IGP), Remote Sensing Unit (RSU), Lisboa, Portugal*

H. Costa, H. Carrão, F. Bação & M. Caetano
*CEGI, Instituto Superior de Estatística e Gestão de Informação, ISEGI, Universidade Nova de Lisboa, Lisboa, Portugal*

Keywords: Land cover map, AWiFS, Decision trees, Maximum likelihood, Portugal

ABSTRACT: Three AWiFS scenes from three intra-annual time periods were explored for land cover mapping of mainland Portugal in an annual operational basis. Two different classification approaches were tested: a more conventional parametric classifier, the maximum likelihood classifier, and a nonparametric classifier, a decision tree. Several tests were designed to evaluate the most suitable classification approach, training sample size effect on the classification accuracy and the images's ability for land cover mapping of mainland Portugal. Overall accuracy values achieved were low regardless the classification approach and training sample size, which suggests that AWiFS images are not suitable for land cover mapping the Portuguese landscape. Results also reveal that decision tree classifier is advantageous relatively to maximum likelihood classifier concerning training size and its characteristics.

## 1 INTRODUCTION

Land Use Land Cover (LULC) maps production is essential to many scientific areas, such as landscape management, environmental modelling, change detection and others (Foody & Mathur, 2004; Justice *et al.*, 1998; Friedl *et al.*, 2002). In Portugal, the CORINE Land Cover map and the Portuguese Land Cover Map of 1990 are the reference LULC maps for major studies, but they are not updated with the required periodicity (Painho & Caetano, 2006; Caetano *et al.*, 2008a; Caetano *et al.*, 2008b). The reason is that these products are derived from visual interpretation of high-resolution images to be more reliable and accurate, but this is economically and timely expensive. This background is extremely problematic to many national applications that claim for updated information. Owing to the impossibility to frequently update existing LULC products by manual methods, the obvious approach is to automatically produce new land cover products with higher periodicity, but with less spatial and thematic detail. Thus, in the framework of the COSMIC project, an ongoing research project within the Remote Sensing Unit (RSU) of the Portuguese Geographic Institute (IGP), we are developing automatic methodologies to produce LULC maps for mainland Portugal. The aim is to produce in a quickly and inexpensively fashion two LULC products with different technical and cartographic specifications that will fill up the lack of updated information. Multitemporal intra-annual MERIS and AWiFS images have been explored for land cover mapping in an annual operational basis.

Two important groups of classifiers have been discussed in the remote sensing community: parametric and non-parametric. Parametric classifiers have been applied since the first steps of remotely sensed data classification and their behavior is well known. This group of classifiers has been criticized in some points: i) they make statistical assumptions concerning the distribution of data, in most cases, the Gaussian distribution (Pal & Mather, 2003), which may not be reasonable in remote sensing images containing mixed pixels (Xu *et al.*, 2005); ii) the characteristics of the training data have a considerable influence on the accuracy of the resulting classification (Campbell, 1981); iii) the required training sample size must be large and will increase rapidly as the number of features increases to avoid the curse of dimensionality, also known as ''Hughes phenomenon'' (Hughes, 1967). Despite parametric classifiers limitations, they still are used in several works. Among this group, maximum likelihood classifier (MLC) is definitely the most used and one of the most accurate.

Recently, other methods have been developed in order to overcome some problems unresolved by parametric approaches. This type of classifiers does not make statistical assumptions concerning the distribution of data and has been described in literature as more accurate (Friedl & Brodley, 1997; Xu *et al.*, 2005; Pal & Mather, 2003; Rogan *et al.*, 2002). Decision trees (DT) are one of these classifiers and have been described as advantageous due to various characteristics, namely: their relatively simple, explicit, and intuitive classification structure (Friedl & Brodley, 1997); the ability to handle non-linear relationships between classes features (Friedl & Brodley, 1997; Xu *et al.*, 2005), data that are represented on different measurement scales (Pal & Mather, 2003) and both continuous and nominal data (Homer *et al.*, 2004); quickness in training stage (Gahegan & West 1998 *in* Pal & Mather, 2003; Homer *et al.*, 2004), and fast computational processing (Pal & Mather, 2003; Homer *et al.*, 2004).

The effect of training sample size in DT has been inconsistently described. Pal & Mather (2003) state that these classifiers do not require a very large training sample to be effective whereas Joy *et al.* (2003) *in* Sesnie *et al.* (2008) state the contrary. Oates & Jenson (1997) *in* Pal & Mather (2003) suggest that training sample size has no effect on classification accuracy. Another inconsistency exists regarding response to outliers. Friedl & Brodley (1997) state that DT are flexible and robust with respect to noisy relations among input features and class labels, whereas Miller & Franklin (2002) *in* Sesnie *et al.* (2008) state that instability of trees to outliers or small changes in the training data are a limitation.

Some countries developed efforts to produce LULC products through automated processes, but based upon different approaches. Land Cover Map of Great Britain (Fuller *et al.*, 1994) or Sweden's CORINE land cover mapping project (Hagner & Reese, 2007) are examples of conventional approaches for production of LULC products in an operational basis. Both used maximum likelihood classifier, but in the Swedish case, with a prior weights. In Australia, a LULC mapping programme was developed based on course satellite images to map land use at national scale through modeling procedures based on Normalized Difference Vegetation Index (NDVI) time-series profiles. The probability of each land use is estimated by comparing each pixel's spectral profile with spectral profiles of control sites with known land uses (BRS, 2006).

Non-conventional approaches, namely DT, were already adopted in operational programs regardless its relatively recent use with remotely sensed data. For example, in the framework of the National Land-Cover Database project (NLCD 2001) of USA, a LULC map was derived with DT and Landsat 5/Landsat 7 images (Homer *et al.*, 2004). A DT approach has been explored to map LULC in a timely manner in India with AWiFS images (Kandrika & Roy, 2007).

In this paper we report the results achieved with intra-annual AWiFS images for land cover mapping of Portuguese mainland. The goal was to evaluate potential methods and images for an annual, quickly and inexpensively operational program for land cover map.

## 2 METHODOLOGY

Three scenes from three intra-annual key seasons (April, July and October) were used to explore land cover classification with AWiFS images (Table 1). Two different classification approaches were tested. On the one hand, we explored a more conventional approach using a parametric classifier, the maximum likelihood classifier. On the other hand, we explored a nonparametric classifier, a decision tree classifier (DTC). Considering the goals of COSMIC project, we defined an adequate methodological approach, which consisted of four main stages: i) territorial stratification; ii) land cover nomenclature definition; iii) image classification; and iv) accuracy assessment. Details of each stage are described in the next sections.

Table 1. Technical characteristics of AWiFS images and scenes date.

| Spatial resolution (m) | Spectral resolution (μm) | Radiometric resolution (bits) | Swath (Km) | Scenes date |
|---|---|---|---|---|
| 60 | 0.52 – 0.59 (Green)<br>0.62 – 0.68 (Red)<br>0.77 – 0.86 (NIR)<br>1.55 – 1.70 (SWIR) | 10 | 370 | April<br>July<br>October |

## 2.1 *Territorial stratification*

The stratification of the territory into areas of homogeneous landscapes was recommended to improve the gain in classification results attended whereas classifying the whole Portuguese territory. With territorial stratification it is possible to isolate physiographic types, land cover distribution, spectral similarities, etc. This is because closest geographical individuals are more alike than abroad individuals of the same land cover class. So, different areas with the same class information can have different spectral responses in different strata and need to be trained independently in a tailored and more efficient manner. The stratification stage led to a thirteen strata form (Figure 1) and was based in early works of landscape characterization and assessment (e.g. DGOTDU, 2004). These works used features that are important for automatic classification, such as topography, land use patterns, and so on. Here we report the research work related to a single stratum, used as study area. Our study area, pilled with grey in Figure 1, was chosen due to its characteristics, such as land cover variety or landscape heterogeneity. The goal was to guarantee that problems arosed in classification of study area were demonstrative of remain strata. All tests and results presented in this paper are related with this single stratum.

Figure 1. Geographical stratification and study area

## 2.2 *Land cover nomenclature*

Land cover nomenclature was not defined by any land cover classification system. Nevertheless, LANDEO nomenclature was used as beginning framework. LANDEO is a hierarchical, detailed and multiscale nomenclature that was developed for Portugal by the RSU with the purpose of being used in LULC maps derived by automatic methods (Araújo & Caetano, 2006). The original LANDEO nomenclature was modified along work's progress based on images ability for land cover classification of Portuguese landscape characteristics. As a result of this iterative process, 15 land cover classes were defined (Table 2).

Table 2. Land cover nomenclature.

| Code | Land cover classes | Code | Land cover classes |
|---|---|---|---|
| 1 | Continuous Artificial Areas | 9 | Agro-forestry Areas |
| 2 | Discontinuous Artificial Areas | 10 | Shrubland |
| 3 | Rainfed Herbaceous Crops | 11 | Herbaceous Vegetation |
| 4 | Irrigated Herbaceous Crops | 12 | Forest Clearcuts |
| 5 | Rice Crops | 13 | Permanent Wetlands |
| 6 | Vineyards | 14 | Barren |
| 7 | Broadleaved Evergreen Trees | 15 | Water Bodies |
| 8 | Needleleaved Trees | | |

## 2.3 *Image classification*

This stage included various tasks such as land cover nomenclature and training sample definitions as well as assessment of different classification tests performed with independent testing sample and classifiers. Details are given below.

### 2.3.1 *Training sample*

The training observation collection consisted in a deterministically identification of single pixels spread across the study area to avoid autocorrelation effects (Campbell, 1981; Labovitz, 1986). A conventional training design was followed by collecting pure pixels of each land cover class. A total of 1405 observations were collected within the study area. To each observation was given two reference labels. Besides the label related with the nomenclature of the 15 land cover classes (Table 2), a second label was defined for observations that belong to the same class but had different spectral signatures profiles across the intra-annual images. Second label allows the option of training the classifiers with spectral subclasses rather than map land cover classes.

### 2.3.2 *Classification tests*

Several tests were designed to evaluate three main issues: i) the most suitable classification approach, ii) the effect of training sample characteristics in the classification and iii) image's ability for land cover mapping of mainland Portugal. First issue included the selection of best classifier, the best classification strategy and the best set of intra-annual images to be used as input features for classification. Second issue aimed to define the training sample size needed to perform the classifications without reducing overall accuracy. Moreover, it aimed to evaluate differences between training the classifier with the map land cover classes or with spectral subclasses, i.e. some tests used a higher number of classes in training stage (spectrally similar, as mentioned in topic 2.3.1) that were aggregated after classification to their respective map land cover classes. This procedure did not change global training sample size but change sample size per class because training observations had to be split in more training classes. Together with all these issues, that are very important concerning COSMIC project's goals, the third issue is evaluated. To conclude about these three main issues, we tested different classifiers, input features for classification (dates, NDVI and principal components), training classes (map land cover classes or spectral subclasses), training sample sizes, classification strategies (single-stage or multistage classification), and land cover nomenclatures. Classification tests were performed with the commercial software PCI Geomatics© and See5©. In Table 3 we summarize the tests performed with both classifiers.

Classification tests 1-5, presented in Table 3, were thought to evaluate simultaneously the impacts of training both classifiers with land cover classes or with their spectral subclasses, as well as to evaluate the best set of intra-annual images to be used as input features. So, tests 1-5 differed about training classes, map classes or input features. In classifications tests 6 and 7 we intended to evaluate the classification accuracy whereas reducing the dimensionality of input data. Tests 8-11 were design to evaluate two different issues. On the one hand, to evaluate the potential classification improvement by using a less detailed land cover nomenclature. In tests 8 and 10 we used a nomenclature constituted by 10 classes (Artificial Areas, Rainfed Herbaceous Crops, Summer Herbaceous Crops, Vineyards, Forest, Agro-forestry areas, Non-forest Natural Vegetation, Forest Clearcuts, Barren and Water/Wetlands) that derived from the nomenclature presented in Table 2. In the same way, in tests 9 and 11 we used a nomenclature constituted by 6 classes (Artificial Areas, Croplands, Forest, Non-forest Natural Vegetation, Barren and Water/Wetlands). On the other hand, tests 8-11 were also design to evaluate the effect of training both classifiers with the map land cover nomenclature or with more detailed and similar spectral subclasses, likewise tests 1-4. The difference was that in tests 10 and 11, the 15 land cover classes were used as spectral subclasses and then were aggregated in 10 and 6 map classes, respectively. Tests 8-11, together with tests 1-4, are important to evaluate the training effort for major classification results as well as the time expended in training stage. Test 12 (Table 3) was design to deeply explore the training sample size effect in DTC. In this case, five different classifications were design with half sample size (half observations per class) randomly selected to evaluate classification accuracy with lower training effort. In tests 13 and 14 (Table 3), two different classification strategies were explored. A multistage classification was performed in test 13. It included three independent classifications, each one using spectral data from a single date as input features and with a different and reduced land cover nomenclatures from the nomenclature presented in Table 2. Then, each pixel was classified based on the three a priori classifications. For example, a pixel was classified as Rainfed Herbaceous Crops if it was classified as green herbaceous vegetation, dry herbaceous vegetation and dry herbaceous vegetation, in April, July and October, respectively. Another multistage classification was performed in test 14. In the beginning of this test, land cover classes were merged and then were split progressively by different trees. In first stage water and permanent wetlands were distinguished apart from remain classes. Second stage distinguished land cover classes with and without vegetation, and so on.

### 2.4 *Accuracy assessment*

A testing sample was produced using a stratified random sampling design based on an early classification result, with 30 observations per class and the pixel as sampling unit, in order to derived unbiased accuracy measures. As reference data we used airborne orthoimages acquired between 2004 and 2006 and a mosaic of LISS-3 and SPOT 4 images from 2006. The labelling protocol consisted in assign the "ground truth" to each testing observation through visual interpretation of reference data. Each testing observation was labelled with a principal land cover class plus a secondary land cover class when necessary. This necessity is related due to transitional or mixed pixels and uncertainty in reference land cover class assignment to some "ground truth" that comprises characteristics of more than one land cover class. Finally, classification results were evaluated with overall accuracy indexes derived with the testing sample.

## 3 RESULTS AND DISCUSSION

The tests results are presented in Table 3. Tests that used both classifiers shows that MLC can produce a land cover map as accurate as DTC or even more accurate. Thus, these results do not follow the statements of expert literature and the results of another research works (e.g. Friedl & Brodley, 1997; Pal & Mather, 2003). Despite the similar results achieved with both classifiers, theirs behavior is very different and lead to important discussion.

Tests 1-5 show that MLC is very sensitive to the training sample size. All tests used the same training sample but different training classes led to different observations per class. Note that the overall accuracy of test 1 was 62.11% and the overall accuracy of test 2 (classified attending spectral data from October) was 37.72%. Here we can see the effect of the curse of dimensionality. The problem of data dimensionality is also the reason why test 4, using MLC with the same spectral information than test 2 but with less training classes, retrieved a result much better (59.30%). Tacking into account the curse of dimensionality and the differences in observations per class between tests, it was expected that test 3 (trained with 15 classes) would produce a better classification result than test 1 (trained with 37 subclasses). However, MLC behavior assumes data's distribution as normal and training MLC with land cover classes (that do not have a perfectly normal distribution), instead their spectral subclasses, is a drawback. Thus, the accuracy achieved with test 3 did not overcome the overall accuracy of test 1. Anyway, even though the number of observations per class for 15 land cover classes is higher, it remains to be insufficient. Therefore, the results obtained with MLC were decreasing as along as features were added (tests 3 to 5).

On the other hand, in tests 1-5, DTC produced similar classification results besides the use of land cover classes or spectral subclasses, which shows that the trees construction is not affected by the number of observation per class. Also, training DTC with 15 classes produced a slightly improvement. This indicates that DTC does not require training data characterized by normal distribution. Moreover, comparing test 5 with tests 3 and 4 shows that the addition of more features (namely, October scene and NDVI) produced improvements on the results, though slightly.

In consequence of previously results achieved with MLC, Tests 6 and 7, designed to reduce the data dimensionality, allowed to avoid the huge decrease observed between the results of tests 1 and 2 with the MLC. Tests 6 and 7, which differs from test 1 and 2 on input features and training classes, could avoid the effect of the curse of dimensionality and reveals slightly improvements with both classifiers. Principal components decreased data dimensionality and removed redundant information from original bands. But even though, overall accuracies remain low and AWiFS images seem to contain insufficient data to retrieve higher results.

Table 3. Results and characteristics of tests performed with MLC and DTC.

| Test | Training classes | Map classes | Input features for classification | Strategy | Training sample size | Overall accuracy (%) | |
|---|---|---|---|---|---|---|---|
| | | | | | | MLC | DTC |
| 1 | 37 | 15 | April and July | Single-stage | Entire set | 62.11 | 55.96 |
| 2 | 37 | 15 | April, July and October | Single-stage | Entire set | 37.72 | 55.96 |
| 3 | 15 | 15 | April and July | Single-stage | Entire set | 61.58 | 57.54 |
| 4 | 15 | 15 | April, July and October | Single-stage | Entire set | 59.30 | 56.67 |
| 5 | 15 | 15 | April, July and October, NDVI of July and April | Single-stage | Entire set | 53.51 | 58.98 |
| 6 | 15 | 15 | July and April's principal components | Single-stage | Entire set | 62.98 | 60.53 |
| 7 | 15 | 15 | Three date's principal components | Single-stage | Entire set | 60.35 | 58.68 |
| 8 | 10 | 10 | April, July and October | Single-stage | Entire set | 71.58 | 72.1 |
| 9 | 6 | 6 | April, July and October | Single-stage | Entire set | 74.74 | 78.4 |
| 10 | 15 | 10 | April, July and October | Single-stage | Entire set | 67.37 | 63.86 |
| 11 | 15 | 6 | April, July and October | Single-stage | Entire set | 71.40 | 67.54 |
| 12 | 15 | 15 | April, July and October | Single-stage | 5 random sets with half sample size | -<br>-<br>-<br>-<br>- | 58.25<br>58.42<br>56.84<br>55.26<br>59.82 |
| 13 | 15 | 15 | April; July; October | Multistage | Entire set | - | 54.03 |
| 14 | 15 | 15 | April, July and October | Multistage | Entire set | - | 62.11 |

Tests 8 to 11 were thought to evaluate image's ability to produce a land cover map thematically less detailed. Obviously classification results with both classifiers were higher with a less detailed land cover nomenclature. Although the results improvements were insufficient to became these maps useful. This means that exist confusion between very different land cover classes. Likewise previous tests, these show that training sample definition is very important to classification accuracy. Thus, training the classifiers with more classes than the ones in final map nomenclature (tests 10-11) did not produced better results, on the contrary, it worsen (tests 8-9 retrieved better results). In the MLC case, it can be argue that data dimensionality can be the responsible for that, but it is not reasonable in DTC case. Comparing tests 1-4 and tests 8-9 reveals that DTC achieve higher results with training sample more heterogeneous, i.e. when map classes were not split in training subclasses. This fact is very important because to split training samples of the same land cover class in spectral subclasses that are used only in training stage are timely expensive.

Test 12 was design to deeply explore the training sample size effect on DTC. Likewise tests 1 to 5, this test underline that decision tree are not sensitive to training sample size and it has no effect on classification accuracy. In general, the five tests performed with half samples per class did not retrieved lower results than test 4 (trained with the entire training sample size).

The last two tests (13 and 14), which were design to evaluate different classification strategies, retrieved different results. Test 13 produced a worse result than the one produced by single-stage classification with the same classification features and nomenclature (test 2). This result is due to the strategy's complexity. After each date classification, for each pixel it decided the final classification considering spectral profile of each land cover class along the three dates. This strategy became complex when classification errors occur at least in one date. These errors change land cover spectral profiles and cause misclassifications. Thus, this multistage strategy that used fewer training classes in single date classifications did not bring improvements to the final map. On the other hand, test 14 produced the higher result retrieved by DTC with 15 classes, though slightly. This result shows that, in this multistage strategy, each decision tree fits better to classes distinction along each classification stage. However, this strategy is timely expensive comparing with single-stage strategy and the small improvement achieved makes it not worthy.

## 4 CONCLUSIONS

In general, the overall accuracy of LULC maps produced based on AWiFS images were low, regardless the classification approach. Although results obtained with MLC and DTC were similar, responses to training sample characteristics were not. MLC was too dependent of training sample size. This is because MLC require that the number of observations per training class should be at least 10-30 times the number of features (Mather, 1999). The training sample used in classification tests did not respect this requirement because the time required to collect such number of observations would became the methodology unfeasible. On the contrary, DTC was insensible to training sample size and worked better with heterogeneous training data, that is, map land cover classes rather spectral subclasses. This allows a quick training stage without significant losses in classification accuracy and show that DTC is a most advantageous classifier in an operational basis. Thus, results also show that the map accuracies were limited by data comprised in AWiFS images rather than by unsuitability of classification approaches or training sample. AWiFS inability to derive LULC maps for mainland Portugal could be explained by its course spectral resolution (Table 1). As demonstrated by Carrão *et al.* (2008) and Costa *et al.* (2008), the spectral resolution is the most important technical characteristic of remotely sensed data to map Portuguese landscape. Also, the use of the three AWiFS scenes and NDVI did not retrieved considerable improvements in the results, neither principal components. This suggests that two AWiFS images from spring and summer seasons are sufficient to map Portuguese mainland.

## REFERENCES

Araújo, A. & Caetano, M. 2006. *Nomenclatura de Ocupação de Solo LANDEO*, Lisboa: Instituto Geográfico Português (IGP).

BSR 2006. *Guidelines for land use mapping in Australia: principles, procedures and definitions*, 3rd edition, Canberra: Bureau of Rural Sciences.

Caetano, M., Nunes, V. & Araújo, A. 2008b. Concepção e desenvolvimento das especificações técnicas da nova Carta de Ocupação do Solo de Portugal Continental, *Actas do X Encontro de Utilizadores de Informação Geográfica (ESIG 2008), Portugal,14 - 16 Maio*, in press.

Caetano, M., Pereira, M., Carrão, H., Araújo, A., Nunes, A. & Nunes, N. 2008a. Cartografia temática de ocupação/uso do solo do Instituto Geográfico Português. *Mapping - Revista Internacional de Ciências de la Tierra*, in press.

Campbell, J. B. 1981. Spatial correlation effects upon accuracy of supervised classification of land cover. *Photogrammetric Engineering & Remote Sensing* 47: 355-363.

Carrão, H., Gonçalves, P. & Caetano, M., 2008. Contribution of multispectral and multitemporal information from MODIS images to land cover classification. *Remote Sensing of Environment* 112: 986-997.

Costa, H., Araújo, A., Carrão, H. & Caetano, M. 2008. Influência das características técnicas das imagens de satélite na produção de cartografia de ocupação do solo: estudo baseado em imagens MERIS e AWiFS, *Actas do X Encontro de Utilizadores de Informação Geográfica (ESIG 2008), Portugal,14 - 16 Maio*, in press.

DGOTDU, 2004. *Contributos para a identificação e caracterização da paisagem em Portugal Continental*, Volume I, II, III, IV and V. Lisboa: Direcção-Geral do Ordenamento do Território e Desenvolvimento Urbano.

Foody, G.M. & Mathur, A. 2004. A relative Evaluation of Multiclass Image Classification by Support Vector Machines. *IEEE transactions on Geoscience and Remote Sensing* 42 (6): 1335-1343.

Fridl, M. A., McIver, D. K., Hodges, J. C. F., Zhang, X. Y., Muchoney, D., Strahler, A. H., Woodcock, C. E., Gopal, S., Schneider, A., Cooper, A., Baccini, A., Gao, F. & Schaaf, C. 2002. Global land cover mapping from MODIS: algorithms and early results. *Remote Sensing of Environment* 83: 287-302.

Friedl, M. &Brodley, C. 1997. Decision tree classification of land cover from remotely sensed data. *Remote Sensing of Environment* 61: 399-409.

Fuller, R. M., Groom, G. B. & Jones, A. R. 1994. The Land Cover Map of Great Britain: an automated classification of Landsat Thematic Mapper data. *Photogrammetric Engineering & Remote Sensing* 60: 553-562.

Hagner, O. & Reese, H. 2007. A method for calibrated maximum likelihood classification of forest types. *Remote Sensing of Environment* 110: 438–444.

Homer, C., Huang, C., Yang, L., Wylle, B. & Coan, M. 2004. Development of a 2001 National Land-Cover Database for the United States. *Photogrammetric engineering & Remote Sensing*, 70 (7): 829-840.

Hughes, G. F. 1968. On the mean accuracy of statistical pattern recognizers. *IEEE Transactions on Information Theory* IT-14(1): 55– 63.

Justice, C., O., Vermote, E., Townshend, J. R. G., DeFries, R., Roy, D. P., Hall, D. K., Salomonson, V. V., Privette, J. L., Riggs, G., Strahler, A., Lucht, W., Myneni, R. B., Knyazikhin, Y., Running, S. W., Nemani, R. R., Wan, Z., Huete, A. R., Leeuwen, W., Wolfe, R. E., Giglio, L., Muller, L.P., Lewis, P. & Barnsley, M. J. 1998. The moderate resolution imaging spectroradiometer (MODIS): land Remote Sensing for global change research. *IEEE transactions on Geoscience and Remote Sensing* 36 (4): 1228-1249.

Kandrika, S. & Roy, P.S. 2007. Land use land cover classification of Orissa using multi-temporal IRS-P6 awifs data: A decision tree approach. *International Journal of applied Earth Observation and Geoinformation*, in press.

Labovitz, M. 1986. Issues arising from sampling designs and band selection in discriminating ground reference attributes using remotely sensed data. *Photogrammetric Engineering & Remote Sensing* 52 (2): 201-211.

Mather, P. 2004. *Computer processing of remotely-sensed images: an introduction*, 3rd edition. Chippenham: Wiley.

Painho, M. & Caetano, M. 2006. *Cartografia de Ocupação do Solo, Portugal Continental, 1985-2000.* Amadora: Instituto do Ambiente.

Pal, M. & Mather, M. 2003. An assessment of the effectiveness of decision tree methods for land cover classification. *Remote Sensing of Environment* 86: 554-565.

Rogan, J., Franklin, J. & Roberts, A. 2002. A comparison of methods for monitoring multitemporal vegetation change using Thematic Mapper imagery. *Remote Sensing of Environment* 80: 143– 156.

Sesnie, S., Gessler, P., Finegan, B. & Thessler, S. 2008. Integrating Landsat TM and SRTM-DEM derived variables with decision trees for habitat classification and change detection in complex neotropical environments. *Remote Sensing of Environment*, in press.

Xu, M., Watanachaturaporn, P., Varshney, P. K. & Arora, M. 2005. Decision tree regression for soft classification of remote sensing data. *Remote Sensing of Environment* 97: 322-336.

*Remote Sensing for a Changing Europe*
*D. Maktav (Ed.)*
*IOS Press, 2009*

*doi:10.3233/978-1-58603-986-8-364*

# Object oriented detection of scattered vegetation from IKONOS data

Jan Kolar
*Dept. of Applied Geoinformatics and Cartography, Charles University, Czech Republic*

Keywords: object oriented classification, satellite data, Ikonos, scattered vegetation

ABSTRACT: This paper deals with the application of VHR satellite data object-oriented classification for landscape scattered vegetation monitoring. The object-oriented methods of image data analysis gives certain advantages and benefits comparing to more traditional means. These methods were applied on IKONOS data acquired over test site in the Czech Republic, using Definiens Professional software. Primarily tool used were a multiresolution segmentation algorithm. A role of multiresolution segmentation and related parameters has been assessed. Based on that, the most suitable combinations of segmentation parameters and specific features for determination of scattered vegetation units have been specified. Due to high variety of the size of individual elements belonging to the scattered vegetation, the scale parameter cannot be uniquely determined. Along with spectral features, also spatial and textural features have been implemented into classification algorithm. Standard deviation and brightness are two most significant parameters for detection of the scattered vegetation. Two different classification methods were developed based on nearest neighbour classifier and on thresholding. The processes developed have been verified on another area of landscape as well as on the area of entire Ikonos scene. Producer and user accuracy achieved when using both methods have reached about 80% which can be useful for vegetation inventory.

## 1 INTRODUCTION

Scattered vegetation presents an important features in many landscapes both natural landscapes, and recently modified landscapes. The ecological importance of scattered trees is widely acknowledged at the local scale, as well as at the landscape scale. In combination, these ecological functions support the argument that scattered trees, shrubs and other types of vegetation are keystone structures. This role of the small vegetation was often overlooked in past and individual trees were abandon in favour of the production value of the landscape. Any kind of efficient control brings demands to have a updated information on quantity, state and place all individual landscape elements. An effective landscape management requires regular monitoring and registration of scattered vegetation to keep its presence in an appropriate spatial pattern. That is why scattered vegetation monitoring is growingly requested and development of remote sensing tools is appropriate.

Data from Earth observation satellites provide an efficient source of information for detecting and monitoring the change of scattered vegetation, especially when images with high spatial resolution become readily available. However, high spatial resolution with fewer spectral bands does not always lead to more accurate extraction results. As scattered vegetation is generally growing on background covered by another type of vegetation, the spectral similarity underlined by shadow effect result in poor classification results. Therefore, traditional pixel-based method does not usually give satisfactory result because of confusion between the spectral reflectance from individual trees or bushes and grass, meadow or early crop.

To improve classification results a spatial type of features could be of help in addition to traditional spectral characteristics of classifying targets (Blaschke and Strobl, 2001). Spatial relationship between pixels representing various classes like texture and shape provides another information. Typically in high resolution imagery this contribute to better identification of individual objects as opposed to single pixels (Thomas *et al.*, 2003). Combination and usage of both spectral (colour, intensity) as well as spatial features (size, shape, texture, context, topological relationship with neighbouring objects) for image classification is basic approach applied in object-oriented classification. This method emulates the process of visual interpretation as it is applied by any human operators for years in many applications.

A limited attention has been so far given to scattered vegetation. Study of object oriented approach to tree and shrub identification has been focused rather on vegetation in urban then rural areas (Zhang and Feng, 2005; Mathieu and Aryal, 2005). Object-oriented multiscale image analysis method was used for assessing the shrub in desert grassland in southern New Mexico (Laliberte et al., 2004). This approach has been more widely and also successfully applied in detecting urban objects (Hofmann, 2001).

This study should contribute to improvement into process of extracting elements of scattered vegetation areas from high resolution satellite imagery using a hierarchical object-oriented approach.

## 2 SOFTWARE TOOLS AND DATA

The software named Definiens Professional (before eCognition) was used for this study. It allows employ a object-oriented approach to satellite data analysis in two main parts: multi-resolution segmentation and context-based classification.

Multi-resolution segmentation allows generating image objects on an arbitrary number of scales taking into account criteria of homogeneity in colour and shape. Additionally, the created segments are embedded into a hierarchical network in which each object knows its neighbouring objects in horizontal and vertical direction. Image analysis leads to meaningful objects only when the image is segmented in 'homogeneous' areas (Baatz *et al.*, 2000). The segmentation ensures the subsequent object based image processing.

The image objects can then be described and classified by an extensive variety of features that include colour, texture, form, and context properties in several forms. This is usually done with a standard nearest neighbour classifier (NN) or fuzzy membership functions, or a combination of both. The variety of object features can be used either to describe fuzzy membership functions, or to determine the feature space for NN (Baatz and Schape 1999).

In this study, Ikonos data acquired in June 2006 of the testfield in the Czech Republic has been used. Four georeferenced spectral bands all sharpened with panchromatic data were handled during the testing. Supporting ancillary data were obtained from vector topographic database Zabaged related to the 1:10000 scale map. This source of spatial data was used for verification purposes.

Selected test area is close to Vrchlabi town located in the northern mountainous part of the Czech Republic. It covers an area of 10 km x 10 km in rural area covered mostly by meadows, grass and forest. Small settlements are sparse. The area has been selected for an abundance of scattered vegetation of various kind.

The classification procedure development, finding optimal segmentation parameters and classification features was accomplished on a one subset of 2150 x 2150 pixels. The second subset of the same image having 1800 x 1700 pixels has been used for verification and control of the final classification method. Final process has been consequently applied to the total test area.

# 3 OBJECT-ORIENTED CLASSIFICATION

## 3.1 *Preparatory steps*

Before the data classification some categories outside of the scattered vegetation has been located and marked in the data using appropriate mask with support of topographic database Zabaged. To obtain a reference layer of scattered vegetation the satellite image was interpreted manually. The result in the shapefile format served as a comparison standard for verification of the automatic classification accuracy.

Only two classes were to be distinguished in classification: 1)scattered vegetation and 2) all other objects. The first class includes individual trees or groups of trees and shrub while meadows, arable land, paths and other non-vegetation surfaces belong to the class 2.

## 3.2 *Segmentation and NN classification*

Data from all four pansharped spectral bands were included for segmentation process with the same weight of 1.

Segmentation process has been fully undertaken with segmentation parameters defined and used in Definiens software. They are scale parameter, shape parameter and compactness (Definiens Imaging 2004).

In this study, all possible combinations of following values of the scale, shape and compactness parameters respectively, were tested: scale parameter - 20, 25, 30 and 35; shape parameter - 0.1, 0.3, 0.5 and 0.7; compactness factor - 0, 0.5 and 1.

The best combination of both parameters has been defined based on nearest neighbour classification results. Every specific set of segmentation parameters has been assessed using classification accuracy of segments. The classification accuracy can be expressed in a ratio of number of correctly classified pixels of given class to total number of pixels of that class (producer accuracy PA). Another user's accuracy (UA) is calculated by dividing the number of correct pixels for a class by the total pixels assigned to that class. In fact, following combinations of both PA and UA named Hallden coefficient HA and Short coefficient SA, respectively, have been used for segmentation assessment:

$$HA(class_i) = \frac{2}{\frac{1}{PA(class_i)} + \frac{1}{UA(class_i)}} \tag{1}$$

$$SA(class_i) = \frac{1}{\frac{1}{PA(class_i)} + \frac{1}{UA(class_i)} - 1} \tag{2}$$

The highest accuracy value has been reached for shape parameter 0.5 and scale parameter 30. That twin of parameters values were taken for next steps. Similar assessment of the deviation from a compact or smooth shape on the classification results has shown that optimal preference ratio of both homogeneity components is 1:1.

## 3.3 *Thresholding classification*

The features considered for classification process included following parameters computed from pixels belonging to an object:

- mean in every spectral band
- brightness (average of the four means divided by number of bands
- standard deviation in every spectral band

- ratio of standard deviation in every band and brightness
- ratio of a sum of standard deviations in all bands and brightness

Tresholding method was used for separation of two classes of interest: scattered vegetation and others. The ratio of standard deviation in every band and brightness has been found as the most effective feature for very first detection of scattered vegetation. The values of the ratio for scattered vegetation pixels were between 1 and 8, and smaller than 1 for other objects. Figure 1 shows distribution of the feature value over the subset 1 in gray levels.

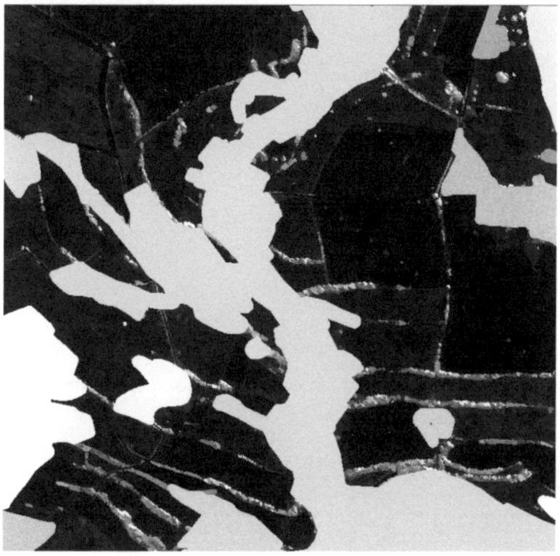

Figure. 1. Values of the ratio of standard deviation in every band and brightness.
High values (light) mark scattered vegetation.

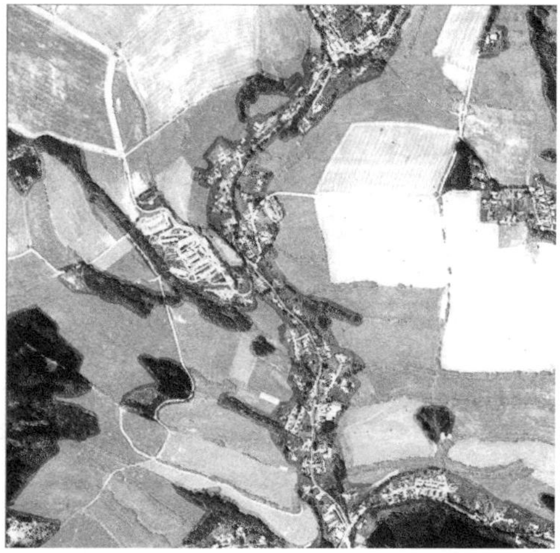

Figure 2. Final result of test area classification by thresholding method.
Green-scattered vegetation, grey-other objects.

As some non-scattered vegetation objects (e.g. paths) were incorrectly included into this class, another feature has been applied on the classification result. The second feature was mean value in red band served as a tool to select non vegetation objects from previous classification result.

Obtained class of scattered vegetation 1 has still contained pixels belonging to grass. Therefore, a ratio of standard deviation in blue band and brightness was used for another division. This feature removed remaining grass objects from the scattered vegetation 1 class.

After that step, the accuracy of the scattered vegetation class has been reduced by missing lighter parts of some trees crowns. To solve the problem a contextual feature of neighborhood to the scattered vegetation 1 class has been applied to the class of other objects 2. This step concluded the hierarchical procedure of tresholding classification method leading to the final classified picture (fig. 2). The procedure is graphically depicted on Fig.3.

Final classification accuracies reached for three areas are presented in Tab.1. The accuracy is similar for both methods with slightly better results for the thresholding classification method. However, the both Hellden and Short coefficient decreased when classification parameters tuned up with subset 1 data were applied on subset 2 and, mainly, on the entire test site. On the other hand, comparison was done with layer of scattered vegetation generated during visual interpretation of the image. Delineation of the vegetation boundaries is generalized which automatically does not correspond to the boundary of the same polygons obtained through classification procedure. On the other hand, the boundary of classification follows pixel boundaries precisely, thus, creating differences with manually drawn boundary.

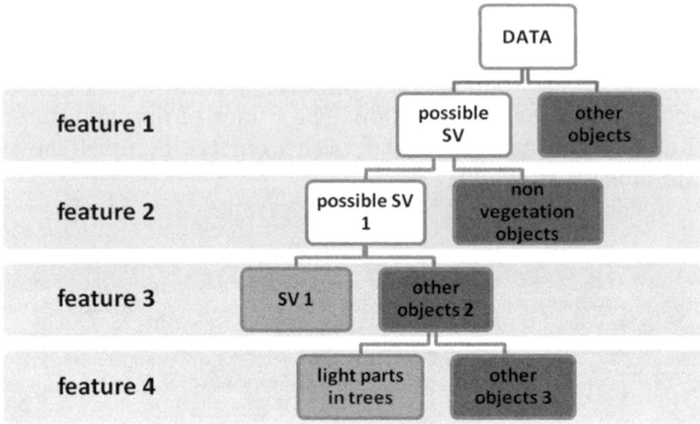


Figure 3. Overview of the hierarchical structure. Class 1 – green, class 2 – gray.

Table 1. Object classification accuracy for the nearest neighbour (NN) and the tresholding method (T).

| Accuracy coefficient | Subset 1 | | Subset 2 | | Test site 10x10 km | |
|---|---|---|---|---|---|---|
| | NN | T | NN | T | NN | T |
| Producer accuracy | 0.823 | 0.839 | 0.895 | 0.876 | 0.766 | 0.718 |
| User's accuracy | 0.887 | 0.948 | 0.777 | 0.840 | 0.782 | 0.838 |
| Hellden coefficient | 0.854 | 0.890 | 0.832 | 0.857 | 0.774 | 0.773 |
| Short coefficient | 0.745 | 0.802 | 0.712 | 0.750 | 0.632 | 0.630 |

## 4 DISCUSSION AND CONCLUSION

Hierarchical object-oriented classification systems provide us a good way to segment arbitrary images into highly homogeneous objects and extract scattered vegetation areas by using rules based on spectral, textural and spatial characteristics. Due to their small size the scattered vegetation objects can be distinguished only on very high resolution satellite data or on aerial images. Moreover, the scattered vegetation features spectral heterogeneity which degrades usage of only spectral-based pixelwise classification methods. Object-oriented classification approached applied in this work through the Definiens software has proved to profit on extension of spectral feature space to spatial and context features.

Two kinds of image classifications were performed in Definiens following multi-resolution segmentation: standard nearest neighbour classifier and threshold classifier based on a fuzzy membership function.

The segmentation has been accomplished with optimal value of the scale parameter for segmentation set to 30, the shape parameter to 0.5 with the same weight for both compactness and smoothness parameters of object homogeneity. It has to be seen as a compromise, though, as elements of the scattered vegetation show wide range of their sizes and shapes. This uncertainty in setting the right segmentation parameters is a weakness of the object-oriented approach. It leads to a time consuming, try-error procedure of finding the optimal parameters.

Altogether fourteen feature parameters were used for the object classification. Among them, the ratio of standard deviation in spectral band to brightness has been identified as the most influential feature for scattered vegetation identification. The ration has given better result than standard deviation in band alone, the feature used in several previous works (Mathieu, 2005; Zhang a Feng, 2005). The explanation for that can be found in specific radiometric characteristics of sparse vegetation. Objects belonging to that category usually shown higher internal heterogeneity (higher variation of pixels) and low reflectance over all optical spectral bands.

Both classification methods bring mostly similar results in terms of classification accuracy measured on every pixel. As a vector layer obtained in visual interpretation was the standard layer used for verification, the classification accuracy is influenced by generalisation effect typical for visual interpretation process. If the concern is identification of the scattered vegetation only, and it is not its exact size, the accuracy of object identification reaches almost 100%.

Object-oriented analysis with Definiens software showed a significant improvement in accuracy. The improvement is due to multiresolution segmentation in which objects' spectral as well as spatial information was involved. Also, the friendly programme environment allows easily modify the hierarchical structure when needed and, thus, to improve the accuracy of the thresholding classification.

Described procedure can serve in efficient manner to get scattered vegetation inventory on local and regional level. Further development of automatic part is planned to improve pixel accuracy allowing proper area measurements and to detect changes of scattered vegetation appearance in time.

## ACKNOWLEDGEMENTS

The work was supported by the Ministry of Education, Youth and Sports of the Czech Republic in frame of the Grant No. MSM 0021620831.

## REFERENCES

Blaschke, T., Strobl, J. 2001. What´s wrong with pixels? Some recent developments interfacing remote sensing and GIS. *GIS Zeitschrift für Geoinformationssysteme* 14(6), 12-17, 2001.

Definiens imaging 2004. eCognition Professional User Guide v4. Definiens Imaging, München, Germany. 485 p.

Hofmann, P. 2001. Detecting urban features from IKONOS data using an object-oriented approach, *RSPS2001 Proceedings*: 79-91.

Laliberte, A.S., et. al 2004. Object-oriented image analysis for mapping shrub encroachment from 1937 to 2003 in southern New Mexico, *Remote Sensing of Environment*, 93: 198-210.

Mathieu, R., Aryal, J. 2005. Object-oriented classification and Ikonos multispectral imagery for mapping vegetation communities in urban areas. In Proceedings: *Seventeenth Annual Colloquium of the Spatial Information Research Centre*. P.A. Whigham (Ed). 24 - 25 Nov, Dunedin, New Zealand. University of Otago, ISBN 1-877139-90-4. pp.181-188.

Zhang, X., Feng, X. 2005. Detecting urban vegetation from IKONOS data using an object-oriented approach. In: *International Geoscience And Remote Sensing symposium, Harmony between Man and Nature*, Seoul, Korea.

*Remote Sensing for a Changing Europe*
*D. Maktav (Ed.)*
*IOS Press, 2009*

*doi:10.3233/978-1-58603-986-8-370*

# CN derivation from LANDSAT ETM+ data and LC changes for rainfall-runoff models

M. Hanzlova, J. Horak
*Institute of geoinformatics, VŠB-Technical University of Ostrava, Ostrava, Czech Republic*

J. Unucka
*Institut of geological ingeneering, VŠB-TU Ostrava, 17. listopadu 15, 70833 Ostrava-Poruba, Czech Republic*

L. Halounova
*Department of Mapping and Cartography, Czech Technical University in Prague, Prague, Czech Republic*

J. Heller
*Cross Czech a.s., Václavské nám. 808/66, Praha 1, Czech Republic*

D. Zidek
*Czech Hydrometeorological Institute - Ostrava, K myslivně 3, 708 00 Ostrava-Poruba, Czech Republic*

Keywords: CN, rainfall-runoff process, LULC, LANDSAT ETM+

ABSTRACT:

Land cover and its characteristics influence actual and long term water balance of landscape segment, particularly evapotranspiration, infiltration, surface and hypodermic runoff. Thus description and modelling of the particular events allow determining basic requirements for land cover classification effective to rainfall-runoff (further in the text as RR) condition assessment. The cooperation "remote sensing – GIS" together with hydrological models proposes solution for comprehension of the investigated phenomenon.

LANDSAT ETM+ data represents a valuable and frequently used data source for description of LULC data as a significant input to rainfall-runoff modelling (further in the text as RRM). LC classified according RRM requirements were compared to the CORINE LC data. The paper shows results comprising very important differences and proves irreplaceableness of remote sensing data for the purpose especially in conditions such as rapid landscape changes (e.g. geological hazards), seasonal changes in vegetation cover, and significant changes in agricultural areas during the year; due to its capability of providing the current LC information.

Land cover data sources (CORINE, LANDSAT ETM+) were use for CN-curve value association with particular land cover and soil conditions and the traditional and new approach of CN-value determination were tested in RRM for 2 small catchment areas (Bela and Olse). Such process requires an evaluation of individual data sources, their processing and evaluation of outcomes from RRM.

## 1 INTRODUCTION

There are many factors that influence rainfall-runoff process in landscape and its numerical modelling. Land use and land cover determine hydrological transformation of atmospheric precipitation to catchment. There are two ways how land use and land cover take part in RR process. First, it is direct land use and land cover participation in RR relationship, such as interception and evaporation. Second and more significant, it is land use and land cover function as soil protecting cover factor where the essential part of precipitation transformation plays role. A lot of analyses have been done on this matter, e.g. for Czech Republic there are works of researchers such as Kantor, Zeleny, Jarabac, Chlebek, for Europe Benecke, Heuveldop, for Russia Rajev, Bitjukov. In Czech Republic there are small experimental catchments (e.g. Cervik and Mala Raztoka in Beskydy mountains). They are mostly devoted to analysis of forest impact (especially differences between Spruce and Beech trees) to RR process.

Land cover and land use data together with other data create information base for hydrological models. Nowadays we use several well-known data sources such as CORINE Land Cover that unsatisfactorily distinguishes agricultural areas, which is necessary to differentiate them according diverse soil infiltration capacity.

Satellite imagery proposes actual information and becomes very efficient source of land use and land cover data. Such data can be then interpreted in diverse ways, according different classification schemas. In result land use and land cover data can cause inefficiency of hydrological modelling.

Land cover data sources (CORINE, LANDSAT ETM+) were use for CN-curve value association with particular land cover and soil conditions and the traditional and new approach of CN-value determination were tested in RRM for 2 small catchment areas (Bela and Olse).

Within ongoing project „**Application of Geoinformation Technologies for Improvement of Rainfall-Runoff Relationships**" (**205/06/1037**) there are selected two pilot areas – Bela catchment (Jeseniky Mountains) and Olse catchment (Beskydy Mountains). The Bela River leaves Czech Republic in Mikulovice and it is tributary of the Nysa Klodzka River in Poland. The Olse River is tributary of the Odra River that flows to Poland. Both rivers drain water from Czech Republic through Poland to the Baltic Sea.

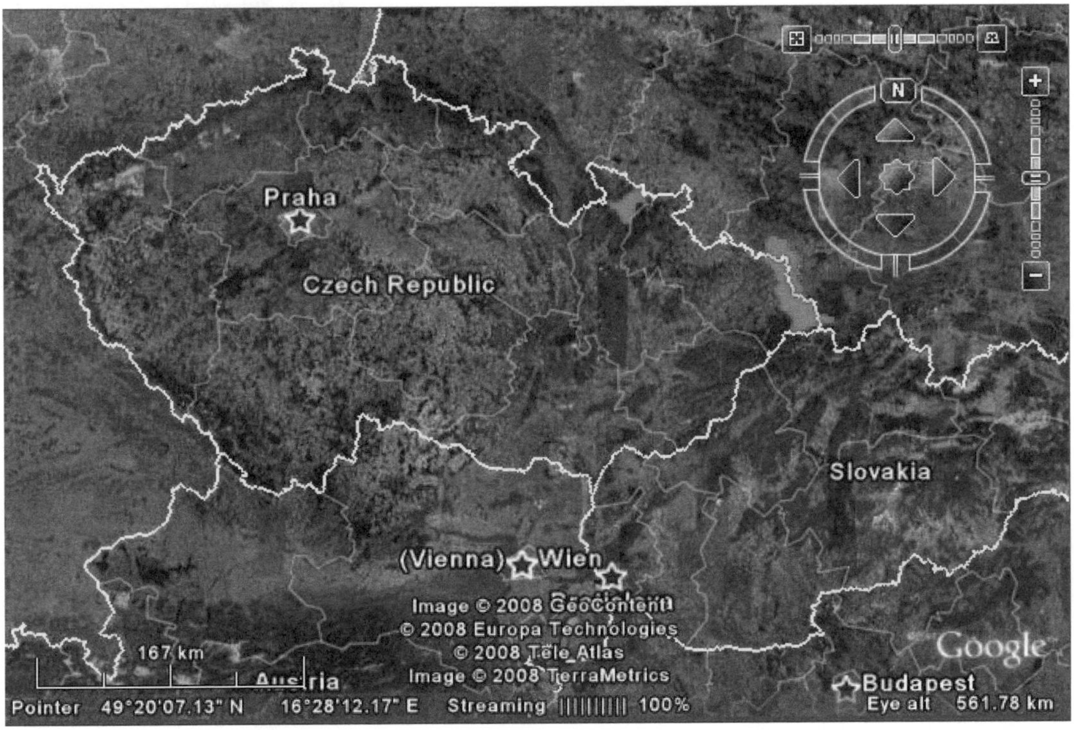


**Figure 1. Pilot areas – Bela and Olse**

## 2 EARTH OBSERVATION DATA AS A LAND COVER SOURCE

Remote sensing proposes one of the significant LULC data sources. EO data were used for CORINE LC database creation (European LC database, http://reports.eea.europa.eu/COR0-landcover/en). CORINE LC data uses specific classification scheme with three levels prepared for certain spatial resolution. Other sources use various other classification schemes such as USGS or UNESCO ones. According necessary scale or size of the studied area there are RS data sources that can be used:

**1 : 25 000 – 1 : 5 000 Very High Resolution Data** (spatial resolution of 1 m) (IKONOS, QuickBird, OrbView 3, Formosat-2),

**1 : 100 000 – 1 : 25 000 High Resolution Data** (spatial resolution of 10 m) (LANDSAT, SPOT, ASTER, IRS, ALI, HYPERION),

**Larger scale than 1 : 100 000** for **regional and global** modelling (e.g. MODIS).

A land cover classification scheme was created to support the investigation of LC effects on r-r-process. Designed schema (hydro-cl) states of four levels, each dedicated to spatial scale, first level copies CORINE LC nomenclature, second level contains 8 classes and follow usage of high resolution EO data (e.g. LANDSAT, ASTER) so as third level (17 classes), fourth level includes 39 classes and is used for more detailed study using very high resolution EO data. Level four needs further ancillary data such as field investigation (mostly for Land use determination).

| LEVEL 1 | LEVEL 2 | LEVEL 3 | LEVEL 4 |
|---|---|---|---|
| ARTIFICIAL AREAS [1] | Urban fabric [11] | Urbanization areas [111] | Residential areas [111.1] |
| | | | Industrial areas [111.2] |
| | | Urban open space [112] | Urban open space [112.1] |
| | Comunications [12] | Paved communications [121] | Paved - tiled, bituminous [121.1] |
| | | | Paved - gravelly [121.2] |
| | | Unpaved communications [122] | Unpaved, clay [122.1] |
| AGRICULTURAL AREAS [2] | Bare soil, pastures, meadows, orchard [21] | Fallow land [211] | After-harvest remains (PZ) [211.1] |
| | | | Spud Areas [211.2] |
| | | Pastures and meadows [212] | Pastures with cover [212.1] |
| | | | Harvested medows [212.2] |
| | | Orchards with grass rows [213] | Orchards with grass rows [213.1] |
| | Line crop and fodder plants [22] | Wide-line crop (root-crops) [221] | Lines in direction of fall-line (SP) [221.1] |
| | | | SP + PZ [221.2] |
| | | | Lines in direction of contourline (VR) [221.3] |
| | | | VR + PZ [221.4] |
| | | | VR + zone-line cultivated crop (PP) [221.5] |
| | | | VR + PP + PZ [221.6] |
| | | Narrow-line crops (cereals) [222] | SP + PZ [222.1] |
| | | | VR [222.2] |
| | | | VR + PZ [222.3] |
| | | | VR + PP [222.4] |
| | | | VR + PP + PZ [222.5] |
| | | Perennial fodder plants, legume [223] | SP [223.1] |
| | | | VR [223.2] |
| | | | VR + PZ [223.3] |
| FOREST AND SEMINATURAL AREAS [3] | Forest [31] | Deciduous forest (LS) [311] | LS heavily damaged [311.1] |
| | | | LS partially damaged [331.2] |
| | | | LS healthy [311.3] |
| | | Coniferous [312] | JL heavily damaged [312.1] |
| | | | JL partially damaged [312.2] |
| | | | JL healthy [312.3] |
| | | Mixed forest (SL) [313] | SL heavily damaged [313.1] |
| | | | SL partially damaged [313.2] |
| | | | SL healthy [313.3] |
| | Shrubbery [32] | Shrubbery [321] | Shrubbery with cover [321.1] |
| | | | Grass and low forest stand [321.2] |
| WATER BODIES [5] | Water course [51] | Water course [511] | Water course [511.1] |
| | Water bodies [52] | Water bodies - shallow, coastal waters [521] | Water bodies - shallow, coastal waters [521.1] |
| | | Water bodies - deep [522] | Water bodies - deep [522.1] |

**Figure 2. HYDRO-LC classification schema**

## 3 CORINE LC AND LANDSAT LC COMPARISON

As the title of this chapter indicates, there were used two LC data sources. One extracted from LANDSAT data, acquisition date is 14 May 2000 and 2 August 2002, and another one represented CORINE LC 1990 and 2000.

Comparisons that were done between datasets as follows:

[comparison-1] CORINE LC 1990 (diss_clc90j) and CORINE LC 2000 (diss_clc00j) – Bela, Olse PA

[comparison-2] LANDSAT LC MAY2000 (diss_Lds00) and LANDSAT LC AUG2002 (diss_Lds02) – Bela PA

[comparison-3] CORINE LC 2000 (diss_clc00j) and LANDSAT LC MAY2000 (diss_Lds00) – Bela PA

Classes that were evaluated are as follows:

[1] ARTIFICIAL AREAS
[2] AGRICULTURAL AREAS
[3] FOREST AND SEMINATURAL AREAS
[5] WATER BODIES

The table below includes investigated LC changes.

| Code | Type of LC change |
|---|---|
| 0 | without change |
| 12 | artificial areas into agricultural areas |
| 13 | artificial areas into forest and seminatural areas |
| 15 | artifical areas into water bodies |
| 21 | agricultural areas into artificial areas |
| 23 | agricultural areas into forest and seminatural areas |
| 25 | agricultural areas into water bodies |
| 31 | forest and seminatural areas into artificial areas |
| 32 | forest and seminatural areas into agricultural areas |
| 35 | forest and seminatural areas into water bodies |
| 51 | water bodies into artificial areas |
| 52 | water bodies into agricultural areas |
| 53 | water bodies and forest and seminatural areas |

**Figure 3. Type of LC changes anticipated in comparison**

### 3.1 *Bela pilot area*

Outputs from comparison CORINE LC data 1990 and 2000 (comparison-1) shows change in artificial area enlarge of 0.1% of total PA area. Such a change cannot project into complex LC conditions in the area. However "comparion-2" indicates changes in 24% of total PA area (higher differences than CORINE comparison). It is important to stress time span for this comparison which is approximately 2 years (2000-2002) further more different season (MAY – AUGUST). The biggest change accrued to artificial areas is that the area increased of 7.0%. There is 4.7% decrease of agricultural areas and the area of forest decreased of 2.4%. In both cases, the $\chi^2$ test proved independence of each LC change.

Comparison of LANDSAT data and CORINE LC (comparison-3) resulted in 18.9% change in total area. It is done thus many aspects (a) CORINE scale 1:100,000 that surely indicate a level of generalization, (b) minimum area for classification is 25 ha. Contrary classification of LANDSAT data were base on per-pixel classification (resolution 30 m).

## 3.2 *Olse pilot area*

The "comparison-1" was done for this pilot area. Statistically, the changes that occurred within the area are 0.8% of total PA area. Based on CAC method we can say there were no changes in the area and as to the change dependency the $\chi^2$ test proves of LC change cohesion.

# 4 LAND COVER AND CN-CURVE NUMBER ASSOCIATION

In terms of LULC changes studies or studies of actual LULC conditions, the GIS implements particular facet, such as data pre-processing, watersheds schematization, meteorological input data processing (e.g. spatial interpolation of precipitation), scenario creation for LU change affect, post-processing.

After initial model calibration based on episode level, there is rising possibility of time series simulation. Thus the aspect of LULC change and its short-term and long-term affect on runoff in the landscape as well as on hydrometeorological characteristics (global climate change) can be simulated according various LULC scenarios.

The essential transparent method of r-r modelling in relation to runoff characteristics and LULC is method based on CN-curve number. CN-curve numbers for LULC classes can be easily found in literature (Hradek, Kurik, HEC-HMS manual). In this case, GIS plays significant role for CN-curve association with LULC where another dimension implies to soil conditions (hydrological soil group, HSG). Using GIS analyses, e.g. UNION, DISSOLVE, and attribute table editing, it is able to create combination of LULC-HSG-CN that proposes more immediate linkage of LULC and r-r process.

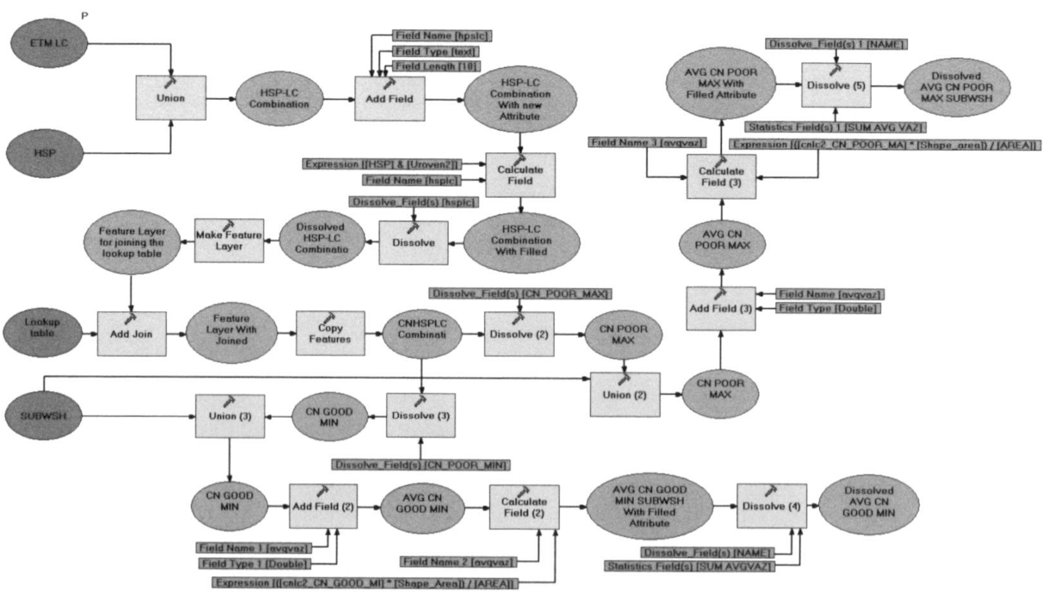


**Figure 4. LC-HSG-CN association**

## 5 HEC-HMS MODEL OUTPUTS

LANDSAT ETM+ imagery (5 scene, track 89 and 90, frame 25) served to LC information extraction analysis. We used per-pixel classification (Maximum likelihood algorithm) according to designed Hydro_LC system that was created with special regard to hydrological effects of land cover. Landsat data cover years 2000-2002 (May, June, August, see Figure 5).

| Acquisition date | RMS [m] (georeferencing) | Spatial resolution | Estimated classification accuracy |
|---|---|---|---|
| 14 MAY 2000 | 10.8 | 25 | 61% |
| 24 MAY 2001 | 2.5 | 25 | 74% |
| 4 MAY 2002 | 31 | 30 | 82% |
| 5 JUNE 2002 | 19 | 30 | 80% |
| 2 AUGUST 2000 | 30 | 39 | 78% |

**Figure 5. Overview of LANDSAT data used.**

Each LANDSAT LC dataset was used for CN-curve number association together with HSG data source. It resulted in 5 CN-number sets for each sub-catchment (minimum and maximum CN value, depends on soil infiltration conditions – poor or good). See Figure 6.

Such data alternates the original used for calibration of hydrological model (HEC-HMS). The effect on run-off is showed in Figure 7 using hydrograph curves.

For PA Olse there were carried out simulated runs with different scenario. Forest LC was enlarged to 100% and 50%. You can see the results in the poster.

| NAME | JUNE 2002 | | MAY 2000 | | MAY2001 | | MAY2002 | | AUGUST 2002 | | CN Original |
|---|---|---|---|---|---|---|---|---|---|---|---|
| | CN POOR MAX | CN GOOD MIN | CN POOR MAX | CN GOOD MIN | CN POOR MAX | CN GOOD MIN | CN POOR MAX | CN GOOD MIN | CN POOR MAX | CN GOOD MIN | BASED on CORINE LC 2000 |
| R100W100 | 70,43 | 61,10 | 69,42 | 60,84 | 69,51 | 60,26 | 68,70 | 59,20 | 70,4 | 59,77 | 68.50 |
| R10W10 | 24,77 | 21,08 | 25,28 | 21,00 | 25,09 | 21,83 | 23,78 | 21,33 | 24,48 | 21,19 | 80.30 |
| R110W110 | 70,21 | 60,37 | 68,39 | 58,31 | 68,68 | 58,60 | 66,28 | 55,86 | 69,91 | 58,98 | 67.60 |
| R120W120 | 75,40 | 67,21 | 74,39 | 66,60 | 74,42 | 66,13 | 73,64 | 65,52 | 74,38 | 65,46 | 75.20 |
| R130W130 | 68,95 | 59,09 | 68,40 | 58,34 | 69,38 | 58,44 | 66,17 | 56,22 | 68,40 | 58,18 | 71.80 |
| R140W140 | 69,30 | 59,99 | 68,81 | 60,31 | 68,91 | 60,10 | 66,33 | 59,06 | 68,80 | 59,88 | 77.40 |
| R150W150 | 77,72 | 69,24 | 76,54 | 68,41 | 76,64 | 68,45 | 74,70 | 66,83 | 76,31 | 67,87 | 76.80 |
| R160W160 | 72,87 | 61,91 | 68,86 | 56,81 | 71,00 | 58,88 | 63,62 | 53,16 | 70,48 | 58,78 | 68.90 |
| R170W170 | 77,68 | 66,79 | 76,94 | 66,52 | 77,66 | 66,40 | 73,59 | 63,30 | 76,00 | 64,91 | 72.30 |
| R180W180 | 71,71 | 61,27 | 68,82 | 56,99 | 68,47 | 57,70 | 64,34 | 54,01 | 70,07 | 58,01 | 70.40 |
| R190W190 | 78,46 | 69,64 | 78,08 | 69,40 | 77,78 | 68,98 | 74,57 | 66,72 | 77,57 | 68,34 | 76.00 |
| R200W200 | 75,13 | 67,65 | 74,30 | 65,87 | 73,60 | 65,10 | 69,34 | 61,06 | 74,30 | 65,00 | 75.70 |
| R20W20 | 72,10 | 54,75 | 71,73 | 55,15 | 71,34 | 55,57 | 60,80 | 54,05 | 71,73 | 55,61 | 81.30 |
| R210W210 | 70,71 | 61,09 | 69,44 | 59,01 | 69,47 | 59,22 | 65,65 | 56,04 | 70,06 | 59,52 | 69.40 |
| R220W220 | 82,54 | 71,16 | 81,39 | 71,60 | 81,57 | 71,81 | 78,74 | 72,37 | 81,34 | 71,33 | 83.20 |
| R230W230 | 77,53 | 70,13 | 76,91 | 69,79 | 77,05 | 69,63 | 75,03 | 68,12 | 77,12 | 69,15 | 77.50 |
| R240W240 | 71,83 | 62,73 | 69,53 | 59,98 | 70,42 | 60,85 | 67,70 | 58,93 | 70,45 | 60,81 | 74.80 |
| R250W250 | 69,97 | 59,79 | 68,50 | 56,45 | 71,08 | 57,97 | 61,63 | 52,44 | 68,52 | 57,91 | 70.30 |
| R260W260 | 67,86 | 56,96 | 69,08 | 54,98 | 72,13 | 56,43 | 58,25 | 48,97 | 67,93 | 56,63 | 71.70 |
| R270W270 | 68,35 | 58,92 | 67,83 | 57,78 | 68,24 | 58,30 | 65,21 | 56,35 | 67,76 | 58,28 | 75.10 |
| R280W280 | 76,70 | 68,56 | 76,23 | 67,95 | 76,53 | 68,25 | 72,55 | 65,37 | 75,95 | 67,42 | 74.30 |
| R290W290 | 69,87 | 59,58 | 68,34 | 57,60 | 69,22 | 58,62 | 64,26 | 54,49 | 69,51 | 58,49 | 68.30 |
| R300W300 | 82,10 | 71,87 | 80,65 | 73,06 | 80,66 | 73,32 | 77,85 | 72,73 | 80,93 | 72,30 | 78.40 |
| R30W30 | 46,76 | 39,76 | 46,62 | 41,05 | 48,22 | 41,60 | 44,84 | 39,94 | 45,38 | 40,19 | 77.20 |
| R310W310 | 69,05 | 58,33 | 69,33 | 57,91 | 70,40 | 58,79 | 64,93 | 55,52 | 69,45 | 58,62 | 75.10 |
| R320W320 | 66,71 | 55,94 | 69,38 | 55,88 | 69,22 | 56,23 | 55,81 | 46,48 | 68,04 | 56,61 | 71.90 |
| R330W330 | 83,15 | 68,37 | 80,94 | 69,30 | 83,74 | 67,99 | 76,56 | 71,63 | 82,16 | 69,49 | 82.60 |
| R340W340 | 81,49 | 73,32 | 80,40 | 73,87 | 80,24 | 74,13 | 77,53 | 73,51 | 81,55 | 71,05 | 82.80 |
| R350W350 | 79,61 | 68,90 | 77,30 | 69,78 | 78,03 | 69,69 | 74,07 | 67,90 | 78,05 | 69,20 | 79.00 |
| R360W360 | 70,41 | 57,47 | 70,55 | 55,86 | 71,79 | 56,91 | 52,07 | 42,79 | 69,71 | 57,15 | 72.30 |
| R370W370 | 67,94 | 57,80 | 67,95 | 56,44 | 68,81 | 57,19 | 63,71 | 54,34 | 67,74 | 57,43 | 74.00 |
| R380W380 | 79,09 | 65,82 | 75,20 | 69,50 | 77,93 | 67,03 | 74,13 | 68,21 | 75,58 | 68,31 | 73.70 |
| R390W390 | 74,96 | 66,15 | 73,17 | 65,01 | 73,46 | 65,49 | 70,62 | 62,18 | 74,75 | 63,72 | 74.50 |
| R400W400 | 79,64 | 68,68 | 77,15 | 69,72 | 76,99 | 69,77 | 72,85 | 66,79 | 76,97 | 68,78 | 73.40 |
| R40W40 | 76,71 | 68,14 | 78,08 | 67,17 | 76,87 | 68,41 | 73,61 | 69,07 | 77,39 | 66,95 | 84.10 |
| R410W410 | 78,04 | 70,77 | 77,50 | 69,91 | 78,49 | 69,39 | 73,80 | 67,38 | 77,01 | 70,15 | 75.20 |
| R420W420 | 67,07 | 55,82 | 68,99 | 55,17 | 69,68 | 56,16 | 51,74 | 43,54 | 67,65 | 56,29 | 72.40 |
| R430W430 | 67,89 | 57,54 | 68,52 | 55,41 | 71,36 | 57,24 | 59,52 | 50,11 | 67,80 | 56,76 | 71.70 |
| R440W440 | 70,14 | 61,43 | 69,72 | 60,15 | 70,73 | 60,93 | 66,81 | 58,18 | 69,94 | 60,98 | 70.50 |
| R450W450 | 67,16 | 57,13 | 67,39 | 55,27 | 68,58 | 56,33 | 61,74 | 52,23 | 67,08 | 56,19 | 70.60 |
| R460W460 | 68,16 | 56,90 | 70,16 | 55,36 | 70,85 | 56,02 | 51,46 | 42,98 | 67,70 | 56,78 | 71.80 |
| R470W470 | 56,24 | 48,05 | 56,51 | 46,83 | 57,79 | 47,70 | 46,50 | 39,70 | 56,70 | 47,85 | 71.70 |
| R50W50 | 68,68 | 58,31 | 69,01 | 58,43 | 71,69 | 59,47 | 65,01 | 56,12 | 67,32 | 57,12 | 72.30 |
| R60W60 | 68,82 | 58,67 | 72,06 | 57,95 | 76,07 | 60,31 | 67,62 | 56,76 | 67,76 | 57,86 | 69.70 |
| R70W70 | 75,49 | 66,76 | 76,13 | 67,17 | 76,86 | 67,39 | 73,22 | 66,80 | 75,01 | 66,34 | 79.40 |
| R80W80 | 69,09 | 59,89 | 68,48 | 59,00 | 69,53 | 59,10 | 65,39 | 56,61 | 68,24 | 58,28 | 71.00 |
| R90W90 | 73,31 | 64,45 | 73,28 | 64,61 | 73,47 | 64,34 | 69,82 | 63,72 | 72,62 | 63,61 | 81.60 |

**Figure 6. CN-curve values for each sub-catchment of PA Bela. Poor soil conditions (pink), good soil conditions (blue) and original values based on CORINE LC data (green).**

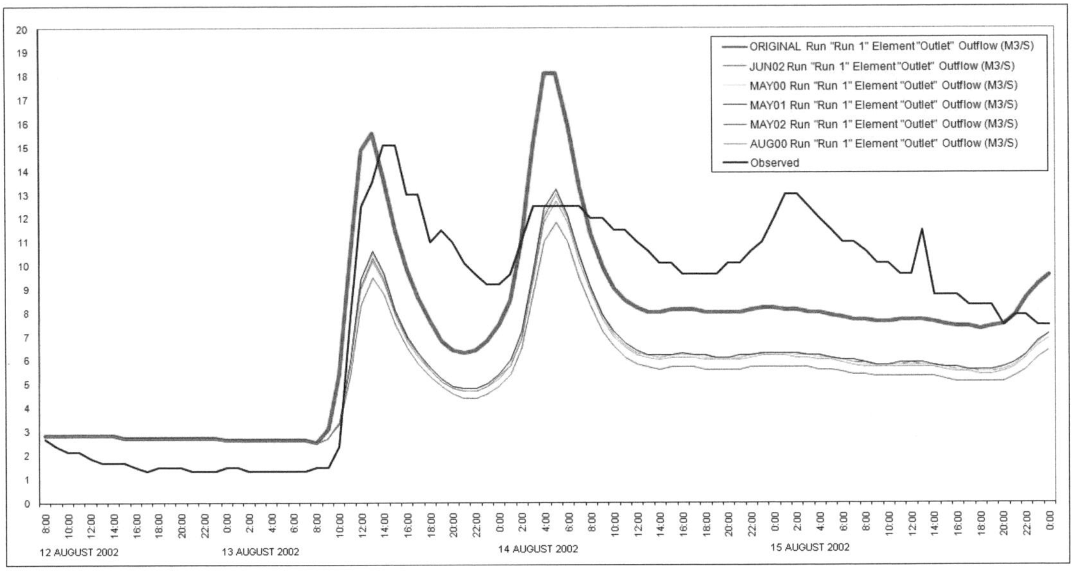


**Figure 7 H dro ra h curves. Observed (black), CN association with CORINE LC (red), CN association with LANDSAT LC alternatives (other colours)**

## 6 CONCLUSIONS

As for CORINE LC and Landsat LC data comparison, there are evident differences. Such variation is caused by data characteristics itself - (a) spatial resolution of basic element (CORINE – 25 ha, Landsat - 30 m), (b) Landsat classification method, (c) CORINE agricultural areas homogeneity.

Such divergence of LC data sources can result in different behavior of hydrological model. In case of local small catchment this would cause huge variation. In case of bigger catchment area this could be taken under consideration in case of LC change bigger than 50% of sub-catchment area.

The problem aspect of LC-CN association stands on soil condition data set together with the fact that both catchments are transboundary ones and the data harmonization needs to be carried out for the best results.

## 7 ACKNOWLEDGMENT

The study has been carried out under grant project „**Application of Geoinformation Technologies for Improvement of Rainfall-Runoff Relationships**“ (**GA205/06/1037**) thanks to Czech grant agency.

## 8 REFERENCES

Used sources but not citited:

ADAMEC, M., UNUCKA, J. (2005): *Vliv dostupnosti a mezinárodní výměny dat na predikci povodňových stavů příhraničních toků na příkladu povodí Olše*. In Transformační procesy 1990 - 2005. 10.10.2005-11.10.2005 Ostrava. Ostrava: Ostravská univerzita v Ostravě, 2005. s. 232 -238.

ADAMEC, M., BREZKOVÁ, L., HANZLOVÁ,M., HORÁK, J., UNUCKA, J. (2006): *Modelování vlivu land use na srážkoodtokové vztahy s podporou GIS*. Příspěvek na konferenci Říční krajina 4. Olomouc, PřF UPOL, s. 335-350

ADAMEC, M., HANZLOVÁ, M., HORÁK, J., UNUCKA, J., ŽIDEK, D. (2007): *Modelování hydrologických extrémů s podporou DPZ a GIS*. In Sborník semináře Změny v krajině a povodňové riziko. Praha 5.6.2007, 12 s.

ADAMEC, M., UNUCKA, J. (2007): *Modelování vlivu land use na erozní procesy s podporou GIS*. In sborník konference Říční krajina 2007. PřF UP, Olomouc (v tisku)

BEDIENT, P.B., HUBER W. C. (2001): *Hydrology and floodplain analysis*. 2nd edition. London, Prentice Hall. 763 s.

BEDIENT, P.B., HUBER, W.C. et VIEUX, B.C. (2007): *Hydrology and floodplain analysis*. 4$^{th}$ edition. Prentice Hall, London, 795 s.

BUZEK, L., KŘÍŽ, V., ŘEHÁNEK, T. (2000): *Hodnocení vodní eroze lesní půdy v povodí horní Ostravice formou plaveninového režimu*. In: Sborník prací PřF OU, 189, 8, Ostrava, s. 37 – 57

CHLEBEK, A., JARABAC, M., HOSEK, A. (1997): *Dlouhodobé odtoky z malých lesnatých povodí*. In Zpravodaj Beskydy. MZLU Brno, p. 51-56

HANZLOVÁ M., UNUCKA J., VOŽENÍLEK V. (2006): *Využití a pokryv krajiny (LULC) ve vazbě na hydrologické modelování*. Příspěvek na XXI. Sjezdu České geografické společnosti, České Budějovice, 30.8.-2.9.2006

HANZLOVÁ, M., HORÁK, J., UNUCKA, J. et al.(2006): *Klasifikace pokryvu území v povodí Bělé pro hodnocení srážko-odtokových poměrů*. In konference GIS ve státní správě, Brno, 8 s.

KANTOR, J et al. (2003): *Lesy a povodně*. Praha, MŽP. 48 p.

UNUCKA, J. (2007): *Modelování hydrologických procesů s podporou DPZ a GIS*. In sborník workshopu Informační technologie pro modelování krizových situací – IT4DM. VŠB-TU Ostrava, 13.9.2007, 15 s.

*Remote Sensing for a Changing Europe*
*D. Maktav (Ed.)*
*IOS Press, 2009*

*doi:10.3233/978-1-58603-986-8-378*

# Hyperspectral image segmentation using FSMLR with Jeffreys prior

J.S. Borges & A.R.S. Marçal
*Faculdade de Ciências, Universidade do Porto, Portugal*

J.M. Bioucas-Dias
*Instituto de Telecomunicações – Instituto Superior Técnico, Lisboa, Portugal*

Keywords: hyperspectral images, image segmentation, feature selection

ABSTRACT: The segmentation of satellite images is a valuable tool to obtain useful information about the spatial distribution of different land cover types. The use of segmentation algorithms instead of the traditional pixel-by-pixel classifiers used to produce land cover maps results on images that exhibit a more homogeneous distribution of classes, showing the piecewise spatial continuity of the real world.
Several segmentation and classification methods are being developed to properly handle the high dimensionality of hyperspectral images. An example is a Bayesian segmentation procedure based on discriminative classifiers with a Multi-Level Logistic Markov-Gibbs prior. This method adopts the Fast Sparse Multinomial Logistic Regression as discriminative classifier, a method that promotes sparsity by including a Laplacian prior. However, the use of this type of prior requires an extensive search to for the best parameter of sparsity. In this work, a modification to this method is introduced. Instead of using the Laplacian Prior to enforce the sparsity of FSMLR classifier, the Jeffreys prior is used. This prior avoids the need to proceed to an extensive search for the best parameter, and also keeps the sparsity of the densities estimators, resulting on a faster and competitive segmentation procedure. The results of the application of this new approach to the benchmarked dataset Indian Pines show the effectiveness of the proposed method when compared with that using the Laplacian prior.

## 1 INTRODUCTION

The classification of satellite images has been used as a powerful tool to access information about the land cover type distribution on a given scene. Until recently, pixel based classification algorithms were the most popular. These classification algorithms base their decisions in the spectral signature of each individual pixel. Recently, there is been a development of classification methods that use information from adjacent pixels. The use of contextual information together with spectral information results on segmentation algorithms that, when compared with pixel based classification algorithms, produce land cover maps that describe better the spatial homogeneity of the real world. This is one reason why segmentation methods are becoming more popular for the production of land cover maps. Other reason is the fact that adjacent pixels are more likely to have the same label, and so, this information can help to improve classification accuracies.

Image segmentation procedures are widely used for many applications such as remote sensing, medical imaging, face and fingertip recognition, machine vision, etc. Theoretically, we might say that, any segmentation algorithm should be able to give a solution to any segmentation problem. Nevertheless, each application has specific characteristics that lead to the development of algorithms somewhat devoted to a specific problem. Regarding the remote sensing application, namely the segmentation algorithms for land cover classification of satellite images, there are different approaches such as thresholding, edge or region segmentations (Pal & Pal 1993). Region segmentation algorithms integrate the spatial and spectral information to take advantage of the complementarities that both sources can provide. Segmentation algorithms based on morphological profiles (Benediktsson *et al.* 2003), endmember extraction (Plaza *et al.* 2006), hierarchical segmentation (Tilton *et al.* 2006) and Markov random fields (Li 2001) have shown good results in segmentation of satellite images. One of the most explored approaches is the Markov Random Fields (MRF). This type of model allows contextual constraints to be incorporated by modeling the spatial neighborhood of a pixel as a spatially distributed random process. Hidden MRF were introduced in image segmentation (Marroquin *et al.* 2003) and (Tso & Olson 2005) with different approaches. However, these types of algorithms showed some problems when dealing with high dimensionality datasets.

The possibility of having an almost continuous spectral signature of the image pixels makes hyperspectral sensors powerful tools to better identify the different materials present in the land cover. Hyperspectral satellite images are characterized for having hundreds of spectral bands, which alongside the advantage of characterizing the different materials in more detail, also have the problem of producing high dimensional datasets. When supervised algorithms are considered, the high dimension of these images, together with limited number of training samples, bring about problems related to Hughes phenomenon (Hughes, 1968) or curse of dimensionality. When conventional pixel based classification algorithms are considered, the Hughes phenomenon can be overcome by increasing the number of training samples, when possible, or by reducing dimensionality. When spatial information is added, the problem of dimensionality increases. A solution to circumvent this problem is the discriminative approach. Discriminative approaches hold the state of the art of hyperspectral image segmentation (Camps-Valls & Bruzzone 2005). In this approach the difficulties in learning class densities are overcome by learning directly the densities of the labels, given the features. One of the most consolidated discriminative supervised classification tools is the Support Vector Machines (SVMs). They have been successfully used for hyperspectral data classification due to their ability to deal with large input spaces efficiently and to produce sparse solutions (Camps-Valls & Bruzzone 2005).

Discriminative approaches incorporate neighborhood interactions in the labels as well as the observed data and have being presented in recent years. The Discriminative Random Fields (DRF) framework proposed by Kumar & Herbert (2006) is an example. In the vein of this approach, we present a Bayesian segmentation approach which improves the classification performance of discriminative classifiers by adding contextual information in the form of spatial dependencies. The major difference of our work from the one presented by Kumar & Herbert (2006) is the way that the parameters are learnt. DRFs learn all the model parameters simultaneously, leading to hard and complex procedures. On the contrary, in the proposed approach, the parameters are learnt in two consecutive, but non simultaneous steps. As a consequence, the proposed method leads to much lighter procedures, still displaying very good results.

The Bayesian segmentation method for hyperspectral images here presented learn the class densities in a supervised fashion with a modified discriminative Fast Sparse Multinomial Regression (FSMLR) (Borges *et al.* 2006). The FSMLR is a fast version of the Sparse Multinomial Regression (SMLR) (Krishnapuram *et al.* 2005). This method is a sparse classification algorithm capable of dealing with high dimensional datasets. The FSMLR uses a Laplacian prior to enforce the sparsity on the class parameters. The degree of sparseness of the class densities estimates is controlled by a sparsity parameter, which has to be tuned by the user. When dealing with high dimensional datasets, such as hyperspectral images, this task may become time consuming. This can be circumvent by using a parameter-free prior. In this work we present a modified version of the Bayesian hyperspectral image segmentation with discriminative class learning by introducing the Jeffreys prior (Bioucas-Dias 2006). The Jeffreys prior keeps the sparsity of the FSMLR without the need to tune any parameter, controlling the complexity of the learned classifier and, therefore, achieving good generalization capabilities.

Having learned the class densities, the spatial dependencies are enforced by a Multi-Level Logistic (MLL) Markov-Gibs prior, which favors neighboring labels of the same class. To compute an approximation to the Maximum A Posteriori probability (MAP) segmentation, we adopt the α-Expansion graph cut based algorithm proposed in (Boykov *et al.* 2001). This tool is computationally efficient and yields nearly optimum solutions.

The segmentation method presented is applied to the benchmarked hyperspectral dataset Indian Pines (Landgrebe 2003).

The paper is organized as follows. In section 2 the methods are presented: FSMLR classifier with Jeffreys prior and the segmentation procedure based on MLL Markov Gibs prior. Section 3 presents the results of the application to the hyperspectral image and section 4 the concluding remarks.

## 2 METHODS

A segmentation can be interpreted as an image of labels $\mathbf{y} = \{y_i\}_{i\in S}$ where $y_i \in L = \{1,2,\ldots,K\}$. Let $\mathbf{x} = \{x_i \in R^d, i \in S\}$ be the observed multi-dimensional image, also known as feature image. The goal of the segmentation is to estimate $\mathbf{y}$, having observed $\mathbf{x}$. Regarding the Bayesian framework, this estimation is done by maximizing the posterior distribution $p(\mathbf{x}\,|\,\mathbf{y}) \propto p(\mathbf{y}\,|\,\mathbf{x})p(\mathbf{y})$, where $p(\mathbf{y}\,|\,\mathbf{x})$ is the likelihood function (or the probability of feature image) and $p(\mathbf{y})$ is the prior over the classes.

The approach here presented, makes use of the discriminative FSMLR classifier (Borges *et al.* 2006) to learn the class densities $p(y_i\,|\,x_i)$. The likelihood is then given by $p(x_i\,|\,y_i) = p(y_i\,|\,x_i)p(x_i)/p(y_i)$. Noting that $p(x_i)$ does not depend on the labeling $\mathbf{y}$ and assuming $p(y_i) = 1/K$, we have

$$p(\mathbf{x}\,|\,\mathbf{y}) \propto \prod_{i\in S} p(y_i\,|\,x_i)/p(y_i) \tag{1}$$

where conditional independence is understood.

In the following sections, the FSMLR method, yielding the density $p(\mathbf{y}\,|\,\mathbf{x})$, the MLL prior $p(\mathbf{y})$ and α -Expansion optimization algorithm are briefly described.

### 2.1 *Class density estimation*

Given the training set $D = \{(x_1, y_1),\ldots,(x_n, y_n)\}$, the SMLR algorithm learns a multi-class classifier based on the multinomial logistic regression. By incorporating a prior, this method performs simultaneously feature selection, to identify a small subset of the most relevant features, and learns the classifier itself (Krishnapuram *et al.* 2005). The goal is to assign to each site $i \in S$ the probability of $y_i = k$, for $k$=1, …, $K$. In particular, if $y_i = \left[y^{(1)},\ldots,y^{(K)}\right]^T$ is a 1-of-$K$ encoding of the $K$ classes, and if $w^{(k)}$ is the feature weight vector associated with class $k$, then the probability of $y_i^{(k)} = 1$ given $x_i$ is

$$P\left(y_i^{(k)} = 1\,|\,x_i, w\right) = \frac{\exp\left(w^{(k)^T} h(x_i)\right)}{\sum_{k=1}^{K} \exp\left(w^{(k)^T} h(x_i)\right)} \tag{2}$$

where $w = \left[w^{(1)^T},\ldots,w^{(K)^T}\right]$ and $h(x) = \left[h_1(x),\ldots,h_l(x)\right]$ is a vector of $l$ fixed functions of the input, often termed features. Possible choices for this vector are linear $h(x_i) = \left[1, x_{i,1},\ldots,x_{i,d}\right]^T$, where $x_{i,j}$ is the $j$th component of $x_i$), and kernel ( $h(x) = \left[1, K(x,x_1),\ldots,K(x,x_n)\right]^T$, where $K(.,.)$ is some symmetric kernel function). The latter nonlinear mapping guarantees that the transformed samples are more likely to be linearly separable. A popular kernel used in image classification is the Gaussian Radial Basis Function (RBF): $K(\mathbf{x}\,|\,z) = -\exp\left(\|\mathbf{x} - z\|^2 / 2\sigma^2\right)$.

The MAP estimate of $w$ is

$$\hat{w}_{MAP} = \arg\max_w L(w) = \arg\max_w \left[l(w) + \log p(w)\right] \tag{3}$$

where $l(w)$ is the log-likelihood function and $p(w)$ is a prior on $w$.

The SMLR presented by Krishnapuram et al. (2005) uses a Laplacian prior on $w$, to control the degree of sparseness of $\hat{w}_{MAP}$. The Laplacian prior is given by $p(w) \propto \exp\left(-\lambda \|w\|_1\right)$ where $\lambda$ acts as a tunable regularization parameter. The process of selecting the optimum $\lambda$ is usually done by cross-validation trough the training process. In the case of high dimensional datasets, such as hyperspectral images, this search often becomes a time consuming task. The process of estimating the class densities must be repeated as many times as the number of $\lambda$ values to be tested. In order to overcome this problem, we introduce a parameter-free prior in the estimation of class densities: the Jeffreys prior (Bioucas-Dias 2006). The Jeffreys prior is given by $p(w) \propto 1/\|w\|_1$ having no longer a sparsity parameter to tune. Adopting this prior, the optimization process is done in a similar way to that done in Krishnapuram et al.(2005).

The weights $w$ are learned using bound optimization tools (Lange 2004), making possible to perform exact MAP multinomial logistic regression, with the same cost as the original iterative reweighted least squares algorithm for maximum likelihood estimation (see (Krishnapuram *et al.* 2005) for details).

The solution of (2) is then given by the iterative equation:

$$\hat{w}^{t+1} = \Gamma^{(t)}\left(\Gamma^{(t)} B \Gamma^{(t)} - I\right)^{-1} \Gamma^{(t)}\left(B\hat{w}^{(t)} - g(\hat{w}^{(t)})\right), \tag{4}$$

where $\Gamma^{(t)}$ is given by

$$\Gamma^{(t)} = diag\left\{\left|\hat{w}_1^{(t)}\right|, \ldots, \left|\hat{w}_{d(K-1)}^{(t)}\right|\right\}, \tag{5}$$

and

$$B \equiv -\frac{1}{2}\left[I - 11^T / K\right] \otimes \sum_{i=1}^{S} x_i x_i^T, \tag{6}$$

and $1 = [1,1,\ldots,1]^T$, $g(w)$ is the gradient of $l(w)$ given by

$$g(w) = \sum_{i=1}^{S} \left(y'_i - p_i(w)\right) \otimes x_i, \tag{7}$$

with $y'_i = \left[y_i^{(1)}, \ldots, y_i^{(K-1)}\right]^T$ and $p_i(w) = \left[p_i^{(1)}(w), \ldots, p_i^{(K-1)}(w)\right]^T$.

As can be observed in equation (4) there is no parameter to be defined by the user. The weights estimation procedure with the inclusion of Jeffreys prior instead of the Laplacian prior becomes independent of the selection of any parameter.

In practice, the computational cost of solving the linear system in equation (4) is often prohibitive. Regarding the application of SMLR to hyperspectral images, this becomes a problem since the cost at each iteration is of the order of $(dK)^3$ where the number of bands $d$ is usually very large.

In order to avoid this problem, a modification to the iterative method used in SMLR can be done. This modification results in a faster and more efficient algorithm: the Fast-SMLR (FSMLR) (Borges *et al.* 2006). FSMLR uses the Block Gauss-Seidel method (Quarteroni *et al.* 2000) to solve the system (4). The modification consists, at each iteration, in solving blocks corresponding to the weights belonging to the same class, instead of computing the complete set of weights. Using this technique, what happens is that, at each iteration, $K$ systems of equal dimension to the number of samples are solved. This results in an improvement in terms of computational effort of the order of $K^2$.

## 2.2 *Including spatial information with the Markov-Gibs prior*

The sparsity enforced by the Jeffreys prior on the estimation of class densities with FSMLR is a key step in the image classification process. However, the estimation of class densities does not include any spatial information about the classes' dispersion in the image. The information of each pixel neighborhood is introduced by the MLL prior. The MLL prior is a MRF that favors neighboring labels of the same class. In this way, we will be able to better model the piecewise smooth of real world images.

The prior over classes $p(w)$ in equation (1) will be defined in a way that allows to model the spatial distribution of neighboring pixels.

According to the Hammersly-Clifford theorem, the density associated with a MRF is a Gibb's distribution (Geman & Geman 1984). Therefore, the prior model for segmentation has the structure

$$p(\mathbf{y}) = \frac{1}{Z}\exp\left(-\sum_{c\in C} V_c(\mathbf{y})\right), \tag{8}$$

where $Z$ is the normalizing constant and the sum is over the prior potentials $V_c(\mathbf{y})$ for the set of cliques $C$ over the image, and

$$-V_c(\mathbf{y}) = \begin{cases} \alpha_{y_i} & if\ |c| = 1 \quad \text{(single clique)} \\ \beta_c & if\ |c| > 1 \quad \text{and } \forall_{i,j\in c}\, y_i = y_j \\ -\beta_c & if\ |c| > 1 \quad \text{and } \exists_{i,j\in c}\, y_i \neq y_j \end{cases} \tag{9}$$

where $\beta_c$ is a non-negative constant.

Equation (8) can be written as

$$p(\mathbf{y}) = \frac{1}{Z} e^{\beta n(\mathbf{y})}, \tag{10}$$

where $n(\mathbf{y})$ denotes the number of cliques having the same label, if we let $\alpha_k = \alpha$ and $\beta_c = \beta/2 > 0$. This choice gives no preference to any label nor to any direction.

The conditional probability $p(y_i = k \mid y_j, j \in S - i)$ is then given by

$$p(y_i = k \mid y_{N_i}) = \frac{e^{\beta n_i(k)}}{\sum_{k=1}^{K} e^{\beta n_i(k)}} \tag{11}$$

where $n_i(k)$ is the number of sites in the neighborhood of site $i$, $N_i$, with label $k$.

## 2.3 *Segmentation algorithm*

All the process was initialized under the Bayesian framework, leading us to the maximization of the posterior distribution $p(\mathbf{x}\mid\mathbf{y}) \propto p(\mathbf{y}\mid\mathbf{x})p(\mathbf{y})$. As described in sections 2.1 and 2.2, the class densities were learned using the FSMLR with Jeffreys prior, while the prior over classes, $p(\mathbf{y})$, was modeled by a MLL Markov-Gibs prior.

This way we have that the MAP segmentation is given by

$$\begin{aligned}\hat{\mathbf{y}} &= \arg\max_{\mathbf{y}} p(\mathbf{x}\mid\mathbf{y})p(\mathbf{y}) = \arg\max_{\mathbf{y}} \sum_{i\in S} \log p(x_i \mid y_i) + \beta n(\mathbf{y}) \\ &= \arg\min_{\mathbf{y}} \sum_{i\in S} -\log p(x_i \mid y_i) + \beta \sum_{i,j\in c} \delta(y_i - y_j), \end{aligned} \tag{12}$$

The minimization of (12) is a hard combinatorial optimization problem. Graph cut techniques from combinatorial optimization are able to find the global minimum for some multi-dimensional energy functions. The minimization problem in (12) presents characteristics that allow us to apply the α-Expansion algorithm (Boykov *et al.* 2001), which achieves very good approximations for this problem.

The α-Expansion algorithm makes use of a min-cut/max-flow algorithm presented by Boykov & Kolmogorov (2004).

## 3 APPLICATION TO A HYPERSPECTRAL IMAGE

To evaluate the performance of the segmentation algorithm presented in this paper, the well-known hyperspectral AVIRIS spectrometer Indian Pines 92 from Northern Indiana was used (Landgrebe, 2003). This benchmarked dataset has been frequently used to test several techniques in the processing of hyperspectral images allowing for a good evaluation.

The ground truth data image consists of 145 x 145 pixels of the AVIRIS image in 220 contiguous spectral bands. Experiments were carried out without 20 noisy bands (Landgrebe, 2003). Due to the insufficient number of training samples, seven classes were discarded, leaving a dataset with 9 classes distributed by 9345 pixels. This dataset was randomly divided into a set of 4757 training samples and 4588 validation samples.

The spatial distribution of the class labels is presented in figure 1.

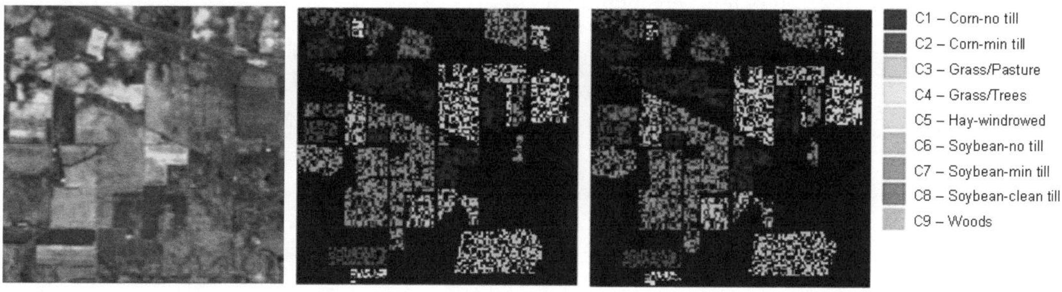


Figure 1. AVIRIS image used for testing. Left: RGB(50, 27, 17) color composite; Centre: training areas; Right: validation areas.

The results presented here are the overall accuracy measured in the independent (validation) dataset with 4588 samples. Experiments were made using 10%, 20% and 50% of the training set with a linear and a RBF kernel to $h(x_i)$. When $h(x_i)$ is set to a linear function the complete training was also used to train the segmentation algorithm. In the GC $\alpha$-Expansion method a $\beta$=1.5 was defined when the complete training set was used, and $\beta$ =4 for subsets of the training data. The overall accuracy results from are presented in table 1.

Table 1. Overall accuracy of Bayesian segmentation using 10%, 20%, 50% and the complete training set, with $h(x_i)$ Linear and RBF, using a Laplacian and a Jeffreys prior.

| | | *Size of training set* | | | |
|---|---|---|---|---|---|
| | | 10% | 20% | 50% | 100% |
| *h* linear | Laplacian prior | 86.05% | 89.45% | 89.69% | 95.60% |
| | Jeffreys prior | 86.18% | 88.58% | 90.43% | 95.66% |
| *h* RBF | Laplacian prior | 92.11% | 94.62% | 97.86% | --- |
| | Jeffreys prior | 89.84% | 95.07% | 96.71% | --- |

By the analysis of table 1, it is possible to observe that the Jeffreys prior achieves competitive results with Laplacian prior. The performance of the classifier was found to be nearly independent of the prior used for all training sets tested, the variations on the overall accuracies are minimal. The increase in the size of training set results in better accuracies for all methods. The results presented in table 1 for the Laplacian prior had already been evaluated in other work (Borges et al., 2007) and proved to be very competitive with the state of the art algorithms for hyperspectral image segmentation. The introduction of Jeffreys prior was able to keep the good performance of the Bayesian segmentation method proposed. It should be noted that with this prior there is no need for searching the parameter that best controls sparsity, something that has to be done with the Laplacian prior. This reduces significantly the time needed to classify the image. The reduction is of the order of the number of sparsity parameters to be tested. Moreover, the sparsity achieved by the FSMLR when using a Jeffreys prior is higher than with the Laplacian prior. This can be observed in table2, which presents the number of significant features selected by each prior, considering $h(x_i)$ as a linear function. In both cases there is a significantly reduction in the number of features (bands) used to produce the land cover map. Recall that the number of bands considered in the experiments were 200. Using the Jeffreys prior and only 10% of the training set it is possible to achieve an overall accuracy of 86.18% using the information of only 18 spectral bands. As the size of training set grows, the number of bands selected by the priors also grows as well as the overall accuracies. The best accuracy achieved with a linear kernel (95.66%) is achieved using only 51 spectral bands from the 200 bands considered initially.

Table 2. Number of significant features selected by each prior, with $h(x_i)$ linear.

| | *Size of training set* | | | |
|---|---|---|---|---|
| | 10% | 20% | 50% | 100% |
| Laplacian prior | 34 | 49 | 71 | 105 |
| **Jeffreys prior** | 18 | 27 | 39 | 51 |

## 4 CONCLUSIONS

In this paper we have presented a Bayesian hyperspectral image segmentation algorithm that uses the non-informative parameter-free Jeffreys prior to compute the class densities with the FSMLR algorithm. This is a first step to the whole segmentation process. After estimating the class densities with the discriminative classifier FSMLR, the segmentation method includes spatial information of the neighborhood of each pixel by adopting a MLL Markov-Gibs prior over the classes. The MAP segmentation is carried out using the CG $\alpha$-expansion algorithm.

The inclusion of the Jeffreys prior instead of the Laplacian prior in the FSMLR method, was able to keep the good performance of the Bayesian segmentation algorithm, while at the same time there was an improvement of the sparsness of the classifier. Moreover, this sparsity improvement was achieved without the need of an extensive search for the parameter that best controls the sparsity, carried out through a cross-validation based model selection. This process, when dealing with high dimensional datasets like hyperspectral images is a challenging problem. The inclusion of the Jefreys prior results on a greatly reduced computational expense mantaining the good performance of the segmentation algorithm

## ACKNOWLEDGEMENTS

The first author would like to thank the Fundação para a Ciência e a Tecnologia (FCT) for the financial support (PhD grant SFRH/BD/ 17191/2004). The authors acknowledge Vladimir Kolmo-

gorov for the max-ow/min-cut C++ code made available on the web, see (Boykov et al, 2001) for more details; and David Landgrebe for providing the AVIRIS data.
This work was supported by the Fundação para a Ciência e Tecnologia, under the project PDCTE/CPS/49967/2003, and by the Instituto de Telecomunicações under the project IT/LA/325/2005, and by CICGE under POCI 2010 programme.

## REFERENCES

Benediktsson, J.A., Pesaresi, M. & Arnason, K. 2003. Classification and feature extraction for remote sensing images from urban areas based on morphological transformations. *IEEE Trans. Geoscience and Remote Sensing* 41:1940–1949.

Bioucas-Dias, J.M. 2006. Bayesian Wavelet-Based Image Deconvolution: A GEM Algorithm Exploiting a Class of Heavy-Tailed Priors. *IEEE Trans. Image Processing* 15(4):937–951.

Borges, J.S., Bioucas-Dias, J. & Marçal, A.R.S. 2006. Fast Sparse Multinomial Regression Applied to Hyperspectral Data. *In: Campilho, A., Kamel, M. (eds.) ICIAR 2006. LNCS* 4142:700–709.

Borges, J.S., Bioucas-Dias, J. & Marçal, A.R.S. 2007. Bayesian Hyperspectral Image Segmentation with Discriminative Class Learning. *In: Pattern Recognition and Image Analysis: 3rd Iberian Conference, IbPRIA 2007, Lecture Notes in Computer Science, Girona, Spain* I:22–29.

Boykov, Y. & Kolmogorov, V. 2004. An experimental comparison of mincut/max-flow algorithms for energy minimization in vision. *IEEE Trans. on Pattern Analysis and Machine Intelligence* 26:1124–1137.

Boykov, Y., Veksler, O. \& Zabih, R. 2001. Fast Approximate Energy Minimization via Graph Cuts. *IEEE Trans. Pattern Analysis and Machine Intelligence* 23(11):1222–1239.

Camps-Valls, G. & Bruzzone, L. 2005. Kernel-based methods for hyperspectral image classification. *IEEE Trans. Geoscience and Remote Sensing* 43(6):1351–1362.

Geman, S. & Geman, D. 1984. Stochastic relaxation, Gibbs distribution, and the Bayesian restoration of images. *IEEE Trans. on Pattern Analysis and Machine Intelligence* 6:721–741.

Hughes, G.F. 1968. On the mean accuracy of statistical pattern recognizers. *IEEE Trans. Information Theory* 14:55–63.

Krishnapuram, B., Carin, L., Figueiredo, M.A.T. & Hartemink, A.J. 2005. Sparse Multinomial Logistic Regression: Fast Algorithms and Generalization Bounds. *IEEE Trans. Pattern Analysis and Machine Intelligence* 27(6):957–968.

Kumar, S. & Hebert, M. 2006. Discriminative Random Fields. *International Journal of Computer Vision* 68(2):179–202.

Landgrebe, D.A. 2003. *Signal Theory Methods in Multispectral Remote Sensing*. New Jersey: John Wiley and Sons, Inc..

Lange, K. 2004. *Optimization*. Springer Texts in Statistics, New York: Springer-Verlag.

Li, S.Z., 2001. *Markov Random Field Modeling in Image Analysis*. Tokyo: Springer-Verlag.

Marroquin, J.L., Santana, E.A. & Botello, S. 2003. Hidden Markov Measure Field Models for Image Segmentation. *IEEE Trans. on Pattern Analysis and Machine Intelligence* 25(11):1380–1387.

Pal, N.R. & Pal, S.K. 1993. A review on image segmentation techniques. *Pattern Recognition* 26(6):1277–1294.

Plaza, A., Benediktsson, J., Boardman, J., Brazile, J., Bruzzone, L., Camps-Valls, G., Chanussot, J., Fauvel, M., Gamba, P., Gualtieri, A., Tilton, J. & Trianni, G. 2006: Advanced Processing of Hyperspectral Images. *IEEE IGARSS 2006 Proceedings* IV:1974–1979.

Quarteroni, A., Sacco, R & Saleri, F. 2000. *Numerical Mathematics* TAM Series (37). New York: Springer-Verlag.

Tilton, J.C., Lawrence, W.J. & Plaza, A. 2006. Utilizing hierarchical segmentation to generate water and snow masks to facilitate monitoring change with remotely sensed data. *GIScience & Remote Sensing* 43:39–66.

Tso, B. & Olsen R.C. 2005. Combining spectral and spatial information into hidden Markov models for unsupervised image classification. *International Journal of Remote Sensing* 26(10):2113–2133.

*Remote Sensing for a Changing Europe*
*D. Maktav (Ed.)*
*IOS Press, 2009*

*doi:10.3233/978-1-58603-986-8-386*

# Impact of topographic parameters on seismic amplification applying Geospatial tools

Muhammad Shafique[1,2], Mark van der Meijde[1], Norman Kerle[1] & Freek van der Meer[1]
[1]*International Institute of Geo-Information Science and Earth Observation (ITC) – The Netherlands*
[2]*National Center of Excellence in Geology, University of Peshawar, Pakistan*

Keywords: Seismic ground shaking, Topographic features, ASTER, SRTM

ABSTRACT: The impact of topography on the uneven distribution of seismic response is well observed synthetically, experimentally and visually during seismic events. Numerical and experimental investigations for predicting topographic effects on seismic shaking are limited to isolated and/or synthetic hills and ridges. Furthermore, most of these studies only focus on one of the many terrain parameters necessary for analyzing the impact of topographic features on amplification or de-amplification of seismic response. Seismic events located in rough terrain, like the Kashmir earthquake (2005) in the western Himalaya, exhibit intensified ground shaking and associated devastation at hill ridges and on inclined slopes. Regional seismic ground shaking maps developed through USGS-ShakeMap ignore this topographic impact. DEM derived topographic attributes and seismic event characteristics can be integrated, to predict the topographic seismic amplification.

Satellite sensor's acquired SRTM (90 m) and ASTER (30 m) DEMs are employed to compute terrain attributes, and also to investigate the impact of DEM random errors and resolution on the topographic attributes, and topographic seismic response. Data disparity of SRTM and ASTER DEMs, and derived topographic attributes, imply the sensitivity of satellite remote sensing DEMs, to rugged terrain and steep slopes. Impact of DEM random errors on derived topographic attributes is quantified through Monte Carlo Simulation that shows higher consistency of SRTM DEM to derived topographic attributes. Impact of slope angle, relative height, wavelength and damping on amplification and de-amplification of seismic response is analyzed, in homogeneous lithological and geotechnical environment. Seismic response is predicted to be sensitive to the slope geometry, among the analysed parameters. DEM resolution and random errors have meager impact on the predicted topographic seismic response.

## 1 INTRODUCTION

Natural disasters are dynamic and uncertain processes that can have adverse impact on, and a threat to, sustainable socio-economic development (Morales, 2002). Forecasting disasters in advance, assists to minimize the potential devastation, but unfortunately predicting earthquakes remains impossible with the current understanding and technology. Earthquakes have proved to be the most devastating natural disaster, with a high mortality rate and a wide spread destruction (Alexander, 1993). The strength and duration of seismic ground shaking, that plays a key role in seismic devastation, depends on the earthquake magnitude, location of epicenter, medium traversed by the seismic waves and the physical characteristics like geology, topography and soil conditions of the site (Kramer, 1996). In rough terrain, like the western Himalaya covering northern Pakistan, topography and soil thickness significantly controls, the intensity of seismic ground shaking. The existing models predicting seismic ground shaking at regional scale such as USGS-Shakemap and (INGV) are ignoring the topographic impact, while the seismic events are usually associated with mountainous regions, therefore limiting their reliability at local scale.

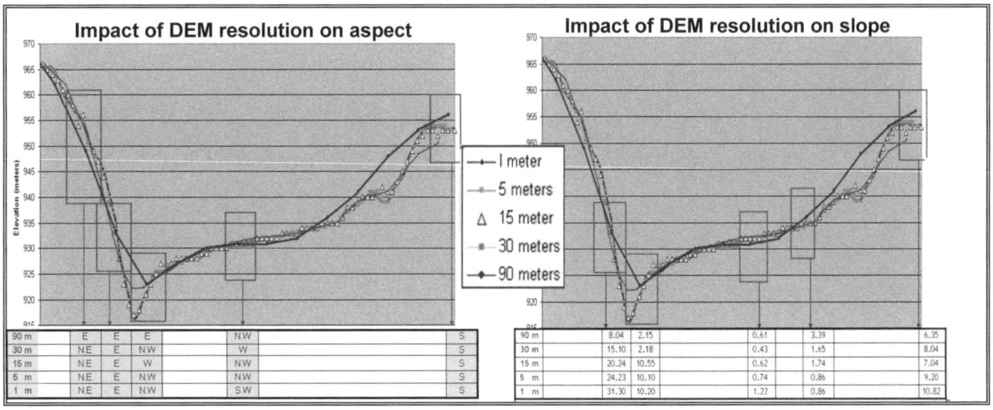


Figure 1. Impact of DEM resolutions on realistic representation of terrain features. At 1, 5, 15 m resolutions, smaller features and edges can be separately identified, while at 30 and 90 m resolution, the terrain appears smoothening and minor features are disappearing.

Extensive numerical, analytical and experimental techniques have been applied to investigate and predict the impact of various terrain features on amplification or deamplification of seismic response. These studies are in agreement in predicting amplification of seismic response at the slope crest, and deamplification at the slope toe (Ashford et al., 1997a; Ashford et al., 1997b; Athanasopoulos et al., 1999; Bard, 1982; Bouckovalas et al., 2005; Chávez-García et al., 2000; Sanchez-sesma et al., 1982; Stamatopoulos et al., 2007). Parametric studies analysing the impact of various topographic features on seismic response, leads to deriving numerical models, that predict topographic aggravation of seismic response. Main limitation of these studies is that they are limited to a synthetic environment and/or isolated hills, not applied at regional scale.

Investigating topographic impact at regional scale, the Satellite Remote Sensing (SRS) acquired Digital Elevation Model (DEM) and their derived terrain information, can potentially be utilized, to predict topographic amplification of seismic response, in real case scenario of rough terrain. With development in remote sensing technology and techniques, SRS acquired DEMs are available at range of resolutions, accuracy and at global coverage. Resolution of DEM significantly affects the realistic representation of the terrain and derived topographic attributes (Fig. 1).

Shuttle Radar Topography Mission (SRTM) DEM at 90 m grid size and Advance Spaceborne Thermal Emission and Reflective Radiometer (ASTER) DEM at 30 m grid size are utilized, to compute the required topographic attributes (slope, aspect, relative height, curvature) for seismic amplification modeling. Consistency of SRTM and ASTER DEMs is investigated, and impact of their respective random errors and resolution on derived topographic attributes is also estimated. The derived topographic attributes were incorporated in synthetically developed numerical model, to predict the topographic aggravation of seismic response. Impact of DEM resolution and data source on predicted topographic amplification is also quantified. The study area is located around the cities of Balakot and Muzaffarabad, located in north Pakistan, that were severely devastated during the 8th October 2005 Kashmir earthquake (Fig. 2).

## 2 METHODOLOGY

This study investigates the applicability, and the impact of DEM random errors and resolution on the predicting topographic amplification of seismic response (Fig. 3). SRS recorded SRTM and ASTER DEMs of study area were acquired from USGS and ERSDAC respectively, with the assumption that blunders and the systematic errors in the DEMs are already adjusted. Due to unavailability of reference data for DEM accuracy assessment, consistency of the utilized SRTM and

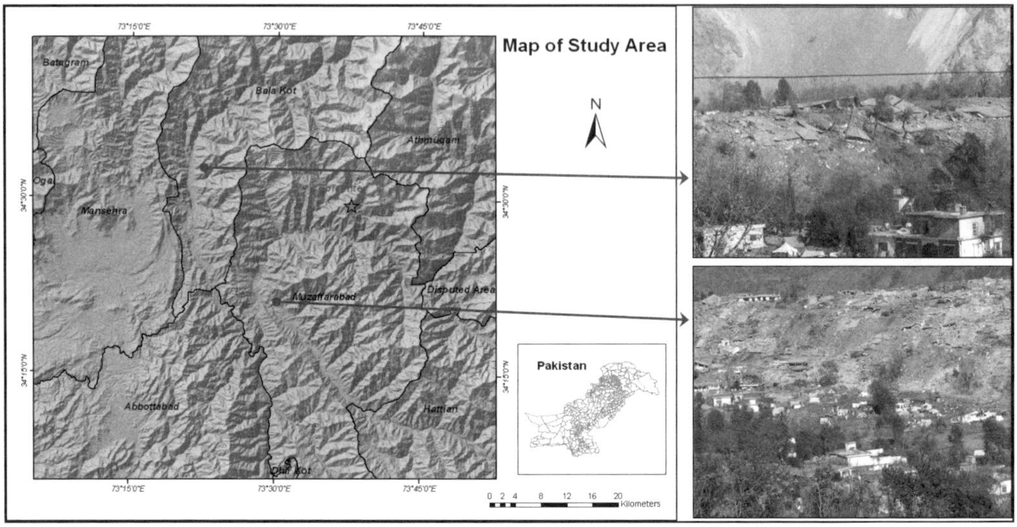


Figure 2. Location map of study area, source and surrounding of 2005 Kashmir earthquake.

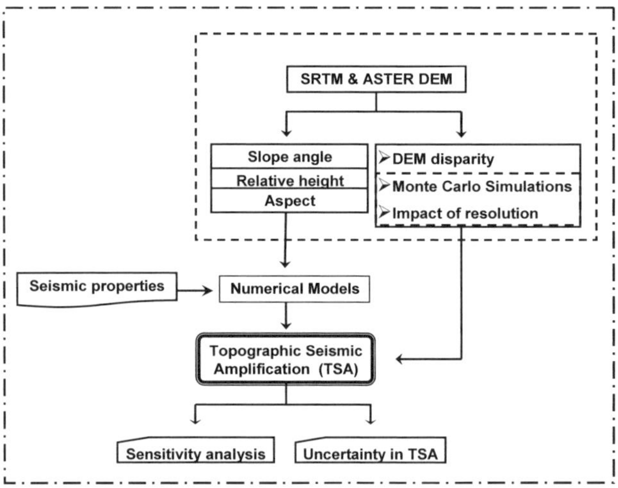


Figure 3. Flow chart of methodology adopted for the study.

ASTER DEMs is investigated through the disparity calculation, of the elevation data and their derived slope and aspect. The impact of DEM resolution on the terrain representation and computed topographic attributes is portrayed through comparing the terrain profile generated from the DEMs of various resolutions.

In contrast to elevation recorded by SRTM and ASTER DEMs, topographic seismic modeling requires relative height of terrain features from the assumed base level. Following the predecessors research such as Bouckovalas et al. (2005), Assimaki et al. (2004), Ashford et al. (1997a) and Pedersen et al. (1994), the nearest drainage network is assumed as base level, for the surrounding terrain features. To implement the theory, local minima are manually selected at the streams confluence and heads, and in the river base. Local minimal spots were assign elevation from the respective DEM, and interpolated to derive a continuous surface, as the assumed base level for the

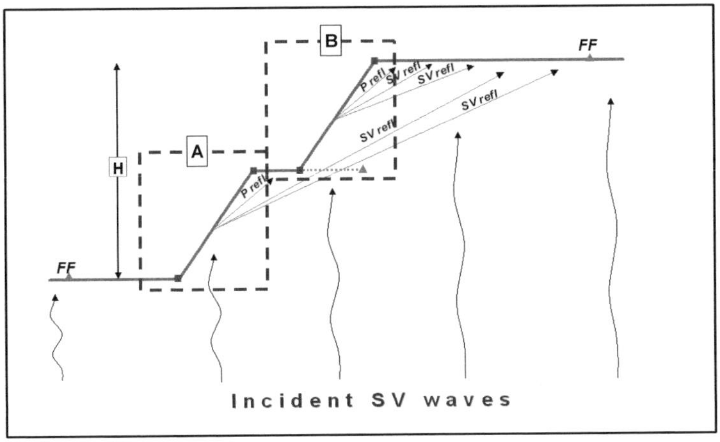


Figure 4. Applicability limitation of synthetic model to real world.

study area. The generated base level surface is subtracted from the original DEM, resulting in surface exhibiting height of terrain from the nearest drainage network.

DEMs acquired directly through the satellite sensors are prone to blunders, systematic and random errors. The blunders and systematic errors were assumed to be adjusted in the utilized DEMs, the presence of un-adjustable random errors injects uncertainty in the derived topographic attributes. Monte Carlo Simulations (MCS) technique is applied to quantify the uncertainty in the slope and aspect computation from SRTM and ASTER DEMs, in the rough terrain of northern Pakistan. MCS is commonly used by many researchers to evaluate errors in GIS data, and specifically to address DEM uncertainty (Heuvelink et al., 1990; Lanter et al., 1992; Oksanen et al., 2005; Wechsler et al., 2006). Elevation values are added to the model in the provided uncertainty (e.g. RMSE), randomly and normally distributed throughout the model to derive the topographic attributes. The 100 simulation are arbitrarily selected, and analysed to quantify the uncertainty in the computed slope and aspect.

Extensive literature is reviewed to search for a numerical model predicting impact of topographic features and seismic wave's properties on seismic response. Bouckovalas et al. (2006) developed and verified numerical model predicting topographic aggravation factor (TAF) taking into consideration slope geometry, height of terrain feature, seismic wave's wavelength and material damping, assuming SV as a incident seismic waves in homogenous soil and lithological environment.

$$A_{h,\max} = a_h / a_{h,ff} \tag{1}$$

Equation 1. Numerical representation of horizontal TAF (Bouckovalas et al., 2006).

where $a_h$ = Peak horizontal acceleration at any point, $a_{h,ff}$ = Peak horizontal acceleration at free field.

The $a_h$ depends on relative height of terrain, slope inclination, wavelength of approaching seismic waves and material damping (Bouckovalas et al., 2006). The applied TAF predicting numerical model is felt to be most comprehensive in incorporating the crucial topographic and seismic parameters, compared to other available models.

The aforementioned numerical model was developed by using the synthetic terrain profile (Fig. 4), ranging from slope toe to crest. Applying this model on the DEM attributes will assume each DEM pixel (Fig. 4, A & B) as separate terrain profile, which limits the applicability of numerical model at regional scale.

Table 1. Disparity of SRTM-ASTER DEMs and derived attributes

| Category | Disparity (SRTM-ASTER) | | |
|---|---|---|---|
| | Minimum | Mean | Maximum |
| Elevation (m) | –1531 | 32.80 | 1621 |
| Slope (degrees) | –82.32 | –0.36 | 74.54 |
| Aspect (degrees) | –360.14 | 1.79 | 360.89 |
| Relative height (m) | –451.77 | 25.84 | 1012.22 |

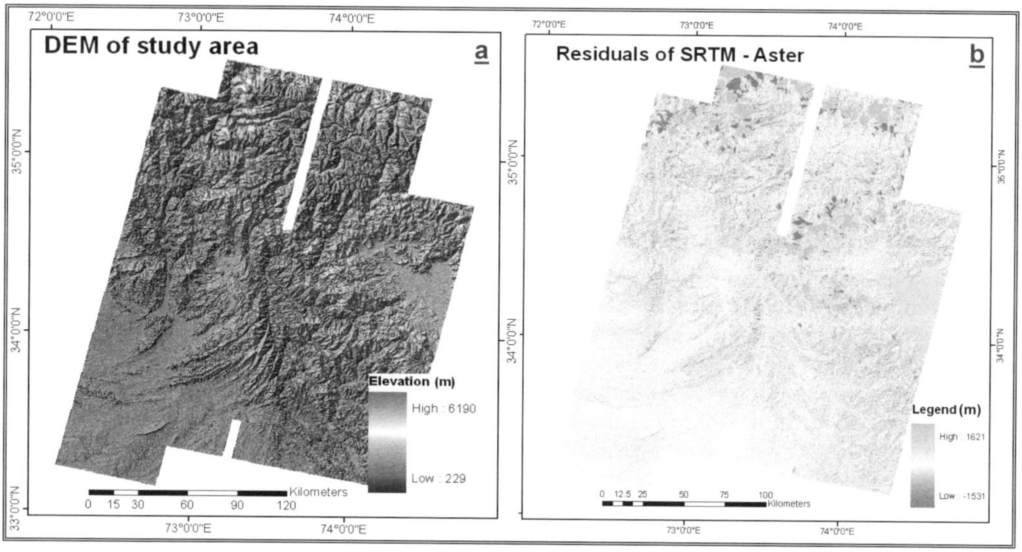


Figure 5. (a) ASTER DEM of the study area, (b) SRTM and ASTER DEMs disparity.

The derived digital topographic parameters and assumed seismic characteristics are integrated in a numerical model, to predict the topographic amplification of seismic response. Sensitivity of the model parameters to seismic response is determined through sensitivity analysis. Uncertainty in the predicted TAF, due to uncertainty in the applied topographic attributes, is quantified, and the impact of DEM resolution is also investigated.

## 3 RESULTS

### 3.1 *Disparity of SRTM and ASTER DEMs*

The disparity of SRTM and ASTER DEMs, predict consistency in recoding elevation and topographic attributes on flat terrain, while large discrepancies are clustered in rugged terrain (Fig. 5, b). On average there are minor differences in slope and aspect estimation at same location from SRTM and ASTER DEMs (Table 1).

Figure 5, reflect clear signs of spatial autocorrelation of residuals, which strengthens the theory of sensitivity of SRS DEMs to rugged terrain.

The relative height of the terrain features estimated from the surrounding drainage network shows that SRTM DEM besides course resolution, present the relative height of terrain feature elevated than ASTER (Table 1). While in geographic space the disparity of SRTM and ASTER DEMs derived relative heights are in agreement with the Fig. 5, presenting higher discrepancy concentrated on steep slopes.

Table 2. Uncertainity in slope and aspect derieved from SRTM and ASTER DEMs

| DEMs | Topographic attributes | Minimum | Mean | Standard deviation | Maximum |
|---|---|---|---|---|---|
| SRTM | Slope (degrees) | 0.03 | 0.93 | 0.17 | 1.67 |
| | Slope (Percentage) | 0.01 | 1.63 | 0.30 | 2.92 |
| | Aspect (Degrees) | 0.2 | 22.69 | 38.68 | 179.61 |
| ASTER | Slope (degrees) | 0.005 | 2.85 | 0.59 | 6.01 |
| | Slope (Percentage) | 0.008 | 4.98 | 1.038 | 10.51 |
| | Aspect(Degrees) | 0 | 41.41 | 45.94 | 179.68 |

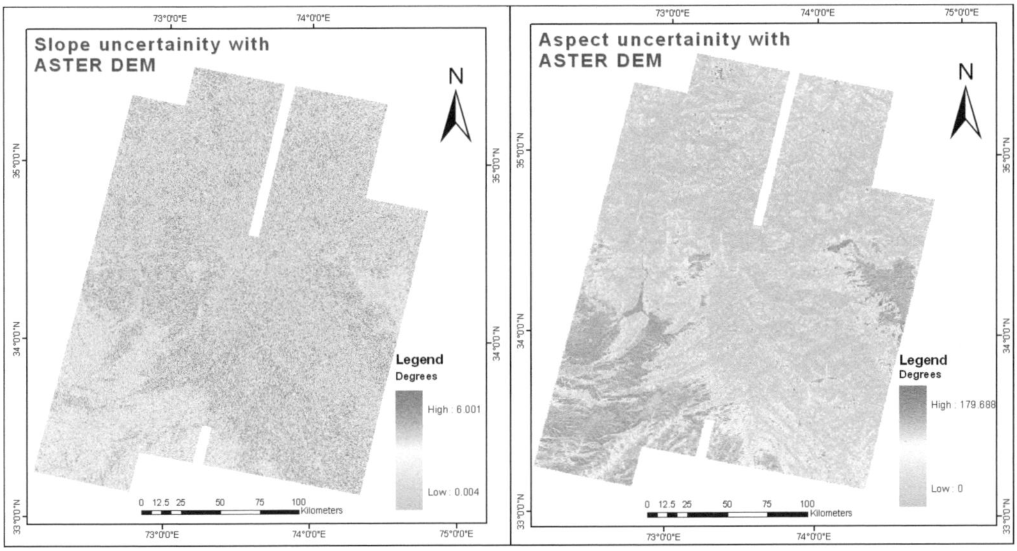


Figure 6. Uncertainty of slope and aspect computed from ASTER DEM in geographic space.

## 3.2 *Impact of DEM resolution and random errors on topographic attributes*

Impact of DEM resolution on the terrain representation and derived topographic attributes is presented in Fig. 1, which shows that as the resolution gets coarser, the features smaller than the DEM grid size are disappearing, due to smoothening effect, and the slope angle also starts reducing. Aspect is affected significantly with varying DEM resolution, particularly in narrow valleys. Figure 1, reflects the sensitivity of aspect estimation to the rugged terrain, while consistency is observed on steep slopes. While slope computation is sensitive to rugged terrain and to steep slopes, while consistent in flat terrain.

The impact of random errors on derived topographic attributes is determined, and uncertainty is quantified through Monte Carlo Simulations (MCS). Statistical analysis of the MCS predicts SRTM DEM more consistent in slope and aspect computation, compared to ASTER DEM. Coarse resolution of SRTM DEM also nullifies the impact of random error on computed slope and aspect, to some extent. Although the applied RMSE of ASTER DEM is just 4 m higher than SRTM DEM, but comparatively fine resolution, magnify the uncertainty in the computed topographic attributes. MCS derived values in Table 2, quantify the possible deviation or the uncertainties in the computed slope and aspect values from the reality.

The spatial distribution of slope and aspect uncertainty in Fig. 6 reflects the slope computation is more sensitive to steep slopes, while the aspect computation is sensitive to flat areas, which are in agreement with the Carter (1992) and Florinsky (1998).

Table 3. Sensitivity of terrain parameters to TAF

| Model Parameter | | Predicted results | | Sensitivity analysis | | | |
|---|---|---|---|---|---|---|---|
| | | TAF parameter | Base-model | Slope 60° | Height 400m | Wavel 200m | Damp 10% |
| Slope | 30° | $A_{h,max}$ | 1.10 | 1.17 | 1.14 | 1.08 | 1.06 |
| Height | 200 m | | | | | | |
| Wavel | 100 m | | | | | | |
| Damp | 5% | | | | | | |

Wavel = Wavelength
Damp = Damping

Table 4. Disparity of TAF computed from SRTM and ASTER DEMs

| | SRTM DEM | | ASTER DEM | | Disparity | | |
|---|---|---|---|---|---|---|---|
| | Min | Max | Min | Max | Min | Mean | Max |
| $A_{h,max}$ | 1 | 1.43 | 1 | 1.66 | –0.260 | 0.003 | 0.227 |

Table 5. Uncertainty in TAF predicted from ASTER and SRTM DEM

| | | | | | |
|---|---|---|---|---|---|
| ASTER DEM | $A_{h,max}$ | Slope (30°) | Slope (32.853°) | Slope (27.147°) | Uncertainty |
| | | 1.104 | 1.103 | 1.105 | 0.002 |
| SRTM DEM | | | Slope (30.935°) | Slope (29.065°) | Uncertainty |
| | | | 1.104 | 1.104 | 0.001 |

### 3.3 *Topographic aggravation of seismic response*

The numerical model (Eq. 1) is applied to predict the impact of topographic features on aggravation of seismic response. The model is applied to synthetic terrain profiles, due to its limitation of applicability, in real environment and at regional scale. The impact of model parameters on derived TAF is determined through base terrain of varying input parameters. The model parameters are doubled gradually, and TAF is computed to explore the sensitivity of each parameter to TAF (Table 3). The sensitivity analysis implies the slope angle as a sensitive parameter for the aggravation of seismic response.

### 3.4 *Impact of DEM resolution and random errors on TAF*

Disparity of TAF computed from SRTM and ASTER reflect meager impact of DEM resolution, on the predicted horizontal TAF (Table 4).

The uncertainty in the computed TAF, due to random errors in ASTER and SRTM DEMs is quantified in Table 5. Due to the high uncertainty in the ASTER DEM derived slope, it shows greater uncertainty in predicted topographic effects than from SRTM DEM.

## 4 DISCUSSIONS AND CONCLUSIONS

The extensive previous efforts on predicting the topographic impact on the variation of seismic response, lead to formulation of numerical models incorporating the topographic features and seismic properties. This study integrates DEM derived topographic features such as slope and relative height, and seismic wave characteristics in a numerical model, to predict topographic amplification or deamplification of seismic response.

Satellite remote sensing DEMs, like SRTM and ASTER DEMs and their derived topographic attributes are observed to be sensitive to steep slopes in terrain representation. The SRTM DEM, with coarse resolution, is predicted more consistent in slope and aspect computation than ASTER DEM. SRTM DEM is observed in representing terrain elevated than the ASTER DEM. The slope computation from SRTM and ASTER DEMs is prone to exaggerated errors on steep slopes, while the aspect computation is sensitive to flat areas. The coarse resolution DEMs were sensitive to the slope computation, especially in steep terrain and for narrow features, particularly when the terrain features were smaller than the DEM grid size.

DEM derived topographic attributes with quantified uncertainty, assumed wavelength of incident seismic waves and material damping are incorporated in numerical model, predict slope angle sensitivity to the predicted TAF. The Slope geometry controls the clustering of reflected seismic waves from terrain features, while height controls the number of incident seismic waves. SRTM and ASTER DEMs random errors and resolution, predict meager impact on the derived TAF.

## REFERENCES


Alexander, D. (1993) "Natural Disasters". USL Press, United Kingdom.

Ashford, S.A. and N. Sitar. (1997a) "Analysis of Topographic Amplification of Inclined Shear Waves in a Steep Coastal Bluff". Bulletin of the Seismological Society of America, 87, 692-700.

Ashford, S.A., N. Sitar, J. Lysmer and N. Deng. (1997b) "Topographic Effects on the Seismic Response of Steep Slopes". Bulletin of the Seismological Society of America, 87, 701-709.

Assimaki, D. and G. Gazetas. (2004) "Soil and Topographic Amplification on Canyon Banks and the 1999 Athens Earthquake". Journal of Earthquake Engineering, Imperial College Press, 8, 1-43.

Athanasopoulos, G.A., P.C. Pelekis and E.A. Leonidou. (1999) "Effects of surface topography on seismic ground response in the Egion (Greece) 15 June 1995 earthquake". Soil Dynamics and Earthquake Engineering, 18, 135-149.

Bard, P.Y. (1982) "Diffracted waves and displacement field over two-dimensional elevated topographies". Geophysical Journal International, 71, 731-760.

Bouckovalas, G. and A.G. Papadimitriou. (2006) "Aggravation of seismic ground motion due to slope topography", First European Conference on Earthquake Engineering and Seismology, Geneva-Switzerland, 1-10.

Bouckovalas, G.D. and A.G. Papadimitriou. (2005) "Numerical evaluation of slope topography effects on seismic ground motion". Soil Dynamics and Earthquake Engineering, 25, 547-558.

Carter, J.R. (1992) "The effect of data precision on the calculation of slope and aspct using gridded DEMs". Cartographica, 29, 22-34.

Chávez-García, F.J., D. Raptakis, K. Makra and K. Pitilakis. (2000) "Site effects at Euroseistest—II. Results from 2D numerical modeling and comparison with observations". Soil Dynamics and Earthquake Engineering, 19, 23-39.

Dunning, S.A., W.A. Mitchella, N.J. Rossera and D.N. Petleya. (2007) "The Hattian Bala rock avalanche and associated landslides triggered by the Kashmir Earthquake of 8 October 2005". Engineering Geology, 93, 130-144.

Florinsky, I.V. (1998) "Accuracy of local topographic variables derived from digital elevation models". International Journal of Geographical Information Science, 12, 47-61.

Heuvelink, G.B.M., P.A. Burrough and H. Leenaers. (1990) "Error propagation in spatial modelling with GIS." Proceedings of the First European Conference on Geographical Information Systems EGIS' 90, Amsterdam, The Netherlands, 453-462.

INGV. (2006) "Istituto Nazionale di Geofisica e Vulcanologia, DPC – S4 Project". Website: http://earthquake.rm.ingv.it/index.php, Accessed: Accessed 15 November 2007.

Kramer, S.L. (1996) "Geotechnical Earthquake Engineering". Prentice Hall International Series, New Jersey, 653 pp.

Lanter, D. and H. Veregin. (1992) "A research based paradigm for propagating error in layer-based GIS". Photogrammetric Engineering & Remote Sensing, 58, 825-833.

Morales, A.L.M. (2002) "Urban Disaster Management, A case study of Earthquake Risk Assessment in Cartago, Costa Rica". PhD-Thesis, Utretch University, The Netherlands. 235 pp.

Oksanen, J. and T. Sarjakoski. (2005) "Error propagation of DEM-based surface derivatives". Computers & Geosciences, 31, 1015-1027.

Pedersen, H.A., F.J. Sanchez-Sesma and M. Campillo. (1994) "Three-Dimensional Scattering by Two-Dimensional Topographies". Bulletin of the Seismological Society of America, 84, 1169-1183.

Sanchez-sesma, F.J., I. Herrera and J. Aviles. (1982) "A boundary method for elastic wave diffraction: Application to scattering of SH waves by surface irregularities". Bulletin of the Seismological Society of America, 72, 473-490.

Stamatopoulos, C.A., M. Bassanou, A.J. Brennan and G. Madabhushi. (2007) "Mitigation of the seismic motion near the edge of cliff-type topographies". Soil Dynamics and Earthquake Engineering, In Press, Corrected Proof.

Wechsler, S.P. and C.N. Kroll. (2006) "Quantifying DEM Uncertainty and its Effect on Topographic Parameters". Photogrammetric Engineering & Remote Sensing, 72, 1081-1090.

*Remote Sensing for a Changing Europe*
*D. Maktav (Ed.)*
*IOS Press, 2009*

*doi:10.3233/978-1-58603-986-8-395*



# An efficient methodology to simulate mixed spectral signatures of land covers through Field Radiometry data

J. M. Vazquez-Sierra
*Dept. of Electr. and Telec. Engineering, Politech. Sch., CEU University – Saint Paul, Madrid, SPAIN*

E. Martinez-Izquierdo & A. Arquero-Hidalgo
*Dept. of Arq. and Tech. of Inform. Systems, FI, Politech. University of Madrid, Madrid, SPAIN*

Keywords: Mixed Spectral Signature, Field Radiometry, Remote Sensing, Sampling

ABSTRACT: An efficient methodology to simulate mixed spectral signatures of land covers, from *endmember* data, using linear statistical modelling based on the least squares estimation approach, is proposed. The optimal set of *endmember* has been obtained by measurements *in situ* with a field spectroradiometer GER 1500. It is proposed the use of new sub-pixel methods based on statistics and certain "units of sampling" to apply to the landscapes. The resultant point estimations for these new units will be the "observations" and all of them will carry out a special role to simulate the final spectral signature. This methodology is used to simulate spectral signatures of a Mediterranean forest landscape near to Madrid (Spain). Furthermore the spectral signature model obtained through Field Radiometry will be correlated with the image data of the same zone provided by the Landsat 7 Enhaced Thematic Mapper Plus (ETM+) sensor once corrected. Promising results about remote and near spectral response correlation have been obtained.

## 1 INTRODUCTION

Traditionally, the production of maps from satellite imagery assumes that each pixel of the image can be assigned to a single land cover class. In the remote sensing imagery, the measured spectral radiance of a pixel is the integration of the radiance reflected from all the objects within the Ground Instantaneous Field Of View, GIFOV, also called Ground Sample Distance, GSD. Mixed pixels are generated if the size of the pixel includes more than one type of land cover.

Obviously, spectral mixing is inherent in any finite-resolution digital imagery of a heterogeneous surface. Solving the spectral mixture problem is, therefore involved in image classification, referring to the technique of spectral unmixing. This technique has been used for analyzing the mixture of components in remotely sensed images. The technique is based on the assumptions that several primitive classes of interest can be selected, that each of these primitive classes has a pure spectral signature, which can be identified and the mixing between these classes can be adequately modelled as a linear combination (Small 2001) of the spectral signatures.

Spectral mixture analysis (SMA) techniques have overcome some of the weaknesses of full pixel approaches by using linear statistical modelling and signal processing techniques (Keshava 2003) (Rand 2001). The key task in linear SMA is to find an appropriate set of pure spectral components which are then used to estimate the fractional abundances of each mixed pixel from its spectrum and the *endmember* spectra by using a linear mixture model (Heinz 2001). The identification of the pure pixel value is often difficult. Nevertheless, the register of land cover spectral response by measurements *in situ* provides a good set of *endmembers*. Field Radiometry (field spectroscopy or field spectrometry) refers to spectroscopic measurements made outdoors, with the sun as the primary source of illumination, and allows to the detection and analysis of the spectral characteristics of the land covers in its natural surroundings.

Previous studies related to the sampling method in the spectral register processes (Vazquez 2004, 2007) exist specifying exhaustively the directives of sampling *in situ*. These investigations contribute to solve a very important aspect, since it is the determination of the optimal number of spectral samples that it is necessary to register for each of the *endmembers*. This number must assure not only that the sampling is effective and the behavior of those samples is representative of the total population they come from.

To simulate mixture spectral signatures of a specific geographical zone from sampling procedures through Field Radiometry, is essential to identify an estimation method optimally adapted to the type and characteristics of the existent land covers for every case. Moreover, among all the most common methods of estimation, such as least squares, maximum likelihood, moments, or the so-called Bayesian, this investigation uses the first one. The fundamental reason is that the last three mentioned above require prior knowledge of the population probability density function. However, the *least squares estimation method* is based on the minimization of distances regardless of the samples density probability function. This is a crucial aspect because usually and from a statistical point of view, *a priori* the samples behavior is not known.

Two fundamental ideas are presented in this paper: on the one hand the assumption of the existing different land covers that can be considered as random variables, and on the other hand, the design of a new statistics generation methodology who will allow to represent in efficient way the typology and variability of each of them. Moreover, we propose new sub-pixel methods based on "ad-hoc" statistics to apply to the landscapes and the use of certain "units of sampling" too. The resultant point estimations for these new units will be the "observations" that will be crucial later to obtain the final landscape spectral signature. Also we present a general solution to the problem of choosing the best sampling method and the optimal number of samples that must be considered in every case. Finally this methodology is applied to simulate spectral signatures of a Mediterranean landscape, correlating the results obtained through Field Radiometry with the existing ones of the same geographical zone, provided by the Landsat ETM+ sensor once atmospheric corrected.

## 2 DATA SET DESCRIPTION

The study zone is situated in the Mountgancedo near Madrid (Spain). This site is located to the south-west of Community of Madrid, and it has a surface area of 125 hectares. The study area selected is located at 40º 24´30´´ N, and 3º49´50´´ W (4473/4474 N, 429/430 W UTM). Several mediterranean species of vegetation can be easily found, such as holm oaks (*Quercus ilex*), spanish lavender (*Lavandula pedunculata*) and the crimson spot rockrose (*Cistus ladanifer*). The surface is also covered by grass and abundant meadows as well as bare soil, and rocks and stones outcrops. The land cover spectral samples have been obtained in the summer by Field Radiometry with the GER 1500, agreement with the date of Landsat ETM+ remote data register. Sampling *in situ* has been carried out according to a methodology (Vazquez 2007) that optimizes the number of samples to pick-up.

Even though the remote sensor (ETM+) has six bands with the same spatial resolution, only the first four ones can be considered according to the spectral register interval of field radiometer; in order to correlate field and remote information.

Since the data registered *in situ*, provides a continuous spectral response, a previous reduction methodology based on the integration of the radiance values (Arquero 2003) into the spectral bands interval selected, was carried out. Thus, the field spectral data had been reduced to field spectral signatures, with four values named R-ETM+1, R-ETM+2, R-ETM+3 y R-ETM+4 in agreement with the first four bands of remote sensor.

## 3 METHODOLOGY

The proposed methodology originally starts with the identification of the landscape unit to be analysed. This unit has equal dimension of the pixel of remote sensor and will be called *Global Unit* (**$G_u$**). This one will be submitted to a process of successive subdivisions, in order to get smaller units or *Intermediate Units*, **$I_u$**, composed, in turn, by indivisible ones called *Elemental Units*, **$E_u$**, (Fig. 2a). From the sampling information of these **$E_u$**, it will be possible to model the mixed spectral response of **$G_u$**. For that, the size of each **$E_u$**, must be such that behaves as a "homogeneous spectral unit". Its spectral content must be as close as possible to the *endmember* in order to represent optimally the cover itself. High land covers variability will involve the existence of a large number of very small ones.

On the other hand, one of the key ideas of this paper is to consider each land cover that make up the different territories as a "random variable". Thus, each one will have its corresponding probability density function, so it is feasible to apply to all of them the principles of the sampling and statistical inference. With this initial hypothesis is possible to consider any landscape as a random variables succession representing the different covers contained therein. So by the *Central Limit Theorem* and the appropriate statistics, it is feasible to model the overall spectral response of the global area analyzed. Thus, each **$E_u$** can be considered as an independent and specific random variable. With this hypothesis, it is possible to design new statistics based on samples that can be considered as random variable successions whose behaviour at the limit provides an adequate representation of the Global Unit, **$G_u$**. Moreover, the size of the **$E_u$** is often small compared with **$G_u$**, especially in the fairly common case when landscapes with great covers variability are analyzed. For this reason, a standard sampling method could not be very efficient from a statistical point of view. To overcome this difficulty and ensure the management of an optimum number of samples it is necessary to create a new sampling unit, called Intermediate Unit, **$I_u$**, whose dimension will be between the Global and the Elementary Units. The criteria to design the more representative statistics for **$I_u$** could be based on a strategy of "considering" more or less the information belonging to the different Elementary Units. In this sense, we propose the "balanced statistic" and the "unbalanced statistic". The first one consists in assigning to each **$E_u$**, identical "weights". So, each of them equally will contribute to the total spectral response of the **$I_u$**. The second one considers weighing "differently" the information contained in the centre than the contained in the peripheral (Fig. 1).

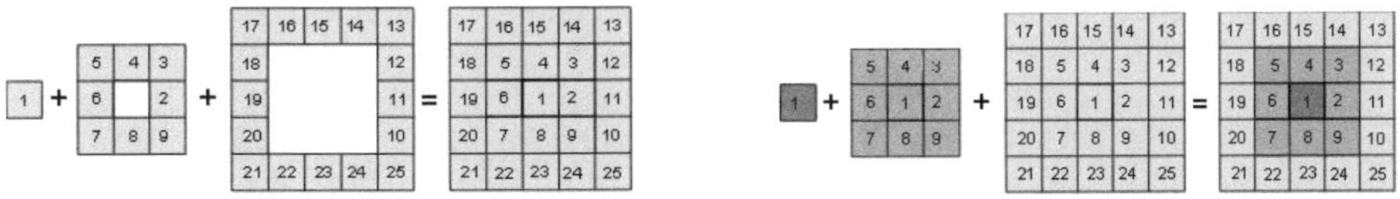

a) b)

Figure 1. Balanced (a) and Unbalanced (b) statistics with an example of Intermediate Unit size equal to 5x5 Elementary Unit.

The analytical expression for these statistics is shown in eq.1 and eq.2.

$$i_{balanced} = \frac{\sum_{i=1}^{i=2n-1} \sum_{j=1}^{j=2n-1} x_{ij}}{(2n-1)^2} \quad (1)$$

$$i_{unbalanced} = \frac{x_{ij} + \sum_{i-1}^{i+1}\sum_{j-1}^{j+1} x_{ij} + \sum_{i-2}^{i+2}\sum_{j-2}^{j+2} x_{ij} + \ldots + \sum_{i-(2n-1)}^{i+(2n-1)} \sum_{j-(2n-1)}^{j+(2n-1)} x_{ij}}{\sum_{i=1}^{i=n} (2n-1)^2} \quad (2)$$

This last strategy compels to consider an **$I_u$** size of odd number of **$E_u$**. Then the problem to be solved focuses on finding the size and the statistical characteristics that best describe this new **$I_u$**. For that it is necessary to consider the best sampling method to gather information for this new **$I_u$** so that, when it is contemplated all together, it ensures an "optimal" representation of the whole **$G_u$**.

For simplicity we suggest that each Global Unit can be considered as a square whose side $N$ consists of a number pair of Elementary Units. This initial square is always possible to divide it into a series of Intermediate Units of side *(2n-1)*, as show in the Figure 2. With the sampling method of unbalanced strategy mentioned above, the samples total number for each Intermediate Unit is the sum of the squares of odd numbers. There is still a key issue to be solved as it is the number of elements that must contain each Elementary Intermediate Unit of sampling. This number can be deduced in terms of "n" and it is the contained in a square of side *N-(2n-1)*, if it is chosen a horizontal sampling method. Accordingly, the general expression of the total number of samples to be analyzed in the Global Unit will be determined by:

$$\textit{Total number of samples} = [N - 2(n-1)]^2 \left[ \frac{n\left(4n^2 - 1\right)}{3} \right] \tag{3}$$

In order to have as much information as possible about the population characteristics, it is desirable to handle the maximum number of samples. This requires finding what is the value of n which maximizes the above expression (eq.1).This value is shown in eq.4.

$$40n^3 - [N+2] \cdot 12n^2 - 6n + N + 2 = 0 \tag{4}$$

In the last expression if the value of *N(n)* is known, it is easy to obtain the value of $n$. Thus, the size of the Intermediate Unit will be a squared integrated by (2n-1) x (2n-1) Elementary Units. In general, the optimum size of the Intermediate Units in terms of the number of Elementary Units that must be considered for different ratios of $N / n$. For example if $\mathbf{G_u}$ is 8x8, $\mathbf{I_u}$ is 5x5; if $\mathbf{G_u}$ is 16x16, $\mathbf{I_u}$ is10x10, ....

On the other hand, the Global Unit sampling method could be as simple as possible. Furthermore, it is crucial to consider the thematic diversity of all the covers existing in each $\mathbf{I_u}$ that conform the $\mathbf{G_u}$. There is not a formal procedure to provide the "best" method of sampling for all cases because, depending on the nature of the terrain to study, the methods can and must be different. The literature recommends following a basic rule: to develop sampling methods depending on the covers type and its variability. One of the most efficient ways to achieve this, is to develop the sampling process along specific lines (usually horizontal, vertical or diagonal). This paper presents a new method for picking up samples: the so-called "double-diagonal" (See Fig.2). This model reflects with acceptable accuracy the variability of the different land covers, and thus achieves a good representation of the Global Unit. This methodology is applied to simulate spectral signatures of a $\mathbf{G_u}$ of a Mediterranean landscape comparing to its spectral response like a mixed pixel. The results of correlation (in reflectance values) study have been obtained through Field Radiometry data with the existing ones of the same geographical zone, provided by the Landsat ETM+ sensor once atmospheric corrected (Vermote 1997).

## 4 RESULTS

The area chosen as Global Unit is a square with one side of N = 32 meters which corresponds, approximately, with the spatial resolution of ETM+ sensor. After making the appropriate subdivision process it founds an elemental square (Elemental Unit) of n = 4 meters side. Thus, the relation between the $\mathbf{G_u}$ and the $\mathbf{E_u}$ will be N = 8n. The $\mathbf{I_u}$ size will be 5x5 Elementary Units (square of 20 meters side). Next, the spectral signatures of near data of the study area, obtained by applying the methodology described with double diagonal sampling method (See Fig. 2b), are presented. To simulate the spectral signatures of $\mathbf{G_u}$ has been used the *least squares method* with abundance constraints. (The constraint considered is the full additivity. This requires that the abundance for a mixed pixel must be one). Once obtained linear mixture algorithms for each band, there has been made a correlation between this information and the provided by the remote sensor in the same period of time. The study area basically contains three relevant land covers: oak, meadow and shrub.

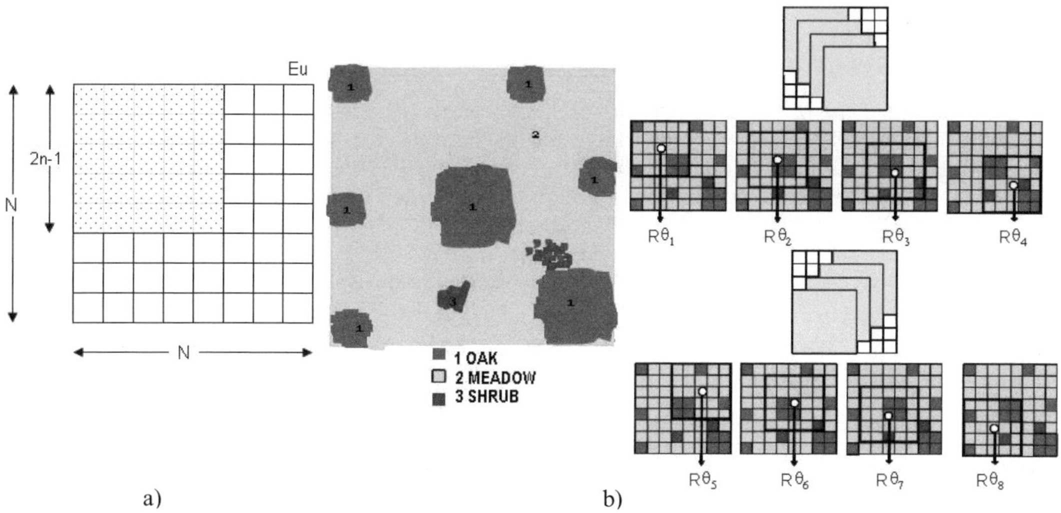


Figure 2. (a) Scheme of example for different sizes $\mathbf{G_u}$, $\mathbf{I_u}$ and $\mathbf{E_u}$. (b) Land covers of Mediterranean Landscape scheme of Unit Global and double diagonal sampling method.

The results of the reflectance values of each Intermediate Unit obtained by applying the balanced and unbalanced statistics are shown in Table 1. These values have been calculated for the interval of bands ETM1+, ETM+2, ETM+3 and ETM+4 and correspond to the point estimations provided by both statistics when a double diagonal sampling method is applied. (See Fig. 2b).

Table 1. Reflectance values obtained by application of balanced and unbalanced statistics to Intermediate Units.

| | Reflectance (%/100) | | | | | | | |
|---|---|---|---|---|---|---|---|---|
| $I_u$ | Balanced method | | | | Unbalanced method | | | |
| | R-ETM+1 | R-ETM+2 | R-ETM+3 | R-ETM+4 | R-ETM+1 | R-ETM+2 | R-ETM+3 | R-ETM+4 |
| $\Theta_1$ | 0.0950 | 0.1340 | 0.1621 | 0.2982 | 0.0935 | 0.1331 | 0.1616 | 0.2981 |
| $\Theta_2$ | 0.0990 | 0.1390 | 0.1702 | 0.2936 | 0.0976 | 0.1383 | 0.1694 | 0.2932 |
| $\Theta_3$ | 0.0910 | 0.1330 | 0.1634 | 0.2987 | 0.0908 | 0.1321 | 0.1631 | 0.2983 |
| $\Theta_4$ | 0.0870 | 0.1250 | 0.1523 | 0.3072 | 0.0866 | 0.1244 | 0.1520 | 0.3056 |
| $\Theta_5$ | 0.0870 | 0.1270 | 0.1542 | 0.3009 | 0.0862 | 0.1266 | 0.1540 | 0.3011 |
| $\Theta_6$ | 0.0960 | 0.1410 | 0.1746 | 0.2957 | 0.0955 | 0.1394 | 0.1731 | 0.2963 |
| $\Theta_7$ | 0.0960 | 0.1370 | 0.1696 | 0.2964 | 0.0956 | 0.1355 | 0.1687 | 0.2966 |
| $\Theta_8$ | 0.0930 | 0.1360 | 0.1687 | 0.2973 | 0.0923 | 0.1351 | 0.1691 | 0.2999 |

The regression models for each band (mixed hyperplanes) obtained by *linear least squares techniques* through balanced (eq. 5-8) and unbalanced (eq. 9-12) statistics are shown in the next equations:

$$R_{R\text{-}ETM+1} = 0.0441p_1 + 0.1036p_2 + 0.0519p_3 \tag{5}$$

$$R_{R\text{-}ETM+2} = 0.0714p_1 + 0.1516p_2 + 0.0843p_3 \tag{6}$$

$$R_{R\text{-}ETM+3} = 0.0882p_1 + 0.1912p_2 + 0.0898p_3 \tag{7}$$

$$R_{R\text{-}ETM+4} = 0.3765p_1 + 0.2877p_2 + 0.2155p_3 \tag{8}$$

$$R_{R\text{-}ETM+1} = 0.0471p_1 + 0.0981p_2 + 0.0273p_3 \quad (9)$$

$$R_{R\text{-}ETM+2} = 0.0722p_1 + 0.1422p_2 + 0.0755p_3 \quad (10)$$

$$R_{R\text{-}ETM+3} = 0.0681p_1 + 0.1952p_2 + 0.0653p_3 \quad (11)$$

$$R_{R\text{-}ETM+4} = 0.3675p_1 + 0.2986p_2 + 0.1784p_3 \quad (12)$$

$R_i$ is reflectance and $p_i$ is abundance for oak (1), meadow (2) and shrub (3).

Finally, the Table 2 presents comparative results between the signature values obtained from simulated models and those obtained through remote sensor ETM+, for de Global Unit. The spectral signature modeled has been obtained through the estimated percentage of abundances: 18.8 of oak class, 78.1 of meadow class and 3.1 of shrub class.

Table 2. Spectral signature values of simulated models and satellite atmospheric corrected data for the $\mathbf{G_u}$.

| | ETM+1 | ETM+2 | ETM+3 | ETM+4 |
|---|---|---|---|---|
| Balanced model | 0.09 | 0.14 | 0.17 | 0.30 |
| Unbalanced model | 0.09 | 0.13 | 0.16 | 0.30 |
| Satellite | 0.08 | 0.11 | 0.15 | 0.31 |

The spectral signatures modeled are very similar that the one provided by the sensor satellite. The simulation predicts signatures values slightly higher than those provided by the satellite sensor in the first three bands, whereas in the fourth band this value is slightly lower. The use of unbalanced statistic provides values closest to the satellite sensor.

## 5 CONCLUSIONS

In conclusion, it has been proposed a methodology to simulate spectral mixed signatures. The methodology derives from the use of new sub-pixel methods based on statistics and certain “units of sampling”. Furthermore, correlation studies have been carried out with near and remote spectral data in order to evaluate this methodology. Their results seem to conclude its efficiency. At the same time, the results open new research guidelines.

## REFERENCES

Arquero, A., Martinez, E. & Gonzalo, C. 2003. Reducción de datos hiperespectrales de Radiometría de Campo”, *Teledetección y Desarrollo regional; X Congreso Nacional de Teledetección, Cáceres (Spain): 439-442.*

Heinz, D & Chang, C. 2001. Fully Constrained Least Squares Linear Spectral Mixture Analysis Method for Material Quantification in Hyperspectral Imagery. *IEEE Transactions on Geoscience and Remote Sensing* 39 (3): 529-545.

Keshava, N. 2003. A Survey of Spectral unmixing Algorithms. *Lincoln Laboratory Journal* 14(1): 55-78.

Rand, R. S. 2001. A Spectral Mixture Process Conditioned by Gibbs-Based Partitioning. *IEEE Transactions on Geoscience and Remote Sensing* 39 (7): 1421-1433.

Small, C. 2001. Estimation of urban vegetation abundance by spectral mixture analysis. *International Journal of Remote Sensing* 22(7): 1305-1334.

Vázquez, J. M., Arquero, A., Martínez, E. & Gonzalo, C. 2004. An efficient methodology for endmembers selection by field radiometry: an application to multispectral mixture model. *Signal and Image Processing for Remote Sensing X* 5573: 435-443.

Vázquez, J. M., Arquero, A. & Martínez, E. 2007. Aplicación de modelos estadísticos a la selección eficiente de cubiertas tipo mediante Radiometría de Campo. *IEEE Latin America Transactions* 5 (8): 561-567.

Vermote E.F., Tanré D., Deuzé J.L., Herman M. & Morcrette J.J. 1997. Second Simulation of the Satellite Signal in the Solar Spectrum: an overview. *IEEE Transactions on Geoscience and Remote Sensing* 35(3): 675-686.

*Remote Sensing for a Changing Europe*
*D. Maktav (Ed.)*
*IOS Press, 2009*

*doi:10.3233/978-1-58603-986-8-402*

# Calibration of land use change models: the potential of spatial metrics

J. van der Kwast & G. Engelen
*Flemish Institute for Technological Research (VITO), Mol, Belgium*

Keywords: remote sensing, landscape metrics, spatial metrics, urban land use, contextual classification, calibration, land use change models

ABSTRACT: Recently, spatial metrics have been introduced in the field of urban land use mapping and modelling to characterize the spatial dynamics of such systems. The question raised in this study is whether spatial metrics can be applied directly on remote sensing images to calibrate and validate dynamic land use models of urban systems. The underlying hypothesis is that a methodology can be developed which enables to calculate metrics on both the remote sensing image and the land use map which quantify the same distinguishing spatial structures at some level of abstraction. A positive answer to this question would simplify and speed up greatly the calibration procedures in so far that the development of time series of land use maps could be avoided. In this study, remote sensing derived land use patterns have been compared with a land use map at four different levels of abstraction: 1) a pixel-based comparison between the classified image and the map, 2) comparison of landscape metrics calculated for each pixel by using a circular moving window, 3) comparison of landscape metrics for all urban classes individually, and 4) comparison of the aggregated urban classes. Results show that the agreement between the two maps depends on the level of abstraction at which they are compared and the spatial metric that has been used. The study concludes by presenting a concept of using landscape metrics in the calibration of urban land use change models.

## 1 INTRODUCTION

More than ever before, planners and policy makers need tools to anticipate and assess the impact of their decisions on the spatial systems that they are to manage. A growing number of high resolution dynamic land use models is currently being developed for this purpose. The calibration of these models remains a major challenge. Typically the required time series of land use maps based on identical and consistent mapping methodologies, legends and scales are missing. As a result, the differences observed in a series of land use maps are often not real land use changes, but mismatches that should not be compared with model predictions of land use changes.

On the other hand, earth observation satellites provide images of the earth's surface since 1972. Their potential temporal availability is relatively high and depends on the revisit time of the satellite: typically between 1 and 16 days. A series of remote sensing images of one sensor is consistent in time and space, with only slight deviations in quality due to changes in gain values of the sensor. However, the spatio-temporal availability and consistency heavily depend on atmospheric conditions. Another problem is that conventional remote sensing based classifications result in land cover maps, based on physical properties of the surface, rather than land use maps representing functional classes.

Recently, landscape metrics or spatial metrics have been introduced in the field of urban land use mapping and modelling to characterize the spatial dynamics of such systems (Herold & Menz 2001). The question raised in the study presented is whether spatial metrics can be applied directly on remote sensing images to calibrate and validate dynamic land use models of urban systems. The underlying hypothesis is that a methodology can be developed which enables to calculate metrics on both the remote sensing image and the land use map which quantify the same distinguishing spatial structures at some level of abstraction. A positive answer to this question would simplify and speed up greatly the calibration procedures in so far that the development of time series of land use maps could be avoided.

In this research, land use patterns have been derived from a Landsat TM5 image of Dublin by applying a contextual classifier, the Optimized Spatial Reclassification Kernel (OSPARK), to an unsupervised classification. The remote sensing derived land use patterns have been compared with a land use map at four different levels of abstraction: 1) a pixel-based comparison between the classified image and the map, 2) comparison of landscape metrics calculated for each pixel by using a circular moving window, 3) comparison of landscape metrics for all urban classes individually, and 4) comparison of the aggregated urban classes.

The study concludes by presenting a concept of using landscape metrics in the calibration of urban land use change models.

## 2 METHODS

First the derivation of land use classes from a Landsat TM5 image of Dublin of June 13$^{th}$ 1988, using the OSPARK algorithm will be described, by introducing the concept of SPARK and its extensions. Next, the theory and application of spatial metrics as a means to compare spatial land use patterns will be explained.

### 2.1 *The concept of SPARK*

The Spatial Reclassification Kernel (SPARK, Barnsley & Bar 1996) is a contextual reclassification method. The conceptual idea behind SPARK is that the land use types of interest can be characterized by the spatial arrangement and size of objects. For urban areas, for example, this can be the spatial arrangement of objects like streets, buildings, bare areas, shade, trees and grass. SPARK examines the local, spatial patterns of objects in a square kernel or moving window. Based on the arrangement of adjacent pixels in the kernel SPARK assigns a new class to the centre pixel. Consequently, the SPARK-method consists of three phases (Barnsley & Barr 1996):

1. Produce a land cover map using any type of pixel-based spectral classifier from a remotely sensed image, further referred to as the 'initial land cover map';
2. Define SPARK decision rules based on local, spatial patterns of objects in typical land use types;
3. Reclassify the initial land cover map into land use types, based on the decision rules of step 2.

For the first phase, the production of the initial land cover map, several methods can be used, e.g. unsupervised clustering techniques and supervised classification techniques.

During the second phase, reference kernels for known land use types are defined. These kernels are referred to as template matrices ($T_k$-matrices) and are based on training datasets in the initial land cover map. Each land use type is identified by one or more $T_k$-matrices. Examples of characteristic template matrices for urban land use can be found in figure 1. Within each $T_k$-matrix the frequency and the spatial arrangement of land cover classes positioned next to each other as well as diagonally are counted. Each pair of neighbouring pixels is called an adjacency event. The results of counting the adjacency events are expressed by an adjacency event matrix (figure 1). Remark that the two template matrices of figure 1 cannot be discriminated based on class frequencies only, ignoring the spatial configuration.

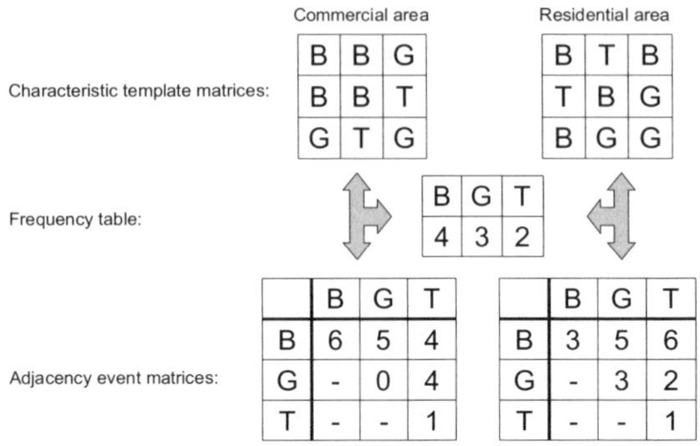


Figure 1. Simulated 3 by 3 kernels of different urban land use types with corresponding frequency table and adjacency event matrix. Only the upper triangle of the matrix is considered because $\mathbf{M}_{ij}$= $\mathbf{M}_{ji}$. **B** = building, **G** = grass and **T** = trees. After Alimohammadi & Shirkavand (2004).

The third phase of the SPARK algorithm involves the counting of the frequency and the spatial arrangement of land cover classes in the initial land cover map. For each pixel, an adjacency event matrix is produced by means of a moving window. The adjacency event matrices of the initial land cover map are called M-matrices.

The M-matrices of the initial land cover map are compared with all $T_k$-matrices using the following equation (Barnsley & Barr 1996):

$$\Delta k = 1 - \sqrt{0.5 \cdot N^2 \cdot \sum_{i=1}^{c}\sum_{j=1}^{c}\left(\mathbf{M}_{ij} - \mathbf{T}_{\mathbf{k}_{ij}}\right)^2} \tag{1}$$

where M is the adjacency event in an $i$ by $j$ matrix, $T_k$ is a template matrix for land use class $k$, $N$ is the total number of adjacency events in the kernel and $c$ is the number of land cover classes in the initial land cover map. The resulting term $\Delta k$ is an index of similarity between the M-matrix and the $T_k$-matrix. The values can range from 0 to 1. If $\Delta k$ equals 0, they are different, while a value of 1 means that they are identical.

Finally, the central pixel in the kernel is assigned to the land use class of the corresponding $T_k$-matrix with the highest $\Delta k$. More template kernels can be defined for each output class. The use of multiple template matrices has the advantage that subtle differences in spatial arrangement of the pixels in the initial land cover map can be accounted for, resulting in a more accurate classification of a particular pixel (Barnsley & Barr 1996). The procedure is summarized in the flowchart of figure 2.

The selection of representative $T_k$-matrices and the determination of the optimal kernel size are the most important factors that determine the quality of the SPARK classification. The optimal kernel size depends on the resolution of the images and the range of the spatial variation in the initial land cover map (Barnsley & Barr 1996). Kernels that are too large in respect to the land use objects will increase the effects of edges, while too small kernels possibly do not include all spatial variation. Different studies (Barnsley & Barr 1996, Sluiter *et al.* 2004) suggest that the kernel size can also be considered like a tuning parameter, which can be iteratively obtained by classifying the initial land cover map with different kernel sizes and evaluating the accuracy of the results.

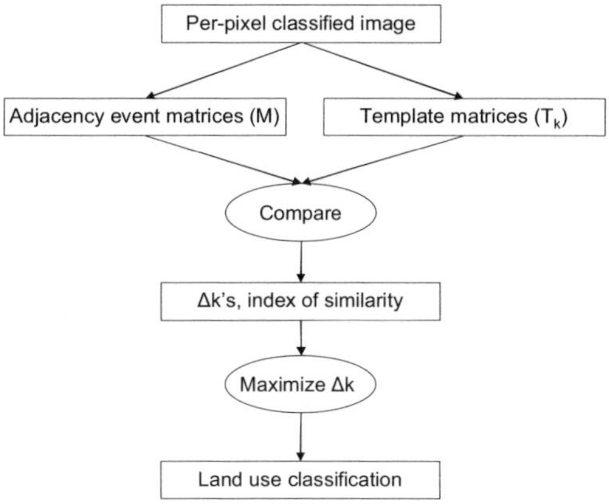


Figure 2. Flowchart of the SPARK algorithm.

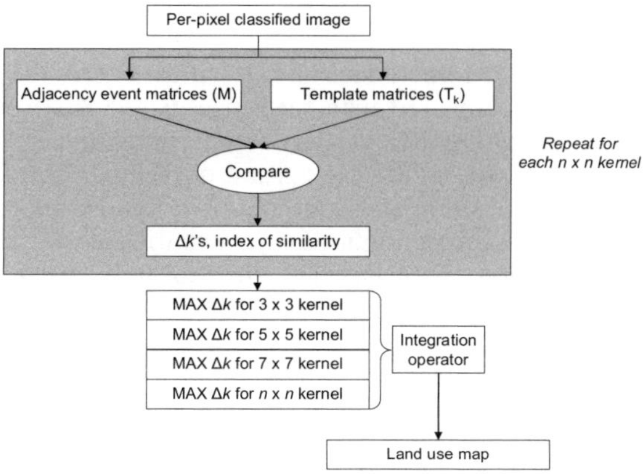


Figure 3. Flowchart of the OSPARK algorithm.

Theoretically, the optimal kernel size may differ for different land use types. For example Barnsley & Barr (1996) indicated that for urban land use mapping purposes, small kernel sizes are more appropriate in non-urban areas, while large kernels should be used to represent the spatial variability in land cover in urban areas. Based on the SPARK algorithm, the following two results are expected (Sluiter *et al.* 2004): (1) complex patterns will benefit less from larger kernel sizes because the chance that a particular M-matrix is similar to one of the $T_k$-matrices decreases, since the variability in the adjacency event matrix increases with increasing kernel size; (2) Pure homogeneous classes will not benefit from larger kernel sizes because edges will influence the classification. Because of these considerations, some extensions to the original SPARK algorithm have been developed.

## 2.2 *Extensions to SPARK*

The Optimized SPARK algorithm (figure 3) is an extension of the original algorithm and differs on two points:

1. OSPARK calculates the $T_k$- and M-matrices for a range of user-defined kernel sizes. The coordinates of the central pixel of the template kernels around which the kernel sizes will be increased, have to be defined by the user;
2. The integration operator assigns to each pixel the land use of the first template kernel for which the $\Delta k$ value is above a user-defined threshold. This is necessary, because larger template matrices are less unique for a particular land use type. Therefore after one or several local maxima $\Delta k$ will increase with increasing kernel size.

## 2.3 *Classification*

First, the initial land cover map has been derived from a Landsat TM5 image of Dublin, Ireland of June 13$^{th}$ 1988. This image has been classified using an unsupervised classification. This resulted in a thematic map of 17 classes with a resolution of 30 m.

Next, centre coordinates of the $T_k$-matrices have been sampled randomly within polygons from a land use map of 1990, which is based on the CORINE (Bossard *et al.* 2000) land use/land cover database. The resulting 288 template centre locations are used in OSPARK to iteratively define the increasing template kernels. In this application, the OSPARK algorithm optimizes the classification using kernel sizes between 3 by 3 pixels and 59 by 59 pixels of 30 m resolution. The minimum $\Delta k$ threshold has been set to 0.7. The resulting classification has been resampled to a resolution of 200 m, corresponding with the resolution of the land use map as used in the MOLAND model (Engelen *et al.* 2002). During resampling the most occurring sub-pixel class within the coarse resolution pixel is assigned to the coarse resolution pixel.

## 2.4 *Spatial metrics*

Traditionally, remote sensing classification results and outcomes of land use change models are pixel-wise compared with reference data. Visual interpretation, however, suggests a better correspondence at higher levels of spatial abstraction.

Landscapes, either urban, rural or (semi-)natural can be described according to two basic properties: composition and configuration. Landscape composition is a non-spatially explicit property, while landscape configuration relates to spatially-explicit properties of a landscape (Rainis 2003). Configuration characteristics, such as polycentrism, dispersion, fragmentation, concentration or linearity, cannot be properly represented by pixel-wise derived confusion matrices (Barredo & Demicheli 2003).

In landscape ecology landscape metrics or spatial metrics are commonly used to quantify landscape composition and configuration in thematic maps (Gustafson 1998). Only recently, scientists concerned with urban land use are using spatial metrics as a tool to quantify the composition and/or structure of urban landscapes, with promising results (Barredo & Demicheli 2003, Herold & Menz 2001, Herold *et al.* 2002, Herold *et al.* 2003, Parker *et al.* 2001, Rainis 2003).

Spatial metrics can be defined as quantitative measures to describe structures and patterns of a landscape at a set point in time and provide information about the contents of the landscape mosaic or the shape of the component landscape elements (Herold & Menz 2001, Rainis 2003). They are derived from digital thematic-categorical data that show spatial heterogeneity at a specific scale and resolution. Therefore a spatial metric always represents spatial patterns at a specific spatial scale, which is determined by the spatial resolution, the extent of the spatial domain and the number and definition of thematic classes of the input dataset (Herold *et al.* 2005). Spatial metrics can be aggregate measures for the whole landscape or spatially explicit.

In this research it will be investigated if spatial metrics can be used to compare patterns in the OSPARK result with patterns in the land use map. Besides the pixel-based comparison, the maps will be compared at three different levels of abstraction: 1) comparison of landscape metrics calculated for each pixel by using a circular moving window (window-level), 2) comparison of landscape metrics calculated for all urban classes individually (class-level), and 3) comparison of the aggregated urban classes (landscape-level).

Many spatial metrics have been included in three public domain GIS-based software packages: FRAGSTATS (McGarigal *et al.* 2002), Patch Analyst (Elkie *et al.* 1999) and the r.le programs within GRASS (Baker & Cai 1992). In this study FRAGSTATS 3.3 has been used. The spatial metrics and the level of abstraction at which they are applied are given in Table 1. For the calculation of these metrics we refer to McGarigal *et al.* (2002).

Table 1. Spatial metrics used in the analysis at different levels of abstraction

| *Spatial metric* | *Abbreviation* | *Window-level* | *Class-level* | *Landscape-level* |
|---|---|---|---|---|
| Class area | CA | | x | |
| Percentage of landscape area | PLAND | | x | |
| Number of Patches | NP | | x | x |
| Patch Density | PD | x | | x |
| Larges Patch Index | LPI | | x | x |
| Edge Density | ED | x | x | x |
| Largest Shape Index | LSI | | x | x |
| Standard deviation of Patch area | AREA_SD | x | x | x |
| Area Weighted Mean Patch Fractal Dimension | FRAC_AM | x | x | x |
| Euclidian Mean Nearest Neighbour Distance | ENN_MN | | x | x |
| Landscape Contagion | CONTAG | x | x | x |

## 3 RESULTS

The result of the OSPARK classification (figure 4, left) has been compared with the land use map (figure 4, right) at four different levels of abstraction: the pixel-level, class-level, landscape-level and moving window-level.

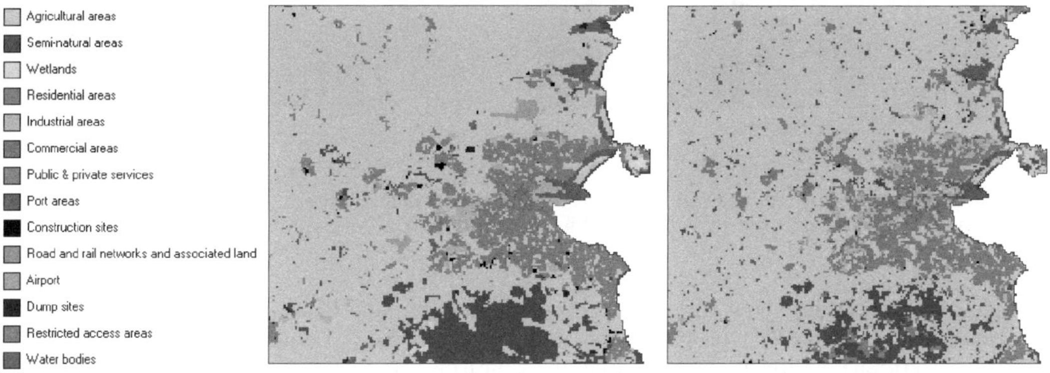


Figure 4. Left: Land use map of Dublin 1990. Right: OSPARK classification of Dublin 1988.

### 3.1 *Pixel-level*

At the pixel-level the OSPARK classification has been compared with the land use map by means of a confusion matrix (aka contingency matrix, error matrix). The overall accuracy of the classification is 79% and the Kappa is 0.55.

When individual land use classes are compared, it can be seen that the *agricultural areas* have the highest user's accuracy (89%), followed by *semi-natural areas* (67%) and *airport* (62%). From the urban classes, in which we are most interested in this research, the *residential areas* have a user's accuracy of 59%. The other urban classes have a low user's accuracy: *industrial areas*, *public & private services* and *commercial areas* have 24%, 12% and 10% respectively. The user's accuracy gives the chance that a class in the map can be found in reality when the map is used.

### 3.2 *Class-level*

At a higher level of abstraction, the class-level, the configuration of different urban classes can be compared by analysing the differences in their metric values. This comparison is not spatially explicit.

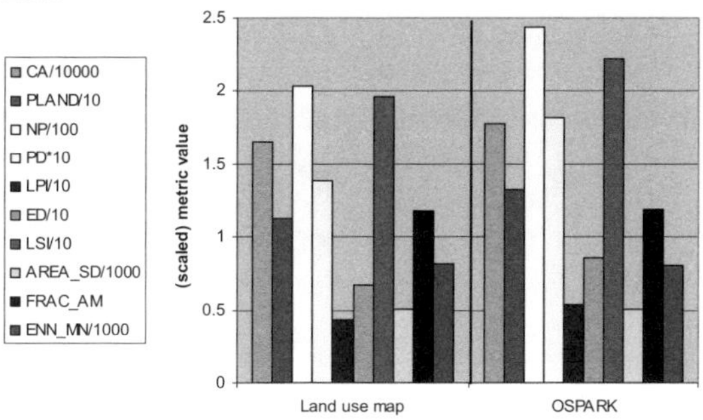


a

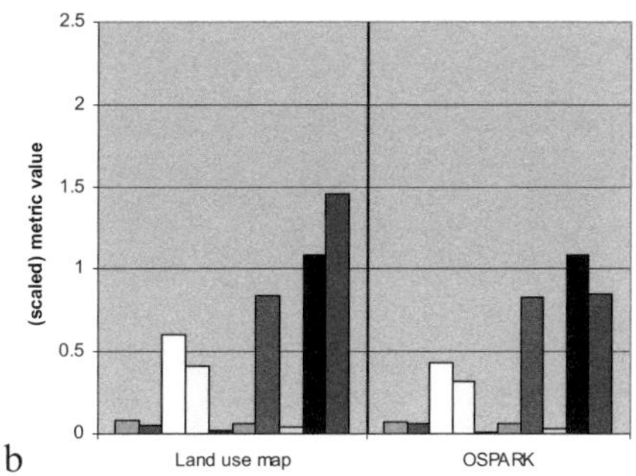


b

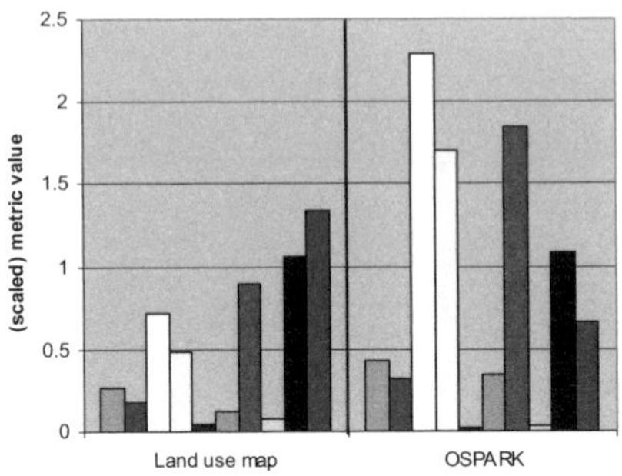


c

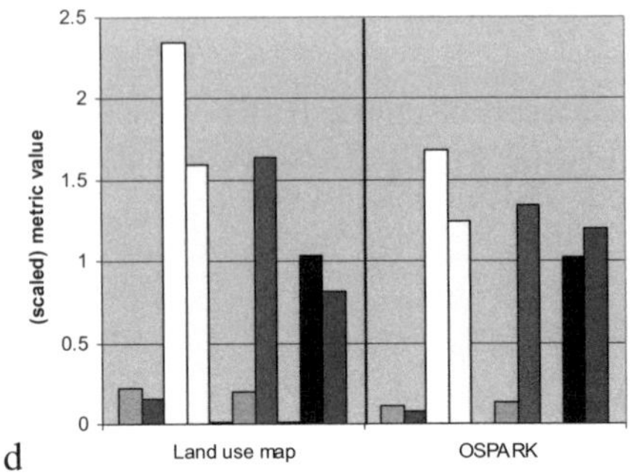


d

Figure 5. A set of class-level metrics calculated from the land use map and the OSPARK classification for: a) Residential areas, b) Commercial areas, c) Industrial areas, and d) Public & private services.

The values of the metrics show large differences between the functional classes (Figure 5). Therefore Herold & Menz (2001) proposed the definition of typical Landscape Metrics Signatures (LMS) to discriminate between classes. Figure 5 confirms the potential of LMS in discriminating functional land use classes based on spatial configuration. In addition, it is proposed here that LMS can be used to compare a land use map and a remote sensing classification at the class-level. Figure 5 shows that for the *residential areas* class the LMS of the land use map and OSPARK are comparable. The other classed show comparable values for some metrics, but differ for other metrics. The challenge for using LMS for an accurate discrimination between classes, is to find the optimal set of spatial metrics.

## 3.3 *Landscape-level*

For the calculation of the non-spatially explicit spatial metrics at the landscape-level, the OSPARK result and the land use map have been reclassified in two classes: urban and non-urban. Figure 6 shows that at the landscape-level the OSPARK classification and the land use map are quite similar in terms of the landscape-level metrics used. Most comparable are the number of patches (NP) and the Area Weighted Mean Patch Fractal Dimension (FRAC_AM).

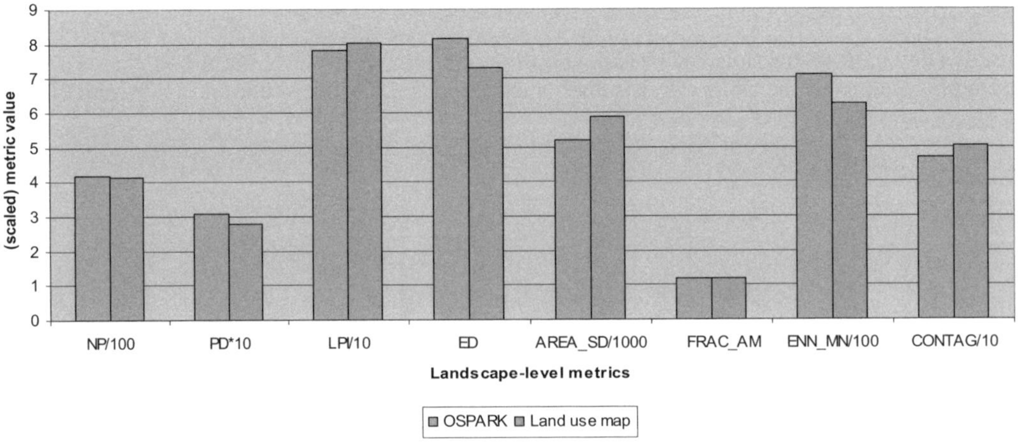


Figure 6. Landscape-level metrics for the land use map and the OSPARK classification.

### 3.4 *Moving window-level*

For the moving window-level a circular moving window with a radius of 1600 m has been used to calculate the metric value of the centre pixel, while moving over the land use map or the OSPARK results, without merging the urban and non-urban classes. The moving window-level gives a spatially explicit result for the metrics that are calculated within the window. The size of this window is chosen similar to the definition of the distance decay functions of the cellular automata model in the MOLAND model (Engelen *et al.* 2002). It is therefore assumed that larger distances have no significant influence on the function of a model gridcell.

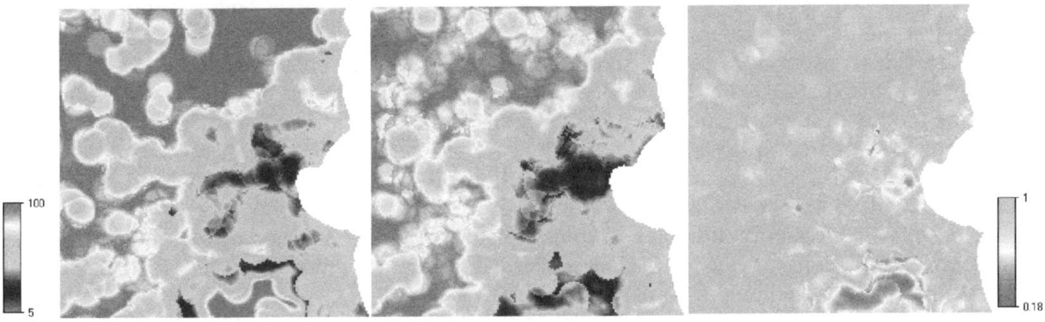


Figure 7. Comparison between moving window-level metrics. Left: CONTAG calculated for the land use map. Middle: CONTAG calculated for OSPARK. Right: Fuzzy Kappa map of the difference between the two CONTAG maps. Average Fuzzy Kappa: 0.875. Note the misclassification in the south (see figure 4).

Figure 7 shows the result for the CONTAG metric, a measure for the interspersion and dispersion of patches (McGarigal *et al.* 2002). The low values for the urban area means that many patch types occur within the window, which are small and dispersed patches. The non-urban area shows the opposite. The maps have been compared by calculating the average Fuzzy Kappa statistic (Hagen-Zanker *et al.* 2005) for the area under consideration. Results are reported in table 2. Also here, some metrics give better results for the comparison than others. FRAC_AM often gives a high Fuzzy Kappa and is therefore less useful in the comparison of maps.

Table 2. The difference between Moving Window-level metrics calculated for the land use map and the OSPARK classification, expressed as the average Fuzzy Kappa (Hagen-Zanker *et al.* 2005) of the area under consideration. For metric names see table 1.

| Moving Window-level metric | Average Fuzzy Kappa |
|---|---|
| AREA_SD | 0.714 |
| CONTAG | 0.875 |
| ED | 0.620 |
| FRAC_AM | 0.982 |
| PD | 0.618 |

## 4 DISCUSSION

There are several possibilities for using spatial metrics in the calibration of land use change models. First, the OSPARK algorithm is an interesting method to create functional land use maps that can be compared with model outcomes in the conventional way of calibrating a model. In fact, the adjacency event matrix used in the OSPARK algorithm is a moving window-level spatial metric. Further research should determine if the replacement of the adjacency event matrix by another spatial metric, or set of metrics (LMS), can improve the OSPARK classification results. Additional modifications to the algorithm that should be evaluated are: the possibility to use circular kernels, the use of different similarity measures than $\Delta k$ and the automatic optimization of the selection of template kernels. This can be accomplished by minimizing the $\Delta k$ value between templates of different classes in order to achieve the highest separability and minimizing the $\Delta k$ value between templates of the same class to prevent data redundancy and reduce calculation times.

Some authors (Barnsley & Barr 2000, Herold *et al.* 2003), however, argue that both pixel and kernel classifiers should be avoided, because they do not consider real objects, but artificial spatial entities. To their opinion it would be a better approach to focus on the analysis of thematically defined, irregularly shaped objects and regions. Current automatic segmentation algorithms, however, have difficulties in delineating real world objects, while traditional visual interpretation approaches are elaborate, time consuming, and they require much expert knowledge.

Furthermore, the spatially-explicit results of the metrics calculated at the moving window-level can be compared with metrics calculated from land use change model outputs. The average Fuzzy Kappa can be used as an objective function for the calibration that should be maximized.

Another option for using metrics in the calibration of land use change models is to compare model results with metrics calculated at a higher level of abstraction, which are not spatially explicit. The goal of the calibration is then to minimize the difference in one or more landscape-level metrics between the remote sensing based classification and the model output. In this case the overall pattern of the model output map is optimized.

When class-level metrics are used, the LMS of the different classes should have a high similarity between the model output and the remote sensing based classification. In this case the model can be optimized for the accurate prediction of patterns of specific classes.

Many spatial metrics have been developed for the analysis of landscape configuration. The choice of which metric is most appropriate for the characterization of a land use class is a point of scientific discussion, since different researchers use different metrics and new metrics are still being developed (Parker *et al.* 2001). Parker et al. (2001) hypothesize "that different metrics are appropriate for different landscapes and that no single set of metrics is appropriate for the characterization of landscape functions in all environments". If that hypothesis is true, the LMS should be trained to the landscape under consideration, like spectral signatures instead of looking for generally applicable landscape metrics signatures. Further research is needed to test this hypothesis.

## 5 CONCLUSIONS

Spatial metrics provide interesting opportunities for the mapping of urban land use and the calibration of urban land use change models. It has been shown that spatial metrics can be spatially explicit or not spatially explicit and can be applied at different levels of abstraction, which makes it useful in many applications.

The OSPARK algorithm provides an interesting framework for the application of moving window-level metrics in the functional classification of remote sensing images. Some modifications have been suggested and need further research.

A framework for the calibration of land use change models using spatial metrics has been suggested. Land use change models can be optimized by comparing spatial metrics calculated from remote sensing based classifications with metrics calculated from maps predicted by the land use change model at different levels of spatial abstraction by using either moving window-level, class-level or landscape-level metrics. This means that, depending on the goal of the model, or the stage in its development, calibration or application, a modeller can choose to optimize the overall pattern in the predicted map, the pattern within the classes or at the moving window-level.

## ACKNOWLEDGEMENTS

We would like to express our acknowledgements to BELSPO (Belgian Science Policy) for funding the MAMUD research project (SR/00/105). The land use data used for Dublin are made available by the MOLAND project of the EU – Joint Research Centre in Ispra, Italy.

## REFERENCES

Alimohammadi, A. & Shirkavand, M. 2004. *Optimization of the SPARK algorithm for land use mapping from remotely sensed data.* Geomatics Conference 83, National Cartographic Center, Tehran, Iran,

Baker, W.L. & Cai, Y. 1992. The r.le programs for multiscale analysis of landscape structure using the GRASS geographical information system. *Landscape Ecology* 7 (4): 291-302.

Barnsley, M.J. & Barr, S.L. 1996. Inferring urban land use from satellite sensor images using kernel-based spatial reclassification. *Photogrammetric Engineering & Remote Sensing* 62 (8): 949-958.

Barnsley, M.J. & Barr, S.L. 2000. Monitoring urban land use by earth observation. *Surveys in Geophysics* 21 269-289.

Barredo, J.I. & Demicheli, L. 2003. Urban sustainability in developing countries' megacities: modelling and predicting future urban growth in Lagos. *Cities* 20 (5): 297-310.

Bossard, M., Feranec, J. & Otahel, J. 2000. *CORINE Land Cover Technical Guide - Addendum 2000. Technical report 40.* EEA.

Elkie, P.C., Rempel, R.S. & Carr, A.P. 1999. *Patch Analyst user's manual: A tool for quantifying landscape structure.* Ontario: Ontario Ministry of Natural Resources.

Engelen, G., White, R. & Uljee, I. 2002. *The MURBANDY and MOLAND models for Dublin. Final report.* Maastricht: RIKS.

Gustafson, E.J. 1998. Quantifying landscape spatial pattern: What is the state of the art? *Ecosystems* 1 143-156.

Hagen-Zanker, A., Straatman, B. & Uljee, I. 2005. Further developments of a fuzzy set map comparison approach. *International Journal of Geographical Information Science* 19 (7): 769 - 785.

Herold, M. & Menz, G. 2001. Landscape Metric Signatures (LMS) to improve urban land use information derived from remotely sensed data. *A decade of trans-european remote sensing cooperation. Proceedings of the 20th EARSeL Symposium. 2001:* 251-256.

Herold, M., Scepan, J. & Clarke, K.C. 2002. The use of remote sensing and landscape metrics to describe structures and changes in urban land uses. *Environment and planning* 34 1443-1458.

Herold, M., Liu, X. & Clarke, K.C. 2003. Spatial metrics and image texture for mapping urban land use. *Photogrammetric Engineering & Remote Sensing* 69 (9): 991-1001.

Herold, M., Couclelis, H. & Clarke, K.C. 2005. The role of spatial metrics in the analysis and modeling of urban land use change. *Comput., Environ. and Urban Systems* 29 369-399.

McGarigal, K., Cushman, S.A., Neel, M.C. & Ene, E. 2002. *FRAGSTATS: Spatial pattern analysis program for categorical maps.* http://www.umass.edu/landeco/research/fragstats/fragstats.html.

Parker, D.C., Evans, T.P. & Meretsky, V. 2001. Measuring emergent properties of agent-based land-cover/land-use models using spatial metrics. *Seventh annual conference of the international society for computational economics. Yale University, 2001:* 24.

Rainis, R. 2003. *Application of GIS and landscape metrics in monitoring urban land use change.* In: Hasmin, N. & Rainis, R. (Eds.), Urban ecosystem studies in Malaysia - A study of change. Singapore: Universal Publishers.

Sluiter, R., De Jong, S.M., Van der Kwast, H. & Walstra, J. 2004. *A contextual approach to classify mediterranean heterogeneous vegetation using the spatial reclassification kernel (SPARK) and DAIS7915 imagery.* In: De Jong, S.M. & Van der Meer, F.D. (Eds.), Remote Sensing image analysis: including the spatial domain. Remote Sensing and Digital Image Processing. Dordrecht: Kluwer academic publishers.

*Remote Sensing for a Changing Europe*
*D. Maktav (Ed.)*
*IOS Press, 2009*

*doi:10.3233/978-1-58603-986-8-412*

# Fire danger assessment using remote sensing and geographic information system Technologies

Emilio Chuvieco
*Department of Geography, University of Alcalá (Spain)*

Juan de la Riva
*Department of Geography and Land Planning, University of Zaragoza (Spain)*

Keywords: Fire risk, Fire Danger, Geographic Information Systems, Remote Sensing, Logistic Regression, Human factors, Fuel moisture content, Lightning

ABSTRACT: Forest fires play a critical role in landscape transformation, vegetation succession, soil degradation and air quality. Improvements in fire risk estimation are vital to reduce the negative impacts of fire. This paper defines a conceptual scheme for fire risk assessment, as well as methods for data generation based on remotely sensed and geographic information system techniques. It presents proposals for their integration into synthetic risk indices, and provides a first validation exercise on several study regions in Spain. All variables were mapped at 1 sq km spatial resolution, and were integrated into a web-mapping service system. This service was active in the summer of 2007 for semi-operational testing of end-users.

## 1 INTRODUCTION

Forest fires are a major factor of environmental transformation in a wide variety of ecosystems (FAO, 2007). Fires have global impacts (Chuvieco, 2008), affecting forested areas and having an important share in greenhouse gas emissions. Additionally, fires have local impacts, associated to soil degradation, soil erosion, lost of lives, biodiversity, and infrastructures (Omi, 2005).

Although fire has been historically used as a tool for land use management and many ecosystems are well adapted to fire cycles, recent changes in both climate and societal factors related to fire can transform traditional fire regimes, increasing the negative effects of fire upon vegetation, soils and human values. Recent changes in land use management in developed countries, with an increasing abandonment of traditional rural uses have implied a remarkable increase of fuel accumulation, which lead to more severe and intense fires, and consequently to higher negative impacts on soils and vegetation resilience. Fire risk evaluation may greatly help to reduce the negative impacts of fire, since pre-fire planning resources require objective tools to monitor when and where a fire is more prone to occur. Traditional fire danger systems rely on meteorological indices, based on variables that are routinely measured by weather stations. However, atmospheric conditions are only one of the components of fire risk, which should also consider human aspects, fuel amount and moisture status, as well as values at stake. This integrated approach was the basis to develop a national project in Spain, being the main objective to develop methods for data generation and integration into synthetic fire danger indices.

# 2 METHODS

## 2.1 *Fire risk scheme*

Although fire risk should be considered a combination of fire occurrence probability and potential damages, only the former was considered for this paper to fit the scope within the paper extension. Fire danger considers the potential that a fire ignites or propagates (figure 1). To obtain an operational assessment of fire risk conditions following the proposed scheme, the following steps were required:

- Generation of risk factors, using a common geographical unit. This phase requires the definition of a target scale and resolution, which should relate to sources of data available.
- Conversion of risk factors to a common risk scale.
- Development of criteria to integrate risk factors.

Since fire danger is a spatial and temporal process, it should be addressed both spatially and temporally. The use of Geographic Information System (GIS) is quite obvious in this regard, since these tools are ideal to manage spatial information, provide adequate spatial processing and visualization of results. For this reason, several previous studies on fire risk estimation have been based on GIS (Chou, 1992; Chuvieco & Congalton, 1989; Nourbakhsh et al., 2006; Yool et al., 1985).

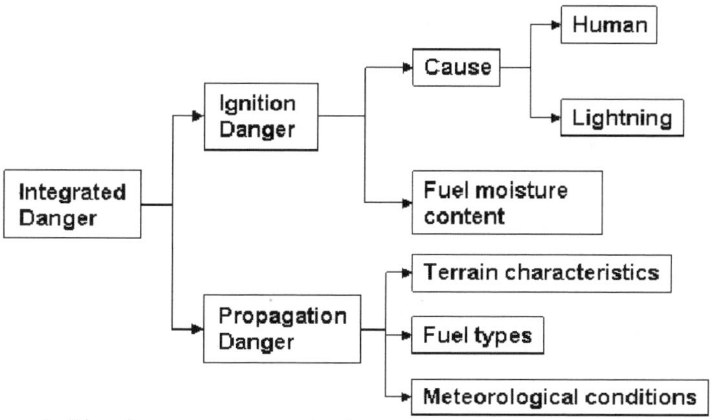


Figure 1. Fire danger assessment scheme

## 2.2 *Study regions*

Several research groups working on Mediterranean conditions participated in developing the fire risk project presented in this paper. Four study areas were selected to develop the methods of data generation and integration. Three of them are autonomous regions of Spain: Aragon with 47,719 $Km^2$; Madrid, with 8,028 $Km^2$, and Valencia with 23,255 $Km^2$, while the fourth is a province of Andalusia (Huelva, 10,148 $Km^2$). Total area covered for the four regions is 89,131 $Km^2$, which accounts for 18% of the total area of Spain. Following end-user recommendations, the minimum mapping unit was fixed at 1 $Km^2$, using as a reference the standard UTM grid. The regions were selected to provide a good assortment of Spanish various fire conditions.

## 2.3 *Generation of risk variables*

The generation of the input variables was based on extensive use of remote sensing and GIS methods. A brief summary of the most critical factors follows, but table 1 includes references for further details (Table 1).

### 2.3.1 *Modeling the Human factors of fire ignition*

In most countries human activities are in one way or the other, the main responsible for fire ignition. Humans have used fire historically for different purposes: light, heat, cooking, land clearing, etc., and still have a critical impact on fire regimes and vegetation distribution. In Mediterranean areas, human factors cause more than 90% of fires (Leone et al., 2003). In Spain, 96,1% of all fires are human-caused (Dirección General de Biodiversidad, 2006).

Table 1: Input factors for the fire danger assessment system

| Factor | Input data | Method |
|---|---|---|
| Human (Vilar, 2006) | Historical occurrence<br>Demographic data<br>Vegetation - DTM | Logistic Regression |
| Lightning (Nieto et al., 2008) | | |
| Dead Fuels Moisture Content (Aguado et al., 2007) | Meteorological Data | Linear Regression Analysis |
| Live Fuels Moisture Content (Chuvieco et al., 2004b; Garcia et al., 2008) | Satellite images | Statistical fitting<br>Inversion of RTM |
| Propagation Danger | Fuel type maps<br>Meteorological Data | Behave Simulation |

In spite of the importance of these human aspects, little work has been devoted to this issue. Most frequently, the studies have focused on variables related to land use or land use-change (rural abandonment, agricultural-forest interface or urban-forest interface), population trends, rural activities, potential conflicts that may lead to vengeances or arson (unemployment, enforcement of conservation areas, reforestation in traditional pastured areas, etc.) (Leone et al., 2003). The approach to consider human factors in fire risk assessment has been commonly based on statistical models, which have tried to explain historical human-caused fire occurrence from a set of independent variables (Martell et al., 1989; Martínez et al., 2004).

For this project, the analysis of human risk conditions were firstly based on selecting the critical variables associated to human-cause fires in Spain, following a detailed reviewed of specialized literature. General factors commonly identified by previous studies needed to be approached using single variables, which should be available for all study sites. In a second phase, variables expressing each factor were mapped at the target spatial resolution of the fire risk assessment system (1 $km^2$), using a wide variety of GIS analysis. Only structural factors were considered at this stage, since the human component was intended to be stable for the whole fire season.

Logistic Regression techniques were used to estimate the probability of occurrence from socio-economic explanatory variables. The dependent variable was the number of fires caused by human activities in the period 1990-2004 derived from official fire statistics, which reference fire records to a UTM 10x10 Km UTM grid and to the municipality where the fire started. Previous to building the statistical model, correlations between independent variables were tested to avoid multicolinearity problems. The model was based on a forward stepwise logistic regression analysis. In each study area, 60% of the input cells were used for model calibration and the remaining 40% for validation.

Table 2 shows the results of the different models for each study region. A 0.5 threshold value was used for classification of the input cases. The number of correctly assigned cells varies from 68.4% (Valencia) to 86.8% (Aragon). The variables included in each model were in agreement with the experience of the forest managers who participated in the project. In the most urbanized regions the urban-wildland interface was the most prominent variable in the model, followed by unemployment rate. Variations in population and agricultural-forest interface were more important in the more rural areas.

### 2.3.2 *Ignition potential from lightning*

In spite of the lower importance of lightning over human factors for fire ignition, lightning strikes are also an important factor to consider in fire danger estimation. They tend to burn larger areas than human-caused fires, because they occur in more isolated and steeper areas and frequently have various simultaneous ignited spots, and therefore are more difficult to control (Wotton & Martell, 2005). Several previous studies have focused on analyzing the geographical variables that are more prone to lightning caused fires, such as the topography, strike polarity and fuel moisture content.

For this project, the structural factors associated to historical lightning-caused fires were analyzed, by comparing spatial patterns of affected and non-affected areas. The dependent variable in this case was the number of lightning-caused fires during the longest possible period of time (when both lightning sources and fire statistics are available), while the independent variables were the total number of light strikes, vegetation and terrain characteristics, and moisture codes derived from the U.S. National Fire Danger Rating System (Bradshaw et al., 1983), and the Canadian Forest Fire Weather Index (Van Wagner, 1987). The daily meteorological database was only available at 3x3 Km resolution and for the period of 2002-2004, to which the analysis was restricted.

Similarly to the human factors, a logistic regression model to predict and explain historical fire occurrence was derived for the lightning-caused fires. In this case, a pure binary variable was taken into account (fire/not fire), since the total number of fires was much smaller than the human-caused fires. The outputs of the models show good classification results, with 70% of the cells correctly classified. The main explicative variable was the number of dry storms (with less than 2 mm).

### 2.3.3 *Ignition potential associated to fuel moisture content status*

Fuel Moisture Content (FMC) is a critical variable to estimate ignition and propagation danger, since the amount of water in the vegetation is inversely related to ignition potential and rate of spread. Following a common approach in forest fire literature, the estimation of FMC was divided in this project between dead and live components. The former were estimated from meteorological variables and the later from satellite images.

The estimation of dead FMC for this project was performed from an empirical approach, based on field sampling developed between 1998 and 2003 in Central Spain (Aguado et al., 2007). The independent variables in this case were two moisture codes routinely used in fire danger estimation: the Fine Fuel Moisture Code (FFMC) and the 10-hour code, the former being part of the Canadian and US fire danger systems, respectively. Similar results were obtained from the two moisture codes, but finally the 10-h code was selected, since it does not require wind speed as an input and therefore it is easier to compute. Once the empirical relations were established, they were extended to a grid of 1x1 Km resolution, interpolated from the data of the European Centre for Medium Range Weather Forecasting (ECMWF)'s using local algorithms. The interpolation algorithm took into account horizontal distances between the grid point and the surrounding stations (quadratic inverse distance algorithm). The effect of altitude of each grid point over the value of the variable (temperature or humidity) was also considered (Aguado et al., 2007). The estimation of dead FMC was computed everyday, based on 12 (noon) forecasted data from the 8 am prediction.

Regarding the estimation of FMC of live species, satellite remote sensing was used as an input. The use of satellite data in live FMC estimation has been discussed by different authors in the last years. In spite of the difficulty of extracting the influence of water absorption over other factors affecting plant reflectance, several studies have found good relationships, especially in grasslands and some shrub species. For this project, empirical models for NOAA-AVHRR images were used, based on our experience from previous projects (Chuvieco et al., 2004b). Previous models were found inappropriate for very dry years, such as 2005, when high overestimations were found. Therefore, a revision of the empirical method was developed. The new functions took into account the rainfall conditions of the Spring season to choose whether a dry or normal year equation should be applied. The outputs provide a more consistent estimation of FMC for contrasting years than a single model (Garcia et al., 2008).

### 2.3.4 *Propagation potential*

Most fire spread simulation models have been designed for local conditions and for active fires that have occurred or have been simulated to occur. For this project, it was intended to produce an estimation of the average propagation potential of each cell, assuming a fire may occur anytime in any cell of the study areas. Another challenge was that fire propagation values should be calculated for coarse grid cells, since our model was addressed to regional scales, which is uncommon in fire behaviour models.

Within these two limitations, average propagation conditions were simulated using the Behave program (Andrews & Chase, 1990). A total of 5525 simulations were run for the 13 fuel types defined within this program, by modifying systematically the slope gradients, from 0% to 90%, and the wind speeds, from 4 Km/h to 20 Km/h. Standard values of FMC were considered: 5% for 1-h moisture fuels, 10%, for 10-h moisture fuels, 12% for 100-h Moisture fuels and 50% for live fuels (Martín Fernández et al., 2002). Those input conditions were selected by considering the worst-case scenario, that is the fire is potentially propagated along the maximum slope gradient and the wind speed is the average of the maximum speeds for the summer time.

The simulated values of flame length and rate of spread were averaged for each fuel type and slope interval, as to generate a potential propagation map of the study sites. Fuel type models were derived from the forest inventory maps, while slope intervals were computed from the 250 x 250 m digital terrain model of the country.

## 2.4 *Model integration*

Once the input risk variables were generated, two additional tasks were required to obtain an integrated fire danger index. On one hand, the input variables needed to be converted to a common risk scale; on the other, they should be properly weighed, so the importance of the different factors was taken into account.

### 2.4.1 *Creating common danger scales*

Several methods have been proposed to find common scales of fire risk, being variable normalization, qualitative categorization and probabilistic approaches the most common (Chuvieco et al., 2003). For this project, all the variables were converted to a 0-1 scale using probability functions. For the consideration of the causes (human and lightning), the estimation models were based on logistic regression analysis and, therefore, the predictions were already expressed in probabilistic terms. For the fuel moisture content, the conversion of FMC to ignition potential (IP, 0-1 scale) was based on a physical model, using the concept of moisture of extinction (ME: (Simard, 1968). This value expresses the maximum moisture value above which a fire is not sustained, and differs for each fuel type. Strictly speaking, the ignition potential of a fuel when FMC equals to ME should be 0. However, in our project a conservative approach was adopted (ME = 0.2) to avoid eliminating areas with mixed fuels. Therefore, the conversion from FMC to IP was based on a linear relation from the FMC minimum value found in the historical data series (PI = 1) to ME (PI=0.2) (Chuvieco et al., 2004a).

Finally, the conversion of the propagation variables to a propagation potential danger was based on a normalization of the cumulative proportion of both rate of spread and flame length in all grid cells of the study areas. For each cell, the maximum probability value between rate of spread and flame length was selected as representative of the worst case conditions.

### 2.4.2 *Integration of risk indices*

The different input variables have different impacts on fire danger conditions. Identifying which are more relevant and how they should be weighed to generate synthetic indices is a critical phase in danger assessment. For this project, the integration of the causative agents (human and lightning) was based on the Kolmogorov probabilistic rule (Tarantola, 2005):

$$P(A \cup B) = P(A) + P(B) - P(A) \cdot P(B)$$

where P(A U B) is the integrated probability, P(A) is the probability of ignition derived from human variables and P(B) is the probability of ignition derived from lightning. The integration of live and dead FMC was performed by averaging both FMC ignition potential values, weighted by the percentage cover of both dead and live fuels.

For the integration of causative agents and FMC a multicriteria evaluation technique was adopted. It was assumed that high risk probability should be associated to situations when both high probability of having causative agents and FMC ignition potential occur. In the case of the integration between ignition and propagation danger (named integrated danger), a similar approach was adopted, although in this case it was assumed that the worst conditions would occur either when the maximum ignition or propagation danger occur.

### 2.5 *Development of a dedicated web-mapping service*

To facilitate the participation of end-users in the proejct, a dedicated web mapping service was developed. The service was based on public domain software. The server was tested during the fire season of 2007 (June to September), and it was well accepted by the end-users. It included all the input risk variables and integrated indices, plus several vector variables as auxiliary information. Zoom, roam, consults, and download facilities were incorporated to the server (http://www.geogra.uah.es:8080/cartofire/index.php, last accessed on April, 23, 2008).

## 3 RESULTS AND VALIDATION

The assessment of a fire risk index typically requires comparing the predictions of the system with the actual occurrence of fires within a particular area and period. The fire danger system proposed in this paper was tested in the summer of 2007. Two types of validation could be considered in this regard, one affecting the input variables, and the other the integrated indices. The former should be associated to the data generation phase. For instance, the assessment of FMC should be based on field measurements of FMC, not on fire occurrence, since fire may not occur even with very low values of FMC when no causative agents are involved, or alternatively fires may occur even with moderate values of FMC if the agents are very active in starting a fire. The validation of the integrated indices, on the other hand, might be done using fire occurrence, since all the dangerfactors should have been considered. However, in this case, validation data from a long time series are more appropriate than from a single season, because short periods may bias some of the theoretical assumptions that are required to build the model.

For this project, the validation exercise focused on evaluating the existence of significant differences between danger values of cells with fires versus those without fires (for all 1 Km cells and periods). For validation purposes, two indices were considered, the ignition danger which included human and lightning factors, and the integrated danger, which included ignition and propagation danger.

The occurrence database that was available for assessment was derived from the fire reports collected by the regional forest fire services. Total ignition points used for validation were 173 in Madrid, 111 in Huelva, 188 in Aragon, and 158 in Valencia. Several statistics to estimate significance of differences between fire and non-fire cells were computed: 1) the distances of Mahalanobis; 2) the Mann-Whitney U test (Mann and Whitney, 1947); and 3) the Nagelkerke R2 coefficient from logistic regression fittings for each integrated index. Processing was done using the R statistical software.

Ignition danger generally showed higher Mahalanobis distances than integrated Danger (table 2). However, the results were very close between the two indices in the Huelva region (with the highest values among the different regions), and Valencia (the worst). The U values confirm those results, since both Ignition danger and Integrated danger provided significant differences in all study regions. The results were poorer for Valencia than for other regions. The two risk components showed higher values for fire cells than for non-fires, showing the potential of these indices to predict fire occurrence in very different regions, although a wide diversity of risk values within non-fire areas was observed (fig. 2).

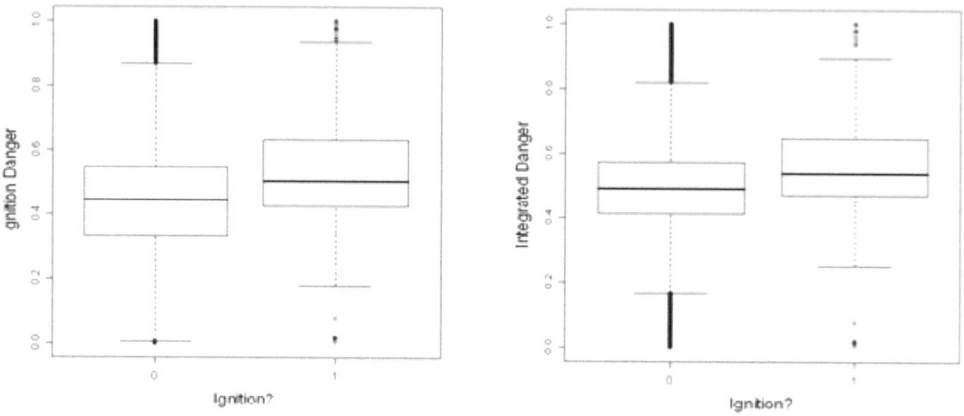


Figure 2. Box graphs showing differences in the integrated danger for cells with and without fires during the summer of 2007

Table 2. Results of the U Mann-Whitney tests. Significance of differences between fire and non-fire cells

| Index | Statistics | Madrid | Huelva | Aragón | Valencia |
|---|---|---|---|---|---|
| Ignition danger | Mann-Whitney U | 73428174 | 56640631 | 504403051 | 223141072 |
| | Z | -6.424 | -6.358 | -6.155 | -4.455 |
| | Significance (2-tailed) | .000 | .000 | .000 | .000 |
| Integrated danger | Mann-Whitney U | 76460728 | 55444348 | 523491786 | 231831563 |
| | Z | -5.967 | -6.609 | -5.489 | -3.781 |
| | Significance (2-tailed) | .000 | .000 | .000 | .000 |

The paper has reviewed the methods to generate critical risk variables and integrate them in a coherent structure for fire risk assessment. Although further work is required to improve procedures of data integration, and extent length of the validation period, the proposed scheme should provide a sound procedure to obtain synthetic and spatially-explicit assessment of fire risk conditions.

## ACKNOWLEDGEMENTS

This paper is based on the Firemap project, which was funded by the Spanish Ministry of Science and Education (CGL2004-060490C04-01/CLI) through the Environment and Climate program. Very useful comments were received from end-users of the project: Civil Protection, Forest services in Madrid, Aragon, Valencia and Andalusia regions.

## REFERENCES

Aguado, I., Chuvieco, E., Boren, R., & Nieto, H. 2007. Estimation of dead fuel moisture content from meteorological data in Mediterranean areas. Applications in fire danger assessment. *International Journal of Wildland Fire* 16: 390-397.

Andrews, P.L., & Chase, C.H. 1990. The BEHAVE Fire Behavior Prediction System. *The Compiler* 8(4): 4-9.

Bradshaw, L., Deeming, J., Burgan, R.E., & Cohen, J. 1983. The 1978 National Fire-Danger Rating System: Technical Documentation. *In* Ogden, Utah: USDA, Forest Service, 44.

Chou, Y.H. 1992. Management of Wildfires with a Geographical Information System. *International Journal of Geographical Information Systems* 6(2): 123-140.

Chuvieco, E. 2008. Satellite observation of biomass burning: implications in global change research. *In* Chuvieco, E. *ed. Earth Observation and Global Change.* New York: Springer: 109-142.
Chuvieco, E., Aguado, I., & Dimitrakopoulos, A. 2004a. Conversion of fuel moisture content values to ignition potential for integrated fire danger assessment. *Canadian Journal of Forest Research-Revue Canadienne De Recherche Forestiere* 34 (11): 2284-2293.
Chuvieco, E., Allgöwer, B., & Salas, F.J. 2003. Integration of physical and human factors in fire danger assessment. *In* Chuvieco, E. *ed. Wildland Fire Danger Estimation and Mapping. The Role of Remote Sensing Data.* Singapore: World Scientific Publishing: 197-218.
Chuvieco, E., Cocero, D., Riaño, D., Martín, M.P., Martínez-Vega, J., de la Riva, J., & Pérez, F. 2004b. Combining NDVI and Surface Temperature for the estimation of live fuel moisture content in forest fire danger rating. *Remote Sensing of Environment* 92: 322–331.
Chuvieco, E., & Congalton, R.G. 1989. Application of remote sensing and Geographic Information Systems to Forest fire hazard mapping. *Remote Sensing of Environment* 29: 147-159.
Dirección General de Biodiversidad. 2006. *Los incendios forestales en España.* Madrid: Ministerio de Medio Ambiente
FAO. 2007. *Fire management - global assessment 2006. A thematic study prepared in the framework of the Global Forest Resources Assessment 2005.* Rome: FAO.
Garcia, M., Aguado, I., & Chuvieco, E. 2008. Combining AVHRR and meteorological data for estimating live fuel moisture content in forest fire danger rating. *Remote Sensing of Environment* en prensa.
Leone, V., Koutsias, N., Martínez, J., Vega-García, C., Allgöwer, B., & Lovreglio, R. 2003. The human factor in fire danger assessment. *In* Chuvieco, E. *ed. Wildland Fire Danger Estimation and Mapping. The Role of Remote Sensing Data.* Singapore: World Scientific Publishing: 143-196.
Martell, D.L., Bevilacqua, E., & Stocks, B.J. 1989. Modelling seasonal variation in daily people-caused forest fire occurrence. *Canadian Journal of Forest Research* 19(12): 1555-1563.
Martín Fernández, S., Martínez Falero, E., & Pérez, J.M. 2002. Optimization of resources management in wildfire combat. *Environmental Management* 30(3): 352-336.
Martínez, J., Chuvieco, E., & Martin, M.P. 2004. Estimating human risk factors in wildland fires in Spain using logistic regression. *II International Symposium on Fire Economics, Planning and Policy: A Global Vision.* Córdoba: Univ. of Cordoba, CD-Rom: 15 pp.
Nieto, H., Aguado, I., & Chuvieco, E. 2008. Lightning-caused fires in Central Spain: development of short-term and long-term probability models of occurrence in two regions of Spain. *International Journal of Wildland Fire* in preparation.
Nourbakhsh, I., Sargent, R., Wright, A., Cramer, K., McClendon, B., & Jones, M. 2006. Mapping disaster zones. *Nature* 439(7078): 787-788.
Omi, P.N. 2005. *Forest fires : a reference handbook.* Santa Barbara, Calif.: ABC-CLIO.
Simard, A.J. 1968. The moisture content of forest fuels - A review of the basic concepts. *In* Ottawa, Ontario: Forest Fire Research Institute, 47.
Tarantola, A. 2005. *Inverse Problem Theory and Methods for Model Parameter Estimation.* Philadelphia: Society for Industrial and Applied Mathematics.
Van Wagner, C.E. 1987. Development and structure of the Canadian Forest Fire Weather Index System. *In* Otawa: Canadian Forest Service, 48.
Vilar, L. 2006. Empleo de regresión logística para la obtención de modelos de riesgo humano de incendios forestales. *XII Congreso Nacional de Tecnologías de la Información Geográfica.* Granada: CdRom
Wotton, B.M., & Martell, D.L. 2005. A lightning fire occurrence model for Ontario. *Canadian Journal of Forest Research* 35: 1389-1401.
Yool, S.R., Eckhardt, D.W., Estes, J.E., & Cosentino, M.J. 1985. Describing the brushfire hazard in southern California. *Annals of the Association of American Geographers* 75(3): 417-430.

*Remote Sensing for a Changing Europe*
*D. Maktav (Ed.)*
*IOS Press, 2009*

*doi:10.3233/978-1-58603-986-8-420*

# Detecting changes in mountainous terrain from time series of remote sensing data

Markéta Potůčková, Eva Štefanová, Jan Kolář, Vít Vilímek
*Charles University in Prague, Faculty of Science, Czech Republic*

Keywords: archive images, image orientation, QuickBird, natural hazards, Huascaran

ABSTRACT: Processing of time series of imagery for the purpose of monitoring changes of terrain in mountainous area over a long period of time from archive aerial images as well as from very high resolution satellite images is discussed. The information eventually obtained can be beneficial to studies of some geomorphological phenomenas like land slides, avalanches or rock slumping. Although the remote sensing technology provides efficient tool for the purpose, the quality of data available is often far from the optimum forcing to look for methods modifying the standard procedure accordingly. In the presented paper an area nearby Huascaran Mt. in Peru is taken as an example. The data available for the processing include aerial images from the years 1967 and 1970 and a QuickBird satellite image from 2003.

## 1 INTRODUCTION

Archive aerial and satellite images are an important source of information for various natural hazards and environmental studies. They are also very useful in quantification of changes in landscape evolution after significant natural disasters - like earthquakes, large slope movements or volcanic activities. To verify the methodology an example from Cordillera Blanca (Peru) was chosen. On $31^{st}$ May 1970 towns Yungay and Ranrahirca nearby Huascaran Mt. (6 655 m a.s.l.), the highest mountain in Peru, were wiped out by earthquake-triggered avalanche. Several authors described the rock and ice avalanche just after the event (e.g. Lomnitz 1971, Plafker et al. 1971) or later with respect to the prevailing geomorphological processes (Vilímek 1995). Recent investigation is focused on the risk assessment for the population at the foothill of Huascaran Mt. (e.g. Vilímek et al. 2000). There was created big pressure on the local authorities to settle again the free area of the accumulation from 1970 disaster.

For better understanding of landscape evolution in this area an investigation of prehistoric avalanches were initiated. In order to have a complete study of the area field survey as well as remote sensing data are needed - thus time series of aerial and satellite images were collected.

The purpose of this paper is to show a potential but also limitations of usage of archive image data for a documentation of an extent of events such as land slides or rock slumping that have happened several years ago as well as development of the area after a disaster. The presented results document the first part of this research which main goal was to evaluate whether any reasonable outputs could be obtained from the available data set and to estimate its quality.

## 2 DATA DESCRIPTION

The data set available contains imagery from three time periods, namely 1967, 1970, and 2003. The images from 1970 were taken very soon after the avalanche event. The selected time series therefore represents a very valuable dataset enabling monitoring the extent of the disaster and the development of the area thirty years afterwards. Unfortunately, there was not enough overlap between all images available. Thus, one stereopair from 1967 images and one from 1970 images was only chosen. An overlap between both stereopairs is about 53% (27 $km^2$) and covers the town of Yungay and its surroundings.

There were no topographic maps that could be used for data processing. The maps containing only contour lines were not of a big use because they missed any additional topographic content that could help with an orientation in the maps. A coordinate grid was missing as well.

### 2.1 *Aerial Images from 1967 and 1970*

Aerial images were taken with Wild RC8 camera using wide-angle objective lens (the principal distance close to 153 mm) at the approximate scales of 1:33 000 and 1:42 000 for 1967 and 1970 images, respectively. Quality of aerial images appeared to be one of the main problems of the data set. Original photographs were not available. Obtained digital copies were produced on an office scanner with a resolution of 400 dpi (64 μm) which corresponds to the ground sampled distance of 2.1 m in 1967 images and 2.7 m in 1970 images. Moreover, the size of the copies was limited by scanner's A4 format and did not cover the photographs' standard 23 cm x 23 cm. At least three fiducial marks were therefore missing on each scan. This fact made processing of the images quite difficult and caused some problems later. The calibration report of the camera was not available as well. Only the camera constant was readable on some of the images and the type of the camera was judged from a typical pattern of fiducial marks. It helped to search a camera report of the same camera type. Being aware of incorrectness and with regard to further processing described later, this camera calibration report was used as the best approximation at the moment.

### 2.2 *QuickBird Image from 2003*

A 2003 image was acquired from the QuickBird satellite. Both panchromatic and multispectral images with the resolution of 0.6 m and 2.4 m, respectively, were delivered as a part of the data set. The level of processing corresponds to "standard imagery", which according to the DigitalGlobe specifications means absolute positional accuracy (RMSE) of 14 m (Digital Globe 2008). There was no correction to the terrain except of shift to the plane of an average height of the area covered by the image (3 738 m a.s.l.). Only the panchromatic image was used for further processing.

### 2.3 *Elevation data*

The maps containing only contour lines with a basic interval of 50 m were available. But as it was difficult to find corresponding points in these maps and in the images and it was not possible to read coordinates from the maps, they were not further used in processing. On the other hand, height information was needed for orientation of aerial images. Thus SRTM height model with the 3 arc second resolution was downloaded. According to USGS ( 2006), its horizontal accuracy is 20 m (circular error at 90% confidence) and its vertical accuracy is 16 m (linear error at 90% confidence, which corresponds to the standard deviation $\sigma_{SRTM}$ = 10 m). The data for SRTM model were collected in February 2000.

## 3 DATA PROCESSING

The first goal of the investigation was to determine the extent of area covered by the avalanche in Yungay. For this purpose, images from all time periods were georeferenced into the same coordinate system (UTM/WGS84) and orthorectified. The orthoimages were also used for evaluation of the development of the landscape after disaster. The second goal was to determine the height of the accumulated material but it was not possible fully accomplished due to low data quality.

This project suffers from a complete lack of ground control points (GCPs). Therefore it was necessary to choose a reference frame. After placing all date into this frame, the relative relations can be kept although the absolute position will be inaccurate. If the relative solution is correct then as soon as GCPs are available, the whole data set can be transformed into its proper absolute position. All image processing was carried out in PCI Geomatica. The QuickBird image was chosen as a reference frame. Its georeferencing was taken from the metadata file obtained together with the imagery. The scale of the satellite image was considered to be correct and the absolute shift in the position of the image in the reference coordinate system is not of high importance at this stage of the research. The image was orthorectified using SRTM elevation model. Assuming positional correspondence between the QuickBird image and the SRTM elevation model, a positional shift in the satellite orthoimage can be estimated from SRTM height accuracy using the formula $dR = dh \tan(\alpha/2)$, where dR is the estimated height error at the edge of the swath, dh is the height error and $\alpha$ is the field of view. In case of QuickBird image with viewing angle of 0 degrees (nadir look) and $dh = 3\sigma_{SRTM} = 30$ m, $dR = 0.6$ m which corresponds to the size of a pixel and is therefore within a generally accepted estimation of orthoimage accuracy of 2 pixels.

Processing of aerial images consisted of three steps – restoration of parameters of inner orientation, determination of parameters of exterior, and orthorectification.

### 3.1 *Inner orientation*

The camera calibration report was not accessible for the processed images. The following approximations were therefore necessary:

- the principal distance was set equal to the value printed in the photograph
- the principal point identified with the fiducial centre
- the coordinates of the fiducial marks were taken from the calibration report of the same type of the camera
- the radial lens distortion was neglected

At least two fiducial marks had to be graphically reconstructed because of incompleteness of the images. The reconstruction was done in a CAD software package (MicroStation) assuming symmetry of the image. Measurement of the fiducial marks was done in the OrthoEngine modul of PCI Geomatica. Residuals after affine transformation from the pixel to the image coordinate system was about 2.5 pixels for 1970 images and 1.5 pixel for 1967 images.

### 3.2 *Exterior orientation*

Exterior orientation was calculated by means of bundle adjustment in the modul OrthoEngine. Ten GCPs, ten check points (CPs) and six tie points were measured in each stereopair. Both GCPs and CPs were selected from the QuickBird orthoimage and their heights were interpolated from SRTM DEM. Due to changes in the landscape, the control points had to be chosen in spots that have not been affected by the avalanche. The CPs check the model orientation only by relative means because they were not measured independently. The measured points were not signalized. Points such as road or path crossings, edges of fields or crossings of line vegetation were chosen. Accuracy of definition of such points can be assumed about 1 to 2 pixels (up to 4 m and 5 m in the 1967 and 1970 stereopairs). Adding an influence of an expected satellite orthoimage positional accuracy of 2 pixels (1.2 m), it leads to the rough estimation of horizontal accuracy of orientation of 5 m. The accuracy of the SRTM elevation model of 10 m had the highest influence of the vertical accuracy of the orientation of the stereoscopic models that could not therefore be much better than 10 m. Table 1 summarizing root mean square errors on GCPs and CPs after bundle adjustment shows that the orientation of the stereomodel from 1970 fulfilled expectations while the residuals in the 1967 model exceeded the expected values twice. Because the GCPs and CPs in both stereomodels were identical or close to each other and from the same source, the observed differences in the RMSE values indicate problems with the inner orientation of the 1967 images, especially the camera constant.

Table 1. Residuals on ground control points and check points after bundle adjustment.

| | Ground Control Points | | | Check Points | | |
|---|---|---|---|---|---|---|
| | RMSE_X [m] | RMSE_Y [m] | RMSE_Z [m] | RMSE_X [m] | RMSE_Y [m] | RMSE_Z [m] |
| 1967 | 7 | 5 | 18 | 7 | 5 | 24 |
| 1970 | 3 | 2 | 8 | 6 | 8 | 13 |

## 3.3 *Orthorectification*

Images from 1970 were orthorectified using the STRM elevation model. A DEM had to be derived from the stereopair from 1967 as the SRTM was not possible to use due to distinct landscape changes caused by the avalanche. A correspondence between the satellite orthoimage and aerial orthoimages was evaluated by means of 20 check points depicted in Figure 1. The mean, RMSE and standard deviation of differences between aerial orthoimages and the satellite orthoimage (a reference frame) are summarized in Table 2.

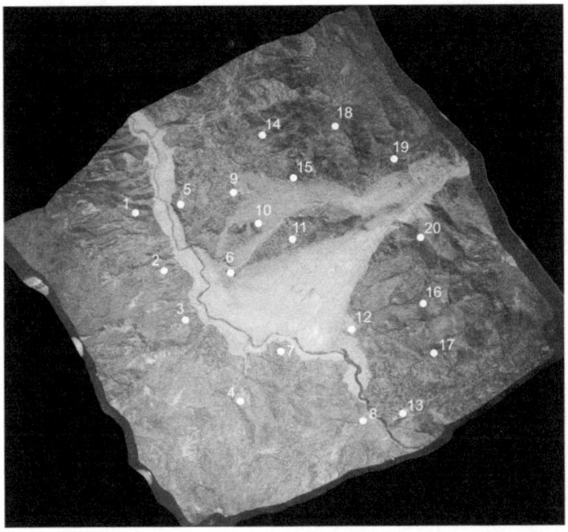


Figure 1. Overview on check points used for evaluation of the positional correspondence between derived aerial orthoimages and the satellite orthoimage.

Table 2. Statistical measures based on horizontal differences between derived aerial orthoimages and the satellite orthoimage.

| | 2003-1967 | 2003-1970 |
|---|---|---|
| RMSE_E [m] | 10 | 3 |
| RMSE_N [m] | 7 | 4 |
| Mean_E [m] | 3 | 0 |
| Mean_N [m] | -3 | -2 |

Similarly to the exterior orientation, the correspondence between the orthoimages 1967 and 2003 is about twice worse than the correspondence between orthoimages 1970 and 2003. The explanation is following. The orthoimage from 1970 was derived using the STRM elevation model as in the case of the satellite orthoimage. In case of 1967 imagery, a DEM from the stereopair had to be derived for the calculation of the orthoimages of that year and this DEM was influenced by inaccurate orientation as that was described above.

# 4 RESULTS

Georeferencing of all imagery into the same reference frame (satellite orthoimage, UTM/WGS84) made mapping of the extent of the accumulation and mapping of land use/cover of the area of interest possible. As this paper is aimed at geometrical part of the image processing, possible outputs from mapping are demonstrated in Figure 2. The extent of the accumulation after the avalanche in 1970 is delineated. Most of the town of Yungay is covered by mad and ice. The change of land cover and land use during after more than 30 years is visible in the 2003 orthoimage.

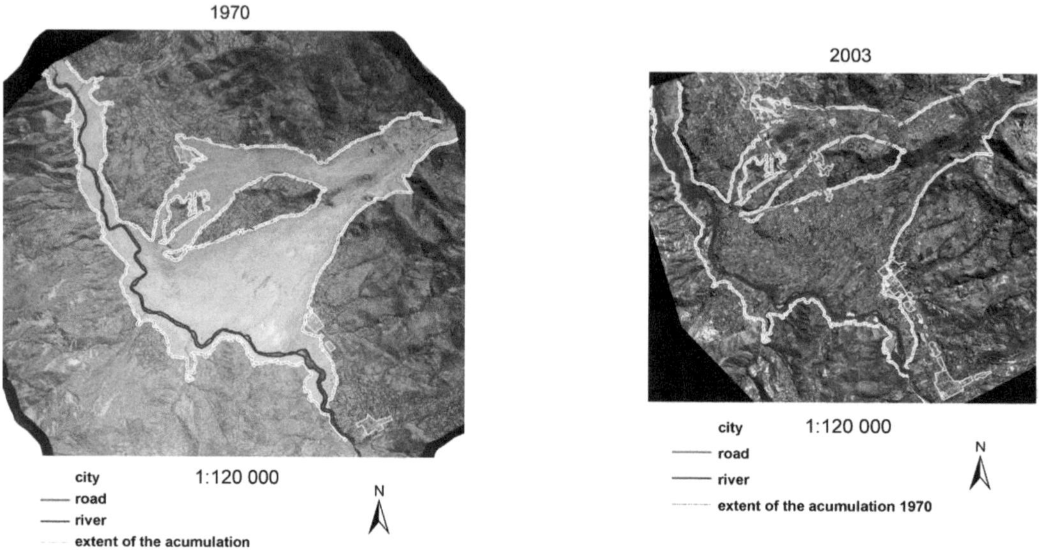


Figure 2. Delineation of the extent of the accumulation in the 1970 orthoimage and its superimposition in the 2003 orthoimage. Some of the main topographic features of the area are added.

Only 2D outputs are presented at this stage of research. Their horizontal accuracy (standard deviation) is estimated to 4 m with respect to the reference frame in case of imagery from 1970. 3D evaluation was not accomplished because low accuracy of the orientation of the images due to lack of calibration information and GCPs with sufficient accuracy.

## 5 DISCUSSION

As mentioned in the previous sections, the accuracy of input data was rather poor and a lot of simplification had to be accepted. In this paragraph, some ideas how to improve the results will be given.

Having both parameters of inner orientation and GCPs, processing of aerial images could be accomplished using a standard processes with following rough estimation of horizontal and vertical accuracy (Kraus 2007):

$$\sigma_{XY}^2 = \sigma_{XY(sig)}^2 + \sigma_{XY(def)}^2 + \sigma_{XY(ori)}^2 \qquad \sigma_Z^2 = \sigma_{Z(sig)}^2 + \sigma_{Z(def)}^2 + \sigma_{Z(ori)}^2$$

where

$\sigma_{(sig)}$ accuracy of a measurement of signalized points in the images,
$\sigma_{Z(sig)}$=m*c*0.3 pixel / b', $\sigma_{XY(sig)}$=0.3 pixel for manual measurement

$\sigma_{(def)}$ uncertainty of definition of natural points

$\sigma_{(ori)}$ accuracy of orientation

m is the scale of the image, c is the principal distance and b' is the length of the base in the image.

In the given dataset, if it was possible to scan the images from original negatives on a photogrammetric scanner with resolution of e.g. 21μm, a camera calibration report was available and GSPs were determined with accuracy of 0.5 m or better in all three coordinates, then accuracy of 1 meter from stereo restitution of the aerial images would be reachable in flat or mildly undulating terrain. In case of mapping of hilly or mountainous terrain, contributions of $\sigma_{Z(sig)}$ and $\sigma_{Z(def)}$ should be replaced by $\sigma_{Z(ter)}$ given by a formula that takes into account the slope of the terrain (Kraus 2007):

$$\sigma_{Z(ter)} = \pm\left(0.00015h + \frac{0.15}{c}h\tan\alpha\right)$$

where

his the flying height, c is the principal distance of the camera [mm] and $\alpha$ is the slope of the terrain.

Example: $m_s = 40\ 000$, $c = 153$ mm, $h = 6210$ m

$\alpha = 10°$, $\sigma_{Z(ter)} = 2$ m, $\alpha = 50°$, $\sigma_{Z(ter)} = 8$ m

With lower image quality or accuracy of GCPs, the accuracy of orthoimages or DEM derived from a stereopair decrease accordingly. If at least 6 GCPs are available per one image, parameters of both inner and exterior orientation can be calculated by means of direct linear transformation (DLT, Kraus 2007).This calculation procedure is useful when camera calibration reports are not reachable which is often the case of processing archive images.

## 6 CONCLUSION

A continuous study of areas exposed to natural hazards by means of both remote sensing and in situ measurements is important in order to understand processes connected to the origin of the hazards and possibly give early warnings in case of rapid development of new events endangering inhabitants and their properties. The presented work showed an example of processing time series of remote sensing images for monitoring changes in landscape due to avalanches. Although the data set contained a lot of unique topographic and thematic information, the geometric quality was poor and therefore all possible information could not be derived from it. Deriving changes in 3D, which would be valuable from the geomorphologic point of view, was not possible due to lack of ground control points of sufficient accuracy. Nevertheless, the presented 2D outputs are of value for study of landscape development and land use/cover changes.

## ACKNOWLEDGEMENTS

The work was supported by the Ministry of Education, Youth and Sports of the Czech Republic in frame of the Grant No. MSM 0021620831.

## REFERENCES

Lomnitz C. 1971: The Peru earthquake of May 31, 1970. Some preliminary seismological results. *Bulletin of the Seismological Society of America 61(3), June*: 535-542.

Plafker G., Ericksen G. E., Concha J. F. 1971. Geological aspects of the May 31, 1970, Perú earthquake. *Bulletin of the Seismological Society of America 61(3), June*: 543-578.

Vilímek V. 1995. Natural hazards in the Cordillera Blanca Mts., Peru. *Acta Montana, Series A, Geodynamics 8 (97)*: 71 - 86.

Vilímek V., Zapata M. L., Stemberk J. 2000. Slope movements in Callejón de Huaylas, Peru. *Acta Univ. Carol., Geographica, 35, Supplementum*: 39-51.

USGS 2006. Shuttle Radar Topography Mission (SRTM). http://seamless.usgs.gov/website/seamless/products/srtm3arc.asp (visited 30th May, 2008)

Digital Globe 2008. Quick Bird Imagery product guide http://www.digitalglobe.com/file.php/511/QuickBird_Imagery_Products-Product_Guide.pdf (visited 30th May, 2008)

*Remote Sensing for a Changing Europe*
*D. Maktav (Ed.)*
*IOS Press, 2009*

*doi:10.3233/978-1-58603-986-8-426*

# Landcover degradation analysis of Mediterranean forest by means of hyperplanes obtained from mixture linear algorithms (MLA)

E. Martinez-Izquierdo & A. Arquero-Hidalgo
*Dept. of Arq. and Tech. of Inform. Systems, FI, Politech. University of Madrid, Madrid, SPAIN*

J. M. Vazquez-Sierra
*Dept. of Electr. and Telec. Engineering, Politech. Sch., CEU University – Saint Paul, Madrid, SPAIN*

Keywords: Linear Mixture, Regression Hyperplanes, land cover degradation estimation

ABSTRACT: The percentage alteration of the Mediterranean forest landscape is one of the primary indicators for its degradation. In this sense, the land cover abundances change analysis by using mixture linear algorithms (MLA), is presented like a good alternative to study this degradation. This research analyzes the use of two information sources like Remote Sensing (Landsat-ETM+) and Field Radiometry (GER 1500) to obtain mixture hyperplanes. These are calculated by models based on *least square estimations*, assuming that each pure land cover (*endmember*) belonging to any geographic area, behaves as a random variable which distribution function is known. The mixture hyperplanes provide spectral signatures with a suitable correlation level with regard to the supplied from remote satellite sensors once corrected, for the same geographical zone. These established hyperplanes can be used in future researches about Mediterranean forest landscape changes, because they can represent the different levels of its degradation. In this sense, it is proposed that they will feed a land cover spectral library with free accessibility.

## 1 INTRODUCTION

The uncontrolled urban development brings about not expected and not wished environmental changes. Particularly, the Mediterranean landscape is a good example because it has been under serious and continuous anthropogenic pressure since historical times. Severe degradation of soil and vegetation resources can be observed as a result of such pressure. An analysis of the "observed" changes of the development of vegetation cover may hence serve as a key indicator of its degradation. In this sense, we addressed the problem researching jointly remote (Remote Sensing) and near (Field Radiometry) sensing data. We feel that both techniques can offer us an adequate source of information about land cover degradation.

In remotely sensed data, pixels containing mixed spectral information about the objects under study are commonly found. This is due to the limitations of the spatial resolution of the remote sensor and the heterogeneity of features on the ground. As a result, most detected surfaces within the instantaneous field of view (IFOV) of the remote sensing instrument are spectrally complex and therefore create a heterogeneous spectral mixture rather than one spectrally "pure" signal within the pixel. In spectral mixture theory, the spectral signal of a pixel can be represented as a mixture of signals contributed by all spectrally "pure" features, or *endmembers* (Lillesand 2000), within the IFOV of the sensor at a given time (Chen 2006).

The identification of the pure pixel value is often difficult (Olthof 2007). Nevertheless, the register of land cover spectral response by measurements *in situ* by Field Radiometry can be a good alternative for this task. This technique is able to provide a good set of *endmembers* and in addition refers to spectroscopic measurements made outdoors, with the sun as the primary source of illumination, and allows to the detection and analysis of the spectral characteristics of the land covers in its natural surroundings.

Previous studies exist, related to the sampling method in the spectral register processes (Vazquez 2004, 2007), specifying exhaustively the directives of sampling *in situ* (Field Radiometry). These researches solve a very important aspect, that is the determination of the minimum and optimal number of spectral samples (so large sample) that it is necessary to register for each land cover *endmember*.

In this paper, we have hence employed a mixture linear approach to cope with the different levels of forest landscape degradation. We show an approach to analyze the question above: the use of mixture linear algorithms (MLA) obtained from mixed models based on *least square estimations*. It is assumed that each of the pure existing cover (*endmember*) belonging to any geographic area behaves as a random variable which distribution function is known (Vazquez 2007). Later, the zone at issue can be subdivided in smaller territorial units until its intrinsic behavior be homogeneous from a spectral point of view. These units will be the basic elements to build different "ad-hoc" statistics that can bear in mind the characteristics and variability of real existent land covers. In the limit, the values of these statistics will be the "observations" that will be needed for the *least squared estimation* method. The obtained mixture hyperplanes (Vazquez 2008) will provide spectral signatures with a suitable correlation level with regard to the supplied data from remote sensor ETM+ with a atmospheric correction (Vermote 1997), for the same geographical zone.

The results of applying this approach to a Mediterranean forest zone with different degradation levels are shown and the different mixed hyperplanes considered are able to represent efficiently the different levels of degradation.

## 2 DATA SET DESCRIPTION

For the present work there has been chosen one of the most typical places of the Madrid Community (Spain), which corresponds to a Mediterranean ecosystem forest principally composed by oaks, bushes and meadows. The study zone is situated in the Mountgancedo near of Madrid City. This site is located to the south-west of the Community, and it has a surface of 125 hectares. The study area selected is located at 40º 24´30´´ N, and 3º49´50´´ W (4473/4474 N, 429/430 W UTM) (Fig. 1). Several Mediterranean species of vegetation coexist besides natural meadows, rocky outcrops, bare soils and a set of generated covers by anthropogenic effects like asphalted paths of some new little cities with commercial buildings and sports center.

The land cover spectral samples have been obtained in the summer by Field Radiometry (GER 1500), in accordance with Landsat ETM+ remote data register. Even though the remote sensor (ETM+) has six bands with the same spatial resolution, only the first four ones can be considered according to the spectral register interval of field radiometer; in order to correlate both sources of information. Sampling *in situ* has been carried out according to a methodology (Vazquez 2007) that optimizes the number of samples to pick-up.

Since the data registered *in situ*, provides a continuous spectral response, a previous reduction methodology based on the integration of the radiance values (Arquero 2003) into the spectral bands interval selected, was carried out. Thus, the field spectral data had been reduced to field spectral signatures, with four values named R-ETM+1, R-ETM+2, R-ETM+3 y R-ETM+4, compatibles with the first four bands of the remote sensor.

## 3 METHODOLOGY

The used sampling methodology starts with the identification of the landscape unit to be analysed. We select geographical units that represent mixed pixels with different levels of Mediterranean forest degradation. With the aim of doing correlation studies between near and remote sensing data, the selected units have equal dimension that the remote sensor spatial resolution and will be called

*Global Unit* ($\mathbf{G_u}$). This one will be submitted to a process of successive subdivisions, with the purpose of get smaller units or *Intermediate Units*, $\mathbf{I_u}$, composed, in turn, by indivisible ones called *Elemental Units*, $\mathbf{E_u}$. These smaller units have a spectral response corresponding to a land cover *endmember*. This methodology has been proposed and justified in a previous research (Vazquez 2008) and a small summary appears later.

Figure 1. Visualization of coloured (RGB) aerial image of Mountgancedo area. The zone of study is marked with a square.

As selected $\mathbf{G_u}$, it has been chosen five squares that represent five levels of degradation, belonging to a Mediterranean forest placed in the surroundings of Madrid (Spain). They have 30x30 meters of side and fit with the spatial resolution of the remote sensor (Landsat 7 ETM+). Each one of them is subdivided in 64 $\mathbf{E_u}$ (4x4 meters). In this way it is possible to assure their good spectral representation. Usually the size of the $\mathbf{E_u}$ is small compared with $\mathbf{G_u}$, especially in the fairly common case when landscapes with great covers variability are analyzed. To assure the management of an optimum number of samples it is necessary to create a new unit, called $\mathbf{I_u}$, whose dimension will be between the Global and the Elementary Units. In our case its dimension will be equal to 20x20 meters that corresponds to 25 $\mathbf{E_u}$ (Fig.1). The criteria to design the more representative statistics for $\mathbf{I_u}$ could be based on a strategy of "considering" more or less the information belonging to the different Elementary Units. In this sense, we propose the “balanced statistic” that consists of assigning to each $\mathbf{E_u}$, identical "weight". So, each of them equally will contribute to the total spectral response of the $\mathbf{I_u}$. This idea is showed in the next expression:

$$R_{I_u} = \frac{\sum_{i=1}^{i=5}\sum_{j=1}^{j=5} x_{ij}}{5^2} \qquad (1)$$

For the representation of the Global Unit in the sampling process, it is proposed the accomplishment of a diagonal double sampling to obtain 8 point estimations corresponding to each of the $\mathbf{I_u}$ included in it (Fig.2).

As for the estimation procedure for the Global Unit reflectance there has been chosen the *least squares method* since this one does not need the previous knowledge of the joint density probability function to the analyzed area. It has been used the *least squares method* with abundance constraints. The constraint considered is the full additivity. This requires that the abundance for a mixed pixel must be one. Otherwise if we suppose that the land cover samples come from normal populations and, in addition, it is possible to demonstrate that the mistakes random variable that represent to them has null average and identical variance, the previous method of adjustment acquires a new perspective from inferential-statistic viewpoint.

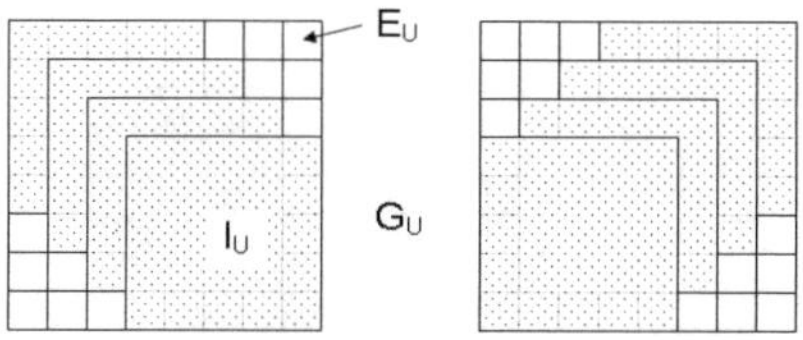


Figure 2. Sampling Units: Global Unit ($G_u$), Intermediate Unit ($I_u$) and Elemental Unit ($E_u$).

This methodology is applied to simulate spectral signatures for the selected geographical $\mathbf{G_u}$ and a set of mixed hyperplanes are been obtained. Once obtained linear mixture results, it is made a correlation between this information and the provided by the remote sensor (once corrected from atmospheric point of view) in the same period of time. The study area basically contains three natural covers: oak, meadow and shrub; and two anthropogenic cover: sport court and fallow.

## 4 RESULTS

The regression models for each band (mixed hyperplanes) obtained by *linear least squares techniques* through balanced statistics applied on $\mathbf{I_u}$ double diagonal sampling in $\mathbf{G_u}$, and are shown in the next table (Table1):

Table 1. Obtained mixed hyperplanes to Global Unit for the five choose levels of degradation

| | |
|---|---|
| Lower Degrad. | |
| $R_{R\text{-}ETM+1} = 0.0266\ p_1 + 0.0364\ p_2 + 0.0802\ p_3$ | ***Endmembers*** |
| $R_{R\text{-}ETM+2} = 0.0281\ p_1 + 0.1333\ p_2 + 0.1831\ p_3$ | $p_1$ : oak |
| $R_{R\text{-}ETM+3} = 0.0153\ p_1 + 0.1717\ p_2 + 0.2301\ p_3$ | $p_2$ : meadow |
| $R_{R\text{-}ETM+4} = 0.2204\ p_1 + 0.3320\ p_2 + 0.9263\ p_3$ | $p_3$ : shrub |
| Medium Degrad. | |
| $R_{R\text{-}ETM+1} = 0.0441\ p_1 + 0.1037\ p_2 + 0.0519\ p_3$ | ***Endmembers*** |
| $R_{R\text{-}ETM+2} = 0.0714\ p_1 + 0.1516\ p_2 + 0.0843\ p_3$ | $p_1$ : oak |
| $R_{R\text{-}ETM+3} = 0.0822\ p_1 + 0.1911\ p_2 + 0.0897\ p_3$ | $p_2$ : meadow |
| $R_{R\text{-}ETM+4} = 0.3765\ p_1 + 0.2877\ p_2 + 0.2155\ p_3$ | $p_3$ : shrub |
| Higher Degrad. | |
| $R_{R\text{-}ETM+1} = 0.1672\ p_1 + 0.0304\ p_2 + 0.3493\ p_3$ | ***Endmembers*** |
| $R_{R\text{-}ETM+2} = 0.3342\ p_1 + 0.0522\ p_2 + 0.4047\ p_3$ | $p_1$ : oak |
| $R_{R\text{-}ETM+3} = 0.4996\ p_1 + 0.0778\ p_2 + 0.4987\ p_3$ | $p_2$ : meadow |
| $R_{R\text{-}ETM+4} = 0.9845\ p_1 + 0.1471\ p_2 + 0.7114\ p_3$ | $p_3$ : shrub |
| With Sport Court | |
| $R_{R\text{-}ETM+1} = 0.1119\ p_1 + 0.1671\ p_2 + 0.0416\ p_3$ | ***Endmembers*** |
| $R_{R\text{-}ETM+2} = 0.1507\ p_1 + 0.2350\ p_2 + 0.0790\ p_3$ | $p_1$ : oak |
| $R_{R\text{-}ETM+3} = 0.1691\ p_1 + 0.2797\ p_2 + 0.1214\ p_3$ | $p_2$ : meadow |
| $R_{R\text{-}ETM+4} = 0.4350\ p_1 + 0.4087\ p_2 + 0.2108\ p_3$ | $p_3$ : sport court |
| With Fallow | |
| $R_{R\text{-}ETM+1} = 0.0283\ p_1 + 0.1717\ p_2 + 0.0607\ p_3$ | ***Endmembers*** |
| $R_{R\text{-}ETM+2} = 0.0355\ p_1 + 0.1954\ p_2 + 0.1348\ p_3$ | $p_1$ : oak |
| $R_{R\text{-}ETM+3} = 0.0270\ p_1 + 0.2338\ p_2 + 0.1913\ p_3$ | $p_2$ : meadow |
| $R_{R\text{-}ETM+4} = 0.2769\ p_1 + 0.2406\ p_2 + 0.3324\ p_3$ | $p_3$ : fallow |

$R_i$ is reflectance and $p_i$ is abundance of *endmember*: oak, meadow, shrub, sport court and fallow.

| *Endmembers* | Low degrad. % |
|---|---|
| $p_1$ : oak | 75.6 |
| $p_2$ : meadow | 10 |
| $p_3$ : shrub | 14.4 |

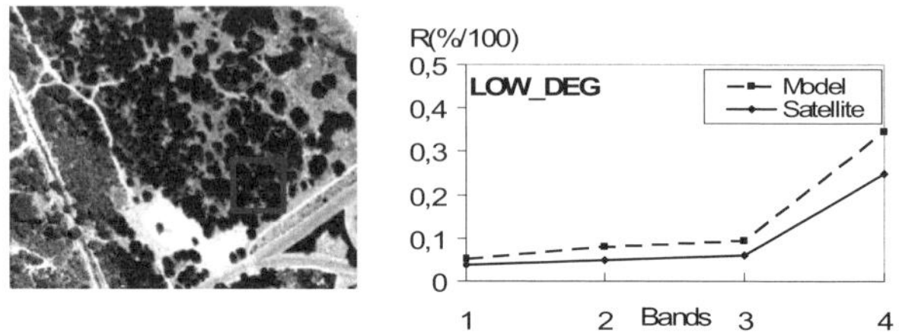


| *Endmembers* | Med. degrad. % |
|---|---|
| $p_1$ : oak | 18.7 |
| $p_2$ : meadow | 78.1 |
| $p_3$ : shrub | 3.2 |

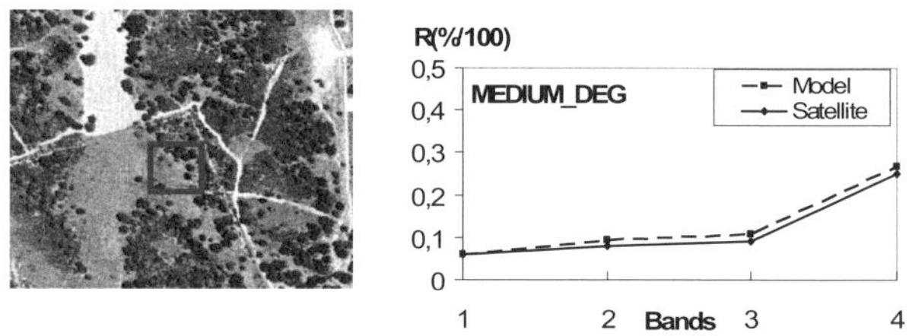


| *Endmembers* | High degrad. % |
|---|---|
| $p_1$ : oak | 9.4 |
| $p_2$ : meadow | 84.3 |
| $p_3$ : shrub | 6.3 |

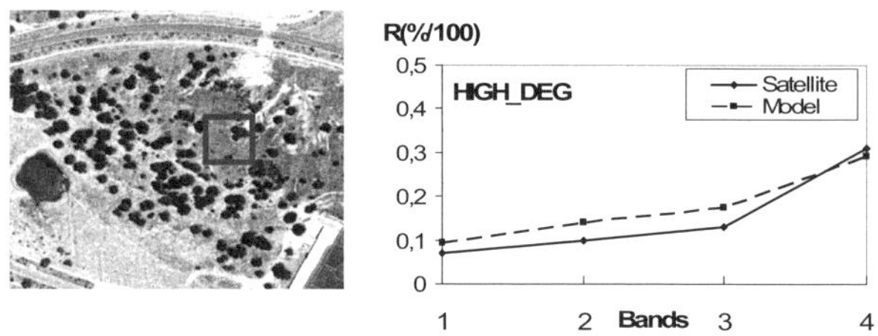


| *Endmembers* | With sport court % |
|---|---|
| $p_1$ : oak | 18 |
| $p_2$ : meadow | 68 |
| $p_3$ : sport court | 14 |

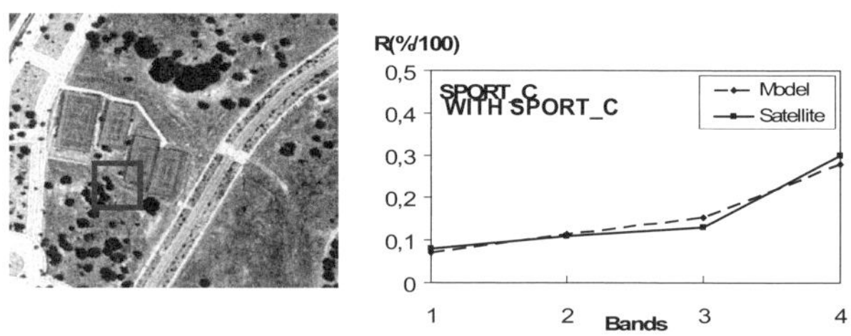


| *Endmembers* | With fallow % |
|---|---|
| $p_1$ : oak | 25 |
| $p_2$ : meadow | 20 |
| $p_3$ : fallow | 55 |

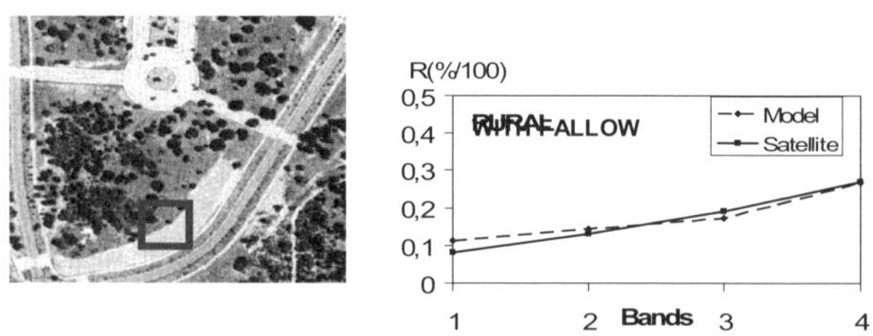


Figure 3. Comparison results between spectral signature values of simulated models and satellite atmospheric corrected data for the five selected Global Units.

The Figure 3 presents comparative results between the signature values (reflectance) obtained from simulated models and those obtained through remote sensor ETM+, for the five Global Units.

The spectral signature modeled has been obtained through the estimated percentage of abundances and are showed in the same figure.

A global analysis of the results showed in the Fig. 3, denote that the spectral signatures modeled are very similar that ones provided by the sensor satellite once atmospheric corrected. Particularly, in the case of Global Units of Lower and Higher Degradation, the simulation predicts signatures values slightly higher than those provided by the satellite sensor but in the other cases it is observed highly correlation. We suggest that the atmospheric correction method of the satellite images, produces an overcorrection in the fourth band (ETM+4), in the cases of high percentage of vegetation, land cover that presents a maximum value in this spectral interval. In the case of Higher Degradation with the higher abundance of class meadow that presents certain spectral variability makes some slightly differences.

## 5 CONCLUSIONS

The procedure to modelling the spectral response in areas composed by mixed land covers through the successive subdivision of the study area, presents satisfactory results. For that, it turns out indispensable to handle spectral reliable values for each of the present *endmembers*. In consequence it is crucial to have information with a great statistical solidity extracted from adequate multispectral information libraries. With this basic information it is possible to conform signature models to a reasonable adjustment degree to their provided by the satellite images previously corrected atmospherically. Different sets of mixed hyperplanes had been obtained for the different examples. In this work the case of the Mediterranean forest with different levels of degradation and areas deeply transformed by the human action are showed.

## REFERENCES

Arquero, A., Martínez, E. & Gonzalo, C. 2003. Reducción de datos hiperespectrales de Radiometría de Campo", *Teledetección y Desarrollo regional; X Congreso Nacional de Teledetección, Cáceres (Spain)*: 439-442.

Chen, X. & Vierling, L. 2006. Spectral mixture analyses of hyperspectral data acquired using a tethered balloon. *Remote Sensing of Environment* 103: 338-350.

Lillesand, T. M. & Kiefer, R. W. 2000. *Remote Sensing and Image Interpretation. John Wiley & Sons, Inc. New York.*

Olthof, I. & Fraser R. H. 2007. *Remote Sensing of Environment* 107: 496-509.

Vázquez, J. M., Arquero, A., Martínez, E. & Gonzalo, C. 2004. An efficient methodology for endmembers selection by field radiometry: an application to multispectral mixture model. *Signal and Image Processing for Remote Sensing X* 5573: 435-443.

Vázquez, J. M., Arquero, A. & Martínez, E. 2007. Aplicación de modelos estadísticos a la selección eficiente de cubiertas tipo mediante Radiometría de Campo. *IEEE Latin America Transactions* 5 (8): 561-567.

Vazquez, J. M., Martínez, E. & Arquero, A. 2008. An efficient methodology to simulate mixed spectral signatures of land covers through Field Radiometry data. *EARSEL08*. In press.

Vermote E.F., Tanré D., Deuzé J.L., Herman M. & Morcrette J.J. 1997. Second Simulation of the Satellite Signal in the Solar Spectrum: an overview. *IEEE Transactions on Geoscience and Remote Sensing* 35(3): 675-686.

*Remote Sensing for a Changing Europe*
*D. Maktav (Ed.)*
*IOS Press, 2009*

*doi:10.3233/978-1-58603-986-8-432*

# The effects of terrain illumination correction on automatic classification

Annamaria Bernardini, Eva S. Malinverni
*Università Politecnica delle Marche, DARDUS, 60131 Ancona, Italy*

Andrea Galli, Ernesto Marcheggiani
*Università Politecnica delle Marche, DiSASC, 60131 Ancona, Italy*

Primo Zingaretti
*Università Politecnica delle Marche, DIIGA, 60131 Ancona, Italy*

Keywords: Remote sensing, IKONOS, Land cover, Algorithms, Correction, Test, Classification

ABSTRACT: In Remote Sensing analysis it was pointed out that the reflectance of many land covers is influenced by solar illumination, that differs in dependence of the shape of the terrain. These terrain illumination effects can affect automatic classification methods, due to the fact that similar land covers produce different reflectance values. In the present study we compared some radiometric image correction methods applied to reduce differential illumination effects to an IKONOS image dataset. In literature *Cosine Correction* (1) based on the assumption of an isotropic (Lambertian) reflectance, is widely argued. However several authors showed that this method overcorrects the images (2, 3, 4) mainly in faintly illuminated areas. In our research we applied the *C-correction* method (1, 3) and a modified version of the *Normalization* method (5). Finally we performed an automatic classification to assess if the proposed DEM based altimetry effect correction improves the results.

## 1 INTRODUCTION

This research is part of a wider project whose main objective is to develop a Corine Land Cover class based automatic classification methodology starting from high resolution Multispectral IKONOS images. The images were provided by Regione Marche Institution thanks to a Research Agreement signed together with the Università Politecnica delle Marche jointed Departments (DARDUS, DIIGA and DiSASC).

The study case refers to an area belonging to the Ancona Province in Italy (Figure 1). The IKONOS dataset covers an extension of approximately 150 km$^2$, with both coastal and mountainous aspect, urban and rural landscapes, natural environment, among which the Conero Natural Park has to be mentioned.

According to a hierarchical approach, different phases are applied for land cover classification based on digital image analysis. At the lowest level the analysis is merely based on the pixel spectral values, while moving toward higher levels, the segmentation will carry out the spatial feature patterns. The present research regards the lower level of the analysis. We analyzed the different solar illumination, due to the irregular shape of the terrain, because it causes a variation of the spectral response into the same cover: shaded areas show lower reflectance values than sunny areas. Although it is possible to take into account this effect in multiband classification process by including aspect or illumination maps as ancillary variables, the process of the solar illumination correction of image Digital Number (DN) is more accurate. This correction is called in literature *Topographic Normalization*. In this work we applied different topographic corrections to IKONOS images testing different methodologies. Successively the corrected images were classified and the results compared by means of the Confusion Matrix.

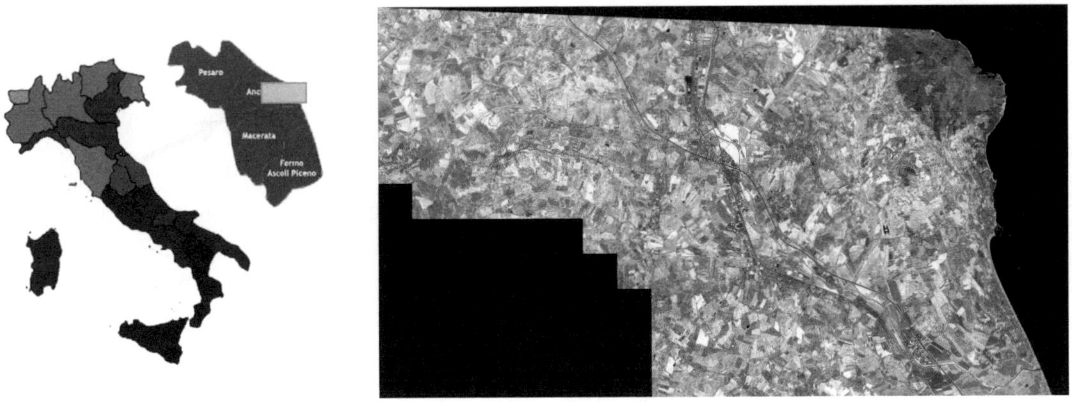


Figure 1 Location of the study area: the surroundings of the Ancona Province, Italy

The work was divided into five stages (Figure 2) which can be summarized as:

- Dataset creation: the image dataset were orthorectified in a chosen coordinate system and the generated DTM was used to create aspect and slope maps.
- Definition of training and control samples: an exhaustive number of samples, based on a hierarchical classification nomenclature, was chosen to derive calibration values for the topographic correction models and successively to train and check the automatic supervised classification.
- Topographic Correction: several methodologies were used to normalize the illumination effects.
- Automatic Classification: the corrected images were applied in an automatic supervised classification.
- Accuracy assessment: the assessment of the classification accuracy was carried out comparing corrected and not corrected images with the training sample set.

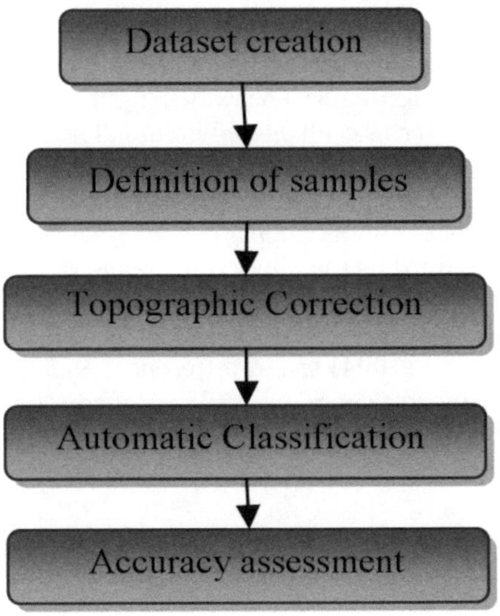


Figure2. The sequence of the 5 classification process stages

Paper is organized in the following seven sections: "Background" section introduces the methodological approach; "Pre-Processing" section provides a description of orthorectification method and DTM processing; "Sample Dataset" section points out the necessary procedures to obtain a sufficient number of training samples which are prerequisites for a successful classification. "Topographic Correction" section describes different methodologies applied for illumination correction. Last two sections, "Assessment of the Classification Accuracy" and "Conclusions", present and comment the results.

## 2 BACKGROUND

In sloped terrains, the radiance depends of the sun's incidence angle $\gamma_i$ in relation to the surface normal direction, , that can be calculated by means of the relation (1) using some DEM information (aspect and slope):

$$\cos\gamma_i = \cos\theta_p \cdot \cos\theta_z + sen\theta_p \cdot sen\theta_z \cdot \cos(\phi_a - \phi_0) \quad (1)$$

Where $\gamma_i$ is the sun's incidence angle; $\theta_p$ is the terrain slope; $\theta_z$ is the sun zenith angle; $\phi_a$ is the sun azimuth and $\phi_0$ is the terrain aspect.

The $IL = \cos\gamma_i$ parameter is in a range of ± 1 and can be computed for the whole image to obtain the illumination map, modelling the illumination conditions. The topographic correction procedure calculates the theoretical reflectance of an horizontal surface $\rho_H$ from observed radiance over sloped terrain $\rho_T$ on the base of the IL coefficient. The method called *Cosine Correction* was proposed by Teillet et al, and it is simpler to apply (2): $\rho_H = \rho_T \cdot \cos\theta_z / IL$ (2)

In this case the method assumes that the reflectance was independent from the observation direction and from the wavelength. The correction is based on the same factor for every direction and for each spectral band, not considering the diffuse irradiance. For these reasons, the method over-corrects the reflectance of the image where the IL parameter is low, as several authors showed (2, 3). Teillet et al (1982) introduced the *C-correction* coefficient in the previous equation to take into account the influence of wavelength, considering the non Lambertian irradiance and the diffuse irradiance (3): $\rho_H = \rho_T \cdot (\cos\theta_z + C) / I(L + C)$ (3)
The *C-correction* coefficient increases the denominator, modelling the effect of the diffuse irradiance, and weakening the over-correction of faintly illuminated data. The wavelength effect is taken into account, by the estimation separately of the parameter in each specific band. The calculation of *C-correction* coefficient is based on a statistical empirical function (4) that assumes a linear correlation between the reflectance of each band and IL results. The significant trend in the observed data is significatively illustrated by the parameters *b* (the intercept) and *m* (the inclination) of the regression equation. The parameter *C* can be obtained as the quotient of *b* and *m* of the regression model (*C*=*m*/*b*); $\rho_T = m \cdot IL + b$ (4)

Several researches (1, 3), applied on Landsat imagery, demonstrated that the *C-Correction* is the most effective illumination correction method. Law and Nichol (2004) tested different topographic correction algorithms on IKONOS images, performing a changed version of the *Normalization* method proposed by Civco that considers the average *IL* conditions ($IL_m$) (5):

$$\rho_H = \rho_T + \rho_T \cdot \frac{IL_m - IL}{IL_m} \quad (5)$$

In this case Law and Nichol introduced a calibration coefficient $C_\lambda$ related to each band (6):

$$\rho_{Hj} = \rho_T + \rho_T \cdot \frac{IL_m - IL}{IL_m} \cdot C_\lambda \quad (6)$$

The $C_\lambda$ calibration coefficient can be derived from statistical measurements of radiance and from IL coefficient referred to a particular sample set, facing to and away from the sun (7):

$$C_\lambda = \frac{S_\lambda - N_\lambda}{\left[\left(N_\lambda \cdot \frac{IL_N - IL_m}{IL_m}\right) - \left(S_\lambda \cdot \frac{IL_S - IL_m}{IL_m}\right)\right]} \quad (7)$$

Where $N_\lambda$ is the mean on the slope away from the sun for a chosen cover class; $S_\lambda$ is the mean on the slope facing to the sun for a chosen cover class; $IL_m$ *is* the mean value for the entire scaled shaded relief model (0, 255); $IL_N$ is the mean of the illumination of a chosen cover class on the slope away from the sun; $IL_S$ is the mean of the illumination of a chosen cover class on the slope facing to the sun.

Choosing training samples set referred to different cover classes you can obtain different coefficients, producing, furthermore, different terrain corrected images.

## 3 PRE-PROCESSING

### 3.1 *Digital Terrain Model (DTM) processing*

Digital contour lines of Technical Regional Map at the scale 1:10.000 were used to generate a DTM with a grid spacing of 4 meter. From the DTM we derived the information about the slope, the aspect and the illumination according to the sun angle and elevation, useful, successively, for the topographic corrections algorithms.

### 3.2 *Image orthorectification*

IKONOS images, acquired on July 2006, with a 29 degrees solar zenith angle, are in Panchromatic mode (0.45-0.90 μm) with 1 meter ground resolution and in Multispectral mode with 4 meter ground resolution

The data are available in 11-bit radiometric resolution. They are delivered georeferenced according to the UTM projection and WGS84 Datum. The whole image dataset was orthorectified by means of the third order rational function model implemented in the software PCI - Orthoengine. We selected 15 Ground Control Points (GCPs) on the Technical Regional Map at the scale 1:10.000 and identified on the panchromatic image. Using also the DTM information we obtained an RMS error below 1 pixel (1 meter ground resolution). The radiometric interpolation, applied by means of nearest neighbour resampling method, preserves the original image values. Next, a sub map of the original image was extracted to define the study area.

## 4 SAMPLE DATASET

A suitable classification system and a sufficient number of training samples are prerequisites for a successful classification. The system should be informative, exhaustive, and separable (6, 7).

At first it is necessary the definition of the hierarchical classification structure that is principally based on the user's necessity and on the spatial resolution of remotely sensed data. On the base of this structure, a significant number of training samples have to be defined.

### 4.1 *Classification structure*

The necessity of a standard classification system suggests the use of Corine Land Cover nomenclature, derived from an European Project, led by the European Environmental Agency (EEA) in coordination with the member countries. The aim of the project was to define land cover inventories for all European Countries, at the scale 1:100.000, based on a standard methodology and nomenclature, useful for remote sensing applications. The legend has a hierarchical structure on three levels, containing 44 land cover classes, grouped into five major categories: Urban Fabric (1), Agriculture Areas (2), Forest and Semi-Natural Areas (3), Wetlands (4), Waterbodies (5). With respect of these structures, the high ground resolution of the IKONOS sensor, suitable for a map at the scale 1:10.000, needs the introduction of others categories of fourth and fifth level. Regarding the Agriculture, the Forest and Semi-Natural Areas, we adopted the extended legend provided by the Italian

Agency for Environmental Protection and Technical services (APAT). For Urban Areas we introduced new levels based on the spectral response of different materials.

### 4.2 *Construction of sample set*

For each class a sample test set was constructed with labelled polygons. Urban Fabric and Water bodies samples were detected by means of the classical photo interpretation technique or basing on specific surveying campaign for "ground truth" collection. Regarding the categories of Agriculture, Forest and Seminatural Areas, to minimize errors in land use class assessment it was necessary to search other information in dedicated databases. The principal cause of error was the high intra annual land use variation for rotations and short cycle agricultural productions. Approximately 18.000 training pixels were extracted, lying to 27 classes. After a separability analysis, based on histograms of the different bands, some classes were merged into single class. Finally the number of classes was set to 15.

## 5 TOPOGRAPHIC CORRECTION

The analysis was performed with the Open Source software ILWIS (Integrated Land and Water Information System). The software, developed in 1984 by the Netherlands ITC (Institute for Aerospace Survey and Earth Sciences), has been granted open source since 1st July of 2007. It's an integrated Remote Sensing – GIS software that allows to use some internal functions and mathematical calculations in a sequence of instructions organized in script format, to make more than customizable the procedures.

In this environment we tested the *C-correction* methodology, that produced, as documented in literature, the best result applied to LANDSAT images, and the *Normalization* method, used successfully by Law on IKONOS images.

### 5.1 *Application of the* C-Correction *method*

The C-correction coefficient evaluation is based on the regression analysis of a selected land cover class. In this research the choice of the Forest class is based on its representatively respect to the pixel amount (resulting from the automatic classification on uncorrected images) and the DN mean value related to each single band. Furthermore the chosen class is the main cover type on the Conero mountain and in the image dataset it presents mainly shadow areas. Figure 3a shows a positive linear relationship between the DN of NIR band and the incidence angle.

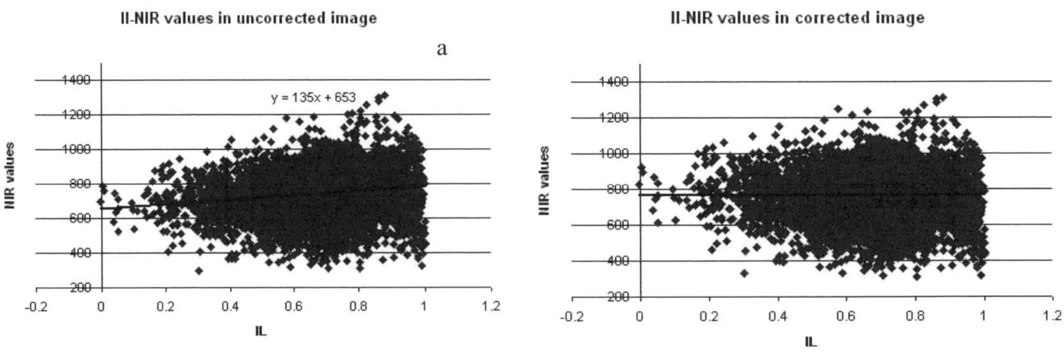


Figure3. Linear regression of incidence angle in NIR band related to uncorrected data (a) and corrected data (b)

The equation for the trend line gives a value for the y intercept of 653 and for the slope of 135. The C-correction coefficient is equal to 4.8. Figure 3b shows the effect applying the C-correction model on the same dataset shown in figure 3a.
The relationship between the DN and the incidence angle is removed from the data. The equation for the trend line gives a value for the y intercept of 770 and for the slope of 2.

## 5.2 *Application of the Normalization method*

We applied the wavelength independent model proposed by Civco using the corrected version coming from Law, that introduces a correction factor $C_\lambda$ estimated separately for each band and here called *C-Normalization* method. In order to perform this solution each sample class was grouped on the base of the surface aspect respect to the sun azimuth. It is well known that in all proposed methodologies the class choice influences the correction. In the graph of the Figure 4 each point represents for a class the percentage variation of the mean values of slopes facing to the sun respect to the mean values of slopes away from the sun.

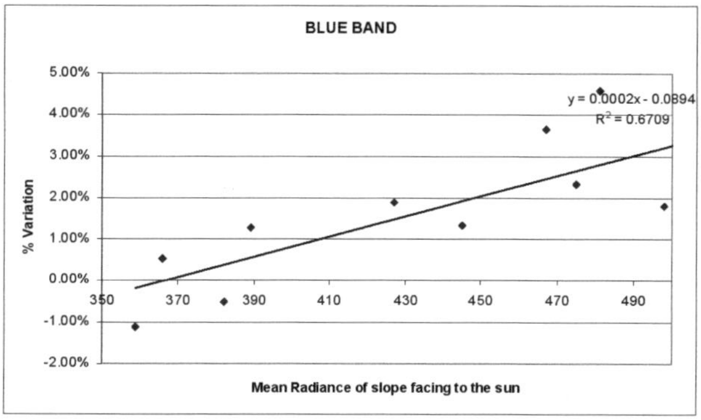


Figure 4. Percentage variation of spectral response of the slopes facing to the sun respect to the slopes away from the sun

This variation is different for each class but it seems that there is a linear dependence from the wavelength inside a specific band. So the $C_\lambda$ coefficient, see equation (7), can be expressed as linear function of the observed radiance values (8): $C_\lambda = m \cdot \rho_T + b$ (8)

Where *m*, *b* are estimated by means of regression analysis on $C_\lambda$ values calculated for the different classes of the sample set. This approach allows to obtain a model that is not influenced by the choice of the sample class and gives better results on the whole image. It is better to apply this model by images with heterogeneous land covers and where not even one is predominant.

# 6 STATISTICAL COMPARISON

In order to evaluate the accuracy of the topographic correction, a statistical analysis was performed. Mean values and standard deviations before and after the topographic correction were calculated both for the whole image and for the training areas of the Forest class. All the proposed methods preserve the radiometric content of the entire image, and the correction is poor probably for the effect of diffuse irradiance. The *C-Normalization* presents a lower standard deviation on the entire image, while *C-Correction* gives better results on the selected Forest class (Table 2, Table 3).

| | **Original Image** | | **Normalization** | | **C-Correction** | | **C-Normalization** | |
|---|---|---|---|---|---|---|---|---|
| **Band** | μ | σ | μ | σ | μ | σ | μ | σ |
| Red | 278.67 | 201.5 | 276.86 | 199.65 | 279.03 | 201.58 | 277.67 | 199.89 |
| Green | 368.84 | 221.33 | 367.04 | 220.95 | 369.1 | 221.41 | 367.83 | 220.18 |
| Blue | 328.38 | 184.38 | 327.06 | 185.5 | 328.49 | 184.87 | 327.59 | 184.12 |
| NIR | 576.98 | 381.86 | 574.32 | 380.93 | 578.5 | 382.39 | 576.19 | 380.68 |

Table 2. Mean and standard deviation of the whole image before and after the correction

| Band | Original Image | | Normalization | | C-Correction | | C-Normalization | |
|---|---|---|---|---|---|---|---|---|
| | μ | σ | μ | σ | μ | σ | μ | σ |
| Red | 234.46 | 13.19 | 281.69 | 38.73 | 239.09 | 12.96 | 239.39 | 13.52 |
| Green | 376.74 | 14.02 | 453.09 | 62.5 | 380.45 | 13.87 | 366.3 | 18.44 |
| Blue | 364.1 | 6.45 | 438.16 | 60.89 | 365.54 | 6.39 | 343.58 | 19.79 |
| NIR | 744.95 | 124.45 | 893.54 | 177.32 | 771.01 | 126.9 | 754.55 | 125.11 |

Table 3. Mean and standard deviation of the Forest class before and after the correction

## 7 ASSESSMENT OF THE CLASSIFICATION ACCURACY

A supervised classification was carried out by Maximum Likelihood (ML) algorithm using both original dataset and the corrected images by the *C-correction* method. Although the User Accuracy (UA) differs inside the classes, the topographic correction doesn't produce improvement in the Overall Accuracy (OA), that is the same for both datasets (Table 4, Table 5).

| | **1101** | **1102** | **1103** | **1201** | **1202** | **1203** | **1301** | **1302** | **1303** | **1401** | **1402** | **1403** | **1501** | **1502** | **1503** | **PA** |
|---|---|---|---|---|---|---|---|---|---|---|---|---|---|---|---|---|
| **1101** | 147 | 8 | 61 | 1 | 16 | 0 | 3 | 0 | 0 | 1370 | 14 | 2 | 5 | 2 | 379 | **7%** |
| **1102** | 0 | 1413 | 45 | 0 | 0 | 0 | 0 | 0 | 0 | 0 | 70 | 6 | 0 | 0 | 65 | **88%** |
| **1103** | 13 | 362 | 731 | 0 | 2 | 0 | 8 | 0 | 0 | 88 | 6 | 89 | 2 | 0 | 1 | **56%** |
| **1201** | 1 | 9 | 16 | 903 | 749 | 1 | 8 | 2 | 34 | 1211 | 107 | 6 | 1 | 0 | 0 | **30%** |
| **1202** | 0 | 3 | 9 | 251 | 2474 | 14 | 0 | 2 | 54 | 16 | 0 | 0 | 2 | 0 | 0 | **88%** |
| **1203** | 2 | 2 | 1 | 7 | 118 | 14 | 1091 | 37 | 61 | 0 | 0 | 3 | 0 | 0 | 0 | **1%** |
| **1301** | 0 | 0 | 0 | 0 | 0 | 0 | 805 | 343 | 10 | 0 | 0 | 0 | 0 | 0 | 0 | **70%** |
| **1302** | 0 | 0 | 0 | 256 | 179 | 19 | 216 | 3840 | 77 | 0 | 0 | 0 | 0 | 0 | 0 | **84%** |
| **1303** | 2 | 0 | 4 | 47 | 2374 | 438 | 115 | 285 | 518 | 0 | 0 | 5 | 0 | 4 | 0 | **14%** |
| **1401** | 15 | 0 | 12 | 28 | 95 | 0 | 0 | 0 | 0 | 5199 | 0 | 1 | 2 | 0 | 0 | **97%** |
| **1402** | 4 | 38 | 2 | 50 | 287 | 0 | 0 | 0 | 0 | 1054 | 5833 | 0 | 2 | 0 | 0 | **80%** |
| **1403** | 1 | 22 | 212 | 15 | 6 | 0 | 110 | 0 | 4 | 2 | 4 | 994 | 0 | 0 | 0 | **73%** |
| **1501** | 0 | 0 | 13 | 18 | 84 | 0 | 0 | 0 | 0 | 22 | 0 | 0 | 79 | 1 | 0 | **36%** |
| **1502** | 0 | 0 | 0 | 0 | 0 | 0 | 18 | 0 | 2 | 0 | 0 | 0 | 0 | 200 | 0 | **91%** |
| **1503** | 0 | 0 | 0 | 0 | 0 | 0 | 0 | 0 | 0 | 0 | 0 | 0 | 0 | 0 | 78542 | **100%** |
| **UA** | **79%** | **76%** | **66%** | **57%** | **39%** | **3%** | **34%** | **85%** | **68%** | **58%** | **97%** | **90%** | **85%** | **97%** | **99%** | |

Table 4. Confusion Matrix of the ML classification carried out on the uncorrected dataset

| | 1101 | 1102 | 1103 | 1201 | 1202 | 1203 | 1301 | 1302 | 1303 | 1401 | 1402 | 1403 | 1501 | 1502 | 1503 | PA |
|---|---|---|---|---|---|---|---|---|---|---|---|---|---|---|---|---|
| **1101** | 141 | 5 | 25 | 0 | 8 | 0 | 3 | 0 | 0 | 1286 | 11 | 1 | 1 | 17 | 258 | **8%** |
| **1102** | 0 | 1517 | 45 | 0 | 0 | 0 | 0 | 0 | 0 | 0 | 66 | 7 | 0 | 0 | 59 | **90%** |
| **1103** | 2 | 298 | 750 | 0 | 0 | 0 | 2 | 0 | 0 | 157 | 4 | 54 | 4 | 0 | 1 | **59%** |
| **1201** | 0 | 5 | 19 | 894 | 841 | 2 | 6 | 2 | 31 | 1336 | 147 | 23 | 0 | 0 | 0 | **27%** |
| **1202** | 0 | 0 | 12 | 252 | 2472 | 16 | 1 | 2 | 88 | 22 | 0 | 0 | 2 | 0 | 0 | **86%** |
| **1203** | 0 | 0 | 0 | 8 | 108 | 18 | 1148 | 26 | 56 | 0 | 0 | 6 | 0 | 0 | 0 | **1%** |
| **1301** | 0 | 0 | 0 | 1 | 0 | 0 | 786 | 322 | 2 | 0 | 0 | 2 | 0 | 0 | 0 | **71%** |
| **1302** | 0 | 0 | 0 | 266 | 185 | 27 | 181 | 3861 | 78 | 0 | 0 | 0 | 0 | 0 | 0 | **84%** |
| **1303** | 0 | 0 | 1 | 38 | 2419 | 471 | 164 | 308 | 524 | 0 | 0 | 6 | 0 | 3 | 0 | **13%** |
| **1401** | 14 | 0 | 5 | 15 | 76 | 0 | 0 | 0 | 0 | 5308 | 0 | 3 | 2 | 0 | 0 | **98%** |
| **1402** | 3 | 43 | 9 | 44 | 289 | 0 | 0 | 0 | 0 | 937 | 5912 | 1 | 4 | 0 | 0 | **82%** |
| **1403** | 0 | 24 | 235 | 10 | 3 | 0 | 44 | 0 | 1 | 1 | 2 | 1022 | 0 | 0 | 0 | **76%** |
| **1501** | 0 | 0 | 0 | 0 | 0 | 0 | 0 | 0 | 0 | 7 | 0 | 0 | 62 | 0 | 0 | **90%** |
| **1502** | 0 | 0 | 0 | 0 | 0 | 0 | 2 | 0 | 0 | 0 | 0 | 0 | 0 | 199 | 0 | **99%** |
| **1503** | 0 | 0 | 0 | 0 | 0 | 0 | 0 | 0 | 0 | 0 | 0 | 0 | 0 | 0 | 78880 | **100%** |
| **UA** | **88%** | **80%** | **68%** | **59%** | **39%** | **3%** | **34%** | **85%** | **67%** | **59%** | **96%** | **91%** | **83%** | **91%** | **100%** | |

Table 5. Confusion Matrix of the ML classification carried out on the corrected image dataset by C-correction method

## CONCLUSIONS

By means of different methodologies we carried out an empirical evaluation of the illumination effects on the radiance values. Analysis has been made inside a specific land cover class, that presented a spectral signature characterized by a mean and a standard deviation in radiance values. We found a correlation between the radiance and IL values, that is wavelength dependent, varying on different bands, but also inside a band on different classes. However the variations produced by the illumination conditions in a land cover class is very limited inside the spectral signature of the class, so the topographic correction doesn't produce meaningful improvements in the automatic classification results. Although the IL model for the chosen dataset has had very low values, the effects on the radiance were not so meaningful, as well the visual analysis showed, probably for the effect of diffuse irradiance. Other tests could be carried out by using different dataset, chosen on the base of visual analysis of the effects due to the illumination. Regarding the methodologies comparison, the *C-Normalization* method seems to perform better results inside the chosen class. Anyway the model is influenced properly by the choice of the class. A correlation could be searched inside a specific band between the $C_\lambda$ coefficient and the mean wavelength of different classes, in order to obtain a model for images which have heterogeneous land covers.

## 8 REFERENCES


1 Teillet P. M., Guindon B., Goodeonugh D. G., 1982, On the slope-aspect correction of multispectral scanner data, Can. Journal Remote Sensing, vol. 8, pp. 84–106.

2 Duguay C. R., LeDrew E. F., 1992, Estimating surface reflectance and albedo from Landsat-5 TM over rugged terrain, Photogrammetric Engineering and Remote Sensing, vol. 58, pp. 551–558.

3 Meyer P., Itten K. I., Kellenbenberger T., Sandmeier S., Sandmeier R., 1993, Radiometric corrections of topographically induced effects on Landsat TM data in an alpine environment, ISPRS Journal Photogramm. Remote Sensing, vol. 48, pp. 17–28.

4 Law K.H., Nichol J., 2004, Topographic correction for differential illumination effects on IKONOS satellite imagery, in Proceedings of XXth ISPRS Congress: Geo-imagery bridging continents, Istanbul, Turkey, Volume XXXV, Part 3B.

5 Civco D. L., 1989, Topographic Normalization of Landsat Thematic Mapper Digital Imagery, Photogrammetric Engineering and Remote Sensing, 55(9), pp. 1303-1309.

6 Jensen J .R., 1996, Introduction to Digital Image Processing: A remote sensing perspective, 2nd edition, Piscataway, NJ: Prentice Hall.

7 Landgrebe D. A., 2003, Signal Theory Methods in Multispectral Remote Sensing, ed. Hoboken, NJ: John Wiley and Sons.

*Remote Sensing for a Changing Europe*
*D. Maktav (Ed.)*
*IOS Press, 2009*


*doi:10.3233/978-1-58603-986-8-441*

# Analysis of Antarctic Sea Ice Extent based on NIC charts and AMSR-E data

Burcu Ozsoy-Cicek, Steve Ackley, Hongjie Xie & Penelope Wagner
*University of Texas at San Antonio*

Keywords: QuikSCAT, AMSR-E, NIC

ABSTRACT: The extent of the Antarctica sea ice is not accurately defined only using low resolution passive microwave data, such as The Advanced Microwave Scanning Radiometer-Earth Observing System (AMSR-E). Due to the varied ice types at the ice edge, it is necessary to use additional resources that will have better results. Therefore sea ice edge data provided by National Ice Center (NIC) were used for better understanding. Sea ice extent for summer time period between January 1$^{st}$ and March 1$^{st}$, and winter time period between August 1$^{st}$ and December 31$^{st}$ were calculated from both AMSR-E and NIC products. The calculated summer ice extent from AMSR-E underestimated the extent by 30% for summer season comparing to the results from NIC dataset. Also sea ice extent calculated from AMSR-E underestimated the extent by 10% for winter season again comparing to the results from NIC data. For the quality purpose and validation NIC data were compared with NASA's Quick Scatterometer (QuikSCAT) to see the detection of ice edge differs or agrees between these dataset. Comparison covered the entire Antarctic continent for the middle of summer and middle of winter. NIC and QuikSCAT data showed very good agreement. However, AMSR-E data did not agree with the other dataset and basically underestimated total ice concentration at the edge during the summer season.

## 1 INTRODUCTION

Sea ice simply is frozen sea water and forms as saline ocean water freezes on the surface in polar regions. Sea ice is considered to be a sensitive indicator of global climate change. It is also used as an input to global weather and climate models. Sea ice in the Antarctic is seasonal and can vary from a minimum in the austral summer month of February to a maximum in the early spring month of September. Sea ice extent varies from annual minimum of 4 to maximum of $18 \times 10^6$ km$^2$. However current climate models suggest that global warming will be felt most acutely in the Polar Regions (IPCC, 2007). That is why mapping the extent of sea ice in the earth's polar regions is of great interest to the scientific community. Researchers have already observed many changes in the Arctic, including the warmest temperatures in the last 400 years and a decline in the extent of spring and summer sea ice. While dramatic changes are occurring in the Arctic region, we still do not have a full picture of how Antarctic sea ice as a whole, is responding to climate change.

Researchers already have utilized satellite measurements to study changes in Antarctic sea ice (Cavalieri et al., 1997, Zwally et al., 2002, Cavalieri, et al., 2003, Kwok 2005, Kwok et al., 2007). However, it is still questionable which satellite data gives better result comparing to others. Comiso et al, 2007 examined multi-sensor characterization of the Antarctic sea ice cover to provide general survey of satellite observations on sea ice cover. Comiso et al, 2007 and Comiso, 2004 mentioned the advent of passive microwave remote sensing since it is able to monitor the entire sea ice cover on a day to day, day/night almost all weather basis. They also pointed that scatterometer data provides complementary information to the passive systems. On the other hand NIC charts provide consistent integration of the various satellite sources includes different type of sensors such as passive microwave, active microwave, and scatterometer to detect the sea ice edge precisely.

We used satellite remote sensing data to detect changes in the ice to be examined over a large area. We compared data sets in terms of deriving sea ice extent to be able to look at the changes for a long time period. Our research included comparing microwave data derived from AMSR-E, ice edge charts generated by the NIC and sea ice extent derived from QuikSCAT to determine how the ice edges differ or agrees for each satellite observations for further geophysical retrievals and interpretations.

## 2 DATA

AMSR-E is an instrument launched on NASA's Aqua satellite that is able to study the Earth's atmospheric, oceanic, cryospheric, and land processes and their relationship to global change (NSIDC). The Aqua Satellite travels in a polar, sun-synchronous orbit, in which it wraps the earth pole to pole daily. This instrument is useful for our purposes because it helps to detect general ice concentration around Antarctica that we can use to establish seasonal trends. The AMSR-E instrument provides passive microwave data including horizontally and vertically polarized brightness temperature (Tb) from 6.9, 10.7, 18.7, 23.8, 36.5, and 89 gigahertz (GHz) frequencies. The spatial resolution of the data (pixel size) is 12.5 km by 12.5 km. The passive microwave has a clear advantage that it provides day/night and almost all weather observation of the sea ice cover. The data product [AMSR-E/Aqua Daily L3 12.5 km Tb, Sea Ice Conc., & Snow Depth Polar Grids] obtained from the NSIDC (Cavalieri and Comiso, 2004) is updated daily and is provided in digital HDF format.

The NIC data provides the daily ice edge products which extracted mostly manually, and visually by trained analysts. Satellite and ancillary data normally used consists of AVHRR (The Advanced Very High Resolution Radiometer), QuikSCAT, Envisat (Environmental Satellite), SSM-I (The Special Sensor Microwave-Imager), buoy data, and foreign ice charts. The use of two or more sensors from the same satellite has the distinct advantage of observing the same surface at the same time. This is especially important when co-registration of sea ice data from different satellites is sometimes difficult. NIC basically uses Quicksat, high resolution microwave radar data, which is used in conjunction with AMSR-E.

QuikSCAT is a polar orbiting satellite with an 1800 km wide measurement swath on the earth's surface. Ku-Band scatterometer data have been used to detect polar sea ice on QuikSCAT. Generally, this results in twice per day coverage over a given geographic region. This data set provides sea ice extent for the Arctic (60–90° N) and Antarctic (52–90° S) in Scatterometer Image Reconstruction (SIR) binary image format, along with ASCII text files containing latitude and longitude coordinates along the sea ice edge. QuikSCAT obtains 12 individual radar normalized backscatter (sigma-0) measurements, called 'slices,' for each footprint as it scans over a 1800 km wide swath. Slices are typically 4 to 6 km long by 20 km wide. The summed measurements of the slices are called 'egg' measurements. The effective resolution and shape of each egg measurement is approximately 20 by 30 km, depending on the antenna beam and instrument mode. This data set contains both slice and egg images for each day. Passive microwave and scatterometers can be used jointly to improve our interpretation of the data and enhance our understanding of the global sea ice cover.

## 3 METHOD

The objective of this study was to examine different satellite observations to obtain the sea ice extent precisely on Antarctic sea ice. 3 different data set were used to derive sea ice extent. First daily AMSR-E images between January 1$^{st}$ and March 1$^{st}$ for 2007 and 2008 were downloaded from http://nsidc.org/. Data then was imported directly to ENVI software and was opened as an external HDF file. The daily files were saved in ENVI standard format and georeferenced as polar stereographic coordinates. Using the ENVI software, Regions of Interest (ROIs) were defined using threshold criteria to obtain only sea ice concentration between 1% and 100%. All pixels were screened out other than the given range. The statistics for sea ice were computed to get pixel number that has ice on it for each day. These statistics were imported to an Excel spreadsheet for further analysis. Since the pixel size is 12.5 km, the total number of pixel is multiplied by the total area of pixel to calculate the sea ice extent (in $km^2$) for each day.

On the other hand, NIC data was directly processed in ArcGIS. National ice center provide the data as shape file. NIC sea ice edge data for Antarctica data between January 1$^{st}$ and March 1$^{st}$ for 2008 were downloaded from http://www.natice.noaa.gov/ and imported directly to ArcGIS software. Each day of NIC data was separately georeferenced as polar stereographic coordinates. And again each day was converted from shape file to coverage file to get the total area of ice extent. Then daily comparison was made between NIC and AMSR-E to see the agreement of detecting the sea ice extent.

The third dataset, QuikSCAT, was also imported directly to ArcGIS software. The data is obtained from http://www.scp.byu.edu/data/Quikscat/Ice/Quikscat_ice.html. The website data set contains both slice and egg images for each day. And the data has been provided as in ASCII format. ASCII files containing latitude/longitude pairs which represent the contour points of the estimated sea ice edge. Each line entry in the file consists of two values: a longitude and latitude. The longitude values range from –180 to +180. Unfortunately the latest day of the provided data is 6$^{th}$ of July, 2006. Because of the lack of the dataset, just only January 1$^{st}$ of 2006 and July 6$^{th}$ of 2006 were downloaded to validate NIC data. The two days of data were converted to dbf file first and added to ARCGIS. QuikSCAT data then was projected as polar stereographic coordinates. Finally all the points at the edge manually connected to generate polygon shapefile to compare with NIC and AMSRE data.

## 4 RESULTS AND CONCLUSION

It is apparent on Fig. 1 that the QuikSCAT backscatters and AMSR-E brightness temperatures provide consistent information both inside the ice pack and in the marginal ice zone for winter season. They also indicate very similar features along the Antarctic ice edge and divergence regions. Backscatters and brightness temperatures signals go in opposite directions usually. The brightness temperature is usually low when the backscatter is high, as with multiyear ice. And the brightness temperature is high when the backscatter is low, as with first year ice. On the other hand, the backscatter and brightness temperature can both be low in some areas inside the pack because of the open water regions. Also NIC data provides coherent information at the edge with the other dataset for winter time.

In January, the QuikSCAT map shows very enhanced backscatter in practically the entire Antarctic ice cover. Because of that high backscatter, the brightness temperature observed as relatively low by AMSR-E. On the other hand the ice concentration has agreement in some regions. Summer is the time when the sea ice breaks into individual ice floes and creates saturated sea ice region at the marginal zone. Because of the atmospheric and oceanic effects, they occasionally collide causing elevated edges for these floes. The surface is thus generally rough and hence the high backscatter as observed in the QuikSCAT data. The rough edges of the floes provide very high backscatter because of the effectively reflected radar signal back from the surface. Because the surfaces are generally covered by wet snow or slush during the summer, the brightness temperature as observed by AMSR-E is generally low (Fig. 2). Since the resolution is lower, it is unable to read the finer details regarding sea ice concentration. Also AMSR-E generally underestimates sea ice concentrations due to the thinning of the sea ice toward open water. However the NIC ice charts consist of a combination of microwave and radar data to compensate for the inaccuracy of AMSR-E. Overall NIC and QuikSCAT comparison shows better agreement than the agreement between QuikSCAT and AMSR-E.

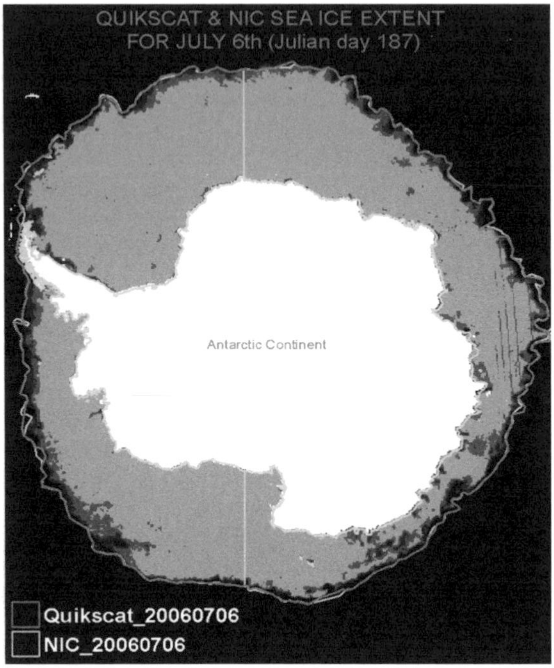


Figure 1. NIC, Quickscat and AMSR-E comparison for winter season.

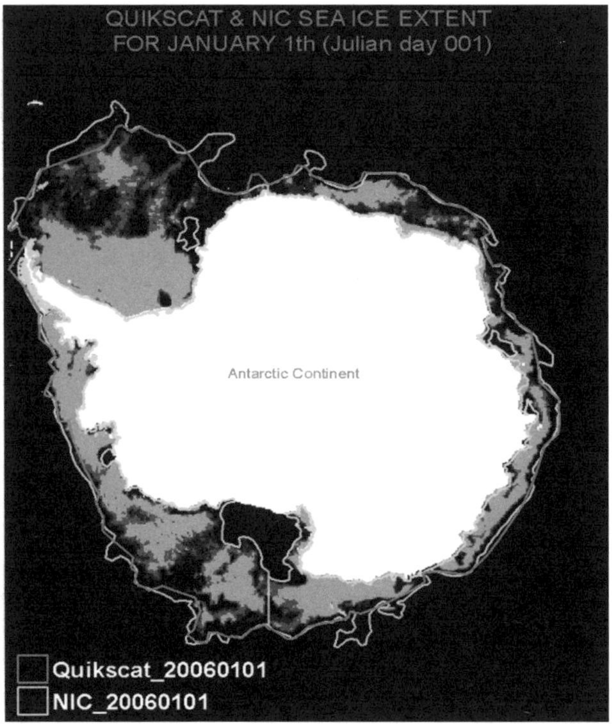


Figure 2. NIC, Quickscat and AMSR-E comparison for summer season.

Table 1. Areal summer sea ice extent calculation for two datasets, NIC and AMSR-E

| DATE | DATA | PIXEL | EXTENT (km^2) | DATA | PIXEL | EXTENT (km^2) | DATA | EXTENT (km^2) |
|---|---|---|---|---|---|---|---|---|
| 1-Jan | AMSR-08 | 61417 | 9596406.25 | AMSR-07 | 46605 | 7282031.25 | NIC-08 | 12265336.53 |
| 2-Jan | AMSR-08 | 60275 | 9417968.75 | AMSR-07 | 44823 | 7003593.75 | NIC-08 | 11936184.47 |
| 3-Jan | AMSR-08 | 59138 | 9240312.50 | AMSR-07 | 44035 | 6880468.75 | NIC-08 | 11533855.49 |
| 4-Jan | AMSR-08 | 58440 | 9131250.00 | AMSR-07 | 42988 | 6716875.00 | NIC-08 | 11182194.35 |
| 5-Jan | AMSR-08 | 56810 | 8876562.50 | AMSR-07 | 42350 | 6617187.50 | NIC-08 | 11165300.89 |
| 6-Jan | AMSR-08 | 55660 | 8696875.00 | AMSR-07 | 41098 | 6421562.50 | NIC-08 | 11268900.65 |
| 7-Jan | AMSR-08 | 54038 | 8443437.50 | AMSR-07 | 39817 | 6221406.25 | NIC-08 | 10654553.75 |
| 8-Jan | AMSR-08 | 52374 | 8183437.50 | AMSR-07 | 38525 | 6019531.25 | NIC-08 | 10450472.18 |
| 9-Jan | AMSR-08 | 51098 | 7984062.50 | AMSR-07 | 37316 | 5830625.00 | NIC-08 | 10250853.38 |
| 10-Jan | AMSR-08 | 49607 | 7751093.75 | AMSR-07 | 36607 | 5719843.75 | NIC-08 | 9904500.64 |

It is also very interesting that the passive microwave provides really critical information is distinguishable from the NIC data but not distinguishable from the QuikSCAT data that is the polynyal region. For example, the ice concentration map clearly shows the occurrence of a summer polynya which is a feature that is not so easy to infer from the QuikSCAT data. It is questionable that maybe the provided data set only includes the ice edge or QuikSCAT does not basically provide the polynya.

The other comparison was made between the sea ice extent from NIC and AMSR-E. It was shown that the NIC data and QuikSCAT data agrees very well during the winter time. And these datasets has better agreement than what the AMSR-E data shows for summer time. It is all because of that the NIC data provides sea ice edge derives from multi satellite imagery. So it is reasonable to consider the NIC as a precise dataset. The table below (Table 1) shows the areal sea ice extent calculation for NIC and AMSR-E data that there is about 2.5 million km$^2$ maximum differences between the datasets. NIC and AMSR-E sea ice extent differs from each other for each season. The maximum different was obtained between January 1$^{st}$ and March 1$^{st}$ 2008 as 2825852.45 km$^2$. The minimum difference was obtained as 761890.02 km$^2$ for the same period. The maximum different was obtained between August 1$^{st}$ and December 27$^{th}$ 2007 as 2424525.62 km$^2$. The minimum difference was obtained as 324133.53 km$^2$ for the same period.

The Fig. 3 shows that within one year (2006–2007) summer sea ice extent increased by over 1 million km$^2$ for the entire continent. Figure 4 indicates that the winter sea ice extent for the entire continent.

## 5 DISCUSSION

We were looking at sea ice conditions in Antarctica. It had been found is that the NIC ice charts give a much better representation of ice extent than the passive microwave. Also this study showed that the QuikScat has agreed better with NIC charts and also disagreed with AMSR-E for summer season. The major reason seems to be that passive microwave influenced by particularly in low concentration conditions at the ice edge, while the NIC ice charts rely on active radar more these days which is probably similar to the scatterometer, so are less troubled in defining the ice edge. The under estimates of AMSR-E mostly coincide with the regions of the sea ice edge where usually ice conditions are highly variable. Even though AMSR-E provides continuous record of sea ice extent, NIC charts clearly shows seasonal effects on the estimation of the sea ice edge.

Results also showed that the summer minimum (area bounded by the ice extent) this year in Antarctic sea ice exceeds last summer's Arctic summer minimum. The two graphs in the AMSR-E file plot the ice extent area for 2007-08 National Ice Center Charts, and AMSR-E (total pixels with ice times pixel area) for both 2007-08 and 2006-07 for comparison. Particularly in summer, the

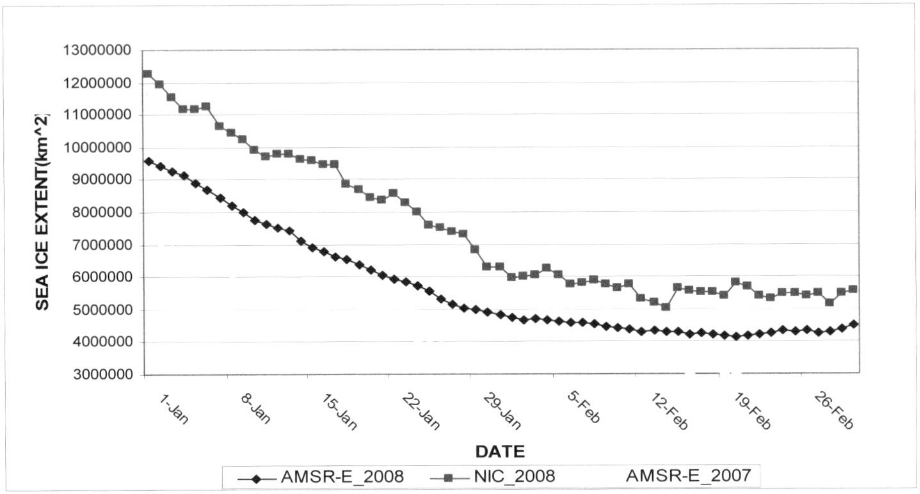


Figure 3. NIC and AMSR-E comparison for summer season.

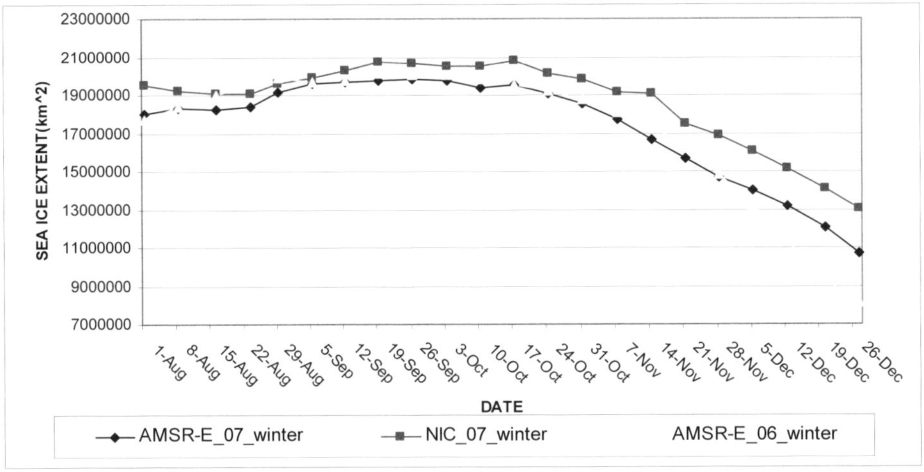


Figure 4. NIC and AMSR-E comparison for winter season.

NIC area exceeds the AMSR-E area, but the other figure with QuikSCAT shows the NIC and Scatterometer (QuikSCAT) generally agree, with both exceeding the AMSR-E. Based on the NIC, the ice extent area bottoms out at 5 million sq km (12 Feb). It must be also noted that NIC areas include the large polynyas in the Ross and Amundsen Seas, similar to the AMSR-E area.

The 2006 to 2007 comparison is particularly interesting as the winter 06 sea ice extent exceeds the 07 value until late November when the decline crosses over (see Fig. 4) so that the summer minimum 07 ends up exceeding the summer 06 value. The idea is that the winds control the ice edge variations and that higher (lower) winds give a higher (lower) winter maximum followed by a lower (higher) summer minimum. The main concept is that when the summer minimum has a high dependence on transport, this will show for the Antarctic, the future of the Arctic summer ice in a different wind regime under climate change than present may be subjected to similar variability, making its disappearance perhaps less predictable than a linear extrapolation of the current trend.

## REFERENCES

Cavalieri, D., C. Parkinson, P. Gloerson, and H.J. Zwally. 1997, updated 2005. Sea ice concentrations from Nimbus-7 SMMR and DMSP SSM/I passive microwave data, June to September 2001. Boulder, CO: National Snow and Ice Data Center. Digital media and CD-ROM.

Cavalieri, D.J., Parkinson, C.L. and Vinnikov, K.Y. 2003, 30-Year satellite record reveals contrasting Arctic and Antarctic decadal sea ice variability. Geophysical Research Letters 30: 10.1029/2003GL018031.

Cavalieri, D. and J. Comiso, 2004, updated daily. AMSR-E/Aqua Daily L3 12.5 km Tb, Sea Ice Conc., & Snow Depth Polar Grids V001, March to June 2004, Boulder CO, USA: National Snow and Ice Data Center. Digital media.

Comiso, J.C., Sea ice algorithm for AMSR-E, Rivista Italiana di Telerilevamento (Italian 718 Journal of Remote Sensing), 30/31, 119-130, 2004.

Comiso, J.C., Kwok, R., Steffen K., and Lobl Elena, 2007, Multi-sensor Characterization of the Antarctic Sea Ice Cover, Submitted to JGR Oceans on 13 April 2007 as part of the Special Section on "Large Scale Characteristics of the Sea Ice Cover".

Kwok, R., Ross Sea ice motion, area flux, and deformation, J. Climate, 18, 3759-3776, 2005.

Kwok, R., J.C. Comiso, S. Martin, and R. Drucker, Ross Sea Polynyas: Response of ice 744 concentration retrievals to large areas of thin ice, J. Geophys. Res., 2007.

nsidc.org/

natice.noaa.gov/

Parkinson, C.L., 2002, Trends in the length of the Southern Ocean sea-ice season, 1979-99, Annals Glaciology, Volume 34, Number 1,1, pp. 435-440(6).

scp.byu.edu/data/Quikscat/Ice/Quikscat_ice.html

The Intergovernmental Panel on Climate Change (IPCC), 2007 reports.

Zwally, H.J., Comiso, J.C., Parkinson, C.L., Cavalieri, D.J., 2002, Variability of Antarctic Sea Ice 1979–1998, Journal of Geophysical research, Vol. 107, No. C5, 10.1029/200JC000733.

*Remote Sensing for a Changing Europe*
*D. Maktav (Ed.)*
*IOS Press, 2009*

*doi:10.3233/978-1-58603-986-8-448*

# Estimation of daily actual evapotranspiration in Algerian semiarid environment with satellite ASTER

A. Hamimed, A. Khaldi & K. Mederbal
*Laboratoire de recherches sur les systèmes biologiques et la géomatique, University of Mascara, Algeria*

A. Seddini
*Dept of Hydraulics, Faculty of Engineering Sciences, University of Tlemcen, Algeria*

Keywords: remote sensing, ASTER, daily evapotranspiration, SEBAL, energy balance, evaporative fraction

ABSTRACT: The main aim of this study is the use of the satellite ASTER data for assessing the daily actual evapotranspiration by solving the energy balance equation at the surface. The energy balance model used is SEBAL (surface energy balance algorithm for land). It is a one-source model that calculates the net radiation, soil heat flux and sensible heat flux at each remote sensing pixel and estimates the latent heat flux as the residual term in the energy balance. It involves in input the NDVI, surface temperature and albedo. The study area encompassed by ASTER image contains extremes in surface albedo, vegetation cover and surface temperature values. It corresponds to Ksar Chellala plains in Algeria.
To estimate the sensible heat flux, which represents the crucial variable in solving the energy balance equation, we opted for a method based on the use of profiles in the surface layer, with the coupling taking place in fluxes at the convective boundary layer base. We showed that a simplification of this method by ignoring the atmosphere thermal stability functions can give the same results approximately. It is then possible to estimate the sensible heat flux and calculate the latent heat flux as the residual term of the energy balance. We can then asses the actual daily evapotranspiration and some moisture indices: evaporative fraction, Priestley and Taylor parameter and surface resistance to evaporation. These indices allow quantitative monitoring in pixel basis of water status degree.

## 1 INTRODUCTION

Monitoring the transfers of mass and energy at a surface is crucial for hydrological and vegetation resources management. It is also necessary for a better comprehension of hydrological and climatic systems as well as their predictions.

Remote sensing is an excellent tool for this monitoring as it provides information related to mass and energy transfers and particularly to evapotranspiration fluxes.

Evapotranspiration is one of the fundamental processes controlling the equilibrium of our planet. It constitutes the link between the hydrological and energetic equilibrium at the soil-vegetation-atmosphere interface and its knowledge is crucial for climatic and agro-meteorological studies.

Depending on the geographical location on the earth's surface, evapotranspiration represents between 60 to 80 % of the precipitation return to the atmosphere (Balaji *et al.* 2000). Consequently, it constitutes one of major phenomena in the hydrological budget, especially in arid and semi-arid regions.

Furthermore, estimation of actual evapotranspiration using satellite data in the visible and infrared has been at the centre of several methodological approaches during the last years (Courault *et al.* 2005).

We can mention three such methods:

- Those using simple empirical relationships relating daily evapotranspiration to an instantaneous surface temperature measurement (Seguin & Itier 1983; Lagouarde 1991; Vidal *et al.* 1987). These methods are inconvenient as they can only give a daily value of actual evapotranspiration, and are essentially statistical in nature.
- Those using atmospheric boundary layer methods do not present the above limitation (Taconet *et al.* 1986). They are however more cumbersome and use many important, difficult to measure, parameters which make them unsuitable to spatial integration in models that are very sensitive to such parameters (Jacob 1999).
- Those avoiding the limitations of the previous two, while still preserving their advantages. They take into account physical laws where possible and empirical relationship otherwise. SEBAL (*Surface Energy Balance Algorithm for Land*) is considered as a third category. It has been developed to solve the energy budget equation using a spatial approach assuming the existence of sites in extreme hydrological conditions (very dry and very wet) (Bastiaanssen *et al.* 1998). Properties from these sites are used to determine some variables at the soil-vegetation-atmosphere interface not possible to obtain through remote sensing.

In this study we propose a methodology based on SEBAL to solve the energy balance and map the actual daily evapotranspiration from the Advanced Space born Thermal Emission and Reflection Radiometer (ASTER) aboard the NASA's Terra Spacecraft.

## 2 THEORETICAL BACKGROUND

Currently, the energy balance equation at the surface is expressed as:

$$Rn = G + H + \lambda E \tag{1}$$

expresses the partition of the net radiation Rn between soil heat flux G, sensible heat flux H and latent heat flux (corresponding to evaporation for a bare soil and transpiration for vegetation canopy).

The net radiation is found from the various components of radiations exchanges:

$$Rn = (1 - r_o).Rg + L\downarrow - L\uparrow \tag{2}$$

where Rg is the incoming global radiation, partly reflected depending on the albedo $r_o$, $L\downarrow$ and $L\uparrow$ are the downwelling and the up-welling long wave radiation, respectively.

The soil heat flux G is usually low in comparison with the other terms. Also, we tend to neglect it, or to give it a fixed proportion of the net radiation (0.1 for example). The sensible heat flux H is expressed as a function of the near-surface air temperature difference ($T_o$-$T_a$) as:

$$H = \frac{\rho.Cp}{r_{ah}}(T_o - T_a) \tag{3}$$

This expression involves (outside constants ρ and Cp, air density and specific heat, respectively) the aerodynamic résistance to heat transport $r_{ah}$ which is a function of wind velocity, thermal stability effects of the atmosphere and surface roughness.

To estimate the aerodynamic resistance to heat transport, some theoretical approaches have been used particularly by Paulson (1970). They are essentially based on the use of logarithmic profiles of mass and energy transfer in the surface boundary layer and the coupling surface/boundary layer which occurs at the fluxes in the convective boundary layer. The integration of speed profiles leads to two similarities functions $\psi_m$ and $\psi_h$ parameterized by the turbulent regime of momentum and heat.

$$u^* = K.u/\left[\ln(za/z_{om}) - \psi_m(za/L)\right] \tag{4}$$

$$r_{ah} = \frac{1}{K u^*}\left[(\ln za/z_{oh}) - \psi_h(za/L)\right] \tag{5}$$

where u is wind speed at the reference height (usually equal to 2 m), $\psi_m$ and $\psi_h$ are respectively stability correction for momentum transport and stability correction for heat transport and L is the Monin-Obukhov's length, expressed as:

$$L = -\frac{\rho.Cp.T_a.u^{*3}}{K.g.H} \tag{6}$$

where u* is the friction velocity, K is the von Karman constant (≈ 0,41), g is the acceleration due to gravity, $z_{om}$ is the roughness length for momentum transport and $z_{oh}$ is the roughness length for heat transport.

$\lambda E$ is the latent heat flux, expressed as:

$$LE = \frac{\rho.Cp}{\gamma}\frac{e_{sat}(T_o) - e_a}{r_{ah} + r_s} \tag{7}$$

where $\gamma$ is the psychrometric constant, $r_s$ is the surface resistance to evaporation, $e_a$ is the water vapour pressure at reference height (za) and esat($T_o$) is the saturated vapour pressure at surface temperature.

## 3 STUDY FRAMEWORK

### 3.1 *Study area and climate*

The study area corresponds to the Ksar Chellala plain. It is located in the east of Tiaret (in the west of Algeria) between latitudes 35°12' and 35°20' N, and longitudes 2°13' and 2°21' E. It covers an area of 126 km2 (Fig. 1).

The selected area is marked by a steppe plateau at the piedmont of Ben Hamad Mountain which constitutes a barrier to the sirocco winds very frequent in summer. The average altitude is about 850 m. The terrains levelling are of miocene formation, consisted of sandy clay, poudings and conglomerates. We also note the presence of some alluvial deposits which cover bed watercourses.

The study area is reported on semi-arid climatic floor enjoying a continental climate. Two principal periods characterize this area: one rainy and sinks during months of autumn, winter and early spring and another dry and torrid in summer. The winter is rather cold. The minimum absolute temperature goes down to -6 °C. The summer is usually dry and hot. The maximum absolute temperature is equal to +47 °C . Annual average precipitations are around 350 mm.

Because of the intense evapotranspiration and wind, only a small portion of annual precipitations, which takes place except estival season, is useful for cultivable soils.

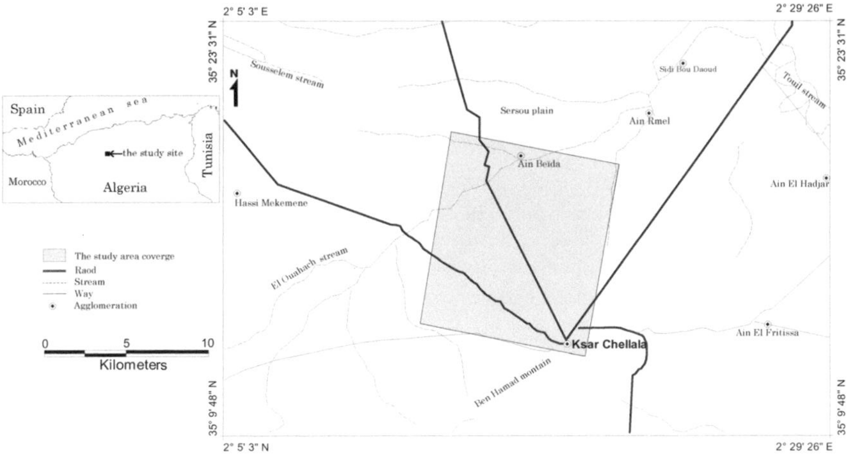


Figure 1. Location of the study site

## 3.2 *Data*

The data set used in this work consists of 14 bands atmospherically corrected of ASTER satellite, acquired on October 25, 2000 at 10 H 48' GMT, and some ground data acquired at the moment of satellite overpass from the Ksar Chellala weather station localised in the selected study area.

# 4 ESTIMATING OF SURFACE ENERGY FLUXES

In order to obtain the actual evapotranspiration, SEBAL is based on the estimation of latent heat flux as the residual term of the energy balance equation:

$$\lambda E = Rn - G - H \tag{8}$$

Indeed, the ASTER sensor measures exoatmospheric spectral radiance in the visible, near infrared, middle infrared and thermal infrared. These radiances can be converted into albedo ($r_o$), vegetation index (NDVI) and surface temperature ($T_o$).

The albedo, defined as the hemispherical surface reflectance in short wavelengths rang (0.3-3.0 μm), is obtained from weighted surface reflectances in the first nine ASTER channels.

The normalized difference vegetation index is calculated from reflectances in the red ($r_R$) and near infrared channels ($r_{NIR}$) as:

$$NDVI = \frac{r_{NIR} - r_R}{r_{NIR} + r_R} \tag{9}$$

The surface temperature ($T_o$) is obtained by conversion according to Planck law of the thermal infrared radiance in brightness temperature which is then corrected for emissivity effects using TES (Temperature/Emissivity Separation) model (Gillepsie *et al.* 1998).

The estimation of net radiation requires evaluation of:

- incoming global radiation Rg, obtained from meteorological observations; this data allows us to compute the atmospheric transmittance ($\tau_{sw}$) which is in our case equal to 0.72;

- the upwelling longwave radiation L↑ from Stephan-Boltzman law using surface temperature;

- the downwelling longwave radiation L↓ using air temperature and atmosphere emissivity. This latter is estimated following to (Brutsaert 1975):

$$\varepsilon' = 1.24 \times \left( \frac{e_a}{T_a} \right)^{1/7} \tag{10}$$

The soil heat flux G is empirically estimated using the expression suggested by Bastiaanssen *et al.* (1998):

$$\frac{G}{Rn} = \frac{T_o}{r_o} \left[ (0.0032 \times r_o{}^{moy} + 0.0062 \times (r_o{}^{moy})^2 \right] \left[ 1 - 0{,}978 \times (NDVI)^4 \right] \tag{11}$$

involving albedo, vegetation index, surface temperature and net radiation.

It is then possible to estimate the sensible heat flux according to the near-surface air temperature difference and the aerodynamic résistance to heat transport ($r_{ah}$). This later takes into account the atmosphere thermal stability effects, which represents a delicate point. A current approximation for estimating this resistance is to overlook the stability correction for momentum and heat transport functions ($\psi_m$ and $\psi_h$) introduced into equations 4 and 5 (Allen *et al.* 1998; Balaji & Raghavan 2000).

The original approach suggested in SEBAL (Bastiaanssen *et al.* 1998) for estimating $r_{ah}$ is based on the use of averages profiles in the atmospheric boundary layer in which vertical fluxes (momentum flux, friction velocity, wind speed) can be considered constant even over heterogeneous terrains. It is therefore possible to assume that, for dry pixels, the latent heat flux is zero and the energy balance equation is written then:

$$H^{dry} = Rn - G \tag{12}$$

The identification of dry pixels is done by analyzing relation between albedo and temperature surface (Hamimed *et al.* 2001). On these pixels, aerodynamic resistance to heat transport ($r_{ah}{}^{dry}$) is determined according to:

$$\frac{\partial H}{\partial T_o} = \frac{\rho . C_p}{r_{ah}^{dry}} \tag{13}$$

$\partial H / \partial T_o$ can be calculated as:

$$\frac{\partial H}{\partial T_o} = -Rg \frac{\partial r_o}{\partial T_o} + \frac{\partial L^*}{\partial T_o} - \frac{\partial G}{\partial T_o} \tag{14}$$

corresponding to a combination of equations 1 and 2 followed by differentiation with respect to surface temperature $T_o$. L* is the net longwave radiation (L* = L↓ - L↑ ).

The determination of $\partial H/\partial T_o$ for dry pixels requires quantification of various terms on the right-hand side of equation 14 (Tab. 1)

Table 1. Partials derivatives of albedo and energy fluxes with respect to surface temperature

| Paramètre | Unité | Valeur |
|---|---|---|
| $\partial r_o/\partial T_o$ | $K^{-1}$ | -0.0495 |
| $\partial G_o/\partial T_o$ | $W.m^{-2}.K^{-1}$ | 0.88 |
| $\partial L^*/\partial T_o$ | $W.m^{-2}.K^{-1}$ | -6.09 |
| $\partial H/\partial T_o$ | $W.m^{-2}.K^{-1}$ | 26.06 |
| $\partial LE/\partial T_o$ | $W.m^{-2}.K^{-1}$ | 0.0 |

Equation 13 enables us to assess the effective aerodynamic resistance to heat transport for dry pixels at blending height (generally set to 100 m) (Bastiaanssen *et al.* 2000). Using $r_{ah}^{dry}$ into an iterative process allows extracting the effective friction velocity $u^*_{100}$ at this height (Tab. 2). We can then obtain the wind speed (u) at the reference height using the logarithmic profiles. The integration of aerodynamic momentum roughness $z_{om}$ into a second iterative process enables us mapping the friction velocity u* at the reference height. Finally, the use of Monin-Oubukhov's similarity hypothesis allows us mapping aerodynamic resistance to heat transport following expression 5.

The near-surface air temperature difference $(T_o-T_a)$ is computed using a linear relation between air and surface temperatures. The slope and offset are estimated inverting the sensible heat flux expression over wet areas ($T_o = T_a$) and dry areas (H = Rn - G).

Table 2. Momentum flux parameters at the surface boundary layer height (100 m)

| Parameter | Unit | Value |
|---|---|---|
| u* | $m.s^{-1}$ | 0.54 |
| L | m | -56.796 |
| $\psi_h$ | - | 2.616 |
| $\psi_m$ | - | 0.069 |
| $r_{ah}$ | $s.m^{-1}$ | 42.55 |
| $T_p$ (potential temperature) | K | 281.84 |

The use of equation 3 enables us computing the sensible heat flux in pixel basis and solving the energy balance equation. This step leads mapping the latent heat flux. This should help interpreting more accurately the behaviour of a surface with respect to water stress. However, its values strongly depend on conditions of its estimate. It is therefore preferable for an easier interpretation to deduce indicators such as evaporative fraction, Priestley and Taylor parameter and the surface resistance to evaporation.

## 5 MAPPING OF DAILY ACTUAL EVAPOTRANSPIRATION

In agriculture, the knowledge of daily energy fluxes particularly the actual evapotranspiration is very significant compared to instantaneous fluxes. It allows appreciating the crop moisture status and the definition of irrigation modes.

Among the existing approaches on the estimate of actual evapotranspiration from remotely sensed data, we can retain that based on evaporative fraction assumption. This one is estimated from instantaneous measurements according to:

$$\Lambda = \frac{LE}{Rn - G} \tag{15}$$

which is considered similar to its daily counterpart (Tasumi & Allen 2000; Kustas *et al.* 1994), and is used to compute the daily actual evapotranspiration $ETR_d$ from instantaneous latent heat fluxes.

Ignoring the soil heat flux on a day basis, the daily actual evapotranspiration can be computed from daily net radiation and instantaneous evaporative fraction:

$$ETR_d = \Lambda \, Rn_d \tag{16}$$

The daily net radiation can be computed according to slob (de Bruin 1987) where the surface albedo is assumed to be similar to the surface albedo during the morning overpass, and the daily global radiation and atmospheric transmittance are taken from the field measurements:

$$Rn_d = (1 - r_o)\, Rg_d - 110\, \tau_{swd} \tag{17}$$

## 6 RESULTS AND DISCUSSION

Through the SEBAL energy balance equation modeling, we showed that the surface parameters obtained from the satellite imagery namely the albedo, vegetation index and surface temperature lead determining the latent heat flux as the residual term of this equation.

The surface temperature is indirectly linked to the latent heat flux through the energy balance equation. It varies on the image between 295.4 and 308.8 K. The higher values correspond to areas where bare soil are dominant, while low values are associated with areas where vegetation cover is dense. Similarly, the dry pixels temperature is more than that of wet pixels (Tab. 3).

The evaluation of rah without correction for atmosphere thermal stability can have values of latent heat flux roughly similar to those obtained by introducing the functions of stability correction for momentum and heat transport (Fig. 2). It is therefore possible to simplify the approach proposed in SEBAL for calculating rah by ignoring contribution of these two functions, but this hypothesis needs to be verified on other sites.

Note also on the table 3 that rah for dry pixels is low, causing the release of sensible heat to the atmosphere. This is justified by high differences between surface and air temperatures. However, wet pixels present high values of rah because the net available energy (Rn - G) is mainly used in evapotranspiration.

As shown in table 4, estimating the daily actual evapotranspiration fairly accurately represents the logic of evapotranspiration spatial variations without having to introduce a large number of input parameters easily accessible to routine.

On figure 3, daily actual evapotranspiration varies from 0 to 5.13 mm with a clear dominance surfaces with a water deficit more or less strong. By against, a good water status is represented only a small area. The dominance of water deficit is clearly observed on frequency distributions of daily evapotranspiration and evaporative fraction (Fig. 4 a & b).

Table 3. Instantaneous and surface fluxes above dry and wet pixels

| Parameter | Unit | Dry pixels | Wet pixels |
|---|---|---|---|
| NDVI | - | 0.11 | 0.48 |
| Albedo | - | 0.35 | 0.10 |
| Emissivity | - | 0.95 | 0.98 |
| Surface temperature | K | 305.6 | 296.4 |
| Net radiation | $W.m^{-2}$ | 281 | 477 |
| Soil heat flux | $W.m^{-2}$ | 49 | 42 |
| Friction velocity | $m.s^{-1}$ | 0.44 | 0.64 |
| Monin-Oubukhov lenght | m | -62.37 | -362.78 |
| Surface resistance to evaporation | $s.m^{-1}$ | 12.33 | 42.35 |
| Sensible heat flux | $W.m^{-2}$ | 232.31 | 3.4 |
| Latent heat flux | $W.m^{-2}$ | 0 | 431 |
| Evaporative Fraction | - | 0 | 0.98 |
| Near-surface air temperature difference | ° C | 7.25 | 0.48 |

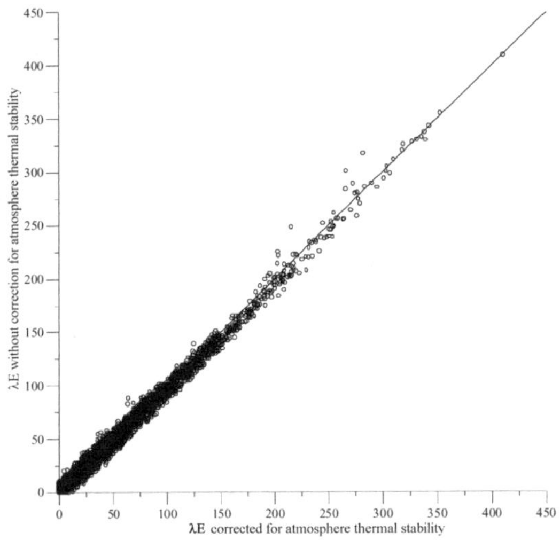


Figure 2. Comparison between corrected and uncorrected latent heat flux for atmosphere thermal stability

Table 4. Variation of daily actual evapotranspiration and moisture indicators with land use

| Land use units | $ETR_d$ (mm) | Λ (-) | $r_s$ ($s.m^{-1}$) | α (-) |
|---|---|---|---|---|
| Bare soil, fallow and rangelands | 0.54 | 0.11 | 323.7 | 0.17 |
| Sparse vegetation cover | 0.86 | 0.17 | 285.4 | 0.27 |
| Moderate vegetation cover | 1.62 | 0.31 | 217.4 | 0.49 |
| Dense vegetation cover | 2.17 | 0.45 | 148.1 | 0.71 |

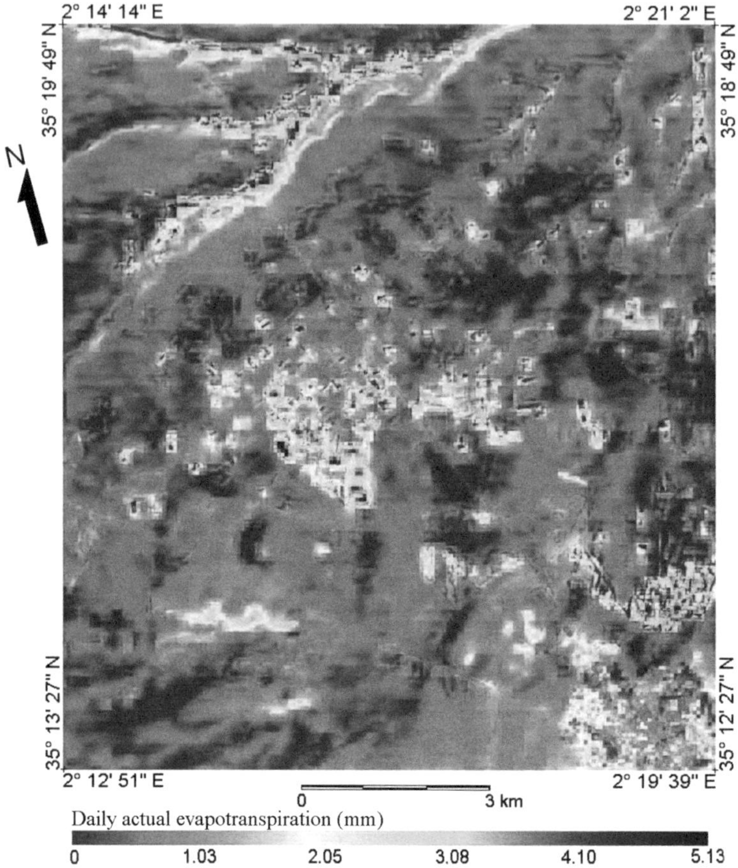


Figure 3. Daily actual evapotranspiration map

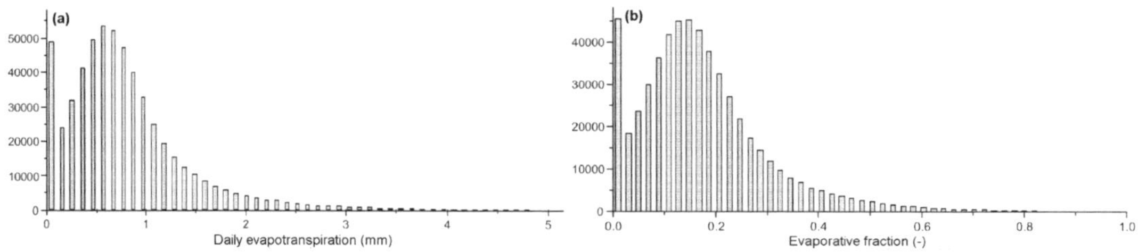


Figure 4. Frequency distribution of daily evapotrnspiration (a) and fraction evaporative (b)

The confrontation of actual daily evapotranspiration values obtained with SEBAL and those with Priestley and Taylor method (Jiang & Islam 2001) shows a good agreement (Fig. 5), which justifies the validity of the used approach.

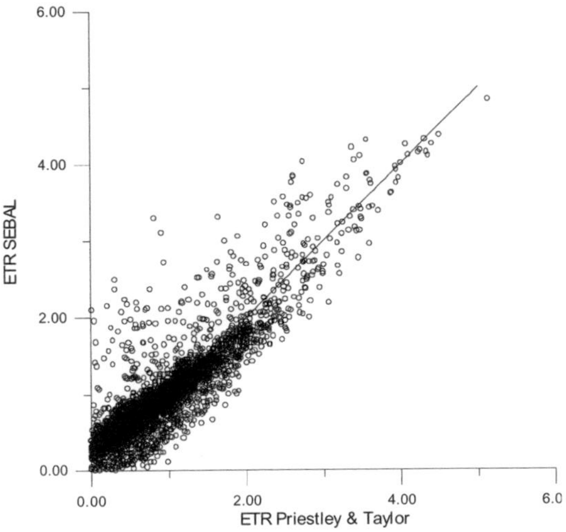


Figure 5. Comparison between SEBAL and Priestley & Taylor evapotranspirations

Indeed, actual evapotranspiration values so estimated are to be treated with caution and have only a relative value, because there is no a validation method as wide at time and space scales.

One way may be seen by considering two climatic data: the instantaneous wind speed and daily pan evaporation. The wind speed measured at the moment of satellite overpass (8 $m.s^{-1}$), does not differ much from the average wind speed simulated on selected site with logarithmic profiles (7.1 $m.s^{-1}$). The used approach gave a maximum value of daily evapotranspiration 5.13 mm (corresponding to the surface resistance to evaporation value of 23.2 $s.m^{-1}$), which is slightly less than 6.04 mm of open water evaporation value, estimated in the same conditions, that corresponds to a surface resistance to evaporation almost zero.

An analysis of the surface resistance to evaporation frequency distribution is considered another way to validate the results. They showed that for most crops covering entirely soil, this resistance values varies between 10 and 400 sm-1 with generally peaking around 80 $s.m^{-1}$. The result shown in figure 6 is roughly consistent with this indication.

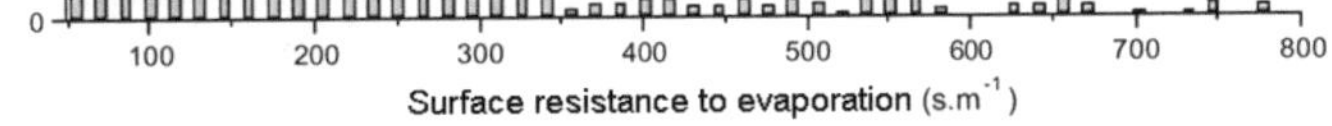


Figure 6. Frequency distribution of the surface resistance to evaporation for pixels with NDVI more than 0.6

## 7 CONCLUSION

The results outlined above confirm the opportunities presented by ASTER satellite data to solve the equation of energy balance, assess the water stress degree, clearly differentiate the parcels subject to different water systems.

The model used is quite suitable for a real exploitation of satellite data to estimate a number of parameters at the soil-plant-atmosphere interface. These parameters have the advantage of being spatialized and offer a spatio-temporal coverage more satisfactory than the data points measured operationally.

They pose a problem against it to be indirect and need using radiative transfer models within the atmosphere and at the surface for their interpretation in terms of physical parameters.

## ACKNOWLEDGEMENTS

This study is supported by the research contract CNEPRU: D2901/51/05.

## REFERENCES

Allen R. G., Periera L. S., Raes D., and Smith M., 1998. Crop evapotranspiration: guidelines for computing crop water requirements. *FAO Irrigation and drainage 56*, FAO - Food and Agriculture Organization of the United Nations, Rome.

Balaji N., Raghavan S., 2000. Determination of regional scale evapotranspiration of Texas from NOAA-AVHRR satellite. Report, *Texas Water Resources Institute*, Texas, 15 p.

Bastiaanssen W. G. M., 2000. SEBAL-based sensible and latent heat fluxes in the irrigated Gediz basin, Turkey. *Journal of Hydrology* 229 : 87-100.

Bastiaanssen W. G. M., Menenti M., Feddes R. A., Holtslag A. A. M., 1998. Remote sensing surface energy balance algorithm for land (SEBAL) : 1. Formulation. *Journal of Hydrology* 213 : 198-212.

Bastiaanssen W. G. M., Van Der Wall T., Visser T. N. M., 1996. Diagnosis of regional evaporation by remote sensing to support irrigation performance assessment. *Irrigation and Drainage Systems* 10 : 1-23.

Bougeault P., Noilhan J., Lacarere P., Mascart P., 1991. An experiment with an advanced surface parameterization in a meso-beta-scale model. Part I: Implementation. *Monthly Weather Review* 119: 2358-2373.

Brutsaert, W., 1975. On a derivable formula for longwave radiation from clear skies. *Water Resources Research* 11: 742-744.

Courault D., Seguin B., Olioso A., 2005. Review on estimation of evapotranspiration from remote sensing data: From empirical to numerical modelling approaches. *Irrigation and Drainage Systems*, 19: 223-249.

De Bruin H, A, R., Holtslag A. A. M., 1982. A simple parameterization of the surface fluxes of sensible and latent heat during daytime compared with the Penman–Monteith concept. *Journal of Applied Meteorology* 21: 1610-1621.

Gillepsie A., Rokugawa S., Matsunaga T., Cothern J. S., Hook S., Kahle A. B., 1998. A temperature and emissivity separation algorithm for advanced spaceborne themal emission and reflection radimeter (ASTER) Images. *IEEE Transactions on Geoscience and Remote Sensing* 36 : 1113-1126.

Hamimed A., Mederbal K., Khaldi A., 2001. Utilisation des données satellitaires TM de Landsat pour le suivi de l'état hydrique d'un couvet végétal dans les conditions semi-arides en Algérie. *Télédétection* 2 : 29-38.

Horiguchi I., Tani H., Motoki T., 1992. Accurate estimation of 1.5 m-height air temperature by GMS IR data. *24th International Symposium of Remote Sensing of environment* 6 :1223-1242.

Jacob F., 1999. Utilisation de la télédétection courtes longueurs d'onde et infrarouge thermique à haute résolution spatiale pour l'estimation des flux d'énergie à l'échelle de la parcelle agricole. Thèse de Doctorat, Université de Toulouse III, 1999, 268 p.

Jiang L., Islam S., 2001. Estimation of surface evaporation map over southern Great Plains using remote sensing data. *Water Ressource Research* 37(2) : 329-340.

Kustas W. P., Perry E. M., Doraiswamy P. C., Moran M. S., 1994. Using satellite remote sensing to extrapolate evapotranspiration estimates in time and space over a semiarid rangeland basin. *Remote Sensing of environment* 49 : 275-286.

Lagouarde J. P., 1991. Use of NOAA-AVHRR data combined with an agrometeorological model for evaporation mapping. *International Journal of Remote Sensing* 12 : 1853-1864.

Paulson C. A., 1970. The mathematical representation of wind speed and temperature profils in the unstable atmospferique surface layer. *Journal of Applied Meteorology* 9 : 857-861.

Seguin B., Itier B., 1983. Using midday surface temperature to estimate daily evaporation from satellite IR data. *International Journal of Remote Sensing* 4 : 371-383.

Taconet O., Bernard R., Vidal-Madjar D., 1986. Evpotranspiration over agricultural region using surface flux/temperature model based on NOAA-AVHRR data. *Journal of Climate and Applied Meteorology* 25 : 284-307.

Tasumi M., Allen R. G., 2000. Application of the SEBAL methodology for estimating consumptive use of water and stream flow depletion in the Bear River Basin of Idaho through Remote Sensing. Final Report, *The Raytheon Systems 10 Company EOSDIS Project*, 107 p.

Vidal A., Kerr Y., Lagouarde J. P., Seguin B., Télédétection et bilan hydrique : utilisation combinée d'un modèle agrométéorologique et des données thermique du satellite NOAA-AVHRR. *Agricultural and Forest Meteorology* 39 : 155-175.

*Remote Sensing for a Changing Europe*
*D. Maktav (Ed.)*
*IOS Press, 2009*

*doi:10.3233/978-1-58603-986-8-460*

# Mapping burned areas and assessing short-term fire effects with the use of object-based analysis and high resolution satellite imagery

T. Katagis*, A. Polychronaki & I. Z. Gitas
*Laboratory of Forest Management and Remote Sensing, Aristotle University of Thessaloniki, P.O. Box 248, University Campus, Greece. E-mail*:thkatag@for.auth.gr*

Keywords: forest fires; object-based classification; post-fire regrowth

ABSTRACT: The mapping of burned areas and the estimation of post-fire forest regeneration is very important in order to examine the vulnerability of Mediterranean forest to the increased occurrence and destructiveness of fires induced by climatic warming and the changes in the traditional land use patterns in Southern Europe. Satellite remote sensing has been widely applied in forest fire-related research and has provided satisfactory results for burned area mapping, as well as valuable information concerning the post-fire condition of affected areas. Moreover, advanced classification techniques, such as object-based methods have been successfully used for the accurate mapping of burned areas and the estimation of the vegetation recovery in fire-altered landscapes.

The aim of this study was to investigate the use of remote sensing in mapping a burned area in Kassandra peninsula, Greece and to assess the short term fire effects on the landscape. The specific objectives were: a) to develop an object-based classification model for mapping the burned area by employing a LANDSAT image and to evaluate the classification accuracy, and b) to develop an object-based classification model to monitor vegetation regrowth one year after fire by employing a SPOT image.

The use of object-based classification resulted in very accurate mapping of the burned area when compared to field and aerial photography data. Moreover, the results of the analysis showed that about 58% of the burned area was covered by vegetation one year after the fire event.

## 1 INTRODUCTION

Natural fires are common in many parts of the world and they are an integral part of many terrestrial ecosystems, including the Mediterranean ones. Fire is a natural factor in such climates and consists an important ecological process that is affecting the natural cycle of vegetation succession and the ecosystem's structure and function (Koutsias and Karteris 2003). During the recent decades, the number of wildfires has increased in the countries of the Mediterranean region and this increase can be attributed to (Pausas and Vallejo 1999): a) land use changes, and b) climatic warming leading to the reduction of fuel humidity and increase in fire risk.

In general, impact of fires on forest ecosystems can be divided to short term and long term. According to Karteris (1996), short term impact assessment involves the quantification of the severity of burn, derivation of the affected land cover types and the approximation of the wood volume lost, while the long term impact estimation includes the evaluation of vegetation recovery, forest natural regrowth, land degradation, hydrological response, atmospheric pollution and ecological impact in terms of modifying or affecting plant and animal species.

Therefore, detailed and current information concerning the location and extent of the burned areas and post-fire vegetation situation is important in order to assess economic and ecological effects (Gitas 1999). Accurate assessments of burned location, fire severity and post-fire vegetation regeneration, provide the necessary information to evaluate the effectiveness of measures taken to rehabilitate the fire damaged area, and allow forest managers to identify and target areas for intensive or special restoration. (Jakubauskas et al. 1990).

Satellite remote sensing has been widely applied in forest fire-related research and has supplied an alternative to conventional techniques that is especially suitable for monitoring burned areas (Chuvieco 1999). Indeed, satellite data have been effectively used in fire research studies, since they provide synoptic information over wide areas with high acquisition frequency (Richards 1993; Myneni et al. 1997) and complement field work measurements.

New classification techniques, such as object-based classification, have proven to be quite accurate for mapping burned areas when utilizing satellite remote sensing imagery. In contrast to traditional image processing methods, the basic processing units of object oriented image analysis are image objects or segments, and not single pixels and the classification acts on image objects (eCognition user guide 2004). Such classification techniques, incorporating contextual and semantic information, can be performed using not only image object attributes, but also the relationship among different image objects (Kosko 1992; Kruse et al. 1993; Pierce et al. 1994). Object-based classification models have been successively developed for Landsat-TM, NOAA-AVHRR and IKONOS images for the mapping of different burned areas in the Mediterranean (Mitria and Gitas 2002; Mitri and Gitas 2006; Gitas et. al. 2004).

The **aim** of this study was to investigate the use of Remote Sensing and GIS in mapping a burned area in Kassandra peninsula in Greece and assessing the short term fire effects on the landscape. The specific objectives can be summarized in developing an object-based classification model for mapping the burned area and estimating vegetation regrowth one year after the fire by using Landsat and SPOT imagery.

## 2 STUDY AREA AND DATASETS

The study area is located in North Greece in the southern part of the Kassandra peninsula and belongs to the prefecture of Chalkidiki. The Kassandra peninsula extends approximately from $25^0$25' to $25^0$35' East to 25035' East (longitude) and $39^0$90' to $40^0$10' North (latitude) and its area is estimated to be about 350 km$^2$. The elevation ranges from sea level to 335 meters and has relatively low slopes and smooth relief.

According to Tsitsoni (1991), forest areas occupy 140.02 km$^2$ (about 40%) of the total area and agricultural lands about 199.69 km$^2$ (about 57%). Private landholders own 40.42 km$^2$ of forest land, while the rest 99.6 km$^2$ belong to the state.

A major fire ignited on August 21st, 2006 and burned until the 25th of August. The intensity and extent of the fire event caused the destruction of thousands of hectares of vegetated land, residences, and even caused casualties among civilians. Over one hundred residences and over one hundred vehicles were destroyed during the first two days of the fire event, as it burned its way, with a front of approximately 20 km, through the villages of Polychrono, Hanioti, Pefkohori and Kriopigi. Moreover, hundreds of houses, hotels and childrens' camps in the wider area were evacuated, while there was a cut off in electricity in the wider area. The development of multiple fronts along with the great intensity of the winds blowing in the area were the major factors that hindered the fire fighting efforts.

The datasets that were used in this research were:

- A Landsat-5 TM satellite image captured on September 2nd 2006, one week after the fire. The multispectral image, which covered the study area, has a spatial resolution of 30m and was used for the development of the object-oriented classification models.
- A SPOT-4 satellite image of Kassandra captured on September 24th 2007, one year after the fire event. The SPOT scene has a spatial resolution of 20 m and was employed for the

development of an object-oriented classification model, in order to assess the short-term impact of the fire on the area.

- A total of 65 field points collected with a GPS device after the fire were employed to facilitate the accuracy assessment of the classification results. The field survey was carried out six days after the fire and the points referred to burned and unburned areas. Additionally, 63 points were gathered in a field survey that was conducted one and a half year after the fire. The latter reference data was used to estimate the accuracy of vegetation recovery classification results.
- An aerial photograph of the burned area was also used for the geometric correction of the images and, furthermore for assisting in validating the burned area mapping classification results.

## 3 DEVELOPMENT OF AN OBJECT-BASED CLASSIFICATION MODEL FOR BURNED AREA MAPPING USING A LANDSAT IMAGE

The object-oriented approach takes the form, textures and spectral information of image objects into account. Its classification phase starts with the crucial initial step of grouping neighbouring pixels into meaningful areas, which can be handled in the later step of classification. Such segmentation and topology generation must be set according to the resolution and the scale of the expected objects. In this method, single pixels are not classified but homogeneous image objects are extracted during a previous segmentation step. This segmentation can be done in multiple resolutions, thus allowing differentiation of several levels of object categories.

Prior to developing the model, the Normalized Burn Ratio was calculated in order to be used as additional information feature in the model. The NBR is widely used nowadays for fire mapping, since after fires the NIR reflectance decreases and mid-infrared (MIR) reflectance increases in comparison to pre-fire values. Therefore, its use could assist in the better discrimination of burned areas. The NBR was computed (Figure 1) by using the following equation:

$$NBR = \frac{TM4 - TM7}{TM4 + TM7} \quad (1)$$

where TM4 and TM7 are the near-infrared and mid-infrared Landsat bands, respectively.

Figure 1. NBR of the Landsat image

The initial spectral bands of the non-corrected atmospherically image along with the NBR layer were used as input for the development of the classification model. Before performing the classification, image segmentation was performed in order to construct a hierarchical network of image objects that represented image information in different spatial resolutions. Afterwards, several processes and algorithms were implemented on different domains and segmentation levels so as to finally classify the burned area. All these processes and algorithms that were used in a certain sequence finally comprised the rule set of the model.

In this study, the strategy before classifying the burned area was to create a three-level graded scale of segmentation. The multiresolution segmentation was a bottom-up region-merging technique and the resulting objects were automatically linked to a network in which each image object knows its neighbours, thus affording important contextual information for later analysis. Several criteria concerning scale, layer weights and homogeneity had to be set. The final classification took place in the smallest scale level following a top-bottom approach by using super-object information.

## 4 DEVELOPMENT OF AN OBJECT-BASED CLASSIFICATION MODEL FOR POST-FIRE IMPACT ASSESSMENT USING A SPOT IMAGE

Prior to developing the model, the Normalized Difference Vegetation Index (NDVI) was computed for the SPOT4 geocorrected image, in order to be used as an additional information layer. Since NDVI has been extensively applied in vegetation studies and is related to biophysical parameters, such as green cover and biomass, its use can assist in estimating vegetation recovery. The NDVI was computed by using the following equation:

$$NDVI = \frac{B1 - B2}{B1 + B2} \tag{2}$$

where B1 and B2 are the near-infrared and red SPOT bands, respectively.

Afterwards, both the SPOT scene and the NDVI layer were subset to fit the fire perimeter that was extracted by the object-based classification of the Landsat image. The subset scene that contained the initial spectral bands along with the vegetation index layer were used as input for the development of the SPOT model

The strategy before classifying the region of interest was to create a two-level graded scale of segmentation. The multiresolution segmentation was a bottom-up region-merging technique and the resulting objects were automatically linked to a network in which each image object knows its neighbours. The final classification was applied in the larger scale level following a bottom-up approach by using sub-object information (Figure 2).

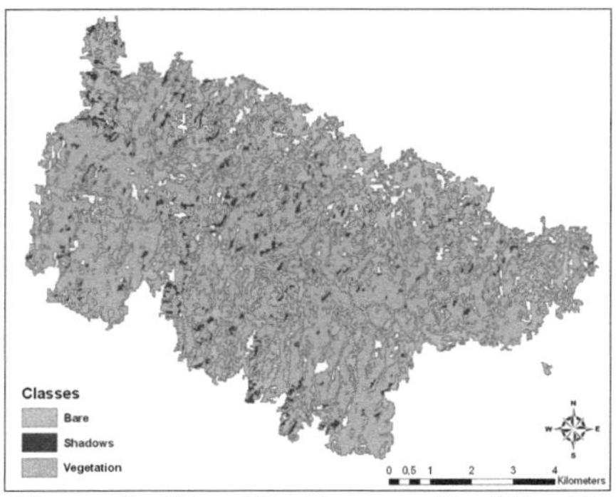

Figure 2. Classification result of the SPOT image one year after the fire in Kassandra peninsula

## 5 RESULTS AND DISCUSSION

For the estimation of the accuracy of the object-based model in mapping the burned area, reference data were selected in the aerial photograph by means of photo-interpretation. After careful examination, two hundred points were randomly chosen in away so that they would be as much dispersed in the photograph and referred to burned and unburned areas. Accuracy assessment is considered more reliable if a large number of randomly and widely spread reference data is used. Due to restricted road access in many parts of the study area, field data were limited in number and not as dispersed as the 200 chosen points. Therefore, it was decided that these 65 ground points would be only used for assisting and validating the photo interpretation.

The results of the descriptive statistics generated in order to estimate the classification accuracy of the model are shown in the following table (Table 1). From the 178 pixels classified as burned, 13 were erroneously classified as unburned. The overall classification was estimated to be 90.50%. Users' accuracy and Producers' accuracy for burned area mapping were estimated at 92.7% and 96.49%, respectively.

| *Class Name* | *Reference Totals* | *Classified Totals* | *Number Correct* | *Producers Accuracy* | *Users Accuracy* |
|---|---|---|---|---|---|
| Unburned area | 29 | 22 | 16 | | |
| Burned area | 171 | 178 | 165 | 96.49% | 92.7% |
| Totals | 200 | 200 | 181 | | |

**Overall Classification Accuracy = 90.50%**

Overall Kappa Statistics = 0.7342

Table 1. Error matrix generated for the classification accuracy assessment of the object-based model using the LANDSAT image

After the classification of the SPOT image, the resulting vector layer was inserted in a GIS and arithmetic results were produced (Table 2). Non-vegetated areas occupy a total area of 2525 ha and represent 37% of the study area, while more than half of the burned area appears to be vegetated. More specifically, about 60% (4109 ha) of the area was classified as vegetation and three vegetation levels were distinguished: low, moderate and dense, which cover 37, 17 and 6 percent of the total area. Shadows were also distinguished and were found to be 3% of the area.

| **Land cover classes** | **Area (ha)** | **Area (%)** |
|---|---|---|
| **Bare/No vegetation** | **2525** | **37** |
| **Dense vegetation** | **399** | **6** |
| **Moderate vegetation** | **1198** | **17** |
| **Low vegetation** | **2512** | **37** |
| **Shadows** | **198** | **3** |
| **Vegetation (total)** | **4109** | **60** |
| **Total area** | **6832** | **100** |

Table 2 Size of the land cover classes in Kassandra, one year after the fire

The proportion of vegetation appears to be quite high, given that one year after the fire seems a very short period for significant vegetation recovery. However, the largest percentage (61%) is categorized as low vegetated by the classification model, while moderate and dense classes occupy 39% of the total amount. The high percentage of low vegetation can be attributed to the substantial recovery of shrubs and poaceous plants in general and pine regeneration. This direct ecosystem recovery is typical of Mediterranean ecosystems after a fire event and these regrowth processes were indeed verified after the field survey, where a significant amount of pine seedlings was detected among the general regrowth of the *maquis* plant community.

For the estimation of the accuracy of the object-based model that was created in order to investigate the potential vegetation regrowth in the burned area, reference data were selected after conducting a field survey. A total of 63 points were selected with a GPS device and referred to vegetated and non-vegetated plots inside the fire perimeter. In addition, a large number of digital photographs was captured and showed the significant recovery of vegetation. As already mentioned, ground relief limitations and poor maintenance of the existing forest roads prohibited a more widespread point collection that would facilitate a more robust assessment of the classification output.

| *Class Name* | *Reference Totals* | *Classified Totals* | *Number Correct* | *Producers Accuracy* | *Users Accuracy* |
|---|---|---|---|---|---|
| No vegetation | 10 | 22 | 9 | | |
| Vegetation | 53 | 41 | 40 | 75.47% | 97.56% |
| Totals | 63 | 63 | 49 | | |

**Overall Classification Accuracy = 77.78%**

Overall Kappa Statistics = 0.6489

Table 3 Error matrix generated for the classification accuracy assessment of the object-based model that was developed with the SPOT image

Despite this fact, the accuracy results were considered satisfactory and are presented in Table 3. The error matrix shows that out of 41 pixels classified as vegetated, only one was erroneously classified, whereas from the 22 pixels classified as non-vegetated only 9 were correct. The overall classification was estimated to be 77.78%. Users' accuracy and Producers' accuracy for burned area mapping were estimated at 97.56% and 75.47%, respectively and the Kappa coefficient was found to be 0.648.

## 6 CONCLUSIONS

The aim of this study was to investigate the use of object-based classification and high resolution satellite data in mapping burned areas and assessing the short term fire effects on the landscape. The main conclusion drawn from this study was that advanced classification techniques such as object-oriented image analysis can accurately map the extent of a burned area and furthermore to assess the short-term impact of fire, thus being a powerful tool in forest fire management.

According to the specific objectives of this work, the major findings can be summarized as follows:

- the object-based classification model that was developed by using a LANDSAT image resulted in the mapping of the burned areas with very high overall accuracy (90.50%), and
- the overall accuracy of the classification resulted from the object-based model to monitor vegetation regrowth one year after the fire by employing a SPOT image was estimated to be 77.78%, when the classification results were compared with field survey reference data, and it was considered satisfactory in mapping the vegetation recovery of the study area.

## REFERENCES

Chuvieco E. 1999. Foreword. A review of remote sensing methods for the study of large wildland fires. E. Chuvieco. Alcalα de Henares, Departamento de Geografva, Universidad de Alcala: 35.

eCognition 2004. User Guide (Definiens Imaging GmbH: München).

Gitas I.Z. 1999. Geographical information systems and remote sensing in mapping and monitoring firealtered forest landscapes. PhD thesis, University of Cambridge, Cambridge, UK.

Gitas I., Mitri G. and Ventura G. 2004. Object-oriented image analysis for burned area mapping using NOAA-AVHRR imagery in Creus Cape, Spain. *Remote Sensing of Environment* 92, 409–413.

Jakubauskas M. E., Lulla K. P. and Mausel P. W. 1990. Assessment of vegetation change in firealtered forest landscape. *Photogrammetric Engineering and Remote Sensing* 56(3): 371-377.

Karteris M. A. 1995. Burned land mapping and post-fire effects. Remote Sensing and GIS applications to Forest Fire Management. E. Chuvieco, University of Alcala de Henares, Alcala de Henares, Spain: 35-44.

Kosko, B. 1992. Neural networks and fuzzy systems, a dynamical systems approach to machine intelligence. Englewood Cliffs (NJ 07632)7 Prentice Hall International.

Koutsias N. and Karteris M. 2003. Classification analyses of vegetation for delineating forest fire fuel complexes in a Mediterranean test site using satellite remote sensing and GIS. I*nternational Journal of Remote Sensing* Vol. 24(15): 3093–3104.

Kruse R., Gebhardt J., and Klawonn F. 1993. Numerical and Logical Approaches to Fuzzy Set Theory by the Context Model. In R. Lowen, & M. Roubens (Eds.), *Fuzzy logic: state of the art* (pp. 365–376). Dordrecht7 Kluwer Academic.

Mitri G. and Gitas I. 2002. The development of an object-oriented classification model for operational burned area mapping on the Mediterranean island of Thasos using Landsat-TM images. In D. X. Viegas (Ed.), *Forest fire research and wildland fire safety* (pp. 79). Rotterdam7 Millpress.

Mitri G. and Gitas I. 2006. Fire type mapping using object-based classification of Ikonos imagery. *International Journal of Wildland Fire*, 15, 457–462.

Myneni R. B., Keeling C. D., Tucker C. J., Asrar G. and Nemani R. R. 1997. Increased plant growth in the northern high latitudes from 1981– 1991. Nature, 386, 698–702.

Pausas J. G. and Vallejo R. V. 1999. The role of fire in European Mediterranean ecosystems. In: E. Chuvieco (Ed.) *Remote Sensing of large wildfires*: (pp.316).

Richards J. A. 1993. Remote sensing digital image analysis: An introduction (2nd ed.).Heilderberg: SpringerVerlag, pp 340.

Pierce E., Ulaby F. T., Sarabandi K. and Dobson M. C. 1994. Knowledgebased classification of polarimetric SAR images. *IEEE Transactions on Geoscience and Remote Sensing*, 32, 1081– 1086.

Tsitsoni T. K. 1991. Structural analysis and physical conditions of regeneration after a fire on forest of Pinus halepensis in peninsula Kassandra of Halkidiki (pp 13-24), Phd Dissertation, Aristotle University of Thessaloniki, Department of Forestry and Natural Environment.

*Remote Sensing for a Changing Europe*
*D. Maktav (Ed.)*
*IOS Press, 2009*


*doi:10.3233/978-1-58603-986-8-467*

# Urban land cover changes assessment from satellite remote sensing data

M. Zoran
*National Institute of R&D for Optoelectronics, Remote Sensing Dept., MG5 Bucharest -Magurele, 077125 Romania, mzoran@inoe.inoe.ro*

Keywords: satellite remote sensing, urban land cover, Spectral Mixture Analysis, spatio-temporal changes, Constantza, Romania.

ABSTRACT: Remote sensing is a key application in global-change science, being very useful for urban climatology and land use/land cover dynamics and morphology analysis. Multi-spectral and multi-temporal satellite imagery (LANDSAT TM, ETM, IKONOS) provide the most reliable technique of monitoring of different urban structures regarding the net radiation and heat fluxes associated with urbanization at the regional scale. Multiple Endmember Spectral Mixture Analysis (MESMA) technique was used to map the physical components of urban land cover for the city of Constantza, Romania, using Landsat Thematic Mapper (TM), Enhanced Thematic Mapper (ETM+) and IKONOS imagery during period of 1989 and 2007 years. The main objectives of this investigation targeted: to develop and validate new techniques for mapping and monitoring land cover and land use within and around Constantza urban area using satellite sensed images and new digital framework data; to analyze the spatial pattern of land cover and the detailed morphology of urban land use across the study area, and quantify the structure of land use parcels ; to devise a methodology for updating of digital urban land-use maps ; to develop an improved information base on urban land-cover and land-use change for land-use/transportation models, urban development planning, urban ecology and local plans.
This study attempts to provide environmental awareness to urban planners in future urban development. The land cover information, properly classified, can provide a spatially and temporally explicit view of societal and environmental attributes and can be an important complement to in-situ measurements.

## 1. INTRODUCTION

Urban zone represents a highly complex area containing a continuum variety of many different spatial, temporal and spectral scales : spatial variability due to the varied landscape; temporal variations are attributed to periodic seasonal changes over the year; spectral variability is due to the great variety of materials and structured associated with the urban area (Biging et al, 1998). Urban areas are currently among the most rapidly changing types of land cover on the planet.
Urban land cover/land use mapping have been relying on data provided by different sensors, but during last years are related to very high resolution (VHR) in both the spatial and the spectral domain (Dell'Acqua et al., 2004), where urban objects may be recognized as distinct blocks, and algorithms based on "per-object" segmentation rather than "per-pixel" classification are feasible.
Many algorithms are considering area-based geometrical and spectral/texture properties in order to recognize "objects" in the original VHR images. Objects are spatial clusters of pixels consistently "homogeneous" with respect to the chosen features and characterized by a set of geometrical, i.e., shape, properties.In the boundary zones between objects, pixels belonging to the boundary are

mixed pixels, and misclassifications may result in inadequate shape recognition, or imprecise data segmentation.

In Romania, land use/land cover patterns of Constantza town on the Romanian Black Sea coastal zone have undergone a fundamental change due to accelerated economic development under its economic reform policies since 1989 and experienced a rapid urban expansion. A large amount of forest and agricultural land has been converted into housing, infrastructure and industrial estates. The resultant impervious urban surface alters the surface energy balance and surface runoff, which in turn could pose serious environmental problems for its inhabitants (e.g., urban waterlogged and thermal pollution). Land-use and land-cover change are significant to a range of themes and issues central to the study of urban environmental change. Alterations in the earth's surface hold major implications for the global radiation balance and energy fluxes, contribute to changes in biogeochemical cycles, alter hydrological cycles, and influence ecological balances and complexity having environmental impacts at local, regional levels through alterations in surface runoff dynamics, lowering of groundwater tables, impacts on rates and types of land degradation, and reduced biodiversity. It is therefore widely recognized that an understanding of changes in urban land and water use over a time span of the next 30–50 years is central to the debate of sustainability.

## 2. SATTELITE DATA FOR URBAN LAND COVER CHANGES ASSESSMENT

Satellite remote sensing imagery can provide a timely and synoptic view of urban land cover, as well as a means to monitor change in urbanizing landscapes and to compare urban environments globally. Due to the great spectral and spatial variability of the materials that compose urban land cover, deriving accurate, quantitative measures from remote sensing imagery over urban areas means a fundamental research challenge (Lu et al.2004).

The highly heterogeneous nature of urban surface materials is complex at multiple spatial scales, resulting in a high percentage of mixed pixels in moderate resolution imagery and even limiting the utility of high spatial resolution imagery (Small, 2005).

Satellite remote sensing data offer great potential of urban areas mapping, being used as an important tool for spatial dynamics and urban morphology assessment. A difficulty in using remote sensing technology for urban and peri-urban studies is the diversity of features found in the environment, including different targets like concrete, asphalt streets and avenues, roofs of different materials, exposed soil, grass, trees, and water .Some of these targets are smaller than the pixel resolution. A landscape is composed of ever-changing elements. Their spatial and temporal patterns distinguish a landscape to an observer; at the same time they inform us of the complexity of dynamic processes at various scales. The changing pattern of the landscape, including the changing biophysical properties of that landscape, is a central theme in the fields of landscape ecology and environmental quality management and planning. Satellite remote sensing instruments provide measurements at a variety of pixel resolutions, spatial extents and temporal scales. However, due to variability in illumination, atmospheric effects, and instrument calibration, conventional supervised or unsupervised classification techniques have difficulty providing pixel to pixel comparisons between images from different times. Different mathematical approaches were applied to modeling land use dynamics. Change detection is the process of identifying differences in the state by observing it at different times.Various change detection techniques have been developed for analyzing of spatio-temporal modifications in natural and anthropogenic surface features (Batty et al., 2001). Digital change detection comprises the quantification of temporal phenomena from multi-date imagery.

Two of the most common uses of multispectral and multitemporal satellite images are mapping land cover via image classification and land cover change via change detection and derived surface bio-geophysical parameters. The recently available Very High Resolution (VHR) satellite images holds a potentially rich source of useful information for the observation of urban features (air, soil, vegetation, water), changes and functioning.

## 3. METHODOLOGY

A continuously increasing amount of multiple sensor data offers opportunities to address new questions and to solve existing problems more effectively in urban remote sensing. A full potential exploitation of these data depends on: data sets selection, use of the proper algorithms and procedures originally developed for single sensor data processing, data fusion modalities. The key to understanding the urban's dynamics and system complexity is to integrate observations at local, regional and global scales, over a broad portion of the electromagnetic spectrum with increasingly refined spectral resolution, spatial resolution and over time scales that encompass phenomenological lifecycles with requisite sampling frequency.

### 3.1 *Subpixel analysis*

The traditional hard classifiers (e.g., minimum distance, Mahalanobis distance, maximum likelihood) can label each pixel only with one class. Information on the fractional amount of spatially mixed spectral signatures from different ground-cover features is not possible with the per-pixel classifiers (hard classifiers). Hence, the traditional classification of mixed pixels may lead to information loss, degradation of classification accuracy, and degradation of modeling quality in successive applications. Subpixel analysis that can provide the relative abundance of surface materials within a pixel is a

potential solution to per-pixel classifiers especially when dealing with medium to coarse spatial resolution satellite sensor images.

### 3.2. *Linear Spectral Mixture Analysis (SMA)*

Spectral mixture analysis (SMA) (Adams et al., 1993) is one of the techniques proposed to provide a solution to the problem of mixed pixels in urban satellite imagery with medium spatial resolution (i.e. 20m or lower). Linear spectral mixture analysis (SMA), uses the multiple linear regression to define the endmembers abundance in the image, which provides subpixel endmember abundance information, being probably the most commonly used approach of all subpixel analysis techniques. The approach is based on the assumption that the spectrum at each pixel is a linear combination of the spectra of all ground components within the pixel, and that the linear mixture coefficients are equal to the fractional area of each ground component in a pixel. The mathematical model of linear spectral mixture analysis can be defined as:

$$X_i = \sum_{k=1}^{n} f_k X_{ik} + e_i \qquad (1)$$

Where, $X_i$ = Total spectral reflectance of band $i$ of a pixel;
$k$ = number of endmembers; $f_k$ = fraction of an endmember $k$ within a pixel; $X_{ik}$ = known spectral reflectance of endmember $k$; within the pixel in band $I$; $e_i$ = error term for band $i$.
The root mean square (RMS) error is given by:

$$RMS = \frac{\left( \sum_{i=1}^{m} (e_i)^2 \right)^{0.5}}{m} \tag{2}$$

where $e_i$ are the error terms for each of the m spectral bands considered. The above constrained least-squares estimate assumes the followings:

$$\sum_{k=1}^{n} f_k = 1 \quad \text{and} \quad 0 \leq f_k \leq 1 \tag{3}$$

Findings from recent studies indicate that although SMA provides superior results to traditional per-pixel classification techniques when applied to urban imagery, a considerable degree of error may be associated with SMA models (Small, 2001). This is because the standard SMA model implements an invariable set of endmembers to model the spectra in all the pixels within an image. This assumption fails to account for the fact that, due to the diversity of urban materials, the number and type of components within the satellite sensor's field of view are variable.
Limitations of linear SMA consist of: (1) Linear spectral mixture classifier does not permit number of representative materials (endmembers) greater than the number of spectral bands.
(2) An invariable set of endmembers to model the spectra in all pixels. This assumption could potentially fail to account for the fact that the number and type of land cover components within each pixel are highly variable. The endmembers used in SMA are the same for each pixel, regardless of whether the materials represented by the endmembers are present in the pixel.

### 3.3. *Multiple Endmember Spectral Mixture Analysis (ESMA)*

An extension of SMA approach that allows the number and type of endmembers to vary for each pixel within an image. MESMA has been proven to be effective in identifying different types of materials in a variety of environments. The MESMA adopts as the best model the one that has smaller root mean square (RMS) error when compared to the spectral curve of the pixel. The algorithm produces the RMS error and the shade information in each pixel as separate layers. It was proposed a solution to this problem by developing a modified SMA algorithm that allows the number and type of endmembers to vary for each pixel in an image. This technique is referred to as multiple endmember spectral mixture analysis or MEMSA. Thus, MESMA can be described as a modified linear SMA approach in which many simple SMA models are first calculated for each pixel in the image. The objective is then to choose, for every pixel in the image, which model amongst the candidate models provides the best fit to the pixel spectrum while producing physically reasonable fractions.

## 4. STUDY AREA AND DATA ANALYSIS

Study area (Fig.1), urban zone Constantza, on North Western Romanian Black Sea coastal zone with complex morphological patterns that are rapidly changing due to a range of complex, interrelated forces of urbanization is defined by Latitude 44.142280, Longitude of 28.657680.
The data utilized in this research included subsets (3113 lines X 4801 samples) from two Landsat TM and ETM+ images acquired in 20/08/1989 ,16/07/2000 and 26/07/2005 and three IKONOS images 29/07/2005, 16/09/2006 and 25/07/2007 images .

During last decades, urbanization has transformed natural landscapes into anthropogenic impervious surfaces, urban land use has become a major driving force for land cover and land use change in Constantza town at Black Sea coast.
The investigations were focused on estimating of urban land cover/use changes. In situ-monitoring additional data were used. ENVI 4.4, IDL 6.3 and ILWIS 3.1 softwares were used.

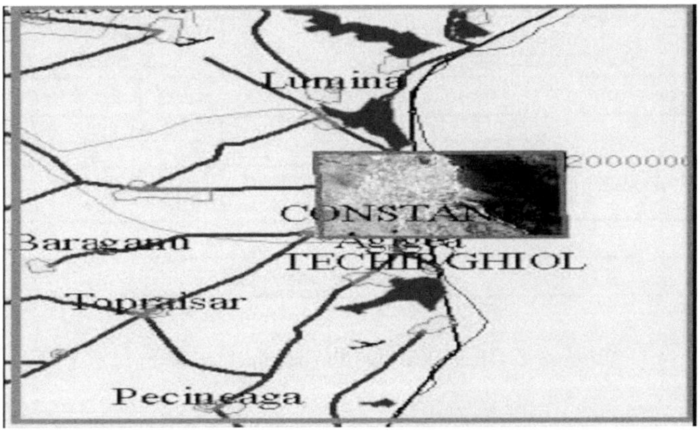

Figure 1. Test site position of Constantza town.

## 5. RESULTS

The main objectives of this paper consisted in developing and validating new techniques based on satellite imagery for mapping and monitoring urban dynamics in urban area Constantza. Linear and non-linear combinations of channels, Principal Component Analysis (PCA) and Relative Channel Analysis, HIS (Intensity-Hue-Saturation) transforms, spectral mixture analysis and unsupervised classification using Maximum Likelihood algorithm assisted by an unsupervised clustering procedure were applied on the available Landsat TM, ETM, and IKONOS data. Because urban development is not generally replaced by agricultural or forest land over time, we reclassified as urban all pixels that were labeled urban in the rasterized 1989 land-cover dataset, but not classified as urban in the 2000 unsupervised classification.
The procedure of applying SMA and MESMA to urban satellite imagery to a single-date image starts by selecting a set of candidate endmembers believed to represent a relatively pure spectral response of the target materials in the scene. A classification of urban surface materials is presented in Figure 2. As a simplification, urban areas may be described in terms of proportions of Vegetation (V), Impervious surfaces (I), and Soil (S). The process of endmember selection is commenced by applying the Pixel Purity Index (PPI) method to screen all the pixels in the image in terms of their relative purity. In the next step, a series of standard SMA models are applied based on a variety of possible combinations of the selected endmembers.
The performance of all models is evaluated so that a smallest subset of candidate models can be selected for every pixel in the image. A reliable candidate model is one that produces physically realistic fractions (i.e. 0-100% range) and does not exceed a certain threshold of error. From the selected candidate models, an optimal model is then identified for each pixel based on the classical maximal covering problem. Finally, the fraction values produced by these optimal models are utilized to map the abundance of general land cover components in the urban scene at a given point of time. The proposed methodology for quantifying the ecological patterns of urban densification consists of three sequential phases. In the first phase, the MESMA technique is separately applied

to individual images in order to derive per-pixel physical measures of urban land cover abundance at a given point of time. Land cover fractions of individual dates are then validated against test data to determine the accuracy of MESMA-derived measures.

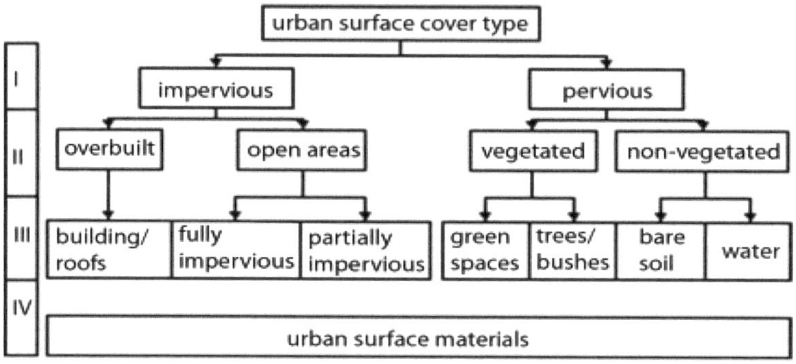


Figure 2. Classification of urban surface materials

Once acceptable, multi-date fractions of corresponding land cover materials are used to calculate per-pixel temporal differences in these fractions. Hence, the resultant fractional differences will represent a direct measure of the changes that take place in the composition of urban morphological patterns over time due to such processes as urban densification and urban sprawl.
This study employs thirty endmembers and six hundred and sixty spectral models to identify soil, impervious, vegetation, and shade in the Constantza area. The mean RMS error for the selected land use land cover classes range from 0.004 to 0.020. The Pearson correlation between the fraction outputs from MESMA and reference data from IKONOS 1m panchromatic resolution data for soil, impervious, and vegetation were 0.7052, 0.7249, and 0.8184 respectively. The results can be summarized in the Table 1.
In this analysis, we apply MESMA to LANDSAT TM and ETM satellite data IKONOS hyperspectral data and focus on 3 key issues: spectral endmember characterization, subpixel component abundance quantification, and error magnitude and distribution for band residuals and total RMS errors.
Figure 3 is presenting a map of the subpixel abundance of generalized urban materials (impervious surfaces, vegetation, and soil), of Constantza urban and periurban areas based on Landsat TM 20/08/1989 satellite data. Figure 4 is showing Landsat ETM 16/07/2000, landcover classification map for Constantza town. This study explored the utility of multiple endmember spectral mixture analysis (MESMA) to capture patterns of change in Constantza urban and periurban land cover through time. MESMA models measured spectra as the linear sum of spectrally pure endmembers and allows endmembers to vary on a per-pixel basis. Constantza urban test area have been analysed using Landsat TM/ETM and IKONOS imagery corresponding to the years 1989, 2000, 2005, 2006, 2007 . Table 1. illustrates mean fraction values of soil, impervious, vegetation, shade, and RMS error of the selected land use/ land cover classess for Constantza urban area derived from IKONOS imagery.
MESMA was applied to each sample, and two categories of maps were generated: (a) maps of the subpixel abundance of generalized urban materials (impervious surfaces, vegetation, and soil), and (b) maps of model complexity (i.e., the number of endmembers required to adequately model each pixel). Model complexity was found to be highly correlated with degree of human impact on the

landscape. The relationships between model complexity, urban growth, and changes in the periurban landscape were explored in the context of this Black Sea coastal zone frontier environment.

Table 1. Mean fraction values of soil, impervious, vegetation, shade, and RMS error of the selected land use/ land cover classess for Constantza urban area, Romania

| | Mean Fraction Values | | | RMS error | |
|---|---|---|---|---|---|
| Land use /Land cover classes | Soil | Imper - vious | Veget a-tion | Shade | |
| Agriculture | 0.024 | 0.015 | 0.796 | 0.078 | 0.020 |
| Harbor | 0.229 | 0.339 | 0.012 | 0.113 | 0.006 |
| Airport | 0.158 | 0.384 | 0.009 | 0.362 | 0.010 |
| Commercial | 0.227 | 0.338 | 0.047 | 0.302 | 0.016 |
| Exposed soil | 0.517 | 0.170 | 0.085 | 0.151 | 0.08 |
| Forest and vegetation | 0.146 | 0.047 | 0.312 | 0.409 | 0.003 |
| Residential | 0.261 | 0.279 | 0.092 | 0.119 | 0.007 |
| Rugged terrain | 0.254 | 0.096 | 0.080 | 0.524 | 0.004 |

Figure 3. IKONOS 29/07/2005 Constantza map of subpixel abundance of generalized urban materials

Figure 4. Landsat ETM 16/07/2000, landcover classification for Constantza town.

In urbanizing landscapes, the first type of change corresponds with urban expansion, the conversion of periurban land cover, such as crops or forest, to built-up land cover. The second type of change is characterized by internal modification of urban land-cover, for example, in-filling of open spaces with high-density buildings, paving of roads, regrowth of vegetation, etc. Most conventional methods of assessing land-cover change only identify transitions between classes, neglecting change within classes due to land-cover modification. This can result in significant error, underestimating the total area experiencing land-cover change, while overestimating the magnitude of change. Additionally, identifying change between classes may not be appropriate in an environment where most change occurs at scales finer than the resolution of the imagery, or where land-cover types are continuous. This work aimed to provide a comprehensive characterization of the Constantza urban landscape in terms of physically meaningful, continuous variables using moderate resolution remote sensing imagery of Landsat TM/ETM and high resolution IKONOS imagery. *Spatial* variability of Constantza urban environment was addressed by mapping the subpixel components of land cover using spectral mixture analysis (SMA), which models each pixel as a linear sum of spectrally 'pure' endmembers. *Spectral* variability of urban land cover was addressed by applying multiple endmember spectral mixture analysis (MESMA), a methodology that allows the number and type of endmembers to vary on a per-pixel basis. The products of this work include a set of maps representing the per-pixel fractional cover of the primary components of urban land cover (i.e., vegetation, impervious surfaces, and soil), as well as maps of spectral complexity (i.e., the number of endmembers necessary to model each pixel). The results are locally specific,capturing the spectral variability that is distinct to the region, yet globally representative of urban land cover, allowing comparison of urban composition across regions and through time.

## 6. CONCLUSIONS

Remote sensing is very useful for urban landcover/use changes assessment, especially in the context of rapid increasing of urbanization in Constantza, Romania. Since LANDSAT TM and ETM contain complementary information, environmental quality and landuse/lancover changes of urban areas mapping is more efficient when the images are used in synergy with high resolution imagery IKONOS. Results from this study demonstrated that the MESMA approach is reliable and the subpixel processor picked out the signatures effectively. It should be noted that a careful selection of endmembers that represent all land covers under study play an important role in the MESMA approach. It was noticed that there is some signature confusion between dry exposed soil/sand bars vs. bright impervious surface and water vs. tar roads/parking lots. It is important as all possible models (combinations of all surface materials) to be considered in the analysis. It is also important to note that number of surface features and all possible combinations of endmember models are increased and generate fraction layers repeatedly until a satisfactory result is received.

## REFERENCES

Adams J.B, Smith M.O., and Gillespie A.R., 1993, Imaging Spectroscopy: Interpretation based on Spectral Mixture Analysis.In: C.M. Pieters and P. Englert (Editors), *Remote Geochemical Analysis: Elemental and Mineralogical Composition.* Cambridge University Press, Cambridge, pp. 145-166.

Biging G., Colby D., Congalton R.G., 1998, Sampling systems for change detection accuracy assessment in *Remote Sensing*, S. Lunetta, & C. D. Eldvidge (Eds.), MI: Ann Arbor Press.

F. Dell'Acqua, P. Gamba, A. Ferrari, J. A. Palmason, J. A. Benediktsson, and K. Arnason, 2004, Exploiting spectral and spatial information in hyperspectral urban data with high resolution, *IEEE Geosci. Remote Sens.Lett.*, 1(4), 322–326.

Lu, D., & Weng, Q., 2004, Spectral mixture analysis of the urban landscape in Indianapolis with Landsat ETM+ imagery. *Photogrammetric Engineering and Remote Sensing*, 70, 1053−1062.

Small C., 2001, Estimation of Urban Vegetation Abundance by Spectral Mixture Analysis. *International Journal of Remote Sensing,* 22(7): 1305-1334.

Small, C., 2005, A global analysis of urban reflectance. *International Journal of Remote Sensing*, 26, 661−681.

*Remote Sensing for a Changing Europe*
*D. Maktav (Ed.)*
*IOS Press, 2009*


*doi:10.3233/978-1-58603-986-8-475*

# Seismic hazard in Vrancea area, Romania, inferred from satellite, GPS and geophysical data

M. Zoran
*National Institute of R&D for Optoelectronics, Remote Sensing Dept., MG5 Bucharest -Magurele, 077125 Romania*

Keywords: seismic hazard maps, geospatial data, Vrancea, Romania.

ABSTRACT: Placed at conjunction of four tectonic blocks which lie on the edge of the Eurasian plate, Vrancea zone in Romania is considered one of the most seismically active area in Europe, being characterized by strong intermediate depth seismicity in a very limited and well defined hypocentral region.The joint analysis of geodetic, seismological , geological and time-series satellite data (LANDSAT MSS, TM and ETM, SAR-ERS 1/2, ASTER, MODIS) is revealing new insights in the understanding of the kinematics and dynamics of the complex plate boundary system and long-term deformation in relation with earthquake activity. Multispectral and multitemporal satellite images over a period 1975-2005 have been analyzed for recognizing the continuity and regional relationships of active faults as well as for geologic and seismic hazard mapping. Seismic hazard maps and classifications of different geologic features and Digital Elevation Models (DEM) generation from SAR ERS1/2, Landsat TM, ETM, and ASTER data are highly correlated with in-situ ground data. GPS measurements provide information on the rate of strain accumulation on active faults (coseismic, postseismic, and interseismic deformation; plate motion and crustal deformation at plate boundaries). Based on GPS Romanian network stations data was revealed a displacement of about 5 or 6 millimeters per year in horizontal direction relative motion, and a few millimeters per year in vertical direction. Future use of long-term InSAR data will be a useful tool in active tectonic investigation for this region.

## 1. INTRODUCTION

Seismic hazard for almost half of the territory of Romania is determined by the Vrancea seismic region, which is situated beneath the southern Carpathian Arc in Romania. Several studies have been carried out to evaluate the seismic hazard in Romania using the probabilistic or deterministic approaches. Earthquakes occur as a result of sudden displacements across a fault within the Earth. The earthquake releases part of its stored strain energy as seismic waves. These waves propagate outward and along the Earth's surface. It is the motion of the ground as these waves move past that is perceived as an earthquake. With most earthquakes, ground shaking is the direct and principal cause of damage to buildings and infrastructure. Considerable damage can be caused by fault rupture at the surface, but this is generally limited to places near the fault.

Space-time anomalies of Earth's emitted radiation (radon in underground water and soil , thermal infrared in spectral range measured from satellite months to weeks before the occurrence of earthquakes etc.), ionospheric and electromagnetic anomalies have been interpreted, by several authors, as pre-seismic signals.Satellite remote sensing provides a systematic, synoptic framework for advancing scientific knowledge of the Earth complex system of geophysical phenomena which often lead to seismic hazards. Geospatial information derived from this research increasingly are addressing the operational requirements of decision support systems used by policy makers, emergency managers as the basis for comprehensive risk and disaster assessments.

Space-based geodetic measurements of the solid Earth with the Global Positioning System, combined with ground-based seismological measurements and satellite remote sensing information are yielding the principal data for modeling lithospheric processes and for accurately estimating the distribution of potentially damaging strong ground motions which is critical for earthquake engineering applications. Moreover, integrated with interferometric synthetic aperture radar, these measurements provide spatially continuous observations of deformation with sub-centimeter accuracy. Seismic and in situ monitoring, geodetic measurements, high-resolution digital elevation models (e.g. from InSAR, IKONOS and digital photogrammetry) and imaging spectroscopy (e.g. using ASTER, MODIS and Hyperion) are contributing significantly to seismic hazard risk assessment.

## 2. DIGITAL REMOTE SENSING IMAGERY

Remote sensing and geospatial information tools and techniques, including numerical modeling, have advanced considerably in recent years, enabling a greater understanding of the Earth as a complex system of geophysical phenomena. The information derived from such systems and analyses are beginning to make their way into operational use by decision makers through a number of information products and decision-support capabilities that inform community based mitigation, preparedness, and/or relief and recovery activities.

Satellite remote sensing represents an important investigation tool of Earth's geologic features being an integrated part of the advanced Information Technology and Telecommunication Infrastructure. Is based on building spectral databases, global large datasets, refining validation, calibration procedures in multi-source, multi-temporal environment. Atmospheric influences on satellite data also increase the difficulty of comparing broadband remote sensing data with laboratory or field spectra. Major improvements are especially in the hyperspectral opportunities, data fusion experiences, storage, management and retrieval of large datasets. The accelerating impact of the available enabling technologies is very important in Earth's features extraction, interpretation by digital image processing, pattern recognition and features identification. The practical value of remotely-sensed data for geologic and seismic hazard assessment has increased significantly in this context with the advent of new, very high spatial resolution optical sensors (*e.g.* the current Indian IRS-1C and US Space Imaging IKONOS, QuickBird and OrbView systems) and Interferometric Synthetic Aperture Radar (InSAR) (Cakir, 2003; CEOS, 2003).However, these new sensors demand new information-extraction methods. Digital map data products can provide information on various geologic and geomorphologic aspects of the seismic areas.

The joint analysis of geodetic, seismological and geological information on the spatial distribution of crustal deformations as well as the analysis of some earthquake precursors is revealing new insights in the field of hazard and risk approach for Vrancea area (Pohl, 1998; Ranchin, 2000).

## 3. GEOLOGICAL FRAMEWORK AND DATA USED

The Vrancea seismic region in Romania is located at the sharp bend of the Southeastern Carpathian orogenic belt (Figure 1) being characterized by the continental collision between the East European plate and the Intra-Alpine to the North and the Moesian subplate to South-West and South-East, respectively . It is considered to be one of the well-defined seismo-active areas of Europe. This collision process is now in its final stage of evolution (Popescu et al., 2001).The largest earthquakes in this region are of intermediate focal depth (60–170km) but shallow smaller earthquakes also occur.The strong seismic events originating here have caused the most destructive effects experienced on the Romanian territory. Seismicity beneath Vrancea is characterized by the occurrence of intermediate depth earthquakes in a narrow epicentral and hypocentral region. The epicentral area is confined to about 30 km x 70 km (Figure 2). Deeper and shallower events have also been recorded but only with small magnitudes.The depth interval of the strong events is bounded by zones of low seismicity between 40 and 60 km and beneath 180 km. Four major events struck within this century: Nov. 10, 1940 , 7.7 M at depth of 150 - 180 km; March 4, 1977 , 7.5 M at depth 90 - 110 km; Aug. 30, 1986 , 7.2 M, depth 130 - 150 km; May, 30, 1990, 6.9 M, depth 70 – 90 km. The main goal is to investigate the way the Vrancea subcrustal earthquakes control the seismic hazard in Romania (Mandrescu, 1990). The numerical synthesis of the ground motion, taking into account the source, the seismic waves propagation and the local site geology represents the key element for the ground motion predictions. Placed at the border of the great East-European Platform , Romanian territory is a region of a complex geological structure dominated by the presence of the Alpine Orogenic Belt of the Carpathian Mountains. This mountains arc has a spectacular change of direction just above the well-known Vrancea seismic region. Subduction of the Black Sea Sub-Plate under the Pannonian Plate produces faulting processes. Figure 2 is shpwing major faults in Vrancea seismic area.The Vrancea earthquakes are produced by the release of stresses that is accumulating in time along some fracture lines belonging to one of the first or second fault-planes (Bazacliu, 1999). The most important factors influencing the level of earthquake ground motion at a site are: the magnitude and distance of earthquake as well as site effects which can amplify or decrease the level of shaking. Seismic hazard maps can be produced on the basis of the regional geomorphologic maps which combines information on rock types, soil types and slopes for seismically active areas.

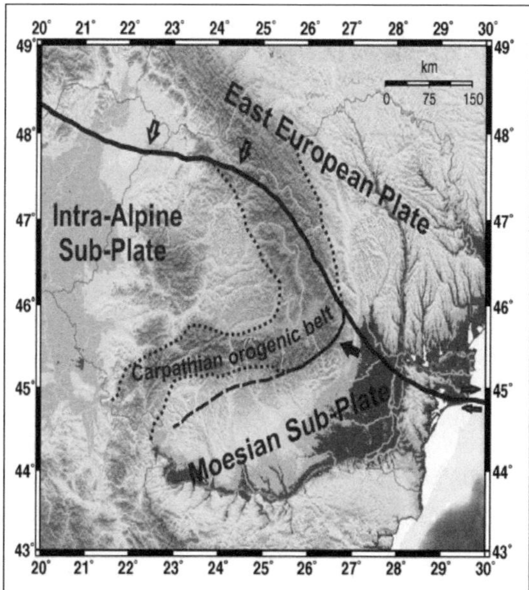


Figure 1. Continental plates and sub-plates in Carpathians. Open arrows show the Neogene, ceased subduction front. Black arrow depicts the active subduction while grey arrows correspond to strike–slip faulting.

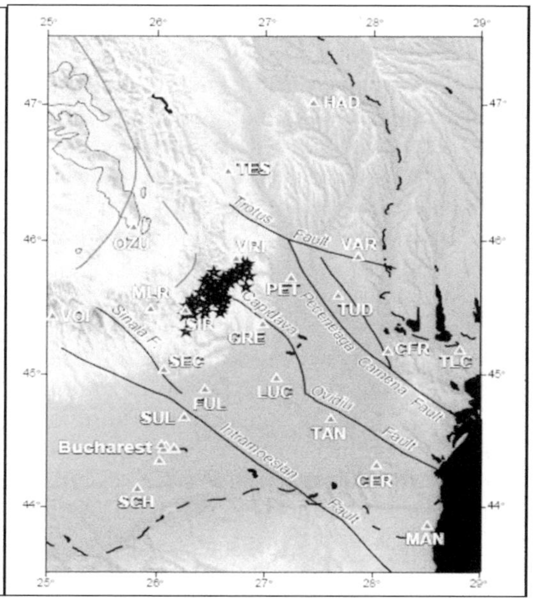


Figure 2. Major faults in Vrancea seismic area and main GPS stations.

The study covers active seismic area Vrancea placed in the South-Eastern part of Carpathians

Romania. The research aims at establishing the changes in geophysical parameters due to geologic activity. Data from optical and microwave satellite images have been analyzed and compared with different ground measurements conducted in conjunction with the satellite data used.
Satellite images used are following: Landsat MSS 18/9/1975; Landsat TM: 2/8/1979; 23/9/1987; 14/10/1989, Landsat ETM 10/5/2002; Landsat ETM 14/09/2004, Landsat ETM 06/09/2005; SAR ERS-1:11/06/1992; ASTER: 30/4/2001; 20/03/2003;14/04/2005; 18/03/2006; MODIS 20/9/2003, MODIS 16/09/2004, MODIS 12/08/2005. Data were digitally processed and classifieds with ENVI 4.4, ILWIS 3.1 and IDL 6.3 software.

## 4. GEOPHYSICAL INFORMATION

Satellite images provide a comprehensive view of the Earth where global structural features can be envisaged in the form of lineaments. These are mostly the traces of discontinuities such as bedding planes, faults, joints and fractures with the ground surface. The analysis of lineaments, especially on maps, provides not only a method for detecting past tectonic trends but also helps in the exploration for minerals, oil and groundwater and for preparing seismic risk maps necessary for construction and mining work.The extraction of lineaments directly from digital images is done by diferent algorithms. The analysis of lineaments as lineament density maps (LDM) is less frequently used, although it permits compilation with other land-use maps using GIS systems. Lineaments and faults are the sources of neotectonic activities which may lead to an increase of seismic hazards in the Vrancea region. Satellite remote sensing has emerged as a powerful tool for seismic hazard assessment and mapping. Synoptic overviews of a wide area and digital remote sensing data are being utilized to map the active tectonic features.
Progress registered in satellite remote sensing technique over the last decade allows its use as a tool for ground deformation studies of seismic areas. The technique of interferometry on satellite SAR images made remote sensing methods very useful for the study of ground deformations. However, this very powerful technique is presently limited by the periodicity of the passes over a given area (35 days in the case of ERS satellites) and by the pixel size (20m at best), both often inadequate for detailed studies on active faults areas. Nevertheless, optical observations made from low altitude airborne platforms as well as the future generation of civil high resolution satellites could still significantly improved ground deformation studies and monitoring on several active seismic areas.

## 5. RESULTS

In order to identify active faults from different sources of remote sensing Images, was done digital image processing approaches used for information enhancement and feature extraction related to faults. Those methods include band ratio, PCA (Principal Components Analysis), Tasseled Cap Transformation, filtering and texture statistics, etc.
Main steps for images processing for seismic zone Vrancea analysis and lineaments detection can be represented as in the Figure 3.
As many of the earthquakes that ever happened on our planet are located in active faults zones it is of most important lo obtain detailed information on regional tectonic structures. The main approach of active faults survey at present is to use geological and geophysical methods, such as in situ measuring, drilling and analysis of gravity and magnetic fields. Remote sensing technique is a complementary tool to map active faults quickly at a large scale, because it can view a side range of area at a time. Meanwhile, remote sensing images cover the spectrum from visible to microwave wavelengths of electromagnetic wave, which provides much more useful information. However, there exist invisible faults underground, which cannot be detected clearly from remote sensing images. In this paper, we used remotely sensed data to identify active faults both at a large spatial

scale and small key study areas. In this paper, we used remotely sensed data to identify active faults both at a large spatial scale and small key study areas.

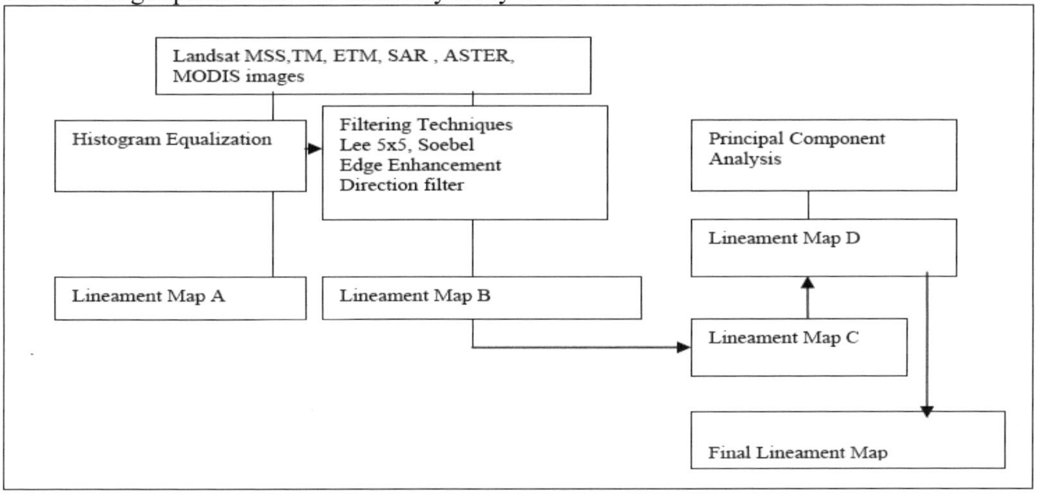


Figure 3. Satellite data processing and analysis for lineament map with filtering techniques

Major faults and lineaments were mapped by visual interpretation of standard False Colour Composite (FCC) images, no features such as lithological dislocations,joints and fracture traces, truncation of outcrops, alignment of streams, sudden bending of streams, etc. Enhanced edges were considered to be related to differences in micro-topography, local vegetation, soil type, soil moisture, or micro-fracture, joints, fault scarps, slopes, etc. The edges reflect the micro-lineament patterns of the area, which were suitably mapped by on-screen digitization of the edges. All possible lineaments were digitized on-screen from the edge enhanced images in four directions, i.e. vertical, horizontal and two diagonal directions. Finally, they were merged with the visually interpreted lineaments to yield total lineament patterns of the Vrancea area.

For estimation of tectonic stress was considered that the orientation of a fault plane is primarily controlled by the direction of principal stress. Thus, the principal stress direction (s1 ) can be determined from the orientation of the fault plane. Quantification of fault plane orientation with respect to principal stress direction is of paramount importance in structural analysis. The work performed by tectonic forces on fault zones is completely consumed and this imposes an important restriction on the sense of displacement along a fault. By definition, the mechanical work done per square centimetre of its surface, is the product of shear stress (*S*) on the fault plane and the relative displacement of the fault blocks. The shear stress in this product is the component parallel to the line of the relative displacement. Since mechanical work is supplied to the fault zone proper, the product is considered to be positive; otherwise, mechanical work would be extracted from the fault. Both the shear stress and the fault plane must therefore have same sign and hence point in the same direction of the maximum compressive stress (i.e. the maximum principal stress s1 has to make an acute angle (h) with the direction of the fault displacement). To make a more accurate determination of angle *Gh*, Mohr's theory of failure for granular material is taken into consideration. Based on Mohr's theory, the angle between the failure plane and the s1 direction is given by (Cernica 1995):

$$Gh = 45 - \phi/2 \qquad (1)$$

where w is the internal frictional angle. The direction of stress acting on each element of failure was estimated using equation (1). We have estimated the stress direction from the lineament pattern using different standard $\phi$ values of the rocks present in the area.

Based on ASTER data we determined active lineaments, i.e. those that show maximum stress direction. For the selection of active lineaments, we considered the orientation of all lineaments with respect to north and their respective maximum stress direction.

A geodynamic model was proposed based on the regional and local tectonic features of the area and GPS measurements.

Figure 4 is ilustrating a Landsat ETM 14/09/2004, 1/5/7 PCA analysis over Vrancea region

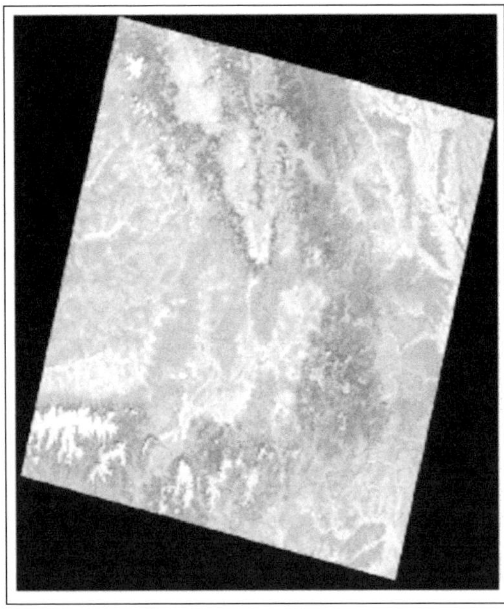

Figure 4. Landsat ETM 14/09/2004, 1/5/7 PCA over Vrancea.

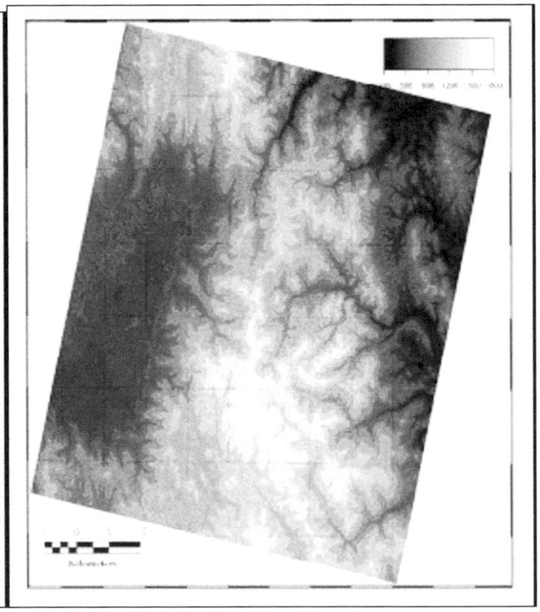

Figure 5. DEM driven from ASTER 18/03/2006 data For Vrancea region.

Figure 5 is showing a DEM driven from ASTER data 18/03/2006 for Vrancea area. High accurate digital elevation models (DEM) acquired periodically over Vrancea seismic area can be used for monitoring crustal deformations. Digital elevation models (DEMs) may be used to determine morphological characteristics such as slopes, volumes, or drainage patterns and integrated with multi-spectral remote sensing imagery for a large number of applications (geology, agriculture, urban planning, etc). The comparison of DEMs relative to different epochs is potentially useful for the detection and description of morphological changes on a three dimensional basis, such as ground deformations related to seismic activity. Expected high resolution DEMs can be controlled through the results of kinematic GPS surveys. This technique consists of the simultaneous operation of a fixed and a rover GPS station, the latter carried either on a vehicle or by a walking operator.

Until a few years ago, deformation studies on seismic active areas were based on geodetic techniques such as spirit levelling, electronic distance measurements (EDM), Global Positioning System (GPS) and on continuous monitoring of sensors like tiltmeters and strainmeters. In a given seismic active area, these various methods allow sampling of the deformation field at only a few ten to a few hundred points, and local or complex deformation fields could not be accurately observed nor modelled.

The Vrancea region is part of an old subduction zone around the Carpathian arc in Romania. While the biggest part of this subduction-system is not active anymore, the tectonic situation in the Vrancea region seems to be different. Is very important to establish if Vrancea slab is still attached to the upper lithosphere, or already detached.This is a the key factor for developing a meaningful geodynamic model for the Vrancea region and therefore an important task of GPS monitoring network (consisting of 50 stations of which 4 permanent) and several GPS campaigns during 1997/2004, as can be seen in Figure 6. GPS monitoring displacement rates field regarding Eurasian Plate and Figure 7. Vertical displacements vectors.

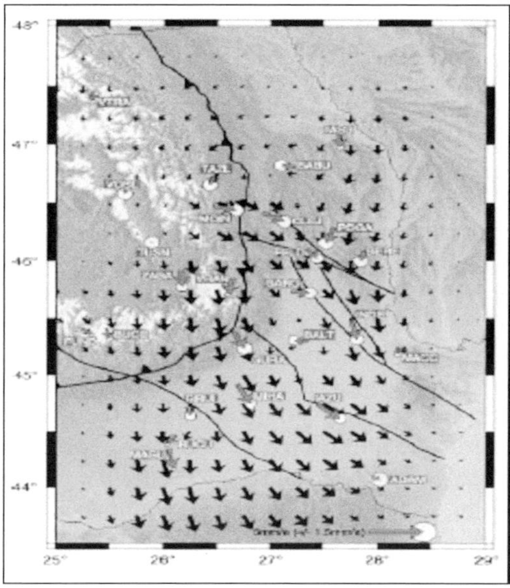


Figure 6. GPS monitoring displacement rates field regarding Eurasian Plate for Vrancea area

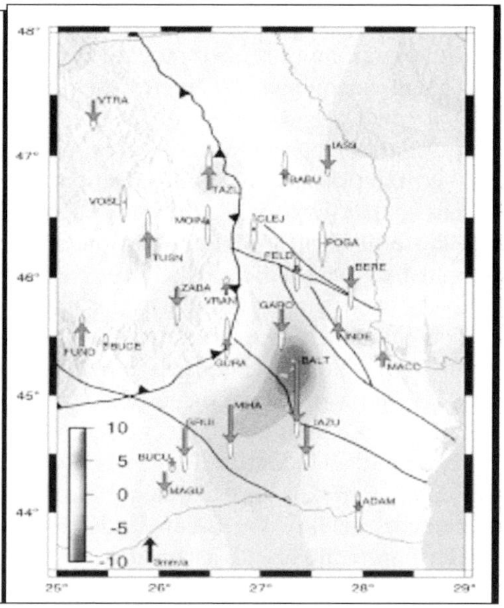


Figure 7. Vertical displacements vectors in Vrancea

Including many points which were already occupied in previous campaigns.Integration of these data with crustal seismicity, surface geology, and topography through a Geographic Information System (GIS) approach places critical constraints on the geodynamic settings for identifying the distribution, geometry, and type of active crustal faults, and (2) elucidating the spatial relationship between the crustal structures and mantle seismicity. The long-wavelength geodetic strain rate field has been computed and compared with the strain field obtained from the analysis of earthquake focal mechanisms.Significant coherence, in terms of style of deformation, between geodetic and seismic strain fields have been obtained in most of the study area. An integrated approach based on satellite monitoring, geophysical modeling, geodetic and seismological methodologies has revealed the major geophysical processes which are active in the Vrancea area.

The combination procedure to integrate satellite data and several geodetic solutions into a single and consistent reference frame with in-situ geophysical information over seismic area Vrancea, requires the availability of the associated statistical information on seismic events .

## 6. CONCLUSIONS

Vrancea region in Romania presents a relatively high potential of seismic hazard mainly due to the subcrustal earthquakes located at the sharp bend of the Southeast Carpathians. Is assumed to be placed at conjunction of four tectonic blocks which lie on the edge of the Eurasian plate. Integrated satellite,GPS and field data of Vrancea area provide a better monitoring of different geophysical parameters and long-term deformation in relation with earthquake activity. Multispectral and multitemporal satellite images (LANDSAT, TM and ETM, SAR-ERS 1/2, ASTER) over a period 1989-2006 have been analyzed for recognizing the continuity and regional relationships of active faults as well as for geologic and seismic hazard mapping. GPS measurements can serve as a reference to these results.
The key to understand the Earth's dynamics and system complexity is to integrate satellite observations at local, regional and global scales, over a broad portion of the electromagnetic spectrum with increasingly refined spectral resolution, spatial resolution and over time scales that encompass phenomenological lifecycles with requisite sampling frequency. Recent advances in computational science and numerical simulations are allowing the study of correlated systems, recognition of subtle patterns in large data volumes, and are speeding up the time necessary to study long-term processes using observational data for constraints and validation.Integrating remotely sensed data into predictive models requires measurements at resolutions substantially superior to those made in the past when the observational systems and the discipline of natural hazards research were less mature than they are today. Furthermore, assimilation of data and model outputs into decision support systems must meet operational requirements for accuracy, spatial coverage and timeliness in order to have positive impact on disaster risk management. Complementarity of different information in their nature, bring new evidence of the complexity and randomness of the earthquake occurrence process, and thus of the difficulty of predicting earthquakes.The continued development and application of a broad spectrum of satellite remote sensing systems and attendant data management infrastructure will contribute needed baseline and time series data, as part of an integrated global observation strategy that includes airborne and in situ measurements of the solid Earth. Understanding the tectonics of the Vrancea earthquakes is mandatory for improving hazard assessment.

## REFERENCES

Bazacliu, O. ,Radulian, M, 1999, Seismicity Patterns in Vrancea (Romania) Region, *Natural Hazards* 19, 2–3, 165–177.
Cakir, Z., Chabalier, J.B., Armijo, R., Meyer, B., Barka, A., Peltzer, G., 2003,Coseismic and early post-seismic slip associated with the 1999 Izmit earthquake (Turkey), from SAR interferometry and tectonic field observations. *Geophys. J.Int.* 155 (1), 93– 110.
CEOS, 2003, *The Use of Earth Observing Satellites for Hazard Support: Assessments and Scenarios. Final Report of the CEOS Disaster Management Support Group* (DMSG), November 2003..
Cernica, J. N., 1995, *Geotechnical Engineering Soil Mechanics* (Chichester: Wiley), pp. 259–271.
Gvirtzman, Z.,2002,Partial detachment of a lithospheric root under the southeast Carpathians:Towards a better definition of the detachment concept, *Geology,*30, 51–54.
Mandrescu N., 1990, Data Concerning Seismic Risk Evaluationin Romania, *Natural Hazards* 3,249-259.
Pohl, C., Van Genderen, J.L., 1998, Multisensor image fusion in remote sensing: concepts, methods and applications, *International Journal of Remote Sensing,* 19 (5), 823–854.
Popescu, E., Radulian, M., 2001. Source characteristics of the seismic sequences in the Eastern Carpathians foredeep region (Romania).*Tectonophysics* 338, 325–337.
Ranchin, T., Wald, L.,2000, Fusion of high spatial and spectral resolution images: the ARSIS concept and its implementation. *Photogrammetric Engineering and Remote Sensing* ,66 (1), 49–61.

*Remote Sensing for a Changing Europe*
*D. Maktav (Ed.)*
*IOS Press, 2009*


*doi:10.3233/978-1-58603-986-8-483*

# Novel vector-based preprocessing of MODIS data

Daniel Kristof
*Dept. of GIS Development, Institute of Geodesy, Cartography and Remote Sensing, Budapest, Hungary*

Robert Pataki
*Dept. Environmental Applications of Remote Sensing, Institute of Geodesy, Cartography and Remote Sensing, Budapest, Hungary*

Keywords: MODIS, preprocessing, vector data model, quality enhancement

ABSTRACT: The Moderate Resolution Imaging Spectroradiometer (MODIS) instruments onboard NASA's Terra and Aqua satellites provide continuous observations and have been increasingly used in different disciplines since 1999. A whole suite of derived products is also produced in a systematic manner. However, these data have some intrinsic characteristics that make their processing a delicate issue when used on pixel- or subpixel-scale studies. Due to the wide field of view, the actual ground pixel size changes continuously off-nadir both in along-track and along-scan directions. All kinds of resampling carried out due to the fixed cell size implemented in the raster data model inevitably result in artifacts in the data. The MODIS gridding process also has a large influence. As a result, the average overlap between grid cells and "real" observations is less than 30%. Therefore, the accuracy of current MODIS data at its original resolution is not sufficient for carrying out pixel- or subpixel-level studies such as spectral unmixing or pixel-level change detection. Spatial or temporal aggregation is widely used to cope with these issues, but may lead to a considerable loss of information.

In this work, we propose and demonstrate an alternative solution based on the vector data model. By using MODIS swath products and geolocation datasets, a polygon layer is created, in which each polygon represents the real pixel footprint sensed during image acquisition. To test the increase of accuracy yielded by the new methods, correlation with same-day high resolution (HR) SPOT images is used as an indicator. The results are encouraging: the original 250-m MODIS reflectance products in the red and near infrared domains show $r^2$ values of 0.4-0.6, whereas the new method yielded values over 0.8 over the same study area.

## 1 INTRODUCTION

The Moderate Resolution Imaging Spectroradiometer (MODIS) instruments mounted on NASA's Terra and Aqua satellites provide observations on 36 spectral bands between 0.405 and 14.385 micrometers, with three spatial resolutions: 250m, 500m and 1km nominal pixel sizes at nadir. The large swath width and the orbit of the two satellites make it possible to acquire imagery over the entire surface of the Earth every one to two days (depending on the latitude: at mid-to-high latitudes, multiple observations per day are possible. For more information, visit http://modis.gsfc.nasa.gov/).

MODIS data are made available free of charge to the scientific community. Besides the original observations, a whole suite of derived products is produced in a systematic manner and can be used for research purposes (Tan et al., 2006). Atmospheric and BRDF correction algorithms based on MODIS observations and other inputs are used to derive surface reflectance (Vermote and Vermeulen, 1999) and more complex algorithms are used to compute other surface properties and processes.

The MODIS processing chain algorithms are improved in a continuous manner, and all the archived datasets are reprocessed regularly. The different versions are referred to as "collections". Starting from January 2007, newly acquired MODIS data are processed into Collection 5, and all archived datasets are being reprocessed with the new algorithms.

MODIS data have some intrinsic characteristics that make their processing delicate. Wolfe et al. (1998) and Tan et al. (2006) give a detailed description of these characteristics and assess their effects for the MODIS sensor. Here we give a brief overview of the main sources of possible distortions, with our own comments and opinions.

1. Due to the wide field of view, the across-track scan angle of MODIS ranges from 0 to 55 degrees. The curvature of the Earth elongates the scan line to 2340 km, and makes the view zenith angle larger than the scan angle. With the view zenith angle increasing, the observation dimensions also increase both in along-track and along-scan directions, and an increasing overlap can be observed among the successive scan lines, among others (Wolfe et al., 1998; Tan et al., 2006).

2. In the light of the above, the raster data model itself has some important limitations. A raster dataset by definition cannot store observations with varying dimensions without the need of resampling and thus modifying the data itself. All kinds of resampling carried out due to the fixed cell size inevitably result in artifacts in the data. Moreover, both the orientation and the size of the cells (pixels) is fixed and cannot change within one raster dataset, which is incompatible with the varying observation dimensions mentioned in point 1. As a consequence, in a MODIS raster image with nominal nadir pixel size, the pixels do not represent the actual Earth surface where the values come from.

3. Although geometric accuracy is an important issue in remote sensing in general, it is even more important when the size of the observation targets and the pixel size are close to each other. The geolocation error of MODIS has been quantified, and was found to be relatively modest: 50 m at 1 sigma at nadir (Wolfe et al., 1998).

4. For MODIS "grid" products, the MODIS gridding process also has a large influence on the resulting grids of MODIS data. The MODIS Data Processing System (MODAPS) uses predefined grids for storing and processing MODIS observations. The predefined MODIS grid cells have a size corresponding to the nominal observation dimensions at nadir. In the gridding process, all observations (image pixels) are stored in a grid cell based on a nearest neighbour resampling algorithm. Therefore, close to the swath edge, several adjacent grid cells will share the same observation. Moreover, the average overlap between grid cells and "real" observations (referred to as "observation coverage") is less than 30% (Tan et al., 2006). The typical values vary between 5% (at swath edge) and 65% (at nadir) for the reflectance products used in our study.

5. In remote sensing, we generally assume that the information content of a pixel originates from its footprint. However, in an actual remotely sensed image, a substantial portion of each pixel comes from its surrounding. Atmospheric effects, instrument optics and electronics and image resampling are the main factors, which can be characterised by the point spread function (PSF) for each sensor (Huang et al., 2002). This becomes extremely important when the size of the studied objects is close to or below pixel size. The image motion PSF, caused by the motion of the scan mirror during the measurement time integration, is the most important component of the MODIS PSF. For a MODIS pixel, approximately 25% of the signal is originating from adjacent pixels while 75% comes from the nominal observation area; the shape of the MODIS PSF can be represented by a Gauss curve (Huang et al., 2002) or approximated by a triangular function (Tan et al., 2006).

## 2 DATA AND METHODOLOGY

### 2.1 *High-resolution SPOT-5 images*

Altogether five multispectral SPOT-5 images were used for this study (acquisition dates: 29/07/2005, 05/05/2006, 25/05/2006, 07/07/2006, 21/07/2006). All images are multispectral (green, red, near infrared, middle infrared) with a ground resolution of 10 meters.
All SPOT images were orthorectified with sub-pixel precision using rigorous transformation with the sensor parameters, a 90 m resolution SRTM digital elevation model and ground control points taken from digital topographic maps of scale 1 : 10 000. UTM projection was chosen in order to avoid the MODIS data to be resampled.

### 2.2 *MODIS: Raster-based approach*

MODIS Level 1 swath products corresponding to the acquisition dates of the SPOT images (see below) were downloaded from the LAADS Web server of NASA (http://ladsweb.nascom.nasa.gov/data/). The datasets include MODIS Geolocation and MODIS Calibrated Radiances (for more details see Table 1.)

Table 1. MODIS swath products used for the observation-to-polygon calculations.

| ***Data type*** | ***Code root (MOD=Terra, MYD=Aqua)*** | ***Relevant contents*** |
|---|---|---|
| MODIS Geolocation | MOD/MYD03 | Latitude, Longitude, Height, Sensor Zenith Angle, Sensor Azimuth Angle, Slant Range |
| MODIS Calibrated Radiances | MOD/MYD02 | Calibrated Radiances, Reflectances (swath format) in original and aggregated form |

The 250-m resolution bands (red and NIR) and their 1000-m aggregated versions from all MODIS swath images were transformed to UTM projection using the standard MODIS Swath Reprojection Tool (available at http://lpdaac.usgs.gov/landdaac/tools/mrtswath/index.asp). The pixel size was chosen accordingly to the nominal nadir pixel size of the bands (250 m and 1000 m, respectively). The resampling was carried out using a simple nearest neighbor method, because the same method is applied in MODAPS during the standard MODIS gridding procedure.

### 2.3 *MODIS: The proposed new vector-based approach*

An overview on the possible sources of error was given in the Introduction section of the present paper. Among these, resampling is one with considerable effects: first, MODIS images are resampled according to the nominal at-nadir pixel size, then gridded to predefined grid cells, both with nearest neighbour method. The low observation coverage values also reflect this effect. Due to this, the pixels of the gridded reflectance products do not represent the Earth surface actually sensed during image acquisition. Tan et al. (2006) suggest that the pixels should be aggregated at a coarser resolution or over longer periods to increase reliability; MODIS data are also available in a composited form. However, this leads to a significant loss in spatial and/or temporal resolution.

In the light of the above, we decided to develop a new method to cope with these problems. The next section gives a brief description of the method.

Instead of using gridded MODIS data, we used swath products (MOD/MYD02) and geolocation datasets (MOD/MYD03) with a new processing scheme. All necessary information is available in these datasets to determine the accurate ground location, dimensions and orientation of each MODIS pixel (see Table 1 for more details). Pixel centroid coordinates (latitude/longitude), slant range, view zenith angle, view azimuth angle are used to calculate observation dimensions and exact position for each observation (the pixel footprint). In the current implementation we assume that the footprints are rectangular in shape, thus the convergence of the sides of each observation footprint (within the range of several meters) is neglected.

First, observation dimensions are calculated for each pixel with the following equations:

$$d_t = \frac{r \cdot d_n}{H} \tag{1}$$

$$d_s = \frac{r \cdot d_n}{\cos\alpha \cdot H} \tag{2}$$

where

$d_t$ is the along-track observation dimension in meters,
$d_s$ is the along-scan observation dimension in meters,
$r$ is the slant range (i.e. the distance of the sensor from the currently sensed surface) in meters,
$H$ is the nominal height of the satellite (i.e. the nominal distance from the earth surface at nadir) in meters (905 000 was used in the calculations),
$d_n$ is the nominal at-nadir pixel size in meters (250, 500 or 1000 depending on the band),
$\alpha$ is the view zenith angle (i.e. the angle between the normal of the observed surface and the scan angle).

Centroid latitudes and longitudes are known for each sample from the MOD03 file. It is also necessary to take into account the orientation of each pixel footprint, which is given by the sensor azimuth angle; the rectangles should be rotated accordingly. (Note that when working in projected coordinate systems such as UTM, the convergence of meridians should also be taken into account and the rotation angles should be recalculated according to the "true north"). Finally, a polygon layer is created by using the appropriately placed and rotated rectangles. Each polygon represents the real Earth surface (pixel footprint) sensed during image acquisition. Thus, the image is stored in polygon instead of raster format, with region topology (permitting overlapping objects to allow scan line overlaps). Reflectance values are stored as attributes in the vector dataset.

### 2.4 *Quality assessment*

To assess and compare the quality of the different representations of MODIS images, the high-resolution SPOT satellite images of the same area acquired on the same day were used.

The principle behind is that the quality of any remotely sensed image can be measured by calculating the correlation of the pixel vales with known reference values. With the general assumption that radiometric effects (including atmospheric and angular effects) are strongly linear in nature over a limited area (e.g. the size of a SPOT image), the reference values should not necessarily be actual surface reflectance measurements. Radiometric rectification methods are based on this well-proven assumption and use linear transformations to cope with the totality of disturbing effects (Hall et al., 1991; Du et al., 2002; Schott et al., 1988). Digital number values of the high-resolution satellite images, aggregated in an appropriate manner within each low-resolution pixel, can thus also serve as a basis for quality assessment. Moreover, if the time difference is negligible between the reference and the image to be assessed (i.e. some hours), it is reasonable to suppose that no significant land cover changes occur. In the case of our MODIS and same-day SPOT 5 images, the time difference does not exceed some hours. Correlation between the aggregated SPOT pixel values and the MODIS pixel values is therefore used as a quality measure: higher correlation indicates a better representation of the real Earth surface.

In the light of the above, we proceeded the following way. MODIS data were subset to the area covered by the SPOT images. Then, based on the 250-m MODIS grid cells, zonal mean was calculated on the red and NIR bands (corresponding to those of the MODIS sensor). All SPOT pixels having their center in a 250-m cell were taken into account and provided one mean value per grid cell; as a result, SPOT data were aggregated to the resolution of the MODIS grid. Then, linear regression was calculated between the SPOT and the MODIS data, and the values of the coefficient of determination ($r^2$) were used as a quality indicator.

## 3 RESULTS AND DISCUSSION

### 3.1 *Results with the original raster-based approach*

The results of the raster-based approach illustrated on the following figures.

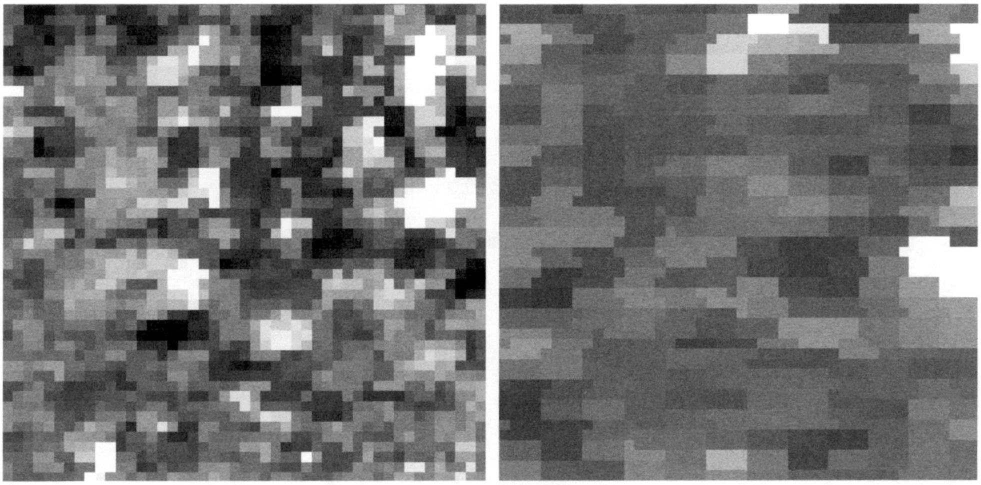

a) b)

Figure 1. Raster-based reprojection results for the MODIS images, created with the MODIS Swath Reprojection Tool at nominal nadir pixel size (250 m) and nearest neighbor resampling for a close-to-nadir subset (a) and an off-nadir subset close to the swath edge (b).

Figure 1 illustrates the effects of resampling artifacts close to swath edge. The nominal nadir pixel size is largely different from the real observation dimensions in this region of the image, and this along with the nearest neighbor resampling results in contiguous blocks of pixels with the same value.

On Figure 2, an example of the regression with the high-resolution SPOT images taken on the same day is shown for the near-infrared bands ( described in 2.4 Quality assessment). The coefficient of determination ($r^2$) values are 0.4702 for the 1-km and 0.6117 for the 250-m MODIS data. Our results indicate the rather poor fit between the raster-based MODIS reflectance products and the SPOT images. This is especially true for off-nadir situations with low observation coverage values.

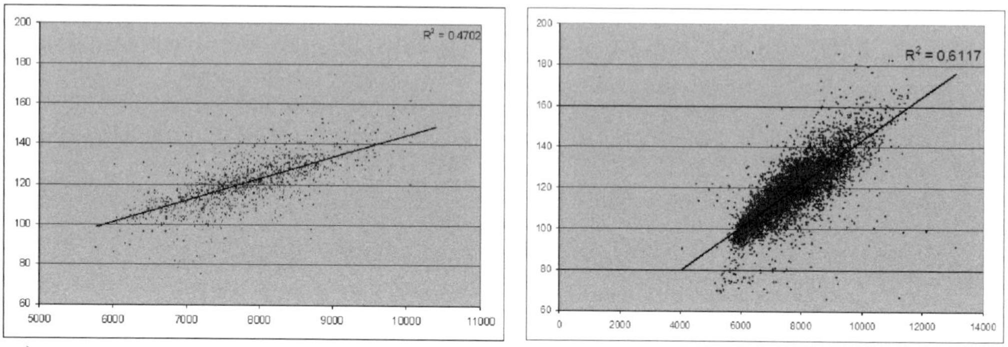


a) b)

Figure 2. Regressions between the NIR bands of the original MODIS reflectance grid (horizontal axis) and that of the SPOT satellite image (vertical axis), both acquired on 05/05/2006, for 1-km (a) and the 250-m resolution (b). The coefficient of determination is 0.4702 for the 1 km and 0.6117 for the 250 m grid.

### 3.2 *Results with the proposed new vector-based approach*

The polygon-based representation of the MODIS images is presented on Figure 3. It can be clearly seen that the original MODIS raster with nominal nadir pixel size and fixed orientation contains blocks of contiguous pixel values not representing the calculated observation footprint.

a) b) c)

Figure 3. Results of the proposed method: polygon-based representation of the MODIS swath reflectance data at a resolution of 1 km (a) and 250 m (b) over the SPOT image as background; and comparison with the original raster representation of the 1-km MODIS data (c). Note the difference in size and orientation between the MODIS raster and the polygon model, and the overlap of the scan lines.

To test the correctness of the new approach, we have recalculated the correlations with the high resolution SPOT images, as described above and in 2.4 Quality assessment. The results are shown on Figure 4. It is evident that the only fact of changing the data model from raster to vector and using the polygon representation yielded a significant increase in the fit between MODIS and SPOT values. The coefficient of determination ($r^2$) increased from 0.4702 to 0.7910 for the band 1-km band and from 0.6117 to 0.7918 for the band with 250 m of nominal resolution. This is due to the elimination of the errors induced by the raster model with a more correct and accurate localization and orientation of the observation footprint of each pixel.

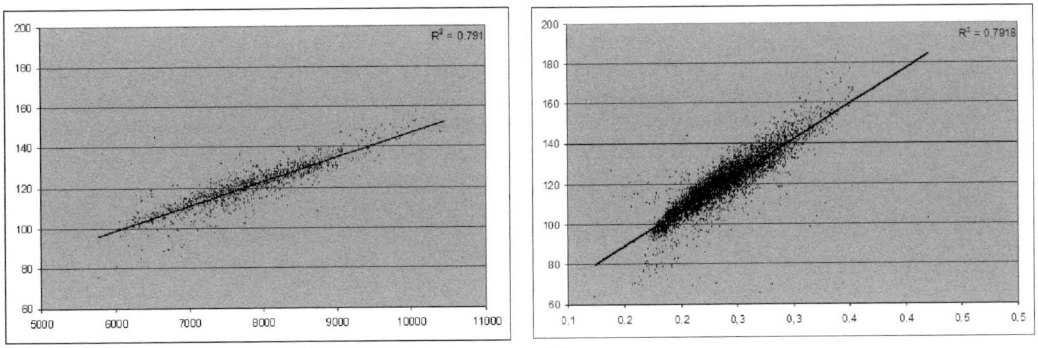


a) b)

Figure 4. Regression between the NIR bands of the new MODIS polygon representation (horizontal axis) and that of the SPOT satellite image (vertical axis), both acquired on 05/05/2006, for 1-km (a) and the 250-m resolution (b). The coefficient of determination has increased to 0.7910 for the 1 km and to 0.7918 for the 250 m grid.

## 4 CONCLUSIONS AND OUTLOOK

Our results show a that with the polygon-based preprocessing of MODIS data, a significant increase in correlation with high-resolution data can be reached. Although the processing in polygon format is more computation intensive, it can offer a good solution e.g. for smaller study areas where sub-pixel accuracy has to be reached. The first results show that with the proposed method, the MODIS reflectance swath products can be efficiently used e.g. for the radiometric rectification of close-date high-resolution satellite images.

Work is still in progress. We are now carrying out a thorough investigation on the usefulness of the observation-to-polygon conversion. Its effect on an entire MODIS reflectance time series is being assessed, along with elaborating a method for the object-based interpolation of values based on individual intersection at each date. Moreover, we are still working on resolving the mixture caused by the triangular point spread function with a deconvolution-based method.

## ACKNOWLEDGEMENTS

The first author (D. Kristof) was supported in his work by the Öveges József Research Grant of the Hungarian National Office for Research and Technology (NKTH).

## REFERENCES

Du, Y., Teillet, P. M., & Cihlar, J. 2002. Radiometric normalization of multitemporal high-resolution satellite images with quality control for land cover change detection. *Remote Sens. Environ.* **82**:123-134.

Hall, F. G., Strebel, D. E., Nickeson, J. E., & Goetz, S. J. 1991. Radiometric Rectification: Toward a Common Radiometric Response Among Multidate, Multisensor Images. *Remote Sens. Environ.* **35:**11-27.

Huang, C., John R.G. Townshend, Shunlin Liang, Satya N.V. Kalluri, Ruth S. DeFries 2002. Impact of a sensor's point spread function on land cover characterization: assessment and deconvolution. *Remote Sensing of Environment*, 80(2002): 203-212.

Schott, J. R., Salvaggio, C., & Volchok, W. 1988. Radiometric scene normalization using pseudoinvariant features. *Remote Sens. Environ.* **26**:1-6.

Tan, B., C.E. Woodcock, J. Hu, P. Zhang, M. Ozdogan, D. Huang, W. Yang, Y. Knyazikhin and R.B. Myneni 2006. The impact of gridding artifacts on the local spatial properties of MODIS data: Implications for validation, compositing, and band-to-band registration across resolutions. *Remote Sensing of Environment*, 105(2006): 98-114.

Vermote, E. and Vermeulen, A. 1999. *Atmospheric correction algorithm: Spectral reflectances (MOD09).* MODIS Algorithm Technical Background Document (ATBD) Version 4.0, April 1999, NASA

Wolfe, R.E., D.P. Roy and E. Vermote 1998. MODIS land data storage, gridding and compositing methodology: Level 2 grid. *IEEE Transactions on Geoscience and Remote Sensing,* (36): 1324-1338.

*Remote Sensing for a Changing Europe*
*D. Maktav (Ed.)*
*IOS Press, 2009*


*doi:10.3233/978-1-58603-986-8-491*

# Semi-automated analysis of time series satellite imagery to assess changes in water storage capacity in a lake in Northern Greece

G. Mallinis & I.Z. Gitas
*School of Forestry and Natural Environment, Lab of Forest Management and Remote Sensing, Aristotle University of Thessaloniki, Box 248, GR-54124, Thessaloniki, Greece. Tel. +302310992731 gmallin@for.auth.gr*

M. Strati-Tsakiri & I Apostolakakis
*Department of Cadastre, Photogrammetry and Cartography, Faculty of Rural and Surveying Engineering, Aristotle University of Thessaloniki, GR-54124, Thessaloniki, Greece. Tel. +302310996146 martsaki@topo.auth.gr*

Keywords: object based, ndwi, mndwi, medium resolution, environmental degradation

ABSTRACT:

During the last decades, the shoreline of lake Koroneia near Thessaloniki in Northern Greece, has significantly changed, resulting to a dramatical decrease of the quality and the quantity of its water reserves.

A semi-automated object based approach was adopted to analyze a series of medium-high resolution satellite images in order to determine the extent of the environmental degradation.

Satellite imagery including LANDSAT, TM, ETM+ and MSS images were co-registered and their Digital Numbers (DN) values were converted to surface reflectance.

A segmentation routine was undertaken to produce appropriate objects in each image. Following that an object based model was developed to delineate the shoreline using orthophotos as reference data. The same model was transferred and applied to the whole dataset after minor adjustments in the classification rules per case.

Bathymetric data of the lake was obtained from a topographic map of the Hellenic Military Geographical Service. This information, along with the shorelines extracted, were transferred within a GIS environment. A spatial analysis procedure was used to estimate changes over time in water storage capacity of the lake.

Results of this study can be used in order to develop restoration plans or to introduce protective measures against any further environmental degradation.

## 1 INTRODUCTION

The delineation and extraction of coastline or shoreline of rivers and lakes is an important task that has application in different fields such as lake/coastline erosion monitoring, lake/coastal-zone management, watershed definition, flood prediction and evaluation of water resources. This task is difficult, time-consuming, and sometimes impossible for a huge region such as an entire country or continent, when using traditional ground survey techniques. This is because water bodies can be very large, have very complex shorelines, may be fast moving as in floods, tides and storm surges, or may be inaccessible. In addition, automatic and replicable techniques are required to update coastline/shoreline maps, evaluate the spatial and temporal evolutions and sensitivities of alterations due to natural and anthropic events (Bagli & Soille 2003, Ouma & Tateish 2005).

Satellite systems for remote sensing play an ever larger role in environmental monitoring and security, they are appropriate to provide the monitoring capacity to monitor changes in water body area (Harris 1994, Peng *et al.* 2006, Yashon & Tateishi 2006) and to estimate water quality (Vincent *et al.* 2004, Kutser *et al.* 2005). In these studies, the most frequently applied remote sensed data are Landsat TM and ETM+ images (Ma *et al* 2007).

Beyond manual digitization, several techniques have been reported in the literature for the derivation of the coastline position from satellite images (Bagli & Soille 2003). The most common are density slice using single or multiple bands and multispectral classification, both supervised and unsupervised as ISODATA-unsupervised classification, principal components analysis (PCA), Tasseled Cap Wetness (TCW), Normalized Difference Water Index (NDWI), maximum-likelihood supervised classification (Ouma & Tateish 2005).

Regarding, the analysis and processing of remotely sensed data, a shift has been observed in recent years from the pixel-based analysis towards object-based analysis with a lot of motivations on that (Hay *et al.* 2005). Object-based analysis has been recently introduced with exceptional success in applications where the classification scheme is binary such as burned area mapping (Gitas *et al.* 2004, Gitas *et al.* 2008, Mallinis et al. 2008). Deficiencies in pixel based approaches, such as confusions between water and built up areas (Xu 2006), could be been minimized by the adoption of objects such as the primary unit of classification for delineating water bodies extent.

The aim of this work is to develop a semi automated approach for the delineation of water bodies in an accurate and consistent manner. Additional objective is the accurate estimation of the change in the extent of the Lake Koroneia in northern Greece.

## 2 STUDY AREA

Lake Koroneia (23°-04'- 23° 14' E, 40° 70' – 40° 43' N) (Fig. 1) is located near the city of Thessaloniki in N Greece, at an altitude of 75 m above sea level. It is a highly eutrophic lake, characterised by low transparency (0.20-0.60 m) and high phytoplanktonic biomass (max. dry weight 5.6 gm−3), and dominated by blue - green algae and diatoms (Kilkilidis *et al.* 1984). It receives water from small streams and torrents within a drainage area of about 780 $km^2$.

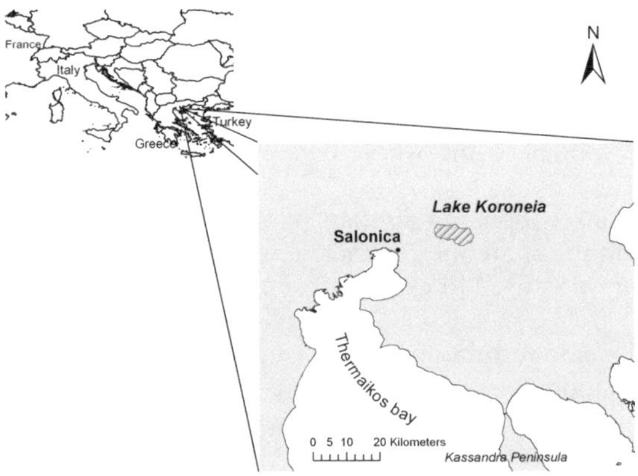


Figure 1 Location of the study area.

The lake and the areas around it are protected by the RAMSAR convention and have been proposed to be included in the protected areas of NATURA 2000. However, water quality of the lake has been declining due to negative water balance and huge amounts of pollutants received from point and diffuse sources.

In the 1970's the lake occupied an area of about 46.2 $km^2$ with a maximum depth of 8.5 m (Psilovikos 1977), while it used to be one of the most productive lakes in Greece concerning fisheries production. During recent decades its water volume has decreased dramatically, due to the overexploitation of water for agricultural and industrial purposes and changes in climatic conditions (in particular lack of rain). In August 1995, an acute change in environmental parameters in combination with the low water volume (depth 1 m, area 30 $km^2$) killed all the fish in the lake, while any efforts since then to re-establish fish populations have failed. Currently the lake has a maximum depth < 1 m, has no macrophytes, no fish and is under a restoration program (which has not started yet) for an increase in its water volume and improvement of water quality, although it still receives agricultural and industrial effluents as well as domestic sewage from the surrounding area (Michaloudi & Kostecka 2004).

## 3 MATERIALS AND METHODS

### 3.1 *Satellite imagery and ancillary data*

A time series of satellite data was used the study (Table 1). All images were acquired during summer. In addition, orthorectified aerial photography of 1:5000 scale, acquired on summer 1996, was used in order to develop the object based approach.

Table 1 Imagery used in the study.

| | **Sensor** | **Period of acquisition** |
|---|---|---|
| 1 | Landsat 2 MSS | July 1977 |
| 2 | Landsat 5 TM | June 1991 |
| 3 | Landsat 5 TM | July 1996 |
| 4 | Landsat 5 TM | July 1997 |
| 5 | Landsat 7 ETM+ | August 200 |
| 6 | Landsat 5 TM | September 2006 |

A Digital Elevation Model (DEM) was generated from 20 interval contours. Finally, the extent of the lake at 1970 as well as bathymetric data was obtained from 1:50000 topographic maps (Fig. 2).

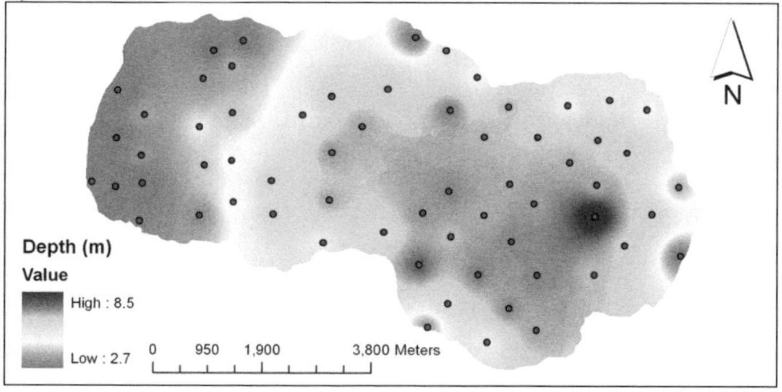


Figure 2 Lake level during 1970

### 3.2 *Image pre-processing*

Pre-processing of multi-date data before change detection analysis is essential to help minimise effects that obscure links between the image data and the biophysical phenomena being studied. The objective is to remove errors associated with data acquisition including sensor effects, atmospheric and illumination effects and misregistration (Galatsiatos *et al.* 2007).

All images were normalized for atmospheric and illumination differences independently and converted to reflectance values using a dark-object subtraction technique (Chavez 1996).

Many approaches, ranging from simple relative calibration and dark-object subtraction to complex model-based calibration approaches (e.g., the 6S radiative transfer code for atmospheric correction), have been developed for radiometric and atmospheric normalization or correction (Chavez 1996, Song *et al.* 2001). Because the dark-object subtraction approach was strictly an image-based procedure, and corrected for the effects caused by sun zenith angle, solar radiance, and atmospheric scattering (Chavez 1996, Lu *et al.* 2002), this approach was used in our study for atmospheric correction of the Landsat images.

Furthermore to enhance spectral separability between water bodies and other land cover/use categories three indices were estimated:

-The familiar Normalized Difference Vegetation Index -NDVI (Tucker 1979)

NDVI = NIR -Red/ NIR +Red.

-The Normalized Difference Water Index-NDWI (McFeeters 1996):

NDWI = Green- NIR / Green + NIR and

-The Modified Normalize Difference Water Index-MNDWI (Xu 2006)

MNDWI = Green-MIR/ Green + MIR

### 3.3 *Water body discrimination*

The segmentation algorithm applied in our work is a component of the multi-scale object-oriented FNEA concept (Baatz and Shape, 2000). Through this segmentation technique, embedded within the commercial software Definiens Developer (ver. 7), individual pixels are perceived as the initial regions, which are sequentially merged pairwise into larger ones with the intent of minimizing the heterogeneity of the resulting objects. The sequence of the merging objects, as well the size and shape of the resulting objects, are empirically determined by the user.

## 4 RESULTS AND DISCUSSION

The Landsat TM image acquired on July 1996 was used to develop an object based classification process-the training stage relied on the use of the orthophotographs, acquired on the same period of year.

Figure 3 Segmentation of the Landsat TM image

The Modified NDWI proved to be the most efficient spectral container both for the segmentation (Fig. 3) and the discrimination of the lake's water body as also noted from Xu (2006). Two levels of segmentation were derived-the classification took place in the upper level whilst the lower was used as source of texture information.

The model was transferred to the whole set of images (Fig. 4). Major modifications were deemed necessary only in the 1977 image where the NDWI was used due to the limited spectral resolution of the sensor.

The results revealed the extent of the environmental degradation occurred in lake's ecosystem during the past 3 decades.

Figure 4 Evolution of the coastline

The most dramatic decrease in the volume and the extent of the lakes's surface occurred after the 1997 (Fig. 5-6).

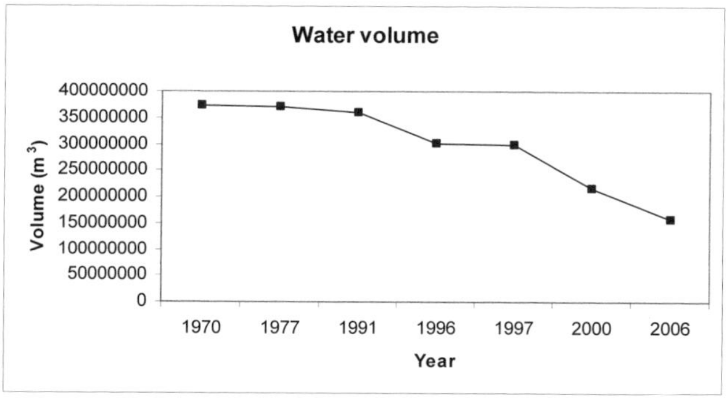


Figure 5 Total volume of the Koroneia lake for individual years

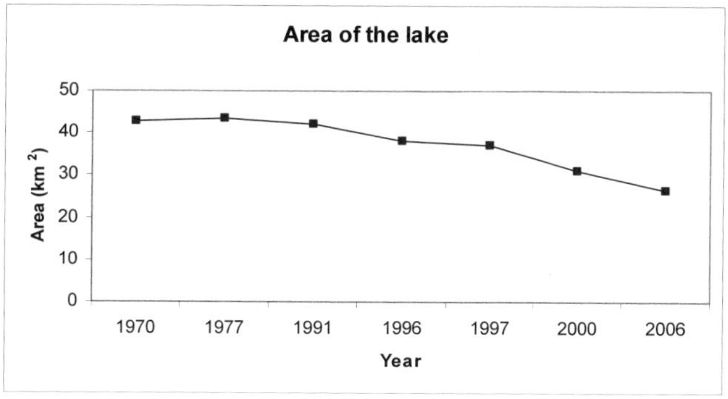


Figure 6 Extent of the Koroneia lake for individual years

## 5 CONCLUSIONS

Object based classification was applied in order to delineate the extent of a water body in different years. The approach was robust and reliable, allowing the estimation of the environmental degradation throughout the study period. Future research , will transfer the model to a national level

## REFERENCES

Baatz, M., Schäpe, M., 2000. Multiresolution Segmentation – an optimization approach for high quality multi-scale image segmentation. In: Strobl, J., Blaschke, T., Griesebner, G. (Hrsg.), Proc. Angewandte Geographische Informationsverarbeitung XII. Beiträge zum AGIT-Symposium, Salzburg, 5-7 July 2000, Herbert Wichmann Verlag: 12–23.

Bagli, S. & Soille, P. 2003. Morphological automatic extraction of coastline from pan-European Landsat TM images. *Proc. Fifth International Symposium on GIS and Computer Cartography for Coastal Zone Management*, Genova, October 2003: 58-59.

Chavez, P.S. Jr., 1996. Image-based atmospheric corrections Revisited and improved. *Photogrammetric Engineering and Remote Sensing* 62(10): 1025–1036.

Galiatsatos, N., Donoghue, D.N.M., Knox, D. & Smith, K. 2007. Radiometric normalization of multi-sensor / multi-temporal satellite images with quality control for forest change detection. *Newcastle, UK: RSPSoc2007: Challenges for earth observation – scientific, technical and commercial.*

Gitas, I.Z., Mitri, G.H. & Ventura, G. 2004, Object-Oriented Image Analysis for Burned Area Mapping using NOAA-AVHRR Imagery in Creus Cape, Spain. *Remote Sensing of Environment* 92: 409-413.

Gitas, I.Z., Polychronaki, A., Katagis, T. & Mallinis, G., 2008. Contribution of remote sensing to disaster management activities: A case study of the large fires in the Peloponnese, Greece. *International Journal of Remote Sensing* 29 (6): 1847-185

Harris, A.R. 1994. Time series remote sensing of a climatically sensitive lake. *Remote Sensing of Environment* 50: 83-94

Hay, G.J., Castilla, G., Wulder, M.A. & Ruiz, J.R. 2005, An automated object-based approach for the multiscale image segmentation of forest scenes. *International Journal of Applied Earth Observation and Geoinformation* 7: 339-359.

Kilikidis, S., KamarianoS, A., Fotis, G., Kousouris, T., Karamanlis, X. & Ouzounis, K. 1984. Ecological research in lakes of North Greece (Lakes Ag. Vassiliou, Doirani and Vistonida). Presuppositions for the establishment of fish reproductive station. *Scientific Annals, Aristotle University* 22: 281-309. [in Greek]

Kutser, T., Pierson, D.C., Kallio, K.Y., Reinart, A. & Sobek, S. 2005. Mapping lake CDOM by satellite remote sensing. *Remote Sensing of Environment* 94 (4): 535-540

Lu, D., Mausel, P., Brondízio, E. & Moran, E., 2002. Assessment of atmospheric correction methods for Landsat TM data applicable to Amazon basin LBA research. *International Journal of Remote Sensing* 23:2651–2671.

Ma, M., Wang, X., Veroustraete, F. & Dong, L. 2007. Change in area of Ebinur Lake during the 1998-2005 period. *International Journal of Remote Sensing* 28 (24): 5523-5533

Mallinis, G., Karamanolis, D., Karteris, M. & Gitas, I. 2008. An object based approach for the implementation of forest legislation in Greece using very high resolution satellite data. In: Blaschke T, Lang S, Hay G (eds.) Lecture Notes in Geoinformation and Cartography. Object-Based Image Analysis Spatial concepts for knowledge-driven remote sensing applications, Springer-Verlag (under publication)

McFeeters, S.K. 1996. The use of normalized difference water index (NDWI) in the delineation of open water features. *International Journal of Remote Sensing* 7: 1425–1432.

Michaloudi, E. & Kostecka, M., 2004. Zooplankton of Lake Koroneia (Macedonia, Greece). *Biologia - Section Zoology* 59 (2): 165-172

Nellis, M.D., Harrington Jr., J.A. & Wu, J. 1998. Remote sensing of temporal and spatial variations in pool size, suspended sediment, turbidity, and Secchi depth in Tuttle Creek Reservoir, Kansas: 1993. *Geomorphology* 21 (3-4): 281-293

Ouma, Y. & Tateishi, R. 2006A water index for rapid mapping of shoreline changes of five East African Rift Valley lakes: An empirical analysis using Landsat TM and ETM+ data. *International Journal of Remote Sensing* 27: 3153-3181.

Peng, D., Guo, S., Liu, P. & Liu, T. 2006. Reservoir storage curve estimation based on remote sensing data. *Journal of Hydrologic Engineering* 11 (2): 165-172.

Psilovikos, A. 1977. Paleogeographic development of Lake Mygdonia (Lagada-Volvi) and the catchment. Doctorate Thesis, Aristotle University, 156 pp.

Song, C., C.E. Woodcock, K.C. Seto, M.P. Lenney & Macomber, S.A. 2001. Classification and change detection using Landsat TM data: When and how to correct atmospheric effect. *Remote Sensing of Environment* 75:230–244.

Tucker, C.J. 1979. Red and photographic infra-red linear combinations for monitoring vegetation. *Remote Sensing of Environment* 8 (2): 127–150.

Vincent, R.K., Qin, X., McKay, R.M.L., Miner, J., Czajkowski, K., Savino, J., Bridgeman, T. 2004. Phycocyanin detection from LANDSAT TM data for mapping cyanobacterial blooms in Lake Erie. *Remote Sensing of Environment* 89 (3): 381-392

Xu, H., 2006. Modification of normalised difference water index (NDWI) to enhance open water features in remotely sensed imagery. *International Journal of Remote Sensing* 27 (14): 3025-3033

Yashon, O. & Tateishi, R. 2006. A water index for rapid mapping of shoreline changes of five East African Rift Valley lakes: an empirical analysis using Landsat TM and ETM+ data. *International Journal of Remote Sensing* 27: 3153–3181.

*Remote Sensing for a Changing Europe*
*D. Maktav (Ed.)*
*IOS Press, 2009*

*doi:10.3233/978-1-58603-986-8-498*

# Analysis of plant condition of the Bystrzanka catchment

B. Zagajewski & A. Jarocinska
*University of Warsaw, Faculty of Geography and Regional Studies, Department of Geoinformatics and Remote Sensing, Krakowskie Przedmiescie 30, 00-927 Warszawa, Poland.*

*E-mail: bogdan@uw.edu.pl, ajarocin@gmail.com*

Keywords: Vegetation indices, hyperspectral imagery, DAIS 7915, NDVI, SAVI, LAI, fAPAR, plant condition, Bystrzanka catchment

ABSTRACT: In natural and semi-natural environments, where the anthropopression is relatively low and natural processes are not disturbed, vegetation can be used as an indicator of other ecosystem components (soils, microclimate etc.). The main role of plant condition investigation plays remote sensing and supportive techniques within plant physiology.
The test area constitutes a mountain zone of the Low Beskid Mts. located at the altitude of 400-750 m a. s. l. The seminatural region is agriculturally, but extensively used; with a domination of natural meadows, cereal and potato crops.
Research algorithm included: field measurements of biometrical indices (LAI, fAPAR, NDVI, temperature index ts-ta, spectral measurements and chlorophyll content measurements); geometric and atmospheric correction of the DAIS 7915; LAI, SAVI, NDVI and fAPAR maps validation according to each land cover unit by field collected data; comparison of min-max values and optimal ranges of biometrical indices (this step was made using maps derived from airborne images).
Results can be outlined: the qualitative and quantitative analysis of pigments showed optimal range of values; spectral signatures of most researched communities are typical for plants in good condition; LAI index oscillates around an optimal values; increase of LAI leads to an increase of absorption of PAR; fAPAR for most researched communities was very high and shows that as much as 90% of sun radiation within the photosynthetic spectrum was absorbed by plants to sustain their growth; temperature differences between plant and air were negative, which indicates good performance of the process of evapotranspiration of the plant species.

## 1 INTRODUCTION

Remote sensing data give many opportunities. For instance, it is possible to analyze biometrical properties of vegetation in different wavelength of electromagnetic spectrum. This method is getting more popular, because these measurements take shorter time and could be extrapolated for large areas. It can be also used to modelling and simulation of biophysical processes. What is more, hyperspectral data have high spectral, spatial and radiometric resolutions. This data can be used to the interpretation of vegetation, land cover and also for analyzing plant condition, forecast biomass and crops, because vegetation cover is very good indicator of environmental condition.
All plant characteristics can be measured in different bands of electromagnetic spectrum. Vegetation indices can be used to measure different characteristic in the same time. The most used regions of spectrum are visible (especially green and red), near and middle infrared. Many kinds of biophysical variables can be measured using hyperspectral data: chlorophyll content, vegetation fresh

or dry biomass, water content, internal structure of leaves, soil moisture and plant surface temperature. It has practical application to crop monitoring and forecasting.

The studies took place in the Low Beskid Mts., which constitute one of the most natural ranges in Carpathian Mts. in Poland. The area extends from 49o34' - 49o 41'N to 21o01'- 21o09'E at the altitude range of 400-750 m a. s. l. The study site focuses on the Bystrzanka catchment around town of Szymbark, where Research and Measurement Station of the Institute of Geography and Spatial Management of the Polish Academy of Sciences and Polish State Environmental Monitoring Station is located. The Bystrzanka catchment encompasses the border zone between the Beskid Mts. and the Carpathian Foothills. The catchment has area around 13.5 $km^2$. The biggest part of area – 40% is covered by forests (especially Carpathian beech with domination of beech and fir). Meadows and pasture are on 28% area. Small fragment of area is covered by arable land.

In this study four of the vegetation indices have been analysed: Normalized Difference Vegetation Index (NDVI), Soil Adjusted Vegetation Index (SAVI), Leaf Area Index (LAI) and fAPAR – fraction of Absorbed Photosynthetically Active Radiation (APAR). These indices measure condition of plants and estimate quantity of biomass.

One of the oldest and best known, widely used is the Normalized Difference Vegetation Index (Rouse et al. 1973; Gamon et al. 1995; Griffith et al. 2002). NDVI is calculated as the difference between near-infrared and red light reflectance values over the sum of these two. Internal structure of green leafs strongly reflects near-infrared radiation, while leaf chlorophyll absorb red visible radiation, that is why this index is related to photosynthetic activity. NDVI is commonly uses to estimate productivity primary production and also biomass of plants. It can be used to forest and crop monitoring. Values of NDVI are from -1 to 1. High values are related with high photosynthetic activity (Rouse 1973; Gamon et al. 1995; di Bella et al. 2004). Typical values for vegetation are from 0.2 to 1, for plants in good condition are above 0,6 (Wang et al. 2004; di Bella et al. 2004). Values of NDVI are influenced by: soil reflection, phonological phase, structure of canopy or optical leaf properties (Myneni & Williams 1994; Epiphanio & Huete 1995; Moreau & Li 1996; Kammerud 1996; Carlson & Ripley 1997; Ridao et al. 1998; Turner et al. 1999; Soudani et al. 2006).

One of the modifications of NDVI is Soil Adjusted Vegetation Index (Huete 1988; Huete & Tucker 1991). SAVI has canopy background correction and it minimized soil influences. It is modified by the parameter L, which is generally constant and equals 0.5. Index is less sensitive to chlorophyll content and is also less dynamic.

Photosynthetic Active Radiation (PAR) is the solar radiation in the wavelength between 400-700nm (Myneni & Williams 1994; Moreau & Li 1996). Absorbed photosynthetically active radiation is the amount of radiation that can be used by plants to photosynthesis. The PAR is increasing, the plants grow well. When the value is getting smaller, plants are in bad condition. The fraction of photosynthetically active radiation intercepted by the canopy – fAPAR – is the amount of radiation, which is actually used by the plant (Epiphanio & Huete 1995). That is why it is closely related to the productivity and net primary production. With this index plant condition can be measured. Values are from 0 to 1 or from 0 to 100%.

Another biometrical index is Leaf Area Index – LAI (Surlock 2001). It is defined as amount of leaf area in a vegetation canopy per 1 square meter. It is a key structural characteristic of vegetation and it is strongly related to photosynthesis, transpiration, productivity and also condition of plants (Serrano et al. 2000). It is a necessary factor to understand and analyze vegetation dynamics due to rapid response to climatic and environmental changes. The optimum values for plants are between 3 and 5; for forest 6-8 and for corn 2-4. The highest value are for forests (LAI>15). LAI can be used to estimate biomass, dynamic of vegetation or to forecast crops (Spanner et al. 1994; Haboudane et al. 2004). Index shows the condition of plants, because in vegetation in stress the values are lower.

In conclusion, vegetation indices are broadly used, especially to vegetation monitoring. Correctly calculated indices give much information about the area. Using indices saves time for time-consuming field measurements. They measure many different biophysical variables, predict biomass and crops.

## 2 MATERIALS AND METHODS

The hyperspectral images were acquired on 29 July 2002 by the airborne scanner DAIS 7915, which was installed on a deck of the plane Dornier Do-228 of the German Aerospace Centre (DLR). The images were acquired during the HySens PL02_05 campaign. The radiometric resolution of the imagery was 15 bit; it has 79 spectral bands from visible, near and middle infrared to thermal IR. The spatial resolution of the scanner is 3meters. Three lines were covering the key areas: Wiatrowki, Taborowka and Biesnik, but only two of them were used for these analyses. The pre-processing was made in January 2003 at the DLR Oberpfaffenhoffen; the geometric correction was made in the PARGE software, the atmospheric correction and the creation of vegetation indices (SAVI, LAI and fAPAR) were made in the ATCOR 4 environment, developed by the DLR and ReSeL laboratories. The image of NDVI was made in ENVI 4.3. Imagery-acquired vegetation indices and results of field measurements were collected and saved in databases of the MS Excel.
Terrain investigation of plant condition using remote sensing field techniques and supportive techniques within plant physiology (including measurements of LAI, fAPAR, plant biometric properties, plant pigments content, plant and soil humidity, plant transpiration and evapotranspiration, soil-vegetation-atmosphere heat exchange) were used to the verification, validation, statistical analysis and making map of vegetation indices. During these measurements, which were made in July and August 2002, data from 47 polygons were collected. They represent different kind of land use: meadows, corn, stubble, clover and potatoes. They were collected by facilities analyzing strictly stated intervals of spectrum using: field spectrometer ASD FieldSpec Pro (NDVI, SAVI), LAI-2000 Plant Canopy Analyzer (LAI) and AccuPAR 80 (fAPAR).
In the end statistical analysis between vegetation indices from ground- and airborne- level have been made. The equations of these correlations were used to make maps of spatial distribution of vegetation index in adequate units. For the relationship between NDVI measured from the ground and airborne levels the calculated equations was $y = 1.2769x - 0.1968$ and $R^2 = 0.91$ (y – field measurement, x – airborne measurement). SAVI ($y = 0.0027x - 0.1596$ and $R^2 = 0.91$), fAPAR ($y = 0.0038x - 0.2081$ and $R^2 = 0.80$) and LAI ($y = 0.0044x + 0.5738$ and $R^2 = 0.80$).

## 3 RESULTS – CONDITION OF CANOPY IN BYSTRZANKA CATCHMENT

The analysis of vegetation was made based on the spatial distribution of four vegetation indices: NDVI, SAVI, LAI and fAPAR. All vegetation indices have high values in the analyzed area, which means that plants are in good and very good condition.
The values of NDVI are the most heterogeneous comparing to the others three indices (Fig. 1a). The average value is high (around 0.6). About 65% of area has values between 0.6 and 0.8, which means that plants are in good condition. Vegetation in average condition is on 26.6% of area. Also forests are in good condition and they have values more than 0.6. On this area values are rather homogenous, because people do not interfere in state of the vegetation. Values are more heterogeneous in corn crops and meadows. They are from 0.2 to 0.8, but most of the area has average plant condition. Only very small (0.6%) area – mostly urbanized area or without plants at all, has small values and plants on this area are in bad condition. High NDVI values means high chlorophyll content and further, high photosynthetic activity, and good plant condition. Vales of NDVI can be perturbed by other elements, for instance soil.
The values of SAVI are much more heterogeneous (Fig. 1b). On particular fields and natural form of land use are les homogenous. It is especially clear in forests. Average values for whole image are around 0.7, which means that most of the soil is covered by plants. Minimum values are equal 0.2, maximum – 1.5. Plants of quite high values between 0.5 and 0.75 are covering the biggest area. Generally, in forest values of SAVI are higher than in corn. The biggest values (more than 1) are on small forest area, mostly in south-eastern part of catchment. Minimum vales (less than 0) are only on two stubbles. Such low values are caused by significant soil reflectance.

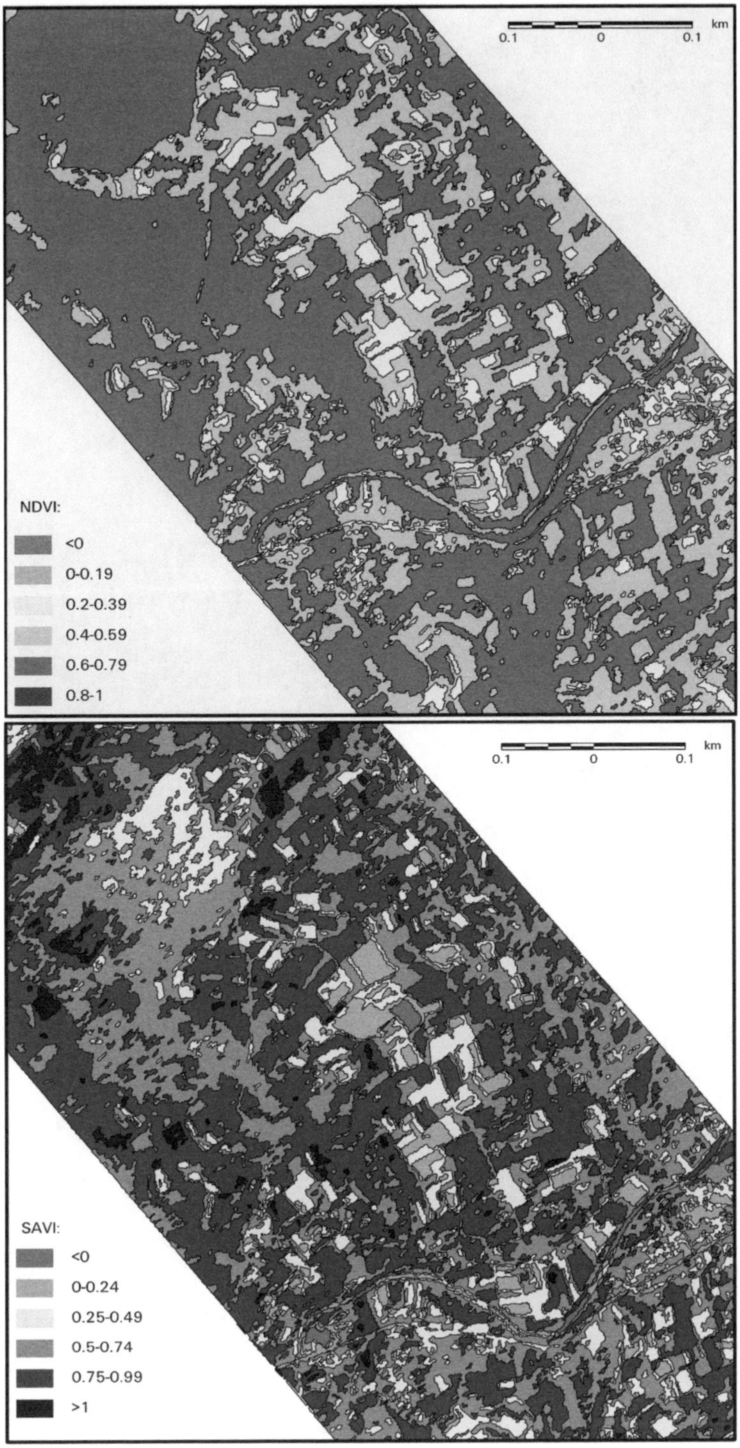


Fig. 1. Spatial distribution for the Wiatrowki key polygon of: a) NDVI and b) SAVI indices

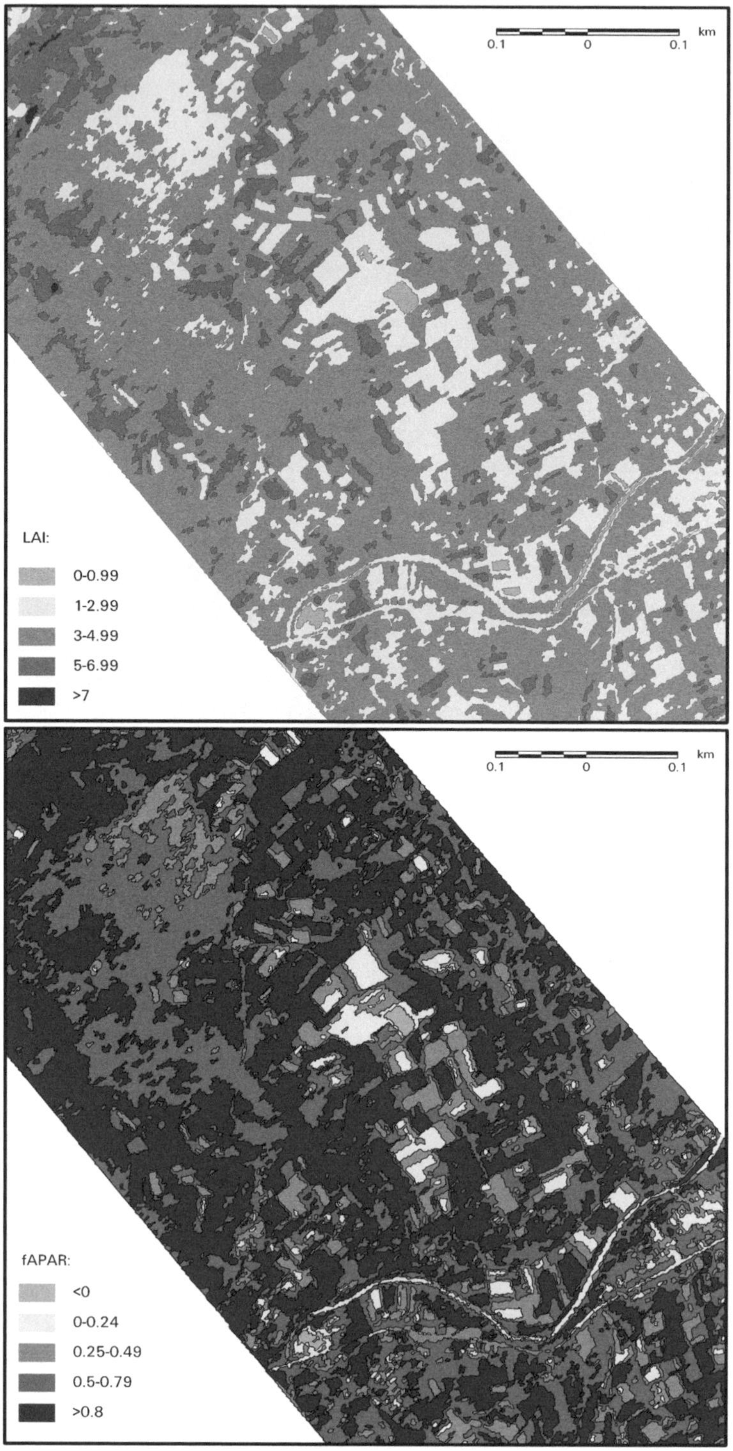


Fig. 2. Spatial distribution for the Wiatrowki key polygon of: a) LAI and b) fAPAR indices

On corn fields values are rather small and canopy is not very compact and this is shown in smaller values of SAVI in this type of land use. On edge of polygons values of analyzed index are different than in the centre of fields. This may be caused by differences in using fertilizers. Low values, between 0 and 0.25, are on very small area, mainly on stubbles, pastures and meadows. On this area the canopy is less compact. When people interfere in quantity of biomass values of SAVI are lower than in natural habitat.

Spatial diversity of LAI (Fig. 2a) is less significant than SAVI (Fig. 1b). Average value is around 3.9, which means that there are almost 4 square meters of leaves on 1 square meter of land. Maximum values are above 10 and minimum 0.6. The lowest values are located on anthropogenic areas and river. The biggest area – about 61%, has values between 1 and 3. On this area plants are covering soil, but canopy is not compact. This is connected with high part of crop fields and meadows, where quantity of biomass is smaller than in natural habitat. Almost 12% of the catchment is covered by dense canopy with big quantity of biomass, values here are above 5. Such values are mostly located in area of forests. Less than 1% of area is covered by plants with very high values – more than 7. This area is located in south-eastern area with forests. On corn crops and meadows values on LAI are rather homogeneous, because the area is rather homogeneous. Values in natural habitat are more heterogeneous.

Values of fAPAR are the most heterogeneous and generally very high (Fig. 2b). Average value is around 0.75, which means that 75% of visible light is used to product biomass and plants has very good condition. About 80% area has values above 0.5 and more than 40% values are higher than 0.8. This means, that 80% of visible light is used in photosynthesis. High values are both in forests and crop fields, which means high crops. Only few corn crops and stubbles have smaller values.

Generally, plants are in good condition of the Bystrzanka catchment. Higher values, apart from fAPAR, are on natural habitants, especially in forests. Values are lower on anthropogenic areas, which are connected with human activity. People are interfering in quantity of biomass, for example mowing meadows. This especially refers to indices: LAI, SAVI and NDVI. SAVI shows the cover of soil with plants, for instance in meadows values of index are quite small after the mowing, LAI also is getting smaller, but values of fAPAR are still high, because big quantity of light is used in photosynthesis.

Indices measured in July 2003 had high values on whole area of the catchment, which means that condition of all plants were good.

## 4 CONCLUSION

Using the remote sensing techniques allows analyzing spatial condition of plants. That kind of methods can be used for automated and objective plant monitoring after all necessary corrections. Different vegetation indices measured plant condition, quantity of biomass and pigments or estimate crops. Because of possibility of repeating the measurements, it is possible to monitor condition and state of vegetation. High values of four analyzed vegetation indices (NDVI, SAVI, LAI and fAPAR) shows good condition of plants of the Bystrzanka catchment in July 2002.

A large number of biometrical input parameters and specific substances and adaptations of plants could be related to reflectance, which can be quantified using hyperspectral data. It is necessary to have support data, which could be acquired using joint research methods of field remote sensing and plant physiology techniques allow differentiating plants in different state and condition.

According to different phenological strategies (flowers, plant's colour, leaves compactness, dry steams) and human activity for plants, condition analysis should be conducted according to each land cover type.

For detailed condition analysis very important are relationships between green pigments, because Chlorophyll a/b and carotenoids index is the best indicator for short term stress, and relationship between chlorophyll b and carotenoids is the best indicator for long term stress.

fAPAR index approximates on the best way a real condition of plants.

## 5 REFERENCES:

Carlson T. N., Ripley D. A., 1997. On the Relation between NDVI, Fractional Vegetation Cover and Leaf Area Index, Remote Sensing of Environment, nr. 46, 241-252.

Di Bella C. M., Paruelos J. M., Becerra J. E., Bacour C., Baret F., 2004. Effect of senescent leaves on NDVI-based estimates of fAPAR: experimental and modelling evidences, International Journal of Remote Sensing, t. 25, nr. 23, 5415-5427.

Epiphanio J. C. N., Huete A. R., 1995. Dependence of NDVI and SAVI on Sun/Sensor Geometry and Its Effect on fAPAR Relationships in Alfalfa, Remote Sensing of Environment, nr 51, 351-360.

Gamon J. A., Filed Ch. B., Goulden M. L., Griffin K. L., Hartely A. E., Joel G., Peňuelas J., Valentini R., 1995. Relationships between NDVI, canopy structure, and photosynthesis in three Californian vegetation types, Ecological Applications, t. 5, nr 1, 28-41.

Griffith J. A., Martinko E. A., Whistler J. L., Price K. P., 2002, Interrelationships among landscapes, NDVI, and stream water quality in the U.S. Central Plains, Ecological Applications, t. 12, nr 6, 1702-1718.

Haboudane D., Miller J. R., Pattey E., Zarco-Tajeda P. J., Strachan I. B., 2004. Hyperspectral vegetation indices and novel algorithms for predicting green LAI of crop canopies: Modeling and validation in the context of precision agriculture, Remote Sensing of Environment, nr 90, 337-352.

Huete A. R., 1988. A soil vegetation adjusted index (SAVI), Remote Sensing of Environment, nr. 25, 295-309.

Huete A. R., Tucker C. J., 1991. Investigation of soil influences in AVHRR red and near-infrared vegetation index imaginery, International Journal of remote Sensing, t. 12, nr. 6, 1223-1242.

Kammerud T. A., 1996. Soil impact on satellite based vegetation monitoring in Sahelian Mali, Georg. Ann., t 78 (A), nr 4, 247-259.

Moreau L., Li Z., 1996. A New Approach for Remote Sensing of Canopy Absorbed Photosynthetically Active Radiation. II: Proportion of Canopy Absorption, Remote Sensing of Environment, nr 55, 192-204.

Mynieni R. B., Williams L. D., 1994. On the Relationship between FAPAR and NDVI, Remote Sensing of Environment, nr 49, 200-211.

Ridao E., Conde J R., Mínguez M I., 1998. Estimating fAPAR from Nine Vegetation Indices for Irrigated and Nonirrigated Faba Bean and Semileafless Pea Canopies, Remote Sensing of Environment, nr 66, 87-100.

Rouse, J.W., Jr., R.H. Haas, J.A. Schell, and D.W. Deering. 1973. Monitoring the vernal advancement and retrogradation (green wave effect) of natural vegetation, Prog. Rep. RSC 1978-1, Remote Sensing Center, Texas A&M Univ., College Station, nr. E73-106393, 93. (NTIS No. E73-106393)

Scurlock J. M. O., Asner G. P., Gower S T., 2001. Worldwide Historical Estimates of Leaf Area Index, 1932-2000, Oak Ridge National Laboratory, UT-BATTELLE.

Serrano L., Gamon J. A., Peňuelas J., 2000. Estimation of canopy photosynthetic and nonphotosynthetic components from spectral transmittance, Ecology, t. 81, nr 11, 3149-3162.

Soudani K., François Ch., la Maire G., Le Dantec V., Dufrêne E., 2006. Comparative analysis of IKONOS, SPOT, and ETM+ data for leaf are index estimation in temperate coniferous and deciduous forest stands, Remote Sensing of Environment, nr 102, 161-175.

Spanner M., Johnson L., Miller J., McCreight R., Freemantle J., Runyon J., Gong P., 1994. Remote sensing of seasonal Leaf Area Index across the Oregon transect, Ecological Applications, t. 4, nr 2, 258-271.

Turner D. P., Cohen W. B., Kennedy R. E., Fassnecht K S., Beriggs J. M., 1999. Relationships between Leaf Area Index and Landsat TM Spectral Vegettion Indices across Three Temperate Zone Sites, Remote Sensing of Environment, nr 70, 52-68.

Wang J., Rich P. M., Price K. P., Kettle W. D., 2004. Relations between NDVI and tree productivity in the central Great Plains, International Journal of Remote Sensing, t. 25, nr. 16, 3127-3138.

*Remote Sensing for a Changing Europe*
*D. Maktav (Ed.)*
*IOS Press, 2009*


*doi:10.3233/978-1-58603-986-8-505*

# SAM and ANN classification of hyperspectral data of seminatural agriculture used areas

B. Zagajewski & D. Olesiuk
*University of Warsaw, Faculty of Geography and Regional Studies, Department of Geoinformatics and Remote Sensing, ul. Krakowskie Przedmiescie 30, 00-927 Warszawa, Poland*

Keywords: ANN, SNNS, SAM, classification, hyperspectral data, plant communities, crops, land use, land cover

ABSTRACT: The goal of this paper is methods and results presentation of artificial neural networks land cover classification based on DAIS 7915 hyperspectral data. Preparing the reference and training data were used SAM classification. To ANN classification experiments using variable pattern size were chosen two data sets included 13 bands obtained after MNF and 40 original bands. The best accuracy were achived using 13 MNF band with the 3x3 pixel subpattern size, what show useful of hyperspectral data extraction methods.

## 1 INTRODUCTION

Vegetation cover is a perfect indicator of all other components of biosphere and should be well researched and mapped. This is possible, using hyperspectral data (which provides very good spectral, spatial and radiometric resolution), because plant species develop specific adaptations (pigment content, plant tissue structure etc.). These adaptations have direct impact on reflectance, which can be quantified using hyperspectral imagery. Application of remotely sensed techniques allows for vegetation research and mapping. One of the characteristics of vegetation is its variable reflectance: identification of some communities may be difficult using only spectral properties. Vegetation reflectance registered by remote sensing instruments is the average of the reflectance of photosynthetic active parts, non-photosynthetic active parts (i.e. branches, dry leaves), shadow and ground. These elements, begin an integral part of plant communities, impede their recognition in case of assuming their spectral properties.
Thus, it is prepared obvious, that vegetation can be characterized by a high variability of the signal and therefore the statistical distribution of reflectance differs from the normal distribution. These relationship is frequently happen over the seminatural and agritulturally used areas, and traditional classification, that uses parametrical approaches does not show satisfying results. A method that uses artificial neural networks does not only depend on statistical parameters of particular class and hence makes it possible to include texture information as additional data. This method may be especially useful to separate and classify vegetation communities (Zagajewski *et al.* 2005).

## 2 STUDY AREA AND DATA SOURCES

The study was conducted in the Low Beskid Mountains, which constitute one of the most natural ranges in the Carpathian Mts. in Poland (Fig. 1). The area is the Bystrzanka catchment extend form 49°34'-49°41'N to 21°01' - 21°09'E. The area constitutes a mountain zone located at the altitude range of 400-750 m a. s. l. The region is agriculturally, but extensively used; with a domination of natural meadows and cereal and potato crops.

DAIS 7915 hyperspectral data used in this study was acquired on 29 July 2002 with cooperation with the German Space Agency (DLR). This instrument is a 79-channel imaging spectrometer operates in the wavelength range 0,4-12,5μm with 15 bit radiometric resolution. After the preprocessing the resulting pixel size was 3 meters. During the overflight 3 lines of DAIS images were acquired (Fig. 1).

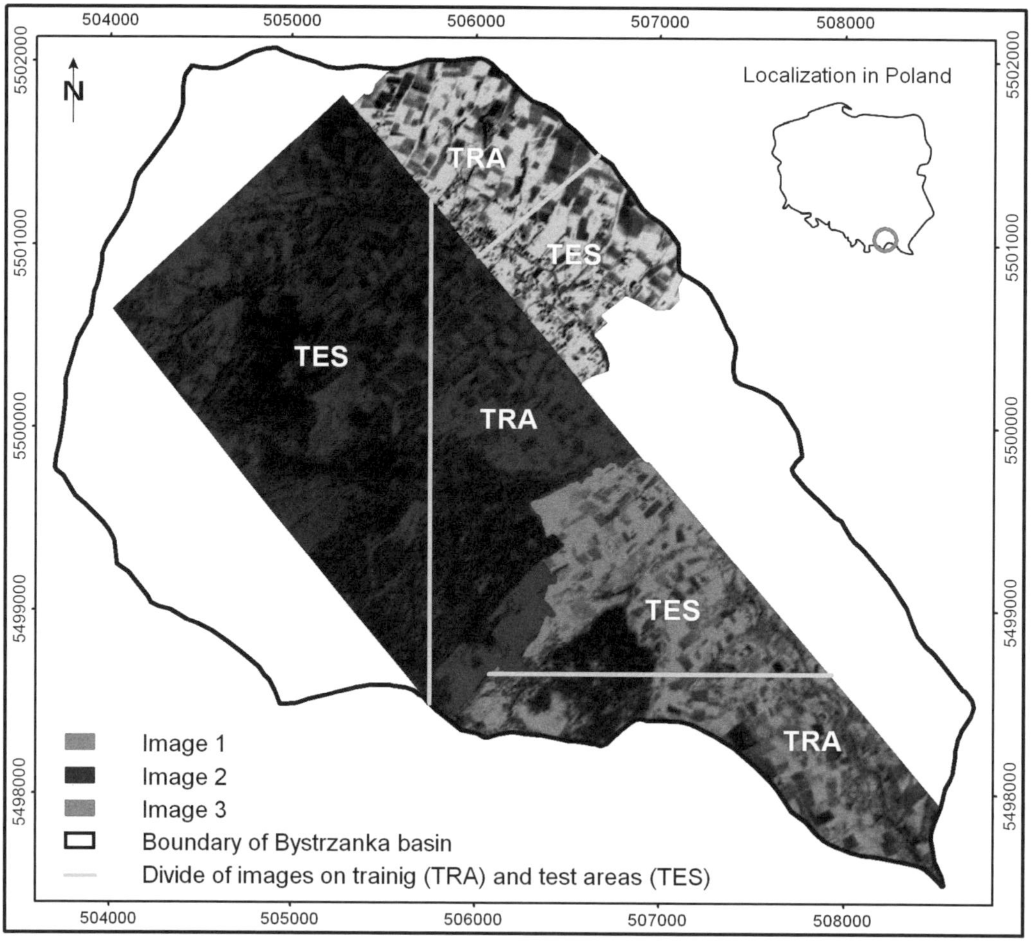


Figure 1. DAIS 7915 images covering the study area.

## 3 METHODS

SAM classification was used to verification land cover map performed in 2002 during the terrain mapping. Endmembers were gained from DAIS imagery (corresponding to the key polygons from the ground measurements). All the polygons represents each class on training area (Fig. 1) were used to learn ANN.

Two data sets using in classification were extracted from 72 bands covering the optical region of the spectrum. First step was reduction by visual (with histogram) inspections bands with severe striping problems, receiving 50 bands. To create first data set, from this 50 bands were choosed 40

(1, 2, 3, 4, 7, 9, 10,11, 12, 14, 15, 16, 17, 19, 20, 21, 22, 24, 25, 26, 27, 29, 30, 31, 32, 34, 35, 36, 37, 38, 39, 40, 42, 46, 47, 50, 52, 53, 55, 56, 58). Second step was compressing the information contained in originally 50 bands using MNF transformation. Thus, second data set was created using the 13 first bands after the MNF transformation.

To classify land cover types were applied a multilayer, one-directional network, trained using a supervised method of back-propagation. In experiments were used two variable pattern size, what is available in the Stuttgart Neural Network Simulator (SNNS) software. To perform classification were applied per-pixel and 3x3 subpattern size (Fig. 2). In per-pixel process number of input nodes was depend of number of bands, but in 3x3 subpattern size window it must be augment by 9. To defining the number of hidden nodes were used the formula 3Ni+1. Every each class was classified separately, so output nodes was 1, and the value of training land cover polygons takes 1, and rest 0. The learning parameters was obtain from Kavzoglu and Mather (2003) papers: the initial weight range [-0.25, 0.25], learning rate 0.2, number of training samples 2500. Trained neural nets was testing on area showing on figure 1 as TES.

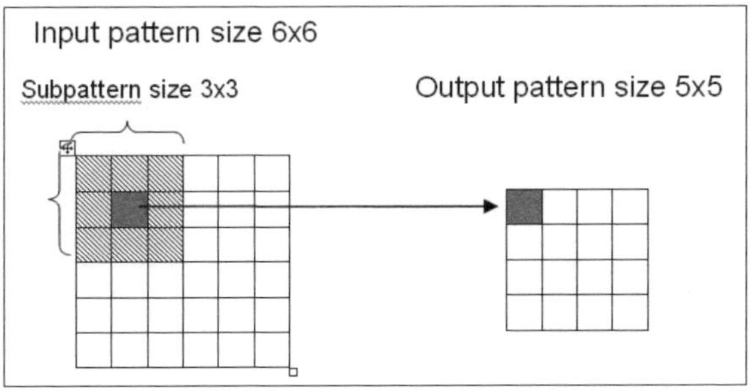


Figure 2. Variable size pattern (source: SNNS user manual; modified).

## 4 RESULTS

Overall accuracy assessment for three images are showed in table 1. Because 3x3 window size (including textural information) process gives better results than per-pixel, image 1 and 2 was classified only with this approaches. Table 2 and 3 showing training and test classification accuracy for image 3 in per-pixel and 3x3 subpattern size process for separately land cover types. Table 4 including confusion matrix of the best results achieved for image 1 using 13 MNF bands with 3x3 subpattern size window (Fig. 3).

Table 1. Overal classification accuracy

| Accuracy / Image | | 13 band MNF 1x1 | 13 band MNF 3x3 | 40 band 1x1 | 40 band 3x3 |
|---|---|---|---|---|---|
| | | Overall accuracy [%] / Kappa | | | |
| TRAINING | Image 3 | 80.1 / 0.7563 | 88.1 / 0.8546 | 85.3 / 0.8204 | 93.1 / 0.9159 |
| | Image 2 | - | 80.7 / 0.7492 | - | 86.6 / 0.8310 |
| | Image 1 | - | 93.8 / 0.9071 | - | 96.8 / 0.9567 |
| TEST | Image 3 | 74.9 / 0.6813 | 75.9 / 0.6942 | 74.5 / 0.6635 | 73.7 / 0.6480 |
| | Image 2 | - | 63.7 / 0.4909 | - | 59.6 / 0.4586 |
| | Image 1 | - | 89.1 / 0.8308 | - | 86.1 / 0.7913 |

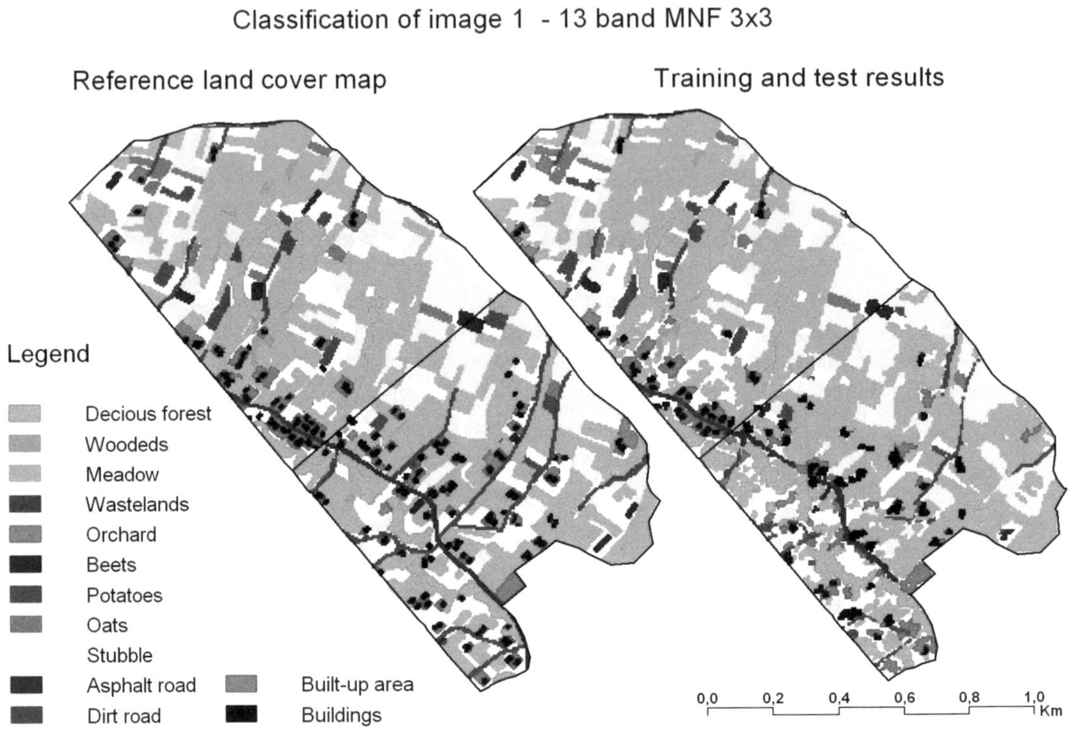


Figure. 3. Results of the DAIS 7915 data classification of the key poygon Biesnik N (1st flightline).

Table 2. Classification accuracy of training area

| Accuracy [%] | 13 k MNF 1x1 | | 13 k MNF 3x3 | | 40 k 1x1 | | 40 k 3x3 | |
|---|---|---|---|---|---|---|---|---|
| | Overall : 74.9 | | Overall : 75.9 | | Overall : 74.5 | | Overall : 73.7 | |
| Land cover | Prod. | User | Prod. | User | Prod. | User | Prod. | User |
| Meadow | 87.6 | 91.2 | 94.3 | 92.4 | 93.1 | 92.2 | 96.6 | 95.1 |
| Wastelands | 81.5 | 53.9 | 81.7 | 89.3 | 83.9 | 75.6 | 91.7 | 90.1 |
| Coniferous fo. | 81.6 | 89.6 | 87.3 | 92.0 | 84.8 | 89.2 | 90.7 | 94.2 |
| Mixed forest | 80.7 | 83.0 | 91.3 | 88.1 | 86.3 | 84.2 | 94.1 | 91.5 |
| Decidous fore. | 83.6 | 74.3 | 84.9 | 88.3 | 85.8 | 80.6 | 92.0 | 92.7 |
| Woodeds | 33.9 | 33.9 | 60.4 | 53.0 | 43.6 | 39.8 | 67.2 | 72.5 |
| Orchard | 37.1 | 28.4 | 70.2 | 76.9 | 44.3 | 56.9 | 86.0 | 91.0 |
| Beets | 19.8 | 70.3 | 74.0 | 85.2 | 75.0 | 84.5 | 96.7 | 100.0 |
| Potatoes | 61.0 | 40.6 | 64.1 | 88.7 | 81.7 | 75.3 | 91.6 | 91.9 |
| Oats | 97.2 | 93.8 | 99.2 | 98.3 | 98.9 | 96.6 | 99.6 | 99.3 |
| Stubble | 97.3 | 96.9 | 97.5 | 97.7 | 98.8 | 94.8 | 98.7 | 98.6 |
| Plough soil | 93.9 | 99.3 | 100.0 | 99.9 | 98.7 | 100.0 | 100.0 | 100.0 |
| Asphalt road | 38.9 | 23.3 | 39.5 | 42.2 | 4.6 | 40.6 | 70.8 | 72.4 |
| Dirt road | 60.5 | 27.3 | 95.6 | 67.4 | 55.5 | 72.6 | 98.6 | 91.2 |
| Built-up area | 5.6 | 46.0 | 63.0 | 65.0 | 43.2 | 57.5 | 72.6 | 83.6 |
| Buildings | 69.8 | 61.7 | 68.8 | 75.1 | 75.3 | 68.4 | 94.4 | 86.3 |

Table 3. Classification accuracy of the test area

| Accuracy [%] Land cover | 13 k MNF 1x1 Overall : 74.9 | | 13 k MNF 3x3 Overall : 75.9 | | 40 k 1x1 Overall : 74.5 | | 40 k 3x3 Overall : 73.7 | |
|---|---|---|---|---|---|---|---|---|
| | Prod. | User | Prod. | User | Prod. | User | Prod. | User |
| Meadow | 89.4 | 85.1 | 93.4 | 78.6 | 89.4 | 85.1 | 93.4 | 78.6 |
| Wastelands | 38.1 | 29.3 | 8.4 | 33.3 | 38.1 | 29.3 | 8.4 | 33.3 |
| Coniferous fo. | 85.0 | 93.6 | 96.2 | 93.4 | 85.0 | 93.6 | 96.2 | 93.4 |
| Mixed forest | 59.2 | 70.7 | 66.5 | 86.2 | 59.2 | 70.7 | 66.5 | 86.2 |
| Decidous fore. | 78.4 | 62.1 | 70.3 | 70.4 | 78.4 | 62.1 | 70.3 | 70.4 |
| Woodeds | 4.8 | 30.9 | 18.7 | 21.9 | 4.8 | 30.9 | 18.7 | 21.9 |
| Orchard | 12.5 | 7.6 | 38.4 | 19.6 | 12.5 | 7.6 | 38.4 | 19.6 |
| Beets | 33.3 | 64.3 | 49.0 | 77.4 | 33.3 | 64.3 | 49.0 | 77.4 |
| Potatoes | 52.1 | 59.7 | 43.5 | 47.6 | 52.1 | 59.7 | 43.5 | 47.6 |
| Oats | 63.4 | 82.5 | 38.7 | 99.1 | 63.4 | 82.5 | 38.7 | 99.1 |
| Stubble | 98.1 | 87.8 | 98.1 | 95.3 | 98.1 | 87.8 | 98.1 | 95.3 |
| Plough soil | 66.3 | 93.3 | 87.8 | 93.7 | 66.3 | 93.3 | 87.8 | 93.7 |
| Asphalt road | 26.3 | 39.1 | 52.9 | 31.9 | 26.3 | 39.1 | 52.9 | 31.9 |
| Dirt road | 15.4 | 17.4 | 63.0 | 49.2 | 15.4 | 17.4 | 63.0 | 49.2 |
| Built-up area | 6.0 | 34.8 | 41.4 | 50.4 | 6.0 | 34.8 | 41.4 | 50.4 |
| Buildings | 70.8 | 53.0 | 73.1 | 54.5 | 70.8 | 53.0 | 73.1 | 54.5 |

Table 4. Confusion matrix of the test classification of the image data set. Expanation: Mea – meadow, Dec – decidous forest, Woo – Woodeds, Orc – orchard, Bee – beets, Pot – potatoes, Oat – oats, Stu – stubble, Asp – asphalt road, Dir – dirt road, B-u – built-up area, Bui – buildings.

| Class | Mea | Dec | Woo | Orc | Bee | Pot | Oat | Stu | Asp | Dir | B-u | Bui | User accu. |
|---|---|---|---|---|---|---|---|---|---|---|---|---|---|
| Mea | 94.8 | 1.2 | 18.4 | 11.2 | 1.4 | 17.7 | 0.0 | 1.4 | 0.0 | 11.2 | 14.3 | 0.3 | 95.8 |
| Dec | 0.7 | 93.9 | 23.0 | 6.8 | 0.0 | 0.0 | 0.0 | 0.0 | 3.8 | 0.2 | 0.2 | 0.0 | 85.4 |
| Woo | 0.3 | 0.6 | 38.8 | 5.5 | 0.0 | 0.0 | 0.0 | 0.0 | 0.0 | 0.0 | 0.0 | 0.0 | 59.5 |
| Orc | 0.1 | 0.1 | 5.2 | 44.7 | 0.0 | 0.0 | 0.0 | 0.0 | 0.0 | 0.0 | 0.0 | 0.1 | 83.0 |
| Bee | 0.0 | 0.0 | 0.0 | 0.0 | 86.0 | 0.2 | 0.0 | 0.0 | 0.0 | 0.0 | 0.1 | 0.3 | 93.2 |
| Pot | 0.1 | 0.0 | 0.0 | 0.0 | 0.0 | 63.6 | 0.0 | 0.6 | 0.0 | 0.0 | 0.0 | 0.0 | 81.6 |
| Oat | 0.0 | 0.0 | 0.0 | 0.0 | 0.0 | 0.0 | 98.4 | 0.1 | 0.3 | 0.0 | 0.0 | 0.0 | 94.7 |
| Stu | 0.0 | 0.0 | 1.2 | 0.0 | 12.6 | 15.4 | 1.6 | 96.6 | 0.0 | 6.0 | 0.0 | 0.3 | 97.6 |
| Asp | 0.0 | 0.4 | 0.0 | 0.0 | 0.0 | 0.0 | 0.0 | 0.0 | 61.3 | 3.7 | 1.9 | 1.7 | 76.6 |
| Dir | 0.5 | 0.0 | 0.9 | 0.0 | 0.0 | 3.2 | 0.0 | 0.4 | 0.3 | 56.1 | 1.2 | 1.3 | 69.6 |
| B-u | 3.3 | 3.8 | 12.6 | 31.9 | 0.0 | 0.0 | 0.0 | 0.8 | 31.0 | 22.1 | 66.8 | 18.7 | 52.3 |
| Bui | 0.2 | 0.1 | 0.0 | 0.0 | 0.0 | 0.0 | 0.0 | 0.1 | 3.4 | 0.7 | 15.5 | 77.3 | 72.2 |
| Prod. Accu. | 94.8 | 93.9 | 38.8 | 44.7 | 86.0 | 63.6 | 98.4 | 96.6 | 61.3 | 56.1 | 66.8 | 77.3 | |

## 5 CONCLUSIONS

Artificial neural network show potential for discriminating land cover types from hyperspectral imagery. Long training time is the most uncomfortable aspect of this classification. but data transformation methods like MNF can short the way to satisfied results.
The best results are observed for: oat crops (99.8 %). stubbles (96.6 %). grasslands (94.8 %). deciduous forest (93.9 %). and the worst for tree clumps (38.8 %). orchards (44.7 %) and side roads (56.1 %). Textural window (3x3) and MNF compressed data (based on 50 bands) increase classification accuracy.

## ACKNOWLEDGEMENTS

Interdisciplinary Centre for Mathematical and Computational Modelling – Warsaw Uniwerity (ICM UW). for open access to high power computers.

## REFERENCES

Kavzoglu T. & Mather P. M. 2003. The use of backpropagating artificial neural networks in land cover classification. *International Journal of Remote Sensing. vol. 24. nr 23, 2003*: 4907-4938

Zagajewski B, Kozlowska A., Krowczynska M., Sobczak M. & Wrzesien M., 2005. Mapping high mountain vegetation using hyperspectral data. *EARSeL eProceedings, 4(1) 2005*: 70-78.

*Remote Sensing for a Changing Europe*
*D. Maktav (Ed.)*
*IOS Press, 2009*

*doi:10.3233/978-1-58603-986-8-511*



# On the evaluation of vegetation resilience after two successive fires in Southern Italy by using SPOT-VGT NDVI time series

R. Lasaponara, A. Lanorte & L. Telesca
*Istituto di Metodologie per l'Analisi Ambientale, CNR, C.da S. Loja 85050 Tito (PZ) Italy*

Keywords: fire resilience, NDVI time series, SPOT-VGT

ABSTRACT: The evaluation of vegetation recovering after fire disturbance (resilience) is challenging in the framework of vegetation monitoring. This paper is focused on the response of vegetation to two successive fires in a test site of southern Italy, using the 1998 to 2005 time series of Normalized Difference Vegetation Index (NDVI) from SPOT-VEGETATION sensor. The detrended fluctuation analysis (DFA) was applied to quantify this response. The obtained results reveal that vegetation dynamics is characterized by a certain degree of resilience after the first fire, which is lower after the second fire. This result shows that the vegetation recovery mechanism is less efficient after the second fire.

## 1 INTRODUCTION

Vegetation has important ecological impacts. The characterization of vegetation reaction to fires is of primary importance since changes in the status or types of vegetation play an active role in ecological processes (such as productivity level, creation of altered patches, modification in vegetation structure and shifts in vegetation cover composition), as well as in land surface processes (such as surface energy, water balance, carbon cycle) (Huemmrich et al. 1999).

Satellite technologies can be profitably used for investigating the dynamics of vegetation regrowth after fire disturbance at different temporal and spatial scales. Nevertheless, fire-induced dynamical processes are very difficult to study since they affect the complex soil-surface-atmosphere system, due to the existence of feedback mechanisms involving human activity, ecological patterns and different subsystems of climate. Therefore, the vegetation patterns constrain fires and at the same time are constrained by the fire processes that influence them.

The remote sensing of vegetation has been traditionally carried out by using vegetation indices, which are quantitative measures, based on vegetation spectral properties, that attempt to measure biomass or vegetative vigor. The vegetation indices operate by contrasting intense chlorophyll pigment absorption in the red against the high reflectance of leaf mesophyll in the near infrared. The simplest form of vegetation index is simply a ratio between two digital values from these two spectral bands. The most widely used index is the well-known normalized difference vegetation index NDVI = [NIR-R] / [NIR+R]. The normalization of the NDVI reduces the effects of variations caused by atmospheric contaminations. High values of the vegetation index identify pixels covered by substantial proportions of healthy vegetation (Campbell 1987). NDVI is indicative of plant photosynthetic activity and has been found to be related to the green leaf area index and the fraction of photosynthetically active radiation absorbed by vegetation. Therefore variations in NDVI values become indicative of variations in vegetation composition and dynamics.

In this study, we analyze the temporal series from 1998 to 2005 of NDVI satellite SPOT VEGETATION data acquired for a shrubland test site in Sardinia (Southern Italy), affected by two consecutive fires. Our objective is to characterize quantitatively the resilient effect of vegetation cover.

## 2 METHODOLOGY

Signal fluctuations are generally investigated using the power spectral density. A $1/f\alpha$-power spectrum, where f is the frequency and $\alpha$ the scaling exponent is typical of systems characterized by long-range correlations. By estimating the scaling exponent we can have quantitative information on the type and strength of correlations in the signal and the kind of mechanisms that may be responsible of its generation. The strength of these correlations provides useful information about the inherent memory of the system (Miramontes & Rohani 2002). The detrended fluctuation analysis (DFA) (Peng et al. 1994) avoids spurious detection of correlations that are artifacts of trends and nonstationarity, that often affects experimental data. Such trends have to be well distinguished from the intrinsic fluctuations of the system in order to find the correct scaling behavior of the fluctuations. Very often we do not know the causes and the scales of these underlying trends (Kantelhardt et al. 2001). The DFA method is based on the scaling behavior of the fluctuation function

$$F(n) = \sqrt{\frac{1}{N}\sum_{k=1}^{N}\left[y(k) - y_n(k)\right]^2} \qquad (1)$$

where $y(k)$ is the integrated function of the NDVI time series and $y_n(k)$ the trend of the data in the box of duration $n$. Calculating $F(n)$ for all the available scales n, one obtains a relationship between $F(n)$ and the box size n, which for long-range correlated signals is a power-law

$$F(n) \sim n^{\alpha}. \qquad (2)$$

The scaling exponent $\alpha$ quantifies the strength of the long-range power-law correlations of the signal: if $\alpha = 0.5$, the signal is uncorrelated; if $\alpha > 0.5$ the correlations of the signal are persistent, where persistence means that a large (small) value (compared to the average) is more likely to be followed by a large (small) value; if $\alpha < 0.5$ the correlations of the signal are antipersistent, which indicates that a large (small) value (compared to the average) is more likely to be followed by a small (large) value.

## 3 RESULTS

We studied the satellite time series of 1998–2005 NDVI data for Bolotana site (Sardinia, southern Italy), in order to evaluate the resilient behaviour of vegetation. This site was affected by two consecutive fires: one occurred in July 1, 1998 and the other occurred in June 28, 2002. Figure 1 shows the test site. The selected site is homogeneous in terms of morphological features and land cover type. This was carefully verified by overlapping to the satellite data both a Digital Terrain Model (considering slope, aspect and elevation) and a Land use map. In particular, the natural shrubland cover was recognized by using the Corine land cover map provided by the European Topic Centre on Land Cover at Environmental satellite data Center in Kiruna, Sweden. The Corine map was recoded and re-sampled at the same spatial resolution as satellite data.

We analyzed the SPOT-VGT NDVI time series of 3 pixels completely affected by the two fires and 3 adjacent pixels unaffected by any fire. Each pixel has a spatial resolution of 1 $km^2$. Such data are available free of charge (http://www.vgt.vito.be). In particular, we analysed the ten-day (decadal) maximum value of daily NDVI maps. The temporal evolution of decadal NDVI composition is regarded as an effective time window able to show the natural seasonal variations, the conse-

Figure 1. Location of Bolotana test site.

quences of extreme climatic events and the man-induced damage suffered by ecosystems. The data were subjected to atmospheric corrections performed by CNES on the basis of the Simplified Method for Atmospheric Corrections (SMAC). The considered NDVI composition also allows for reducing the contamination effects due to residual clouds, atmospheric perturbations, variable illumination and viewing geometry that are generally present in daily NDVI maps.

In order to eliminate the phenological fluctuations, for each decadal composition of each pixel, we focused on the departure $NDVI_d = [NDVI - \langle NDVI \rangle]/\sigma$, where $\langle NDVI \rangle$ is the decadal mean and $\sigma$ is the decadal standard deviation. The decadal mean $\langle NDVI \rangle$ and the standard deviation were calculated for each decade, e.g. 1st decade of January, by averaging over all years in the record. Figure 2 shows the time variation of $NDVI_d$, corresponding to two pixels one unburned (Fig. 2a) and the other burned (Fig. 2b).

Figure 3 shows the results of the DFA performed on the time variation of $NDVI_d$ of pixels shown in Fig. 2: Fig. 3a shows the fluctuation function for the unburned pixel, Fig. 3b the DFA result for the subset of $NDVI_d$ series of the burned pixel between the two fires, and Fig. 3c the results of DFA for the $NDVI_d$ of the subset after the second fires.

We observe that all the pixels show a value of the scaling exponents, estimated by the slope of the line that fits in a least square sense the curves plotted in log-log scales, larger than 0.5. This indicates that the temporal fluctuations of all time series are positively correlated or persistent. Persistence means that the investigated ecosystems are governed by positive feedback mechanisms, which tend to destabilize the system under external forces. The feedback mechanisms express a positive circular causality that acts as a growth-generating phenomenon and therefore drives unstable patterns. Therefore the vegetational processes take memory of external shocks, which drive the time dynamics of the vegetational covers. After the first fire, the scaling exponent is larger than that for the unburned pixel, indicating a larger persistent behaviour of vegetation after fire to recover. This suggests that fires play an important role in the temporal evolution of the vegetation, increasing the persistence of its dynamics. This seems to express the inherent character of the fire-related vegetation recovery processes, which indicate the existence of positive relation between the amounts of burned and regenerated biomass. This result highlights that fires drive more unstable

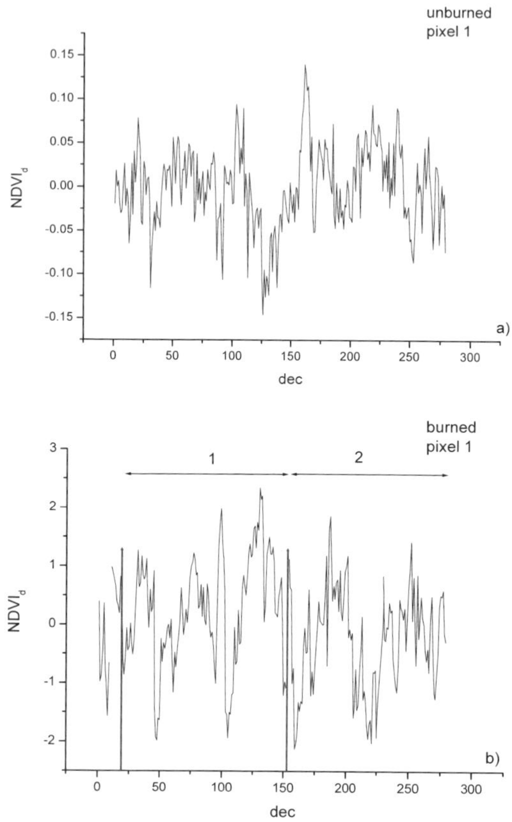


Figure 2. $NDVI_d$ time series for an unburned (a) and burned (b) pixel in Bolotana site.

vegetational patterns in shrubland covers, and this indicates an efficient fire-induced vegetation recovery processes.

But, after the second fire, the scaling exponent becomes slightly lower, thus indicating a less efficient recovery mechanism of vegetation, which is interrupted by a successive stress, like the second fire.

Figure 4 shows the mean scaling exponent calculated for the three NDVI time series of pixels not affected by any fire, those included within the two fires, and those after the second fire. It is clear the discrimination between the three stati of vegetation, with the lowest persistent degree for the unburned pixels, the highest persistent degree for the pixel time series between the tow fires. After the second fire vegetation decreases its recovery capacity, thus reducing its resilience.

## 4 CONCLUSIONS

The detrended fluctuation analysis of NDVI data was applied to quantify the behavioral dynamical trends in vegetation dynamics of a shrubland test site in Italy, affected by two successive fires. The estimated scaling exponents are larger than 0.5, suggesting that vegetation is persistent, but the first fire occurrence increases the degree of persistence of vegetation. This indicates that more unstable patterns have been induced in vegetation dynamics, after fire disturbance. But after the second fire, the efficiency of the recovery mechanism lowers, due to a more stable response of vegetation to the

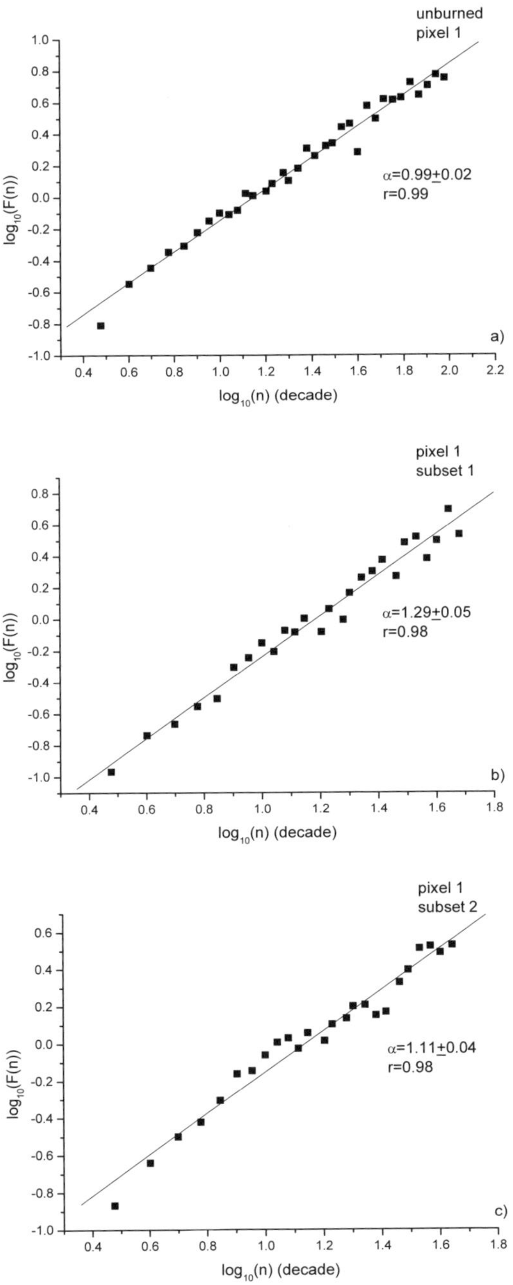


Figure 3. DFA results for the pixel time series shown in Fig. 2: (a) unburned, (b) between the two fires), c) after the second fire.

second fire stress. The value of $\alpha$, after the fire occurrence, can be used as a quantitative indicator of the resilience, which the capability of vegetation recovery after the fire occurrence. The $\alpha$ exponent larger than 0.5 in unburned sites probably is related to the successional status of ecosystem: in situation of dynamical equilibrium, the variation among years respond to the meteorological oscil-

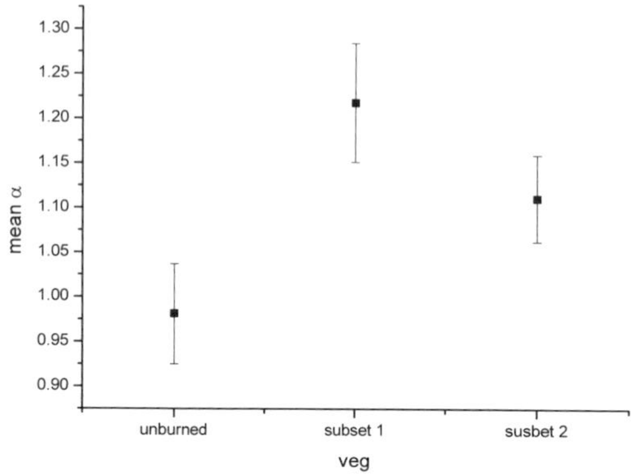


Figure 4. Mean and standard deviation of the α exponents calculated for the Bolotana pixels.

lations and the changes in year (intra-annual) respond to the seasonal dynamics of biomass (phenologycal changes). The perturbation caused by a fire increases the vegetation persistence (larger α), thus indicating the trend of ecosystem in recovering after fire; but the second fire interrupts this recovery, leading to a less efficient resilient behaviour.

## REFERENCES

Huemmrich, K. E., Black, T. A., Jarvis, P. G., McCaughey, J. H. & Hall, E. G. 1999. Remote sensing of carbon/water/energy parameters: High temporal resolution NDVI phenology from micrometeorological radiation sensors. *Journal of Geophysical Research* 104: 27935-27944.

Campbell, J. B. 1987. *Introduction to Remote Sensing* (New York, Guildford).

Miramontes, O. & Rohani, P. 2002. Estimatine $1/f^{\alpha}$ scaling exponents from short time-series. *Physica D* 166: 147-154.

Peng, C.-K., Buldyrev, S. V., Havlin, S., Simons, M. Stanley, H. E. & Goldberger, A. L. 1994. Mosaic organization of DNA nucleotides. *Physical Review E* 49: 1685-1689.

Kantelhardt, J. W., Konscienly-Bunde, E., Rego, H. H. A., Havlin, S. & Bunde, A. 2001. Detecting long-range correlations with detrended fluctuation analysis. *Physica A* 295: 441-454.

*Remote Sensing for a Changing Europe*
*D. Maktav (Ed.)*
*IOS Press, 2009*


*doi:10.3233/978-1-58603-986-8-517*

# Evaluation of urban shape variability in Southern Italy by using LANDSAT data

R. Lasaponara, R. Coluzzi & L. Telesca
*Istituto di Metodologie per l'Analisi Ambientale, CNR, C.da S. Loja 85050 Tito (PZ) Italy*

Keywords: Urban morphology, box counting, LANDSAT

ABSTRACT: This study deals with the spatial characterization of urban expansion of a small town in southern Italy by using spatial fractal analysis applied to multidate Multispectral Scanner (MSS) and Thematic Mapper (TM) satellite images. The investigation was focused on Altamura town (southern Italy), for which the border was extracted from NASA Landsat images acquired in 1976 (MSS), in 1987 (TM) and 1998 (TM). The border was analyzed using the box counting method, which is a well-know technique to estimate the spatial fractal dimension. The fractal dimension is a parameter used to quantify the shape irregularity of an object. The obtained results show a certain increase of the fractal dimension indicating an increase of the irregularity degree of the border of the investigated town from 1976 to 1998. This variation is also connected with the urban expansion and the population growth.

## 1 INTRODUCTION

Understanding the size distribution and dynamic expansion of urban areas is a key issue for the management of city growth and mitigation of negative impacts on environment and ecosystems. Although urban growth is perceived as necessary for a sustainable economy, uncontrolled or sprawling urban growth can cause various problems such as loss of open space, landscape alteration, environmental pollution, traffic congestion, infrastructure pressure, and other social and economical issues. To face these drawbacks, a continuous monitoring of the urban growth evolution in terms of type and extent of changes over time are essential for supporting planners and decision makers in future urban planning. The analysis of the city size distribution deals with different disciplines such as geography, economy, demography, ecology, physics, statistics because the evolution of a city is a dynamic process involving a number of different factors. The main issue of great importance in modelling urban growth includes spatial and temporal dynamics, scale dynamics, man-induced land use change.

A critical point for understanding and monitoring urban expansion processes is the availability of both (i) time-series data set and (ii) updated information relating to the current urban spatial structure and city edges to define and locate the evolution trends. In such a context, an effective contribution can be offered by satellite remote sensing technologies, which are able to provide both historical data archive and up-to-date imagery. Satellite technologies represent a cost-effective mean for obtaining useful data that can be easily and systematically updated for the whole globe. The use of satellite imagery along with spatial analysis techniques can be used for the monitoring and planning purposes as these enable the reporting of ongoing trends of urban growth at a detailed level.

Nevertheless, the exploitation of satellite Earth Observation in the field of the urban growth monitoring is a relatively new tool, although during the last three decades great efforts have been addressed to the application of remote sensing to detecting land use and land cover change.
Over the years, satellite time series such as Landsat TM, MSS images were used to assess the urban growth, mainly for several cities (Masek et al. 2000; Yang & Lo 2002; Yuan et al. 2002).

Many recent researches have also explored the way of measuring the dynamics of urban morphology. Shen (2002) for example compared the morphology of 20 urban areas in USA obtaining a wide range of results due to the different size and character of each case study. Frankhauser (1998) has also used the fractal dimension in the examination of outskirt areas in European cities trying to obtain a typology of urban agglomerations. Finally Benguigui et al. (2000) by examining the built-up settlement of Tel Aviv concludes that the fractal dimension tends to increase through time.

This paper is focused on the fractal analysis of the border of one small town in southern Italy (Altamura), using multidate NASA Landsat images acquired in 1976, 1987 and 1998. The fractal method used was the box counting, well suited to analyse the spatial properties of fractal objects.

## 2 DATA SET AND STUDY AREA

Since 1972, the Landsat satellites have provided repetitive, synoptic, global coverage of high-resolution multispectral imagery. The characteristics of the MSS and TM bands were selected to maximize each band's capabilities for detecting and monitoring different types of land surface cover characteristics. In the present study, the spatial resolution of MSS is 57 m, while that of TM is 30m. The data were downloaded freely from NASA satellite archive (www.landsat.org).

The investigation was performed by using one MSS image acquired in 1976 and two TM images acquired in 1987 and 1998 respectively (Figs 1–3). The increase of the urban area observed from 1976 to 1998 was connected with the urban expansion due to economic and demographic factors. Altamura town is located on the Murge plateau that is a characteristic area of Apulia (southern Italy) composed of limestone hills sparsely covered by vegetation but with a very high ecological values so that in the in 2004, a national park named Parco Nazionale dell'Alta Murgia was instituted. The town is characterized by an active and dynamic local economy mainly based on small and medium enterprises operative in the commerce, industry and services.

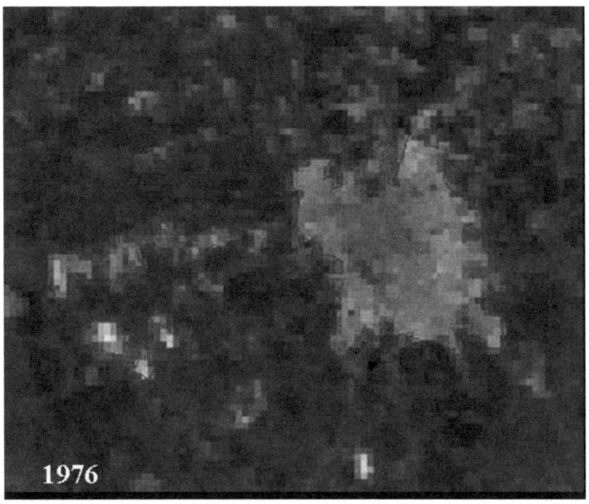


Figure 1. Border of Altamura town in 1976 (MSS).

Figure 2. Border of Altamura town in 1987 (TM).

Figure 3. Border of Altamura town in 1998 (TM).

## 3 METHOD

The use of fractal dimension as a measure of shape irregularity is well known. For a regular 2-dimensional curve the fractal dimension is 1. Spatial fractal curves are characterized by fractal dimension larger than 2. In this paper, the fractal dimension is computed with the box-counting method. The box-counting method computes the degree of irregularity of borders; therefore higher the fractal dimension, more irregular the border. The image of the town under study is divided in non-overlapping cells of size L. Then the number of cells occupied by at least one point belonging to the border is computed. The procedure is iterated for different sizes L and the number of cells as a function of the size L behaves as a power-law for fractal borders:

$$N(L) \approx L^{-df_{BOX}} \tag{1}$$

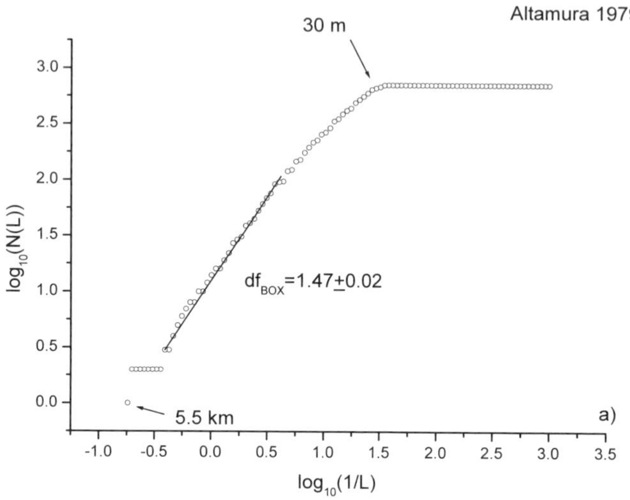


Figure 4. Box-counting results of the town borders in the years 1976, 1987 and 1998.

where $df_{box}$ is the box-counting fractal dimension:

$$df_{box} = \lim_{L \to 0} \frac{\ln(N(L))}{\ln(1/L)} \tag{2}$$

The estimate of the fractal dimension is performed calculating the slope of the line fitting the Eq. 1 plotted in log-log scales in its linear part.

## 4 RESULTS

The border of the investigated town was analysed by comparing the patterns of development of the years 1976, 1987 and 1998. Figure 4 shows the number of cells N(L) as function of 1/L, where L is the size of the box. In all the plots the length scale varies between approximately 30 m and 5.5–6.5 km. The higher length scale is consistent with the maximum linear dimension of the border; the lower is consistent with the spatial resolution of the images. The fractal dimension was calculated in the linear range scales between 300m and 2.5 km.

It must be noted that the reliability of the calculated $df_{Box}$ value is examined by the correlation coefficient r, which in our case ranges around 0.99. The value of the coefficient can be considered good if we take into account the suggestions of Batty (2005) and Benguigui et al. (2000). It can be therefore stated that the urban pattern under investigation displays clearly a fractal structure.

It is visible an increase of the fractal dimension from 1.47 in 1976, to 1.51 in 1987, and 1.52 in 1998, indicating a tendency of the urban expansion to a more irregular shape.

Taking into account the obtained results of fractal analysis, we can observe that the distribution of the built-up area was more homogeneous and less fragmented in the year 1976. During the period up to 1998 changes led to an increase of density on every scale leading into an increase in the value of dfBox from 1.47 to 1.52. Furthermore the fragmentation and heterogeneity of the urban development were increased, because the urban boundary has been enlarged to areas that in 1976 were empty.

## 5 CONCLUSIONS

In the present letter, fractal analysis was considered as a method of examining the transformations taking place in one urbanized area located in southern Italy. The built-up pattern of the town was examined and the main objective was the determination of whether its structure can be described as more or less regular. Analysing and comparing three different years, the process of urban intensification was observed, and the increase of the value of the fractal dimension of the urbanized area was revealed. This change shows the transformation that took place in the town under investigation and the transformation from a quite regular to a more fragmented peripheral settlement. Moreover the box-counting analysis showed the spatial fractal character of the urban pattern where similar structures are observed at many scales, according with results obtained in other researches (Frankhauser 1998; Benguigui et al. 2000; Batty &Longley 1994).

The relevance of the technique used here is that it provides a reliable way of quantifying the urban structure and its transformation through time.

However this study is preliminary and quite suggestive and its main objective was to present a way of applying the fractal analysis to the monitoring of urban area evolution. The need for the analysis of more time periods and a comparative analysis between many urban areas would be fruitful, and the application of fractal analysis and the focusing on the variation of fractal dimension across space and through time constitutes a major challenge of further investigation.

## REFERENCES

Batty, M. & Longley, P. 1994. *Fractal cities, a Geometry of Form and Function*. Academic Press.

Batty, M. 2005. *Cities and complexity: Understanding cities with Cellular Automata, Agent-Based models, and Fractals*. MIT Press, Cambridge, Massachusetts, London, England.

Benguigui, B., Chamanski, D. & Marinov, M. 2000. When and where is a city fractal?. Environ. Planning B 27: 507-519.

Frankhauser, P. 1998. The Fractal Approach, a new tool for the spatial analysis of urban agglomerations, Population: An English Selection, Vol. 10, No. 1, New Methodological Approaches in the Social Sciences, 205-240, Institut National d' Etudes Démographiques.

Masek, J. G., Lindsay, F. E. & Goward, S. N. 2000. Dynamics of urban growth in the Washington DC metropolitan area, 1973-1996, from Landsat observations. Int. J. Rem. Sensing 21: 3473-3486.

Shen, G. 2002. Fractal dimension and fractal growth of urbanized areas, *Int. J. Geogr. Inf. Sci.* 16: 419-437.

Yang, X. & Lo, C. P. 2002. Using a time series of satellite imagery to detect land use and land cover changes in the Atlanta, Georgia metropolitan area. Int. J. Rem. Sensing 23: 1775-1798.

Yuan, F., Sawaya, K., Loeffelholz, B. C. & Bauer, M. E. 2005. Land cover classification and change analysis of the Twin Cities (Minnesota) Metropolitan Area by multitemporal Landsat remote sensing. Rem. Sensing Environ. 98: 317-328.

*Remote Sensing for a Changing Europe*
*D. Maktav (Ed.)*
*IOS Press, 2009*

*doi:10.3233/978-1-58603-986-8-522*

# A review of satellite data processing methods applied in archaeological context

R. Lasaponara, [1] & N. Masini [2]

[1] *Istituto di Metodologie per l'Analisi Ambientale, CNR, C.da S. Loja 85050 Tito (PZ) Italy Tel. ++39 0971 423214 – fax ++39 0971 427 271– e_mail: lasaponara@imaa..cnr.it*

[2] *Istituto Beni Archeologici e Monumentali, IBAM-CNR, C.da S. Loja, 85050 Tito Scalo (PZ), Italy. Tel. ++39 0971 427321 – fax ++39 0971 427 222– e_mail: n.masini@ibam.cnr.it*

Keywords: Remote sensing of archaeology, VHR satellite, PCA, HIS, Data Fusion, edge detection

ABSTRACT: During the last century, aerial photos have been the most common remote sensing data source used for archaeological investigation. Beginning from the eighties multi-spectral satellite images,such Landsat TM, ASTER, have been successfully applied in the field of cultural heritage with particular reference to landscape archeology and palaeoenvironment studies.
More recently the higher spatial resolution of Very High Resolution (VHR) satellite images, such as, IKONOS (1999), QuickBird (2002) and Worldview (2007), offers improved capability in locating potential buried archaeological sites. Nevertheless, the effectiveness of satellite needs to be tested in relation to the specific archaeological problems and individual regions. The satellite-based identification of traces of ancient human activity is highly dependent on many factors, among them. (i) the image spatial resolution, (ii) the extension of buried sites, (iii) specific surface characteristics, (iv) illumination conditions, and (v) data processing adopted to enhance archaeological features.

This paper provides a review of satellite data processing methods, applied to support the archaeological research, such as RGB, IHS, PCA, TCT, Spectral indices, pixel-based and object oriented classification techniques, data fusion and edge detection.

## 1 INTRODUCTION

The identification of traces of human activity on ancient environments is one of the greatest challenges not only in the field of archaeology, but, also for botany, forestry, hydrology, soil and geoscience. Traces of ancient human activity can provide suitable information for improving the knowledge and the understanding of the environmental interaction, climate change, and the Earth's system.

The detection of archaeological features can be fruitfully performed by using remote sensing technologies, available from ground, air and space. Over the years, aerial photos and geophysical prospection have been the most common remote sensing data source used by the archaeologists. More recently, satellite remote sensing data from active and passive sensors have been adopted for searching larger areas in order to acquire accurate quantitative information on ancient landscape, and to detect archeological features in a more rapid manner compared to other traditional survey techniques. In particular, compared to other remote sensing techniques, satellite images are able to provide both: (i) archaeological evidence linked to traces of past human activity and (ii) a synoptic view of the landscape with its geographic attributes and natural environment context.

The higher spatial resolution of Very High Resolution (VHR) satellite images, such as (i) IKONOS (1999) with panchromatic at 1 m, and multispectral at 4m, (ii) QuickBird (2001) with panchromatic at 0.6m and multispectral at 2.4, and (iii) Worldview (2007) panchromatic at 0.5m, offers improved capability in locating potential buried archaeological sites. Moreover, compared to aerial photos the VHR satellite imagery offers a very large coverage, and a powerful multispectral capability. Moreover, they are geo-referenced and, therefore, ready to be integrated with external data sources. This makes VHR satellite imagery an ideal tool for searching large areas and for supporting study performed in areas where aerial photography could be restricted because of military or political reasons. Nevertheless, the possibility of locating unknown buried archaeological sites as well as large scale cultural features is highly dependent on many factors, among them, the image spatial resolution and extension of buried sites, the specific archaeological marks and ground characteristics, the acquisition time (year, season, day) and illumination conditions.

This paper provides a review of satellite data processing methods widely applied in the archaeological context to enhance spatial features linked to traces of ancient human activity.

## 2 SATELLITE DATA AND ARCHAEOLOGICAL FEATURE IDENTIFICATION

Over the years, different satellite sensors have been used for archaeological investigations (see references listed below) addressed to the: (i) characterization of spectral signatures of archaeological marks, (ii) mapping of subsurface remains, (iii) geo-statistical analysis, (iv) management and protection of archaeological sites.

(i) Landsat, (http://landsat.gsfc.nasa.gov/about/L7_td.html ) (see, for example, Parry, 1992, Fowler, 2002, Behnaz & Samani 2006,),

(ii) SPOT HTTP://WWW.SPOTIMAGE.FR/WEB/EN/171-PRODUCTS-AND-SERVICES.PHP) (Clark et al., 1998),

(iii) ASTER http://lpdaac.usgs.gov/aster/AST14OTH.asp, (Altaweel, 2005)

(iv) CORONA http://msl.jpl.nasa.gov/Programs/corona.html (see; for example,Kennedy, 1998, Goossens et al., 2006, Beck et al. 2007),

(v) IKONOS http://www.landinfo.com/satprices.htm (De Laet et al. 2007, Beck et al. 2007),

(vi) QUICKBIRD http://www.digitalglobe.com/index.php/6/DigitalGlobe+Products (Lasaponara &Masini, 2005,2006, Masini & Lasaponara, 2007)

The satellite-based identification of spatial features linked to the presence of buried archaeological remains is one of the most complex and challenging tasks faced by computer vision and photogrammetry communities This is due to the fact that archaeological deposits tend to induce various kinds of marks, such as, soil, crop and shadow marks (Dassie 1978; Wilson 1982; Bewley 2003) generally characterized by spatial and spectral anomalies. As an example, Figure 1 shows the typical phenological behaviour and spectral characteristics of crop marks.

Archeological marks are generally not evident on the ground, but, they could be only recognized from air. The visibility of such marks strongly depend on (i) vegetation cover, (ii) soil type, (iii) illumination conditions, (iv) view geometry, (v) specific (day, year) time of data acquisition, etc. The enhancement of the different type of archeological marks can be performed by using several data processing methodologies (see section 3).

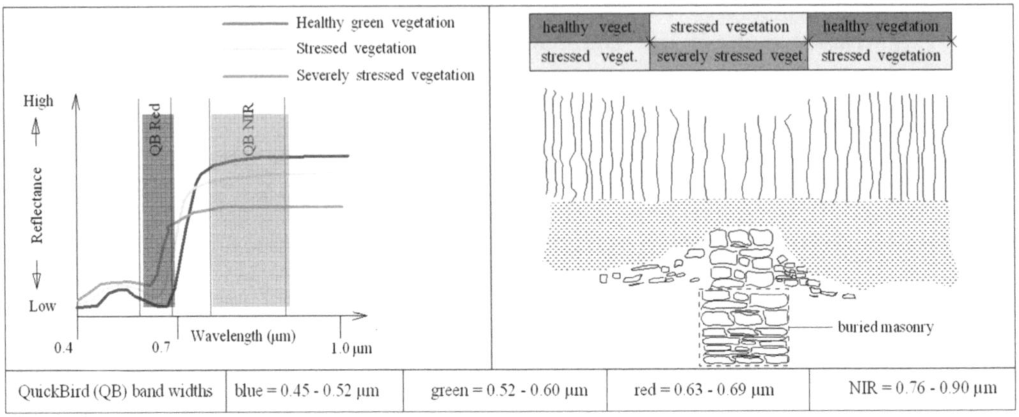


Fig. 1 Spectral characteristics of vegetation (left) ; crop-marks: phenological behavior (right)

## 3 METHOD

Traditional methods generally adopted for enhancing archaeological feature are: (i) RGB, (ii) IHS, (iii) PCA, (iv) TCT, (v) Spectral indices, (vi) Classification (Supervised, MLC, SAM, Object ; oriented). Recently other methods have been used, such as: (i) data fusion, (ii) edge- detection, fourier and wavelet transform.

### 3.1 RGB and IHS

Over the years, RGB has been widely used for enhancing archaeological features in satellite images as a unique tool for data analysis (see, for example, Fowler,2002) or along with other (). RGB can be regarded as a three-dimensional image in which each pixel is characterized by a vector red, green, and blue (RGB). Figure 2 shows the RGB colour cube.

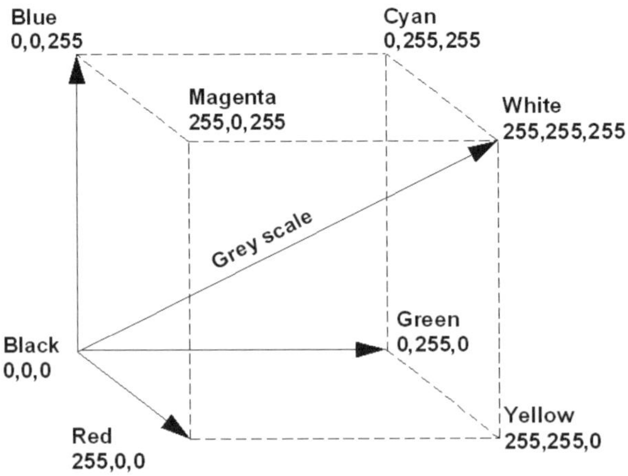


Fig. 2 RGB colour cube, which shows the interrelationships between the colours and is defined by the brightness levels of each of red, green and blue.

The RGB visualization is generally used to merge multiple co-registered images in order (i) to improve the visual and analytical quality of the data and (ii) to overcome the difficult for the human eye to select different features of the earth's surface in a greyscale image

In the RGB visualization, 3 greyscale satellite images, generally red (red spectral band), green (green spectral band) and blue (blue spectral band) are combined into one RGB colour image. Using such combination, the RGB image close resembles a colour photograph. The colours of the different features depend on the selected RGB bands The use of different spectral combinations provides different renderings of the same feature.

An alternative way to describe colours by their RGB components is the Intensity-Hue-Saturation (IHS), also used in archaeological context (see, for example Campana & Forte, 2001). The IHS colour space is much more closer than RGB to what the human eye can perceive. The IHS colour transformation is based on the color sphere in which the vertical axis represents Intensity, the radius represents saturation, and the circumference represents hue.

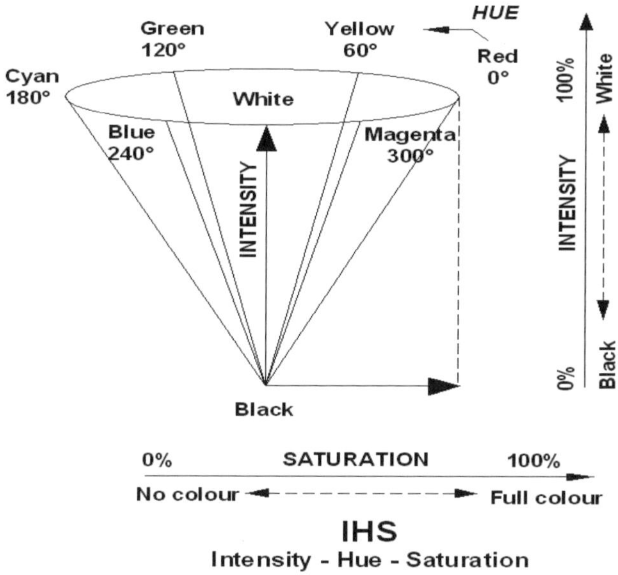


Figure 3 The IHS colour transformation

The intensity represents brightness variations, saturation represents the purity of color or amount of white, and hue represents the dominant wavelength of color. IHS can improve image sharpness and edge extraction. That why is commonly used as a basic data fusion technique.

## 3.2 Principal Component Analysis.

The Principal Component Analysis (PCA) has been widely adopted to identify the small spectral signals linked to archaeological features (see for example, Campana and Forte, 2001). The PCA (see, for example, Richards, 1986, 2004) is a linear transformation which decorrelates multivariate data by translating and/ or rotating the axes of the original feature space. In this way the data can be represented without correlation in a new component space. In order to do this, the process firstly computes the covariance matrix (Unstandardized PCA) or the correlation (standardized PCA) among all input spectral. Then eigenvalues and eigenvectors are calculated in order to obtain the new feature components.

The PCA transforms the input multispectral bands in new components whose number is equal (or less) to the input channels. The first component contains the major portion of the variance, each successive component contains less of the total dataset variance. The first component provides a sort of average of all the input channels, whereas the later components may represent information variance for a small area or essentially noise. In the latter case it must be disregarded. Some problems can arise from the fact that eigenvectors can not have a general and universal meaning since they are extracted from the data set itself. Therefore, a part for the first PCA component, it is not possible to give indication on the meaning of the later PCA components.

The PCA should be able to make the identification of distinct features and surface type easier. This is a direct result of different facts: (i) only the meaningful low correlated data can be considered, (ii) the effect of noise can be easily identified and strongly reduced because it is in the later components. Moreover, the PCA can be used to obtain a new color enhancement technique, "Decorrelation Stretching", based on the following steps: (i) PCA transformation, (ii) each PCA component is contrast stretched, (iii) stretched components are rotated back using the inverse PCA transformation (iv) finally, the resulting components are shown in RGB. The removal of the contribution of noisy is very important, because if noise is a major component of the scene variation, the algorithm will enhance it. The decorrelation stretch is best suited to the channels with Gaussian (or near Gaussian) distribution, although, the algorithm is quite insensitive to substantial deviations from the ideal Gaussian form.

### 3.3 *Tasseled-Cap Transformation (TCT)*

The TCT, also known as Kauth-Thomas technique, was devised for enhancing spectral information content of satellite data ( see for example Richards, 1986). The reader is referred to Crist and Kauth (1986), for an introduction. The TCT is a linear affine transformation substantially based on the conversion of given input channel data set in a new data set of composite values. The transformation depends on the considered sensor. The original TCT was derived (Kauth and Thomas, 1976) for the four bands of the Landsat MSS sensor. Later, the TCT was extended to the Landsat TM (Crist and Cicone, 1984), ETM (as available, for example, in a routine of geomatica, PCI software) and IKONOS sensor (Horne 2003). All the existing TCTs are performed on a pixel basis to best show the underlying structure of the image by using weighted sums of the input channels. Usually there are just three composite variable tasseled cap transform bands that are used: (i) TCT-band 1; TCT-band 2; TCT-band 3.

TCT-band 1 is a weighted sum of all spectral bands and can be interpreted as the overall brightness or albedo at the surface. TCT-band 2 primarily measures the contrast between the visible bands and near-infrared bands and is similar to a vegetation index. The TCT-band 3 can be interpreted as a measure of soil and plant moisture. The weighted sums, adapted from the ETM input channels, are listed in equations 5 to 7 :

$$TCT_{ETM}\text{-band } 1 = 0.1544 \text{ BLUE} + 0.2552 \text{ GREEN} + 0.3592 \text{ RED} + 0.5494 \text{ NIR} \quad (5)$$

$$TCT_{ETM}\text{-band } 2 = -0.1009 \text{ BLUE} - 0.1255 \text{ GREEN} - 0.2866 \text{ RED} + 0.8226 \text{ NIR} \quad (6)$$

$$TCT_{ETM}\text{-band } 3 = 0.3191 \text{ BLUE} + 0.5061 \text{ GREEN} + 0.5534 \text{ RED} + 0.0301 \text{ NIR} \quad (7)$$

The weighted sums developed by Horne (2003) for IKONOS input channels are listed in equations 8 to 11:

$$TCT_{IKONOS}\text{-band } 1 = 0.326 \text{ BLUE} + 0.509 \text{ GREEN} + 0.560 \text{ RED} + 0.567 \text{ NIR} \quad (8)$$

$$TCT_{IKONOS}\text{-band } 2 = -0.311 \text{ BLUE} - 0.356 \text{ GREEN} -0.325 \text{ RED} + 0.819 \text{ NIR} \quad (9)$$

$$TCT_{IKONOS}\text{-band } 3 = -0.612 \text{ BLUE} - 0.312 \text{ GREEN} + 0.722 \text{ RED} -0.081 \text{ NIR} \quad (10)$$

$$TCT_{IKONOS}\text{-band } 4 = -0.650 \text{ BLUE} + 0.719 \text{ GREEN} -0.243 \text{ RED} -0.031 \text{ NIR} \quad (11)$$

As in the case of PCA, the tasseled cap transformation also corresponds to a linear transformation of the data to a new set of orthogonal axes. In PCA, the axes are dependent on statistical relationships among the spectral bands of the particular image under investigation, whereas in the tasseled cap transformation the axes are defined a priori, "wetness", "greenness" and "brightness", from physical-based statistically analyses. The TCT is particularly useful in defining soils (wetness) and vegetation (greenness). The TCT should be able to make the identification of distinct archaeological features easier, because (i) archaeological remains that appear on the surface may influence surface brightness and therefore they could be enhanced from the TCT-band 1, (ii) buried remains may influence the state of vegetation cover or soil moisture content, and therefore they could be enhanced from the TCT-band 2 and 3 respectively.

### 3.3 *Spectral indices*

Spectral indices are combinations of different bands which are widely used to emphasize the spectral changes caused by the presence/type/status of vegetation cover, as well as soil types and characteristics.

The widely used index is the Normalized Difference Vegetation index (NDVI) obtained by using the following formula:

$$NDVI=(\rho NIR-\rho RED)/(\rho NIR+\rho RED) \qquad (12)$$

The NDVI operates by contrasting intense chlorophyll pigment absorption in the red against the high reflectance of leaf mesophyll in the near infrared. The normalization of the NDVI reduces the effects of variations caused by atmospheric contamination.

The NDVI is has been successfully used to enhance crop marks for both homogeneous vegetation covers (Lasaponara & Masini) and sparse herbaceous plants (Lasaponara & Masini) for satellite images acquired during different period of the year spring to summer, characterized by green and dried vegetation respectively.

NDVI a quantitative measure, based on the vegetation spectral properties, that attempt to measure biomass or vegetative vigor. NDVI is indicative of plant photosynthetic activity and has been found to be related to the green leaf area index and the fraction of photosynthetically active radiation absorbed by vegetation. High values of the vegetation index identify pixels covered by substantial proportions of healthy vegetation, whilst disease or stressed vegetation exhibits lower NDVI values.

### 3.4 Classification

The Image classification techniques allow us to identify and group pixels according to their similar scattering (within a single image) or spectral characteristic (within multiple images). Digital image classification procedures can be differentiated into (i) unsupervised and (ii) supervised. The unsupervised classification techniques is performed without any prior knowledge of the image. Pixels are grouped into a pre-defined number of classes, according to their reflectance features. Supervised classification is a semi-automatic technique where the training phase is the basic step, to identify the spectral characteristics and group pixels according to the spectral signatures of the training areas.

Supervised and unsupervised classification algorithms are methods based on pixel analysis. Both of them could fail when applied to VHR satellite images, producing inconsistent classification results. To cope with this issue new classification techniques, based on object-oriented approach have been developed. The object-oriented classification techniques take the form, textures and spectral information into account.

In an object-oriented classification, instead of processing and classifying single pixels, homogenous image objects are extracted during the basic step of segmentation, which groups neighboring pixels into meaningful areas. Such segmentation and topology generation must be performed accordingly. It must be set considering both the resolution and the scale of the expected objects, and can be done in multiple resolutions, to differentiate several levels of object categories. The object-oriented classification has been successfully applied in archaeological context by De Laet et al. 2007.

### 3.5 Data Fusion

The enhancement of archaeological marks can be fruitfully improved by using data fusion techniques which allow us to jointly exploit the higher spatial resolution of the satellite panchromatic image and the spectral capability of the multispectral images. This can make the detection of archaeological features easier. Moreover, another advantage of using data fusion products is that the increased spatial resolution can fruitfully provide a more accurate localization of the archaeological features. This more accurate localization, from the initial spatial resolution of multispectral data to the spatial resolution of panchromatic can be very helpful during in situ survey, such as GPS (Global Position System) campaigns, geophysical prospection or excavation trials.

The data fusion can be performed using various satellite data such as Landsat ETM, SPOT, IKONOS and Quickbird which provides both panchromatic and multispectral images acquired at the same time. The widely used algorithms applied to enhance the archaeological features in VHR satellite images are those based on HIS, PCA transformation, Zhang, for a comprehensive review the reader is referred to (Lasaponara and Masini 2007).

### 3.6 Edge detection

The edge technique is commonly used to (ii) enhance spatial pattern to be recognized easier by a visual interpretation and (ii) extract features using local statistics. A different edge detection approaches have been developed over the years, see for example Richards (1986 and later editions).

In satellite data processing for archaeological prospection, the edge detection has been performed by applying a multiscale approach based on the scale-space theory (see Lasaponara & Masini 2006a,b , Masini & lasaponara 2006) that uses Gaussian smoothing kernels. The selection of scale was performed by keeping in mind that, in the case of archaeological features detection it is necessary to focus on structures having small sizes and signal amplitudes as expected in the case of surface anomalies due to buried walls, buildings and roads.

## 4 CONCLUSIONS

This paper provides a brief review of the most widely used data processing techniques in the field of satellite- archaeology. The availability of VHR satellite images has open new perspective for searching large areas to identify traces of past human activity. It is expected that the potentiality of satellite images for identifying and analyzing exposed and subsurface archaeological remains will strongly increase according to the improved spectral and spatial capability of next satellite sensors (http://www.satimagingcorp.com/satellite-sensors/geoeye-1.html) such as (i) the GeoEye-1 (whose launch has been anticipated in late 2008) that will be capable of acquiring image data at 0.41 meter panchromatic and 1.65 meter multispectral resolution; or the WorldView-2 (whose launch has been anticipated in late 2008) that will provide a high resolution Panchromatic (at 0.46 meter at nadir) eight (8) Multispectral bands; four (4) standard colors (red, green, blue, and near-infrared 1) and four (4) new bands (coastal, yellow, red edge, and near-infrared 2), at 1.8 m at nadir view.

The improved capability of next satellite data, in terms of both spectral and spatial resolution, along with the development/application of effective and robust data processing techniques will provide ideal tools to be used in different phases of archaeological heritage management from detection of unknown of sites, detailed reconstruction, protection, conservation and valorisation.

## REFERENCES

Altaweel, M., 2005. The use of aster satellite imagery in archaeological contexts. Archaeological Prospection 12, pp. 151–166.

Beck A., G. Philip, M Abdulkarim. D. Donoghue, 2007 Evaluation of Corona and Ikonos satellite imagery for archeological prospection in wesrtern Syria, 81 161-175

Behnaz A. & F. Samani 2006, Identifying the boundaries of the historical site of Persepolis using remote sensing Remote Sensing of Environment 102, -252-62

Bewley R H 2003 Aerial survey for archaeology *Photogrammetric Record* **18** (104) 273-292.

Campana S., Forte M., Remote Sensing in Archaeology, XI Ciclo di Lezioni sulla Ricerca Applicata in Archeologia (Certosa di Pontignano, Siena, 6-11 novembre 1999), Quaderni del Dipartimento di Archeologia e Storia delle Arti – Sezione Archeologia, n.51-52, Firenze 2001.

Clark C D, Garrod S M and Parker Pearson M 1998 Landscape archaeology and Remote Sensing in southern Madagascar International Journal of Remote Sensing 19(8) 1461-1477.

De Laet V., Paulissen E., Waelkens M Methods for the extraction of archaeological features very high-resolution Ikonos-2 remote sensing imagery, Hisar (southwest Turkey). Journal of Archaeological Science 34 (2007) 830e841

Dassie J 1978 *Manuel d'archeologie aerienne* (Paris : Editions Technip).

Fowler M J F 2002 Satellite Remote Sensing and Archaeology: a comparative study of satellite imagery of the environs of Figsbury Ring, Wiltshire Archaeological Prospection 9 55-69.

Goossens R, De Wulf A, Bourgeois J., Gheyle W.*, Willems T. Satellite imagery and archaeology: the example of CORONA in the Altai Mountains, Journal of Archaeological Science 33 745-755.

Lasaponara R and Masini N 2006a On the potential of Quickbird data for archaeological prospection International Journal of Remote Sensing, Vol. 27, No. 16, 3607–3614

Lasaponara R and Masini N 2006b Identification of archaeological buried remains based on Normalized Difference Vegetation Index (NDVI) from Quickbird satellite data IEEE Transactions on Geoscience and Remote Sensing Letters,

Masini N and R Lasaponara 2006 Satellite-based recognition of landscape archaeological features related to ancient human transformation Journal of Geophysics and Engineering 3, 230-235.

Parry J T 1992 The investigative role of Landsat TM in the examination of pre-proto-historic water management sites in northeast Thailand Geocarto Int. 4 5–24.

Richards J. A. 1986 ( and following editions) Remote Sensing Digital Image Analysis Springeer-Verlag

Wilson D R 1982 Air photo interpretation for archaeologists (London: St. Martin's Press).

*Remote Sensing for a Changing Europe*
*D. Maktav (Ed.)*
*IOS Press, 2009*

*doi:10.3233/978-1-58603-986-8-530*

# Architectural survey for documentation of cultural heritage with new sensor technologies

Cemal Özgür Kivilcim
*Istanbul Metropolitan Municipality, Historical Sites Protection*

Keywords: new sensors and instruments, LIDAR, remote sensing for archaeology and heritage, world heritage

ABSTRACT: Ancient Istanbul, settled on top of seven hills, was the capital of Ottoman and Eastern Roman Empire. Today, with 12.5 million inhabitants, the metropolis is still a major centre for business, culture and arts of the modern Turkish Republic. Furthermore, the old city is continuing to its mission for centuries; offering the daily basic needs for citizens. This unique ancient city is in the UNESCO World Heritage List more, Istanbul has been announced as the European Capital of Culture for the year of 2010. On the other hand, the pressure of high dense urban-commercial areas and the effects of the nature are pushing the old city into its limits. With, these challenging tasks and motivations, the expectations from the local, national and international authorities are highly increased to protect the historical areas for the next generations. Currently, numerous historical and cultural heritage documentation, restoration, reconstruction, renewal and rehabilitation projects are in progress. To provide a good base for such projects, architectural surveys are highly demanded within the limited time periods. In this paper, the needs and provided solutions for such projects with the combination of different remote sensing technologies such as photogrammetry and laser scanning are studied and example work algorithms are discussed from the point of design, needs and results.

## INTRODUCTION

Clues from the past life styles and habits of the mankind has always been interesting and valuable to people who are living on the same land at different time. Documentation and protection of the historical places and structures is not only a local or national issue but also a global interest to keep the memory of the past of the mankind. In order to achieve these challenging aims there are numerous works and needs from different scientific and technical backgrounds. Among these, architectural surveys play a very important part for the restoration, renovation and conservation works. Due to the fact that the demand for fast, accurate architectural survey in realistic visualisation is becoming a highly followed trend and it is greatly appreciated by the society. As Turkey has a rich historical background and culture, architectural survey of the cultural heritage structures such as houses from different cultures and civilizations, palaces, military and civil architecture examples dating back to centuries needed to be documented for special conservation and restoration projects. However, lack of detailed documentation strategy standardization and the limited traditional methods to capture the structure elements and details of façades is a general problem. More, as the monuments and structures differ a lot from each other, proposals to obtain a standard documentation for architectural survey have some gaps. Today, possibilities achieved with the technological developments in Survey Techniques and Computer Science, the demand of highly detailed and accurate documentation, survey and realistic visualisation is possible. On the other hand, the possibilities offered with these technologies are new to the end users and the using new sensor technologies for the documentation is only in use for the last few years.

Figure 1. Terrestrial Laser Scanner.

## CURRENT TECHNIQUES IN USE

In Turkey, the traditional techniques for architectural survey are widely used by Conservatives and Architects. Basically; levelling, distometers, meters and non-calibrated images are commonly used for documenting elements of a façade for any type of structure. Some fine details are preferred to be recorded by drawings, directly copied with special kind of papers. With the developments, the reflectorless total stations formed a good way of using metric techniques in the production fields.

However, the need to record thousands of points and some difficulties to provide the final output due to the fact that total stations mostly record only points can be also time consuming. Terrestrial photogrammetry is also a very well-known technique by the users. It is an efficient way to work computed 3D coordinates in the object environment and it is widely preferred. However, the users in the practical world usually deal with a high number of projects in a limited time, therefore they seek for any alternative methods to create fast, accurate and solutions to present results easily in a 3D realistic environment within an independent scale.

## NEW TRENDS AND TECHNOLOGIES

With the developments in laser signalling technology during the last few decades, Light Detection and Ranging Technology known as LIDAR, has broaden a new area for data capturing. Laser scanner devices, are able to acquire to measure and record the outgoing and reflected laser signal from environment back to the device. These instruments provide a rapid 3D point cloud of the surrounding area and objects usually with support of integrated digital cameras. There are numerous fields where laser scanners are used from airborne, terrestrial or mobile platforms.

## 1-ARCHITECTURAL DRAWINGS WITH POINT CLOUDS

The nature of the laser scanners is to generate point clouds from reflected surfaces. Thus within a workflow of Target Selection, Laser Scanning, Registration of Point Clouds and Digitalization it is easy to produce architectural surveys with the help of an internal camera and/or external non-calibrated images for interpretation of the elements. Nevertheless, with this technique it is not enough to interpret all the point cloud within the parts where for instance; reflected laser points and the low contrast areas in the point clouds which have influence on determining the shapes. In Fig. 2 this work algorithm can be seen with the represented screenshots of a case study.

## 2-TLS WITH RECTIFIED DIGITAL IMAGES

After the first generation of the TLSs, new versions mostly equipped with digital camera features. The digital cameras provide an easy way for capturing and viewing field of interest and this can be

1. Gathering Point Cloud with TLS   2. Describing Surfaces and Digitalization   3. Final facade output as a CAD file

Figure 2.Working With Point Clouds for Generating and architectural Survey.

Figure 3. Istanbul University Main Portal produced with combination of laser scanning and image rectification.

very useful not only for interpretation of the details of a façade but also to provide an enhanced texture for the structure. However, only a few brands in the commercial market can use high resolution calibrated cameras. With the provided software by the producers, the approaches are usually for mapping the texture to the point clouds with the information gathered from the cameras. As, the benefits of the calibrated images have been recognized, free hand calibrated digital images are used with laser point clouds for easy recognize of the building elements and accurate drawings of the façades.

## 3-USING LASER SCANNERS IN A NON-STATIC PLATFORM

Due to time periods and suitable conditions such as traffic, light and crowded public areas, it is eager to collect the data in a very short time intervals. Therefore, under these conditions the use of laser scanners can be provided with highly sophisticated systems such as GNSS, INS, Gyroscopes, video cameras with high speed hardware supported software. Although satellilte techniques for location based solutions is quite impressive, enough control stations and/or CORS should be considered.

Figure 4. Intergrated GPS, INS, TLS and video camera components of a mobile TLS.

## EVALUATION OF SYSTEMS

- Beside the geometrical accuracy of the TLSs used in the projects, many of them provide the basic needs of the demands from the conservatives and architects. However, work algorithms regarding the structures vary on different problems and results. The shape and the structure of the target object is the mature limit on the quality of the output. On the other hand, a number of key factors are determined to be examined carefully before planning a TLS methodology.
- With the examined works and sample targets, it is seen that some basic principles should be always kept in mind. For instance, there can be applications where simple shaped structures which are not historical itself but in an historical area and therefore they may be needed for the documentation. In such cases the observation with total station has significantly a better solution for optimization of the workflow. With an enough number of points mathematical surfaces can be defined successfully. From point of work design, this is a time convenient and enough accurate application.
- User inexperience is a critical issue with the quality of the final product and time. For most of the users, drawing in a real time 3D environment is a new term and this takes time to adapt.
- Data storage and management; without proper facilitates, the data storage can easily turn into a trap due to the fact that each point cloud dataset contain huge number of data. Currently, software provided by the producers of TLSs have problems to fully handle the optimum conditions for handling data and produce architectural survey.
- Independent from any work planning mentioned above, there are always missing gaps in the point clouds especially with the roof and fine detailed parts of the structures. In order to produce a full survey for the necessary structures this also shows that additional systems such as high resolution aerial images or airborne LIDAR Systems can be good to complete some missing elements. Structured light could be another possible solution for the fine details.
- From point of abilities however Laser Scanners provide fast and accurate solutions for documentation, it is still expensive for the single users.

## PERSPECTIVES AND CONCLUSION

New development and possibilities of using TLS is deeply attracted the ideal architectural survey in minds. As today still the CAD drawings of the façades are used for analyzing the change detection, deformation and documentation, the trend has rapidly changed to products combined with photogrammetry and laser scanning. Another key issue for successful projects is the determination of the optimized workflow for generating the façade plan.

## REFERENCES

IBB, Projects Department Historical Sites Protection Archive of Projects.
C.Ö.Kivilcim, "An undergraduate project with terrestial laser scanner for purpose of architectural survey", ISPRS 2nd Summer School on Theory and Application of Laser Scanning. Young Authors Session. 2007, Ljubljana.
C.Ö. Kivilcim, "3-D Modelling with Terrestial Laser Scanner", Yildiz Technical University Graduation Project, Unpublished, 2005, Istanbul.
N. Demir, T. Vögtle, "Laser Scanning for Terrestial Photogrammetry, Alternative System or Combined with Traditional System?", 2004, ISPRS Congress Istanbul.
N. Hadda, T. Akasheh, "Documentation of Archaeological Sites and Monuments: Ancient Theatres in Jersah" 2005, CIPA 2005 XX International Symposium.
Wolfgang Boehler, Andreas Marbs, 3D Scanning Instruments, 2001.

*Remote Sensing for a Changing Europe*
*D. Maktav (Ed.)*
*IOS Press, 2009*

*doi:10.3233/978-1-58603-986-8-535*


# A new index for the evaluation of land management in the framework of land degradation assessment

V. Imbrenda[1], T. Simoniello [1,2], T. Carone [3], R. Coppola [2,4], M. D'Emilio [1,2], M. Lanfredi [1,2], M. Liberti [2,4], M. Macchiato [1,5]

*1 Consorzio Interuniversitario per la Fisica della Materia (CNISM) UdR-NA, Naples, Italy*
*2 Istituto di Metodologie per l'Analisi Ambientale (IMAA-CNR), Tito (PZ), Italy*
*3 Agenzia Regionale per la Protezione dell'Ambiente (ARPAB), Potenza, Italy*
*4 Università della Basilicata-Dipartimento di Ingegneria e Fisica dell'Ambiente (DIFA), Potenza,Italy*
*5 Università degli Studi "Federico II"-Dipartimento di Scienze Fisiche (DSF), Naples, Italy*

Keywords: Landsat-TM, land management, land degradation, mechanization level, land cover

ABSTRACT:
Soil degradation phenomena are induced by both natural (soil characteristics, climate conditions) and anthropogenic (overgrazing, deforestation, improper agricultural practices) factors. Among the man-induced factors, the intensive agriculture is one of the widest diffused causes of soil deterioration. Particularly, persistent passes of operative machineries produce a dangerous alteration of soil chemical-physical properties. In the framework of Interreg IIIB MILDMAP-MEDIA project, we developed a new index based on the integration of land cover classifications and slope for assessing the level of mechanization that combined with information on soil types allows for estimating the soil vulnerability to compaction.
Two Landsat-TM images were classified with a hybrid unsupervised/supervised approach to obtain land cover maps. On the basis of cultivation type and slope values, derived from a DEM, we associated the required number of machinery passes and evaluated the different use between tyred and tracked vehicles. Results obtained form the comparison of the implemented index with other commonly used indices enhanced its capability to provide a more detailed classification of vulnerability to soil compaction representing precious information for land degradation assessment.

## 1 INTRODUCTION

It is widely recognized that soil degradation phenomena are induced by both natural (e.g. soil characteristics, climate conditions) and anthropogenic (e.g. overgrazing, deforestation, improper agricultural practices) factors. Among the man-induced factors, the intensive agriculture is one of the widest diffused causes of soil deterioration, especially in developed countries (EEA, 2005). Persistent passes of operative machineries produce a dangerous alteration of soil chemical-physical properties. In particular, the formation of the so-called "ploughsole" layer, which is a compressed stratum characterized by a very low permeability, can brings out very dangerous effects on soil production. The compression of soil causes both direct and indirect problems for vegetation growth. In the first case, the increase of the penetration resistance and the shear stress limits the growth of vegetation roots. In the second case, as indirect result, the induced porosity decrease reduces the water and air penetration causing drought and asphyxiation conditions (see e.g., Hoque et al., 2000; Rusanov, 1991). In turn, vegetation stress implies a yield reduction and then an economic loss.

Due to the complexity of the phenomenon, the definition of an index capable of describing such mechanisms and identifying the threatened areas is difficult to determine in analytic terms. There are no parameters directly correlated to machinery passes and so to soil compaction. In situ mea-

surements of penetration resistance are very accurate but they provide very local information and an extensive investigation over large areas is very expensive.
Therefore, in order to identify areas more vulnerable to soil compaction usually proxy indices are adopted accounting for the level of mechanization of the considered territory.

A direct index of mechanization level was adopted by Italian Agency for Environment Protection and Technical Services in inventory of indicators of physico-chemical soil degradation (APAT, 2000). It considers the number of farm property machineries per area utilized for agriculture. Bazzoffi et al. (2003) elaborated another version of this index that takes into account also the machinery weight per unit of area and on the average passes number per year. The principal deficiency of such indices is the absence of information related to actual exploitation of agricultural areas, i.e. the kind of the cultivation actually present. Since the number of machinery passes is strictly dependent on the type of cultivation, neglecting such a variable implies a uniformity of the compaction factor over areas with different vulnerability levels.

In order to obtain information with a high spatial detail and to account for the different cultivation types, we devised a new index based on the integration of slope data and land cover classifications for assessing the mechanization level.

## 2 MATERIALS AND METHODS

The analysis was performed in two study sites located in Basilicata region in Southern Italy (Figure 1). The selected areas are characterized by different climatic and environmental conditions. The first one is prevalently plain and lies from the Ionian coast (east) to the inland territory northeastward, where there are many areas threatened by land degradation processes, the so called "calanchi" landscape (Piccarreta et al., 2006). Climate is typical of arid and semi-arid areas having precipitation less than potential evapotranspiration. Such an area is strongly exploited by agriculture both inland territories and along the coast. In particular, in the area behind the coast (Metapontum areas – Fig. 1a), there are the largest fruit tree cultivation of the region. The second site corresponds to high valley of the Agri River (Fig. 1a) that is characterized by a large altitude variation: the central alluvial plane, intensively cultivated, is surrounded by a mountainous landscape having a rich faunal and vegetation biodiversity.

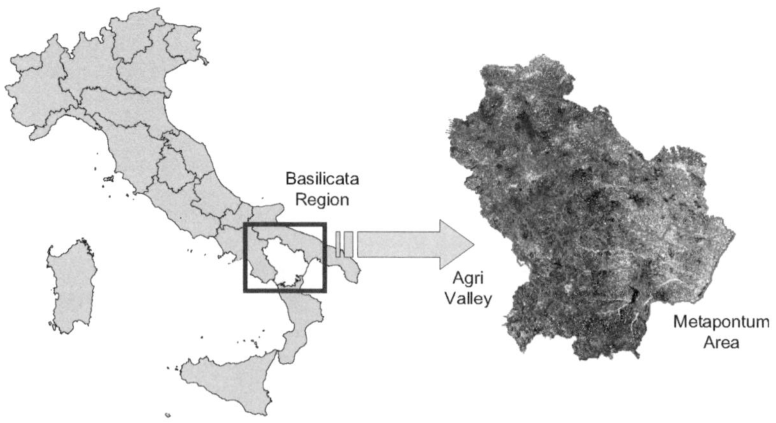


Figure 1. Location of study areas.

In order to obtain information on the type of cultivation for the selected areas, we elaborated two Landsat – TM imagery (path 188 – row 032): one acquired in June 1998 and the second in May 2004. In particular, the first image was elaborated for the Agri basin that is mainly characterized by complex cultivation pattern having the maximum of productivity in June; whereas for the

Matapontum area we used the image acquire in May since the large presence of corn cultivation is not detectable after the harvesting period, i.e. the first weeks of June.

After the preprocessing (DN-radiance-reflectance conversion, cloud detection, etc.), the imagery were classified by adopting a hybrid unsupervised/supervised approach. Clustering was applied to define the signature separability and to identify the principal classes. Different classifications were elaborated by using the Isodata algorithm (see, e.g., Richards & Jia, 2006) with a large number of required classes. Then, on the basis of the territory knowledge they were merged. Finally, the identified clusters, integrated with training areas of peculiar signatures, were used as input for the supervised classification. For the badland identification within the Metapontum area, we adopted an integrated approach that combines the spectral responses with morphological information (Liberti at al., 2006). The final maps were obtained with a maximum likelihood classifier.

In order to associate to each cover type an average number of required machinery passes, we revised data provided by ENAMA (Italian National Agency of Agricultural Mechanization) and obtained a land cover recoding as shown in Equation 1. Since the induced compaction level is different if a soil is pressed by tyred and tracked vehicles and their use is mainly determined by territory morphology (Keller et al., 2002; Servadio et al., 2001; Rusanov et al., 1991), we used DEM (Digital Elevation Model) having a spatial resolution of 20m to obtain a slope map for the investigated territories. Finally, data from agricultural census (ISTAT, 2000) were acquired to derive information on agricultural machinery and on the extension of UAA (Utilized Agricultural Area) at municipal level.

The new mechanization level index, developed in the framework of Interreg IIIB MILDMAP-MEDIA project, evaluates the territory from the point of view of the coverage actually present and the needed machinery passes for each cultivation type. In analytical terms, at municipal level it is calculated as follows:

$$ML = \sum_{i=1}^{nc} N_P \frac{A_i}{UAA} f \tag{1}$$

where, NP is the average number of machinery passes for each cover class; nc is the number of small areas within the given municipality; Ai is the single small area within the municipal boundaries; UAA is the Utilized Agricultural Area at municipal level; f is the correction factor according to tyres or tracks use.

In order to exploit the spatial detail provided by satellite-derived land covers, we computed the index by excluding the weighting step at municipal level and, thus, by considering the following formula:

$$ML = N_P\, f \tag{2}$$

On this basis, to each land cover the number of machinery passes was associated by using the code in Equation 1. The link to the correction factor for tyres or tracks use was derived from the slope map by taking into account that tracked vehicles are mainly used for cultivation in areas with a steep slope, generally higher than 18-20%; whereas plain areas are managed with tyred vehicles. The areas of application is shown in Figure 2. The correction assigns a higher level of vulnerability to tyres use since its compaction effects can reach medium-deep subsoil layers; whereas, even if the compaction induced by tracked vehicles is higher, they mainly affect the superficial soil, which are easier recoverable (Pagliai et al., 2003; Keller et al., 2002; Servadio et al., 2001).

In addition, we also computed the indices by following the first formula in order to compare results with two indices available in literature. In particular, we evaluated the APAT (2000) and Bazzoffi at al. (2003) indices that are computed from data of agricultural census as follows. The APAT index considers the number of agricultural machineries N every 100 hectares of Utilised Agricultural Area:

$$ML_{APAT} = \frac{\text{N}}{100\ ha\ of\ UAA} \quad (3)$$

The second index differentiates the effects of vehicles by including information on their weights as reported in the formula:

$$Ws = \frac{kW * W * N * 5}{S} \quad (4)$$

where, Ws is the weight-summation, W is the average weight and N is the number of the agricultural machineries, 5 is the average passes number per year, and S is the sum of arable land and permanent crops.

In order to compare the computed indices and to make the new index integrable within land degradation models to assess the vulnerability due to land management practices, they were classified it in a common range (1-2), from very low to very high sensitivity (e.g., Genesio et al., 2004; Motroni et al., 2004).

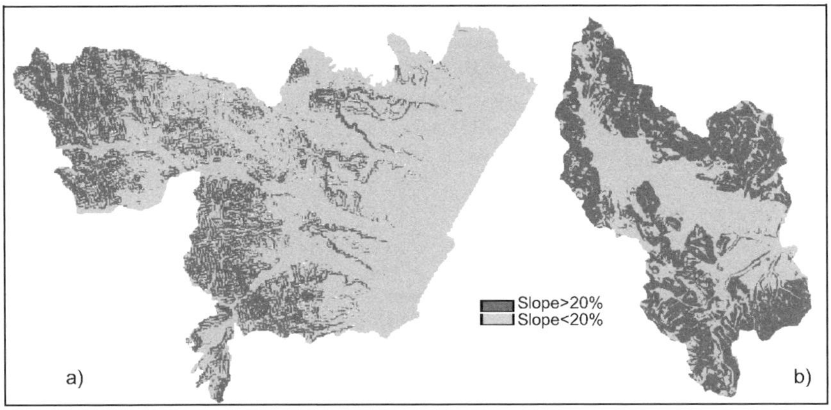


Figure 2. Areas of application of tyred (green) and tracked (red) vehicles derived by thresholding the slope map for the two considered study sites:Metapontum areas (a) and Agri basin (b).

## 3 RESULTS AND DISCUSSION

Land cover maps obtained from the Landsat-TM classification are shown in Figure 3. In the Metapontum areas we found the predominance of cultivated areas of both herbaceous and fruit tree plantation. They are mainly clustered behind the coniferous forests of coastal areas. The badlands class, mainly located in the centre of the investigated areas, includes wheat and maize cultivation and olive groves. Areas of natural vegetation such as broad-leaved and beach forests represent a small part of the territory at north-west and south-west. On the contrary, such classes are dominant in the landscape of Agri basin and cultivated classes are grouped at low altitude with a prevalence of annual crops and mixed cultivation including fruit trees, vineyards and olive groves.

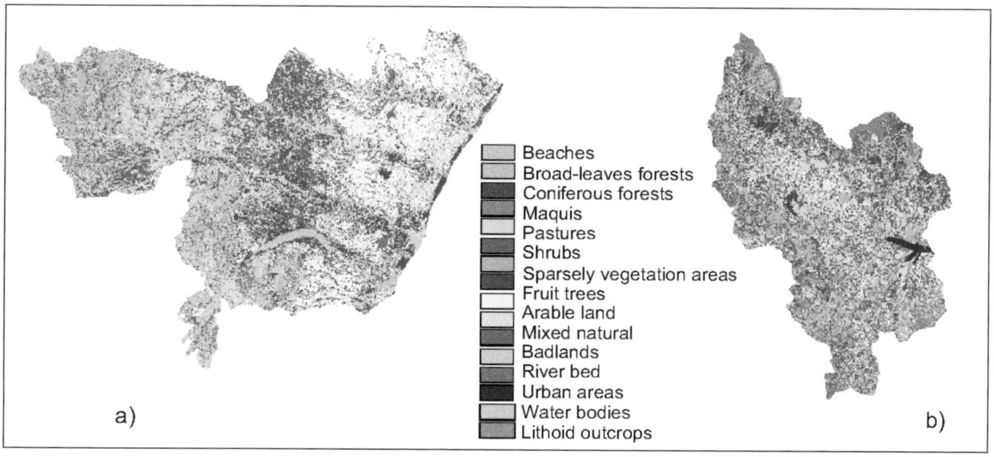


Figure 3. Land cover maps obtained from the classification of Landsat-TM imagery for the Metapontum area (a) and Agri Basin (b).

The mechanization level index elaborated on the basis of such land cover maps (Figure 4) showed a very high level of soil sensitivity to a heavy use of agricultural machineries for large areas in the eastern part of Metapontum region. Also along the principal rivers a medium high level of sensitivity was found. According to the area characteristics, there is a lower percentage of high vulnerable soils in the Agri basin. They are mainly located in the upper part of the valley.

Results obtained from the other two selected mechanization level indices are shown in Figure 5. For the first area, they showed quite similar results enhancing only four municipalities with slightly higher vulnerability compared to the reaming part of the territory. Slight differences were found for the Agri basin where the Bazzoffi et al. index enhanced the northernmost municipality as the more sensible within the basin.

To compare such indices with the new one, we considered the index at municipal level by weighting the values for the areas of each class within the UAA. The obtained maps are reported in Figure 6. In absolute terms, the vulnerability level is higher than the previous maps, but there are consistent differences also by considering the relative level among the municipalities. For the Metapontum area, Policoro and Scanzano showed the highest vulnerability level as in the map of APAT index; whereas the number of municipalities with intermediate level is increased in the map of new index. The comparison for the Agri basin is more explicative since the new index enhanced the central part of the basin as the more vulnerable in accordance of territory exploitation. On the contrary, for both the other two indices the relative highest vulnerability was found for municipalities mainly covered by broad-leaved and beach forest. Such erroneous interpretation of territory can be linked to evaluation of machinery number is based on the property. Instead, there are many agricultural areas managed with machineries of a third party property or join ownership. Therefore, the evaluated number can be not representative of the actual use of machinery within a municipality. Such a misestimation can be particularly evident where contiguous municipalities have similar cultivation (i.e. exchangeable machineries) with different sizes of cultivated area.

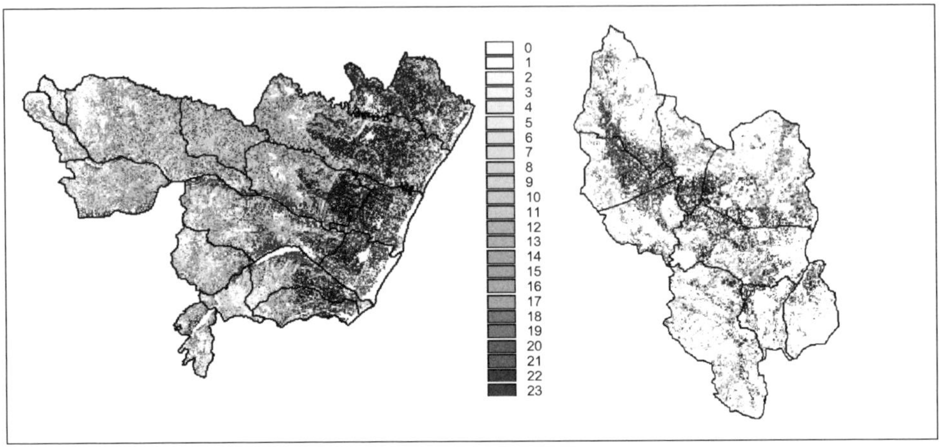


Figure 4. Maps of satellite-based Mechanization Level Index obtained by integrating the recoded land cover maps and the correction factor derived from the slope threshold.

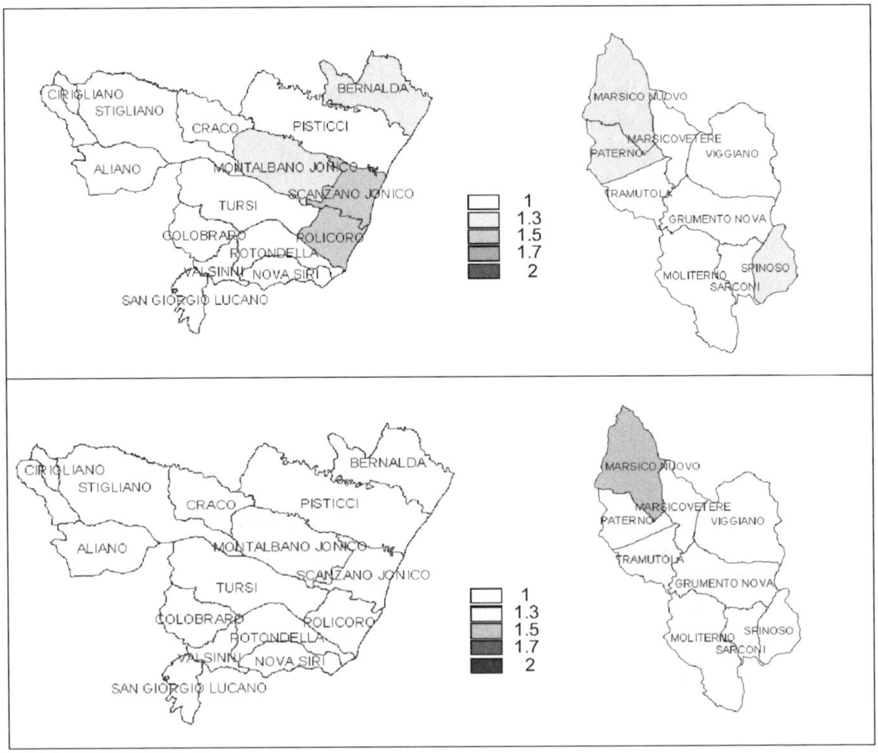


Figure 5. Maps of Mechanization Level Index obtained by applying the APAT index (on the top) and the Bazzoffi et al. index (on the bottom).

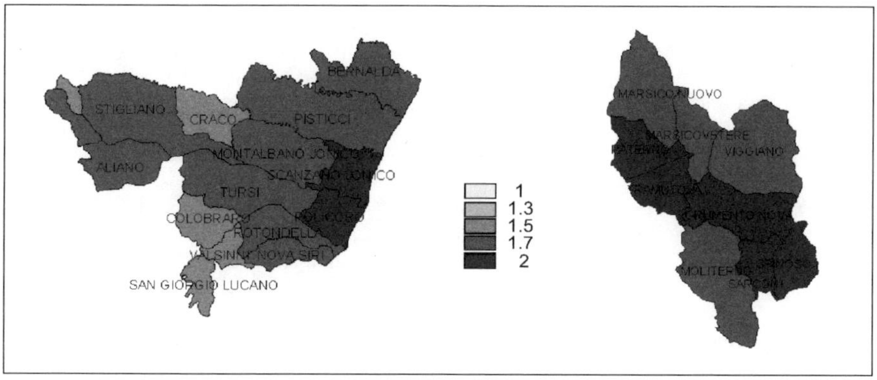


Figure 6. Maps of satellite-based Mechanization Level Index averaged at municipal level.

## 4 CONCLUSION

We introduced a new assessment perspective in evaluating soil vulnerability to compaction by estimating a satellite-based mechanization level. The new index takes into account the actual land exploitation by considering the required passages for each land cover and the used type of vehicles for territorial morphology. The comparison with commonly used indices based on agriculture census data showed that the new index provide a more realistic picture of the investigated landscapes. Moreover, it is cost effective compared to the efforts needed to gather data from agriculture census and can be more frequently updated. Another relevant peculiarity is the high spatial detail since it avoids the average of values on wide areas, such as the municipalities. The spatial detail and updating frequency are strictly linked to the resolution of available satellite data.

## ACKNOWLEDGEMENTS

This work was financed by EU funds in framework of MILDMAP project (Interreg III B – ArchiMed) and was also supported in part by grants from the University of Basilicata (PhD grant) and the Basilicata Region (Val d'Agri funds). We thank Prof. Paola D'Antonio and Dr. Francesco Genovese at University of Basilicata (DITEC) for their valuable discussions on agricultural mechanization in Southern Italy.

## REFERENCES

APAT (formely ANPA) 2000. Sviluppo di indicatori per il suolo e i siti contaminati, *RTI CTN_SSC 1/2000*, Italian National Environmental Protection Agency, 85.

Bazzoffi et al. 2003. In, *Annuario dei Dati Ambientali*, Italian Agency for Environment Protection and for Technical Services - APAT, section: Geosfera, 278-279.

Blunden, B.G., McBride, R.A., Daniel, H., Blackwell, P.S. 1994. Compaction of an earthy sand by rubber tracked and tyred vehicles. *Aust. J. Soil Res.*, 32, 1095-1108.

EEA 2005. The European environment, state and outlook 2005. European Environmental Agency.

Genesio, L., Magno, R., Capecchi, V., Crisci, A. , Bottai, L., Ferrari, R., Angeli, L., Gardin, L. 2004. *Final Report of Desert-Net project – Pilot Action WP10*, 128.

Hoque, M. M., Kobata T. 2000. Effect of soil compaction on the grain yield of rice (Oryza sativa L.) under water-deficit stress during the reproductive stage. *Plant production science* 3 (3), 316-322.

Keller, T., Trautner, A., Arvidsson, J. 2002. Stress distribution and soil displacement under a rubber-tracked and a wheeled tractor during ploughing, both on-land and within furrows. *Soil Till. Res.* 68 (1), 39-47.
Liberti, M., Simoniello, T., Carone, M. T., Coppola, R., D'Emilio, M., Lanfredi, M., & Macchiato, M. 2006. Badlands area mapping from Landsat ETM data. *Proceedings of the 2th Workshop of the EARSeL SIG on Land Use and Land Cover*: 434-440.
Motroni A., Canu, S., Bianco, G., Loj, G. 2004. Environmentally sensitive areas to desertification in Sardinia, *Techical Report*, 46.
Piccarreta, M., Capolongo, D., Bonzi, F., & Bentivenga, M. 2006. Implications of decadal changes in precipitation and land use policy to soil erosion in Basilicata, Italy. *Catena*, 65, 138-161.
Pagliai, M., Marsili, A., Servadio, P., Vignozzi, N., Pellegrini, S. 2003. Changes in some physical properties of a clay soil in central Italy following the passage of rubber tracked and wheeled tractors of medium power. *Soil Till. Res.*, 73(1-2), 119-129.
Richards, J.A., & Jia, X. 2006. *Remote Sensing Digital Image Analysis: An introduction*, 439. Fourth Edition. Berlin: Springer.
Rusanov, V.A. 1991. Effects of wheel and track traffic on the soil and crop growth and yield. *Soil Till. Res.* 19, 131-143.
Servadio, P., Marsili, A., Pagliai, M., Pellegrini, S., Vignozzi, N. 2001. Effects on some clay soil qualities following the passage of rubber-tracked and wheeled tractors in central Italy *Soil Till. Res.* 61(3-4), 143-155.
ISTAT 2000. 5th Agriculture Census. Italian National Institute of Statistics. http://www.census.istat.it/wibdsi/default.asp

*Remote Sensing for a Changing Europe*
*D. Maktav (Ed.)*
*IOS Press, 2009*

*doi:10.3233/978-1-58603-986-8-543*



# Analysis of persistent trends in vegetation activity in Mediterranean areas affected by land degradation phenomena

Lanfredi M.[1,2], Simoniello T.[1,2], Liberti M.[1,3], Coppola R. [1,3], and Macchiato M.[2,4]

1 Istituto di Metodologie per l'Analisi Ambientale (IMAA-CNR), Tito (PZ), Italy
2 Consorzio Interuniversitario per la Fisica della Materia (CNISM) UdR-NA, Naples, Italy
3 Università della Basilicata-Dipartimento di Ingegneria e Fisica dell'Ambiente (DIFA), Potenza, Italy
4 Università degli Studi "Federico II"-Dipartimento di Scienze Fisiche (DSF), Naples, Italy

Keywords: land degradation, vegetation, AVHRR, NDVI

ABSTRACT: The characterization of vegetation cover dynamics plays an important role for both land degradation and climate change studies. In such a context AVHRR/NDVI time series provide a powerful tool for improving knowledge on processes involving vegetation and consequently for supporting restoration policies.

For studying vegetation patterns in the framework of "Methodology Integration of EO Techniques as Operative Tool for Land Degradation Management and Planning in Mediterranean Areas" (Mildmap project –Interreg IIIB Archimed), we analyzed NDVI trends from GIMMS (Global Inventory Monitoring and Modeling System) dataset (1982-2003) available at 8 km resolution over the involved regions (South Italy and Greece).

For such areas, we investigated long-term stationary properties of vegetation dynamics of different covers by estimating the persistence probability according to a mechanical-statistical approach. The followed procedure, considers the times during which vegetation deviates from a given reference status before returning back to it. Our reference map kept the signs of NDVI trends related to the 10-year period (1982-1992); then, the signs of trends concerning longer periods up to 2003 were considered. From the persistence probability, we derived the mean life time of positive and negative trends for the selected covers, by fitting the number of surviving trends per year with the exponential decay law.

Our results show a general healthy level of vegetation and some differences between the two areas mainly for the natural covers (Forests and Maquis). They confirm the valuable use of NDVI time series for capturing recovery and/or endurance signs of vegetation cover against natural or human-induced stresses.

## 1 INTRODUCTION

The history of land degradation in Mediterranean regions is very long and it is extensively studied and analyzed in many countries (Brandt & Thornes 1996, Rubio & Calvo, 1996, van Der Leeuw 1996). In spite of the high diversity in their ecosystem, the Mediterranean areas are extremely fragile currently and the potential changes due to climatic and human causes may drive the start of land degradation and desertification processes. Climatic adverse characteristics, such as drought, and ecologically unbalanced human interventions are held accountable for desertification processes in Mediterranean area. Vegetation, as a major fraction of land cover plays a fundamental role in degradation processes by preserving and maintaining the soil from destructive rainfall impacts as well as by protecting against gullies and rills erosion, surface wash and crusting (Lal et al. 1997).

Vegetation distribution, on the other hand, is linked to the regional climate, so change in cover might indicate a climate change at regional scale (Pilke, 2002).

On the basis of such considerations, the characterization of vegetation cover dynamics can be crucial for both land degradation and climate change studies. Indeed, it is fundamental in models devoted to land degradation/desertification assessment (Enne & Zucca 2000) and represents also key information for understanding vegetation responses to climate change (White et al. 2005) and, more in general, complex feedbacks between vegetation and climate variables.

At present, the usefulness of NOAA-AVHRR satellite time series, which is the main source of global information over land, is widely accepted and even more investigated for such purposes (Los 2001, Lucht et al. 2002; Zhou et al. 2003, Piao et al. 2003, Ji & Peters 2005). Long-term analyses on the derived Normalized Difference Vegetation Index (NDVI), allowing a reliable monitoring of the photosynthetic activity over wide areas, can be a reliable tool for addressing adequate recovery policies and any efforts to preserve vegetation and the natural resources.

We analyzed trends derived from GIMMS time series, one of the most widely used AVHRR-NDVI global dataset (Myneni et al. 1997, Nemani et al. 2005), since we were interested in using this dataset for evaluating long-term vegetation dynamics at Mediterranean scale.

In particular, trends from GIMMS-NDVI dataset were adopted to assess vegetation persistence according to a methodology proposed by Lanfredi et al. (2004) and applied to the AVHRR/NDVI time series at full resolution (1.1 km). This method allows for studying the stability properties of vegetation dynamics through the analysis of persistence of NDVI trends and can be used for assessing indirectly resilience and resistance properties of vegetation, i.e. respectively the ability of vegetation to recover from stresses and to go up against them.

## 2 MATERIALS AND METHODS

In order to study vegetation patterns in the framework of "Methodology Integration of EO Techniques as Operative Tool for Land Degradation Management and Planning in Mediterranean Areas" (Mildmap project –Interreg IIIB Archimed), we analyzed NDVI trends from GIMMS (Global Inventory Monitoring and Modeling System) time series (1982-2003), at 8 km resolution over the involved Mediterranean regions (South Italy and Greece) and obtained from the free ftp website of Global Land Cover Facility (GLCF) of University of Maryland, where data were preprocessed (Pinzon et al. 2004, Tucker et al. 2005) (Figure 1). We composed the starting 15-day Maximum NDVI into annual MVC maps. MVC allowed for removing the influence of seasonal pattern so to be able to estimate inter-annual variations of vegetation activity. We also attempted a preliminary validation of such a dataset, which showed to be suitable for performing multitemporal analyses useful for pointing out anomalies in vegetation patterns and for studying long-term characteristics of vegetation covers (Liberti et al. 2008). NDVI trends were generated per pixel by means of linear regressions (least squares on first order polynomial) for obtaining slope values, on which the analysis of the trend persistence were performed (Lanfredi et al. 2004, Simoniello et al. 2008).

Persistence analysis is able to capture possible characteristic time scales from the persistence probability and/or highlight possible non-stationary dynamics. In the first case, vegetation that efficiently recovers from shock describes long-term stationary patterns. On the contrary, long lasting anomalies in vegetation activity may suggest non-stationary dynamics, which might be linked to climate change and or anthropic stress (Lanfredi et al, 2004). The followed approach, based on the study of trend signs, considers the times during which vegetation deviates from a given reference status before returning back to it. As a reference period for vegetation activity, the map of the signs of the trends related to the 10-year period 1982-1992 was selected; then, we evaluated for how many time these signs last by considering longer periods at one year rate up to 2003.

The aim of this analysis was to investigate the persistence property of negative and positive trends related to different vegetation covers, as vegetation resilience and resistance depend on the type of vegetation itself. Persistence probability was therefore estimated for different vegetated classes obtained from Corine Land Cover map available for both regions (Figure 2). Subsequently we fitted the number of the negative and positive trends that survive per year and for each covers and found that the exponential decay law was the best ($R^2$>0.90). Finally we derived the mean life time of positive and negative trends for the investigated areas. From a mathematical point of view, the persistence probability, which is the probability that a trend is still being undecayed at time t, can be estimated as:

$$q(t) = \frac{N(t)}{N(0)} \quad (1)$$

where N(t) is the number of trends having a same sign at time t and N(0) is their total number. We estimated such a ratio for positive and negative trends separately. The best fits for the persistence probability were obtained by using the exponential decay law:

$$q(t) = e^{-\frac{t}{\tau}} \quad (2)$$

where $\tau$ is the characteristic parameter (mean life time), given by the average time <T> during which a trend survives before it decays:

$$\langle T \rangle = \int_0^\infty t\ e^{-\frac{t}{\tau}} dt = \tau \quad (3)$$

This law was successfully used for fitting persistence probability of 1 km AVHRR data (Lanfredi et al. 2004). The time scale $\tau$ provides a measure of the time autocorrelation length: the higher the value of $\tau$, the more persistent the trends are on average.

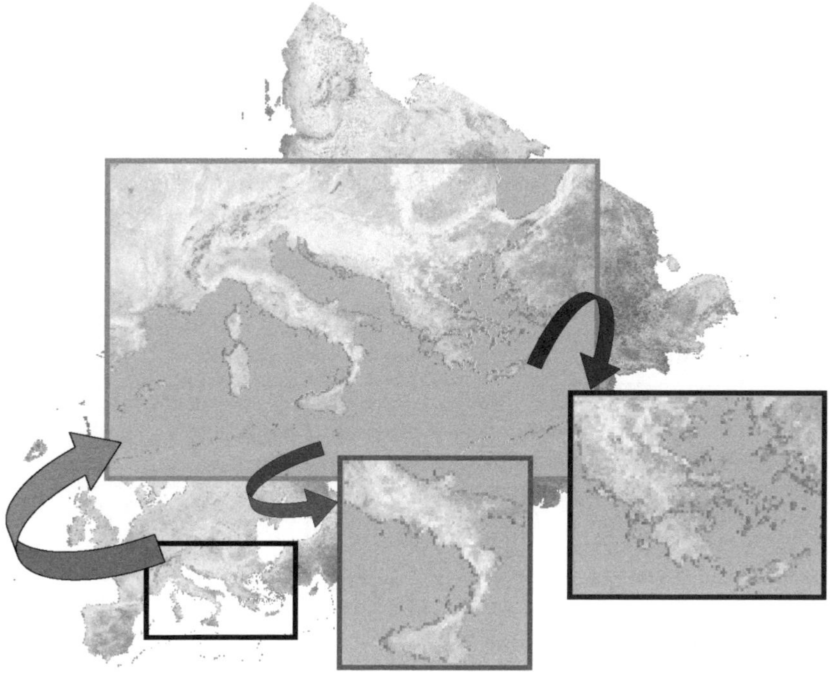

Figure 1. Images of GIMMS related to 15day NDVI composite of August 1998: Subset on Mediterranean area (red box), zoom in for South Italy and Greece (green and blue boxes), Eurasia continent in the background.

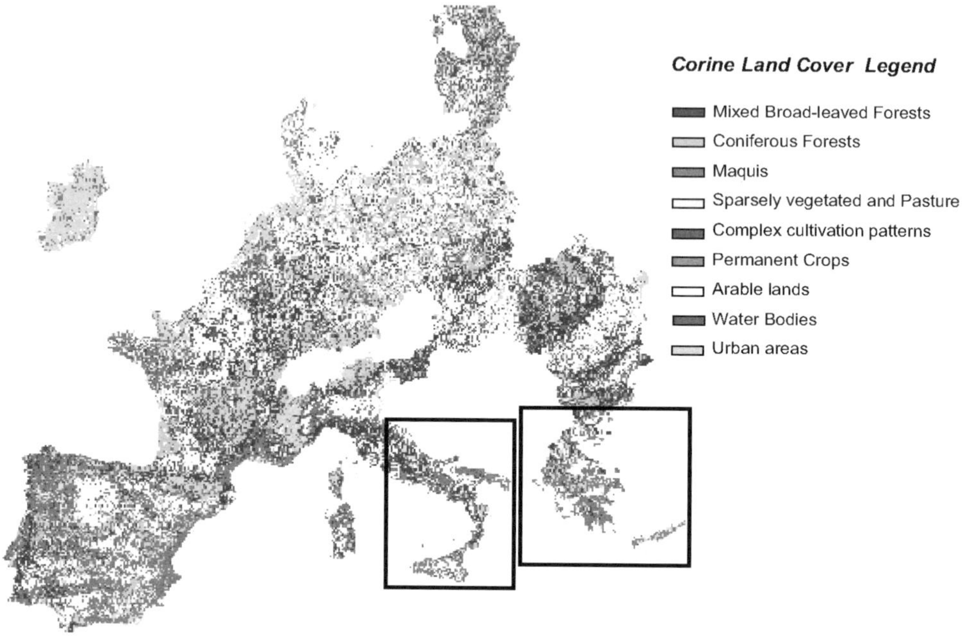


Figure 2. Corine Land Cover (CLC) Map recoded in 9 classes and rescaled at 8-km spatial resolution. Black rectangles delimit the study sites.

## 3 RESULTS

We selected from the CLC map 6 classes: Forests (Mixed Broad-lived and Coniferous), Maquis, Sparsely vegetatated and Pasture merged, Permanent Crops, Complex cultivation patterns, Arable lands. Table 1 shows the mean lifetimes τ (years) as well as the determination coefficients $R^2$ of the exponential best fits obtained for the different land covers.

Table 1. Characteristic time scale per vegetation cover classes found for Europe and Italy: mean life times (τ) expressed in years and determination coefficient ($R^2$)

| | Positive trends | | | | Negative trends | | | |
|---|---|---|---|---|---|---|---|---|
| Land cover | South Italy | | Greece | | South Italy | | Greece | |
| | τ | $R^2$ | τ | $R^2$ | τ | $R^2$ | τ | $R^2$ |
| Forests | 39.1 | 0.83 | 12.6 | 0.98 | 12.5 | 0.96 | 11.6 | 0.96 |
| Maquis | 35.7 | 0.86 | 13.2 | 0.98 | 15.7 | 0.93 | 9.5 | 0.96 |
| Sparsely vegetated and Pasture | 18.4 | 0.95 | 11.8 | 0.94 | 16.9 | 0.96 | 12.6 | 0.99 |
| Permanent Crops | 16.3 | 0.93 | 13.8 | 0.98 | 13.3 | 0.95 | 8.6 | 0.92 |
| Complex cultivation patterns | 19.4 | 0.97 | 15.2 | 0.97 | 13.8 | 0.99 | 9.1 | 0.96 |
| Arable lands | 16.5 | 0.97 | 16.7 | 0.97 | 18.5 | 0.96 | 24.8 | 0.93 |

The relative best fits on survival trend distributions are shown in Figures 3-4.

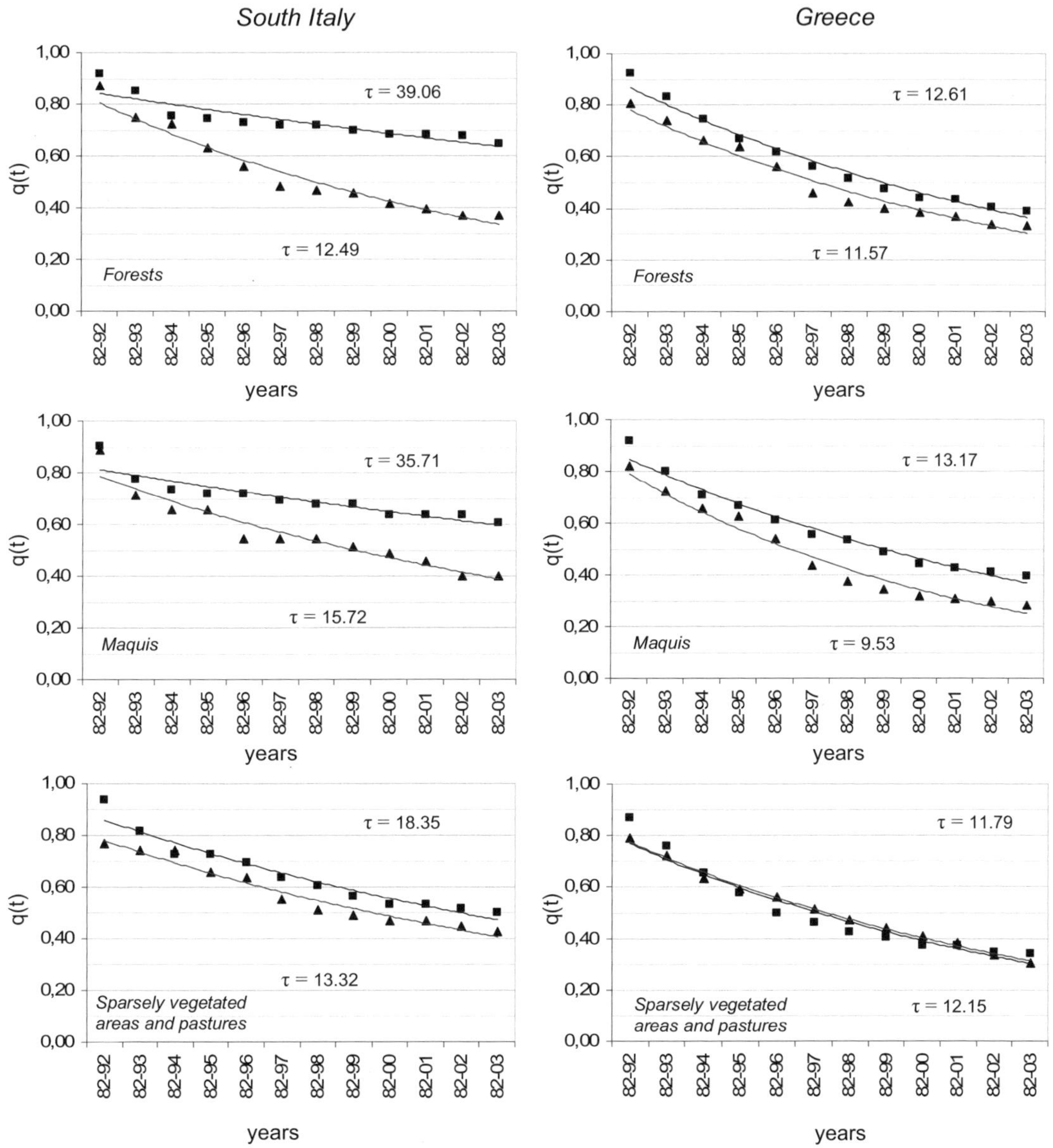


Figura 3. Comparison between the best fits of survived NDVI trend distribution for positive (square) and negative (triangles) over the two regions related to Forests, Maquis, and Sparse vegetation and pastures. The abscissas indicate the sequence of periods for which trend persistence is estimated from the reference period 1982-1991 (the first trend is 82-92, the last one 82-03).

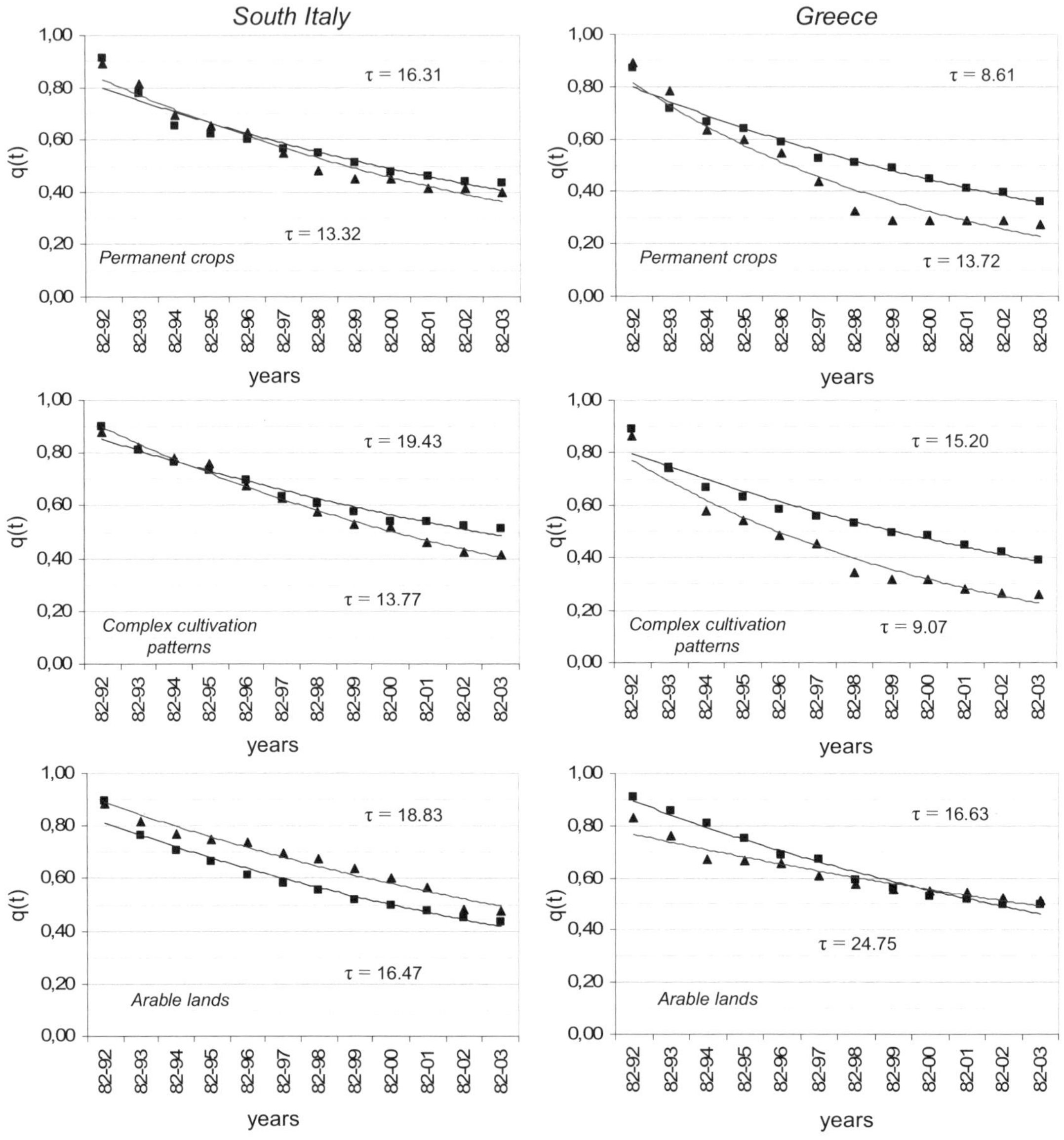


Figura 4. Comparison between the best fits of survived NDVI trend distribution for positive (square) and negative (triangles) over the two regions related to Permanent crops, Complex cultivation patterns, and Arable lands. The abscissas indicate the sequence of periods for which trend persistence is estimated from the reference period 1982-1991 (the first trend is 82-92, the last one 82-03).

Estimation of decay time shows for most of the considered covers that negative trends decay faster than the positive ones. As far as the status of vegetation in Southern Italy we can observe that the mean life times return a rather good scenario, with sparsely vegetated areas and pastures and arable lands, showing similar decay times, and with all the other covers, having mean return times of positive trends longer than the negative ones. In particular, very long decay times for positive trends were estimated for the natural covers, such as Forests and Maquis. If we evaluate the behavior of positive trends for these covers by comparing the two regions, we find the largest differences. In fact, the related mean life times for Southern Italy are much higher than those in Greece. For negative trends, the decay times for Forests were similar whereas for Maquis, Greece is characterized by shorter recovery times, which means more resilience power. Furthermore, for Greece the forests have similar decay times of positive and negative trends, which suggest a more equilibrated state.

Also in the case of Maquis the characteristic times estimated for Greece are closer than those estimated for Italy. Concerning the anthropic covers, both regions have decay times of positive trends longer than the needed recovery times from negative shocks, with the exception of arable lands. In particular, such cover class for Greece returns the highest mean life time of negative trends.

## 4 CONCLUSION

Altogether, the obtained results provided a general healthy level of vegetation covers: for both the regions, positive trends last more than the negative ones. However, some peculiar differences between the two investigated areas were pointed out. For both Maquis and Forests, Greek land covers showed a higher resilience than the Italian's ones, which were characterized by longer extinction times of negative trends. Nevertheless, in Italy the life time of positive trends is really very long: it is twice over the other ones. Italian natural areas are characterized by the highest mean life time difference between positive and negative trends, whereas higher stability seems to characterize the Greek territory. The only case that could evidence vulnerability to land degradation concerns the greek Arable land, which showed the highest recovery times from negative trends.

Information on vegetation covers dynamics through the estimation of their persistence probability can be very relevant for land degradation assessment as well as for climate change studies. In such a context, we showed that NDVI time series can be very useful for capturing recovery and/or endurance signs of vegetation cover against natural or human-induced stresses. By concluding, we highlight the precious contribute of NOAA-AVHRR historical data source for land degradation studies devoted both for providing support for policy makers and for the improvement of our knowledge on the involved processes.

## ACKNOWLEDGEMENTS

This work was financed by EU funds in framework of MILDMAP project (Interreg III B – ArchiMed) and was also supported in part by grants from the University of Basilicata (PhD grant) and the Basilicata Region (Val d'Agri funds).

## REFERENCES

Brandt, C. J. and Thornes, J. B., (Eds.), 1996. Mediterranean Desertification and Land Use. Wiley and Sons. Chichester, 600 pp.

Enne, G. & Zucca, C. 2000. Desertification indicators for the european mediterranean region, State of the art and possible methodological approaches. *ANPA and NRD.* Printed by Sped Srl - Rome (Italy)

Herrmann, S. M., Anyambab, A. & Tucker, C. J. 2005. Recent trends in vegetation dynamics in the African Sahel and their relationship to climate. *Global Environmental Change* 15,(2005) 394-404.

Ji, L. & Peters, A. J. 2005. Lag and seasonality considerations in evaluating AVHRR NDVI response to precipitation. *Photogrammetric Engineering & Remote Sensing*, 71, (2005) 1053-1061.

Lal, R., Blum, W. E. H., Valentin, C. & Stewart, B. A. (Eds) ,1997. Methods for Assessment of Soil Degradation. *Advances in Soil Science*, vol. 9, pp. 576.

Lanfredi, M., Simoniello, T. & Macchiato, M. 2004. Temporal persistence in vegetation cover changes observed from satellite: Development of an estimation procedure in the test site of the Mediterranean Italy. *Remote Sens. of Environ.* 93, 565–576.

Liberti, M., Simoniello T., Lanfredi T., Coppola R., Macchiato M. & Cuomo,V. 2008. Preliminary validation of GIMMS time series on Mediterranean ecosystems. *Geophysical Research Abstracts*,

Vol. 10, EGU2008-A-11987, SRef-ID: 1607-7962/gra/EGU2008-A-11987.

Myneni, R.B., Keeling, C.D., Tucker, C.J., Asrar, G. & Nemani, R.R. 1997. Increased plant growth in the northern high latitudes from 1981 to 1991. *Nature* 386, 698-702.

Nemani, R.R., C.D., Kelling, H. Hascimoyo, W.M. Jolly, S.C. Piper, C.J. Tucker, R.B. Myneni & Running, S.W. 2005. Climate-Driven Increases in Global Terrestrial Net Primary Production from 1982-1999. *Science* 6,vol.300 n°5625, 1560-1563.

Rubio, J. L. & Calvo, A., (Eds.), 1996. Soil degradation and desertification in Mediterranean environments. Geoforma Ediciones. Logroño, 290 pp.

Pielke, R.A.,Sr., Mariland, G.,Betts, R.A, Chase, T.N., Eastman,J.L., Niles,J.O., Niyogi,D.S. & Running, S.W. 2002. The influence of land-use change and landscape dynamics on the climate system: relevance to climate-change policy beyond the radiative effect of greenhouse gases. *Phil.Trans.R. Soc. Lond.* A 360, 1705-1719.

Pinzon, J., M. and Brown, E., et al., 2004. Satellite time series correction of orbital drift artifacts using empirical mode decomposition. *Hilbert-Huang Transform: Introduction and Applications*. N. Huang: Chapter 10, Part II. Applications.

Simoniello T., Lanfredi M., Liberti M., Coppola R., & Macchiato M. Estimation of vegetation cover resilience from satellite time series, in Special Issue “Climate-soil and vegetation interactions in ecological-hydrological processes”, Eds: V. Iacobellis, S. Manfreda, and M. Sivapalan, *Hydrol. Earth Syst. Sci. Discuss.* 5, 511–546, 2008.

Tucker, C. J., Pinzon, J. E., Brown, M.E., Slayback, A., Pak, E.W., Mahoney, R., Vermote, E.,F., & El Saleous, N., 2005. An Extended AVHRR 8-km NDVI Data Set Compatible with MODIS and SPOT Vegetation NDVI Data. *Int. J. Remote Sensing* 26, n°20, pp.4485-4498.

van der Leeuw, S.E. (Ed.), 1996. Understanding the natural and anthropogenic causes of soil degradation and desertification in the Mediterranean basin. *Archaeomedes Project* 6. European Commission.

White, M. A., Hoffman, F., Hargrove, W. W. & Nemani R. R., 2005. A global framework for monitoring phenological responses to climate change. *Geophys. Resear. Lett.*, vol. 32 L04705, doi:10.1029/2004GL021961.

*Remote Sensing for a Changing Europe*
*D. Maktav (Ed.)*
*IOS Press, 2009*

*doi:10.3233/978-1-58603-986-8-551*


# Urban change extraction from aerial photographs and multispectral scanner - an applied study from Graz/Austria

W. Sulzer, K. Kern & St. Eichberger
*Institute for Geography and Regional Science, University of Graz, Austria*

Keywords: land-use/land-cover, sealed/built-up, urban sprawl, Graz

ABSTRACT: Urbanization processes are a major factor of change in Central Europe cities, where urban settlement structures have changed over the past decades, especially since the 2nd World War. Due to economic and political pressures several cities rapidly became regional centers or international nodes. Urbanization causes land-cover and land-use changes, which can lead to deeper social, economic and environmental changes. For planning purposes different methodologies and classification algorithm of remote sensing data are required. The semantic meaning of land-use and land-cover classes and their differences itself will influence the potential of further planning issues. This topic will be documented by the concrete example of the terms sealed and built-up.

## 1 INTRODUCTION

Remote sensed urban change extraction became a valuable tool in various town development planning purposes. Aerial photographs as well as high resolution satellite images (Banzhaf & Netzband 2004, Moeller, 2005) are used to provide information about ongoing changes in urban landscape patterns. The inventory of land cover changes is necessary for the analysis of trends, causes and consequences of natural and artificial processes, impact assessment, maintenance of ecological stability and its observation in decision-making processes. The methodology of information extraction reaches from simple visual interpretation up to sophisticated digital classification routines (cf. Donnay et al. 2001, Lu et al. 2003, Moeller, 2005).

The main difference between these two approaches is basically explained by the terms land-use and land-cover (Comber et al. 2005 a, b). Land-use represents how land is used by humans; it refers to the (e.g. economic) use to which land is transferred. For example the land can be used for private purposes (single family houses), for commercial purposes (stores, office buildings, etc.), for industrial purposes (factories) or for recreational or agricultural purposes. On the other hand, land-cover refers to the vegetation, structures or other features that cover the land. For example, the land is covered by grass, by trees, by water or by artificial surfaces. Two land parcels may have similar land cover, but different land use. For instance, an industrial factory may look very much like a shopping centre from the outside. The first is an example of industrial use, the latter an example of commercial use. But also the reverse case is possible: two land parcels that have similar land-use may have different land-cover. A golf course and a sports ground (a swimming pool with a roof) are both recreation land uses. The former would have a land cover of grass, while the latter would be considered sealed. However, in practice the scientific community is used to describe those aspects under the term land cover (FAO, 1997).

Although the opportunities of context based classifiers are well documented, they cannot substitute human interpretation knowledge in land-use classification. Most of the urban planning topics are not aiming in a simple pixel based classification of spectral urban features (land-cover); urban planning needs the information about the use of certain urban lots. In contrast the cadastral land-use planning of cities and communities defines the future land-use structures, but not exactly the special land cover type. Hence we considered the terms “sealed area” and “built-up area” to be two terms which can describe these different semantic meanings very well. This paper will document several multitemporal classification results, their legends and methodological approaches.

## 2 CASE STUDY GRAZ

With app. 250.000 inhabitants Graz is the second largest town in Austria. Its agglomeration reaches a half million inhabitants and spreads from the south-eastern rim of the Alps to alpine foreland. The town is settled in a wide basin of the Mur valley and on the Tertiary and Palaeozoic ridges besides. The investigation area covers the whole municipality and the accompanied settlements in the outer fringe (14 x 15km), where – especially in the flat southern part - mighty changes in land cover und land use took place in the last decades (Eicherger & Sulzer 2005; Sulzer & Kern 2008). The data source acquired for the study consists of: (i) black and white ortho-rectified aerial photography (1m resolution) and colour ortho-rectified aerial photography (0,25m resolution), (ii) multispectral Daedalus scanner data (2,5m and 1m resolution) and (iii) the digital cadastral map of the city of Graz.

This study is embedded in a long time applied research programme called “Thermal 04”, which deals with climate and environmental change studies for planning purposes financed by the government of the Province of Styria (environmental policies) and the town government of Graz (Environmental Office, Sewage System Office, Surveys Office, Town Planning). This study started in 2004 and is planned to continue until 2009.

## 3 OBJECTIVES AND METHODOLOGY

For the city of Graz (Austria) a long-term growth analysis utilizing digital airborne remote sensor data (DAEDALUS) and historical aerial photographs was conducted. A historic look at urban growth using remote sensing will allow planners and the public to visualize the expansion occurring in and around the city. The purpose of the project will be to assess land-use/land-cover information to detect changes within the urban area and its surrounding. This study will use a multi-tiered strategy where aerial photographs and airborne multispectral scanner data will provide a long-term data set for change analysis.

Change detection methodologies will be analyzed, including a visual interpretation of b/w aerial photographs and a classification comparison between pixel and object based classification methods. It is expected that different change detection techniques (caused mainly by different data and image processing sources and tools) will produce varying results, which could ultimately lead to different planning and management procedures by policy makers. It is hoped that an assessment, such as this, will assist in developing growth management policies and aid in determining environmental priority areas due to urban change.

This paper aims in a comparison of different multitemporal land-use/land-cover (LULC) classification schemata. Traditional visual interpretation of multitemporal aerial photographs and digital classification of airborne multispectral scanner data will provide different LULC data (cf. de Sherbinin 2002). The Federal Office of Metrology and Surveying (BEV) and the government of Styria provided supplemental digital cadastre information which includes LULC information as well. Another information source is provided by CORINE land-cover classification which is available from 1990 and 2000.

This comparison is focused on the terms “sealed” and “built-up”, which have different meaning based on the different methodology of visual and digital classification techniques.

Table 1. Workflow of data integration, data processing and analyses.

| | built-up | | sealed | |
|---|---|---|---|---|
| | **1997** | **2004** | **1996** | **2004** |
| **data acquisitions** | b/w ortho-rectified aerial photography | colour aerial photography | scanner flight *(operated by DLR)* | scanner flight *(operated by DLR)* |
| **geocoding** | ---- | ortho-photo production *(LPS)* | ortho-rectification *(performed by DLR)* | ortho-rectification *(performed by Joanneum Research / RSG)* |
| **mosaiking** | ---- | *(LPS)* | mosaicing of the swaths *(performed by DLR)* | mosaicing of the swaths *(ERDAS Imagine)* |
| **enhancement** | radiometric enhancement *(ERDAS Imagine)* | radiometric enhancement *(ERDAS Imagine)* | ---- | shadow correction *(ERDAS Imagine)* |
| **classification** | visual interpretation *(ArcView, ArcGIS)* | | object-based classification *(Definiens Professional)* | |
| **representation and usage** | - digital and analogue maps (1:10.000)<br>- GIS database<br>- analysis within district boundaries, town community and on regional level<br>- thematic mapping and representation of detailed studies<br>- basis for the development plan of Graz | | - sealing maps<br>- land-cover/ land-use maps<br>- integration of the sealing maps into the town drainage/ sewage water system management of the city of Graz | |
| **change detection** | *(ArcGIS)* | | *(ERDAS Imagine)* | |

**Multitemporal aerial photograph investigation:** In a first step conventional multitemporal (1952/1959/1968/1975/1984/1990/1997/2004) aerial photographs of the investigation area were processed to obtain orthophotographs for a visual land use classification. The main focus for the use of multitemporal interpretation of aerial photographs is the registration of changes (Eichberger & Sulzer 2005). The whole interpretation-process was carried out by means of object-visibility, size of the study area and realization of the objectives on a scale of 1:5.000. The use of a larger scale would lead into the direction of digital cadastral maps or land-use-plans without focusing on the functional view and useful generalization of the dataset. A cartographic representation on a scale of 1:10.000 results from the maximum plot-size and object visibility respectively. The classification used in this study was built up in collaboration with the department of Urban Development Graz and is basically founded on a catalogue of different kinds of land use to detect the actual land-use and land use plans suggested by the Austrian Conference on Spatial Planning (ÖROK, 1992). Because of the different aims and objectives of the different official investigations a common missing of homogenous documentation of land-use and land-wastage in Austria should be pointed in the course of this study.

The classification key (see Table 2) used in this study was generated in collaboration with the Department of Town Planning Graz and is basically founded on a catalogue of different kinds of land-use to detect the actual land-use and land-use plans suggested by the Austrian Conference on Spatial Planning (ÖROK, 1992). For our comparison the class "built-up" consists of the following classes out of Table 2: (i) traffic areas, which include all types of streets as well as parking lots, (ii) built-up areas, all buildings and their immediate surrounding areas (e.g.: single family houses also include doorways, gardens, small fruit tree areas, etc.) as well as trade and industrial areas and (iii) technical supply areas.

Table 2. Classification legend (visual interpretation of aerial photographs).

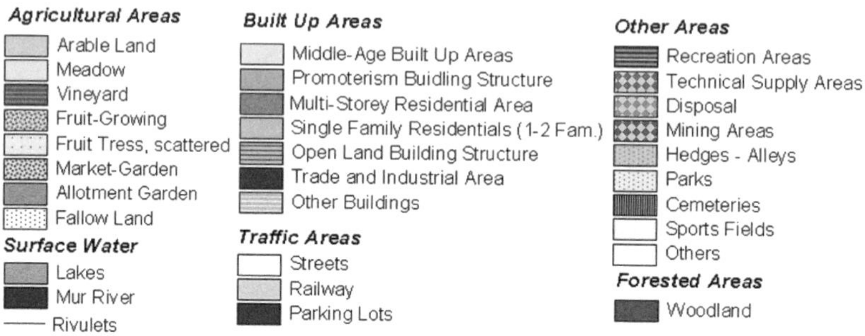


**Multitemporal airborne multispectral scanner investigation:** The main objectives of this study was to provide time-series information to define and locate the urban sprawl trends in sealing processes in Graz/Austria and to document the application and integration of sealing maps in planning purposes of the governmental building authority of the city of Graz (especially for constructing sewerage systems). Multitemporal sealing maps were generated by a digital classification of multispectral DAEDALUS Scanner data recorded in 1996 and 2004 (Sulzer & Kern 2008).

Due to its acquisition in late October the 2004 image was seriously affected by shadow effects. Shadow effects are frequent in high spatial resolution images, particularly in urban areas and depend on daytime, seasons, camera angle and flight altitude. The loss of information caused by shadow effects within the 2004 image had to be minimized by a histogram matching of shadow affected and non affected areas within the image. For the classification of sealed surfaces we generally assumed, that greens are unsealed and sealed surfaces are not covered with vegetation. Hence we considered the normalised difference vegetation index, or NDVI, a suitable tool to create sealing maps. Because of the high information content and the possibility to compare the results of the digital classification with the results of the visual interpretation we decided not just to develop a sealing map, but also to perform a land-cover/land-use classification. For that purpose, an object-oriented approach by using the Definiens Professional software was realized. In both cases (1996 and 2004) a hierarchical bottom-up network consisting of three levels was created. The class hierarchy was constructed to classify from the coarsest to the finest segmentation level. As mentioned before, sealed surfaces were considered to be surfaces without any vegetation. So the NDVI was used to separate sealed from unsealed areas. All segments not assigned to be sealed by using the NDVI were analysed based on their mean grey value, shape, texture, neighbourhood relations and relation to segments of higher and lower levels and classified as one of the following land cover types: wood, green, field, water and mineral extraction sites.

**Digital cadastre map**: From the digital cadastre map provided by the city of Graz be picked the following classes: buildings, built-up area (surfaced), roads, business premises, storage area and technical supply areas.

Table 3. Classification legend of the digital cadastre.

| | | | |
|---|---|---|---|
| **1. agricultural areas** | **3. water bodies** | **5. built-up areas** | **6. others** |
| - arable land | - river | - built-up, surfaced | - mining area |
| - fruit trees, scattered | - lakes | - built-up, green | - fallow land |
| - vineyards | **4. traffic areas** | - buildings | - recreation area |
| - pastures | - airports | - technical supply areas | - rocks and debris |
| - gardens | - railroad | | - inland marsh |
| **2. forested land** | - roads | - business premises | - waste land |
| - woodland | | - storage area | - others |

**CORINE Land Cover:** The objective of the European project CORINE Land Cover (CLC) is the provision of a unique and comparable data set of land-cover for Europe. It is part of the European Union programme CORINE (Coordination of Information on the Environment). The mapping of land-cover and land-use was performed on the basis of satellite remote sensing images on a scale of 1:100.000. The first CLC data base CLC1990, which was finalized in the 1990s, consistently provided land-use information comprising 44 classes. In the project CORINE Land Cover 2000 (CLC2000) an update of the database and a mapping of changes have been accomplished using the year 2000 as reference. The aim of CLC2000 is to update the CORINE land cover map for the year 2000 and to collect information on land cover changes in Europe during the last decade (1990-2000; cf. ECSC - EEC - EAEC, 1997).

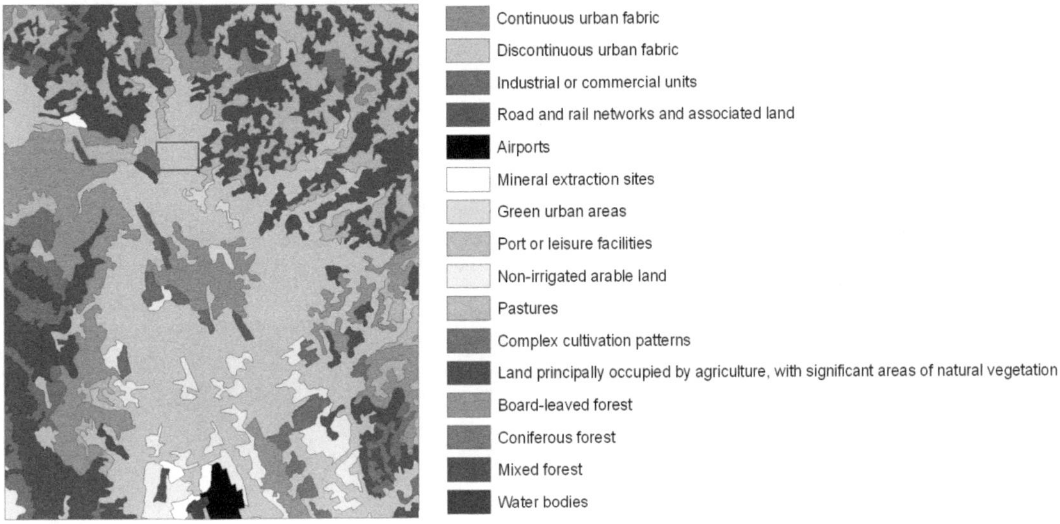


Figure 1. CORINE classification and area of interest (rectangle).

## 4 RESULTS

Table 4: Built-up and sealing processes of the municipality of Graz.

| **built-up area** derived from aerial photographs | | | **sealed area** derived from Daedalus data | | | **built-up area** derived from cadastre map | | |
|---|---|---|---|---|---|---|---|---|
| year | ha | % | year | ha | % | year | ha | % |
| 1997 | 5577,75 | 43,72 | 1996 | 3355,15 | 26,30 | ---- | ---- | ---- |
| 2004 | 6010,08 | 47,11 | 2004 | 3566,96 | 27,96 | 2004 | 2353,96 | 18,45 |

Table 4 shows the results of an overall classification of the city of Graz. The figures derived from the visual interpretation of aerial photographs, the object based classification of DAEDALUS Scanner data and the digital cadastre. The very broad term of built-up shows the highest rate. Approximately 50% of the municipality of Graz is built-up area, whereat approx. 30% is definitely sealed. The built-up area which is derived from the digital cadastre reveals this class differences stronger. The tendency (1996/97 – 2004) of increasing the sealing and built up processes is indicated in both remote sensing data sources. The results of CORINE land-use change (1990-2000) shows only minimal changes in the vicinity of Graz and is therefore not presented in figures.

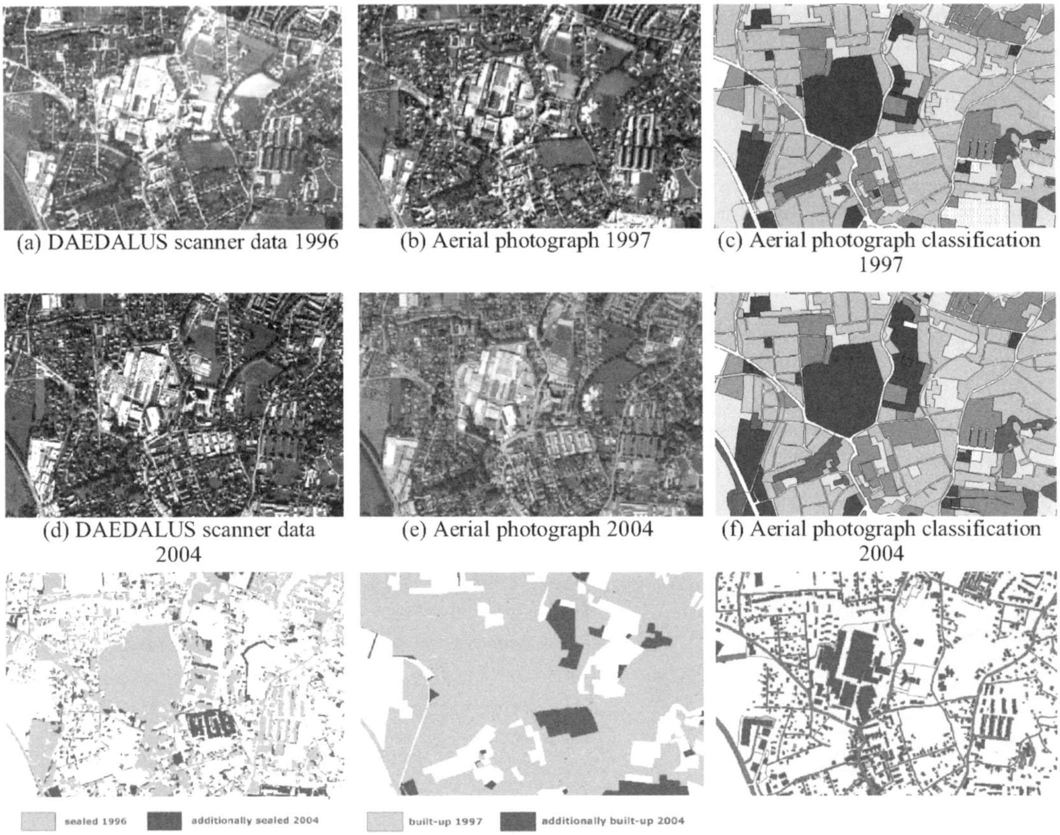


(a) DAEDALUS scanner data 1996 (b) Aerial photograph 1997 (c) Aerial photograph classification 1997

(d) DAEDALUS scanner data 2004 (e) Aerial photograph 2004 (f) Aerial photograph classification 2004

(g) Development of sealed areas (DAEDALUS) (h) Development of built-up areas (aerial photographs) (i) Digital cadastral map

Figure 2. Comparison of Daedalus scanner data (a/d) and aerial photographs (b/e) as well as comparison of the visual classification of 1997 (c), 2004 (f) and of built-up areas (h) the object-oriented classification of sealed areas (g) and the digital cadastral map (i).

(a) Digital cadastre map (streets and buildings in black) (b) DAEDALUS Scanner data (c) Digital cadastre map with sealing layer (in red)

Figure 3: Comparison of digital cadastral map (a) and Daedalus scanner data (b) as well as an overlay of the cadastre with the sealing layer (c).

## 5 CONCLUSION

The assessment of land-use or land-cover change dynamic is driven by the final usage of the data information. Does it leads in an integration e.g. in a sewage manage systems; the term land-cover is more suitable with its product of a sealing map than the heterogonous term built-up.

The used methodology of visual interpretation and digital image analysis will lead to different results, whereby a human induced visual interpretation will tends more in a differentiation of land-use then of land-cover. This mainly caused by the synoptic and analytic interpretation capacity of the human's brain. Object respectively context based classification algorithm are showing the first possibilities of a more sophisticated land-use classification.

## ACKNOWLEDGEMENTS

Special thanks go to Winfried Ganster from the Geodetic Survey Department of the government of Graz for his excellent project coordination and data providing.

## REFERENCES:

Banzhaf, E., Netzband, M., 2004. Detecting urban brownfields by means of high resolution satellite imagery. The International Archives of the Photogrammetry, Remote Sensing and Spatial Information Sciences, XXXV, Part B, Comm. VII. pp. 460-466.

Comber, A.J., Fisher, P.F. & Wadsworth, R.A., 2005a. What is land cover? Environment and Planning B: Planning and Design, 32, 199-209.

Comber, A.J., Fisher, P.F. & Wadsworth, R.A., 2005b. You know what land cover is but does anyone else?…an investigation into semantic and ontological confusion. International Journal of Remote Sensing, 26, 223-228.

de Sherbinin, A., 2002. A CIESIN Thematic Guide to Land Land-Use and Land Land-Cover Change (LUCC). Center for International Earth Science Information Network (CIESIN) Columbia University Palisades, NY, USA Available online at: http://sedac.ciesin.columbia.edu/guides/lu/ CIESIN_LUCC_TG.pdf. Webpage accessed in May 2008.

Donnay, J-P., M. J. Barnsley & P. A. Longley (eds.), 2001. Remote Sensing and Urban Analysis, Taylor & Francis, London & New York.

ECSC-EC-EAEC, 1997. Technical and Methodological Guide for Updating the CORINE Land Cover Data Base, Brussels • Luxembourg, 132p.

Eichberger, S. & Sulzer,W., 2005. Urban development of Graz - A time-series analysis with historical aerial photographs. GGRS2004 - 1st Göttingen GIS & Remote Sensing Days - Environmental Studies - Göttingen, 07. - 08.10.2004, Göttinger Geographische Abhandlungen, Heft 113, pp. 63-70.

FAO. 1997a. Africover land cover classification. FAO, Rome.

Kern, K. 2007. Objektorientierte Klassifikation von Scannerdaten am Beispiel der Thermalbefliegung 2004 Graz. Unpublished diploma-thesis, Inst. of Geography and Regional Science, University of Graz, 86 pp.

Lu, D., Mausel, P., Brondizio, E. & Moran, E., 2003. Change detection techniques. Int. Journal of Remote Sensing, 25 (12), pp. 2365-2407.

Moeller, M., 2005. Remote Sensing for the Monitoring of Urban Growth Patterns. In: The International Archives of the Photogrammetry, Remote Sensing and Spatial Information Sciences, Vol. XXXVI - 8/W27, on CD.

Österreichische Raumordnungskonferenz (ÖROK), 1992. Österreichisches Raumordnungskonzept 1991.Schriftenreihe der Österreichischen Raumordnungskonferenz, Nr. 96, Wien, 224S.

Sulzer, W. & Kern, K., 2008. High resolution airborne scanner data for multitemporal sealing maps of Graz – and their applicability in town planning. In: C. Jürgens (Ed.). Remote Sensing – New Challenges of High Resolution. EARSeL Joint Workshop Bochum, March 5-7, 2008. Selbstverlag des Geographischen Instituts für Geographie der Ruhr-Universität Bochum. ISBN 978-3-952143-79-3, 265-274.

Umweltbundesamt, 2008: CORINE Landcover Nomenklatur. Available online at: http://www.umweltbundesamt.at/fileadmin/site/umweltthemen/raumplanung/1_flaechennutzung/corine/CORINE_Nomenklatur.pdf (Webpage accessed in May 2008).

*Remote Sensing for a Changing Europe*
*D. Maktav (Ed.)*
*IOS Press, 2009*

*doi:10.3233/978-1-58603-986-8-558*

# Landscape structure variation of protected area of Kujawsko-Pomorskie Voivodeship

M. Kunz
*Department of Cartography, Remote Sensing and GIS, Nicolaus Copernicus University, Torun, Poland*

Keywords: landscape indices, CORINE Land Cover, SDI, Landsat

ABSTRACT: Knowledge of a landcover is very essential to understand and know relations between a man and an environment. The landcover is closely related to the rest elements of geographic environment. It also illustrates the variety of the structure of geographic environment. Till the end of the last century, there were no data representing full, reliable and covering the whole country information about the landcover. Maps and studies over this element of landscape were based on other methods of mappings, various scales and spatial ranges. This fact made it impossible to make comparison analysis and to have proper conclusions concerning processes that were taking place. The realization of National Geographic Information Systems (NSIG) and works connected with the creation of CORINE Land Cover (CLC) is a big help in this field. It was impossible without the development of spatial imagery, universality of the application of spatial information systems, and the use of army materials in the civilian, formal cases. Source spatial data used on the paper, mutually supplement and cover one another. Research connected with the analysis of a landscape structure was done in the chosen protected areas in Kujawsko-Pomorskie Voivodeship. All (8) landscape parks were in the Voivodeship. Total area of the preserved territories is almost 12% of the total Voivodeship area, and represent various types of a natural landscape. The structure of the landscape in landscape parks was analyzed with the application of landscape indices. The following indices describing the structure of landscape were calculated and compared: a number of landcover classes, a number of patches, total length of the edge, a shape index, a fractal dimension, and Shannon's diversity index. Comparisons of indices among parks were done in relation to various scales of source materials.

## 1 INTRODUCTION

Existing resources of digital spatial data (state referential data) should be applied in studies over geographical space at any level of specificity. Landscape structure and its variability in the space can be analysed in a few ways applying various research methods and tools. Geographic Information Systems (GIS) are the basic research techniques applied in a landscape ecology. These modern systems make possible to estimate a landscape structure with the help of several ecological measures and indices worked out on the basis of various information sources. This technology makes possible fast and recurrent creation of thematic maps, calculation of some indices, comparison analysis, and forecasting changes in the future. Landscape ecology has been applying geoinformation tools and procedures together with standard methods used for many years with success.

Present image of spatial landscape pattern is a result of all the activities and processes which took place in the analysed area in the past. Almost every activity of man concerning space has its result in landscape mosaic created by various forms of landcover and landuse.

The aim of the study was the analysis and comparison of landcover diversity of landscape parks in Kujawsko-Pomorskie Voivodeship on the basis of reference spatial data and satellite imageries.

## 2 STUDY AREA

Studies have been done in the area of landscape parks situated in Kujawsko-Pomorskie Voivodeship (Fig. 1). Landscape parks are protected territories because of their natural, historical, and cultural values. They are created to preserve, popularize and disseminate these values in the conditions of sustainable development (Walczak et al. 2001). In the areas founded as landscape parks economical activities take place which are slightly limited and under some strictness. These territories are the basic element of a large-space nature preservation system.

Figure 1. Situation of study area.

120 landscape parks were created in Poland till the end of December 2005. Their total area is 2 482 211 ha (about 8% of Polish territory). Within this total number 8 landscape parks were founded in Kujawsko-Pomorskie Voivodeship (over 209 200 ha which is 15% of the Voivodeship area). The following landscape parks were created in Kujawsko-Pomorskie Voivodeship (Fig. 1):

- Brodnicki Landscape Park (BLP) – created in 1985, in the center of the Brodnica Lakeland; the area of the park is 13 674 ha, 9 338 ha is in Kujawsko-Pomorskie Voivodeship; forests (60%) and water (12%) dominate in the structure of the landcover,
- Gostyninsko-Włoclawski Landscape Park (GWLP) – created in 1979 in the area of the Plock Basin; the area of the park is 38 950 ha, 22 200 ha is in Kujawsko-Pomorskie Voivodeship; forests (60%) dominate in the structure of the landcover,
- Gorzniensko-Lidzbarski Landscape Park (GLLP) – created in 1990 in the eastern part of the Chełmno-Dobrzyn Lakeland; the area of the park is 27 764 ha, 13 902 ha is in Kujawsko-Pomorskie Voivodeship; forests (70%) dominate in the structure of the landcover,
- Krajenski Landscape Park (KLP) – created in 1998 in the central part of the Krajen Lakeland; the area of the park is 54 395 ha, woodiness in the area is 30%,
- Nadgoplanski Millennium Park (NMP) – created in 1992 in the area of the Wielkopolska Lakeland; the area of the park is 8 898 ha, water covers 24% of the territory of the park,
- Tucholski Landscape Park (TLP) – created in 1985 in the Pomorze Lakeland in the central part of the Tuchola Forest: the area of the park is 36 983 ha, forests (86%) dominate in the structure of the landcover,

- Wdecki Landscape Park (WLP) – created in 1993 in the central-eastern part of the Tuchola Forest; the area of the park is 19 177 ha, forests (70%) dominate in the structure of the landcover,
- Chełminski and Nadwislanski Complex of Landscape Parks (CNCLP) – created in 2003 (after the connection of two landscape parks) in the Lower Vistula Valley and the Chelmno-Dobrzyn Lakeland.

## 3 MATERIALS AND METHODS

Spatial data which are national reference data for the study area – General Geographic Database (GGD) and CORINE Land Cover (CLC) Database were applied to realise the aim of the study. The applied data are the part of Spatial Data Infrastructure (SDI).

Within the NGIS project for the whole state, according to the same standards, rules and instructions, General Geographic Database (GGD) is created which informational range consists of 8 thematic groups. This is done using 3 scales: 1:250 000, 1:500 000 and 1:1 000 000. According to the scale of elaboration some thematic groups have various thematic layers: from 23 for the scale 1:250 000 to 16 for the scale 1:1 000 000. In each scale of the presence of GGD there was the landcover layer worked out, and it illustrates the spatial structure of this element. This base makes possible to distinguish 19 classes of landcover in the biggest scale and 5 classes of landcover in the smallest one. The main source of information applied during the creation of this basis was Vector Map Level 2 (VMAP L2) made by The Board of Military Geographic General Headquarters of Polish Armed Forces or other state registers (e.g. The State Registry of Boundaries, The State Registry of Roads).

CORINE Land Cover is the second database which consists of information about the landcover in the whole state. The map of the landcover for the state was created for 2000 on the basis of Landsat satellite data. CLC is hierarchically organized on three levels, which differ in their scales and numbers of categories of landcover. The third level (state level) covers 44 forms of landcover, within them 31 exist in the area of Poland. There are main 5 categories of landcover in the first level (European level). Source spatial data described and used on the paper, mutually supplement and cover one another.

The structure of landscape in landscape parks was analyzed with the application of landscape indices. The following indices of size, edges, shapes, and diversity describing the structure of landscape were calculated and compared:

- a number of cover classes (NUCP), which shows a number of distinguished classes of landcover,
- a number of patches (NUMP), which shows a number of distinguished patches of landcover,
- total length of the edge (TE), summing up the length of edges between all patches of each category,
- shape index (MSI), showing the shape of patches; the higher the value of this index the less regular are landscape patches,
- fractal dimension (FD), applied for description of roughness of landscape patches, and it is the index of structure complexity,
- Shannon's diversity index (SDI), shows the share of separate categories in the structure of the study area,
- NDVI index according to the formula proposed by Rouse et al. (1973) [NDVI = (IR-R)/(IR+R), where IR means near-by infrared radiation, and R – red radiation].

The first six indices on the list were calculated on the basis of General Geographic Database and CORINE Land Cover Database. The last index (NDVI) was worked out applying Landsat satellite imagery from May 2000. Comparisons of indices among parks were done in relation to various scales of source materials (from 1:250 000 to 1:1 000 000).

## 4 RESULTS

Maps of landcover were worked out on the basis of referential data (Fig. 2). Measures of the landscape structure were also calculated for each landscape parks in Kujawsko-Pomorskie Voivodeship (see Table 1). NDVI index was calculated for landscape parks on the basis of satellite imagery. Because there was no full range of the analysed satellite scene in relation to Kujawsko-Pomorskie Voivodeship area, this index was calculated only for 5 landscape parks.

Table 1. Landscape indices for landscape parks in Kujawsko-Pomorskie Voivodeship in relation to various reference data.

| Landscape Park | Source of data | Spatial landscape indices | | | | | | |
|---|---|---|---|---|---|---|---|---|
| | | NUCP | NUMP | TE | MSI | FD | SDI | NDVI |
| BLP | CLC 2000 | 9 | 76 | 5528.1 | 2.41 | 1.28 | 0.63 | 0.15 |
| | GGD 250k | 9 | 118 | 6217.4 | 2.19 | 1.29 | 0.65 | |
| | GGD 500k | 5 | 38 | 11051.4 | 2.77 | 1.27 | 0.30 | |
| | GGD 1000k | 3 | 16 | 35135.0 | 4.64 | 1.26 | 0.08 | |
| GWLP | CLC 2000 | 13 | 164 | 19111.1 | 2.54 | 1.28 | 0.98 | no data |
| | GGD 250k | 11 | 266 | 11587.2 | 2.03 | 1.29 | 0.73 | |
| | GGD 500k | 5 | 70 | 32916.6 | 2.65 | 1.27 | 0.17 | |
| | GGD 1000k | 3 | 26 | 91515.4 | 4.49 | 1.24 | 0.05 | |
| GLLP | CLC 2000 | 11 | 126 | 4190.6 | 2.44 | 1.29 | 1.12 | no data |
| | GGD 250k | 11 | 156 | 4408.8 | 2.30 | 1.29 | 1.16 | |
| | GGD 500k | 6 | 50 | 11727.9 | 2.93 | 1.28 | 0.36 | |
| | GGD 1000k | 2 | 19 | 34963.9 | 3.93 | 1.24 | 0.06 | |
| KLP | CLC 2000 | 13 | 298 | 6273.9 | 2.03 | 1.27 | 1.29 | 0.34 |
| | GGD 250k | 11 | 422 | 6714.2 | 1.87 | 1.28 | 1.29 | |
| | GGD 500k | 6 | 157 | 4500.3 | 1.87 | 1.26 | 0.94 | |
| | GGD 1000k | 3 | 67 | 91526.8 | 2.64 | 1.24 | 0.05 | |
| NMP | CLC 2000 | 10 | 41 | 12708.1 | 2.84 | 1.28 | 0.15 | no data |
| | GGD 250k | 10 | 89 | 9919.2 | 2.32 | 1.30 | 0.27 | |
| | GGD 500k | 6 | 29 | 31871.6 | 3.93 | 1.29 | 0.06 | |
| | GGD 1000k | 3 | 10 | 90831.3 | 9.17 | 1.30 | 0.01 | |
| TLP | CLC 2000 | 12 | 102 | 7749.9 | 2.41 | 1.29 | 0.77 | 0.40 |
| | GGD 250k | 10 | 160 | 6179.5 | 2.12 | 1.29 | 0.98 | |
| | GGD 500k | 6 | 80 | 14289.5 | 2.52 | 1.28 | 0.44 | |
| | GGD 1000k | 3 | 40 | 102247.3 | 4.45 | 1.25 | 0.32 | |
| WLP | CLC 2000 | 10 | 75 | 6805.1 | 2.62 | 1.29 | 0.54 | 0.44 |
| | GGD 250k | 10 | 106 | 4919.1 | 2.26 | 1.30 | 0.70 | |
| | GGD 500k | 5 | 40 | 13683.8 | 3.09 | 1.28 | 0.33 | |
| | GGD 1000k | 3 | 28 | 10986.4 | 2.92 | 1.25 | 0.31 | |
| CNCLP | CLC 2000 | 14 | 318 | 20314.5 | 2.59 | 1.29 | 1.50 | 0.30 |
| | GGD 250k | 14 | 509 | 15402.3 | 2.22 | 1.30 | 1.36 | |
| | GGD 500k | 6 | 101 | 30115.0 | 3.16 | 1.28 | 0.88 | |
| | GGD 1000k | 4 | 45 | 49826.9 | 4.08 | 1.27 | 0.59 | |

In the Table 1, marking of landscape parks and landscape indices according to marks presented in chapter 2.

The analysis of spatial structure of the landscape showed that the most substantial diversity was characteristic for: Chełminski Landscape Park, Nadwislanski Complex of Landscape Parks, Krajenski Landscape Park and Gorzniensko-Lidzbarski Landscape Park. The smallest diversity is connected with Nadgoplanski Millenium Park.

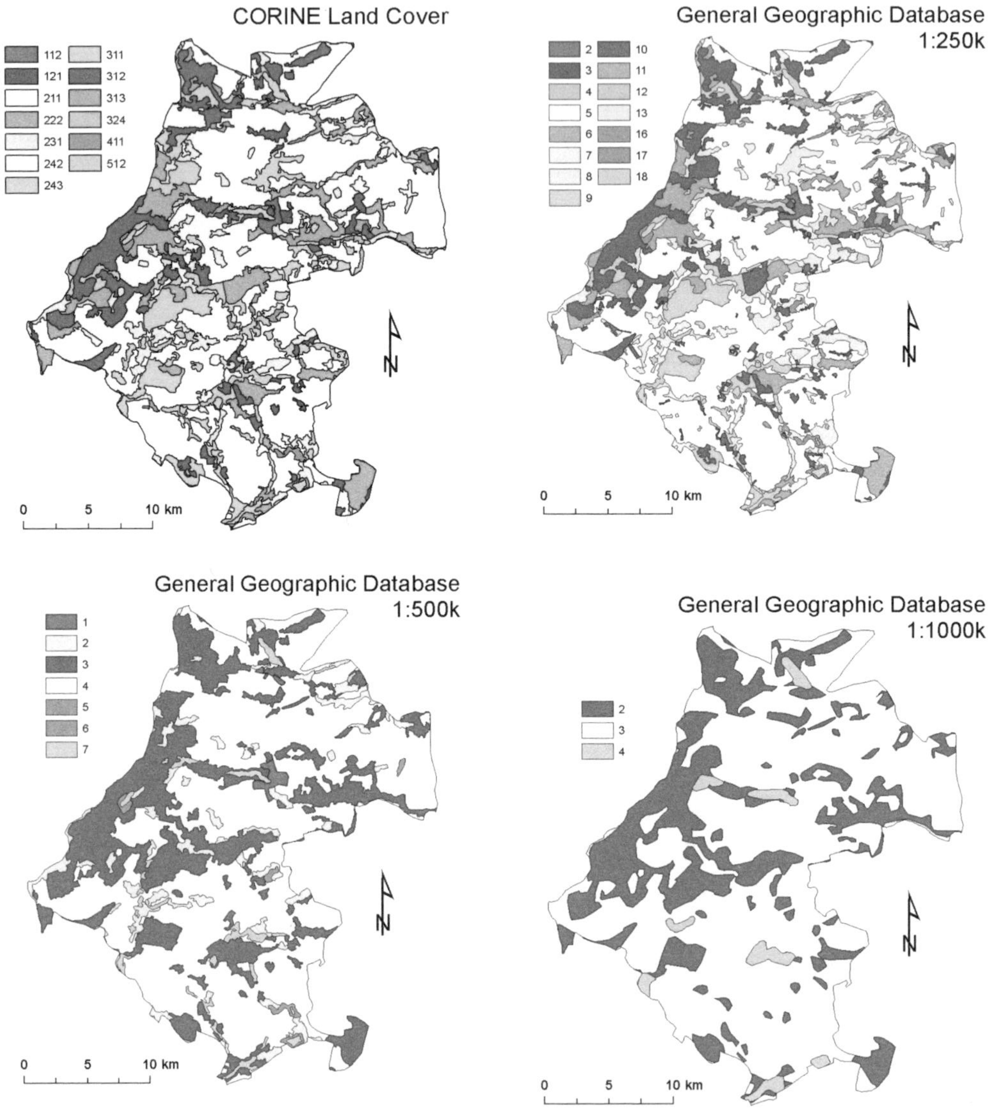


Figure 2. Landcover of the Krajenski Landscape Park (KLP) according to different data sources. Explanation of landcover categories: CLC – according to European standards; GGD250k - 2 (buildings), 3 (industry areas), 4 (city vegetation), 5 (arable fields), 6 (orchards), 7 (meadows, pastures), 8 (other arable fields), 9 (deciduous forests), 10 (coniferous forests), 11 (mixed forests), 12 (other vegetation), 13 (sands, dunes), 16 (swamps, peat-bogs), 17 (rivers), 18 (waterlands); GGD500k - 1 (buildings), 2 (meadows, pastures), 3 (forests), 4 (others), 5 (swamps, peat-bogs), 6 (rivers), 7 (waterlands); GGD1000k – 2 (forests), 3 (others), 4 (waterlands).

Landscape parks with smaller share of forests in the structure of landcover (>30%) had characteristic bigger number of patches of landcover calculated on the basis of CORINE Land Cover data as well as General Geographic Database. The exception was only the smallest in the area Nadgoplanski Millennium Park, where the smallest number of patches was distinguished.

From 9 to 14 classes of landcover were distinguished in the area of landscape parks. The biggest number of types of landcover was characteristic for Chełminski and Nadwislanski Complex of Landscape Parks and Krajenski Landscape Park. This index was correlated with diversity. The number of classes of landcover distinguished on the basis of General Geographic Database in the 1:250 000 scale was identical to the number of categories in CORINE Land Cover. Three parks (TLP, KLP and GWLP), where in CLC distinguished 2 more categories were the exception.

The smallest in the area landscape park (TLP) had characteristic the biggest shape index. It showed its substantial irregularity of distinguished patches of landcover.

No substantial correlation among number of distinguished categories of landcover, the fractal dimension, and the total length of boarders of patches was defined.

Landscape parks with bigger share of forests in landcover structure had higher NDVI index that those which had more arable areas.

## 5 CONCLUSIONS

Analysis of measures of spatial pattern of the landscape worked out on the basis of landcover from database (General Geographic Database and CORINE Land Cover) shows differences in landscape structure. The reason for these differences are specific landscape values of parks (e.g. forests, lakes). The results differentiate landscape parks in an identical way irrespectively of source materials and the scale of data. CLC class seems to be fuller taking into consideration the richness of separated classes. Taking into consideration detailed elements GGD is more precise. The applied source materials may be used complementary for observations and estimation of relationship among various forms of a landcover, a landscape state, and its transformation. On the basis of analysis of the results it may be stated that the presented research approach may show structural differences among landscape fragments.

## REFERENCES

Rouse, J.W., Haas R.H., Schell J.A., Deering D.W. 1973. Monitoring vegetation systems in the great plains with ERTS, ERTS Symposium. *NASA SP-351* I: 309-317.

Walczak, M., Radziejowski, J., Smogorzewska, M., Sienkiewicz, J., Gacka-Grzesikiewicz, E., Pisarski, Z. 2001. *Obszary chronione w Polsce.* Warszawa: Instytut Ochrony Srodowiska.

*Remote Sensing for a Changing Europe*
*D. Maktav (Ed.)*
*IOS Press, 2009*

*doi:10.3233/978-1-58603-986-8-564*

# Landuse changes of Goleniow Forest in 1973-2000 using remote sensing methods

M. Kunz & M. Klimecki
*Department of Cartography, Remote Sensing and GIS, Nicolaus Copernicus University, Torun, Poland*

Keywords: landuse changes, reforestation, GIS, Landsat, Goleniow Forest

ABSTRACT: The Goleniow Forest is a very interesting, compact compact forest complex situated on the eastern coasts of the Szczecinska Bay and Dabie Lake (north-western part of Poland). The study area administratively belongs to Zachodniopomorskie Voivodeship and it is managed by the Goleniow Forest Inspectorate. It covers the area of 53 989 ha, within it over 22 300 ha are forests, mainly pine-trees. Recently, in the study area, there have been clear anthropogenic changes mainly connected with the development of wood industry. There are intensive cuttings in the area of the Forest Inspectorate. 83 500 $m^2$ of wood are cut every year. Clearings are reforested together with non-forest areas. Because of good natural conditions, 4 natural reserves have been created in the area of the Inspectorate, and further 4 are being worked out. The analyzed region is included in the national ecological net ECONET-POLAND as an international junction area, and within it the area of habitat preservation has been created in the scheme of NATURA 2000. The aim of the study was to find out the direction and dynamics of changes in a landuse in the area of the Goleniow Forest in 1975-2000 with the application of accessible satellite imageries LANDSAT and general maps of treestands. Satellite imageries from 1975 and 2000, general maps of treestands from 1973, 1985 and 1997, and topographic maps were used for the analysis. The study was done with the technology of geographic information system.

## 1 INTRODUCTION

The first paragraph. Taking into consideration nature, the Goleniow Forest is a very interesting area situated in the north-western part of Poland (Fig. 1). In spite of the fact that first information concerning Goleniow Forest is from XIV century, management of this area before the WWII is still unknown because of the lack of any written documents. In 1939-1945, there was a very wasteful exploitation of forest (by clear cuttings) without any plans of management or reforestation of the territory. Recently, anthropogenic transformations have been taking place. They were mainly connected with the development of wood industry. Intensive clear cuttings take place in the area of the Inspectorate. The territories of cuttings as well as non-forest areas are reforested again.

The aim of the study was to find out the direction and dynamics of changes in a landuse in the area of the Goleniow Forest in 1973-2000 with the application of accessible Landsat satellite imageries and general maps of treestands.

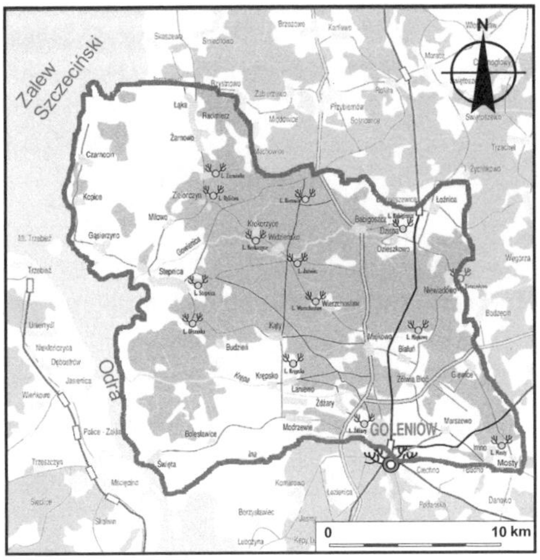


Figure 1. Location of study area.

## 2 STUDY AREA

The Goleniow Forest is a very interesting, compact forest complex situated on the eastern coasts of the Szczecinska Bay and Dabie Lake (north-western part of Poland) (Kondracki 2002). The study area administratively belongs to Zachodniopomorskie Voivodeship and it is managed by the Goleniow Forest Inspectorate. The Inspectorate has been situated within the present boarders since 1972, the year when contemporary existing inspectorates: Goleniow, Stepnica and Widzensko were joined. It covers the area of 53 989 ha. Forest economy takes place on over 22 000 ha, there. Forests cover almost 21 000 ha, and non-forest areas like swamps, waters, and meadows create the rest of the landscape. The area of the Inspectorate has been divided into 14 forestries functioning within one forest district (Kosacki 2001).

There is hardly any species diversity in forests of the Goleniow Inspectorate. Pine-trees dominate on 76% of the area. This species creates homogenous tree stands mainly on a poor, sandy soil. Average age of pine-trees in the study area is 52 years. Moist areas, especially on the west and in river valleys are covered by alder. It grows on 9.7% of the territory. The average age of the species is 58 years. Young tree stands of birch trees (average age 36 years) cover 8.5% and they exist mainly on formerly arable areas. Other interesting tree stands are an oak (1.4% of the area), a beech (1.1% of the area), and a spruce-tree (1.7% of the area). Other species are not so significant. Taking into consideration the habitat, mixed fresh coniferous forest and fresh coniferous forest are predominant in the study area.

83 500 $m^3$ of wood are cut in the area of the Forest Inspectorate every year. Over half of this number are loggings, and improvement cuttings are the rest. Every year, tree clearances take place only on 90-120 ha of the forest, and ripening activities are on about 5 000 ha. Yearly restockings and reforestations take place in the area of about 190 ha.

Taking into consideration nature, the most important regions in the Inspectorate are swamps and moors together with green crops and plantings around them. In the study area, four nature reserves have been created because of their unique character: the "Czarnocin" moor reserve (9.4 ha), "Uroczysko Swiete" floral reserve (9.5 ha), the "Wilcze Uroczysko" moor reserve (61.8 ha), and the "Olszanka" moor-floral reserve (1 291 ha). At present, activities are taking place over the creation of the next four reserves. Other forms of nature preservation in the study area are nature monuments (12), and bird protection zones (22) mainly of a white-tailed eagle, a black stork, and a red kite.

The analyzed region is included in the national ecological net ECONET-POLAND as an international junction area, and within it the area of habitat preservation was created in the scheme of NATURA 2000 (Liro 1995).

## 3 MATERIALS AND METHODS

Landsat satellite imageries and general maps of tree stands were applied (Table 1 and 2) in the analysis of landuse changes in the Goleniow Forest and the area situated within the borders of the Goleniow Inspectorate. Additionally, during calibration of general maps of tree stands, topographic maps in scale 1:25 000 were analysed.

Table 1. List of used satellite imageries.

| Satellite | Scanner | Registration data | WRS |
|---|---|---|---|
| Landsat 5 | MSS | 27.05.1975 | 208-22 |
| Landsat 7 | ETM+ | 20.06.2000 | 192-23 |

Table 2. List of used general maps of tree stands.

| Publication year | Scale |
|---|---|
| 1973 | 1:20 000 |
| 1985 | 1:25 000 |
| 1997 | 1:20 000 |

General maps of tree stands were created by the Office of Forest Management and Forest Geodesy in Gorzow Wielkopolski. They show the division of the Inspectorate into several sections and subsections together with the way of their management, type and age of tree stands, and the characteristics of the habitat. The supplement of maps is a impact assessment for the forest management. This material is very detailed.

Raw Landsat satellite imageries were offered for free by the University of Maryland in the US. Analysis of landuse changes were done on the basis of satellite imageries, and it was created for the whole territory of the Goleniow Inspectorate. In case of general maps of tree stands, the study area was narrowed to the places covered with the map.

All the analogue maps (general maps of tree stands, and topographic maps) were scanned and calibrated to the proper reference system. Satellite imageries were also under the process of rectification.

Geometrically correct source materials were the basis for the production of multitemporal database concerning the exploitation of the Goleniow Forest. After the analysis of source materials 8 categories of landuse were distinguished: a forest, a cutting, a meadow, an arable area, a water, a swamp, a pasture, and a forest nursery. Because of the resolution of spatial imagery materials, last three categories were not mentioned during analysis.

On the basis of interpretation of source materials, spatial databases was created which presents the landuse in this area in 1973, 1985, 1997 on the basis of cartographic data and 1975-2000 on the basis on satellite imageries.

This database makes it possible to define areas covered by separate landuse categories, size and dynamics of changes. On the basis of the first state (1973 or 1975) and the last one (1997 or 2000) the analysis of changes was done which took place in the Goleniow Forest. Changes of more than 25 metres in the position of object were considered significant. As a result of this procedure, 559 transformation areas were defined on the basis of cartographic data and 532 transformation areas on the basis of satellite imageries.

The study was done with the technology of geographical information system with the application of licensed ArcView (ESRI) and MicroStation (Bentley) software.

## 4 RESULTS

Spatial data basis concerning the use of the Goleniow Forest in 1973-2000 make possible complete analysis of transformations which took place in this range, and definition of their size and direction. Because the landuse was interpreted on the basis of two different sources, the results were described independently.

### 4.1 *Changes of landuse in the Goleniow Forest in 1973-1997 on the basis of interpretations of general maps of tree stands*

In 1973-1997 changes in landuse concerned 1575 ha, which is 7% of the analysed area (Fig. 2). Among territories under transformation, two especially large areas may be distinguished. One of them is a swamp covered by alders. It is the area of 310 ha, situated in the north-western part of the Inspectorate. As a result of drainages and plantings it was transformed into the complex of an alder-birch forest. Second area with the most substantial changes of area is the place covering 280 ha, which is managed by the army. In the 80s the area was taken over by the army unit connected with the airport in Goleniow.

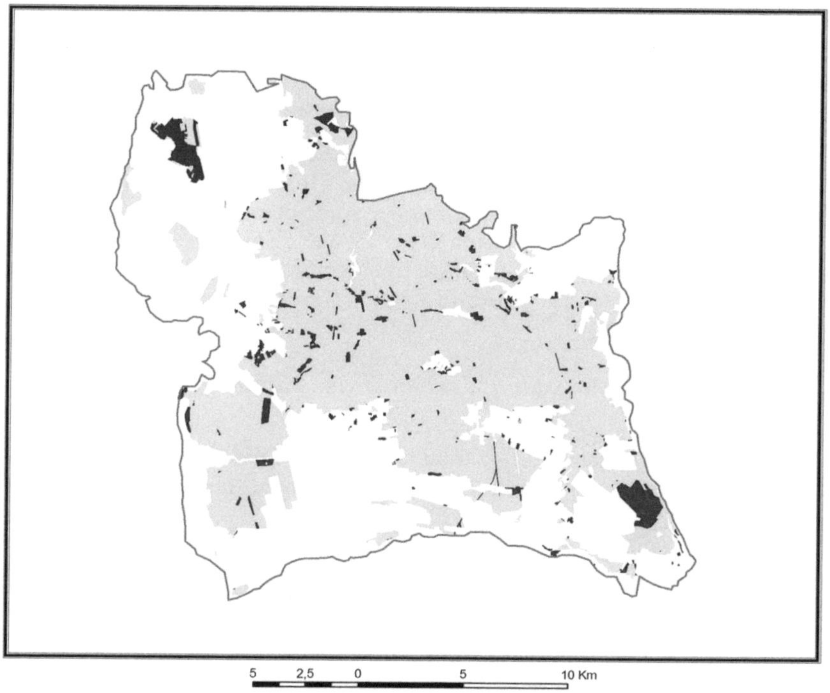


Figure. 2. Landuse changes in the Goleniow Inspectorate in 1973-1997 on the basis of general maps of tree stands.

Except changes in two large areas, smaller ones took place on almost entire territory of the Inspectorate. In general, in the analysed period of time there were 315 changes of one category of landuse into the other one. There were also changes connected with the ownership of land – which was under or out of the management to the Inspectorate. The biggest number of changes took place in the western part of the Inspectorate, on the west from the heads of the Krepa River. However, changes in these areas did not have a homogenous direction. The most stable areas were Olszanka

region, the complex of swamps situated in the south-western part of the analised territory, and the central part of the Inspectorate.
The average size of the area of changes was 5 ha. However, if two largest sample plots were not taken into consideration, this number would decrease to only 1.5 ha.

The most considerable changes in 1973-1997 took place in swamp ecosystems (see Table 3). Total share of swamps under transformations was over 28% of all areas under landuse changes. This number was considerable influenced by transformation of swamps into the alder-birch forest.

Table. 3. Landuse transformations in the Goleniow Inspectorate in 1973-1997 on the basis of general maps of tree stands.

| 1973 \ 1997 | swamps | arable fields | forests | meadows | pastures | waters | cuttings | areas out of Inspectorate |
|---|---|---|---|---|---|---|---|---|
| swamps | 513.8 | - | 416.0 | - | 1.9 | 5.2 | - | 21.4 |
| arable fields | - | 60.5 | 77.0 | - | 4.9 | - | - | 4.3 |
| forests | 66.8 | - | 20 543.0 | - | 7.5 | - | 60.5 | 297.8 |
| meadows | 54.9 | - | 125.5 | 151.4 | 30.8 | - | - | 36.3 |
| pastures | - | - | 72.9 | - | 37.3 | - | - | - |
| waters | 5.8 | - | - | - | - | 90.4 | - | - |
| cuttings | - | - | 41.5 | - | - | - | 0 | - |
| areas out of inspectorate | 16.2 | - | 227.0 | 2.2 | - | - | - | x |

In spite of the fact that considerable number of transformations relates to forests, it is also the result of taking quite a big part of forests out of the management of the Goleniow Inspectorate. In the analysed area only 2% of general territory of forests were under some changes. Some parts of forests, mainly on the fringes, were given for the settlement. Over 15% of changes were on meadows. Half of them were reforested in 1973, the rest of them were transformed into swamps or pastures. Arable areas were mainly changed into pastures. Water reservoirs preserved their shapes and they did not change their areas.

## 4.2 *Changes of landuse in the Goleniow Forest in 1975-2000 on the basis of interpretation of satellite imageries*

In 1975-2000, landuse transformations concerned 8 260 ha, which was 20% of the analysed area (Fig. 3). It may be noticed that changes took place in the area of the whole forest, and the number of changed places with diverse area is considerable.
The biggest area of changes in the analysed period of time was a 400 ha territory situated on the north from Goleniow, where meadows were transformed into arable areas. Next place with the identical transformation was 380 ha territory on the west of Stepnica. There was the opposite direction of changes from arable areas into meadows in the area situated in south-western part of the Goleniow Forest. There were 400 areas where transformations in landuse were observed. The average territory of landuse changes was 21.5 ha, however half of these areas were not bigger than 10 ha. 10 areas were distinguished and their territory of transformations were over 200 ha.
The biggest transformations took place in the areas which in 1975 were meadows (see Table 4). These changes were over 42% of all transformations, and they were connected with changes into arable areas as well as with reforestations. Moreover, part of meadows (especially in the region of the Swiete Range) was flooded and transformed into marshy grounds, and swamps. Changes connected with arable areas constituted 25% of all transformations, and they were connected with their transformations into meadows. In this time, about 500 ha of arable areas were reforested. They were mainly situated on the fringe of the main forest complex. Forests had similar share in transformations as arable areas. Most transformations of forests were connected with their changes into arable areas.

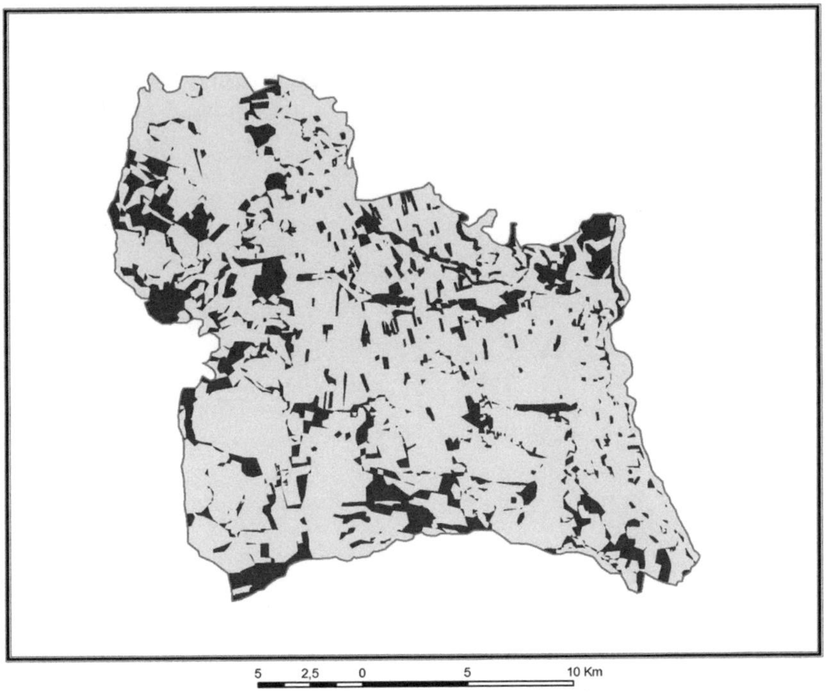


Figure. 3. Landuse changes in the Goleniow Forest in 1975-2000 on the basis of Landsat satellite imageries.

Table. 4. Landuse transformations in the Goleniow Forest in 1975-2000 on the basis of interpretation of Landsat satellite imageries.

| 1975 | 2000arable fields | forests | meadows | waters | cuttings |
|---|---|---|---|---|---|
| arable fields | 5 136.3 | 546.6 | 1631.0 | - | - |
| forests | 384.2 | 19 395.7 | 1305.2 | 34.9 | 438.8 |
| meadows | 2 797.2 | 820.7 | 10 362.9 | 46.9 | - |
| waters | - | 48.7 | 21.1 | 216.6 | - |
| cuttings | - | 626.1 | 3.9 | - | 5.8 |

## 5 CONCLUSIONS

Most areas covered with changes in case of Landsat satellite imageries from 1975-2000 are out of transformational areas interpreted on general maps of tree stands from 1973-1997.

On the basis of studies in the area of the Goleniow Inspectorate, applying the interpretation of Landsat satellite imageries (1975-2000) and general maps of tree stands (1973-1997) it can be stated:

- the forest area of Goleniow Inspectorate is over 90% and it is still increasing,
- the forest area of Goleniow Forest is over 49% and it is decreasing,
- increase of forest areas is at the sacrifice of meadows,
- the biggest increase in the area was on arable areas, which expanded mainly at the expense of meadows,
- transformations covered the total number of 7% of territories managed by the Goleniow Inspectorate, and 20% of the Goleniow Forest,
- in the future, the results of this study will determine the basis for conclusions concerning landuse transformations in this area, and they will be a referential material in further studies and analysis.

This method, based on the analysis of cartographic materials as well as spatial imagery data can well be used for studies over transformations taking place in other Forest Inspectorates.

The aim of the study was to find out the direction and dynamics of changes in a landuse in the area of the Goleniow Forest in 1973-2000 with the application of accessible Landsat satellite imageries and general maps of treestands.

## REFERENCES

Kondracki, J. 2002. *Geografia regionalna Polski.* Warszawa: PWN.

Kosacki, J. 2001. Lesny Kompleks promocyjny Puszcza Bukowa i Goleniowska. *Wedrowiec Zachodniopomorski* 1: 18-21.

Liro, A. (ed.) 1995. *Koncepcja krajowej sieci ekologicznej ECONET-POLSKA.* Fundacja IUCN Poland: Warszawa.

*Remote Sensing for a Changing Europe*
*D. Maktav (Ed.)*
*IOS Press, 2009*

*doi:10.3233/978-1-58603-986-8-571*


# Land cover change assessment in Belek forest based on change vector analysis

A. Akkartal & F. Sunar
*Center of Satellite Communication and Remote Sensing, Istanbul Technical University, Turkey*

Keywords: change detection, change vector analysis, desertification

ABSTRACT: The detection and monitoring of land use / land cover change using satellite multispectral imagery has been a topic of interest in remote sensing. Different techniques, mostly grouped into two general classes which are based on spectral categorization (classification) and based on radiometric change, for accomplishing change detection have been formulated, applied and evaluated in many environmental applications. Change vector analysis is an effective approach for detecting and characterizing land cover change that processes and analyses change in all multi-spectral and multi-temporal data. In this context, the main objective of this paper is to authenticate the land cover change based on a radiometric technique, Change Vector Analysis (CVA) for the Belek forest, Antalya.

## 1 INTRODUCTION

The growth in the population of Earth and the needs of human beings causes unconscious deformation of natural resources more and more. The perishing of natural resources affects the wild life, ecological balance, food chain and finally the life of posterity. As a result, planners and resource managers need reliable mechanism to assess these consequences by detecting, monitoring and analyzing land use changes quickly and effectively.

The monitoring and detection of land use/cover change using satellite multispectral imagery has been an important topic in remote sensing. A number of techniques for realizing the detection of land cover dynamics using remotely sensed imagery have been formulated, tested and assessed with the results varying with respect to the change scenario under investigation, the information required and the imagery applied (Siwe & Koch 2007). Most of these techniques can be grouped into two general classes: (1) those based on spectral categorization (classification) of the input data; and (2) those based on radiometric change between acquisition dates (Johnson & Kasischke 1998).

Monitoring change in vegetation between two time periods can assess the health and vigor of forest and plant species, assess vegetation growth and regrowth following a cataclysmic event, or quantify forest loss caused by deforestation and timber harvesting (Lawrence & Ripple 1999). These changes can be classified effectively using the change vector analysis (CVA) (Lambin & Strahler 1994, Johnson & Kasischke 1998, Allen & Kupfer 2000, Lorena et al 2002, Lunetta et al 2004). CVA is a radiometric technique that examines corresponding pixels of two maps by comparing two bands of each map to produce images of change direction and change magnitude (Kuzera et al 2005).

The SPOT -5 multispectral satellite images of the area covering the forest and destined areas in Belek Antalya are acquired by SPOT Image on the 27th of October, 2005 and 23rd of May, 2007, respectively and analyzed by ITU-CSCRS (Istanbul Technical University – Center of Satellite

Communication and Remote Sensing). As the most critical pre-processing requirements for CVA, the images were registered geometrically and normalized radiometrically. Therefore after the pre-processing step, CVA was applied to the Greenness and Brightness variables that were calculated from three multispectral bands (having 10 m spatial resolution) of SPOT -5 images. The output was the change image indicating the change vector direction and the multispectral change magnitude that gave information about the amount of the change in the forested area.

## 2 STUDY AREA AND DATA USED

The Belek region on the Mediterranean coast is located 30 kilometers on the east side of Antalya province. The region, which was declared a tourism zone in 1990 and promoted as the place with the best and most dense forests in the Antalya region, has been subject to deforestation as a result of the golf boom in the region (Cyprus Observer 2007).

Belek forest, the worlds second great stone pine area, was planted in 1960's and now it's being destined to golf areas and tourism resorts. It was the region's and one of Turkey's most important ecosystem that sheltered 27 species and 1 sub species of endemic plants, 109 birds that are under danger with average of 600 – 700 thousand trees spread over 2300 hectares.

The coastal regions of the forests are the second largest known nesting area for sea turtles after the Zakynthos Island of Greece for Caretta Caretta sea turtles and the vegetation is important for migrating birds that come to the region.

Depending on its environmental importance Belek Forest were chosen as the study area (Figure 1).

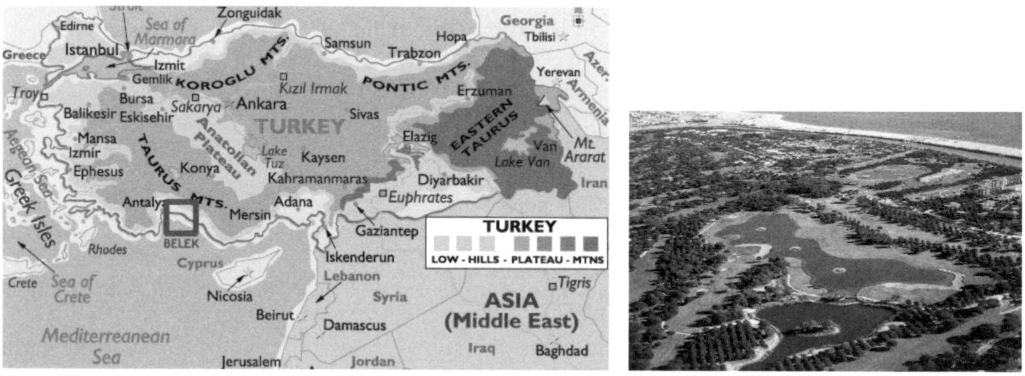


Figure 1. Map and aerial photograph of the study area

SPOT-5 images, with 5m spatial resolution, were used to detect the deforestation of the area between two dates that the images were acquired (Table 1).

Table 1 . The characteristics of the satellite data used

| Satellite | Date | Spectral Res. (nm) | Spatial Res. (m) |
|---|---|---|---|
| SPOT – 5 | 27.11.05<br>23.05.07 | Green: 500 – 590<br>Red: 610 – 680<br>Near IR: 780 – 890 | 5 |

## 3 METHODOLOGY

Classical digital change detection techniques using remotely sensed sequential data taken from the same area are based on the comparison of land cover classifications, multidate classifications, image differencing and ratioing, vegetation index differencing, principal component analysis and change vector analysis (Taberner & Sunar 1998). In this study only NDVI differencing and CVA were used.

### 3.1 *NDVI differencing*

Vegetation indexes depend on the reflectance of vegetation, which is very different in near infrared and red bands. Healthy vegetation should absorb the visible light and reflects most of the near infrared light; on the other hand unhealthy vegetation reflects more visible light and less near infrared light. The reflection on visible band is related with the pigments in the leaves of plants but in the near infrared, it depends on the cell structure.

The algorithm of NDVI is (near infrared band– red band)/(near infrared band + red band). Resulted values change between -1 and +1 regarding to the vegetated area. Such as, if the result is 0,1 or below, it corresponds to an area of rocks; if it is between 0.2 and 0.3, it indicates an area of shrubs or grasslands; if it is between 0.6 and 0.8 it corresponds to an area of tropical rainforests.

### 3.2 *Change vector analyze*

Change vector analysis (CVA) is a robust approach for detecting and characterizing radiometric change in multispectral remote sensing data sets. The CVA algorithm produces two `channels' of output change information: (1) change vector direction; and (2) multispectral change magnitude (Johnson & Kasischke 1998) (Figure 2).

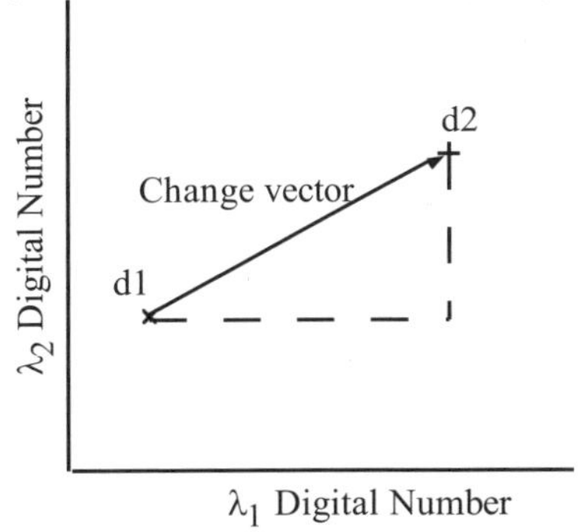


Figure 2. Change vector description for a sample in a particular wavelength band at two different dates

The CVA approach has shown that change vector direction to be a useful aid in discrimination of different phenomenological types of change, and change magnitude to be useful for relative comparisons within and among change types (Johnson & Kasischke 1998).

A drawback in the use of the CVA technique is its susceptibility to noise, mainly caused by the atmosphere, occurring between dates (Taberner & Sunar 1998). The most critical pre-processing requirements for CVA are the accurate geometric registration and radiometric normalization of the input data (Johnson & Kasischke 1998). In registration phase, sufficient ground control points should be used to ensure sub-pixel RMS errors. Radiometric normalization of the input data is critical in order to minimize detection of false changes due to poor normalization.

In this study, data acquired from different bands compose a new axis associated with biophysical properties of targets. These axes are Difference Vegetation Index (DVI), associated with the presence of vegetation, and Soil Brightness Index associated with variations of soil reflectance.

– *Difference vegetation index*

Difference Vegetation Index (DVI), which is also sensitive to the amount of the vegetation. Mathematically, it is in the form of (near infrared band) – (red band). DVI has the ability to distinguish the soil and vegetation but not in shady areas.

– *Soil Brightness Index*

Soil brightness index indicates bare areas such as agricultural fields, beaches, parking lots as the lightest features whereas agricultural fields with harvest and forests as the darkest features within the image. The soil brightness index for SPOT-5 was performed using the following formula (Lau et al 1998);

$$BRIGHTNESS=0.60539*B1+0.61922*B2+0.50008*B3 \quad (1)$$

After DVI and SBI were calculated, the position variation of the same pixel in the multitemporal data within the space, formed by these two axes is used to determine the magnitude and direction of the spectral change vectors.

## 4 APPLICATION & RESULTS

### 4.1 *Geometric and radiometric correction*

Accurate geometric and radiometric corrections are very important for CVA analysis. In the geometric correction phase, image registration was applied to the after image acquired on 23.05.07. In the registration process 14 GCP's were used and the RMS error was obtained as ± 0.78.

Relative radiometric correction is a method of correction that applies one image as a reference and adjusts the radiometric properties of subject images to match the reference. Relative radiometric correction was preferred depending on the fact that this kind of normalization does not require ancillary datasets on for instance atmospheric temperature, relative humidity and/or aerosol back-scatter. Minimum - maximum normalization was chosen and applied to the second image by taking the first image as reference. Each band of the second image was normalized using the formula given below separately.

$$\text{Normalized DN} : A\,x + B$$

$$A: (y_{max} - y_{min}) / (x_{max} - x_{min}) \quad (2)$$

$$B: y_{min} - A\,x_{min}$$

In the formulation of minimum - maximum normalization, $y_{max}$ and $y_{min}$ represents the max. and min. values of the reference image where $x_{max}$ and $x_{min}$ represents the max. and min. values of the image to be normalized.

### 4.2 *NDVI differencing*

NDVI values of both images were calculated. The results derived from the images were extracted from each other (Figure 3).

Figure 3. NDVI differencing image

It is observed that not all the woodcutted areas have low NDVI values like it supposed to be. Since the old woodcutted areas being sprouted up, these areas have higher NDVI values than forested areas due to the different spectral characteristics of grass and pine tress (Figure 4).

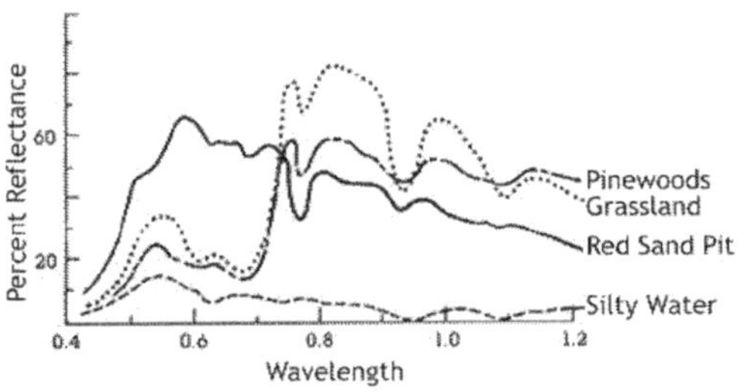


Figure 4. Spectral reflectance characteristics of grasslands and pinewoods

As a result, areas that were sprouted after woodcutting weren't taken in consideration as deforestrated areas in the change detection process.

## 4.3 *Change vector analysis*

The first step of the CVA method was to calculate Soil Brightness Index and DVI values, in order to reduce the amount of redundant information of images to be analyzed. Spectral change vectors were determined by the position variation of the same pixel in multitemporal data within the space formed (Figure 5).

The magnitudes of these vectors were than calculated with the Euclidean Distance indicating the differences in positions of the same pixels.

$$R: \sqrt{(y_2 - y_1)^2 + (x_2 - x_1)^2} \quad (3)$$

Change direction is measured as the angle of the change vector from a pixel measurement at date 1 to the corresponding pixel measurement at date 2 (Kuzera et al 2005).

Angles measured between 90° and 180° indicate an increase in DVI and a decrease in SBI that represent regeneration of vegetation. Angles measured between -90° and 0° indicate a decrease in DVI and an increase in SBI that represent deforestation. Angles measured between 0° - 90° and -180° - -90° indicate either increases or decreases in both bands of DVI and SBI indicating no change (Kuzera et al 2005) (Figure 5).

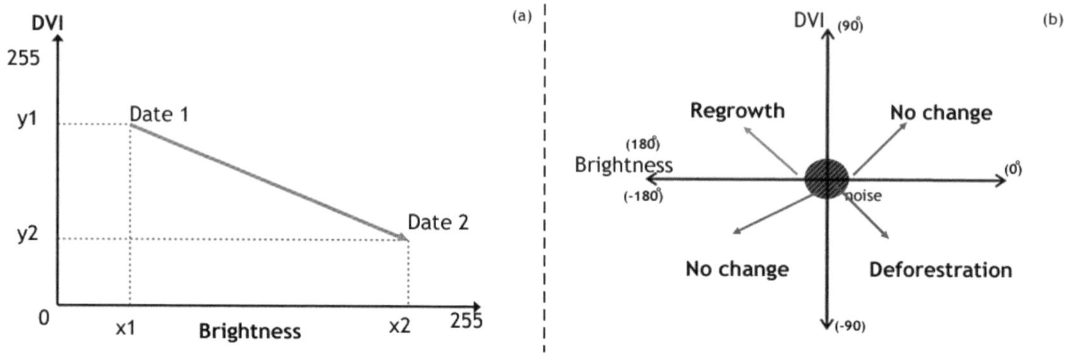


Figure 5. The process for detecting (a) the magnitude of change and (b) the direction of change within CVA

As a threshold value of 26 DN was chosen and the values less than 26 DN were considered as noise or non-efficient normalization (Figure 6). Deforestrated areas that had decrease in DVI and an increase in SBI as the direction of change (i.e. general deforestration) were calculated as 745.8ha.

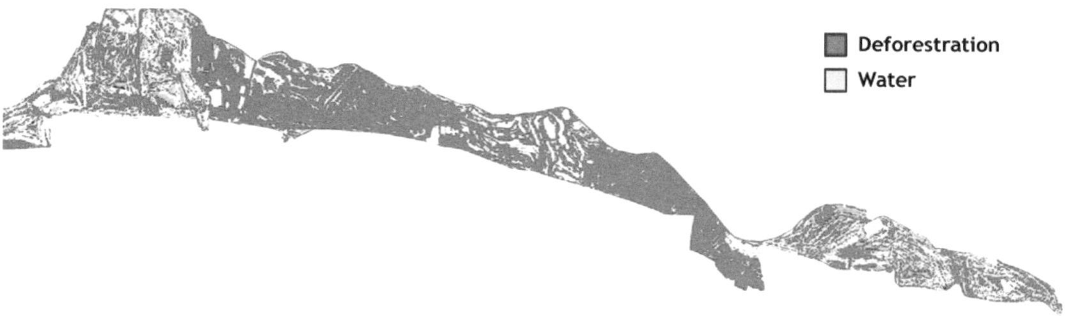


Figure 6. Change Image – Deforestration in the study area with the threshold value of 26 DN

In order to analyze the deforestration amount of denser forest areas the threshold value was taken as 80 DN (Figure 7). The calculated area indicating the denser deforestration was found as 235.6ha.

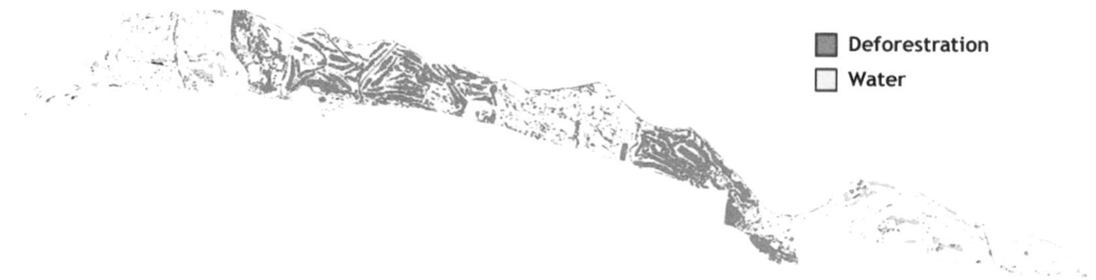


Figure 7. Change Image – Deforestration in the study area with the threshold value of 80 DN

## 5 CONCLUSION

Natural resources of a country should be protected and monitored in order to save them for the benefits of all humankind. In this content, remote sensing techniques are very adequate to analyze the changes in land cover/use with a high accuracy.

As shown in this study, radiometric change techniques such as band differencing and CVA, were used to detect the deforestration due to woodcutting in Belek forests. Band differencing gave information about the general change and its detectability whereas CVA offers qualitative information concerning the direction and the intensity of the change occurred in vegetation. The conversion of the bands to DVI and SBI reduced the dimensionality of the bands and at the same time highlighted vegetative properties of the landscape.

It was seen that, CVA approach of multitemporal data is more sensitive to, and provides more information on, subtle changes seasonality and vegetation phenology than more classic approaches.

## ACKNOWLEDGEMENTS

The authors would like to sincerely thank to the SPOT Image S.A. for providing the satellite images used in this study.

## REFERENCES

Allen, T. R., Kupfer , J.A. 2000. Application of Spherical Statistics to Change Vector Analysis of Landsat Data: Southern Appalachian Spruce-Fir Forests. Remote Sensing of Environment 74: 482-493

Cyprus Observer 2007. Deforestration revealed in Belek. http://www.observercyprus.com/observer/NewsDetails.aspx?id=2450Siwe, R.N., Koch, B. 2007. Change vector analysis to categorise land cover change processes using the tasselled cap as biophysical indicator. Environ Monit Assess, DOI 10.1007/s10661-007-0031-6

Johnson, R. D., Kasischke, E. S. 1998. Change vector analysis: a technique for the multispectral monitoring of land cover and condition. International Journal of Remote Sensing, vol. 19, no. 3, 411- 426

Kuzera, K., Rogan, J., Eastman, J. R. 2005. Monıtorıng vegetatıon regeneratıon and deforestatıon usıng Change vector analysıs: mt. St. Helens study area. ASPRS 2005 Annual Conferance, Baltimore, Maryland, March 7-11

Lambin, E. F., Strahler, A. H. 1994. Change-Vector Analysis in Multitemporal Space: A Tool to Detect and Categorize Land-Cover Change Processes Using High Temporal-Resolution Satellite Data. Remote Sensing of Environment 48: 231-244

Lawrence, R. L., Ripple, W. J. 1999. Calculating Change Curves for Multitemporal Satellite Imagery: Mount St. Helens 1980-1995. Remote Sensing of Environment 67: 309-319

Lorena, R. B., Santos, J. R., Shimabukuro, Y. E., Brown, I. F., Kux, H. J. H. 2002. A change vector analysis technique to monitor land use/land cover change in SW Brazilan Amazon: Acre Site. Proceeding from Integrated Remote Sensing at the Global, Regional, and Local Scale: ISPRS Commission I Mid-Term Symposium in conjunction with Pecora 15/Land Satellite Information IV Conference, November 10-15, 2002, Denver, Colorado, USA. 8 p

Lunetta, R. S., Johnson, D. M., Lyon, J. G., Crotwell, J. 2004. Impacts of imagery temporal frequency on land-cover change detection monitoring. Remote Sensing of Environment 89: 444-454

Taberner, M., Sunar, F. 1998 An implementation of the change vector analysis technique to assess the changes in land cover with multidate remotely sensed images, a case study: Istanbul, Turkey. 27th Int. Symp. on Remote Sensing of Environment – Information for sustainability, pp. 454-457. 8-12 June Tromso, Norway

*Remote Sensing for a Changing Europe*
*D. Maktav (Ed.)*
*IOS Press, 2009*

*doi:10.3233/978-1-58603-986-8-578*

# Pinro Airborne research on Mackerel distribution in the Norwegian sea

Vladimir Zabavnikov & Sergey Egorov
*Knipovich Polar Research Institute of Marine Fisheries and Oceanography (PINRO), Remote Sensing Laboratory, Murmansk, Russia, e-mail: ltei@pinro.ru*

## ABSTRACTS AND INTRODUCTION

To estimate stocks of commercial marine organisms the special surveys are used nowadays. The trawl-acoustic ones are the most popularized. However, this method is characterized by some limitations, which leave certain traces on quality, efficiency and reliability of these surveys and, in the first place, when estimating feeding mackerel biomass by instruments.

This pelagic fish species, as well as the others, may avoid the ambient noises (Chernook et al., 2003) from vessel and trawl when surveying, especially, in the feeding period. Besides, many vessels have low speed, the optimal value of which should be more than 5 m/s (Gavrilov, 1999). The results of acoustic survey are affected by different reflectance of objects depending on their concentration and biology.

Feeding mackerel concentration distinctive mobility and instability in time (further we shall only say about this period of the fish life cycle) required using special mobile means of survey (search), for which in 1982, since June-July, started to be applied was the aircraft IL-18 (Churnside et al., 2001) reequipped to the IL-18 DORR (distant oceanic fish reconnaissance airborne laboratory) afterwards, in the end of 1985. Eyes and experience were the main instruments of the researcher to estimate mackerel distribution during 1982–1985, before IL-18-18DORR to become applicable. But visual aerial surveys were characterized by some limitations. The main of them were subjectivity and impossibility to give instrumental numeral estimation of biomass.

Therefore, in 2001, after having conducted some necessary scientific, methodic, organization and technical activities, in PINRO, aboard specialized and associated aircraft-laboratory AN-26 “Arktika”, the complex of laser location developed and created by specialists from “MULTITECH” Ltd. (St. Petersburg), and named as polarized aerial (PAL-1), was installed and started to be exploited. Later in was modernized to PAL-1M.

Using data of PAL-1M laser location based on the analysis of sea water optical characteristics allows to determine its transparency, the depth of optical heterogeneities (OH) in the sub-surface layer of the sea identifying pycnocline, plankton and schools of pelagic fishes (mackerel, in the first place), as well as the location of oceanographic fronts and boundaries of water masses (Churnside et al., 2002).

## MATERIAL AND METHODS, RESEARCH AREA

As it has been already mentioned, PAL-1M was installed and has been exploited aboard the two-motor aircraft AN-26 since 2002 (Fig. 1). The transmitting and receiving unit (radiator and receiver) of LIDAR is placed in camera window (Fig. 2) closed by flat illuminator made of optical glass. LIDAR optical unit was set at the angle of 17°.

Solid-state pulse laser on YAP:Nd was used as a source of sounding radiation in PAL-1M. Two echo signal components (orthogonal or cross-polarization and normal or co-polarization) are recordered by two identical photo-detectors in the inlet of the optical systems of which the analyzers on the basis of Polaroid film are set. The main specifications of PAL-1M: wave length – 532 nm,

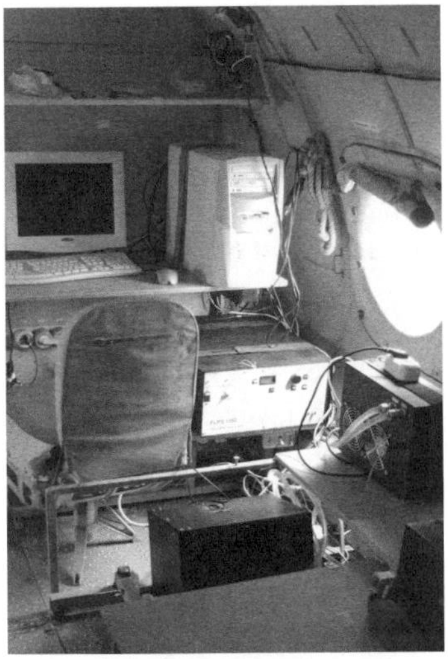

Figure 1. LIDAR aboard "Arktika" airborne laboratory.

Figure 2. Optical LIDAR unit in photo camera hatch of "Arktika" airborne laboratory.

pulse duration – 12.5 ns, pulse energy – 120 MJ, pulse frequency – 1–40 Hz, patch diameter on surface – about 1 m, optical receiver diameter – 10 cm, polarization vector – normal and cross.

There's also a fluorescent channel set in the system of cross-polarization receiving (in cross-channel), where scattered light passing through interference filter with 685 nm wave length gets photomultiplier tube (PMT).

It should be noticed that laser location is one of the elements of complex fishery and oceanographic airborne research including those ones for feeding mackerel, as it has been already mentioned. Therefore, besides LIDAR, board measurement complexes contain that one of Macrowave and Infrared Radiometry (MWR, IR), digital photo- and video cameras, synthetic aperture radar (SAR), satellite navigation system (GPS) and board automatized system (BAS). The aircraft is equipped with four blisters that allow observers to register fish schools, plankton strips, hydrodynamic effects and phenomena on the sea surface (Klochkov et al., 2001).

LIDAR aerial sounding is usually made from 150–170 m altitude of the aircraft flight with the average speed of about 83 m/s and the sea wave of not more than 3. The depth of laser pulse penetration into water depends on natural water transparency and equals to 30–40 m, the pulse frequency is 7 Hz with 1.4 m pitch of stratification using analog-to-digital converter (ADC) "PUPSIK" with sample rate 80 MHz and 0.44 m when applying TDS-3032 digital oscillograph with sample rate 200 MHz.

Each LIDAR pulse relates to the sounding point determined using GPS in space and time. The comments of board observers, sea surface temperature (SST), parameters of MWR, voltage at PMT, laser power and others are also recorded in LIDAR file.

In the course of LIDAR sounding the information obtained by PAL-1M is processed in two stages:

aboard the aircraft, in real time, with the aid of LIDAR board programme so-called data post-processing after the flight, when more detailed analysis is made and air sounding data are interpreted, as well as new algorithms and software are evaluated.

*Light scattering layer – LSL*

By light scattering layer (LSL) the optical heterogeneities in sub-surface layer scattering light waves in the wide range of angles and quite extended in space are meant. In experimental works and calibrations with vessels the probable biological nature of LSL, that are, most of all, concentrations of small marine organisms (zooplankton, phytoplankton, et cetera) was determined. This assumption requires further studying including making nature measurements of LSL optical characteristics, obtaining quantitative data on concentration of marine organisms of different species and size.

LSL provides the possibility of visualization of some hydrological heterogeneties, such as, for instance, the layer of density change (pycnocline), internal waves and etc.

*"Lidarogramma"*

By "lidarogramma" the data from laser location visualized to graphs on which presented are geographical positions of the beginning and end of this or that area of survey, sounding depth and intensity scale of signals in conditional units is understood. Presently, "lidarogramma" allows the depth of different objects including fish schools, plankton, dense concentrations of jellyfish occurrence in surface sea layers and pycnocline to be estimated. The "lidarogramma" gives the opportunity to determine linear dimensions of the mentioned objects vertically and horizontally, and later, after relevant calibrations, their density also. The main stages of algorithm to create lidarogram are described in (Kukhorenko, 1986). Nowadays, lidarograms are plotted both for normal and cross-polarized channel. In the last case more contrast visualization of image is provided that allows us to split found objects into single and LSL.

*Signals reflected (echos) from fish schools – SRFS*

This paper considers SRFS from feeding mackerel, where the greatest experience and knowledge on recording and identification have been acquired. Usually, two kinds of schools – small and large – are singled out for this pelagic fish species. The small ones are the concentrations, which are 0–60 m across, the large ones are more than 60 m across. Feeding mackerel concentrating to small aggregations is very typical for this fish group behaviour (Zabavnikov et al., 1994).

Laser aerial sounding is executed with the frequency of 7 Hz that corresponds to the distance of about 15 m between subsequent pulses, on horizontal, along the flight track. Due to this, after the analysis of mackerel school "lidarogrammas" taking the light scattering in water into account, it was determined that powerful single signals with a typical big amplitude in normal channel (more than 25 units) may be considered as signals of mackerel small school.

Despite the fact mentioned, research on mackerel school recording and identification should be proceeded and, in the first place, in checked conditions, both in joint experiments and calibrations with vessels and in cages.

## RESEARCH AREA

As it has been mentioned, considered LIDAR aerial surveys were carried out during the complex aerial surveys for feeding mackerel in the Norwegian Sea, during July each year usually. Figure 3

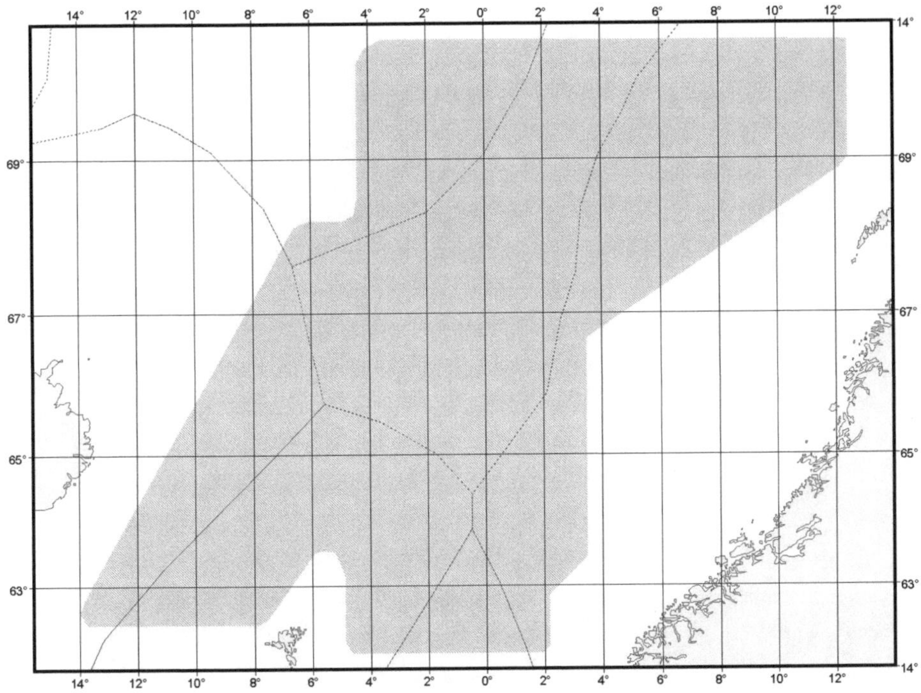


Figure 3. Area of complex aerial surveys.

shows traditional research area. The main tacks of survey tracks were plotted along the latitudes, i.e. across maximal variability of the main oceanographic parameters. For this, the distance between them amounted to not more than 45 miles in the large-scale aerial surveys, and less than 30 miles, in the local parts and mesoscale ground in the aerial surveys reducing down to 10 miles in some cases, when a significant number of mackerel schools is found in the local part and also a high variability of hydrographic parameters is observed.

## RESULTS AND DISCUSSION

Stable LIDAR signals from plankton under the surface were observed, practically, over the whole area of the aerial surveys. The example of such echo is given in Fig. 4. Post-processing integration and software processing of initial data allowed the light scattering layer (LSL) of higher density to be visualized that is shown in Fig. 5 and interpreted as plankton. When analyzing the figure it is shown that this layer is depicted as a stretched part with the length of about 1.7 km, the depth of which location varies (the upper boundary – 10–12 m and the lower one – 15–18 m) that is, in the first place, conditioned by hydrodynamic effects under the surface.

The echo from mackerel surface school is somewhat different than that one visually demonstrated in Fig. 6. The dark blue line shows the level of received echo in normal and the white one – in cross channels. Here it is clearly seen that echo from mackerel school has additional contrasts on both channels, i.e. using two polarizations permits mackerel concentrations to be singled out more correctly.

Figure 7 shows the example of "lidarogramma" for mackerel school with the following parameters: the vertical size – 5–7 m, the horizontal one, along the flight track of aircraft-laboratory – 100 m. Mentioned geometrical parameters additionally corroborate that the registered object may be more probably related to the surface school of mackerel.

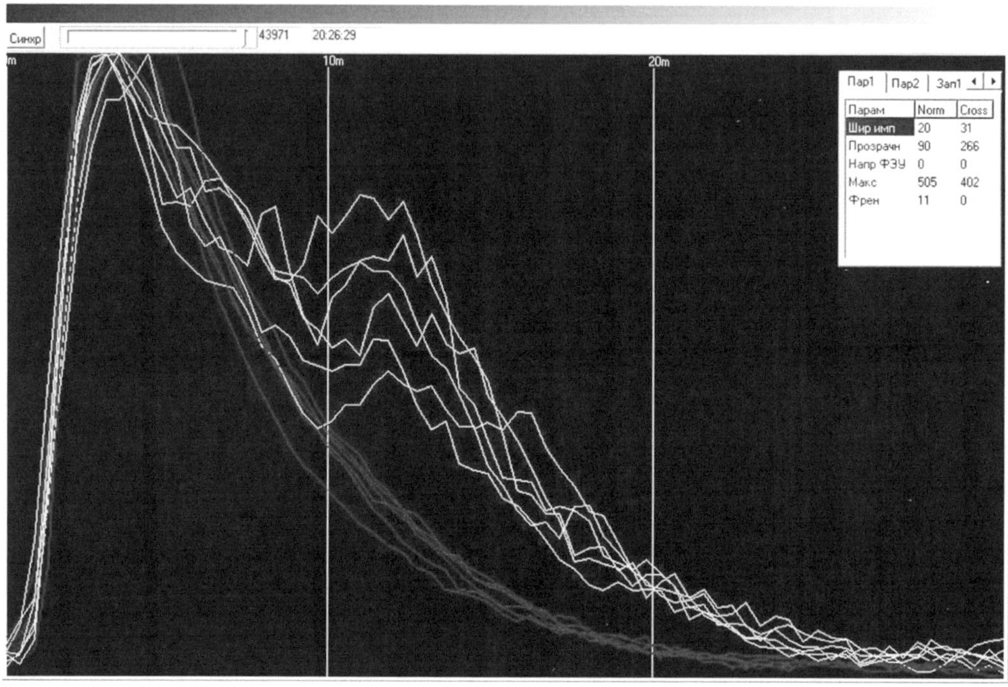


Figure 4. Typical echo-signal from plankton.

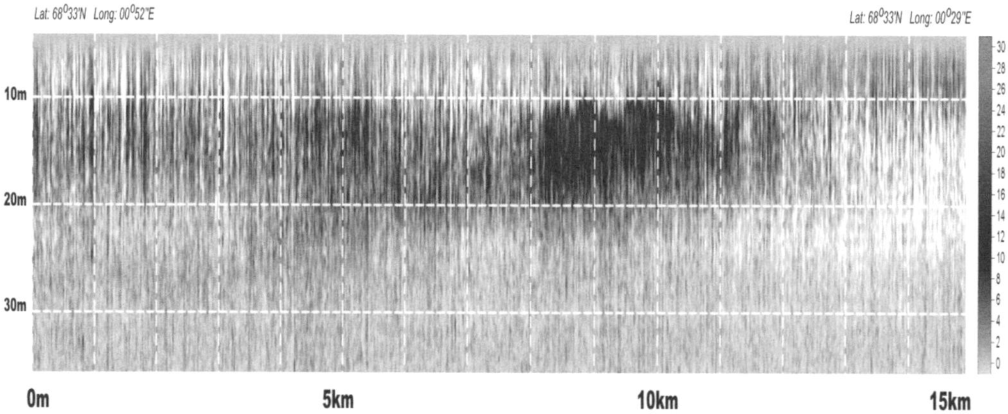


Figure 5. "Lidarogramma" of light scattering layer (LSL).

It should be noticed that the majority of mackerel schools registered when LIDAR aerial sounding was verified during visual observations. In future, after the relevant experimental and calibration works, within the framework of which collecting necessary, statistically reliable, data is planned, there will be the transition to the estimation of density and biomass of single school with the purpose of the following calculations using instrumental estimation of the Norwegian Sea feeding mackerel biomass.

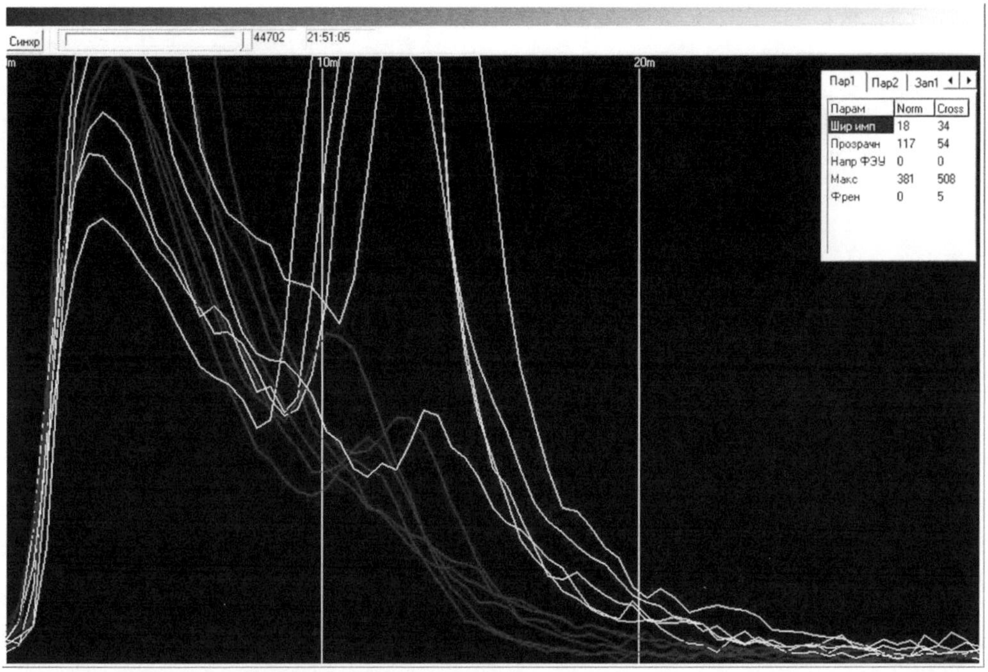


Figure 6. Typical PAL-1M single signal from mackerel school.

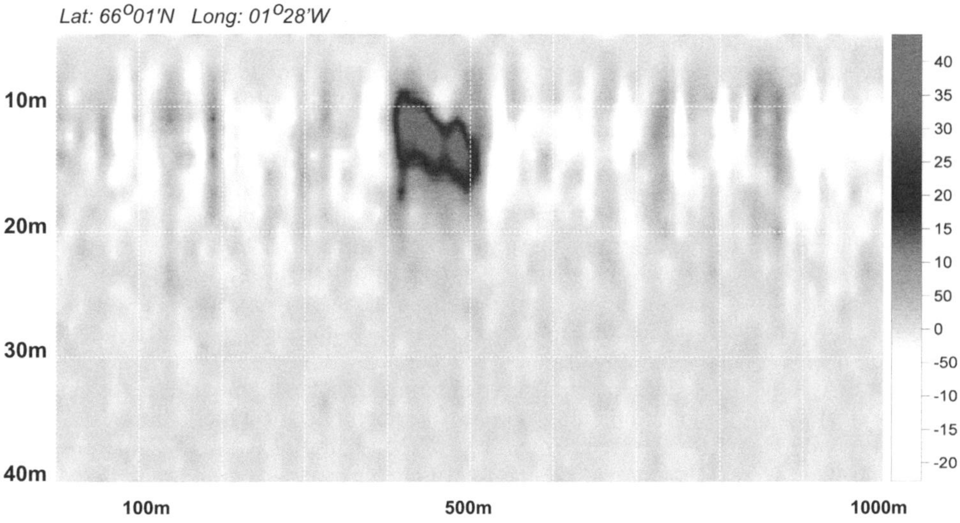


Figure 7. "Lidarogramma" of mackerel school.

The distribution of LIDAR signals in the area of complex aerial surveys for feeding mackerel identified as fish schools is presented in Fig. 8. For this, it should be noticed that they had different intensity that indicated on different density of those concentrations. The distribution of mackerel RFSS shows a good food supply in the Norwegian Sea area, in common, and in the open part of the Norwegian Sea, in particular. However, food organisms, due to the hydrodynamical conditions in summer 2005 (for example), did not form steady dense local concentrations in the Norwegian Sea

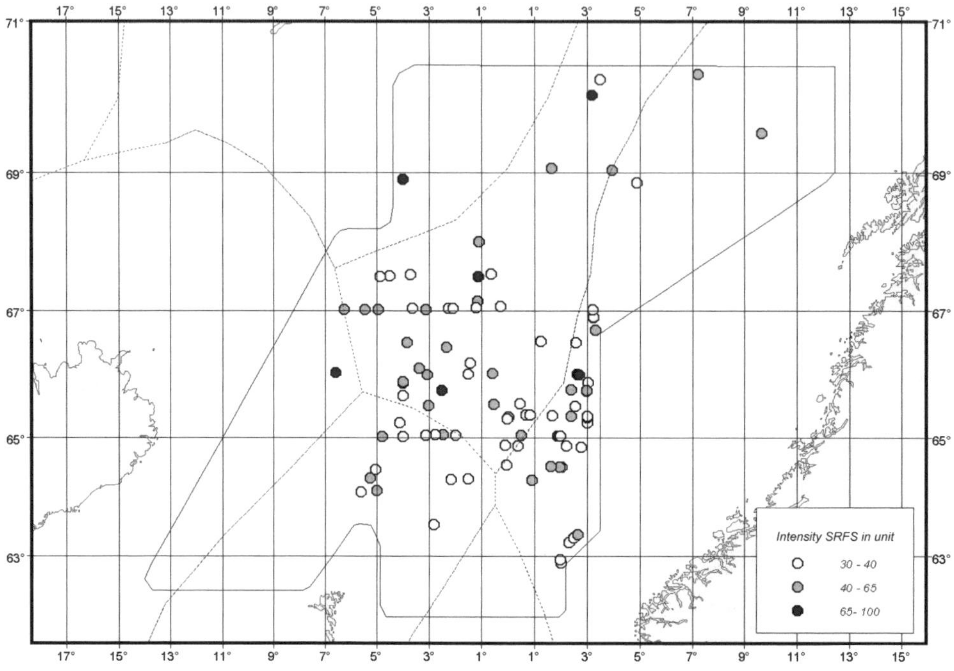


Figure 8. Distribution of SRFS (as example in 2005).

area, but were regularly distributed in the wide area. For this, a great number of echoes with small intensity indicate on small dimensions of mackerel schools.

In addition, mackerel school distribution observed visually and that one identified by the data of laser sounding were compared that is presented in general in Fig. 9. For this, it was found that in the separate areas the position of schools coincided or they were located close by each other that supplemented and corroborated registration of object surveyed. But, there are some areas, where the schools were only identified using LIDAR or they were detected during visual observations.

This situation is caused by the impossibility to register the schools of mackerel by LIDAR in the most upper surface layers, at the depth of not more than 5 m, which is distinctly recorded by observer. Besides, the sounding laser beam goes along the tack of aerial survey route, while the observers register schools not only under the aircraft, but at the distance from board, in the field of vision equal to the flight altitude. At the same time, observers cannot accurately detect and identify mackerel schools in the layer deeper than 6–8 m that is successfully done by LIDAR.

In the period of carrying out considered aerial survey complex for feeding mackerel (as example 2005), in the course of visual observations, 84 schools were recorded, of which about 40% were surface, and when conducting LIDAR aerial sounding, based on the analysis and interpretation described above, detected were 119 schools, i.e. the total number of detected and recordered concentrations of mackerel amounted to 203. Thus, a wide use of the described above approach, which was only partly used in 2005, will allow us to estimate more objectively and correctly distribution and biomass of feeding mackerel, that is planned to do in future.

## CONCLUSION

The results from carried out aerial surveys for feeding mackerel show the prospects and efficiency of using LIDAR aerial sounding of sea surface layer when studying distribution and, in prospect, estimation of feeding biomass mackerel in the Norwegian Sea.

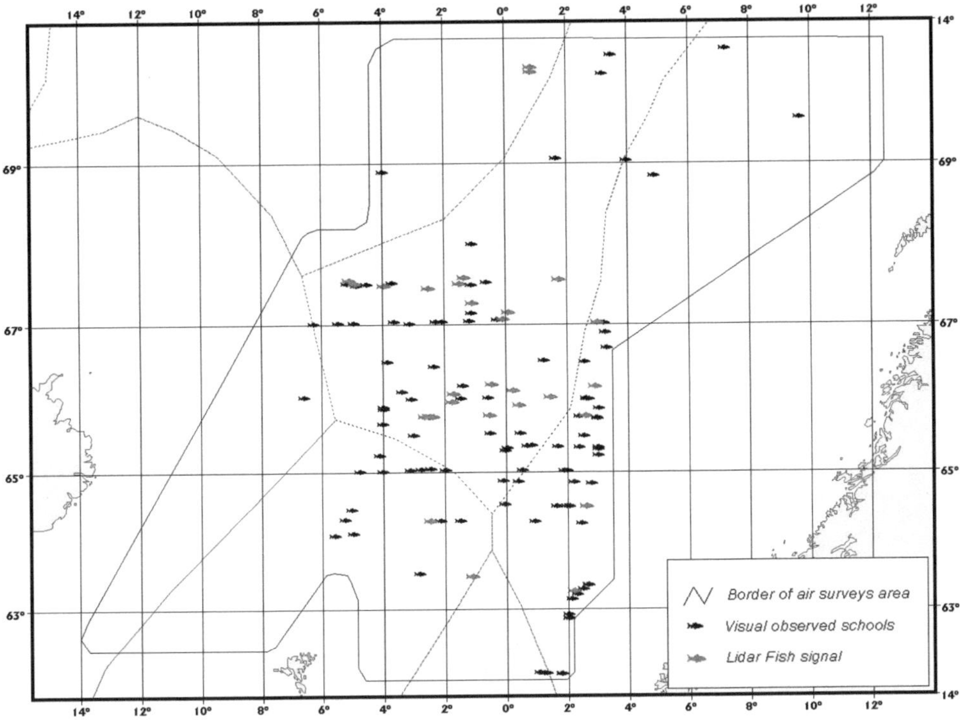


Figure 9. Distribution of visually detected schools of mackerel and RFSS.

In the nearest future it is necessary to carry out a series of experimental works and investigations as using the scheme "airborne laboratrory – vessel", as mounting LIDAR aboard research vessel, as well as in checked conditions at one of the experimental bases of PINRO. As for the two first mentioned, the first steps in that direction in implementing of which good and promising results have been obtained, as well as further prospects and main directions of experimental research and works have been determined were made in previous years.

So, for instance, when carrying out LIDAR surveys, during the technical cruise by RV "Smolensk", it was revealed that to mount and use laser sounding complex aboard the vessel it should be updated, since the principal scheme and "ideology" of sounding, climatic and exploitation conditions, as well as many others changed.

For this, it should be taken into consideration that the principal scheme and "ideology" of acoustic and LIDAR surveys are quite alike and due to this fact it is highly reasonable and useful to consider the long-term experience of using acoustic methods in surveys for fish biomass estimation also including joint acoustic-LIDAR works both applying the scheme "airborne laboratory-vessel", and synchronously, aboard the same vessel.

## REFERENCES

Chernook, V.I., Vasiljev A.N. and Lisovsky A.S. 2003. Using aerial LIDAR PAL-1 to survey sea areas in 2001–2002. Abstracts from the report meeting of PINRO and SevPINRO by the results of the research work in 2001–2002. PINRO. Murmansk: PINRO Press, 2003. P. 60-161 (in Russian).

Gavrilov, E.N. 1999. On the level of underwater noises aboard research vessels conducted hydroacoustic surveys. Development of fishery research technical methods: Selected papers/PINRO. Murmansk: PINRO Press, 1999, P. 5-12 (in Russian).

James H., Churnside, KouchiI S. and Tsuyoshi O. 2001. A comparison of airborne LIDAR and echo sounder performance in fisheries. Marine Acoust. Soc. Jpn. Vol. 28 no. 3, July 2001.
James H. Churnside, James J. Wilson. 2002. Results of an Airborne Lidar Survey of Mackerel in the Norwegian Sea. July 2002.
Klochkov, D.N. and Shamray E.A. 2001. Migration, stock status and prospects for mackerel fishing in the Norwegian Sea. Problems of fisheries, Vol. Issue 4(8), 2001, 736 pp.
Kukhorenko, K.G. 1986. Mackerel behaviour in the Central Eastern Atlantic and accessibility for Fishery. In: Biological resources of the Atlantic Ocean. M: Nauka, 1986. P. 310-329 (in Russian).
Zabavnikov, V.B. and Chernook V.I. Aerial methods of investigating sea environment to estimate distribution of pelagic fishery objects. Murmansk: PINRO Press, 1994. P. 60 (in Russian).

*Remote Sensing for a Changing Europe*
*D. Maktav (Ed.)*
*IOS Press, 2009*

*doi:10.3233/978-1-58603-986-8-587*



# Using of Airborne LIDAR for research carrying out in the interest of fisheries oceanography information providing

Vladimir Zabavnikov & Sergey Egorov
*Knipovich Polar Research Institute of Marine Fisheries and Oceanography (PINRO), Remote Sensing Laboratory, Murmansk, Russia, e-mail: ltei@pinro.ru*

ABSTRACT: At present requirements on truth and accuracy initial data in the interests of fisheries oceanography information providing is increased. Also it is very important to get complex data from great sea area for short time. One way of that, in the first for above second part is using of airborne remote sensing data from different type remote sensing equipments, and next is using new type equipments. Therefore during last several years PINRO uses for fisheries oceanography information providing new remote sensing methods, which is base on using of LIDAR systems. Name of it's come from English words abbreviation – LIght Detection And Ranging.

Ocean air remote sensing LIDAR methods are based on analyze of intensity, spectral, polarization, angular and temporal characteristics of reflected LIDAR echo-signal, which appears to sound of water subsurface layers by short strong laser impulse.

The same LIDAR was designed and manufactured by "MULTITEKH" company from St.- Petersburg in 2001. It was named polarization aviation LIDAR (PAL-1). Later it was modernized to PAL-1M/ This LIDAR system is installed and operate onboard of research aircraft Antonov-26 (An-26) named "Arktika". PAL-1M is operated by PINRO specialist. PAL-1M has not analogues in the Russia in modern stage for carrying out of air research in the interests of fisheries oceanography.

During this time PAL-1M was highly successfully, reliably, effectively and qualitatively used in remote sensing detection of upper sea water layers (penetration depth is dependent on sea transparency), including oceanographic parameters (transparency, picnocline depth, subsurface plankton concentration), detection of fish aggregations, and first mackerel, determination of vertical profiles and optical characteristics for difference water type (masses).

## INTRODUCTION – WHAT IS LIDAR MEHTOD FOR FISHERIES OCEANOGRAPHY USING?

One of optic electromagnetic radiation range peculiarities is capacity to extend at the water environment and air (laser location), that allow to LIDAR to carry out of water sounding through line between air and water, penetrating at the sea closely surface layer. Depth of laser impulse is dependent on water transparency. Besides, if laser sounding impulse has linear polarization and has two receiving and recording channels, which are intended for recording of two echo-signal components with reciprocally orthogonal polarization has possibility to make identification of displayed object in echo-signal analyze.

Therefore laser location (LIDAR) method is based on analyze of intensity, spectral, polarization, angular and temporal characteristics of reflected echo-signal appears to sound of water subsurface layers by short strong laser impulse.

This circumstance showed that laser location data based on the analysis of sea water optical characteristics allows to determine its transparency, the depth of optical heterogeneities (OH) in the subsurface layers of the sea identifying as picnocline, plankton and pelagic fish schools (mackerel, in the first) as well as the location of oceanographic fronts and difference water masses edge.

Above tasks have been realized as was indicated earlier for PAL-1M with success and effectiveness.

## PAL-1M BRIEF TECHNICAL DESCRIPTION

PAL-1M was installed and has been exploited onboard the two engine aircraft "Arktika" (see Figs 1–3). LIDAR transmitting and receiving unit is placed in the photo-camera hatch closed by flat illuminator made of optical glass. LIDAR optical unit was set at the angle of 17°.

Solid-state pulse laser on YAP:Nd was used as source of sounding radiation. Two echo-signal components (orthogonal/cross polarization and normal/co-polarization) are recorded by two identical photo-detectors in the inlet of the optical systems of which the analyzers on the basis of Polaroid film are set. PAL-1M main technical specifications are following:

- wave length – 532 nm;
- pulse duration – 12.5 ns;
- pulse energy – 120 MJ;
- pulse frequency – 140 Hz;
- patch diameter on sea surface – approx. 1 m;
- diameter of optical receiver – 10 cm;
- polarization vector – normal and cross.

There's also fluorescent channel set in the system of cross-polarization receiving (in the cross-channel), where scattered light passing through interference filter with 685 nm wave length gets photomultiplier tube (PMT), that allowed to get information about initial biological productivities (chlorophyll "a" of phytoplankton) concentration on sea surface.

Figure 1. Aircraft-laboratory "Arktika" appearance.

Figure 2. PAL-1M onboard of “Arktika”.

Figure 3. Optical LIDAR unit in photo-camera hatch.

## PAL-1M USING DURING FISHERIES OCEANOGRAPHY COMPLEX AIR SURVEYS CARRYING OUT AND SOME LIDAR DATA

Laser location with PAL-1M using is one element of complex fisheries and oceanographic airborne research in the first for feeding mackerel in the Norwegian Sea. At present the maximum of PAL-1M signal penetration in the Norwegian Sea is 45 meters and it is dependent on sea water transparency.

During airborne complex research usually carry out LIDAR sounding, Microwave and Infrared Radiometry (MWR, IR), including Infrared scanning, digital photo- and video surveys, synthetic aperture radar (SAR) sounding. Also onboard of “Arktika” installed satellite navigation system – GPS and onboard computer system (OCS). The aircraft is equipped with four blisters that allow observers to register fish schools, plankton strips, hydrodynamics effects and phenomena on sea surface, including carrying out observation on marine mammals and sea birds.

LIDAR sounding is the best effective and reliable from 150–200 m altitude of aircraft flight with the average speed about 85 m/s and the sea wave no more than 4 (height < 1.5 m). Each

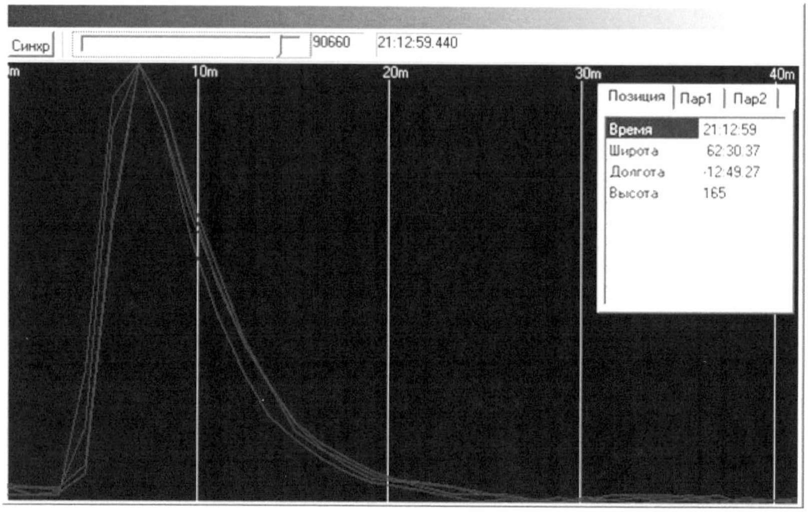


Figure 4. PAL-1M transparency typical single signal.

PAL-1 pulse relates to the sounding point determined using GPS in the space and time. The comments of observers, sea surface temperature (SST), MWR data, PMT voltage, laser power and some others are also recorded in LIDAR file.

PAL-1M information is processed in two stages:

- onboard of "Arktika", in the real time, with the using special LIDAR onboard mathematic calculations and software;
- so-called post-processing data calculation after the each flight, when more detailed analysis is made and air remote sensing data are interpreted, as well as new algorithms and software are evaluated.

After LIDAR (PAL-1M) using in the complex air surveys we get following information as mapping in GIS presentation:

- data about positions of area with high level of subsurface plankton concentration.
- transparency spatial distribution;
- values of chlorophyll "a" fluorescence on sea surface;
- data about picnocline depth;

Also usually after several flights calculate so-named "Lidarogramma" with purpose of pelagic fish schools, in the first mackerel, identification. "Lidarogramma" is the same as acoustic echogrammas.

In Figs 4–10 are presented characterized (conformably) examples of transparency single signal (so named clear signal, without OH) and transparency spatial distribution in the Norwegian Sea; plankton typical single signal and plankton "Lidarogramma"; mackerel typical single signal and mackerel school "Lidarogramma"; fluorescence chlorophyll "a" spatial distribution in the Norwegian Sea.

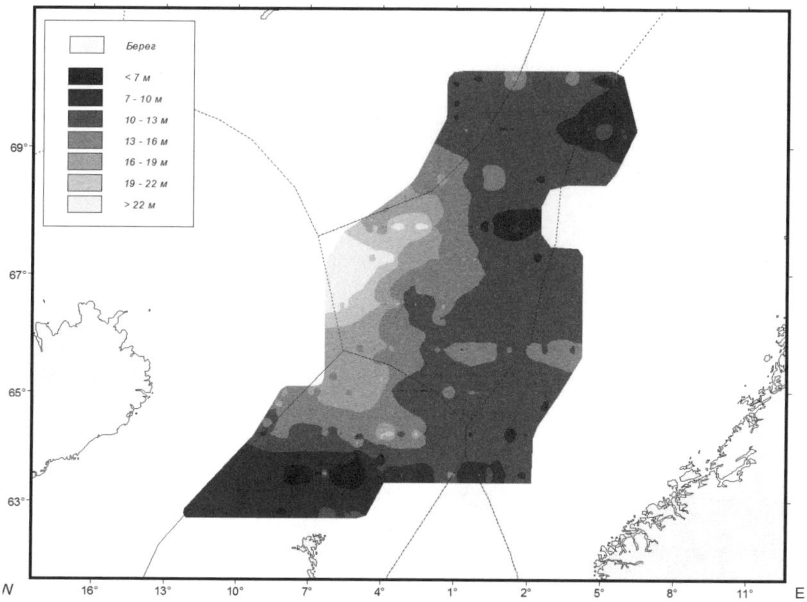


Figure 5. Example of transparency spatial distribution in the Norwegian Sea.

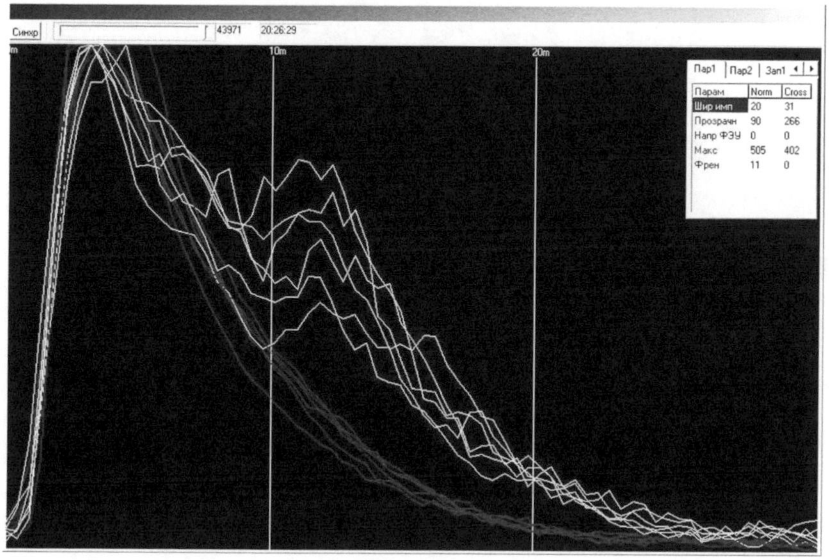


Figure 6. PAL-1M plankton typical single signal.

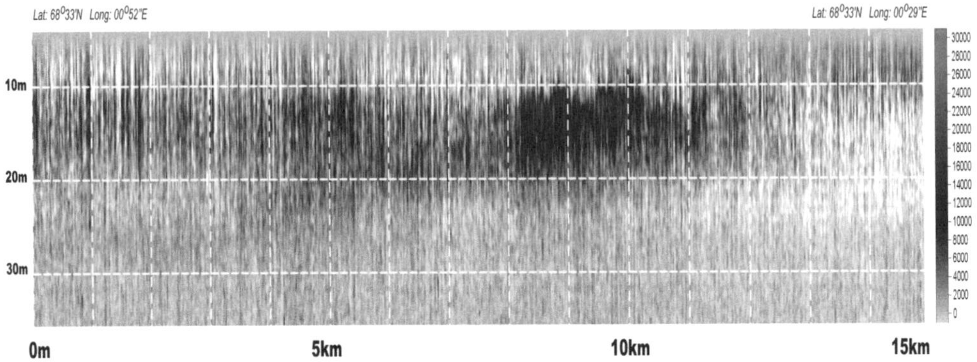


Figure 7. Example of plankton in the subsurface layers "Lidarogramma".

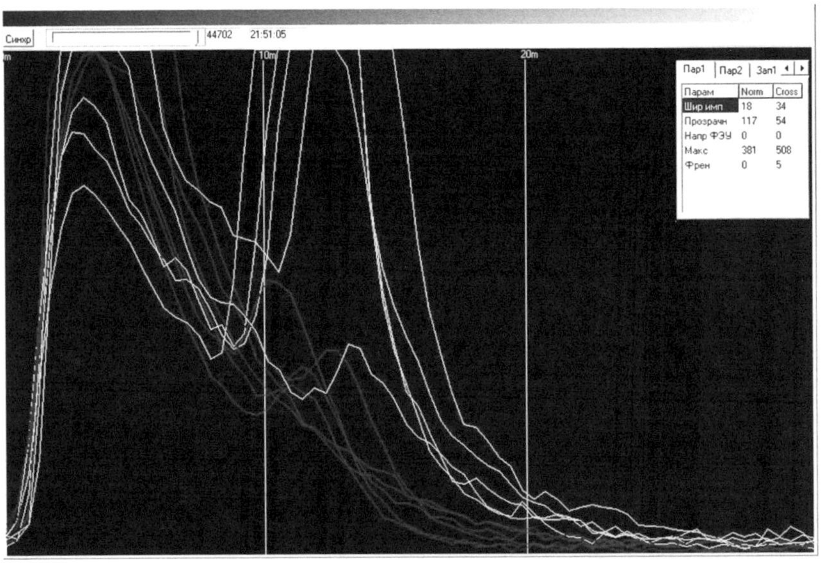


Figure 8. PAL-1M mackerel school typical single signal.

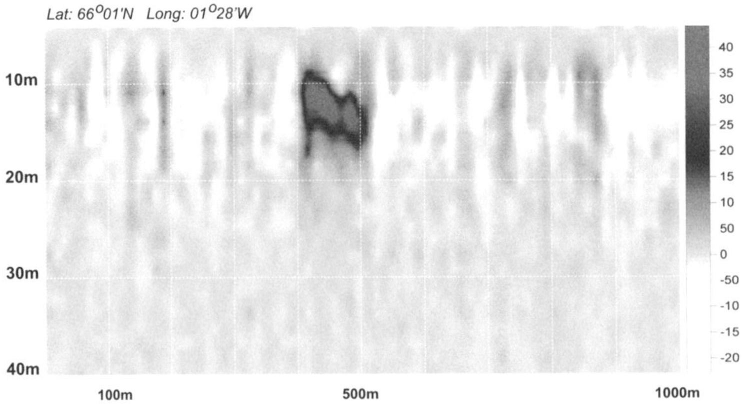


Figure 9. Example of mackerel school in the subsurface layers "Lidarogramma".

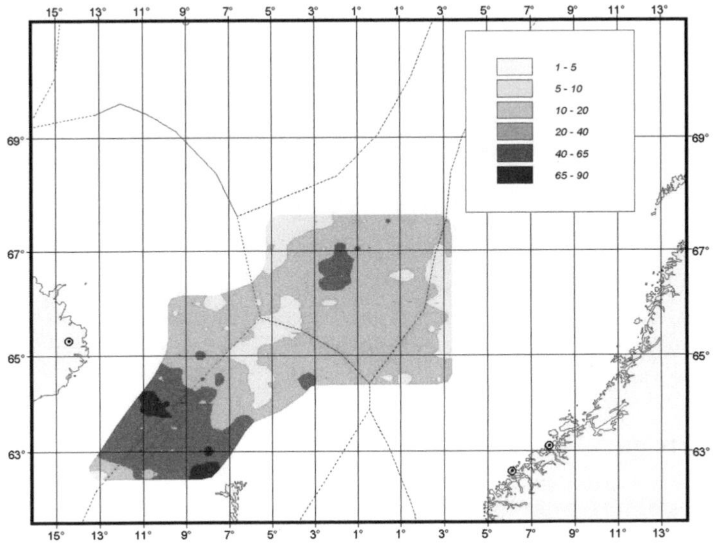


Figure 10. Example of fluorescence chlorophyll "a" spatial distribution in the Norwegian Sea.

## CONCLUSIONS

PAL-1M using in complex airborne fisheries oceanography research carry out showed good results, quality and effective in the several cases higher than during vessels surveys in the first for feeding mackerel in the Norwegian Sea. Nevertheless in the future time is need to carry out series experimental works and research with using the scheme "aircraft – vessel" and special investigations in the checked conditions at one of the PINRO experimental bases.

## ACKNOWLEDGEMENTS

Authors are very grateful to Russian State Fisheries Committee and PINRO Administration for financial supporting these researches. Also its could not be carried out without Flight Company "Arkhangelsk Airlines" – aircraft An-26 "Arktika" owner. And finally we are very grateful to crew of aircraft laboratory "Arktika" for the highest level of aircraft operation, navigation and technical service.

*Remote Sensing for a Changing Europe*
*D. Maktav (Ed.)*
*IOS Press, 2009*

*doi:10.3233/978-1-58603-986-8-594*

# Measurement of multispectral BRF effects of the megacity Cairo, Egypt using CHRIS/PROBA data

C. M. Frey & E. Parlow
*Institute of Meteorology, Climatology and Remote Sensing, Basel University, Switzerland*
*(corinne.frey@unibas.ch, eberhard.parlow@unibas.ch)*

Keywords: BRF, BRDF, urban, CHRIS/PROBA, reflectance

ABSTRACT: In this analysis, a series of images from the ESA imaging spectrometer CHRIS on the technology demonstration satellite PROBA, taken in a row over Cairo on 17th February 2007 and 24th March 2008 at each five different viewing zenith and azimuth angles are analyzed to find the BRF (bidirectional reflection function) which is typical for urban areas in a developing Arab country. The results are only valid for these sun-sensor constellations, since only two scenes were analyzed. CHRIS/PROBA features 18 bands in the shortwave range. All bands were first destriped and then corrected for atmospheric influences using the radiative transfer model MODTRAN. Furthermore the images were geo-referenced using manually set pass points. The accuracy of the geolocation varies across the image. Generally pixels match well; however, in some areas there is a shift of one or two pixels. There is a clear dependency in the reflectances on the viewing geometry of the sensor. The regression coefficients $r^2$ from the regression between the reflectance and the viewing geometry are high for all bands in all urban areas. No distinct difference was found for high or medium dense housing.

## 1. Introduction

Urban surfaces differ from natural environments not only in their surface materials, but also in their diverse geometric forms, presented by blocks, houses, streets and open spaces. Measuring the reflected radiance of an urban area from space, the values of the resulting pixels are probably a mixture between variously illuminated surfaces and possible shading. Additionally, the albedo of roofs is mostly different from the albedo of walls. A side looking sensor is therefore measuring a different surface property than a nadir-viewing sensor. Considering pixel sizes in the mesoscale – like for example LANDSAT or ASTER data - these effects combine to an angular dependent reflection, referred to as 'bidirectional reflection distribution function' (BRDF). This function is not constant, but changes according to the sun zenith and azimuth angle. The BRDF effect was addressed for natural surfaces by many authors: Nicodemus (1970) described the bidirectional reflectance of diffusely scattering, homogenous surfaces. As the BRDF cannot be measured directly in practice (Schiefer et al. 2006), the bidirectional reflectance factor (BRF) is measured over finite solid angles.

Urban areas consist of a variety of materials and structures and don't fulfil the criterion to be homogeneous. A direct measurement of the urban BRF is only possible in the microscale and was done for selected surface materials by Meister et al. (1996, 1998). Schiefer et al. (2006) deducted BRF functions for four pure urban classes from 4 m resolution HyMap images of Berlin. They stress the difficulties with mixed pixels in the classification and their interference in the correction of the brightness gradient. On the macroscale it is assumed that for an urban class with similar characteristics in type of construction and vegetation ratio a large-scale BRF might converge. Meister et al. (1999) analyzed this topic using up scaled airborne data over the city of Nuremberg, Germany and deducted BRDFs for the city of Nuremberg.

In this study the total reflectance and the large-scale BRF effect of Cairo shall be examined to determine the influence of the sensor's view-angle and –direction and the illumination geometry to the urban reflectance.

## 2. STUDY AREA

The study area is located in Cairo, Egypt. Cairo is a strongly growing megacity, facing manifold problems like traffic congestion and air pollution (Robaa 2000). Housing in Cairo can be roughly divided into three classes. Firstly there are the very high density housing areas, where the spaces between the houses often just allow pedestrian walking. The houses are often of bad quality. Further there are other quarters which are built more spacious and belong to the richer parts of the population. Finally, there are huge areas of newly built houses in the outer quarters of Cairo, where the percentage of buildings in construction is high. In this study, only the first two classes will be considered.

## 3. DATA

For this study, two scenes from the CHRIS (Compact High Resolution Imaging Spectrometer) instrument were used. CHRIS is a hyperspectral instrument on the technology demonstration satellite PROBA from ESA. Its objective is the collection of BRDF (Bidirectional Reflectance Distribution Function) data. PROBA was launched on 22nd October 2001 from Sriharikota Island, India. It flies on an altitude between 570 and 670 km in a sun synchronous orbit. Its inclination is 97.9°. In Mode 1 (land) CHRIS features 18 bands between 400 and 1050 nm with a spatial resolution of 17 m at perigee. The swath width at nadir is 13 km. Main characteristic of the CHRIS data sets are the multi-angular acquisitions. The five targeted viewing angles are -55°, -36°, 0°, +36° and +55° (Cutter & Johns 2003, Guanter et al. 2005). Figure 1 shows the actual acquisition geometry of the two used scenes of 17th February 2007 and 24th March 2008 as extracted from the header information of the scenes.

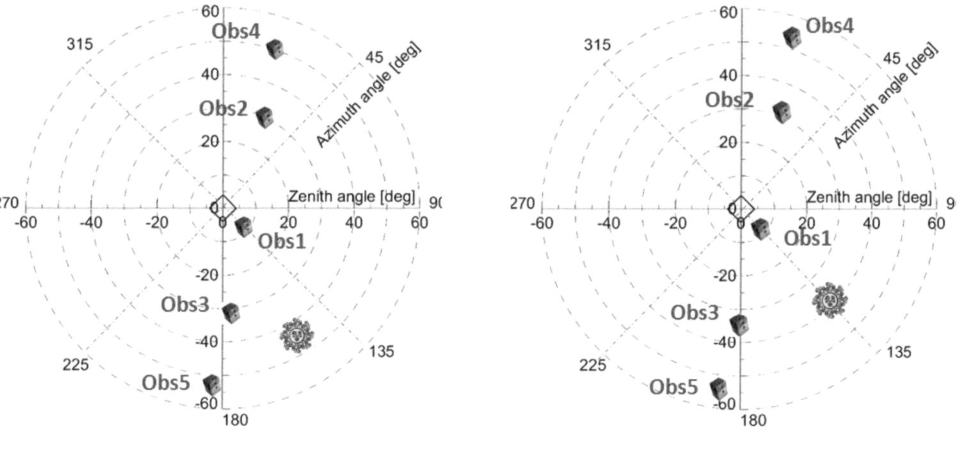


**Figure 1 Actual acquisition geometry of the CHRIS/PROBA scenes.**
**Left: February 17, 2007. Right: March 24, 2008**

All data were provided from the European Space Agency. Unfortunately CHRIS data are affected by horizontal and vertical striping due to errors in the alignment of the sensors in the construction of the instrument and thermal fluctuations during the orbit (Garcia & Moreno 2004). ESA provides a software tool for the correction of these effects (HDFclean V2). However, a faint vertical stripping remains after the usage of the tool in most of the nadir-looking bands.

## 4. Methods

### a. Atmospheric correction

Before analyzing the images, two correction steps were performed. Firstly, an atmospheric correction was performed on all bands using the radiative transfer model MODTRAN (MODerate resolution atmospheric TRANsmission (Berk et al. 1999)). As it is very difficult to obtain an exact estimation of the input atmosphere, the correction procedure was coupled with a statistical approach. Firstly, the path radiance and the transmissivity were estimated using radiosonde data from Helwan, which is south of Cairo, and the standard urban aerosol option as input atmosphere. The output was then convolved for each band width. The obtained path radiance and transmissivity were used to calculate the radiance at surface ($L_{BOA}$) according to formula 1 (Liang 2000).

$$L_{TOA}(\lambda) = L_P(\lambda) + \tau(\lambda) \cdot \rho(\lambda) \cdot L_{BOA}(\lambda) \quad (1)$$

| | |
|---|---|
| $L_{TOA}$ = | At-sensor radiance [$Wm^{-2}sr^{-1}\mu m^{-1}$] |
| $L_P$ = | Path radiance [$Wm^{-2}sr^{-1}\mu m^{-1}$] |
| $L_{BOA}$ = | Bottom of the atmosphere radiance [$Wm^{-2}sr^{-1}\mu m^{-1}$] |
| $\tau$ = | Average atmospheric transmissivity |
| $\lambda$ = | Band width |

The path radiance is almost linear dependent on the reflectance of the Earth surface; therefore a reflectance-dependent path radiance was used for the correction of the image. This term was estimated from a linear relation obtained from two MODTRAN runs with the albedo being 0.1 and 0.3. This linear relation was then applied to a slightly modified top-of-the-atmosphere reflectance.

Subsequently, the surface reflectance was calculated from the radiance at surface and the global irradiance, which was also convolved for each band, using formula 2 (Wang et al. 2000).

$$\rho(\lambda) = \frac{L_{BOA}(\lambda)}{L_G(\lambda)} \quad (2)$$

| | |
|---|---|
| $\rho$ = | Reflectance |
| $L_G$= | Global irradiance [$Wm^{-2}sr^{-1}\mu m^{-1}$] |

Adjacency effects were corrected iteratively using formula 3 and 4 (Del Frate 2007):

$$\rho^{(2)}(\lambda,x,y) = \rho^{(1)}(\lambda,x,y) + \frac{L_P(\lambda,x,y)}{L_G(\lambda,x,y)}(\rho^{(1)}(\lambda,x,y) - \overline{\rho}(\lambda,x,y) \quad (3)$$

$$\overline{\rho}(\lambda,x,y) = \frac{1}{n^2}\sum_{i,j=0}^{n} \rho^{(1)}{}_{i,j}(\lambda,x,y) \quad (4)$$

This procedure resulted in slightly too low reflectances. Therefore, the first guess of aerosol optical depth was itineratively reduced until only a small number of negative pixels were detected in the image. Figure 2 shows the spectral lines of 3 arbitrary chosen pixels of the scene of 17th February 2007 together with spectral lines from the ASTER JPL spectral library of similar land use (http://speclib.jpl.nasa.gov). The lines show a good agreement, considering, that only "similar" materials are compared. However, band 17 shows an overcorrection which might be due to radiometric calibration problems (Guanter et al. 2005). The scene of 24th March 2008 shows an analogue agreement, with exception of the vegetation curve that reaches only 30-40% reflectance in the near infrared. This lower reflection might be due to a different crop or a different growth phase of the crop.

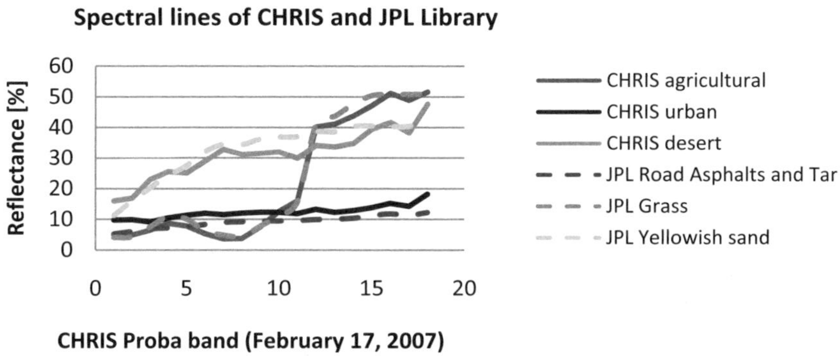


**Figure 2 Spectral curves of three example pixels of the CHRIS scene of 17th February 2007 (nadir viewing)**

b. GEOREFERENCING

After the atmospheric correction, the images were georeferenced to the nadir viewing image with polynomial equations using ground control points. Table 1 shows the RMS (Root Mean Square) and the grade of the polynomials used. RMS is the distance of the georeferenced control point from the specified coordinate. While these RMS values correspond to the whole scene, only a subset of the scenes was used for the analysis. The agreement in the sub scene was even better.

**Table 1 RMS values and grade of the polynomials used for the georeferencing**

| | February 17, 2007 | | March 24, 2008 | |
|---|---|---|---|---|
| Observation | RMS | Grade | RMS | Grade |
| Obs 1 | - | 1 | 0.42 | 1 |
| Obs 2 | 0.755 | 1 | 0.584 | 2 |
| Obs 3 | 0.597 | 1 | 0.639 | 1 |
| Obs 4 | 1.226 | 2 | 1.077 | 2 |
| Obs 5 | 0.861 | 2 | 0.933 | 2 |

### c. Definition of Land Use Classes

The analysis was done separately for five different land use classes: 'high density housing', 'low density housing', 'agricultural fields', 'Nile water' and 'desert'. The classification was done manual, including only 'known' areas.

Even though the georeferencing showed a good agreement, it was not possible to do a pixel-to-pixel comparison. Therefore, the analysis was done using a running average of each pixel. The running average was calculated with a kernel of 9x9 pixels, whereas only pixels with a surrounding of at least 65 valid pixels were considered. For the class 'low density housing', this resulted in 126/126 valid values for both scenes. For the class 'high density housing' the numbers were 137/80. The class 'agricultural fields' showed only 9/9 values, the class 'Nile water' 20/8 values and the class 'desert' 14/4 values. The main purpose of this study was laid on the urban classes, the others were included only for comparison.

## 5. Results

A clear dependence of the reflectance on the viewing geometry of the sensor was found. Figure 3 shows the mean reflectances of three cuts from the classes 'high density housing' and 'medium density housing' of all bands for the different viewing geometries. Observations 3 and 5 have the highest values, which can be explained by their close angular position towards the sun ("hot spot" effect). Lowest values are found for observation 2 and 4, being opposite to the sun. This finding is consistent with Begiebing & Bach (2004). The nadir viewing observation (obs 1) shows similar values like observation 2 and 4.

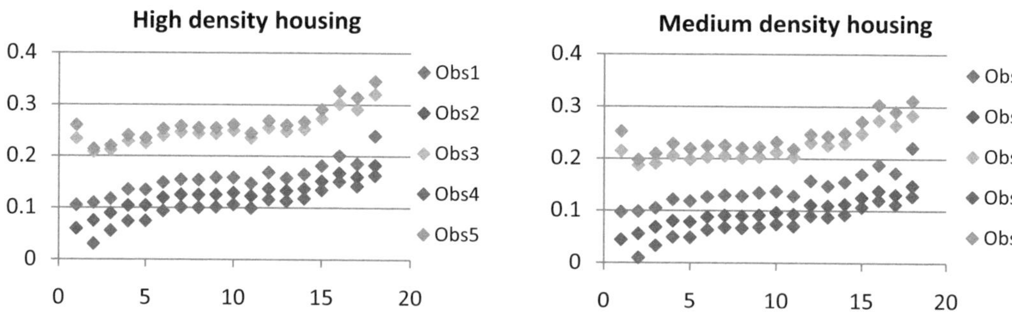


**Figure 3 Angular reflectances of the scene of February 17, 2007**

The class 'Nile water' only shows a slight angular dependence in the first few bands; afterwards the reflectances conform to each other. Furthermore the water spectra are almost flat. This is in agreement with Gatebeck et al. (2005), who analyzed sea water BRDF and found only weak BRDF effects. In case of the agricultural class, the observations almost merge in the red bands. This is the region where the chlorophyll has its absorption maximum. Begiebing & Bach (2004) found a similar behaviour for maize. Figure 4 shows band 3 (531.2 nm) of all land use classes for the different viewing angles on February 17, 2007. The yellow arrow depicts the sun zenith angle. Here again it becomes clear that reflectances are highest, when the sun is "in the back" of the sensor.

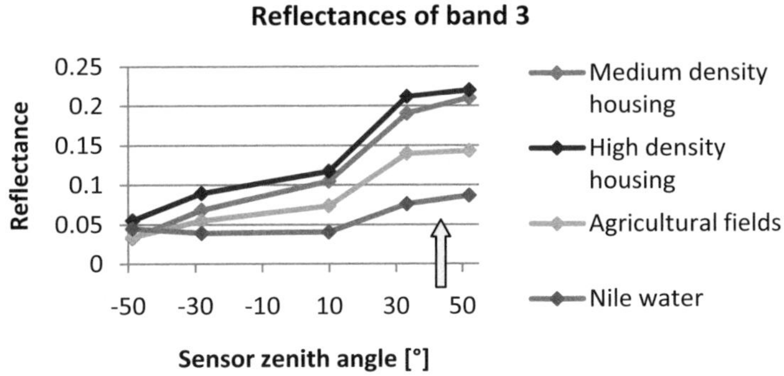


**Figure 4 All land use classes for the different viewing angles on 17th February 2007 (band 3)**

The running average reflectances, as explained in the last chapter, were compared to the zenith and azimuth angles of each acquisition and high correlations were found for both zenith and azimuth angles for both urban classes. The correlation coefficients $r^2$ range between 0.80 and 0.94 for the zenith angle and between 0.82 and 0.90 for the azimuth angle with the class 'high density housing' having slightly higher values than the class 'medium density housing'. The class 'desert' shows similar good correlations like the urban classes. As expected, the class 'Nile water' shows in all cases decreasing $r^2$ values with increasing wavelength. Also the class 'agricultural fields' fails to show good correlations on 24th March 2008, especially in the red bands, where the reflectances almost merge. Due to this merging of values, no slope can be detected and the regression must fail. In case of the scene of 17th February 2008 a good correlation is found for the class 'agricultural fields' for all bands. Figure 5 shows these correlations with the zenith and azimuth angle. However, the results of the class 'desert', 'Nile water' and 'Agricultural fields' must be treated carefully, as the basic sets for these classes are not big enough to deduct robust statistic. Figure 5 shows the $r^2$ values for the zenith and azimuth angles for both scenes.

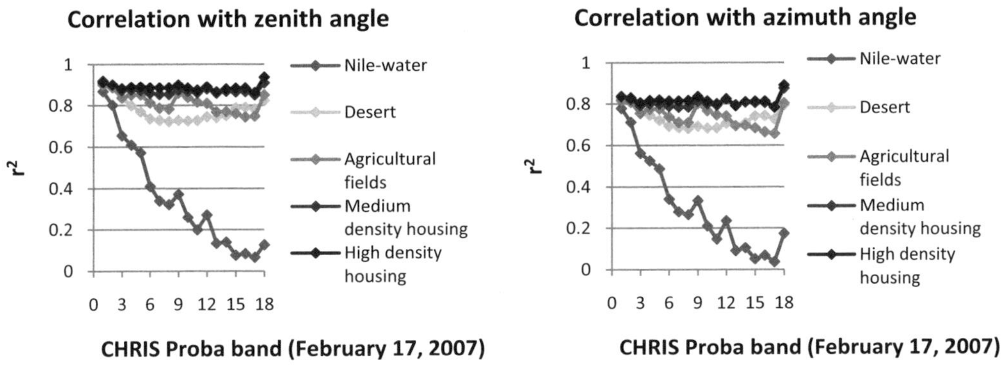


**Figure 5 Correlation of surface reflectance with sensor zenith and sensor azimuth angle for the different land use classes.**

## 6. Discussion

In this analysis, the angular reflectances of two CHRIS Proba scenes were analyzed regarding their dependence on the sensor view geometry. After the necessary atmospheric correction and georeferencing, selected pixels of different land use classes ('high density housing', 'medium density housing', 'agricultural fields', 'desert' and 'Nile water') were compared and regressions with the sensor zenith and azimuth angle performed.

For both urban classes, high correlation coefficients were found. But surprisingly the difference between the two classes was negligible, even though the geometry of the houses differed notably. The class 'high density housing' showed a slightly higher correlation than the class 'medium density housing'. This might be due to the higher proportion of walls visible to the sensor. Walls are variously illuminated according to their exposition towards the sun. The desert class showed similar high correlation coefficients, whereas the classes 'agricultural fields' and 'water' showed weak correlations.

Two limiting factors might limit partly the quality of the results. The first factor is the limited accuracy of the georeferencing process. Even though a running average was used, the results might be skewed slightly. The second factor applies only for the classes 'agricultural fields' and 'Nile water', where the basic set was not big enough to deduct a proper statistic.

## References


Berk, A., Anderson, G.P., Bernstein, L.S., Acharya, P.K., Doethe, H., Matthew, M.W., Adler-Golden, S.M., Chetwynd, J.H., Richtsmeister, S.C., Pukall, B., Allred, C.L., Jeong, L.S. & Hoke, L.S. 1999. MODTRAN 4 radiative transfer modeling for atmospheric correction. SPIE Proceeding, Optical Spectroscopic Techniques and Instrumentation for Atmospheric and Space Research III, Vol. 3756, 6

Cutter, M. & Johns L. 2005. CHRIS Data Products - Latest Issue. Third CHRIS/Proba Workshop 21 – 23 March, ESRIN, Frascati, Italy, 6

Del Frate, F., Duca, R., Solimini, D. 2007. Urban feature retrieved by hyperspectral multi-angle CHRIS Proba images. 2007 Urban Remote Sensing Joint Event, 11-13 April 2007, Paris, France, 5

Garcia, J.C. & Moreno, J. 2004. Removal of Noises in Chris/proba Images: Application to the Sparc Campaign Data. Second Chris Proba Workshop, 28-30 April 2004, ESA-ESRIN, Frascati, Italy, 5

Gatebeck, C.K., King, M.D., Lyapustin, A., Arnold, G.T., Redemann, J. 2005. Airborne spectral measurements of ocean directional reflectance. Journal of the atmospheric sciences, Vol. 62, Nr. 4, 1072-1092

Guanter, L., Alonso, L. & Moreno, J. 2005. A method for the surface reflectance retrieval from PROBA/CHRIS data over land: application to ESA SPARC campaigns. IEEE Transactions on Geoscience & Remote Sensing, Vol. 43, No. 12, 2908-2917

Liang, S. (2000): Narrowband to broadband conversions of land surface albedo I Algorithms. Remote sensing of environment, Vol. 76, 213-238

Meister, G., Lucht, W., Rothkirch, A. & Spitzer, H. 1999. Large Scale Multispectral BRDF of an Urban Area. IEEE International Geoscience and Remote Sensing Symposium, IGARSS'99, Hamburg, Vol. 2, 821– 823

Meister, G., Rothkirch, A., Wiemker, R., Bienlein, J. & Spitzer, H. 1998. Modeling the directional feflectance (BRDF) of a corrugated roof and experimental verification. IEEE International Geoscience and Remote Sensing Symposium, IGARSS'98, Seattle, Vol. 3, 1487 – 1489

Meister, G., Wiemker, R., Bienlein, J. & Spitzer, H. 1996. In Situ BRDF Measurements of Selected Surface Materials to Improve Analysis of Remotely Sensed Multispectral Imagery. Proceedings of the XVIII. Congress of the International Society for Photogrammetry and Remote Sensing ISPRS 1996, Vienna, volume XXXI part B7 of International Archives of Photogrammetry and Remote Sensing, 493 – 498

Nicodemus, F. 1970. Reflectance nomenclature and directional reflectance and emissivity. *Applied Optics*, Vol. 4, 767 – 773

Robaa, S.M. 2000. Giza monuments and local climate change. ICEHM2000, September 2000, Cairo University, Egypt, 187-200

Schiefer, S., Damm, A. & Hostert, P. (2006): Correcting brightness gradients in hyperspectral data from urban areas. *Remote Sensing of Environment*, Vol. 101, 25 – 37

Wang, J., White, K. & Robinson, G.J. 2000. Estimating surface net solar radiation by use of Landsat-5 TM and digital elevation models. International Journal of Remote Sensing, Vol. 21, No. 1, 31 - 43

*Remote Sensing for a Changing Europe*
*D. Maktav (Ed.)*
*IOS Press, 2009*

*doi:10.3233/978-1-58603-986-8-602*

# An operational approach for annual land cover mapping at the national scale with MERIS images

A. Araújo, H. Carrão & M. Caetano
*Remote Sensing Unit (RSU), Portuguese Geographic Institute (IGP), Lisbon, Portugal; antonio.araujo@igeo.pt*

A. Araújo
*Instituto de Engenharia de Estruturas, Território e Construção (ICIST), Instituto Superior Técnico (IST), Universidade Técnica de Lisboa, Lisbon, Portugal*

H. Carrão & M. Caetano
*CEGI, Instituto Superior de Estatística e Gestão de Informação, ISEGI, Universidade Nova de Lisboa, Lisbon, Portugal*

Keywords: Land cover, annual mapping, operational approach, MERIS, Portugal.

ABSTRACT: This study evaluates the effectiveness of a previous developed methodolog , in which was produced a medium scale land cover map for Portugal for 2005, when tested with up-to-date data. If proved successful, this methodology can be used on an operational basis for the establishment of an annual land cover mapping program at the national scale. The main point of this work is to capitalize the investment that was made before, namely in the reutilization of the reference samples (train and test) previously collected, for the development of up-to-date land cover maps. To carry out our study we have produced a land cover map for 2006 using the same methods, nomenclature, reference and auxiliary datasets, and others, that were exploited and established with success before. For map classification we used a supervised Linear Discriminant Classifier (LDC) combined with a vegetation index differentiating technique. Prior to the classification, we submitted the 2005 training database to an outliers detection procedure in order to remove and replace those sample observations that registered class changes from 2005 to 2006. The 2006 map was then rigorously validated and compared with 2005 map. Overall accuracy of the final map was of 70%, showing to be smaller than the one obtained for 2005 map, yet being considerably high. Comparison results demonstrated high agreement between 2006 and 2005 maps. General conclusions suggest that the developed procedure is effective and therefore can be use on an operational basis.

## 1 INTRODUCTION

The Remote Sensing Unit (RSU) of the Portuguese Geographic Institute (IGP) has recently investigated and developed with success a methodological approach for medium scale land cover mapping in mainland Portugal. This method, which is based on the exploitation of the enhanced spectral and temporal characteristics of the ENVISAT-MERIS sensor, through the use of automatic procedures, has conduced to the development of a 16 class land cover map for Portugal, for the year 2005, with an estimated overall accuracy of 80% with an absolute precision of 2% at the 95% confidence level (Carrão *et al.* 2008). Considering that IGP is the national reference centre for cartography, it is of all interest to capitalize the investment and effort that was made by the RSU-IGP in the development of the proposed methodology, and proceed with its improvement aiming for the establishment of an operational procedure for the annual production of medium scale land cover maps in Portugal. The development of such procedure is of greatest importance, principally if considering that the existent operational land cover programs at national scales (e.g. CORINE Land Cover) cannot assure the updated production of land cover knowledge that is required nowadays within many applications. This is because these programs still resort to visual interpretation and manual classification of high spatial resolution data and therefore are time consuming and economically expensive. Moreover, the existent global land cover maps, such as MOD12Q1 or the GLC2000 are also insufficient to meet national user needs in land cover information because, although these are regularly updated, their land cover nomenclatures are not adapted to the landscape characteristics of particular regions. As consequence, global maps overall and per-class accuracies present values below acceptable for their effective appliance at national scales. Supporting these assertions, the work of Cerdeira *et al.* (2006) demonstrated that the GLC2000 and MOD12Q1 present for Portugal overall accuracies of 61% and 57%, respectively. Also, the research of Caetano & Araújo (2006), where was performed a comparison of MOD12Q1 with CORINE Land Cover map for 2000 within the Portuguese region, showed an agreement of only 50% between the two products.

In order to set up the desired operational annual land cover mapping program, we need to evaluate the effectiveness of the referred previously developed methodology (Carrão *et al.* 2008), when applied to up-to-date data. If proved successful, this methodology can be used on an operational basis for annual land cover mapping at the national scale. For this purpose, we propose to take advantage of the methods, procedures and satellite and ancillary data that have already been exploited and previously established with success, for the development of a MERIS based 16 class land cover map for mainland Portugal for 2006. Furthermore, we aim at taking advantage from the existing exhaustive reference sample databases (train and test) that were collected in the framework of the previous experimental land cover classification study, because the collection of such data is highly expensive and time consuming. Automatic production of annual land cover maps, on an operational basis, can only reimburse the economic investment made in the development of methodological classification approaches and outcomes, if these can effectively be used in forthcoming applications. The methodology for map classification consists of two phases: 1) intra-annual stationary classes (e.g. artificial areas) characterization resorts to a supervised Linear Discriminant Classifier (LDC); 2) intra-annual dynamic classes (e.g. burnt areas and clear-cuts) are identified by a vegetation index differentiating technique. Prior to the classification, we submitted the 2005 training database to an outliers detection procedure in order to remove and replace those sample observations that registered class changes from 2005 to 2006.

In the end, 2006 map developed in this study is compared with 2005 map developed in (Carrão *et al.* 2008) and results are analyzed for similarities in terms of agreement and overall accuracy in order to conclude that the methodology is effective and therefore can be used in the operational production of annual land cover maps for mainland Portugal, giving a step forward in responding to the national user needs.

## 2 LAND COVER NOMENCLATURE

The used land cover nomenclature encompasses 16 comprehensive land cover classes (Table 1).

Table 1. Land cover class nomenclature (simplified version of the LANDEO nomenclature).

| Label | Land cover class | Label | Land cover class |
|---|---|---|---|
| 5 | Wetlands | 242 | Vineyards |
| 6 | Bare to sparsely vegetated areas | 34 | Shrubland |
| 7 | Water | 35 | Grassland |
| 11 | Continuous artificial areas | 38 | Burnt areas and clear-cuts |
| 12 | Discontinuous artificial areas | 311 | Broadleaved forest |
| 21 | Non-irrigated herbaceous crops | 312 | Agro-forestry areas |
| 22 | Irrigated herbaceous crops | 321 | Needleleaf forest |
| 23 | Rice crops | 331 | Mixed forest |

This nomenclature is a simplified version of a hierarchical 33 class exhaustive nomenclature, entitled LANDEO, developed by the RSU of IGP (Araújo & Caetano 2006). LANDEO nomenclature was develop through the Land Cover Classification System (LCCS) from Food and Agriculture Organization (FAO) (Di Gregorio & Jansen 2000) according to a three-fold rational: (1) a nomenclature that is adapted to the characteristics of landscapes existent is Portugal mainland, (2) a nomenclature that is compatible with established ones (e.g. CORINE Land Cover, Global Land Cover, and the International Geosphere-Biosphere Programme) in order to turn possible the comparison between our maps and others using different nomenclatures, and (3) a nomenclature that can be derived from satellite images with different spatial resolutions.

## 3 DATASET

The data used to perform this study includes: 1) MERIS L1b satellite images for the year 2006; 2) orthorectified aerial images acquired during the years of 2004, 2005 and 2006, and covering the whole Portuguese territory – these images have the following characteristics: four spectral bands in the blue, green, red and infrared wavelengths, and 50 cm spatial resolution; 3) two reference samples characterizing the reflectance of each land cover class within spectral wavelengths measured by MERIS sensor, which will be used independently for training and testing our classification methodology (reference sample is a set of geographic locations representing the true cover type at Earth's surface in those areas); and 4) other geographical ancillary information.

### 3.1 *Imagery data*

In this study we exploited 19 Level 1b Full Resolution MERIS images for the year of 2006. These images measure the solar radiation reflected by the Earth at a ground spatial resolution of 300m in the range of the visible to the near infrared wavelengths (390nm to 1040nm), and consists of radiance measures in 15 groups of wavelengths (for classification purposes, original bands 11 and 15 were removed since they address O2 content and water vapor, respectively). These images contain additional quality flags provided on a pixel by pixel basis, which address the quality and the validity of the product. Regarding the pre-processing of the imagery data, we used BEAM-VISAT 4.1 software to perform the atmospheric corrections (Rayleigh scattering, ozone, water vapor absorption and aerosol content) and radiance to surface-reflectance conversion. It is important to refer that due to software limitations, the aerosol content was considered as constant and equal for all images. Images were also geo-referenced and co-registered using the same software. For the classification of intra-annual static classes we used 12 images, discarding the other 7 on the basis of presenting significant cloud cover. On the subject of dynamic classes' identification, we resorted to two MERIS L1b FR images from September 2005 and 2006.

### 3.2 *Reference samples*

In this study we used existent reference sample datasets (train and test), developed in previous studies (Carrão *et al.* 2008). These reference samples were collected based on 2004, 2005 and 2006 orthorectified aerial images with the support of several auxiliary datasets (e.g. CORINE Land Cover 2000). Training sample observations were selected for each land cover class (except for the dynamic classes). To guarantee land cover homogeneity within training observations, they were selected and identified for geographical spots that corresponded to the centre pixel of 3x3 MERIS pixels within homogeneous land cover. To assure that training sample observations could be used with success for map production with MERIS images for 2006, a chi2 multivariate analysis of land cover classes was performed (this procedure is further explained in section 4). The objective was to perceive if any of the training observations incurred in some land cover change between 2005 and 2006. If this happened, then the observation corresponds to a class outlier and therefore was excluded and replaced while maintaining the per-class sample sizes necessary for classifier appliance. Testing sample observations, corresponding also to a MERIS pixel area, were collected following a stratified random sampling design, as previously proposed by Carrão *et al.* (2007a). The design of the reference database used for testing and its detailed analysis are presented in Carrão *et al.* (2007b). To avoid confounding classification errors with others, primary and alternate reference land cover labels were identified and characterized for each testing sample observation. A second label was only attributed when image interpreter did not have absolute confidence in the first label. This strategy served also to accommodate the two predominant cover types within a mosaic area. The reference samples (train and test) included some observations that could not be used properly as they were flagged "invalid" by MERIS data, i.e. pixels that were identified has containing clouds, shadows and other inadequate features. Therefore these pixels were excluded from the datasets.

## 4 METHODOLOGY

The methodology used to derive a land cover map for Portugal Continental from MERIS L1b FR imagery of 2006 pursued the research of Carrão *et al.* (2008). This methodological approach follows three phases: 1) Firstly, we identified and removed from training data set all observations within each land cover class that incurred in land cover change between 2005 and 2006, and as a consequence are outlier observations; 2) then we proceeded with the classification of the 15 intra-annual static land cover classes using the supervised Linear Discriminant Classifier (LDC); 3) in the end we produced a map of intra-annual dynamic land cover classes and overlaid it to the intra-annual static land cover map.

The validation of the developed map was carried through a probabilistic accuracy assessment using the test reference database. Besides the accuracy assessment, 2006 map was also compared with 2005 map developed in Carrão *et al.* (2008). If both maps are similar in terms of agreement and overall accuracy we can conclude that the adopted methodology is effective and therefore can be used on an operational basis for medium scale annual land cover mapping in mainland Portugal.

### 4.1 *Detecting outliers and cleaning training data set*

To detect outliers we used a chi2 multivariate analysis of land cover classes (Johnson & Wichern 1998). This outliers detection procedure compares the square distances of observations to the respective land cover class gravity center with the chi2 distribution at some confidence level. If the square distance measured for a given sample observation is greater than the chi2 for some confidence level, than we can consider that the observation incurred in land cover change between 2005 and 2006. In our study, we performed this outliers detection procedure considering a level of confidence of 95% using MERIS images spectral information from April, July and August of 2006 (dates that presented us the best results in uni-temporal classification). These dates were used independently because of the following reasons: 1) to respect the condition/constrain related to the available number of variables (p spectral bands) and number of observations per-class (n), where p – n must be equal or greater then 25 or 30; 2) to increase the confidence in the outliers detection, where only the sample observations found as outliers in at least two of the three analyzed dates were considered as definite outliers, and therefore were discarded and replaced.

### 4.2 *Intra-annual static land cover classes classification*

Intra-annual static land cover classes were classified using LDC with 12 MERIS L1b images of 2006 as input features. This classifier has already proved to be successful. For example, and as already referred, in Carrão *et al.* (2008) the developed map registered an estimated overall accuracy of 80%, with an absolute precision of 2% at the 95% confidence level.

### 4.3 *Intra-annual dynamic land cover classes classification*

The methodology applied to identify areas with intra-annual vegetation decreases (i.e. burnt areas and clear-cuts) was the vegetation index differencing (Lu *et al.* 2004). NDVI maps were produced separately for September of 2005 and 2006, and then the vegetation index map of the earlier date was subtracted from the vegetation index map of the later date. The result was a new map relative to vegetation change. In order to identify true change areas there was a need to define a difference threshold, since not all differences could be consequence of significative changes of vegetation. For the threshold definition it was considered that change pixels are located at a distance higher than 1.25 standard deviations to the mean value of the image resulting from the vegetation index subtraction. These was suggested in Armas & Caetano (2005) for detecting changes in forest and natural vegetation areas in Portugal, using satellite images of medium spatial resolution. Considering that the main goal at this phase was to detect decreases in forest and natural vegetation areas, due to fires and clear-cuts, an analysis mask derived from the intra-annual static land cover classes classification representing these covers was used and changes determined only for these areas. Armas & Caetano (2005) also stated that, for medium spatial resolution images, this methodology should be used with a minimum mapping unit of 10 pixels in order to achieve the best compromise between commission and omission errors. Thus, we have discarded all changes smaller than 90 ha from our final map.

### 4.4 *Accuracy assessment*

The accuracy assessment of the developed map was made through the construction and analysis of an error matrix. In this matter, Card (1982) points out that, for the stratified sampling case, the overall proportion of correctly classified individuals in the map, given the reference land cover categories, should include the known areas of each map category to improve the estimation of the proportion of correctly mapped individuals. If not, bias may be introduced by possible differential sampling rates within map categories. Taking this into account, we decided to develop an accuracy assessment of the produced map considering the area of each map class. For this latter purpose we supported our accuracy assessment methodology on Cochran (1977) work, later adopted by Card (1982). The equations used to derive the specific and overall accuracy indexes were those proposed by Carrão *et al.* (2007a), derived from Cochran (1977).

### 4.5 *Comparison between 2006 and 2005 land cover maps*

The comparison between 2006 and 2005 land cover maps was made according to two different strategies: 1) the first strategy consisted in the direct comparison of the two maps on a per-pixel basis (pixel level comparison); 2) the second strategy tried to eliminate the disagreements that could result from: positional errors between the two maps; problems in classifying mosaic areas (pixels located in the border of land cover classes) where both maps can be correct at the same time, each classifying a different part of the mosaic. For this latter purpose, and as suggested in Araújo *et al.* (2007), before maps comparison we extracted from each map areas that were encompassed by the same land cover class. We called these areas as equal-neighbors regions. Then we proceeded with the comparison assuring that positional errors smaller than one pixel would not affect the agreement result. The class 38 (Burnt areas and clear-cuts) was removed from both comparisons because it is an intra-annual transient class, consequently is logical and expected to be in disagreement.

## 5 RESULTS AND DISCUSSION

In Fig. 1 we present the land cover map produced for the Portuguese territory from MERIS L1b FR images for the year 2006. In Table 2 we introduce the accuracy assessment results of 2006 map, accompanied by the accuracy assessment results of 2005 map (Carrão *et al.* 2008) for later comparison and discussion in this section. Overall accuracy of 2006 final map was estimated at 70% with an absolute precision of 2% at the 95% confidence level. Estimated user's accuracies were all superior to 60%, except for Bare to sparsely vegetated areas (6), Vineyards (242), Broadleaf forest (311) and Mixed forest (331). Absolute precisions for user's accuracies estimates were, for almost all classes, around 20% at the 95% confidence level. Estimated producer's accuracies were all superior to 60%, except for Irrigated herbaceous crops (22), Rice crops (23), Burnt areas and clear-cuts (38), Needleleaf forest (321) and Mixed forest (331), with absolute precisions, for the majority of classes, around 20% at the 95% confidence level.

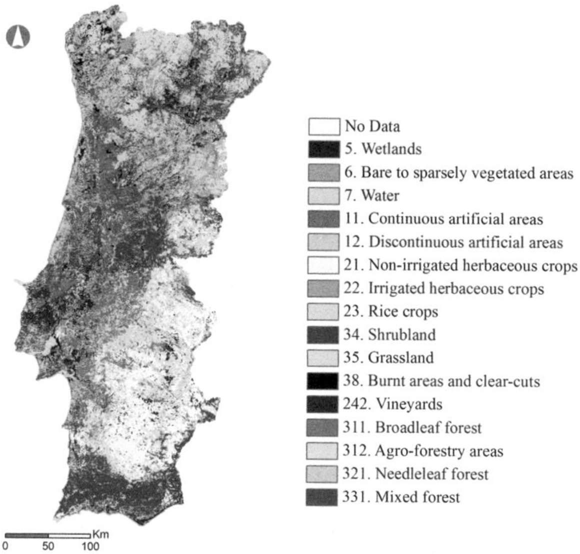


Figure 1. Land cover map for Continental Portugal derived from multi-temporal MERIS images of 2006.

Table 2. Area occupied per-mapped land cover class (Map) and estimated 95% confidence intervals for: area occupied per-reference land cover class (Ref.); user's accuracy (UA); producer's accuracy (PA); and overall accuracy (OA). Results for 2005 map where extracted from Carrão *et al.* (2008).

| | 2006 Map | | | | 2005 Map | | | |
|---|---|---|---|---|---|---|---|---|
| Class | Map (%) | Ref. (%) | UA | PA | Map (%) | Ref. (%) | UA | PA |
| 5 | 0.7 | [0.6, 0.8] | [0.8, 1.0] | [1, 1] | 0.8 | [0.5, 0.9] | [0.7, 0.9] | [0.6, 1.0] |
| 6 | 4.2 | [2.7, 4.1] | [0.5, 0.7] | [0.6, 0.8] | 5.3 | [3.3, 4.7] | [0.5, 0.7] | [0.6, 1.0] |
| 7 | 1.8 | [1.7, 2.1] | [1.0, 1.0] | [0.8, 1.0] | 0.5 | [0.5, 0.7] | [1.0, 1.0] | [0.7, 0.9] |
| 11 | 0.4 | [0.4, 0.6] | [0.8, 1.0] | [0.6, 1.0] | 0.3 | [0.4, 0.8] | [0.8, 1.0] | [0.3, 0.7] |
| 12 | 3.6 | [3.4, 4.4] | [0.7, 0.9] | [0.7, 0.9] | 3.9 | [3.2, 4.2] | [0.7, 0.9] | [0.7, 1.0] |
| 21 | 12.2 | [13.9, 16.1] | [0.9, 0.9] | [0.6, 0.8] | 12.9 | [13.6, 15.8] | [0.9, 1.0] | [0.5, 0.9] |
| 22 | 1.0 | [1.5, 2.7] | [0.6, 0.8] | [0.2, 0.4] | 0.7 | [1.8, 3.2] | [0.8, 1.0] | [0.1, 0.4] |
| 23 | 0.3 | [0.3, 0.7] | [0.9, 1.0] | [0.3, 0.7] | 0.3 | [0.6, 1.6] | [0.8, 1.0] | [0.1, 0.4] |
| 34 | 17.3 | [16.9, 19.7] | [0.8, 0.8] | [0.7, 0.7] | 17.8 | [16, 18.6] | [0.8, 0.8] | [0.7, 1.0] |
| 35 | 17 | [16.7, 19.7] | [0.6, 0.8] | [0.7, 0.7] | 18.3 | [16, 19] | [0.6, 0.8] | [0.5, 1.0] |
| 38 | 0.7 | [0.6, 1.2] | [0.6, 0.8] | [0.3, 0.7] | 2.8 | [2.2, 3.2] | [0.6, 0.8] | [0.5, 1.0] |
| 242 | 6.1 | [2.4, 3.8] | [0.3, 0.5] | [0.7, 0.9] | 4.2 | [2.5, 3.9] | [0.5, 0.7] | [0.5, 1.0] |
| 311 | 9.3 | [5.7, 7.7] | [0.4, 0.6] | [0.6, 0.8] | 7.1 | [5.7, 7.7] | [0.6, 0.8] | [0.5, 1.0] |
| 312 | 15 | [12, 14.6] | [0.6, 0.8] | [0.7, 0.9] | 15.2 | [12.2, 15] | [0.6, 0.8] | [0.6, 1.0] |
| 321 | 3.5 | [5.0, 6.6] | [0.7, 0.9] | [0.4, 0.6] | 3.5 | [4.6, 6.2] | [0.8, 0.8] | [0.3, 0.8] |
| 331 | 6.8 | [4.8, 6.6] | [0.4, 0.6] | [0.5, 0.7] | 6.6 | [4.9, 6.5] | [0.6, 0.8] | [0.6, 1.0] |
| | OA = [0.68, 0.72] | | | | OA = [0.78, 0.82] | | | |

When comparing these results with those registered for 2005 map, its easy to perceive that the used methodology is effective when applied to up-to-date data. The overall accuracy value of 2006 map, although inferior than the 80% value registered for 2005 map, is considerably high. Per-class accuracies where similar for user's and producer's indices, even though confidence intervals values of 2006 map are at times lower than those of 2005 map. A major difference would exist if any of the class confidence intervals did not intersect each other. This didn't happen. Even the four worst cases (Irrigated crops (22), Vineyards (242), Broadleaf forest (311) and Mixed forest (331)) presented confidence intervals that intersected each others in one of their limits. This evidences that maps have identical classes' accuracies. Moreover, for example looking at user's accuracy, both maps presented difficulties in discriminating properly the same classes (Bare to sparsely vegetated areas (6), Vineyards (242), Mixed forests (331)), as demonstrated by the fact that these presented worst accuracies in both maps, while also discriminated with superior accuracy the same classes (Water (7), Continuous artificial areas (11), Non-irrigated crops (21), Rice crops (23)) as demonstrated by the fact that these presented better accuracies in both maps. The same behavior is observable for producer's accuracy as demonstrated in classes 22 and 23 and in classes 5, 7, 12, 242 and 312, respectively. Mapped class areas for 2006 and 2005, and estimated confidence intervals for area occupied per-reference land cover class, are similar with exception for Water (7) and Burnt areas and clear-cuts (38) classes. Although these classes attained good accuracy results for both maps, they presented for these cases significantly different results. The explanation for this apparent strange result is simple. Class 38 is an intra-annual dynamic class, therefore the area occupied by it varies as a result of the size and number of fires and forest cuts of the year. Class 7 was conditioned by the use of an updated boundary when masking the area of analysis in 2006 map. This boundary included more water areas (ocean and estuaries) then the one used for 2005 map. Looking at the differences in both maps, we believe that these are mainly due to the fact that the MERIS L1b images present geometric problems, as confirmed by several authors (e.g. Clevers *et al.* 2004). This fact creates strong difficulties in the correct co-registration of the imagery and consequently has implications in the multi-temporal classification performance for map generation. Moreover, we have used a less sophisticated process to perform the atmospheric corrections, in which the aerosol content was considered as constant and equal for all images, and it may have contributed to the lower accuracies obtained for 2006 map. On the other hand, 2005 map was derived from MERIS Ionia images. These images where submitted to enhanced procedures in order to be positionally and atmospherically corrected, as they are the base images for the development of the GLOBCOVER product for 2005. Therefore, the classification results of 2005 map were not conditioned or diminished by any image constraints, justifying the achieved better results.

Besides evaluating and comparing individual maps accuracy assessments, we also compared the two maps using two different map comparison strategies (pixel level comparison and equal-neighbors regions comparison) (see chapter 4.5) (Fig. 2).

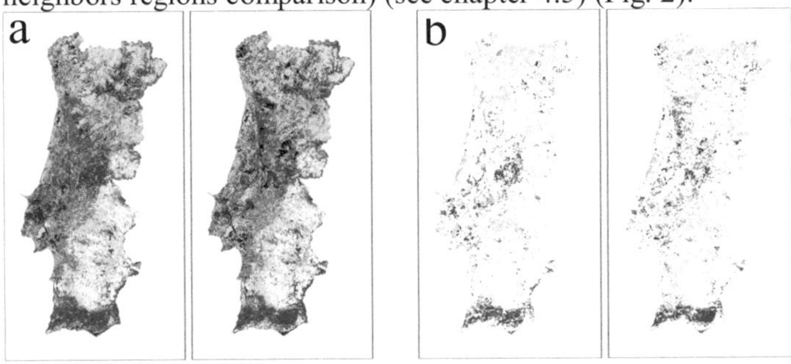


Figure 2. Map areas analyzed for each of the comparison strategies: a) pixel level comparison (2006 map on the left and 2005 map on the right); b) equal-neighbors regions comparison (2006 map to the left and 2005 map to the right).

Visually the maps are much alike. The determined agreement between both maps was of 65% using the pixel level comparison and 99% using the equal-neighbors regions comparison. These results were determined based on the permitted intersections (class direct correspondence – diagonal cells of the comparison matrix) between the two products, as a percentage of the total study area. Results confirm visual similarities between 2006 and 2005 maps, especially if considering the equal-neighbors regions comparison. The difference attained with the two comparisons is related to the fact that equal-neighbors regions have a minimum area of 3x3 MERIS pixels, i.e. 90 hectares. Areas as big as these are unlikely to change their land cover during one year period, with exception for the intra-annual dynamic classes that were not considered in both comparisons. To test this theory, we compared both maps considering not only the direct correspondence of classes, but also the correspondences between those classes that may have change from 2005 to 2006. This strategy increased maps agreement from 65% to 71% at pixel level. After this, we performed another comparison where we added to the previous permitted correspondences, those between classes that are spectrally and visually similar, and because of that could be classified differently in both maps. These situations would be in disagreement but both maps could be correctly classified, because reference data used for accuracy assessment is uncertain for these pixels. With this approach, the agreement result increased from 71% to 79%. Analyzing these results and having in mind that at pixel level comparison some unsolved disagreements can be due to positional errors or due to the fact that mosaic pixels (pixels located in the border of land cover classes) are classified differently in both maps, we can say that maps are very similar.

## 6 CONCLUSION

Considering the accuracy assessment results for 2006 map and their comparison with the accuracy assessment results of 2005 map, and considering the agreement obtained between 2006 and 2005 maps, we can conclude that the used methodology is effective for medium land cover maps production in Portugal.

## REFERENCES

Araújo, A., Caetano, M. 2006. *Nomenclatura de Ocupação de Solo LANDEO.* Lisbon: Portuguese Geographic Institute (IGP), Portugal, 56 pp.

Araújo, A., Caetano, M., Cerdeira, C. & Raposo, D. 2007. Avaliação da incerteza de mapas globais de ocupação do solo para portugal Continental. In J. Casaca & J. Matos (Ed.), *Cartografia e Geodesia 2007* (pp. 147-156). Lisboa: Lidel.

Armas, R., Caetano, M. 2005. Cartografia de áreas ardidas com imagens multirresolução do sensor MODIS. In J. Casaca & J. Matos (Ed.), *Cartografia e Geodesia 2005* (pp. 293-303). Lisboa: Lidel.

Caetano, M., Araújo, A. 2006. Comparing land cover products CLC2000 & MOD12Q1 for Portugal. In A. Marçal (Ed.), *Global Developments in Environmental Earth Observation from Space* (pp. 469-477). Rotterdam: Millpress.

Card, D. 1982. Using known map category marginal frequencies to improve estimates of thematic map accuracy. *Photogrammetric Engineering and Remote Sensing*, 48: 431–439.

Carrão, H., Caetano, M. & Coelho, P. 2007a. Sample Design & Analysis for Thematic Map Accuracy Assessment: An Approach Based on Domain Estimation for the Validation of Land Cover Products. *In Proceedings of the 32nd International Symposium on Remote Sensing of Environment* (unpaginated CD-ROM), 25-29 June 2007, San Jose, Costa Rica.

Carrão, H., Araújo, A., Cerdeira, C., Sarmento, P., Capão, L. & Caetano, M. 2007b. A reference sample database for the accuracy assessment of medium spatial resolution land cover products in Portugal. *In Proceedings of the IEEE International Conference on Geoscience & Remote Sensing Symposium* (IGARSS´2007), 20-29 July 2007, Barcelona, Spain.

Carrão, H., Araújo, A., Gonçalves, P. & Caetano, M. 2008. Multitemporal MERIS images for land cover mapping at national scale: the case study of Portugal. *International Journal of Remote Sensing*, in press.

Cerdeira, C., Araújo, A., Carrão, H. & Caetano, M. 2006. Validação das Cartografias Globais de Ocupação do Solo, GLC2000 e MOD12Q1, para Portugal Continental. *In Actas do IX Encontro de Utilizadores de Informação Geográfica* (unpaginated CD-ROM), 15 – 17 November 2006, Oeiras, Portugal.

Clevers, J.G.P.W., Zurita Milla, R., Schaepman, M.E. & Bartholomeus, H.M. 2004. Use of MERIS Data for Land Cover Mapping. *Proc. ENVISAT Symposium*, 6-10 September, Austria.

Cochran, W. 1977. *Sampling Techniques*, 3rd Ed. New York: Wiley.

Di Gregorio, A., Jansen, L.J.M. 2000. *Land Cover Classification System.* FAO, Rome.

Johnson, R.A., Wichern, D.W. 1998. *Applied Multivariate Statistical Analysis*, 4th Ed. New Jersey: Prentice Hall, Upper Saddle River.

Lu, D., Mausel, P., Brondízio, E. & Moran, E. 2004. Change detection techniques. *International Journal of Remote Sensing*, 25: 2365-2401.

*Remote Sensing for a Changing Europe*
*D. Maktav (Ed.)*
*IOS Press, 2009*

*doi:10.3233/978-1-58603-986-8-610*

# Open source GRIFINOR platform for 3D spatial data presentation

J. Kolar
*Center for 3D Geoinformation, Aalborg University, Aalborg, Denmark*

J. Kolar, T. Bayer & S. Grill
*Faculty of Science, Charles University in Prague, Prague, Czech Republic*

Keywords: 3D visualization, software platform, open source, spatial data modeling

ABSTRACT: The paper describes practical use of the research software platform GRIFINOR developed as a open-source tool for comprehensive presentation of spatial data in 3D format. It presents a supporting tool to project-oriented type of education in geoinformation technology (geomatics). Being developed in Aalborg University, the GRIFINOR platform is now subject of common development inside a group of interested researchers and software programmers spread around the world. The GRIFINOR platform development is based on vision of a product of modern media technology which allows data management, visualization, and exploration of geographic information in three-dimensions. The platform opens opportunities to implement concrete focused on particular concepts of geomatics and studying individual components of geomatics technology. The paper discusses some important skills that students can practice and also presents GRIFINOR research potentials. Some examples are given when the platform has been applied in teaching master students in geomatics study program.

## 1 INTRODUCTION

Spatial data represents such data, for which connection between information and location on the Earth is an important consequence. During last twenty years methods of spatial data acquisition and processing have dramatically developed and spatial analysis, spatial models and spatial structures take a crucial role of research activities. Implementation of the third dimension into data represents an important step enabling particularly different perception of the Earth. Three dimensional model has less constraints than traditional planar projection. Those trends have been reflected by cartographers and software engineers and brought new visualization techniques based on 3D models. There are a lot of platforms processing and visualizing spatial data. This article presents GRIFINOR platform as a new open-source solution which allows advanced data management, visualization, and exploration of geographic information in three-dimensions.

## 2 GRIFINOR PLATFORM

GRIFINOR is an open-source research software platform developed for comprehensive presentation, exploration, publishing and analysis of 3D spatial data. The name GRIFINOR arise as an acronym for "Geographic Reference Interface For Internet Networks" (GRIFIN). This paper describes GRIFINOR as a project, platform and media technology suitable for education of students in geomatics and bringing opportunities to understand, explain, implement tool and procedures of data geoprocessing and data management. This project is primarily addressed to students, researchers in geoinformatics, cartography, 3-D computer graphics, geovisualization, landscape, construction and

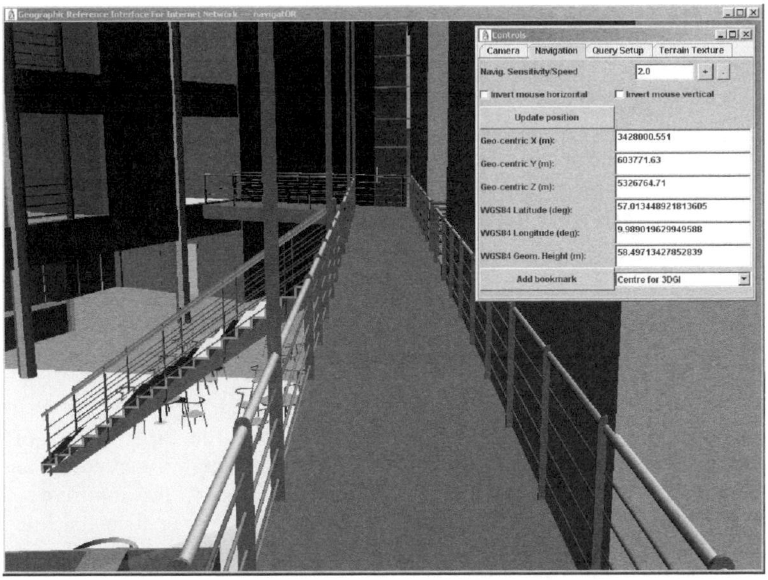


Figure 1. Sample scene illustrating functionality of GRIFINOR platform.

urban planning, architecture. GRIFINOR has been developed at the Aalborg university by the group of software engineers and researchers since 2001 and dynamically spreading around the world.

GRIFINOR platform is an open-source 3D virtual globe project written in Java with graphical user interface like Google Earth. It is freely available under LGP license, source codes are suitable for public usage, study and further development. Scene rendering and all graphics operations are implemented using OpenGL. GRIFINOR uses object oriented model and technology, all components are hierarchically organized, data representation, tools and communication protocols are defined in term of objects. The main software development tools used by GRIFINOR developers are NetBeans and Eclipse. GRIFINOR graphic user interface was projected to be intuitive, its functionality can be easy explored and gives users software product with simple and powerful structure.

GRIFINOR project is not only scientific and research platform, but also a tool suitable for education of students illustrating principles of digital cartography and other geosciences. Role of GRIFINOR platform in educational process are introduced at the Chapter 3.

## 3 ROLE OF GRIFINOR PLATFORM IN EDUCATION

Role of the GRIFINOR in education process could be divided in several consequently steps important for explanation and understanding background of geoprocessing and data management. During last 20 years geosciences has been radically changed in many ways. New data acquisition models are replacing traditional terrestrial measurement techniques, two dimensional maps are becoming insufficient for representation of the Earth. Huge amounts of data are putting accent on geoprocessing, data adjustment or data analysis. The research platform GRIFINOR is in compliance with those trends. Fully supports education of five main activities (see Table 1) and it is suitable for teaching master students of geomatics at Aalborg University and Charles University in Prague.

The main part of the article presents some educational concepts important in teaching geoinformatic technology and illustrates how GRIFINOR could be used as an integral part of the educational process.

Table 1. Applications of GRIFINOR platform in education process

| Activity | Description |
|---|---|
| Application development | Identify and develop tools and instruments to satisfy specific requirements. |
| Data analysis and interpretation | Process data and obtain information required for efficient decision-making process. |
| Data management | Catalogue, archive, retrieve and distribute geospatial data. |
| Visualization | Render data and information into visual geospatial representations. |

### 3.1 *Interoperability concept and GRIFINOR*

Interoperability is a property referring the ability of various systems to exchange, use and share information enabling work together (=interoperate). Interoperability express the capability of different geographic software tools to exchange geodatasets in such formats enabling reading and writing data independent on operation system. According to Open GIS consortium requirements interoperability contributes to the standardization of programs and data exchanging formats. Interoperability brings an important economic benefit and improves usability. An example of interoperability solution is Java language, that products the same byte code across the software platforms. Full interoperability belongs to unsolved problems.

Network interoperability could be comprehended as an ability to provide map via network using such format, that could be addressed and used by another system (Kolar, 2006). There has to be an unified message mechanism called protocol enabling independently sharing data (for example http). This concept is important for GRIFINOR platform.

GRIFINOR project was developed as a platform independent of any closed data standard or proprietary solution. Interoperability concept is realized on the level of programing language instead of data representation. GRIFINOR is coded in Java, both the communication protocol and the data representation of features are defined in terms of objects. The important part of the protocol is implemented in the class `grifinor.godb.GRIFINMessage`. Instances of this class represent query messages that are sent between system components over network (Kolar, 2006).

The data representation of any feature that can be viewed in GRIFINOR must extend the class `grifinor.datarep.GO.GO` is a Java abstract class and ensures the scalability for custom definitions of geographic features. It allows de-facto arbitrary definitions of features, their meaning, the relationships between them and their functionality. With this in mind, GRIFINOR must be regarded as a platform that supports the definitions of custom data representations (Kolar, 2006).

The interoperability within GRIFINOR platform is provided by implementing a custom class loader. This means that even if a system component receives an unknown object defined by a custom application it will be able to obtain the executable definition of the class from the provider and modify itself at runtime. This ensures that both data and processing methods can be exchanged in GRIFINOR. More about this is described in (Kjems and Kolar 2005).

Currently GRIFINOR works in a client-server manner. Nevertheless the design is such that both client and server components are tightly coupled. This means that a GRIFINOR instance can behave as both client and server, which allows to build a peer-to-peer network. However, a specification of such network is currently missing. This concept illustrates some interoperability features and attemps students to explain benefits of the presented scheme with accent to importance of the presented concept (Kolar, 2006).

### 3.2 *Cartography concept in GRIFINOR*

The concepts of cartography were presented by Claudius Ptolemaeus in his famous work books Introduction into Geography. Cartography is over two millenia old science and art. Adding mathematical fundaments and analytical methods to the process of mapping and creation of cartographic materials resulted in the birth of earth sciences like geodesy or mathematical cartography. Cartography hundred years ago represents first of all handly work. Today cartographers and map makers prefer scientific work with automated or semiautomated data acquisition and processing using computers. Earth has not been perceived and represented as a flat land for a long time, planar coordinate systems are becoming more and more a historical relict depending and associating with a paper media. Due to new knowledges in mathematics, physics, geometry, statistics and informatics the methods of creating maps have been rapidly changed and enforced.

Position of the point with respect to the earth center expressed using spatial coordinates could be located by global positioning systems. Present trend and future expectations in cartography could be simplified to phrase "going from two dimensions to three dimensions". Local two dimensional maps are changing to global georeferenced models. This process brings some advantages and disadvantages. Representation of features on the ellipsoid (or geoid), basic calculations are more complicated than in traditional 2D representation, however in many cases planar representation is (and will be) sufficient. The big advantage is elimination of cartographic projection constrains, first of all geometric distortions, limited area range and numerical imprecisions.

Those trends are reflected by cartographers and map makers in production of 3D maps and models. Virtual models of buildings and cities are useful for presentation, help people to orient or make some human decisions easier and faster. Because of missing geo-referenced model they are not used by professionals. Three dimensional geocentric approach we can find in GRIFINOR platform. GRIFINOR could be used as a power tool which describes fundaments and explains concepts of mathematical cartography in geoiformatics. This approach helps students to recognize relationship between geocentric and non-geocentric models and illustrate advantages and disadvantages of both solutions. Students could explore fundaments of mathematical cartography like ellipsoid flattening, excentricity, stretch, length, areal distortion ant other relationships between data on the earth, data in the plane and data in geo-referenced model.

GRIFINOR uses three dimensional coordinate system analogous to World Geographic System (WGS). Origin is identical to Earth centre of mass, z-axis is defined by axis of rotation, x-axis lies in equatorial plane, y-axis is perpendicular to x,y. Reference pole and meridian are defined by International Earth Rotation Service. GRIFINOR does not use cartographic projections in the typical consequence, for demonstration supports Mercator projection and UTM projection including inverse variants (see Fig. 2). It works with three systems of coordinates: carthesian coordinates $[x, y, z]$ for data representation, visualization, spherical coordinates $[\varphi, \lambda, r]$ for indexing and representation geopotential model of Earth, geographic coordinates $[\varphi, \lambda, h]$ for reporting position on the Earth to the user. Source code analysis could help to explain various coordinates representation, basic coordinates transformations and cartographic projections.

### 3.3 *Visualization concept in GRIFINOR*

Visualization of geospatial data has a long tradition. Graphic approach helps to recognize and explain spatial relationship between features and synthesize new knowledges. Implementation of the third dimension to cartography emphase on visualization process. Three dimensional model has less constraints than traditional planar projection of the space to paper, but visualization techniques became more difficult than techniques for flat world approximation. The are a lot of graphic frameworks supporting geovisualization. Open Geospatial Consortium has published specifications, recommendations and standards necessary for graphical interoperability.

Swanson, 1996, defined geographic visualization as a process of creation third-dimensional scene on two dimensional area represented by computer monitor. Third dimensional model brings better simulation of reality, allows easier orientation in terrain, creates more corresponding model of the real world, it is more suitable for presentations and from psychological aspect is more innate. It allows cartographer to perform continuous movement throw the interactive world. Important part of the geovisulization represents terrain visualization. Using computational geometry algorithms continuous terrain has to be reconstructed from discrete measurements, from topographic data we obtain digital terrain model.

```
public class TransverseMercator {
    private double a,b, e, ee, ee1, e4, e6, ep, FE, FN, k, phi0, lam0, M0;
    private double s,t,c,N,T,TT,C,CC,A,M,x,ro,D,DD;
    private double M_1, mi_1, e_1,e2_1,e3_1,e4_1, phi_1 ;

    public TransverseMercator(GlobalReferenceEllipsoid grs) {
//      default constructor defines UTM zone 32 on WGS84 ellipsoid
            a = grs.a; //WGS84
            b = grs.b; //WGS84
            e = grs.e;
            ee = grs.e2;
            ee1 = 1 - ee; //System.out.println("ee: " + ee);
            e4 = ee*ee;
            e6 = e4*ee;
            ep = ee/ee1; //System.out.println("e'^2: " + ep);
            FE = 500000.0;
            FN = 0.0;
            k = 0.9996;
            phi0 = 0;
            lam0 = Math.toRadians(9);
            if(phi0 == 0){
                M0 = 0; //System.out.println("M0: " + M0);
            } else {
                M0 = a*(     //THIS IS NECESSARY FOR GENERAL TM, UTM has this = 0 on north hemisphere.
                    (1 - ee/4 - 3*e4/64 - 5*e6/256)*phi0 -
                    (3*ee/8 + 3*e4/32 + 45*e6/1024)*Math.sin(2*phi0) +
                    (15*e4/256 + 45*e6/1024)*Math.sin(4*phi0) -
                    (35*e6/3072)*Math.sin(6*phi0)); // System.out.println("M0: " + M0);
            }
    }
```

Figure 2. Implementation of Mercator projection class in GRIFINOR.

Visualization of the third-dimensional data is considered to be a difficult task synthesizing common knowledges from various branches. Modeling techniques for the terrain geometry reconstructions are using set of primitives (cylinders, cubes, spheres) acombinated with sets of non-primitives entities (solids, iso-surfaces). Fast visualization of the whole terrain stored in memory is due to time complexity of calculation almost impossible, only parts of the scene can be rendered. In GRIFINOR platform suitable areas, nearby the virtual visitor, are found very fast using global index method.

As mentioned above, GRIFINOR platform is a 3D virtual globe project with graphical user interface like Google Earth. Visitor can move freely through the virtual world represented by georeferenced model and use perspective look on the scene. He is able to define start position visually, using geographic coordinates or bookmarks. An interesting fact is the possibility of moving inside buildings with high level of details. GRIFINOR is using an object oriented model, every piece of terrain is a separate object with own properties and methods. This feature illustrates properties of object oriented programming concept and it is an important difference to other systems. Student can create own georeferenced virtual scene with buildings located on the terrain writing a short script in Java (see Fig. 3). Graphic kernel of GRIFINOR platform is based on OpengGL libraries using Java language. Demonstration maps bundled to GRIFINOR illustrate features and uses of the system, other examples could be downloaded from webpages.

## 3.4 *Computational geometry concept in GRIFINOR*

Computational Geometry studies features of geometry algorithms ands their computer implementation. Computational Geometry is trying to find optimal solution for geometry problems. Earth surface in cartographic representation could be divided into the planar geometric structures like points, lines, polylines, curves or polygones. Therefore in digital cartography and geomatics computational geometry plays a fundamental role. Spatial data operations, like analysis of the relationships of geometric primitives, discrete 2D or 3D geometry tasks, combinatorics tasks, graph analysis, visualization techniques are the most frequent problems that could be partially transformed into geometric solutions.

Figure 3. Visualization of Aalborg city center in GRIFINOR.

Problem of the geographic data organization and geographic data management is important for positioning features around the globe. There is a demand on global indexing method for 2D or 3D datasets enabling fast access of huge amounts of data, supporting levels of details and enabling visual navigation through virtual worlds. Concepts of vector and raster data management, data organization, have been elaborated by a large community of researchers in many conferences an journals. This effort leads to the applications of global grids suitable for plane or space subdivisions and resulted in a birth of the global indexing techniques.

There are irregular grids (Quaternary Triangulated Meshes) dividing sphere using recursive subdivision octahedral faces into triangles (Dutton, 1989). Another approach presented by Lukatella, 1987 is based on Voronoi diagrams on the sphere. Each cell has the similar size, position around the globe can be coded into position index (integer number). GRIFINOR uses a non regular grid based on tessellation of the surface of the sphere. Cell is an essential entity and spatial unit in geoindex defined by spherical latitude and longitude of the circle. Edges are given by vectors from the origin with the same radial distance to two centroids.

An important operation in GRIFINOR is classification of the point in the scene to corresponding cell. This process results in proximity centroid retrieval. It is necessary to choose candidate centroids from all centroids using geoindex. Solution is based on geometric predicates and using two parallels going northern and southern to the point. Tessellation of geoindex projected to sphere results in semiregular global grid. Dimensions and geometry of cells are not unified, but they can be considered as hexagons. The number of grids is not unique, it is different for each resolution of geoindex, this concept is called as concept of levels. Presented solution is suitable for building the hierarchical structure through levels, that could be accessed regardless the required resolution (see Fig. 4).

This concept model helps students to understand dependence between digital cartography and computational geometry. It leads to appreciation of basic geometric principles and techniques in two or three-dimensional space, first of all Delaunay triangulation, Voronoi diagrams, convex hulls, geometric searching problems, point, line and polygone locations, intersection algorithms, etc. There is an assumption of basic knowledge of math, geometry, data structures or data management. Exploration of the algorithms in GRIFIONOR is supported by frequent usage of comments and brings students a new look to geometric and cartographic problems.

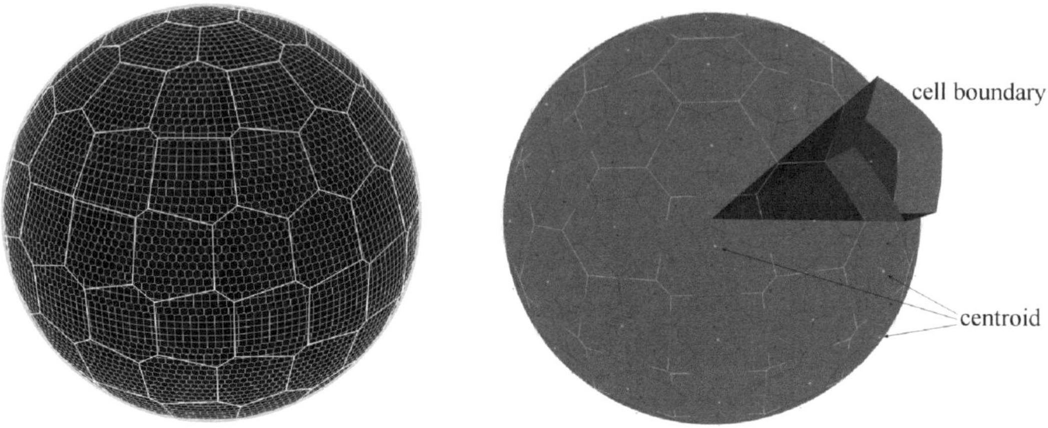


Figure 4. Geoindex tessellation with various resolutions.

### 3.5 *Algorithms implementation concept in GRIFINOR*

The process of the implementation of cartographic, geometric or visualization algorithms leads to robust routines and calculations. The purpose of algorithms analysis and coding is to motivate students to solve specific problems and in result increases their knowledge of computational geometry, programming skills and affects their logical thinking and cognitive abilities. It is not important to find maximum effectively solution, but acquaint students with various techniques like greedy algorithms, divide and conquer algorithms, randomized algorithms on geoinformatic problems, which are easy to understand.

Algorithm implementation in GRIFINOR project we could illustrate on process of surface reconstruction. For online topographic surface reconstruction in GRIFIONOR Delaunay triangulation is used. Presented method of incremental insertion is described in Garland and Hecbert, 1995. Two dimensional Delaunay triangulation is performed inside each cell in specified level of detail, that is necessary for visualization part of the scene. GRIFINOR presents multiresolution triangulation concept (see Fig. 5). First step of the triangulation begins with approximation of the points set by two triangles. In next steps are points with highest importance inserted into Delaunay triangulation. Presented algorithm illustrates divide and conquer solution of the problem, and it could be explored and implemented by students. GRIFINOR is written in Java, this language is well-arranged, clear and suitable also for beginning developers. Developing GRIFINOR using open source tools, libraries and IDE helps people to collaborate and share results without any additional costs.

GRIFINOR has been designed as a software available for anybody interested in implemented solutions. It eliminates disadvantages of proprietary software binded by restrictive software licenses. Those software solutions are usable for creating current "business" operations but often unavailable for research and constrained for usage. With proprietary software is not possible to explain internals of the system, deeper aspect are hidden for talented students and represents a secret black box processing output data from input data. University education is focused on eplanation understanding of facts and not for proprietary software education. Another important reason is an economic criterion, that results benefits in saving money. GRIFINOR is a project from a large software family distributed under LGPL license that guarantees freedom for all users and developers through its license.

## 4 CONCLUSIONS

This article describes software platform GRIFINOR as a comprehensive tool for the presentation of spatial data and illustrates opportunities of usage of this system on the educational process. It also

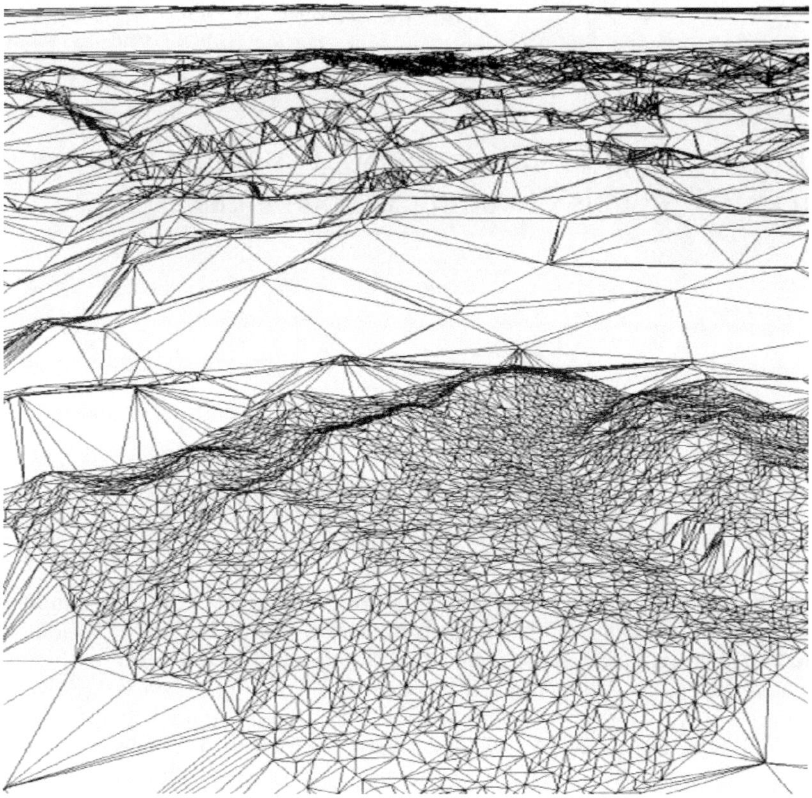

Figure 5. Delaunay triangulation used for online topographic surface reconstruction inside a cell, a concept of multiresolution triangulation.

shows GRIFINOR research potentials, and presents the background of geodata processing routines. Giving accent to cartographic visualization allows users to move freely through the virtual world, explaining fundaments of mathematical cartography enables construction of geo-referenced models. Understanding of basic geometric principles and techniques in two or three-dimensional space motivates students to solve specific problems. GRIFINOR is not a proprietary software, it is independent to existing standards and its functionality could be extend. Everybody is invited to try GRIFINOR or joining the development tool.

## REFERENCES

Kjems E. & Kolář J., From Mapping to virtual geography, 2005 Aalborg, Denmark.
Kolář J., Education in Geomatics with GRIFINOR, 2007, Aalborg, Denmark.
Kolář J., Managed Interoperability for Digital Earth Technology, 2007, Aalborg, Denmark.
Kolář J., Global Indexing of 3D Vector Geographic Feautres, 2004, Aalborg, Denmark.

*Remote Sensing for a Changing Europe*
*D. Maktav (Ed.)*
*IOS Press, 2009*

*doi:10.3233/978-1-58603-986-8-618*

# Investigation of the potential of multiscopic VHR satellite imagery for the production of 3D models of complex urban areas

F. Tack & R. Goossens
*Dept. of Geography, Ghent University, Ghent, Belgium*

G. Buyuksalih
*Bimtas, Istanbul, Turkey*

Keywords: photogrammetry, city DSM generation, tri-stereoscopy, Ikonos, urban

ABSTRACT: In this treatise the discussion of a methodology and results of semi-automatic city DSM extraction from an Ikonos triplet, is introduced. Built-up areas are known as being complex for photogrammetric purposes, partly because of the steep changes in elevation caused by buildings and urban features. To make DSM extraction more robust and to cope with the specific problems of height displacement, concealed areas and shadow, a multi-image based approach is followed. For the VHR tri-stereoscopic study an area extending from the centre of Istanbul to the urban fringe is chosen. Research will concentrate, in first phase on the development of methods to optimize the extraction of photogrammetric products from the bundled Ikonos triplet. Optimal methods need to be found to improve the radiometry and geometry of the imagery, to improve the semi-automatically derivation of DSM's and to improve the postprocessing of the products. Secondly we will also investigate the possibilities of creating stereo models out of images from the same sensor taken on a different date, e.g. one image of the stereo pair combined with the third image. Finally the photogrammetric products derived from the Ikonos stereo pair as well as the products created out of the triplet and the constructed stereo models will be investigated by comparison with a 3D reference. This evaluation should show the increase of accuracy when multi-imagery is used instead of stereo pairs.

## 1 INTRODUCTION TO THE MAMUD PROJECT

The high level of detail and geometric accuracy of very high resolution satellite imagery as Ikonos, has made this kind of imagery suitable for DSM generation at feature level of urban environments. As urban areas are known as complex for photogrammetric purposes, a lot of research is done to cope with the specific problems of such areas during 3D modeling from standard stereopairs. As a multi-image based approach can make the 3D modeling more robust, in this treatise the discussion of a methodology and results of semi-automatic city DSM production from an Ikonos triplet, is introduced. Comparison of the benefits of a multi-image approach with DSM extraction from standard stereo couples will be highlighted. Only a few investigations have been published dealing with the concerning subject. Research published in (Baltsavias et al. 2006) and (Raggam 2006) can be referred.

Research is conducted within the framework of the MAMUD project (Measuring And Modelling of Urban Dynamics) funded by the STEREO (Support to The Exploitation and Research of Earth Observation data) program of Belgian Science Policy.

Urban change processes are affecting the human and natural environment in a not unimportant way. This enlarges the need for more effective urban management approaches based on sustainable development. A sustainable urban management needs sufficiently detailed and reliable base information on the urban environment and its dynamics. The objectives of the MAMUD research project is to investigate the possibilities of earth observation for a better monitoring, modelling and understanding of urban dynamics. Five research teams, each with its own background and know-how, join their strengths to accomplish the objectives. The Ghent university team has the following main objectives:

- City surface model time-series generation from across track, multi temporal imagery: creation of stereo models out of two images of the same sensor (e.g. SPOT) taken of the same area but at a different date. (multi temporal approach).
- City DSM generation from multi-sensor images: investigation of the possibilities of DEM generation from non stereo VHR images from two different sensors (e.g. Ikonos and Quickbird). (multi sensor approach).
- City DSM generation from image triplets: investigation of the advantages of an image triplet, compared with a stereo pair. An image triplet might be constructed from multi-orbit images (e.g. one stereo pair combined with another image from more or less the same orbit) (multi scopic approach).

Multi scopic, multi sensor imagery research is an interesting challenge and, if successful, will increase the flexibility of producing 3D city models from VHR archive data (Ikonos, Quickbird, SPOT), which may be very useful in the future for studying urban dynamics.

## 2 COMPLEXITY OF URBAN AREAS

A Digital Surface Model is a digital representation of the terrain and topographic object height in a grid structure. Interpolation of the discrete height values is needed to approximate the continuity of the ground surface. Urban environments are experienced as complex for 3D modelling purposes because of the steep changes in elevation and the discrepancy between the smoothness of the ground surface and abrupt discontinuities caused by buildings and other urban features. Without manual editing or filter techniques it's difficult to reconstruct vertical walls out of VHR satellite imagery. Kriging creates a smoothed surface and causes that individual buildings will have the shape of a bell instead of the rectangular geometry in an automatic derived surface model. A second consequence of steep changes in elevation is the occurrence of shadow and concealed areas. Due to the convergent viewing angle of VHR sensors like Ikonos, terrain features with certain height above the surface are geometrically displaced in the imagery, leading to dissimilarities between the stereo images. By this distortion of their true position, parts of the ground surface can be hidden in the satellite image, the so called occlusion areas. Shadow areas, which have poor contrast, and occlusion areas lead to mismatches during the image matching algorithm and errors in the resulting surface model. Manual editing of these errors leads to a high accuracy and more detailed results but is not cost effective, so this process step must be minimized. Methods need to be find to get a maximum accuracy for the digital surface model but with the lowest effort.

As discussed in (Buyuksalih & Jacobson 2007) the first problem can be approached by applying a median filter. The filter removes noise and enhances edges. Errors in the surface model caused by the presence of shadow, occlusion and noise (clouds, moving vehicles, etc.) in the image data can be reduced by getting the image information out of more then two images or by the so-called multi scopic record. As the stereo case is the minimum case for 3D mapping, the redundancy of an image triplet gives better constraints. The redundancy of the tri-stereoscopic approach will allow matching with a higher success rate as a correct match can be made if a point in object space is visible in at least two images. With the redundant information of a third image, the effect of occlusion and random noise as clouds and moving vehicles can be reduced. This approach strengthen also the geometric reconstruction because points in object space are calculated by the intersection or best fit of three image rays instead of two. This results in a more optimal image orientation and makes the model more reliable.

Table 1. Characteristics of the three VHR satellite images acquired over the study field

| Image ID | Acquisition date | Elevation angle | Collection azimuth | Sun elevation angle | Projection & datum |
|---|---|---|---|---|---|
| A (Forward) | 1/03/2002 | 67.59° | 1.6° | 39.1° | Epipolar – WGS84 |
| B (Backward) | 1/03/2002 | 75.59° | 214.1° | 39.1° | Epipolar – WGS84 |
| C (Nadir) | 16/05/2005 | 80.93° | 23.5° | 65.5° | UTM – WGS84 |

## 3 DATA SET AND STUDY AREA

The satellite Ikonos is able to rotate the CCD Linear Array sensor up to an angle of 26° off-nadir so the satellite can take images of the same location from two different view points on the same orbital track. Next to along track stereo pairs, it is also possible to create stereo couples out of images from the same area but taken from a different orbit at a different date. These are so called across track stereo pairs. This approach to form couples has some disadvantages as radiometric differences and changes of the ground surface due to the time gap between acquisition of the imagery. A triplet is constructed out of an along track Ikonos stereo pair taken in March 2002 and a third image taken in May 2005. The third image can be considered as a nadir image. Selection criteria for the near vertical image were multiple: overlap with stereo couple, cloud-free acquisition, stereo constellation, minimal time interval and optimal stereo constellation. Despite the big time interval, the 2005 Ikonos image was chosen to be the most optimal candidate.

The Ikonos STEREO product imagery, which comprises of a forward and backward image acquisition and the GEO Ortho Kit 2005 image are panchromatic, resampled to a spatial resolution of 1 m by the image provider and provided with the Rational Polynomial Coefficients (RPC) camera model file. Further characteristics of each image of the triplet can be found in Table 1.

Parts of the megacity Istanbul, Turkey are chosen as test field for the project, partly because it's a city characterized by an intense urban growth. Despite the enormous size, it's a very compact city concentrated along the Bosphorus strait. The high resolution test area covers the overlapping area between the Ikonos 2002 stereo pair and the 2005 image and covers an area of approximately 60 km$^2$, containing Istanbul's historic peninsula and going up to the north to the urban fringe. It concerns a densely built-up area with a height range of 220 m with the lowest point at sea level and geo-morphologically characterized by a hilly landscape.

## 4 PREPROCESSING THE SATELLITE DATA

Before processing the VHR imagery a contrast enhancement is executed as this lead to a more reliable image matching. Especially between images of the same area but taken at different dates from different orbits large radiometric differences can occur due to different illumination and atmospheric conditions, leading to poor matching results. To enhance the contrast for each image individually and to equalize the radiometric differences between the imagery, a Wallis filter was applied (Wallis 1976). The Wallis filter performs a non-linear, locally adaptive contrast enhancement. Actually a large kernel divides the image in different sections and within each section the local contrast is optimized. Applying a Wallis filter on the original images does not only result in a good local contrast and equal overall contrast but normalizes also the radiometry between the images of the across-track stereopairs.

Next to the radiometric enhancement a method for geometric normalization was devised. The Ikonos 2002 stereo couple is epipolar projected by the image vendor and suitable for stereo applications. As the 2005 image is taken from a different orbit, the images are displaced to each other and the internal geometry will be slightly different because of the different scan direction.

Figure 1. High resolution study area: the green polygon covers the Ikonos 2002 stereopair, the blue polygon covers the extent of the high resolution test field or overlap between the Ikonos 2002 stereopair and the 2005 image. (Source: Google earth)

Geometric normalization of the 2005 image with the 2002 imagery is done by image coregistration in ENVI. A first-order polynomial transformation is performed to geometrically align the multitemporal imagery. A first-order polynomial transformation corrects for rotation, translation, scaling and shearing.

## 5 DSM PRODUCTION

Digital surface model extraction from high resolution satellite imagery requires the availability of a database of ground control points. During two field trips to Istanbul the necessary GCP's for photogrammetric processing of the DSM's were collected in close collaboration with Dr. Gurcan Buyuksalih from the Istanbul Metropolitan Planning Centre (IMP-Bimtas). Because accurate large-scale maps were available for the study area and because of the difficulties of GPS measurements in the narrow streets of the densely built-up area, there is chosen for an approach to derive the GCP from maps. 37 GCP were derived from 1:5000 scale topographic maps, even distributed over the high resolution study area.

### 5.1 *Standard stereoscopy*

DSM results from stereopair and triplet need to be compared with each other to get an idea of the possible improvements of the tri-stereoscopic approach. In first place, the along track stereopair is processed. DSM extraction from the Ikonos stereopair was performed with the photogrammetric

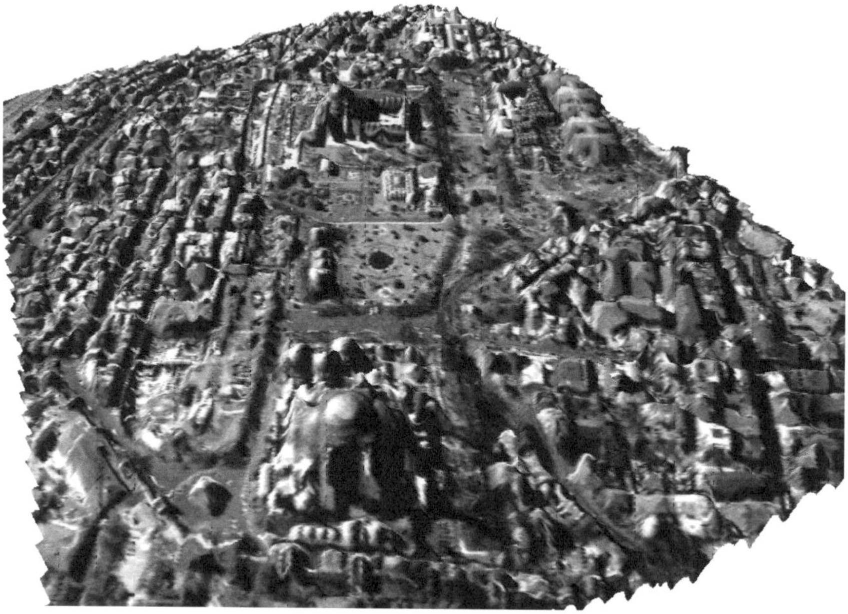

Figure 2. Part of DSM generated from Ikonos stereopair with ortho-image draped over it for 3D photorealistic visualization.

workstation VirtuoZo by Supresoft inc. After importing the imagery and GCP, the image orientation can be performed. During relative orientation conjugate points are searched to relocate position and orientation of the images of a stereopair relatively. In VirtuoZo this step is done in a fully automatic way according to a feature-based matching algorithm, but it needs a manual check for mismatched points, e.g. points on moving vehicles or vegetation. During the absolute orientation, the mathematical relationship between image coordinates and object coordinates is fixed by adding 17 ground control points. On both images of the stereopair the points with known 3D coordinates are located manually. The GCP's have a homogeneous distribution over the test field. The DSM is processed by measuring the disparity between corresponding pixels of the images of the stereocouple in epipolar format. For each pixel to be matched in the first image, the matching algorithm searches for the conjugate pixel in the second image that correlates the most by shifting a kernel of certain size along the epipolar line. The image correlation matching algorithm is a mixture of an area-based and feature-based approach. The DSM is calculated at a grid size of 3 meters. The RMS for the GCP residuals were for X, Y and Z respectively 0.57, 0.72 and 1.92 m. A true ortho-image, which can be draped over the wire-frame for visualization purposes, is produced with a resolution of 1 meter. In a final phase the DSM is manually edited for major blunders (water bodies, clouds, multi-temporal differences, etc.).

### 5.2 *Tri-stereoscopy*

From a theoretical point of view the redundancy of a third image should lead to a more reliable photogrammetric processing. First of all a more optimal image orientation is possible because of the redundancy in the geometric reconstruction. Points in object space can be calculated by spatial intersection or best fit of three convergent image rays instead of two. Secondly, image matching can be performed with a higher success rate because a successful match can be made if an object point is clearly identifiable in at least two images. With the information of a third image, the effect of occlusion and random noise as clouds and moving vehicles could be reduced.

Figure 3. Subimage of ikonos A: high buildings leading to huge image displacement, long shadows and occluded areas.

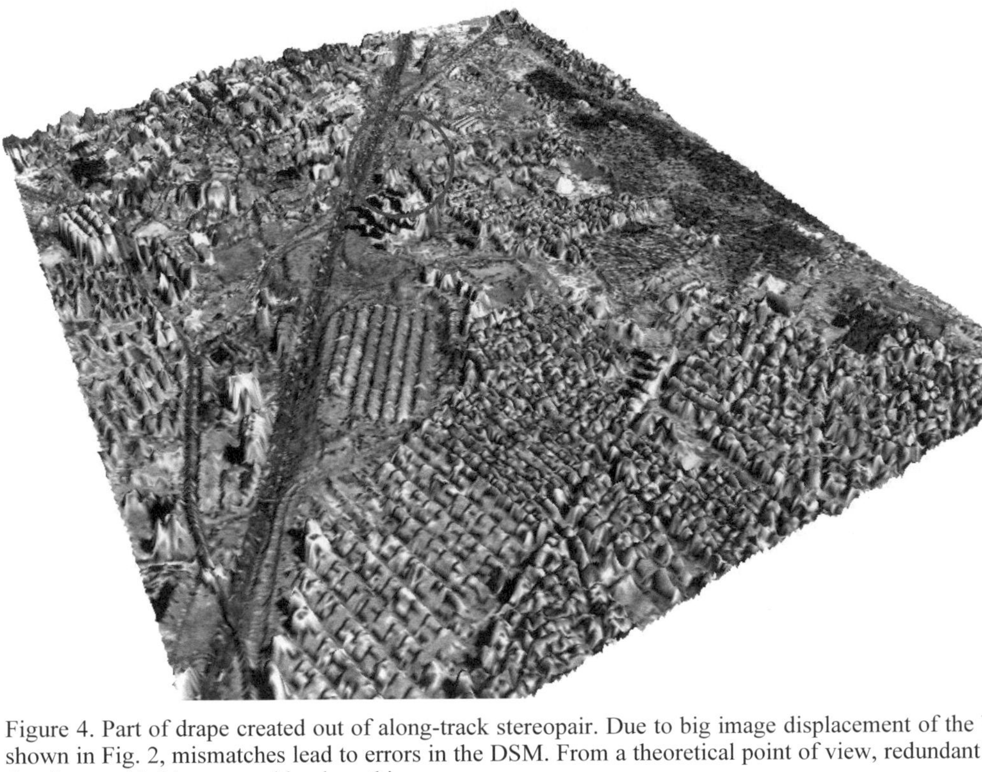

Figure 4. Part of drape created out of along-track stereopair. Due to big image displacement of the buildings shown in Fig. 2, mismatches lead to errors in the DSM. From a theoretical point of view, redundant information from a third image could reduce this error.

Processing of the Ikonos triplet will be performed with SAT-PP. SAT-PP, commercially released by ETH Zurich in April 2008, is able to perform image matching on more then two images simultaneously (Zhang & Gruen 2006). This is in contrast to most commercial photogrammetric software (also Virtuozo) which is able to match only two images at the same time. After processing the Ikonos triplet, DSM results can be compared with a DSM reference to describe the geometric accuracy and to quantify the possible improvements of the tri-stereoscopic approach.

## 6 FUTURE WORK

We introduced an approach to reduce noise and blunders in DSM by using the information from VHR images taken from different angles. However, only an initial step of our planned research is presented in this treatise. The future work deals in first phase with the photogrammetric processing of the bundled Ikonos triplet with the newly acquired photogrammetric workstation SAT-PP, released by ETH Zurich in April 2008. After processing the Ikonos triplet, the resulting DSM from stereopair and triplet will be compared with a DSM reference to describe the geometric accuracy and to quantify the possible improvements of the tri-stereoscopic approach. A DSM reference was already derived from 1:5000 digital topographic maps at the Istanbul Metropolitan Planning centre.

## ACKNOWLEDGEMENTS

The Belgian Science Policy Office is gratefully acknowledged for funding the work presented in this paper (SR/00/105). The authors also wish to thank the other teams of the MAMUD project. More information about the MAMUD project can be found on www.mamud.be

## REFERENCES

Baltsavias E., Pateraki M., Zhang L., 2001. Radiometric and geometric evaluation of Ikonos GEO images and their use for 3D building modeling, In: Proc. Joint ISPRS Workshop High Resolution Mapping from Space 2001, Hannover, 19-21 September, 2001.

Baltsavias E., Zhang L., Eisenbeiss H., 2006. DSM generation and interior orientation determination of ikonos images using a testfield in Switzerland, In: Photogrammetrie, Fernerkundung, Geoinformation, (1), pp. 41-54.

Bethel J.S., McGlone J.Ch., Mikhail E.M., 2001. Introduction to Modern Photogrammetry, John Wiley & Sons, Inc., New York, 477 p.

Buyuksalih G., Jacobson K., 2007. Digital surface models in build up areas based on very high resolution space images. In: ASPRS 2007 Annual Conference, Tampa, Florida, 7-11 May, 2007.

Devriendt D., Goossens R., Dewulf A., Binard M., 2003. Improving spatial information extraction for local and regional authorities using Very-High-Resolution data – geometric aspects. In: High Resolution Mapping from Space 2003.

Jacobson K., 2005. Analysis of Digital Elevation Models based on space information. In: New strategies for European Remote Sensing, Rotterdam : Millpress, pp. 439-451.

Raggam H., 2006. Surface mapping using image triplets: Case studies and benefit assessment in comparison to stereo image processing, In: Photogrammetric engineering and remote sensing, vol. 72, n° 5, pp. 551-563.

Taillieu K., Goossens R., Devriendt D., De Wulf A., Van Coillie S., Willems T., 2004. Generation of DEMs and orthoimages based on non-stereoscopical IKONOS images, pp. 453-460. In: Proceedings of the 24th symposium of the European association of remote sensing laboratories, Dubrovnik, Croatia, 25–27 May 2004.

Wallis R., 1976. An approach to the space variant restoration and enhancement of images, In: Proceedings of Symposium on Current Mathematical Problems in Image Science, Naval Postgraduate School, Monterey, CA.

Zhang L., Gruen A., 2006. Multi-image matching for DSM generation from Ikonos imagery, In: ISPRS Journal of Photogrammetry and Remote Sensing 60 (3) (2006), pp. 195-211.

Remote Sensing for a Changing Europe
D. Maktav (Ed.)
IOS Press, 2009



# Subject Index

# Author Index